13th International Conference on Gravitation, Astrophysics and Cosmology (ICGAC-XIII) and 15th Italian-Korean Joint Symposium on Relativistic Astrophysics (IK15)

EPJ Web of Conferences Volume 168 (2018)

Seoul, South Korea
3 - 7 July 2017

Editors:

Bogeun Gwak
Gungwon Kang
Chanju Kim
Hyeong-Chan Kim
Chang-Hwan Lee
Jaeweon Lee
Seokcheon Lee
Wonwoo Lee

ISBN: 978-1-5108-5574-8

Printed from e-media with permission by:

Curran Associates, Inc.
57 Morehouse Lane
Red Hook, NY 12571

Some format issues inherent in the e-media version may also appear in this print version.

Printed by Curran Associates, Inc. (2018)

For additional information, please contact EDP Sciences – Web of Conferences at the address below.

EDP Sciences – Web of Conferences
17, Avenue du Hoggar
Parc d'Activité de Courtabœuf
BP 112
F-91944 Les Ulis Cedex A
France

Phone: +33 (0) 1 69 18 75 75
Fax: +33 (0) 1 69 28 84 91

contact-edps@webofconferences.org

Additional copies of this publication are available from:

Curran Associates, Inc.
57 Morehouse Lane
Red Hook, NY 12571 USA
Phone: 845-758-0400
Fax: 845-758-2633
Email: curran@proceedings.com
Web: www.proceedings.com

TABLE OF CONTENTS

Preface

Dear Colleagues

I would like to invite you on behalf of the International Organization Committee.

I am pleased to invite you to the conference in Seoul, Korea, based on the accomplishments of the ICGAC community in the Asia-Pacific region over the past 26 years and the academic friendship that has been achieved through relativistic astrophysics between Italy and Korea for 30 years.

The evidence of the gravitational wave released in February of last year, have made the belief to Einstein's relativity theory even stronger, and have contributed greatly to the creation of a new observation field of gravitational wave physics, so called "New Window to our Universe". In addition, many forms of observation and discovery of our universe have played a role in astrophysics and cosmology to expand the horizon. In the coming decade, these new compasses will direct the direction of our knowledge advancement and will do important work in human history. As a result, our understanding of the universe will be expanded and applied to the world, and a new era of physics will emerge.

The 13th International Conference on Gravitation, Astrophysics, and Cosmology (ICGAC-XIII) and the 15th Italian-Korean Joint Symposium on Relativistic Astrophysics (IK15) are combined at the prestigious Ewha Womans University with a history of more than 130 years and an integrated conference will be held. I hope that you will have valuable time to make good conclusions by learning, discussing and sharing recent issues during the conference. I also hope that many young scholars attending this conference learn and acquire a lot of skills and design to carry on future astrophysics.

I hope you will take time to enjoy the values of Ewha Womans University as well as to have a meaningful time to explore Seoul, Korea.

Thank you to all IAC, IOC, LOC committee members and organizations that helped make this conference happen. I am deeply grateful to all of you, especially the Korean Physics Society Astrophysics Division, International Center for Relativistic Astrophysics Network (ICRANet), Asia-Pacific Center for Theoretical Physics (APCTP), Division of Astrophysics, Cosmology and Gravitation (DACG) of Asia Pacific Physics Society (AAPPS), National Research Foundation of Korea, Consiglio Nazionale delle Ricerche (CNR, Italy), Institute for Basic Science (IBS), Ewha Womans University, Konkuk University, Kunsan National University, Physics Department and Professor Leung Memorial Fund at University of Massachusetts Dartmouth, and Springer Verlag.

Without their supports, this conference will not be held here, at this time.

Welcome to ICGAC-XIII & IK15 !

Sung-Won Kim
Chair of IOC, ICGAC-XIII & IK15

International Organizing Committee

Sung-Won Kim (EWU, Chair), **Remo Ruffini** (ICRANet, Co-Chair), **Sang-Pyo Kim** (Kunsan Nat'l U., Co-Chair), **Gungwon Kang** (KISTI, Co-Chair), **Jonghyuk Yoon** (Konkuk U., Co-Chair), **Warrick Couch** (AAO), **Bruce Dawson** (U. Adelaide), **Yungui Gong** (HUST), **Zhong-Hong Zhu** (BNU), **Yipeng Jing** (SJTU), **Pascal Chardonnet** (CNRS), **Jutta Kunz** (U. Oldenburg), **Claus Lammerzhahl** (U. Bremen), **She-Sheng Xue** (ICRANet), **Gregory Vereshchagin** (ICRANet), **Takahiro Tanaka** (Kyoto U.), **Jun'ichi Yokoyama** (U. Tokyo), **Jiro Soda** (Kobe U.), **Kiwoon Choi** (KAIST), **Hyung Won Lee** (Inje U.), **M.H.P.M. van Putten** (Sejong U.), **V.N. Melnikov** (MSU), **A.A. Starobinsky** (ITP Russia), **Pisin Chen** (NTU), **Hoi-Lai Yu** (Sinica)

International Advisory Committee

David Blair (UWA), **John Webb** (UNSW), **Valeri Frolov** (U. Alberta), **Don Page** (U. Alberta), **Rong-Gen Cai** (ITP, CAS), **Jun Luo** (HUST), **Jean-Michel Alimi** (LUTH), **Thibault Damour** (IHES), **Roland Triay** (CNRS), **Varun Sahni** (IUCAA), **Vladimir Belinski** (ICRANet), **Salvatore Capozziello** (U. Naples), **Alexander Dolgov** (U. Ferrara), **Remo Ruffini** (ICRANet), **Kei-ichi Maeda** (Waseda U.), **Masakatsu Kenmoku** (NWU), **Misao Sasaki** (Kyoto U.), **Medeu Abishev** (KNU), **Ratbay Myrzakulov** (ENU), **Yongmin Cho** (Konkuk U.), **Chul Hoon Lee** (Hanyang U.), **Hyun Kyu Lee** (Hanyang U.), **Hyung Mok Lee** (SNU), **Il Hung Park** (SKKU), **A. Garcia** (Cinvestav), **Hernando Quevedo** (UNAM), **J. Buchbinder** (TSPU), **D. Gal'tsov** (MSU), **S.V. Chervon** (UlSPU), **D.S. Gorbunov** (INR), **Chiang-Mei Chen** (NCU), **James M. Nester** (NCU), **W.-T. Ni** (NCU), **Yu. Shtanov** (BITP), **A. Zhuk** (ONU), **Dieter Brill** (UMD), **Jong-Ping Hsu** (UMass), **Ho Jung Paik** (UMD), **Bobomurat Ahmedov** (INP), **Nguyen Suan Han** (VNU), **Pierre Darriulat** (VATLY)

Local Organizing Committee

Hyeong-Chan Kim (KNUT, Chair), **Inyong Cho** (SNUT), **Bogeun Gwak** (Sejong U.), **YuRi Kang** (EWU), **Chanju Kim** (EWU), **Hee Il Kim** (SNU & ZettaHPC), **Kyoung Yee Kim** (Inje U.), **Seoktae Koh** (Jeju Nat'l U.), **Chang-Hwan Lee** (Pusan Nat'l U.), **Hyunjoo Lee** (Incheon Nat'l U.), **Jae-Weon Lee** (Jungwon U.), **Seokcheon Lee** (KIAS), **Wonwoo Lee** (Sogang U.), **John J. Oh** (NIMS), **Sang Hoon Oh** (NIMS), **Mu-In Park** (Sogang U.), **Edwin J. Son** (NIMS)

Host Organization

- Research Institute of Curriculum Instruction,
  Ewha Womans University

Managing Organizations

- Astrophysics Division of Korea Physical Society
- International Center for Relativistic Astrophysics Network (ICRANet)

Supporting Organizations

- Asia Pacific Center for Theoretical Physics
- Division of Astrophysics, Cosmology and Gravitation in the Association of Asia Pacific Physics Society (DACG-AAPPS)
- National Research Foundation of Korea
- Consiglio Nazionale delle Ricerche (CNR, Italy)
- Konkuk University
- Institute for Basic Science (IBS)
- Kunsan National University
- Physics Department and Professor Leung Memorial Fund at University of Massachusetts Dartmouth
- Springer Verlag

Program Timetable

Place: Lee Sam Bong Hall (B4F, Main Rm), B147, and B149 in ECC, Ewha Womans University

Time (Mon.)	3 Mon.	4 Tue.	5 Wed.	6 Thu.	7 Fri.	Time (T/W/T/F)
08:00–09:00	**Registration**					08:50–09:00
09:00–09:45	**V. Frolov** (Main Rm)	**Yungui Gong** (Main Rm)	**Bum-Hoon Lee** (Main Rm)	**Y. Neiman** (Main Rm)	**Jong-Ping Hsu** (Main Rm)	09:00–09:35
09:45–10:30	**R. Ruffini**	**Hyung Mok Lee**	**J. Yokoyama**	**Rong-Gen Cai**	**D. Brill**	09:35–10:15
10:30–10:50	*Coffee Break*	*Coffee Break*	*Coffee Break*	*Coffee Break*	*Coffee Break*	10:15–10:45
10:50–11:10	**Opening Ceremony**	**S. Miyoki**	**Il Hung Park**	**D. Grumiller**	**Yong-Seon Song**	10:45–11:20
11:10–11:55	**M. Sasaki**	**Wei-Tou Ni**	**K. Maeda**	**Jeong-Hyuck Park**	**Eunil Won**	11:20–11:55
11:55–12:40	**Yongmin Cho**	**Ho Jung Paik** ***Photo Session***	**J. A. Rueda**	**Chiang-Mei Chen**	**M. Muccino**	11:55–12:30
12:40–14:00	*Lunch*	*Lunch*	*Lunch*	*Lunch*		12:30–14:00
14:00–15:30	**P1-1** Classical Gravity (Main Rm) **P2-1** Quantum Gravity (B147)	**P3-1** Astro physics (Main Rm) **P4-1** Gravita tional Waves (B147) **P5-1** Inflation and Dark Matter (B149)	Free Time for Excursions, Discussions, Informal Meetings, etc.	**P3-3** Astrophysics (Main Rm) **P6-1** Gravity in String Theory (B147)	**P7-1** Cosmology (Main Rm) **P8-1** Black Holes (B147)	14:00–15:30
15:30–16:00	*Coffee Break*			*Coffee Break*		15:30–16:00
16:00–18:00	**P1-2** Classical Gravity (Main Rm) **P2-2** Quantum Gravity (B147)	**P3-2** Astro physics (Main Rm) **P4-2** Gravita tional Waves (B147) **P5-2** Inflation and Dark Matter (B149)		**P3-4** Astrophysics (Main Rm) **P6-2** Gravity in String Theory (B147)	**P7-2** Cosmology (Main Rm) **P8-2** Black Holes (B147) **Closing Remark**	16:00–18:00
		Banquet				

Plenary Sessions

※ Place: Lee Sam Bong Hall (B4F, Main Rm)

3 July, Monday

Chair: Sang-Pyo Kim (Kunsan Nat'l U.)
08:00 - 09:00 Registration
09:00 - 09:45 Valeri P. Frolov (U. Alberta) Remarks on non-singular black holes
09:45 - 10:30 Remo Ruffini (ICRANet) From Supernovae, to Hypernovae to Binary Driven Hypernovae

10:30 - 10:50 ***Coffee Break***

10:50 - 11:10 Opening Ceremony

Chair: Gungwon Kang (KISTI)
11:10 - 11:55 Misao Sasaki (Kyoto U.) Inflationary massive gravity
11:55 - 12:40 Yongmin Cho (Konkuk U.) Cosmic Production of the Electroweak Monopole

12:40 - 14:00 ***Lunch***

4 July, Tuesday

Chair: Misao Sasaki (Kyoto U.)
09:00 - 09:35 Yungui Gong (Huazhong U. Sci. Tech.) On the polarization of gravitational waves in scalar-tensor theories of gravity
09:35 - 10:15 Hyung Mok Lee (Seoul Nat'l U.) What have we learned from the detection of gravitational waves?

10:15 - 10:45 ***Coffee Break***

Chair: Medeu Abishev (Al Farabi Kazakh Nat'l U.)
10:45 - 11:20 Shinji Miyoki (ICRR) Current Status of KAGRA Gravitational Wave Telescope
11:20 - 11:55 Wei-Tou Ni (Nat'l Tsing Hua U.) Gravitational Wave Classification and the Space GW Detection Sensitivities
11:55 - 12:30 Ho Jung Paik (U. Maryland) Superconducting Tensor Gravitational-Wave Detector and Mitigation of Newtonian Noise
12:30 - 12:40 ***Photo Session***

12:40 - 14:00 ***Lunch***

18:30-20:30 ***Banquet*** **(B403 ECC Theater)**

5 July, Wednesday

Chair: Yongmin Cho (Konkuk U.)
09:00 - 09:35 Bum-Hoon Lee (Sogang U.) Higher Curvature Gravity
09:35 - 10:15 Jun'ichi Yokoyama (U. Tokyo) Creation of the inflationary universe from a black hole?

10:15 - 10:45 ***Coffee Break***

Chair: Jun'ichi Yokoyama (U. Tokyo)
10:45 - 11:20 Il Hung Park (Sungkyunkwan U.) Status and Perspectives in GRB observations
11:20 - 11:55 Kei-ichi Maeda (Waseda U.) Inflation and Disformal Transformation
11:55 - 12:30 Jorge Armando Rueda (ICRANet) On the rate and on the gravitational wave emission of short and long GRBs

6 July, Thursday

Chair: Dieter Rudolf Brill (U. Maryland)
09:00 - 09:35 Yaakov (Yasha) Neiman (Perimeter Institute) Towards a working model of causal-patch physics in dS/CFT
09:35 - 10:15 Rong-Gen Cai (Chinese Academy of Science) Standard sirens and dark sector with Gaussian process

10:15 - 10:45 ***Coffee Break***

Chair: Valeri P. Frolov (U. Alberta)
10:45 - 11:20 Daniel Grumiller (Vienna U. of Technology) Soft Heisenberg Hair
11:20 - 11:55 Jeong-Hyuck Park (Sogang U.) Stringy Gravity & solution to dark matter problem
11:55 - 12:30 Chiang-Mei Chen (Nat'l Central U.) Pair Production of Scalar Dyons in Kerr-Newman Black Holes

12:30 - 14:00 ***Lunch***

7 July, Friday

Chair: Wei-Tou Ni (Nat'l Tsing Hua U.)
09:00 - 09:35 Jong-Ping Hsu (U. of Massachusetts Dartmouth) Experiments on the CMB Spectrum and Their Implications for the Missing Half of the Universe
09:35 - 10:15 Dieter Rudolf Brill (U. Maryland) The revival of General Relativity at Princeton.

10:15 - 10:45 ***Coffee Break***

Chair: Rong-Gen Cai (Chinese Academy of Science)
10:45 - 11:20 Yong-Seon Song (KASI) Cosmological Implications of RSD
11:20 - 11:55 Eunil Won (Korea U.) Status of the GroundBIRD experiment
11:55 - 12:30 Marco Muccino (ICRANet) What can we learn from gamma-ray bursts?

12:30 - 14:00 ***Lunch***

17:50 - 18:00 ***Closing Remark***

Parallel Sessions

※ All talks are for 20 minutes unless being marked with (15 mins)

3 July, Monday- Parallel Sessions 1, 2

P1.* (Main Rm) *Classical Gravity – Classical gravity, Modified gravity, etc.

***P1-1.* 14:00-15:30 (10 minutes for overall discussions)**
Chair: Jong-Ping Hsu (U. of Massachusetts Dartmouth)
Jong Hyuk Yoon (Konkuk U.) Hamiltonian reduction of Einstein\'s gravity without isometries
Seung Hun Oh (Konkuk U.) Solution-generating methods of Einstein's equations by (2+2) Hamiltonian reduction
Chulmoon Yoo (Nagoya U.) Gravitational collapse of massless fields in an expanding universe
Hyeong-Chan Kim (Korea Nat'l U. Transportation) Matter Equation of State in General Relativity

15:30-16:00 ***Coffee Break***

***P1-2.* 16:00-18:15**
Chair: Seung Hun Oh (Konkuk U.)
Jia-Rui Sun (Sun Yat-Sen U.) Analogue gravity in BEC and its gravity dual
Alexey Toporensky (Sternberg Astron. Inst.) A viable compactification scenario in Gauss-Bonnet gravity
Jong-Ping Hsu (U. of Massachusetts Dartmouth) Experiments on the Frequency Dependence of the Deflection of Light in Yang-Mills Gravity
Inyong Cho (Seoul Nat'l U. of Science & Technology) Black Holes in S3 and H3
(15 mins) Laura Marcela Becerra Bayona (Sapienza University of Rome and IcraNet) SPH simulations of the Induced Gravitational Collapse
Atsushi Naruko (Tohoku U.) Extended vector-tensor theories
J. Rodriguez (ICRANet) Strong-field gravitational-wave emission in Schwarzschild and Kerr geometries: some general considerations

P2.* (B147) *Quantum Gravity – Quantum gravity, Black hole thermodynamics, Black hole entropy, etc.

***P2-1.* 14:00-15: 35**
Chair: Chiang-Mei Chen (Nat'l Central U.)
Sang Pyo Kim (Kunsan Nat'l U.) Quantum Gravity Effects in Cosmology
Ehsan Bavarsad (U. Kashan) Effect of a magnetic field on Schwinger mechanism in de Sitter spacetime

Eoin O Colgain (APCTP) Calibrated Entanglement Entropy
Wonwoo Lee (CQUeST) Tunneling decay of false vortices with gravitation
(15 mins) Taishi Ikeda (Nagoya U.) Oscillon in Einstein-scalar system with double well potential and its properties

15: 35-16:00 *Coffee Break*

P2-2. **16:00-18:00**
Chair: Hyung Won Lee (Inje U.)
Hyun Seok Yang (CQUeST) Dark Matter and Dark Energy from Emergent Spacetime Picture
Alex Giacomini (U. Austral de Chile) On the compatibility of thermodynamic equilibrium conditions with lattice propagators
Muhammad Sharif (U. Punjab) Study of Generalized Second Law of Thermodynamics in Curvature-Matter Coupling Gravity
Yoonbai Kim (Sungkyunkwan U.) Entropy in Spacetime with Topological Hair
Hideki Maeda (Hokkai-Gakuen U.) Throat quantization of the Schwarzschild-Tangherlini(-AdS) Black Hole
Wontae Kim (Sogang U.) Origin of Hawking radiation

4 July, Tuesday- Parallel Sessions 3, 4, 5

P3. **(Main Rm)** ***Astrophysics - Relativistic astrophysics, Observation, Cosmic rays, GRB, etc.***

P3-1. **14:00-15:35**
Chair: Myeong-Gu Park (Kyungpook Nat'l U.)
Medeu Abishev (Al Farabi Kazakh Nat'l U.) The orbital stability of a test particle motion in the field of two massive rotating bodies
Giovanni Battista Pisani (Sapienza University of Rome and ICRANet) The 1st ICRANet Catalog of Binary-driven HyperNovae
Gregory Vereshchagin (ICRANet) Cosmic horizon for GeV sources and photon-photon scattering
Naser Ahmadiniaz (Institute for Basic Science) Master formulas for the dressed scalar propagator in a constant field
(15 mins) Mehbub Khan (UMass) Baryonic Force for Accelerated Cosmic Expansion and Generalized Yang-Mills Symmetry

15:35-16:00 *Coffee Break*

P3-2. **16:00-18:00**
Chair: Gregory Vereshchagin (ICRANet)
Myeong-Gu Park (Kyungpook Nat'l U.) Accretion onto Black Holes with Outflow
Sehrish Iftikhar (Lahore College for Women U.) Particle Dynamics Around a Charged Black Hole
Mohammad Wali Hossain (APCTP) Quintessential inflation: A unified scenario of inflation and dark energy

(15 mins) Daria Primorac (ICRANet & Sapienza U. Roma) Analysis of the GRB110731A within the fireshell model
(15 mins) Yongsoo Jho (Yonsei U.) Search for Sphaleron from the Ultra-High-Energy cosmic rays and neutrinos
(15 mins) Rahim Moradi (Sapienza U. of Rome and ICRANet) Charged Cosmological Black holes
(15 mins) Julio David Melon Fuksman (Sapienza U. Rome & ICRANet) Simulation of an electron-positron plasma in the context of the IGC paradigm

18:30-20:30 ***Banquet*** **(B403 ECC Theater)**

P4. **(B147)** ***Gravitational Waves – Gravitational wave astronomy/astrophysics, Numerical relativity, etc.***

P4-1. **14:00-15:30**
Chair: Muhammad Sharif (U. Punjab)
Dong-Hoon Kim (Seoul Nat'l U.) Gravitational waves with effects of radiation reaction,
Toshinori Matsui (KIAS) Gravitational waves from the first order electroweak phase transition in the Z_3 symmetric singlet scalar model
Hyung Won Lee (Inje University) TaylorF2 CBC waveform with eccentricity corrected phase
(15 mins) Luis Gabriel Gómez (Sapienza U. Rome & ICRANet) Gravitational wave emission versus dark matter dynamical friction in the evolution of compact star binaries
(15 mins) Gihyuk Cho (Seoul Nat'l U.) Theory of Gravitational Radiation

15:30-16:00 ***Coffee Break***

P4-2. **16:00-18:00 (10 minutes for overall discussions)**
Chair: Hyung Mok Lee (Seoul Nat'l U.)
Jinn-ouk Gong (APCTP) Induced gravitational waves
Hisaaki Shinkai (*Invited,* Osaka Inst. Tech.) Gravitational waves from merging intermediate-mass black holes
Sunghoon Jung (*Invited,* Seoul Nat'l U.) Angular Resolutions of Mid-Frequency Gravitational Wave Detectors
Gungwon Kang (KISTI) Structural analysis of the SOGRO platform
(15 mins) Dawoo Park (Seoul Nat'l U.) Black Hole Binaries Dynamically Formed in Globular Clusters
(15 mins) HanGil Choi (Seoul Nat'l U.) Finding Quasi-Spherical Orbit Initial Condition of Black Hole Binary using Effective-One-Body Model

18:30-20:30 ***Banquet*** **(B403 ECC Theater)**

P5. **(B149)** ***Inflation and Dark Matter***

P5-1. **14:00-15:40**
Chair: Jae-Weon Lee (Jungwon U.)

Hyun Min Lee (Chung-Ang U) Flattening the inflaton potential beyond the minimal gravity
Aditya Aravind (APCTP) Higgs Portal Inflation with Fermionic Dark Matter
(15 mins) Echal Chang (Chungnam Nat'l U.) Toward a more realistic Randall-Sundrum Brane World
(15 mins) Zhu Yi (Huazhong U. Sci. Tech.) Nonminimal coupling and inflationary attractors
(15 mins) Yongwan Gim (Sogang U.) On the thermodynamic origin of the initial radiation energy density in warm inflation
(15 mins) Dong Woo Kang (Yonsei U.) Electroweak Kaluza-Klein Dark Matter

15:40-16:00 ***Coffee Break***

P5-2. **16:00-18:00 (10 minutes for overall discussions)**
Chair: Yungui Gong (Huazhong U. Sci. Tech.)
Jae-Weon Lee (Jungwon U.) Brief History of Ultra-light Scalar Dark Matter Models
Sichun Sun (National Taiwan U.) New views on dark matter from modified gravity
Yasuhiro Yamamoto (Yonsei U.) Protophobic Light Vector Boson as a Mediator to the Dark Sector
Seodong Shin (Yonsei U.) Non-minimal dark matter search in dark matter colliders
(15 mins) Soo Min Choi (Chung-Ang U) SIMP dark matter and its cosmic abundances
(15 mins) Yoo-Jin Kang (Chung-Ang U) Forbidden Channels and SIMP Dark Matter

18:30-20:30 ***Banquet*** **(B403 ECC Theater)**

6 July, Thursday- Parallel Sessions 3, 6

P3. **(Main Rm)** ***Astrophysics - Relativistic astrophysics, Observation, Cosmic rays, GRB, etc.***

P3-3. **14:00-15:30 (10 minutes for overall discussions)**
Chair: Hyun Kyu Lee (Hanyang U.)
Anna Pollmann (*Invited,* U. Wuppertal) Searches for magnetic monopoles with IceCube
Insik Hahn (Ewha Womans U.) Nuclear astrophysics experiments related to the rp-process using rare isotope beams
Myung-Ki Cheoun (*Invited,* Soongsil U.) Modified Gravity in Nuclear Astrophysics and Big Bang Nucleosynthesis
Chang-Hwan Lee (Pusan Nat'l U.) Strangeness in Neutron Star Cooling

15:30-16:00 ***Coffee Break***

P3-4. **16:00-18:00 (5 minutes for overall discussions)**
Chair: Jorge Armando Rueda (ICRANet)
Hyun Kyu Lee (Hanyang U.) Hadronic matter with emergent symmetries at high density
Hongjun An (*Invited,* Chungbuk Nat'l U.) Studying intrabinary shock emission in pulsar binaries
Dong-Hoon Kim (Seoul Nat'l U.) Pulsar radiation with general relativistic effects

Alexander Yushchenko (Sejong University) Velocity distribution in the clusters of galaxies
Hisaaki Shinkai (Osaka Inst. Tech.) Nonlinear dynamics in Gauss-Bonnet gravity
(15 mins) Andreas Krut (ICRANet) Dark matter and galactic structures

P6.* (B147) *Gravity in String Theory

***P6-1.* 14:00-15:30 (10 minutes for overall discussions)**
Chair: Chanju Kim (Ewha Womans U.)
Jae-Hyuk Oh (Hanyang U.) Phase transition in anisotropic holographic superfluids with arbitrary z and α
Run-Qiu Yang (KIAS) Strong energy condition and complexity growth bound in holography
Yun-Long Zhang (APCTP) Bell inequality from holographic gravity
Sunyoung Shin (CQUeST) Walls of massive Kähler sigma models on SO(2N)/U(N) in three dimensions

15:30-16:00 ***Coffee Break***

***P6-2.* 16:00-18:00**
Chair: Keun-Young Kim (GIST)
Chanju Kim (Ewha Womans U.) Holographic micro state deformations of BTZ black holes
O-Kab Kwon (Sungkyunkwan U.) Exact holography of massive M2-brane theories and entanglement entropy
Kyung Kiu Kim (Sejong U.) Thermodynamic volume in AdS/CFT
Taejin Lee (Kangwon Nat'l U.) Gravitational Scattering Amplitudes and Closed String Field Theory in the Proper-Time Gauge
Hongsu Kim (KASI) 3rd Quantization of Taub Universe
Hongsu Kim (KASI) Callan-Rubakov Effect could answer the baryon asymmetry puzzle

7 July, Friday-Parallel Sessions 7, 8

P7.* (Main Rm) *Cosmology – Cosmology, CMB, Perturbations, Structure formation, Quantum effect, etc.

***P7-1.* 14:00-15:30 (10 minutes for overall discussions)**
Chair: Wonwoo Lee (CQUeST)
Seokcheon Lee (Gyeongsang Nat'l U.) Conformal equality
Dong-han Yeom (Nat'l Taiwan U.) Hartle-Hawking wave function and large-scale power suppression of CMB
Mu-In Park (Sogang U.) On Gauge Invariant Cosmological Perturbations in UV-modified Horava Gravity
Tae Hoon Lee (*Invited,* Soongsil U.) Quintessence in low-energy effective theory

15:30-16:00 ***Coffee Break***

***P7-2.* 16:00-17:50 (10 minutes for overall discussions)**
Chair: Seokcheon Lee (Gyeongsang Nat'l U.)
Ki-Young Choi (Chonnam Nat'l U.) Lower-Bound on the reheating temperature with dark matter
Kyungjin Ahn (Chosun U.) Small-scale structure formation under the large-scale density and CDM-baryon drift velocity environment
Sungwook E Hong (KASI) Small-scale Features of Thermal Inflation: CMB Distortion and Substructure Abundance
Maurice H van Putten (*Invited*, Sejong U.) Fast evolution of the deceleration parameter in surveys of H(z)
Muhammad Sharif (U. Punjab) Stability Analysis of Bulk Viscous Cosmology

17:50-18:00 ***Closing Remark***

P8.* (B147) *Black Holes – Black holes, Wormholes, etc

***P8-1.* 14:00-15:40**
Chair: Jong Hyuk Yoon (Konkuk U.)
Masakatsu Kenmoku (Nara Sci. Academy) Superradiant Phenomena for Spinor Fields in Rotating Black Hole Geometry
Khalid Saifullah (Quaid-i-Azam U.) Thermodynamics of rotating non-Kerr black holes
Bogeun Gwak (*Invited*, Sejong U.) Upper Limit of Radiation from Coalescence of Rotating Hayward Black Holes
Sung-Won Kim (Ewha Womans U.) Wormhole cosmological model
Jin Young Kim (Kunsan Nat'l U.) Wormholes in Einstein-Born-Infeld Gravity

15:40-16:00 ***Coffee Break***

***P8-2.* 16:00-17:50 (10 minutes for overall discussions)**
Chair: Bogeun Gwak (Sejong U.)
Kimitake Hayasaki (*Invited*, Chungbuk Nat'l U.) Detection of gravitational wave emission by supermassive black hole binaries through tidal disruption flare
Medeu Abishev (Al Farabi Kazakh Nat'l U.) Dilatonic dyon black hole solutions in the model with two Abelian gauge fields
Seung Hun Oh (Konkuk U.) Poisson algebra of gauge-invariant quasilocal angular momentum and its asymptotic limit
Deniz Olgu Devecioglu(Sogang U.) Lifshitz black holes in Einstein-Yang-Mills theory
Lunchakorn Tannukij (Hanyang U.) Black holes in dRGT Massive Gravity

17:50-18:00 ***Closing Remark***

Exploring Curved Spacetime---LIU Liao's Adventure

Zheng Zhao[1,*], *Shou-Yong Pei*[1,**], *Jong-Ping Hsu*[2,***]

[1]Department of Physics, Beijing Normal University, Beijing, 100875, China
[2]Department of Physics, University of Massachusetts Dartmouth, North Dartmouth, MA. 02747, USA

In 1999, Prof. LIU, with the help of the Chinese Physical Society and the Chinese Society of Gravitation and Relativistic Astrophysics, among others, organized the Fourth International Workshop on Gravitation and Astrophysics at Beijing Normal University. The purpose was to promote exchange of scientific ideas, which was in line with DENG Xiaoping's open policy after the Culture Revolution. The following year, the Proceedings of the Fourth International Workshop on Gravitation and Astrophysics was published by the World Scientific. During the political turmoil, LIU Liao suffered for 22 years---demoted from "assistant professor" in 1957 to a reference staff in the physics department library until 1979. In spite of these difficulties, he taught himself general relativity, quantum field theory and particle physics. LIU Liao became a pioneer of general relativity in China and established a major center for research of black holes, cosmology and gravitational waves at Beijing Normal University. He is respected among the scientific community in China and abroad.

LIU Liao (12/24/1928 - 4/27/2016) was born in Shen-Yang city. His father LIU Po was a prominent Professor of Chinese Literature at Northeast University. He married Dr. ZHANG Ruiying and had one son LIU Yi. During his student years before 1949, he actively participated in the students' political movement against the Nationalist government. In 1952, LIU Liao graduated from Peking University.

LIU Liao joined Beijing Normal University (BNU) in 1956 to teach physics. In 1957, because LIU Liao honestly expressed his critical opinion with the good intention of helping the Party to improve its operation, he was strongly criticized and denounced repeatedly. Unexpectedly, he became a despised 'rightist' for 18 years. He could no longer teach physics and became reference staff in the physics department library. Then the Cultural Revolution exploded, LIU Liao and many others were rounded up as criminals, "reformed through labor" and subjected to barbarous treatment. As a result, young LIU Liao suffered from lower back pain, which persisted throughout his life. LIU Liao's family also suffered for a long period of time, and his whole family lived in one single room. Every night, he put upz bed boards to sleep. He had no desk to study at and write on, so he used a large bench as a desk. In order for his wife, Dr. Zhang, to sleep better while he studied late at night, she slept in a shower booth.

Through all these years of being a "rightist" and being "reformed through labor," he studied and thought about the profound idea of curved space-time and all that. It was no longer just to satisfy his thirst for understanding the physical universe and for the joy of comprehension; it also helped him to forget the suffering of daily life. In this aspect, Beethoven was a model and hero for him because he also refused to be beaten and crushed

* zhaoz43@hotmail.com ** psy@bnu.cn *** jhsu@umassd.edu

by such a difficult situation of political turmoil. He was also comforted by Beethoven's music and spirit: "I shall seize Fate by the throat; it shall certainly not bend and crush me completely."

During the Cultural Revolution, LIU Liao already had some in-depth understanding of curved space-time based on books such as Einstein's *The Meaning of Relativity*, Møller's *The Theory of Relativity* and Landau and Lifshitz's *The Classical Theory of Fields*. He happened to meet two physics students, YANG Yihon and LIU Zhongzhu. Both of them were labeled as counter-revolutionary students because they posted bigcharacter posters against LIN Biao, who enjoyed Mao's complete trust at that time. These students realized that LIU Liao was not a criminal. Actually, they came to admire LIU Liao's self-study of general relativity. Since they were also interested in general relativity, they secretly came to LIU Liao's home in the night to discuss and to learn general relativity, etc. Fortunately, LIU Liao lived outside the campus at that time. Otherwise, such a secret meeting to learn physics at night would have been a serious crime of 'counter-revolutionary networking', if discovered by those bad people.

Naturally, YANG Yihong and LIU Zhongzhu were the first two bright students of LIU Liao when he became an associated professor in 1979. In 1980, YANG Yihong was one of the first group of students helped by CUSPEA (initiated and managed by T. D. Lee at Columbia University) to pursue a Ph.D. degree in the U.S. He earned his Ph.D. degree in 1984 at the University of Utah. After he returned, Beijing Normal University broke rules for the first time to promote him to be a full physics professor. Unfortunately, YANG Yihong died young in 1988 because of his poor health related to long-term persecution. LIU Zhongzhu, LIU Liao's protégé, also became a professor at Huazhong University of Science and Technology.

When DENG Xiaoping brought the Spring of Reform and Opening to China around 1978, LIU Liao saw hope and future. He made a great effort to spread general relativity theory everywhere in China. He established one of the three centers for general relativity research at Beijing Normal University. The other two centers were at the Institute of Theoretical Physics at the Chinese Academy of Sciences and at the Chinese University of Science and Technology. Their leaders were those who persisted in doing physics research under the very harsh environment of political turmoil.

In 1979, LIU Liao became an Associate Professor and a Member of the Board of Directors of the Chinese Society of Gravitation and Relativistic Astrophysics (CSGRA). From 1983 to 1994, he became a Professor of Physics and the Associate Chairman and the Chairman of CSGRA. From 1991 to 2008, he served as a member of The Chinese Physical Society, an editor of the Russian journal 'Gravitation and Cosmology,' and a member of The International Society of General Relativity and Gravitation.

As a student at BNU in 1978, ZHAO Zhen was surprised that LIU Liao appeared to be the only physicist in China who knew general relativity thoroughly by himself in those years of political turmoil. As far as he knew, others were gifted young men, who formed a group to study and discuss general relativity. LIU Liao also taught himself particle physics and quantum field theories. He enjoyed classical music, literature and philosophy. He also had a good knowledge of languages and could read books in English, German and Russian.

ZHAO Zheng joined LIU Liao's research group and immediately realized that he was easy to approach and very considerate to his students. He stressed that students must follow the process of 'Learning by Doing' and pay attention to the literature for recent physics activities abroad. He also advised students that the most effective way to learn theoretical physics was both to follow a book and derive all equations. Simply reading books without deriving equations may be easier and more enjoyable, but the understanding is only superficial.

LIU Liao was completely dedicated to physics research and education throughout his life. Around 1973, his first research on particle physics resulted in an interesting empirical formula for hadron masses, which gave a better fit than the usual Gell-Mann-Okubo mass formula. In the 1980's, he initiated the teaching and research on black hole physics and quantum field theory in curved space-time in China. Later, he led his group of students into a broad research of the inflationary universe, quantum cosmology, worm holes and time machines. His research interest and results include the following in order of importance:

(a) *Proving that (i) the most general stationary states of black hole emit Dirac particles through thermal radiations. and (ii) future horizons of all stationary space-time must have thermal radiations. (LIU Liao, XU Dianyan--Prof. of Peking Univ.).*

(b) *Hawking thermal radiation and calculations of the temperature and entropy of a non-stationary black hole. (LIU Liao, ZHAO Zheng).*

(c) *The membrane model of black holes* (HUANG Chaoguang, LIU Liao, XU Feng).

(d) A new idea that worm holes could be produced from vacuum fluctuation. (LIU L. Phys. Rev. **D** 1993, 48, R5463).

(e) The cosmological origin of Higgs fields, (LIU Liao, PEI Shouyong).

(f) A new approach to the quantization of evaporating black hole. Any evaporating black hole has a remnant left and a quantized Schwarzschild black hole approaches its ground state.

(g) A new approach to removing the divergence difficulties in QED by using worm holes.

(h) Quantum cosmology, inflationary cosmology, and time machines.

LIU Liao published more than 60 scientific papers in refereed journals with numerous students and colleagues. He also published nine monographs and books with colleagues and students:

[1] Quantum Field Theory in Curved Space-Time and Quantum Cosmology (LIU Liao, HUANG Chaoguang, Beijing, Science Press, 2013).
[2] Black Hole and Properties of Time (LIU Liao, ZHAO Zheng, TIAN Guihua, ZHANG Jingyi, Beijing, Peking University Press, 2008).
[3] Inflationary Cosmology and Cosmic String (JIANG Yuanfang, LIU Liao, QIAN Zhenhua, Shanghai, East China Normal University Press,1993).
[4] Introduction to Lie group and Lie algebra (LIU Liao, CAO Yufang, Beijing, Beijing Normal Univ. Press, 1987).
[5] Special Relativity, second edition. (LIU Liao, FEI Baojun, ZHANG Yunzhong,

Beijing, Science Press, 2008).
[6] General Relativity, second edition (LIU Liao, Zhao Zheng, Beijing, Higher Education Press, 2004).
[7] Gravitation and Astrophysics, the Proceedings of the Fourth International Workshop. (LIU Liao, LUO Jun, LI Xin-Zhou, HSU Jong-Ping. World Scientific, Singapore, 2000).
[8] Quantum Field Theory (Flat Space-time) (LIU Liao, Beijing, Beijing Normal University Press, 2003).
[9] LIU Liao Collected Literary Works, (LIU Liao, Changsha, Hunan Science and Technology Press, 2008).

In 2003, LIU Liao eloquently summarized his life and hopes in the Afterword of his book *Quantum Field Theory (Flat Space-Time)* as follows:

"My life has been a rugged journey, full of unthinkable adversity and calamity. That I was able to survive penitentiary, cowsheds[A] and all other hardship is most fortunate.

In the old days, I loved Beethoven's Symphony #5 and admired him tremendously. At the worst moments of my life, I would quote his "Ich will dem Schicksal in den Rachen greifen...."[B] to help me endure the hardship. His "I will grab fate by the neck, To force fate to comply, Not be crushed or surrender " became my mantra. But a "professional fighter" I am not. During the Anti-Rightist Movement and the Cultural Revolution, I experienced and witnessed all kinds of evil doings and atrocity. I was powerless. I was often thinking of living as a recluse, yet nowhere could I find peace and serenity but tribulations without end. I could only take to reading to find shade "under the greenwood tree" [C] that would shelter me from the extreme torture to my body and soul. What I learned from my study during that period not only gave me some pleasure and relief but helped me persevere.

I spent half a century of my life studying the microscopic and cosmic worlds, keeping company with the creator and all the immortals in the universe. That was critical for my survival and remained a lasting pleasure.

Now I am old and feeble. I wish to pass on to the young the truth and solid laws of nature that I discovered under the shade of the greenwood tree."

Footnotes:

[A] 'Cowsheds refer to flimsy structures with dank floors and reed mats for prisoners of the Red Guards, where starving was magnified by physical exertion and the constant threat of beating.'

[B] "I will grab fate by the neck, To force fate to comply, not be crushed or surrender" is quoted from Ludwig van Beethoven- Famous Quotes and Quotations: "Ich will dem Schicksal in den Rachen greifen, ganz niederbeugen soll es mich grewiB nich ."

[C] Greenwood tree refers to Shakespeare's poem: "Under the greenwood tree/Who loves to lie with me/And turn his merry note/Unto the sweet bird's throat/Come hither, come hither, come hither!/Here shall he see/No enemy/But winter and rough weather.''

We would like to thank L. Haseltine, H. Jiang and A. Parelman for their help.

Seminar at BNU.

Reference room in the physics department library at BNU---LIU Liao's universe. ('.... *keeping company with the creator and all the immortals in the universe.'*)

1999 International Worshop on Gravitation and Astrophysics at BNU.

HSU Jongping, LIU Liao, ZHAO Zheng (from left to right)

LIAN Zhujian, LIU Liao, HSU Jongping, PEI Shouyong, SHI Tianyi(from left to right)

Remembering Vitaly N. Melnikov

Jong-Ping Hsu[1,*], *Sang Pyo Kim*[2,**] and *Sung-Won Kim*[3,***]

[1]Department of Physics, University of Massachusetts Dartmouth, North Dartmouth, MA. 02747, USA

[2]Department of Physics, Kunsan National University, Kunsan 54150, Korea

[3]Department of Science Education, Ewha Womans University, Seoul 12075, Korea

Vitaly Nikolayevich Melnikov (1941-2017) was a major organizer for the Asia-Pacific International Conference on Gravitation, Astrophysics and Cosmology (ICGAC) in the past 24 years. He organized two conferences in 2001 and 2015 at Peoples' Friendship University of Russia (PFUR) in Moscow. His endeavor gave a big boost to the status of ICGAC in the international arena. He commanded respect among the scientific community.

Vitaly was born on January 24, 1941 in Moscow, son of Nikolai Melnikov and Revecca Melnikova. He married Inna Fyodorovna Krutsenko and had three children: Andrey, Oleg and Igor. He completed his undergraduate studies at the Faculty of Physics of Lomonosov Moscow State University (MSU) in 1964. In 1968 he earned a Ph.D. degree at MSU specializing in Theoretical Physics under the supervision of Academician N. N. Bogoliubov. In 1983, he defended at MGU[a] the degree of Doctor of Sciences and passed with flying colors. In 1990, he became a full professor.

From 1967 to 1995, Melnikov was a researcher, a section head and the head of the Theoretical Department at Russian State Committee Standards in Moscow. From 1972 to 1993, he was a professor of theoretical physics at Moscow State University. In 1994, he joined Peoples' Friendship University of Russia (PFUR), Moscow. Since 1995, he served as the head of the Center of Gravitation and Fundamental Metrology VNIIMS. He was the President of the Russian Gravitational Society in Moscow. He was a founder and Editor-in-Chief of the international journal 'Gravitation and Cosmology' (published by Springer).

Melnikov was an Academician of the Russian Academy Metrology and a member of the Scientific Council in Astronomy of the Russian Academy of Sciences.

Melnikov was a visiting Professor in many scientific institutes, including Observatoire Paris-Mendon (1999); LUTH, Observatorie de Paris (2004, 2005, 2006, 2007); Paris Diderot University (2005, 2006); ITP. Universite de Provence, Marseille, France (1996, 1999); Aegean University, Greece (1998, 1999); NWU, Nara, Japan (1999-2000); CINVESTAV, Mexico (2000-2001); CBPF, Brazil (1993,1995,2002,2004, 2006, 2008).

*e-mail: jhsu@umassd.edu **e-mail: sangkim@kunsan.ac.kr ***e-mail: sungwon@ewha.ac.kr

[a] Moscow State University (Moskovskij Gosudarstvennyj Universitet or MGU, in short) is the oldest University of Russia. It was founded on 12th of January 1755 on the initiative of the great Russian scientist M. V. Lomonosov, after whom it was named.

As a research physicist, Melnikov appeared to have had abundant grant support. He was the head of scientific grants Minobrnauki, Russian Foundation Basic Research, since 1995; the German Research Foundation (Deutsche Forschungsgemeinschaft), Germany, 1994-2006; the Conselho Nacional de Pesquisa Brazil.

He also served as the chairman of organizing committees for 12 Russian gravitational conferences (starting from 1989), the director of courses of International School in Gravitation and Cosmology in Erice (Sicily) in 1987, 2003 and 2004, the chairman of joint scientific seminar PFUR-VNIIMS and a co-chairman of the scientific seminar of the Russian Gravitational Society.

Melnikov had seven monographs and proceedings:

(1) Hydrodynamics, Fields and Constants in the Theory of Gravitation (K. P. Stanyukovich and V. N. Melnikov, 1971).

(2) Gravitational Measurements, Fundamental Metrology and Constants (V. de Sabbata, Vitaly N. Melnikov; Springer Science and Business Media, Dec 2012, 552 pages; Proceedings of the NATO Advanced Study Institute, Erice, Italy, May 2-12, 1987).

(3) Multidimensional classical and quantum cosmology and gravitation. Exact solutions and variations of constants, Centro Brasileiro de Pesquisas Fisicas (CBPF) Publ.NF- 051/93 (Rio de Janeiro, 1993).

(4) Multidimensional cosmology and gravitation, CBPF, Publ., MO-002/95 (Rio de Janeiro, 1995).

(5) Exact solutions in multidimensional gravity and cosmology III, CBPF, Publ.MO-03/02 (Rio de Janeiro, 2002).

(6) The Gravitational Constant: Generalized Gravitational Theories and Experiments (Ed. V. de Sabbata, George T. Gillies, Vitaly N. Melnikov, 2004).

(7) Gravitation, Astrophysics, and Cosmology: Proceedings of the Twelfth Asia-Pacific International Conference (Ed. V. N. Melnikov and Jong-Ping Hsu, World Scientific, Feb. 2016, 400 pages).

Melnikov has been completely dedicated to physics research and education throughout his life. He published 289 scientific papers in refereed journals with numerous students and colleagues. [46 papers in Russian (1985- 2009), 192 papers in English (1988-2009) and 51 papers (2010-2016)]. His research interests include (a) Gravitation and cosmology in diverse dimensionalities, exact solutions; (b) Quantum cosmology; (c) Fundamental metrology, fundamental physical constants and their stability; (d) dark energy and dark matter; (e) theory of gravitational experiments, particularly cosmic (satellite) experiments.

Vitaly Melnikov, with his colleagues and students, has created new trends in the field of gravity. He has been one of the founders of a new trend in metrological science, Gravitational-Relativistic Metrology, connected with a thorough study of possible variations of fundamental physical constants and their influence on various physical measurements. Much of his work was devoted to exact solutions to the equations of gravity in four and more dimensions and interpretations of these solutions, describing various cosmological models, black holes and solitons. Together with Prof. A. Sanders (University of Tennessee), he headed the group which developed a theoretical foundation of Project Satellite Energy Exchange aimed at measuring a number of parameters of gravity (including the constant G) with unprecedented accuracy. [1,2]

In a lecture on 'Cosmological Models with Fluids, Scalar Fields and Forms in Diverse Dimensions,' Melnikov explained: The discovery of accelerated expansion of the Universe and the fact that flat Friedmann model with the cosmological constant or quintessence now fits best the set of observational data, created problems of dark matter and dark energy. This is a real revolution in modern physics as we do not know now what really the dark matter (0.20 of 0.30) and what the dark energy (0.70) of the total energy are..... In the lectures I describe our different attempts to solve some basic problems of modern cosmology using exact solutions with different matter sources in diverse dimensions.

In another paper (2000), he expressed his view that 'Gravity as a Key Problem of the Millennium.' Gravitation is a fundamental interaction that governs all phenomena at large and very small scales, but still not well understood at a quantum level, is a missing cardinal link to unification of all physical interactions. Problems of the absolute G measurements and its possible time and range variations are reflections of the unification problem. Integrable multidimensional models of gravitation and cosmology make up one of the proper approaches to studying basic issues and strong field objects, the Early Universe and black hole physics in particular.

His collaborator Kirill Bronnikov (VNIIMS, Moscow) observed: 'He has always been very quiet, positive and friendly, he had many friends not only close around him but throughout the world, just to mention (certainly not all) such countries as the USA, Germany, France, Brazil, Korea and Japan. He had been a very good manager: he was a leader of our research group for about half a century, he was a founder of the Russian Gravitational Society in 1988 and headed it for 29 years; he also founded and headed the Gravitation and Cosmology journal. As a group leader, he always supported his colleagues and was ready to give a good advice. So for us he was always a friend, and we are really missing him.'

Sang Pyo Kim friendship with Melnikov dates back to his surprising invitation to the International School and Conference on Multidimensional Gravity and Cosmology at Yaroslavl, Russia in June 1994. The Conference was organized by the Russian Gravitational Society, which was founded by Melnikov. It was two years after Kim settled down at Kunsan National University, located in his parent hometown. Probably the local organizers would have noticed his papers on quantum cosmology in Physical Review D based on his thesis at Pennsylvania State University in 1991, since he received another letter from Alexander Zhuk asking for hard copies of these papers before the invitation letter arrived. It was an era of airmails rather than emails, and published journals rather than e-journals.

At that international conference Sung-Won Kim and Sang Pyo Kim met many good friends, Kirill A. Bronnikov and Vladimir D. Ivashchuk as local organizers, and Eduardo I. Guendelman, Vladimir N. Lukash, Paulo V. Moniz as speakers, to name a few. Sang Pyo Kim since then oftentimes met Melnikov at ICGA(C) and other conferences. Sang Pyo once chaired a session for Melnikov at the 2nd ICGA, Hsinchu, Taiwan in 1995, which was organized by Wei-Tou Ni and truly the first international conference series of ICGA(C). Melnikov was absorbed in his own talk and enthusiastic about his research and so could not finish the talk on time and continued for additional 15 more minutes.

With Melnikov, Sang Pyo Kim co-chaired ICGAC 10 at Quy Nhon, Vietnam in 2011, ICGAC 11 at Almaty, Kazakhstan in 2013, and then ICGAC 12 in Moscow, Russia in 2015. Kim felt heartbreaking sadness when he heard of Melnikov's passing away from Medeu Abishev, another co-chair of ICGAC 11, since Kim was expecting to see him again at ICGAC 13 in Seoul, Korea in 2017. Melnikov was always kind and gentle in organizing activities with an enthusiasm for science. Those who knew him will remember his enthusiasm and leadership in the field of gravitation and cosmology.

Melnikov and Sung-Won Kim had been good friends since 1994. Kim remembers Melnikov as a good friend to colleagues all over the world, a great leader of the Russian Gravitational Society, and a well-intentioned boss to the members of VNIIMS. Kim organized and chaired the last five ICGAC conferences in 1999, 2003, 2007, 2009, and 2017. He first met Melnikov in 1994 at the Yaroslavl meeting and continued the contact at the 1996 Novgorod and the 1999 Vladimir meetings. Melnikov was then invited to the International Organizing Committee (IOC) and became an active member of The Asia-Pacific International Conference on Gravitation and Astrophysics (formal name prior to ICGAC). Because of their special relationship, Kim served as a member of IAC or IOC at eight international conferences organized by Melnikov and held in Russia. Melnikov's enthusiasm for the international journal "Gravitation & Cosmology" prompted him to invite Kim to be a member of its editorial board. Melnikov also visited Ewha and collaborated with Kim for a month in 2003, publishing two papers as co-author with Vladimir Ivashcuk (JMP 1999, G&C 2004).

The first time Jong-Ping Hsu got in touch with Melnikov was related to the proceedings of the 2015 ICGAC Conference. They corresponded and agreed that it would be nice if there was a record of the conference free for physicists who participated, and then they set out to realize the proceedings. As the organizer, Melnikov was responsible for selecting papers from presentations in planetary and parallel sessions. Hsu dealt with the publisher World Scientific and the in-house editor. Hsu sent the publisher a camera-ready manuscript and made sure that they would send a free copy to each researcher who contributed a paper. Melnikov and colleagues made an effort to make sure that the contents of papers in the proceedings were in line with the contemporary level of world research in gravitation, astrophysics and cosmology. In a draft of the Preface, there was a statement 'these works (including Einstein and Grossmann's 1913 paper) are milestones in our understanding of the universe.' Melnikov revised it to read '.... are milestones in our *present* understanding of the universe.' Hsu was impressed and realized that this was a truly researcher with admirable insight of the evolution of ideas in physics.

In 2001, Melnikov led the Russian Gravitational Society (that he headed for 13 years) and the Center of Gravitation and Fundamental Metrology VNIIMS to organize the Asia-Pacific Conference ICGA-5 in Moscow. In 2015, Melnikov, with the help of RGS, the Institute of Gravitation and Cosmology, PFUR, VNIIMS, BLTP JINR, again organized a most successful, large and exciting Conference ICGAC-12 at the People's Friendship University of Russia, Moscow. More than 130 leading and young Russian participants were joined by 45 distinguished and young guests from Brazil, Estonia, Germany, Italy, Kazakhstan, and Ukraine, in addition to the Asia-Pacific countries. A trip for participants to enjoy the spectacular views of the Moscow River Cruise was incomparable. Melnikov arranged a delightful banquet with bountiful foods and Russian wines, and Dmitry Galtsov spontaneously went to a piano and played Chopin music to entertain colleagues. It was an exciting international conference. Young Min Cho (who initiated the series of ICGAC conference in 1993 in Korea) proposed the idea of expanding the future ICGAC conference from the Asia-Pacific region to the international arena, similar to GRG and the Marcel Grossmann conferences. The idea was discussed and agreed upon in the last-day meeting of the Organizing and Advisory Committees chaired by Melnikov in Moscow. It was gratifying that the first step had been realized in the 2017 conference organized by Sung-Won Kim in Seoul by establishing the Standing Committee to decide policies and to oversee future ICGAC conferences.

Melnikov was a very cheerful, friendly and energetic person, careful and responsive in research to his colleagues and students, with a good sense of humor and intelligence. [1]

Melnikov's life is a symphony of global friendship and collaboration.

With Love from the Asia-Pacific and beyond.

[1] We would like to thank V. Ivashchuk and K. Bronnikov (VNIIMS, Moscow) for materials regarding Melnikov, and A. Parelman for help.

[2] Selected Works:

K. P. Staniukovich and V. N. Melnikov, *Hydrodynamics, Fields and Constants in the Theory of Gravitation* (Energoatomizdat, Moscow, 1983) [book, in Russian].

V. N. Melnikov, *Gravity as a key Problem of the Millennium*, Proc. 2000 NASA/JPL Conf. on Fundam. Physics in MicrKgravity, NASA Document D-21522, 2001, pp. 4.1.17 (CA, USA).

V. D. Ivashchuk and V. N. Melnikov, *Sigma-Model for Generalized Composite p-Branes*, Class. Quantum Grav. **14**, 3001-3029 (1997); **15**, 3941 (1998).

V. D. Ivashchuk and V. N. Melnikov, *Multidimensional Cosmological and Spherically Symmetric Solutions with Intersecting p-Branes*, Lecture Notes in Physics, **537** (2000).

K. A. Bronnikov and V. N. Melnikov, *p-Brane Black Holes as Stability Islands*, Nucl. Phys. B **584**, 43658 (2000).

A. J. Sanders, D Alexeev, W. Allison, Antonov, A. Bronnikov, W. Campbell, R Cates, A. Corcovilos, D. Earl, Gadfort, T. Gillies, J. Harris, I. Kolosnitsyn, Yu Konstantinov, N. Melnikov, J. Newby, G. Schunk and L. Smalley, *Project SEE (Satellite Energy Exchange): an international effort to develop a space-based mission for precise measurements of gravitation*. Class. Quantum Grav. **17**, 2331 (2000).

A. D. Alexeev, A. Bronnikov, I. Kolosnitsyn, Y. Konstantinov, N. Melnikov and J. Sanders, *Measurement of the gravitational constant G in space (SEE project): sensitivity to orbital parameters and space charge effect*. Metrologia **38**, 397 (2001).

MOHREN
BRÄU

First three photos courtesy of Kirill Bronnikov, Center of Gravitation and Fundamental Metrology VNIIMS, Moscow, Russia.

Fourth and fifth photos (2003 ICGAC) courtesy of S. W. Kim, Ewha Womans University, Seoul, Korea.

Sixth photo (2015) courtesy of J. P. Hsu, University of Massachusetts Dartmouth, USA.

Conference Poster (Credit: IK15 Web http://www.icranet.org/ik15)

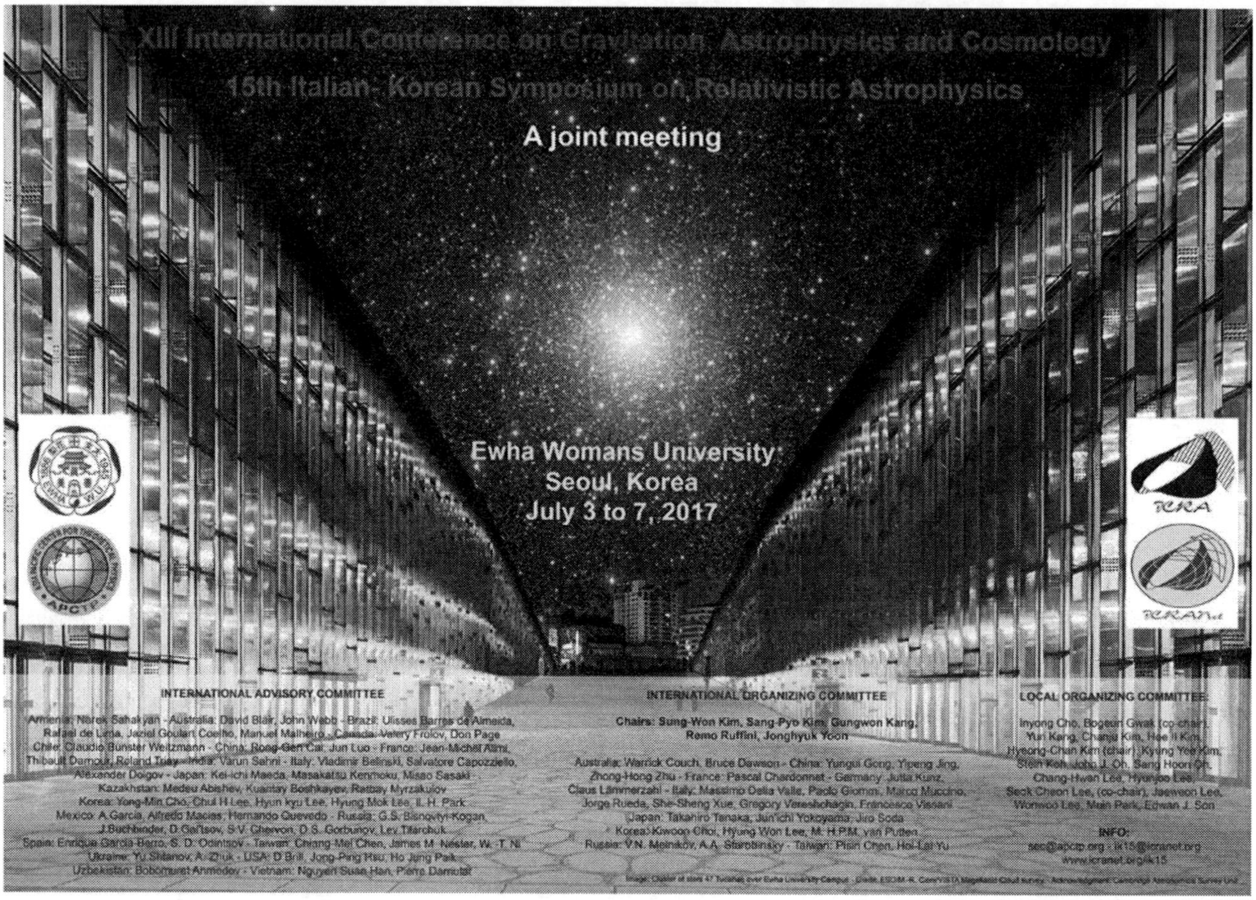

Abstracts

3 July, Monday

Plenary Presentations

Valeri P. Frolov (U. Alberta): "Remarks on non-singular black holes"

We discuss spherically symmetric metrics which represent non-singular black holes in four- and higher-dimensional spacetimes. We first consider static metrics, which obey the following conditions: (i) Regularity at the center and (ii) Validity of the limiting curvature condition. We describe the Hayward model and its generalizations, which obey these properties. After this we discuss quantum radiation of a massless scalar field from an evaporating spherically symmetric non-singular black hole with finite lifetime. We demonstrate that in a general case there exists a huge outburst of the quantum radiation, emitted from the black hole interior from the domains close to the inner horizon. We discuss a relation of this phenomenon with the mass inflation effect. Finally, we shall make remarks on attempts to solve this problem and to provide self-consistency of the non-singular model in the quantum domain.

Remo Ruffini (ICRANet): "From Supernovae, to Hypernovae to Binary Driven Hypernovae"

Our concept of Induced Gravitational Collapse (IGC paradigm) from an incipient Supernova into a companion Neutron Star, has unlocked the understanding of seven different families of GRBs, indicating the path for the formation of Black Holes in the Universe. An authentic laboratory of relativistic astrophysics has been unveiled in which new paradigms have been introduced in order to advance in the knowledge of the most energetic, distant and complex systems of our Universe. A novel paradigm of the Cosmic Matrix has been introduced which parallel, in a relativistic cosmic level, the concepts of an S Matrix introduced by Feynmann, Wheeler and Heisenberg in the quantum world of microphysics. Here the "in" states are represented by a Neutron Star and a Supernova, the "out" states, generated, in less then a second, by a new Neutron Star and a Black Hole. This novel field of research needs very powerful technological observations in all wavelength ranging from Radio, to Optical, to X and Gamma radiation all the way to ultra-high-energy cosmic rays.

Misao Sasaki (Kyoto U.): "Inflationary massive gravity"

Inflation is a natural platform for modified gravity. Here we present a new massive gravity theory in which the gravitational wave (GW) modes become massive during inflation. Then we discuss its observational signatures, which are particularly intriguing in this coming era of GW physics/astronomy.

Yongmin Cho (Konkuk U.): "Cosmic Production of the Electroweak Monopole"

We discuss the cosmological production and the successive evolution of the electroweak monopole in the standard model, and estimate the remnant monopole density at present universe. We confirm that, although the electroweak phase transition is of the first order, it is very mildly first order. So, the monopole production arises from the thermal fluctuations of the Higgs field after the phase transition, not the vacuum bubble collisions during the phase transition. Moreover, while the monopoles are produced copiously around the Ginzburg temperature $T_G\simeq 59.6~{\rm TeV}$, most of them are annihilated as soon as created. This annihilation process continues very long, untill the temperature cools down to about 29.5 MeV. As the result the remnant monopole density in the present universe becomes very small, of 10^{-11} of the critical density, too small to affect the standard cosmology and too small comprise a major component of dark matter. We discuss the physical implications of our results on the ongoing monopole detection experiments, in particular on MoEDAL, IceCube, ANTARES, and Auger.

P1. Classical Gravity – Classical gravity, Modified gravity, etc.

Jong Hyuk Yoon (Konkuk U.): "Hamiltonian reduction of Einstein's gravity without isometries"

I apply the Hamiltonian reduction procedure to general spacetimes of 4 dimensions in the (2+2) formalism and find privileged spacetime coordinates in which the physical Hamiltonian is expressed in terms of the conformal two metric and its conjugate momentum. Physical time is the area element of the spatial cross-section of null hypersurfaces, and the physical radial coordinate is defined by equipotential surfaces on a given spacelike hypersurface of constant physical time. The physical Hamiltonian is local and positive in the privileged coordinates. I present the complete set of Hamilton's equations and find that they coincide with the Einstein's equations written in the privileged coordinates. This shows that the Hamiltonian reduction is self-consistent and respects general covariance.

Seung Hun Oh (Konkuk U.): "Solution-generating methods of Einstein's equations by (2+2) Hamiltonian reduction"

The purpose of this talk is to demonstrate a new method of generating exact solutions to the Einstein's equations obtained by the Hamiltonian reduction. The key element to the successful Hamiltonian reduction is finding the privileged spacetime coordinates in which physical degrees of freedom manifestly reside in the conformal two-metric, and all the other metric components are determined by the conformal two-metric. In the privileged coordinates the Einstein's constraint equations become trivial; the Hamiltonian and momentum constraints are simply the defining equations of a non-vanishing gravitational Hamiltonian and momentum densities in terms of conformal two-metric and its conjugate momentum, respectively. Thus, given any conformal two-metric, which is a constraint-free data, one can construct the whole 4-dimensional spacetime by integrating the first-order superpotential equations. As the first examples of using Hamiltonian reduction in solving the Einstein's equations, we found two exact solutions to the Einstein's equations in the privileged coordinates. Suitable coordinate transformations from the privileged to the standard coordinates show that they are just the Einstein-Rosen wave and the Schwarzschild solution.

The local gravitational Hamiltonian and momentum densities of these spacetimes are also presented in the privileged coordinates.

Chulmoon Yoo (Nagoya U.): "Gravitational collapse of massless fields in an expanding universe"

Gravitational collapse of nonlinear massless fields(scalar fields, gravitational waves) with a periodic boundary condition in a box is considered. We show the existence of the threshold value of the initial amplitude of the local inhomogeneity for black hole formation. The late time expansion law of the universe is also discussed.

Hyeong-Chan Kim (Korea Nat'l U. Transportation): "Matter Equation of State in General Relativity"

We study how a strong gravity affects the equation of state of matters. For this purpose, we employ a canonical ensemble of classical monoatomic ideal gas inside a box in a Rindler spacetime. The total energy decreases monotonically with the increase of the external gravity representing its attractiveness. It is however bounded below, which is different from that of the Newtonian gravity case. As for the entropy, it decreases with the external gravity in the Newtonian regime. However, in the presence of strong gravity or ultra-relativistic high temperature, the entropy increases with the gravity. This result can be a resolution of the negative entropy problem of the ideal gas in the Newtonian gravity. In the presence of strong gravity, the bottom of the box is very close to the event horizon of the Rindler spacetime mimicking a blackhole and the gas behaves as if it is on an effective two dimensional surface located at the bottom of the box. Investigating the equation of state in the strong gravity regime, the temperature of the system is found to be not a free parameter but to approach a fixed value proportional to the external gravity, which is reminiscent of the Unruh temperature.

Jia-Rui Sun (Sun Yat-Sen U.): "Analogue gravity in BEC and its gravity dual"

Analogue gravity appeared in non-gravitational systems show many features resemble to real gravitational systems such as the appearance of curved geometry and horizon. However, deep dynamical connections between the analogy gravity and the real gravity are still lacking. In this paper, we will show that the analogy gravity formed from the Bose-Einstein condensation (BEC) can be dynamically connected with a real gravitational system by using the approach of the gauge/gravity duality. In this sense, the analogy gravity in BEC is no longer just an analogy.

Alexey Toporensky (Sternberg Astron. Inst.): "A viable compactification scenario in Gauss-Bonnet gravity"

We propose a scenario in which a multidimensional anisotropic Universe starting with some non-zero measure set of initial conditions tends to a situation with three big almost isotropic dimensions and inner dimensions with constant radius. The stabilization of compact dimensions occures due to negative spatial curvature. In contrast to earlier work on this type of compactification the present scenario does not require exact isotropy of big and inner dimensions.

Jong-Ping Hsu (UMass): "Experiments on the Frequency Dependence of the Deflection of Light in Yang-Mills Gravity"

Yang-Mills gravity is based on 4-dimensional translational gauge symmetry in flat space-time, can be quantized and is consistent with all known experiments. The eikonal equation for a light ray is derived from a version of Maxwells wave equations modified by gravity in the limit of geometric-optics. One obtains a Hamilton-Jacobi equation for the light ray, $G\mu\nu\partial\mu\psi\partial\nu\psi = 0$. Thus, in Yang-Mills gravity, light rays move as if L they were in a curved space-time with the effective metric tensor $G\mu\nu$. The deflection L angle of a light ray by the sun is about 1.53″ for optical frequencies ≈ 1014Hz. It is roughly 12% smaller than the usual value 1.75″. However, the experimental data for the bending of light by the sun in optical frequencies have uncertainties of 10-20% due to large systematic errors. If one does not take the geometric-optics limit, the eikonal equation is $G\mu\nu\ [\partial\mu\psi\partial\nu\ \psi + (\partial\mu\partial\nu\ \psi)\tan\psi] = 0$. This eikonal equation implies L that the angle of light deflection by the sun should be frequency dependent. All recent accurate measurements of the light deflection angle by the sun have been performed in radio frequencies near 109 Hz, which is too far away from optical frequencies to see this effect. Yang-Mills gravity could be tested by detecting the frequency-dependence of the deflection angle by the sun. We propose carrying out a new experiment using frequencies 1012 Hz which would have an uncertainty of 0.1%, which is technically possible. This experiment could test new implications of Yang-Mills gravity.

Inyong Cho (SeoulTech): "Black Holes in S3 and H3"

The black-hole solutions in closed (S3) and open (H3) spaces are introduced. These black holes are formed in the presence of static perfect fluid. The charged case is also discussed. The spacetime structure, the geodesics, and the stability of the black holes will be discussed.

Laura Marcela Becerra Bayona (Sapienza University of Rome and IcraNet): "SPH simulations of the Induced Gravitational Collapse"

The induced gravitational collapse (IGC) paradigm will be presented. This has been applied to explain the long gamma ray burst (GRB) associated with type Ic supernova, and recently the X-ray flashes (XRFs). The progenitor is a carbon-oxygen core (CO) and a neutron star (NS) tight binary system. The CO core collapses and undergoes a supernova explosion (SN) which triggers the hypercritical accretion onto the NS companion (up to $10^{-2}\,M_\odot s^{-1}$). For the binary driven hypernova (BdHNe), the binary system is enough bound, the NS reach its critical mass, and collapse to a black hole (BH) with a GRB emission characterized by an isotropic energy $E_{iso}>10^{52}$ erg . Otherwise, for binary systems with larger binary separations, the hypercritical accretion onto the NS is not sufficient to induced its gravitational collapse, a X-ray flash (XRF) is produced with Eiso <~

1052 erg. With the SPH code, we're going to simulate the SN ejecta motion under the gravitational atrattion of the NS in order to identify the binary parameters that limits the BdHNe systems with the XRFs systems.

Atsushi Naruko (Tohoku U.): "Extended vector-tensor theories"

Recently, several extensions of massive vector theory in curved space-time have been proposed in many literatures. In this talk, we consider the most general vector-tensor theories that contain up to two derivatives with respect to metric and vector field. By imposing a degeneracy condition of the Lagrangian in the context of ADM decomposition of space-time to eliminate an unwanted mode, we construct a new class of massive vector theories where five degrees of freedom can propagate, corresponding to three for massive vector modes and two for massless tensor modes. We find that the generalized Proca and the beyond generalized Proca theories up to the quartic Lagrangian, which should be included in this formulation, are degenerate theories even in curved space-time. Finally, introducing new metric and vector field transformations, we investigate the properties of thus obtained theories under such transformations.

J. Rodriguez (ICRANet): "Strong-field gravitational-wave emission in Schwarzschild and Kerr geometries: some general considerations"

We show how the concurrent implementation of the exact solutions of the Einstein equations, of the equations of motion of the test particles, and of the relativistic estimate of the emission of gravitational waves from test particles, can establish a priori constraints on the possible phenomena occurring in Nature. Two examples of test particles starting at infinite distance or from finite distance in a circular orbit around a Kerr black hole are considered: the first leads to a well defined gravitational wave burst the second to a smooth merging into the black hole. We notice a difference between our treatment and the one by Ori and Thorne (2000) which will affect the gravitational wave signal. This analysis is necessary for the study of the waveforms in merging binary systems.

P2: Quantum Gravity – Quantum gravity, Black hole thermodynamics, Black hole entropy, etc.

Sang Pyo Kim (Kunsan Nat'l U.): "Quantum Gravity Effects in Cosmology"

Within the geometrodynamic approach to quantum cosmology, we studied the quantum gravity effects in cosmology. The Gibbons-Hawking temperature is corrected by quantum gravity due to spacetime fluctuations and the power spectrum also gets a quantum gravity effect.

Ehsan Bavarsad (U. Kashan): "Effect of a magnetic field on Schwinger mechanism in de Sitter spacetime"

We investigate the effect of magnetic field background on the scalar QED pair production in de Sitter spacetime (dS). We obtained the pair production rate, which agrees with the known Schwinger result in Minkowski spacetime and with the Hawking radiation in the limit of zero electric field in dS. It implies how the cosmic magnetic field affects on the pair production rate. In addition, using the zeta function regularization scheme we calculate the induced current and examine the effect of magnetic field on the vacuum expectation value of the current operator. We find that in the strong electric E or magnetic B field the current responds as E.B, instead in the infrared regime, it responds as B/E, which leads to a phenomenon of infrared hyperconductivity. Those results of the induced current would be important for discussing the cosmic magnetic field evolution.

Eoin O Colgain (APCTP): "Calibrated Entanglement Entropy"

Entanglement entropy can be determined holographically through the identification of minimal surfaces in Anti-de Sitter (AdS) spacetimes. We report on the application of calibrated cycles for identifying minimal surfaces in both unwarped and warped AdS geometries.

Wonwoo Lee (CQUeST): "Tunneling decay of false vortices with gravitation"

We investigate the tunneling decay of vortices initially trapped in the false vacuum of scalar electromagnetic theory with gravitation in three spacetime dimensions. This paper is the extension of our previous paper, tunneling decay of false vortices, with gravitation. In this study, the false vacuum contains metastable vortex solutions instead of the homogeneous configuration at the initial moment. The vortex solution can have inside true vacuum state and the magnetic flux with the thin-wall in the present of gravitation. We present the numerical solutions for the gauge, scalar field, and metric functions.In the limit of vanishing gravitational coupling we smoothly retrieve our previously found results for the decay of the false vortex.

Taishi Ikeda (Nagoya U.): "Oscillon in Einstein-scalar system with double well potential and its properties"

We focus on the Einstein-scalar system with double well potential. In case of a scalar field with double well potential in Minkowsiki background, it is known that there is a longevity localized solution. This solution is called oscillon. Therefore, when we consider the gravitational collapse of the bubble in the Einstein-scalar system with double well potential, it is possible that oscillon appears as an intermediate state. In many previous researches, the oscillon in a scalar field with double well potential in Minkowski background has been examined. In this study, we examined the fundamental properties of the oscillon in the Einstein-scalar system with double well potential, and found the several interesting properties.

Hyun Seok Yang (CQUeST): "Dark Matter and Dark Energy from Emergent Spacetime

Picture"

We emphasize that noncommutative (NC) spacetime necessarily implies emergent spacetime if spacetime at microscopic scales should be viewed as NC. The emergent gravity from NC U(1) gauge theory is the large N duality and the emergent spacetime picture admits a background-independent formulation of quantum gravity. We show that dark matter and dark energy arise as a holographic (UV-IR) manifestation of the coherent vacuum of Planck energy condensate. It turns out that emergent gravity distinguishes dark matter and dark energy according to the causal structure of emergent spacetime.

Alex Giacomini (U. Austral de Chile): "On the compatibility of thermodynamic equilibrium conditions with lattice propagators"

In this paper the compatibility is analyzed of the non-perturbative equations of state of quarks and gluons arising from the lattice with some natural requirements for self-gravitating objects at equilibrium: the existence of an equation of state (namely, the possibility to define the pressure as a function of the energy density), the absence of superluminal propagation and Le Chatelier's principle. It is discussed under which conditions it is possible to extract an equation of state (in the above sense) from the non-perturbative propagators arising from the fits of the latest lattice data. In the quark case, there is a small but non-vanishing range of temperatures in which it is not possible to define a single-valued functional relation between density and pressure. Interestingly enough, a small change of the parameters appearing in the fit of the lattice quark propagator (of around 10~\%) could guarantee the fulfillment of all the three conditions (keeping alive, at the same time, the violation of positivity of the spectral representation, which is the expected signal of confinement). As far as gluons are concerned, the analysis shows very similar results. Whether or not the non-perturbative quark and gluon propagators satisfy these conditions can have a strong impact on the estimate of the maximal mass of quark stars.

Muhammad Sharif (U. Punjab): "Study of Generalized Second Law of Thermodynamics in Curvature-Matter Coupling Gravity"

This paper explores the non-equilibrium behavior of thermodynamics at the apparent horizon of isotropic and homogeneous universe model in f(G,T) gravity (G and T represent the Gauss-Bonnet invariant and trace of the energy-momentum tensor, respectively). We construct the corresponding field equations and analyze the first as well as generalized second law of thermodynamics in this scenario. It is found that an auxiliary term corresponding to entropy production appears due to the non-equilibrium picture of thermodynamics in first law. The universal condition for the validity of generalized second law of thermodynamics is also obtained. Finally, we check the validity of generalized second law of thermodynamics for the reconstructed f(G,T) models (de Sitter and power-law solutions). We conclude that this law holds for suitable choices of free parameters.

Yoonbai Kim (Sungkyunkwan U.): "Entropy in Spacetime with Topological Hair"

Global topological solitons of the hedgehog ansatz are added to Schwarzschild black hole or de Sitter spacetime in arbitrary dimensions larger than three, and thermodynamic law is checked at the horizon without additional horizons. All geometric and thermodynamic quantities are varied in the presence of this interacting matter distribution including pressure, however the area law is satisfied in exact form.

Hideki Maeda (Hokkai-Gakuen U.): "Throat quantization of the Schwarzschild-Tangherlini(-AdS) Black Hole"

Adopting the throat quantization pioneered by Louko and Makela, we derive the mass and area spectra for the Schwarzschild-Tangherlini black hole and its anti-de Sitter (AdS) generalization in arbitrary dimensions. We obtain exact spectra in three special cases: the three-dimensional BTZ black hole, toroidal black holes in any dimension, and five-dimensional Schwarzshild-Tangherlini(-AdS) black holes. For the remaining cases the spectra are obtained for large mass using the WKB approximation. For asymptotically flat black holes, the area/entropy has an equally spaced spectrum, as expected from previous work. In the asymptotically AdS case on the other hand, it is the mass spectrum that is equally spaced.

Wontae Kim (Sogang U.): "Origin of Hawking radiation"

The origin of Hawking radiation will be discussed in connection with the firewall issue in the Unruh vacuum.

4 July, Tuesday

Plenary Presentations

Yungui Gong (Huazhong U. Sci. Tech.): "On the polarization of gravitational waves in scalar-tensor theories of gravity"

According to the Newman-Penrose formalism, in general there are six polarizations for general theories of gravity. In particular, the presence of the longitudinal mode means that the theory has all six polarizations. For massive scalar-tensor theory of gravity or f(R) theory, the longitudinal mode presents, but the theory has only three dynamical degress of freedom, it seems that there is inconsistency somewhere. We point out that the Newman-Penrose formalism is not directly applicable to massive mode, the polarization of the massive scalar field is a mix of breathing and longitudinal mode and the longitudinal mode is proportional to the mass, when the mass becomes zero, the mix mode becomes the pure breathing mode which is consistent with the Newman-Penrose formalism. Since the longitudinal mode is proportional to the mass, so it can be detected only in very low frequency band, like the PTA band.

Hyung Mok Lee (Seoul Nat'l U.): "What have we learned from the detection of gravitational waves?"

The advanced LIGO detected three black hole binary mergers and one candidate during two observing runs. We are still in very early stage of gravitational wave astronomy, but we already learned many new things in astrophysics. The existence of the binary black hole was never established observationally until the LIGO made detections of binary black hole mergers. The binary neutron star merger have been considered to be the most robust source of gravitational waves for LIGO type detectors but such an event has not been detected yet. The masses of the black holes in the binaries before the merger appear to be systematically larger than the typical masses of the black hole candidates in X-ray binaries in the Galaxy. The black hole binary merger detected most recently appear to have spins that are not aligned to the orbital angular momentum. These results pause many fundamental questions regarding the evolution of massive stars, the epoch of formation of the black holes and origin of the compact black hole binaries. We need many more detections with high signal to noise retios in order to answer these questions. In this talk, I will focus on the astrophysical significances of the detected gravitational waves so far and discuss future prospects of gravitational astrophysics with current and future generation of gravitational wave detectors.

Shinji Miyoki (ICRR): "Current Status of KAGRA Gravitational Wave Telescope"

I will talk about the present status and future commissioning plan of KAGRA gravitational wave telescope in Japan.

Wei-Tou Ni (Nat'l Tsing Hua U.): "Gravitational Wave Classification and the Space GW Detection Sensitivities"

After reviewing the gravitational wave (GW) spectral classification. we discuss the sensitivities of GW detection in space aimed at low frequency band (100nHz–100mHz) and middle frequency band (100mHz–10Hz) with the science goals of detecting GWs from (i) Supermassive Black Holes; (ii) Extreme-Mass-Ratio Black Hole Inspirals; (iii) Intermediate-Mass Black Holes; (iv) Galactic Compact Binaries and (v) Relic GW Background. We present an overview on the sensitivity, orbit design, basic orbit configuration, angular resolution, orbit optimization, deployment, time-delay interferometry (TDI) and payload concept of the current proposed GW detectors in space under study. The detector proposals under study have arm length ranging from 1000 km to 1.3×10^9 km (8.6AU) including (a) Solar orbiting detectors
and (b) Earth orbiting detectors.

Ho Jung Paik (U. Maryland): "Superconducting Tensor Gravitational-Wave Detector and Mitigation of Newtonian Noise"

Detection of gravitational waves (GWs) from binary black holes (BHs) by Advanced LIGO has opened a new window of astronomical observation. Many conceivable sources such as intermediate-mass BH binaries and white dwarf binaries, as well as stellar-mass BH inspiral, would emit GWs below 10 Hz. It is highly desirable to open a new window for GW astronomy in the infrasound frequency band. A low-frequency tensor detector could be constructed by combining six magnetically levitated superconducting test masses. Such a detector would be equally sensitive to GWs coming from anywhere in the sky, and would be capable of resolving the source direction and wave polarization. I will present a design concept of a new terrestrial GW detector, named SOGRO, which could reach a strain sensitivity of 10^(-19)-10^(-21) Hz^(-1/2) at 0.1-10 Hz. Seismic and Newtonian gravity noises are serious obstacles in constructing terrestrial GW detectors at frequencies below 10 Hz. I will explain how these noises are rejected in SOGRO. I will also discuss the possibility of mitigating the Newtonian noise for advanced laser interferometers by directly detecting and removing it with mini-SOGROs collocated with the interferometer test masses.

P3: Astrophysics - Relativistic astrophysics, Observation, Cosmic rays, GRB, etc.

Medeu Abishev (Al Farabi Kazakh Nat'l U.): "The orbital stability of a test particle motion in the field of two massive rotating bodies"

We investigate a motion of the test body in the restricted three body problem, where two massive bodies have their own rotation. We consider the problem of orbital stability of circular motion of a test body, where disturbances to motion of the test body (in the plane of motion of the second body) from the circular orbital motion of the second body in the field of central body are in order of relativistic corrections to the motion of a test body from the central body.
In the work [1], the orbital stability of the motion of a test body in the restricted three body problem in mechanics of general relativity (GR) is investigated (all bodies have no proper rotation). As a result[1], it is shown that the motion of the test body in the plane of the orbit of the two body is stable when orbit of the test body is circular. In this work, we consider the case when massive bodies have their own rotation. We found the evolution equation of motion for the test (third) body, which describes the average change of its orbital momentum. To do that, we'll study the evolutionary equations of motion by the asymptotic adiabatic theory, through the process of averaging of the corresponding equations using the vector elements M (the orbital moment) and A (the Laplace vector).
References
[1] Abishev M.E., Toktarbay S., Zhami B.A. On the Stability of Circular Orbits of a Test Body in the Restricted Three-Body Problem in GR Mechanics. Gravitation and cosmology, 20. (2014)

Giovanni Battista Pisani (Sapienza University of Rome and ICRANet): "The 1st ICRANet Catalog of Binary-driven HyperNovae"

In a series of recent publications, scientists from ICRANet led by Remo Ruffini have reached a novel comprehensive picture of "gamma-ray bursts" (GRBs) thanks to their development of a series of new theoretical approaches. Among those, the "induced gravitational collapse" paradigm explains a class of energetic, long-duration GRBs associated with Ib/c supernovae, recently named "binary-driven hypernovae" (BdHNe). Thanks to this novel theoretical and observational understanding, it was possible for ICRANet scientists to build the 1st BdHNe catalog, composed by the 344 BdHNe identified up to the end of 2016.

Gregory Vereshchagin (ICRANet): "Cosmic horizon for GeV sources and photon-photon scattering"

Propagation of ultra high energy particles in the Universe is reviewed. Particular emphasis is given to the photon-photon scattering process.

Naser Ahmadiniaz (Institute for Basic Science): "Master formulas for the dressed scalar propagator in a constant field"

The worldline formalism has previously been used for deriving compact master formulas for the one-loop N-photon amplitudes in both scalar and spinor QED, and in the vacuum as well as in a constant external field. For scalar QED, there is also an analogous master formula for the propagator dressed with N photons in the vacuum. Here, we extend this master formula to include a constant field. The two-photon case is worked out explicitly, yielding an integral representation for the Compton scattering cross section in the field suitable for numerical integration in the full range of electric and magnetic field strengths.

Mehbub Khan (UMass): "Baryonic Force for Accelerated Cosmic Expansion and Generalized Yang-Mills Symmetry"

Based on conserved baryon number (or charge) and the generalized Yang-Mills symmetry, we discuss a new baryonic gauge field and its linear potential for two point-like baryon charges. The force between two baryons is repulsive, extremely weak, but independent of distance. Only in the later stage of cosmic evolution, when two baryonic galaxies are separated by an extremely large distance, the baryonic force can overcome their gravitation attractive force. Thus, this provides a gauge-field-theoretic understanding of the late-time accelerated cosmic expansion. The general forces between two big objects with baryon charges are calculated. The trajectories of two baryonic objects are also numerically calculated and displayed.

Myeong-Gu Park (Kyungpook Nat'l U.): "Accretion onto Black Holes with Outflow"

Recent works on the accretion flow onto black holes show that the accretion rate of the flow can be significantly different from the Bondi mass accretion rate. In addition, hot accretion flows are expected to develop outflows. We discuss the implications of the outflow in terms of the accretion rate and energetics as well as black hole growth.

Sehrish Iftikhar (Lahore College for Women U.): "Particle Dynamics Around a Charged Black Hole"

Black holes are the most important predictions of general relativity, which are the end product of gravitational collapse. The study of geodesics reveals geometrical properties of such a curved spacetime. A physical particle follows either null or timelike geodesics. We explore the particle motion near a charged black hole. In this context, we study the stability of orbits through effective potential and examine the effect of charge on the motion of particles.

Mohammad Wali Hossain (APCTP): "Quintessential inflation: A unified scenario of inflation and dark energy"

In quintessential inflation, inflation and late time acceleration are described by a single scalar field. Such a scenario will be discussed with canonical and noncanonical scalar fields. The scalar field behaves as a inflaton field during inflation and as a quintessence field during late time. Comparison with recent Planck data will also be done.

Daria Primorac (ICRANet & Sapienza U. Roma): "Analysis of the GRB110731A within the fireshell model"

Long gamma-ray burst GRB 110731A was bright enough to cause Autonomous Repoint Request when it triggered the GBM onboard Fermi spacecraft. Swift Burst Alert Telescope was also triggered and immediately slew into the position. Many more observatories responded with prompt follow-up observation. As a result, provided data ranges from optical to gamma rays, covering few hundred seconds. This gives an unique opportunity to study the burst and identify its properties within the IGC paradigm, using the fireshell model. In the fireshell model, description of the GRB acceleration process invokes the dynamics of an optically-thick e^+e^- plasma, which gradual annihilation causes its expansion and self acceleration. The dynamics of the fireshell up to the transparency point is fully described by the total energy of the plasma E and the barion load B. We determine these quantities by performing the time-resolved spectral analysis of the GRB 110731A light curve. We proceed with the light curve simulation from which we deduce the CBM density values and it's inhomogeneities. Finally, time-integrated spectra of the simulated light curve was compared with the spectra observed by NaI and BGO detectors onboard Fermi spacecraft.

Yongsoo Jho (Yonsei U.): "Search for Sphaleron from the Ultra-High-Energy cosmic rays and neutrinos"

We study the Ultra High Energy Cosmic Rays (UHECR) as a probe of the sphaleron processes. Our focuses are on the high energy proton-proton and neutrino-nucleon collisions for the cosmic ray observatories such as TA and Pierre Auger. With the attenuation and regeneration effects due to generic high-multiplicity processes such as the electroweak sphaleron and the microscopic Black hole, constraints from the neutrino telescopes, IceCube and Super-Kamiokande, are also discussed.

Rahim Moradi (Sapienza U. of Rome and ICRANet): "Charged Cosmological Black holes"

The cosmological black holes are black holes living not in an asymptotically flat universe but in an expanding spacetime. They have a rich dynamics in particular for their mass and horizon. In this article we perform a natural step in investigating this new type of black hole: we consider the possibility of a charged cosmological black hole. We derive the general equations of motion governing its dynamics and report a new analytic solution for the special case of the charged Lemaitre-Tolman- Bondi equations of motion that describe a charged cosmological black hole. We then study various relevant quantities for the characterization of the black hole such as the C-function, the effect of the charge on the black hole flux and the nature of the singularity. We also perform numerical investigations to strengthen our results.

Julio David Melon Fuksman (Sapienza U. Rome & ICRANet): "Simulation of an electron-positron plasma in the context of the IGC paradigm"

The BdHN model has been introduced in the last years, to explain a subfamily of gamma-ray bursts (GRBs) with energies Eiso>= 10^{52} erg associated with type Ic supernovae. Such BdHNe have as progenitor a tight binary system composed of a carbon-oxigen core (COcore) and a neutron star (NS) undergoing an induced gravitational collapse (IGC) to a black hole (BH), triggered by the COcore explosion as a supernova (SN). This collapse produces an optically-thick e^+e^- plasma, part of which expands acceleratedly and impacts the SN ejecta at r~10^{10}cm. This process is here considered as a candidate for the production of X-ray flares, which are frequently observed following the prompt emission of GRBs. In this preliminary work we simulate the evolution of the e^+e^- plasma as it interacts with the SN ejecta, and estimate the subsequent photon emission.

P4: Gravitational Waves – Gravitational wave astronomy/astrophysics, Numerical relativity, etc.

Dong-Hoon Kim (Seoul Nat'l U.): "Gravitational waves with effects of radiation reaction"

In General Relativity, it is known that a particle moving in curved spacetime undergoes a force which results from the interaction of the particle with its own field; namely, a self-force or radiation reaction force. In this talk, I discuss the effects of the self-force on the orbital motion of a small object about a black hole: e.g. a test particle orbiting a small black hole (of about a solar mass), a small black hole (of about a solar mass) orbiting a supermassive black hole (of about a million solar masses), etc.. This study can be applied to design accurately the theoretical gravitational waveforms from binary systems consisting of a small black hole and a supermassive black hole (so called extreme-mass-ratio binaries), which are possible target sources of gravitational waves for eLISA detection.

Toshinori Matsui (KIAS): "Gravitational waves from the first order electroweak phase transition in the Z_3 symmetric singlet scalar model"

Among various scenarios of baryon asymmetry of the Universe, electroweak baryogenesis is directly connected with physics of the Higgs sector. We discuss spectra of gravitational waves which are originated by the strongly first order phase transition at the electroweak symmetry breaking, which is required for a successful scenario of electroweak baryogenesis. In the Z_3 symmetric singlet scalar model in which the dark matter candidate is included, the significant gravitational waves are caused by the multi-step phase transition. We show that the model can be tested at future gravitational wave interferometers such as eLISA and DECIGO. This talk is based on a project collaborated with Zhaofeng Kang and Pyungwon Ko.

Hyung Won Lee (Inje University): "TaylorF2 CBC waveform with eccentricity corrected phase"

We developed eccentric corrected TaylorF2 waveform and performed to check determinability for eccentricity usign this waveform.

Luis Gabriel Gómez (Sapienza U. Rome & ICRANet): "Gravitational wave emission versus dark matter dynamical friction in the evolution of compact star binaries"

The measured orbital period decay of relativistic compact star binaries, with characteristic orbital periods ~ 0.1~days, is explained with very high precision by the gravitational wave (GW) emission of an inspiraling binary in vacuum predicted by general relativity. However, the binary gravitational binding energy is also affected by an usually neglected phenomenon, namely the dark matter dynamical friction (DMDF) produced by the interaction of the binary components with their respective DM gravitational wakes. Therefore, the inclusion of the DMDF might lead to a binary evolution which is different from a purely GW-driven one. The entity of this effect depends on the orbital period and on the local value of the DM density, hence on the position of the binary in the Galaxy. We evaluate the DMDF produced by three different DM profiles: the Navarro-Frenk-White (NFW) profile, the non-singular-isothermal-sphere (NSIS) and the fermionic Ruffini-Arg\"uelles-Rueda (RAR) model. We first show that indeed, due to their Galactic position, the GW emission dominates over the DMDF in the NS-NS, NS-WD and WD-WD binaries for which measurements of the orbital decay exist. Then, we evaluate the conditions (i.e. orbital period and Galactic location) under which the effect of DMDF on the binary evolution becomes comparable to, or overcomes, the one of the GW emission. In the case of NFW profile, we find that this occurs for orbital periods longer that 18~days for 1.3--0.2 $M_\odot$ NS-WD binaries and around 30~days for both 1.3--1.3~$M_\odot$ NS-NS binaries and 0.25--0.50~$M_\odot$ WD-WD binaries, located at distances 0.1--1.5~kpc from the Galactic center. For closer distances to the Galactic center, the DMDF effect increases and the above critical orbital periods become interestingly shorter.

Gihyuk Cho (Seoul Nat'l U.) Theory of Gravitational radiation

To detect GW in high precision we need to have highly accurate waveform of GWs. I pay attention to this problem in analytic method, specially using post-Newtonian theory. I will show some results and ongoing projects.

Jinn-ouk Gong (APCTP): "Induced gravitational waves"

We present the induced gravitational waves during matter dominated epoch sourced by not only the quadratic combinations of the linear scalar-scalar perturbations, but also tensor-tensor and scalar-tensor perturbations that were not known before. These contributions can be significant in certain frequency windows.

Hisaaki Shinkai (Osaka Inst. Tech.): "Gravitational waves from merging intermediate-mass black holes"

Based on a dynamical formation model of a supermassive black hole (SMBH), we estimate the expected observational profile of gravitational waves at ground-based detectors, such as KAGRA or advanced LIGO/ VIRGO. We estimate the number density of galaxies from the halo formation model and estimate the number of BH mergers from the giant molecular cloud model assuming hierarchical growth of merged cores. At the designed KAGRA (and/or advanced LIGO/VIRGO), we find that the BH merger of its total mass $M \sim 60$ Msun is at the peak of the expected mass distribution. With its signal-to-noise ratio $r = 10$ (30), we estimate the event rate $R \sim 200$ (20) per year in the most optimistic case, and we also find that BH mergers in the range $M < 150$ Msun are $R > 1$ per year for $r = 10$. Thus, if we observe a BH with more than 100 Msun in future gravitational-wave observations, our model naturally explains its source. [ApJ 835 (2017)276]

Sunghoon Jung (*Invited,* Seoul Nat'l U.): "Angular Resolutions of Mid-Frequency Gravitational Wave Detectors"

Opening up the mid-frequency band of f=0.1-10 Hz is a next important goal of gravitational-wave physics. In particular, this band has an ideal balance between gravitational-wave lifetime and frequency for sub-degree angular resolution. We demonstrate how well a mid-frequency detector can localize compact binary mergers and discuss this ideal balance. Although we use an atom interferometer as an example detector in this band, any space detectors can generally achieve similar performance. But terrestrial detectors in this band would be swamped by gravity gradient noise.

Gungwon Kang (KISTI): "Structural analysis of the SOGRO platform"

SOGRO is a conceptual design for a superconducting low-frequency gravitational wave telescope. Its platform has three arms whose ends support six superconducting test masses. Among many required properties for the platform to work well as a main part of the detector, vibrational noises should be outside of the frequency bandwidth in interest (e.g., 0.1~10Hz). We report preliminary results of structural analyses in finite element method for several platform configurations including 30m and 50 m arm lengths.

Dawoo Park (Seoul Nat'l U.): "Black Hole Binaries Dynamically Formed in Globular

Clusters"

Globular clusters (GCs) are an ideal environment to form black hole (BH) binaries. BHs in GC quickly segregate into central region through dynamical friction, then BH-BH binaries can form under high densiy environment. Subsequent interactions of the binaries with other stars or BHs make the binaries tighter. The tightening of binaries means the release of binding energy in the form of kinetic energies of binaries and intereacting stars. Eventually binaries can be ejected from the GC. We investigate properties of BH binaries using direct N-body simulations. We considered the GC with various BH mass population, and studied properties of binaries that are ejected from the clusters. Our results show higher mass BHs are more likely to form BH binaries. The mass ratios of dynamically formed binaries are found to be close to one. The rate of BH binary merger rate density is estimated to be about 10 per year per cubic Gpc, which is consistent with the current estimation based on the detected events by LIGO so far. We further note that the progenitor of the latest BH merger event GW170104 may have been a dynamically formed binary system because the at least the spin axis of one BHs appeared to be non-aligned with the angular momemtum axis.

HanGil Choi (Seoul Nat'l U.): "Finding Quasi-Spherical Orbit Initial condition of Black Hole Binary using Effective-One-body model"

We are developing a iterative method for eccentricity reduction of compact binary inspiral based on the Effective-One-Body model[M. Pürrer et al. Phys. Rev. D. 85, 124051(2012)]. We improve the original method in using more state-of-the-art version of the EOB model, SEOBNRv4[A. Bohé et al. Phys. Rev. D. 95, 044028(2017)]. To cover spinning, precessing binary black hole configurations by SEOBNRv4, we adopt precessing orthonormal frame when integrating equation of motion of EOB instead of cartesian frame.

P5: Inflation and Dark Matter

Hyun Min Lee (Chung-Ang U): "Flattening the inflaton potential beyond the minimal gravity"

We consider inflation models at large field values, motivated by low energy physics to describe the Higgs phenomena such as the SM Higgs mechanism or the B-L symmetry for generating neutrino masses. We show that non-minimal coupling to gravity and/or modified gravity for large inflaton energy density play a crucial role in flattening the inflaton potential at large field values, being consistent with observed CMB data.

Aditya Aravind (APCTP): "Higgs Portal Inflation with Fermionic Dark Matter"

We discuss an inflationary model involving a gauge singlet scalar field and fermionic dark matter added to the standard model. Either the Higgs boson or the singlet scalar could play the role of the inflaton, and slow roll is realized through its non-minimal coupling to gravity. The effective scalar potential is stabilized by the mixing between the scalars as well as the coupling with the fermionic field. Mixing of the two scalars also provides a portal to dark matter. Constraints on the model come from collider searches, dark matter relic density and direct detection and impose a constraining relationship on the masses of dark matter and scalar fields. Inflationary predictions are generically consistent with current Planck data.

Echal Chang (Chungnam Nat'l U.): "Toward a more realistic Randall-Sundrum Brane World"

In the original version of Randall-Sundrum Brane World scenario, they demonstrated the gravity localization/graviton trappon a "Poincare-invariant" 4-dim. Brane world. This is, however, a too much simplification as the "Poincare-invariant" 4-dim. Brane world can only be a cold, empty universe which obviously lacks its realistic, practical nature. We[Hongsu Kim; KASI], therefore, relax this original set-up to a more realistic/practical set-up Where the graviton still gets trapped on a "Ricci-flat" 4-dim. Brane world that, now, may allow for realistic structures like black holes or homogeneous, anisotropic cosmology [i.e., expanding universe] which are, particularly, solutions to the vacuum Einstein equation.

Zhu Yi (Huazhong U. Sci. Tech.): "Nonminimal coupling and inflationary attractors

We show explicitly how the T model, E model, and Hilltop inflations are obtained from the general scalar-tensor theory of gravity with arbitrary conformal factors in the strong coupling limit. We argue that ξ attractors can give any observables ns and r by this method. The existence of attractors imposes a challenge to distinguish different models.

Yongwan Gim (Sogang U.): "On the thermodynamic origin of the initial radiation energy density in warm inflation"

In warm inflation scenarios, radiation always exists, so that the radiation energy density is also assumed to be finite when inflation starts. To find out the origin of the non-vanishing initial radiation energy density, we revisit thermodynamic analysis for a warm inflation model and then derive an effective Stefan-Boltzmann law which is commensurate with the temperature-dependent effective potential by taking into account the non-vanishing trace of the total energy-momentum tensors. The effective Stefan-Boltzmann law shows that the zero energy density for radiation at the Grand Unification epoch increases until the inflation starts and it becomes eventually finite at the initial stage of warm inflation. By using the above effective Stefan-Boltzmann law, we also study the cosmological scalar perturbation, and obtain the sufficient radiation energy density in order for GUT baryogenesis at the end of inflation.

Dong Woo Kang (Yonsei U.): "Electroweak Kaluza-Klein Dark Matter"

In models with universal extra dimensions (UED), the lightest Kaluza-Klein excitation of neutral electroweak gauge bosons is a stable, weakly interacting massive particle and thus is a candidate for dark matter thanks to Kaluza-Klein parity. We examine concrete model realizations of such dark matter in the context of non-minimal UED extensions. The boundary localized kinetic terms for the electroweak gauge bosons lead to a non-trivial mixing among the first Kaluza-Klein excitations of the $SU(2)_W$ and $U(1)_Y$ gauge bosons and the resultant low energy phenomenology is rich. We investigate implications of various experiments including low energy electroweak precision measurements, direct and indirect detection of dark matter particles and direct collider searches at the LHC. Notably, we show that the electroweak Kaluza-Klein dark matter can be as heavy as 2.4 TeV, which is significantly higher than 1.3 TeV as is indicated as an upper bound in the minimal UED model.

Jae-Weon Lee (Jungwon U.): "Brief History of Ultra-light Scalar Dark Matter Models"

I review the brief history of the scalar field dark matter model also known as fuzzy dark matter, BEC dark matter, wave dark matter, or ultra-light axion. In this model ultra-light scalar dark matter particles with mass $m = O(10^{-22})eV$ condense in a single Bose-Einstein condensate state and behave collectively like a classical wave. Galactic dark matter halos can be described as a self-gravitating coherent scalar field configuration called boson stars. At the scale larger than galaxies the dark matter acts like cold dark matter, while below the scale quantum pressure from the uncertainty suppresses the smaller structure formation so that it can resolve the problems of the conventional cold dark matter model.

Sichun Sun (National Taiwan U.): "New views on dark matter from modified gravity"

We discussed a scenario that the dark matter emerge from higher dimensional embedding of the 4-dimensional spacetime, such that the dark matter density can be determined by the Hubble constant, local gravity and the visible matter content. Our approach may give a new viewpoint for Verlinde's emergent gravity from higher dimensions. We also comment on the some phenomenological implications of this type of models, including gravitational wave solutions and MOND limit.

Yasuhiro Yamamoto (Yonsei U.): "Protophobic Light Vector Boson as a Mediator to the Dark Sector"

The observation of a protophobic 16.7 MeV vector boson has been reported by a ^{8}Be nuclear transition experiment. Such a new particle could mediate between the Standard Model and a dark sector, which includes the dark matter. In this paper, we show some simple models of the dark matter which satisfy the thermal relic abundance under the current experimental bounds from the direct and the indirect detections. In a model, it is found that an appropriate self-scattering cross section to solve the small scale structure puzzles can be achieved.

Seodong Shin (Yonsei U.): "Non-minimal dark matter search in dark matter colliders"

I will discuss the relativistic collisions of dark matter in non-minimal dark sector with the targets in current and future neutrino detectors such as Super/Hyper Kamiokande and DUNE, which induces cascade signals in a scenario like inelastic boosted dark matter.

Soo Min Choi (Chung-Ang U): "SIMP dark matter and its cosmic abundances"

Thermal production of light dark matter with sub-GeV scale mass can be attributed to 3→2 self-annihilation processes. We consider the thermal average for annihilation cross sections of dark matter at 3→2 and general higher-order interactions. A correct thermal average for initial dark matter particles is important, in particular, for annihilation cross sections with overall velocity dependence and/or resonance poles. We apply our general results to benchmark models for SIMP dark matter and discuss the effects of the resonance pole in determining the relic density.

Yoo-Jin Kang (Chung-Amg U): "Forbidden Channels and SIMP Dark Matter"

We consider a thermal production of self-interacting dark matter in models with gauged Z3 symmetry. In particular, the 2-to-2 forbidden channels in the dark sector assist producing a SIMP dark matter from thermal freeze-out and extend the parameter space for a correct relic density. We show that dark photon and/or dark Higgs should be relatively light for unitarity and for forbidden channels to work. We identify the constraints on the parameter space of dark matter self-interaction and mass in the case that forbidden channels are important in determining the relic density.

5 July, Wednesday

Plenary Presentations

Bum-Hoon Lee (Sogang U.): "Higher Curvature Gravity"

We consider the properties of the gravity theories with higher curvature term, especially the Gauss-Bonnet term. The black holes carry the scalar charge, which is determined by the mass. There also exists the minimum mass for the existance of the black hole. We also study the cosmological implication by studying the role of the Gauss-Bonnet term during the inflation.

Jun'ichi Yokoyama (U. Tokyo): "Creation of the inflationary universe from a black hole?"

First I revisit the derivation of the Hawking-Moss transition rate between two de Sitter spacetimes. Using the static coordinates I show that the Euclidean action is entirely determined by the contribution of de Sitter entropy. I then discuss creation of a new inflationary universe from an evaporating black hole which produces high energy radiation around it due to the Hawking radiation. A thin wall bubble created around a black hole may tunnel to produce another universe.

Il Hung Park (Sungkyunkwan U.): "Status and Perspectives in GRB observations"

We present the status of and a few selected topics of GRB observations in space and ground are presented. The important issues for future observations will be discussed as well.

Kei-ichi Maeda (Waseda U.): "Inflation and Disformal Transformation"

The accelerating expansion of the Universe is a big mystery in modern cosmology. To solve it, modified gravity theories have been proposed. In this talk, we show that a disformal transformation is useful to analyze such complicated models, especially a slow-roll inflationary scenario. We present some examples, i.e., three Higgs inflationary models (the original Higgs inflation, the new Higgs inflation and the hybrid Higgs inflation).

Jorge Armando Rueda (ICRANet): "On the rate and on the gravitational wave emission of short and long GRBs"

GRBs, traditionally classified as "long" and "short", have been often assumed, till recently, to originate from a single black hole (BH) with an ultrarelativistic jetted emission. There is evidence that both long and short bursts have as progenitors merging and/or accreting binaries, each composed by a different combination of carbon-oxygen cores (CO core), neutron stars (NSs), BHs and white dwarfs (WDs). Consequently, the traditional long bursts have been sub-classified as (I) X-ray flashes (XRFs), (II) binary-driven hypernovae (BdHNe), and (III) BH-supernovae (BH-SNe). They are framed within the induced gravitational collapse (IGC) paradigm which envisages as progenitor a tight binary composed of a CO core and a NS or BH companion. The SN explosion of the CO core, originating a new NS (ν NS), triggers a hypercritical accretion process onto the companion NS or BH. If the accretion is not sufficient for the NS to reach its critical mass, an XRF occurs, leading to a ν NS-NS system. Instead, when the BH is already present or formed by the hypercritical accretion, a BdHN occurs, leading to a ν NS-BH system. Similarly, the traditional short bursts, originating in NS-NS mergers, are sub-classified as (IV) short gamma-ray flashes (S-GRFs) and (V) short GRBs (S-GRBs), respectively when the merging process does not lead or leads to BH formation. Two additional families are (VI) ultra-short GRBs (U-GRBs) and (VII) gamma-ray flashes (GRFs), respectively formed in ν NS-BH and NS-WD mergers. We use the estimated occurrence rate of the above sub-classes to assess the gravitational wave emission in the merging process and its detectability by Advanced LIGO, Advanced Virgo, eLISA, and resonant bars.

6 July, Thursday

Plenary Presentations

Yaakov (Yasha) Neiman (Perimeter Institute): "Towards a working model of causal-patch physics in dS/CFT"

Quantum gravity with de Sitter (dS) asymptotics is a major open question, both conceptually and technically. One possible avenue is to adapt the holographic framework of AdS/CFT to the de Sitter setup. Working models of dS/CFT are hard to come by, but one such model has been found in 4 dimensions - a duality between higher-spin gravity in the bulk and a free vector model on the boundary. I will describe my attempts, within this model, to bridge the gap between the CFT at future de Sitter infinity and physics in the causal patch of a bulk observer. In the process, I will present some new results on the relationship between holography and twistor theory.

Rong-Gen Cai (Chinese Academy of Science): "Standard sirens and dark sector with Gaussian process"

Gravitational wave from a binary system is a standard siren to probe cosmological evolution history. In this talk, I will first show how to the reconstruction of dark matter and dark energy by using the SINa union 2.1 data, as an application of the Gaussian process. Then I will apply the method to reconstruct the interaction between dark matter and dark energy from mock data from Lisa and to show the constraint ability on the cosmological parameters with the mock data from Einstein Telescope.

Daniel Grumiller (Vienna U. of Technology): "Soft Heisenberg Hair"

Soft hair"" refers to zero energy excitations in the near horizon region of black holes or cosmologies, advocated by Hawking, Perry and Strominger. I review recent results on soft hair in 3 spacetime dimensions. The near horizon symmetry algebra is surprisingly simple, namely infinite copies of the Heisenberg algebra. The results are fairly universal and allow a semi-classical construction of microstates for BTZ black holes, whose degeneracy correctly accounts for their Bekenstein-Hawking entropy.

Jeong-Hyuck Park (Sogang U.): "Stringy Gravity & solution to dark matter problem"

Recent development in string theory has led to the extension of General Relativity, i.e. Stringy Gravity. It postulates the entire closed string massless sector to be geometric and thus gravitational. I will first introduce the mathematical foundation and then discuss a solution to dark matter/energy problems. In terms of R/(MG), i.e. the dimensionless radial variable normalized by mass, Stringy Gravity agrees with General Relativity near infinity, but modifies it at short distance. At far short distance, gravitational force can be even repulsive. These may solve the dark matter and energy problems, as they essentially arise from small R/(MG) observations: far distance divided by much heavier mass.

Chiang-Mei Chen (Nat'l Central U.): "Pair Production of Scalar Dyons in Kerr-Newman Black Holes"

We study the spontaneous pair production of scalar dyons in the near extremal dyonic Kerr-Newman (KN) black hole, which contains a warped AdS$_3$ structure in the near horizon region. The leading term contribution of the pair production rate and the absorption cross section ratio are also calculated using the Hamilton-Jacobi approach and the thermal interpretation is given. In addition, the holographic dual conformal field theories (CFTs) descriptions of the pair production rate and absorption cross section ratios are analyzed both in the \$J\$-, \$Q\$- and \$P\$-pictures respectively based on the threefold dyonic KN/CFTs dualities.

P3: Astrophysics - Relativistic astrophysics, Observation, Cosmic rays, GRB, etc.

Anna Pollmann (*Invited,* U. Wuppertal): "Searches for magnetic monopoles with IceCube"

Particles that carry a single magnetic charge are proposed by various theories which go beyond the Standard Model of particle physics. The expected mass of the magnetic monopoles varies depending on the theory describing its origin, but generally the monopole mass exceeds those possible by direct creation in accelerators by far. In large scale magnetic fields, magnetic monopoles can gain kinetic energy up to relativistic velocities. IceCube is a high energy neutrino detector using the clear ice close to the South Pole as the detection medium. Relativistic magnetic monopoles induce Cherenkov or luminescence light production while passing through the ice. Depending on the monopole kinetic energy, light can also be induced by secondary particles which result from ionization or the proposed catalysis of nucleon decay by magnetic monopoles. A review of possible detection methods and the limits for the monopole flux over a broad parameter range are described for the IceCube Neutrino Observatory, the most sensitive detector available for magnetic monopole searches.

Insik Hahn (Ewha Womans U.): "Nuclear astrophysics experiments related to the rp-process using rare isotope beams"

Nuclear astrophysics experiments using radio isotope beams have been very successful to understand some of the important reactions related to explosive hydrogen burning process. Some reactions such as $^{14}O(a,p)^{17}F$, $^{15}O(a,g)^{19}Ne$, $^{18}F(p,a)^{15}O$ are important for studying the breakout mechanism from the HCNO cycle to the rp process. The above reactions were investigated using radioactive ion beams at the Center for Nuclear Study in RIKEN. We plan to study the structures of neutron-deficient ^{66}Se and ^{56}Zn relevant to the rp-process at RIKEN. Korea will build the radioactive ion beam accelerator facility called RAON. This new facility is expected to carry out nuclear astrophysics experiments for understanding explosive nuclear synthesis in stellar sites such as X-ray bursts and novae as well as the origin of proton-rich elements in the rp-process. Current activities and prospects of nuclear astrophysics in Korea will be discussed.

Myung-Ki Cheoun (*Invited,* Soongsil U.): "Modified Gravity in Nuclear Astrophysics and Big Bang Nucleosynthesis"

We discuss roles of gravitation in the compact cosmological objects like neutron star because equations of state turn out to be so sensitive on the gravity and the modified gravity. In this talk, we also make a brief summary about the equations of state which is now widely and intensively discussed in the nuclear astrophysics. Finally, the Big-Bang nucleosynthesis will be shown to be a useful test of the gravitation as well as the new physics beyond the standard model.

Chang-Hwan Lee (Pusan Nat'l U.): "Strangeness in Neutron Star Cooling"

I this talk, I will review our recent work on the thermal evolution of neutron stars in the resence of hyperons or kaons in the core. Our results indicate that the nucleon and hyperon direct Urca processes play crucial roles for macroscopic cooling behavior of neutron stars. There are possibilities that the neutron star equation of states with hyperons and kaons are consistent with both mass and temperature observations. However, requirement of the fine-tuning of the parameters to explain the observations is still an open problem.

Hyun Kyu Lee (Hanyang U.): "Hadronic matter with emergent symmetries at high density"

The neutron stars with high core density have been considered among the most relevant objects for the formaton of GRB centers and gravitational radiations. It is expected that the detection of GRBs and gravitational radiations provide us the opportunity of constraing the EoS of extreme high dense hadronic matter at the core of neutron stars. The possibilty of emergent scale symmetry along with local flavor symmetry will be discussed in relation to EoS with GRB and gravitational wave detections.

Hongjun An (*Invited,* Chungbuk Nat'l U.): "Studying intrabinary shock emission in pulsar binaries"

We present our studies of intrabinary emission of the gamma-ray binary 1FGL J1018.6-5856. We measure the light curve and phase-resolved spectra, and model them using an intrabinary shock emission model. The model assumes a slow and a fast population of particles accelerated in the shock, and computes the high-energy emission spectra produced by synchrotron and self-Compton processes of the high-energy particles in the shock. Fitting the data with the model allows us to constrain the binary geometry, most importantly the inclination angle. This model can be used for other pulsar binaries and help to determine the inclination angle of the binary hence the mass of the neutron star.

Dong-Hoon Kim (Seoul Nat'l U.): "Pulsar radiation with general relativistic effects"

We consider a magnetic dipole model of a pulsar and investigate general relativistic effects on electromagnetic radiation from the pulsar. The general relativistic modifications should be found applicable to many well-known issues in pulsar astronomy. Among other things, the modifications of Goldreich-Julian model and subpulse drift would be challenging issues of significant interest. The electromagnetic fields in the pulsar magnetosphere are computed by solving Maxwell's equations defined in the strongly curved spacetime around the pulsar, hence containing the properties of the strong gravitational effect. In combination with this effect, the effects from rotation and obliqueness of the pulsar are investigated so that we work out general-relativistic extensions of Goldreich-Julian model and subpulse drift.

Alexander Yushchenko (Sejong University): "Velocity distribution in the clusters of galaxies"

Using SLOAN survey we analyzed the distribution of galaxy redshift in five hundred clusters of galaxies. The mean profiles were constructed for different redshift intervals. It was found that rich clusters have members with discordant redshifts.The possible interpretations are discussed.

Hisaaki Shinkai (Osaka Inst. Tech.): "Nonlinear dynamics in Gauss-Bonnet gravity"

We numerically investigated how the dynamics depends on the dimensionality and how the higher-order curvature terms affect to singularity formation in two models: (i) perturbed wormhole in spherically symmetric space-time, and (ii) colliding scalar pulses in planar space-time. Our numerical code uses dual-null formulation, and we compare the dynamics in 5, 6 and 7-dimensional General Relativity and Gauss-Bonnet (GB) gravity. Both results suggest that GB correction works for avoiding singularity formation in their dynamics. We also found that the existence of the trapped surface in GB gravity does not directly indicates formation of a black hole.

Andreas Krut (ICRANet): "Dark matter and galactic structures"

A self-gravitating system composed of massive fermions in spherical symmetry is considered. We solve the Einstein equation for a thermal and semi-degenerate fermionic gas considered as a perfect fluid in hydrostatic equilibrium. No additional interaction is assumed for the fermions besides their fulfilling of quantum statistics and the relativistic gravitational equation. Within this more general approach a new family of density profiles arises which explains DM halo constraints of the Galaxy and provides at the same time an alternative to the central black hole scenario. Further, we show predictions of that model for galactic structures from dwarf spheroidals to big ellipticals in agreement with observations.

P6: Gravity in String Theory

Jae-Hyuk Oh (Hanyang U.): "Phase transition in anisotropic holographic superfluids with arbitrary z and α"

Einstein-dilaton-U(2) gauge field theory is considered in a spacetime characterised by α and z, which are the hyperscaling violation factor and the dynamical critical exponent respectively. We obtain the critical values of chemical potential μc that is defined on its boundary dual fluid and derives phase transition from spatially isotropic to anisotropic phase for the various values of the α and z. To do so, we first apply Sturm-Liouville theory and estimate the upper bounds of the critical values of the chemical potential. We also employ a numerical method in the ranges of $1 \leq z \leq 4$ and $0 \leq \alpha \leq 4$ to check if the Sturm-Liouville method correctly estimates the critical values of the chemical potential. It turns out that the two methods are agreed within 10 percent error ranges. Finally, we compute free energy density of the dual fluid by using its gravity dual and check if the system shows phase transition at the critical values of the chemical potential μc for the given parameter region of α and z. Interestingly, it is observed that the anisotropic phase is more favoured than the isotropic phase for small values of z and α. However, for large values of z and α, the anisotropic phase is not favoured.

Run-Qiu Yang (KIAS): "Strong energy condition and complexity growth bound in holography"

In this talk, I will discuss a result about strong energy condition and action growth bound in holographic complexity. I will prove that if eternal neutral black holes satisfy some general conditions and matter fields only appear in the outside of the Killing horizon, the strong energy condition is a sufficient condition to insure that the vacuum Schwarzschild black hole has the fastest action growth of the same total energy. This result is consistent with the bound of computational complexity growth rate and gives a strong evidence for the holographic complexity-action conjecture.

Yun-Long Zhang (APCTP): "Bell inequality from holographic gravity"

We show that the manifestation of quantum entanglement in Bell inequality can be reproduced from the gravitational theory in the bulk, and the CHSH formula of Bell inequality can be related with the holographic Schwinger-Keldysh correlator from AdS gravity.

Sunyoung Shin (CQUeST): "Walls of massive Kähler sigma models on SO(2N)/U(N) in three dimensions"

We construct walls of massive Kähler sigma models on SO(2N)/U(N) in three dimensions by using moduli matrices. We discuss the operators generating walls and the root vectors.

Chanju Kim (Ewha Womans U.): "Holographic micro state deformations of BTZ black holes"

We find general deformations of BTZ spacetime and identify the corresponding thermofield initial states of the dual CFT. We deform the geometry by introducing bulk fields dual to primary operators and find the back-reacted gravity solutions to the quadratic order of the deformation parameter.

O-Kab Kwon (Sungkyunkwan U.): "Exact holography of massive M2-brane theories and entanglement entropy"

We obtain an exact holographic relation for the vacuum expectation values of the chiral primary operator with conformal dimension 1 in the large N limit in mass-deformed ABJM theory. Our results involve infinite number of exact dual relations for all possible supersymmetric Higgs vacua and so provide a nontrivial test of gauge/gravity duality away from the conformal fixed point in the large N limit. Using the Ryu-Takayanagi conjucture to obtain the entanglement entropy in gravity side, we calculate the holographic entanglement entropy (HEE) in the reduced 4-dimensional gravity theory. We also obtain constraints for the HEE, which is the counterpart of the perturbed Einstein equations.

Kyung Kiu Kim (Sejong U.): "Thermodynamic volume in AdS/CFT"

The cosmological constant in AdS space can be interpreted as a pressure of the bulk. If the variation of the cosmological constant is allowed, the pressure has a dual thermodynamic volume. We study on the physical meaning of the variation and related quantities through the standard AdS/CFT dictionary.

Taejin Lee (Kangwon Nat'l U.): "Gravitational Scattering Amplitudes and Closed String Field Theory in the Proper-Time Gauge"

We construct a closed string field theory in the proper-time gauge which is the closed analog of the deformed cubic open string field theory and define the general closed string scattering amplitudes. Taking the zero-slope limit, we explicitly evaluate the three-graviton scattering amplitudes and the four-graviton scattering amplitudes. We discuss in the framework of the closed string field theory, the Kawai-Lewellen-Tye (KLT) relations, which relate the tree level closed string scattering amplitudes to those of open string.

Hongsu Kim (KASI): "3rd Quantization of Taub Universe"

In the present work, we address the issue of Quantum Field Theory of Taub Cosmology. To be more concrete, we describe this issue in more detail one at a time: "Taub Cosmology model [A.H. Taub, Am. Math. 53,472 (1951)] can be thought of as a special limit of the Bianchi-type IX (Mixmaster) cosmology. Namely, it is a homogeneous but anisotropic cosmology model. Next, in a physical system consisting of matter and gravity(spacetime), we 1st[quantum mechanics] and 2nd[quantum field theory] quantize the matter fields to begin with. Then we turn to gravity and 1st[universe wave function as a solution to the Wheeler-DeWitt eq.] and 2nd[promote the universe wave function to a field operator a. la. Hosoya-Morikawa] quantize the spacetime. Now the name, "3rd quantization" manifests itself as the gravity is 2nd quantized only after the matter in it gets 2nd quantized beforehand!, Lastly, in order to work it out, the Bogoliubov transformation comes into play...
That is, the choice of mode functions to span the universe field operator is NOT unique as each set of mode functions is separately complete, orthonormal (in the Hilbert space). This is because the physical state of the system (subject to the mode expansions) keeps changing during its (time) evolution. As a result, the Bogoliubov transformations come into play and

result in non-trivial Mode (or, frequency) mixing that, in turn, leads to the notion of vacuum and creation, annihilation of universe(quanta) in its Hilbert space…
To summarize, even the universe itself can be treated as quanta that undergo "birth and destroy".

Hongsu Kim (KASI): "Callan-Rubakov Effect could answer the baryon asymmetry puzzle"

Back in 1982,Callan and Rubakov, independently, pointed out that the "t'Hooft-Polyakov magnetic monopole", a topological soliton solution in Yang-Mills-Higgs theory with SSB in WGS standard model for the electroweak interaction could "catalyze" the proton decay. This is obviously an academic speculation which has NOT been motivated by any precision observational evidence at the time as the proton is known to be definitely stable &secure with its lifetime being as long as the age of the universe itself! As such, we find, after 30 years, that this effect is not just a purely academic speculation, but also a severely wrong-motivated set-up, as now we know the matter-antimatter(or baryon) asymmetry in our present universe… That is, in the present universe, we have severe matter-antimatter(or baryon) asymmetry in which, the number of proton is overwhelmingly greater than that of antiproton… This reality teaches us the lesson that after all, it is "anti proton" rather than proton that should(spontaneously) decay and disappear! To summarize, the age-old Callan-Rubakov Effect has to be modified to its opposite objective; "monopole-catalyzed antiproton decay"!
In the present work, therefore, we briefly revisit and review the original version of the Callan-Rubakov effect and discuss how it should be modified and reorganized for our new perspective and new objective: "monopole-catalyzed antiproton decay"

7 July, Friday

Plenary Presentations

Jong-Ping Hsu (UMass): "Experiments on the CMB Spectrum and Their Implications for the Missing Half of the Universe"

Based on the concept of Lorentz-Poincar\'e invariance and its limiting continuation, we propose an alternative formulation of the generalized Planck distribution that leads to a new physical interpretation of the observed anisotropy of the Cosmic Microwave Background (CMB). We also discuss how data from satellite experiments might be used to search and discover the missing half of the universe, i.e., an extremely distant `antimatter-blackbody,' through accurate measurements of the CMB spectra towards and away from the constellation Leo.

Dieter Rudolf Brill (U. Maryland): "The revival of General Relativity at Princeton."

After General Relativity was established in essentially its present form in 1915 it was celebrated as a great success of mathematical physics. But the initial hopes for this theory as a basis for all of physics began to fade in the next several decades, as General Relativity was relegated to the margins of theoretical physics. Its fate began to rise in the 1950's in a revival of interest and research that over time made gravitational physics one of the hottest research topics it is today. One center of this renaissance was Princeton, where two relative newcomers explored new and different approaches to gravitational physics. Robert Dicke showed that gravity is not as inaccessible to experiment as was thought, and John Wheeler propelled it into the mainstream by proposing highly original and imaginative consequences of Einstein's theory. We will concentrate on these ideas that, in his characteristically intriguing style, Wheeler called ""Daring Conservatism"" -- a term well known to his associates, but one he never menstioned in print. With the aid of unpublished manuscripts and notes we will explore Daring Conservatism's origin and motivation, its successes and failures, and the legacy it left behind. In passing we will encounter a small footnote to the geon problem.

Yong-Seon Song (KASI): "Cosmological Implications of RSD"

The mapping of dark matter clustering from real space to redshift space introduces the anisotropic property to the measured density power spectrum in redshift space, known as the redshift space distortion effect. The mapping formula is intrinsically non-linear, which is complicated by the higher order polynomials due to indefinite cross correlations between the density and velocity fields, and the Finger--of--God effect due to the randomness of the peculiar velocity field. Whilst the full higher order polynomials remain unknown, the other systematics can be controlled consistently within the same order truncation in the expansion of the mapping formula, as shown in this paper. The systematic due to the unknown non--linear density and velocity fields is removed by separately measuring all terms in the expansion directly using simulations. The uncertainty caused by the velocity randomness is controlled by splitting the FoG term into two pieces, 1) the ``one--point" FoG term being independent of the separation vector between two different points, and 2) the ``correlated" FoG term appearing as an indefinite polynomials which is expanded in the same order as all other perturbative polynomials. Using 100 realizations of simulations, we find that the Gaussian FoG function with only one scale--independent free parameter works quite well, and that our new mapping formulation accurately reproduces the observed 2--dimensional density power spectrum in redshift space at the smallest scales by far, up to $k\sim 0.2\hompc$, considering the resolution of future experiments.

Eunil Won (Korea U.): "Status of the GroundBIRD experiment"

The GroundBIRD experiment, a ground-based telescope for the detection of the B-mode of the CMB polarization down to the tensor to scalar ratio of 0.01 is under construction at KEK in Japan. The observation site is Tenerife Island in Spain and the telescope is planned to be moved to the site in early 2018. We discuss present construction details and prospects for the future observation.

Marco Muccino (ICRANet) "What can we learn from gamma-ray bursts?"

Gamma-ray busts (GRBs) are powerful transient sources spanning a wide range in their energy release, duration and redshift distributions. These properties makes GRBs perfect candidates for studying their rates and the possible progenitor systems.

On behalf of a large collaboration, I review the major results obtained in the past two years: 1) the classification of GRBs into seven different sub-classes, each one characterized by distinct set of observational features, different astrophysical systems as progenitors and occurrence rates; 2) the assessment of the gravitational wave emission for some of the above sub-classes and its detectability by Advanced LIGO, Advanced Virgo, eLISA, and resonant bars.

P7: Cosmology – Cosmology, CMB, Perturbations, Structure formation, Quantum effect, etc.

Seokcheon Lee (Gyeongsang Nat'l U.): "Conformal equality"

It is well known that cosmological observables (like Primordial curvature perturbation and Hubble parameters) are frame independent (under conformal transformation) at early Universe. However, late time cosmological observables are frame dependent and issues for physical frame are open. We discuss these issues both in classical level and in quantum one.

Dong-han Yeom (Nat'l Taiwan U.): "artle-Hawking wave function and large-scale power suppression of CMB"

In this talk, I will first describe the Hartle-Hawking wave function in the Euclidean path integral approach. After we introduce perturbations to the background instanton solution, following the formalism developed by Halliwell-Hawking and Laflamme, one can obtain the scale-invariant power spectrum for small-scales. We further emphasize that the Hartle-Hawking wave function can explain the large-scale power suppression by choosing suitable potential parameters, where this will be a possible window to confirm or falsify models of quantum cosmology. Finally, we further comment on possible variations of background solutions, e.g., Euclidean wormholes, which can result in distinct signatures to the power spectrum.

Mu-In Park (Sogang U.): "On Gauge Invariant Cosmological Perturbations in UV-modified Horava Gravity"

We revisit gauge invariant cosmological perturbations in UV-modified, $z = 3$ Horava gravity with one scalar matter field, which has been proposed as a renormalizable gravity theory without the ghost problem in four dimensions.

Tae Hoon Lee (*Invited,* Soongsil U.): "Quintessence in low-energy effective theory"

Considering a theory of scalar-tensor gravity with non-renormalizable couplings of heavy fields, we derive the low-energy effective theory action in the universe of temperature much lower than the heavy field mass. An effective potential of the scalar field is induced through the heavy field interactions in the late-time universe and we find a de Sitter cosmological solution with the potential of the scalar field. We also investigate stability properties of the solution.

Ki-Young Choi (Chonnam Nat'l U.): "Lower-Bound on the reheating temperature with dark matter"

We suggest new lower-bound on the reheating temperature from the cosmological evolution of the dark matter.

Kyungjin Ahn (Chosun U.): "Small-scale structure formation under the large-scale density and CDM-baryon drift velocity environment

The cosmological structure formation, if in the mildly nonlinear regime, cannot be well estimated by the linear approximation. Even in the regime that used to be believed to be linear, Tseliakhovich and Hirata have shown that small-scale structure formation can be affected by the large-scale, CDM-baryon drift velocity environment and thus the linear approximation is invalid. We show that an even stronger effect comes from the large-scale density environment (ApJ 830:68). Therefore, how small-scale structure evolves is determined by the competition between the density environment and the CDM-baryon drift-velocity environment. In terms of the nonlinear objects, minihalos whose mass is about 10^8 solar mass or below, are the ones that are strongly affected. This may also affect the formation of more massive halos that are usually used for galaxy surveys through suppressed merger, so cosmology from galaxy surveys may have to be examined more carefully. We also present BCCOMICS, a cosmological initial condition generator for numerical simulations, which includes this effect and thus surpasses the accuracy of the linear Boltzmann solver such as CAMB.

Sungwook E Hong (KASI): "Small-scale Features of Thermal Inflation: CMB Distortion and Substructure Abundance"

Thermal inflation is an additional inflationary mechanism before the big bang nucleosynthesis, which solves the moduli problem and naturally provides a plausible dark matter candidate. Thermal inflation leaves a slight enhancement followed by huge suppression of a factor of ~50 in the curvature and matter power spectrum, which can be expressed in terms of a single characteristic scale $k_{\rm b}$. Here we describe the observability of the small-scale features of thermal inflation from various observations, such as CMB distortion, satellite galaxy abundance in the Milky-Way-sized galaxies, and 21-cm power spectrum before the epoch of reionization.

Maurice H van Putten (*Invited,* Sejong U.): "Fast evolution of the deceleration parameter in surveys of H(z)"

Redshift evolution Q(z)=dq/dz of the deceleration parameter q(z) is shown to satisfy Q0\simeq 2.8 in recent heterogeneous data on H(z) over $0 < z < 2$. This observation rules out Q0\lesssim1 in LambdaCDM at a level of confidence of 4.36\sigma.

Muhammad Sharif (U. Punjab): "Stability Analysis of Bulk Viscous Cosmology"

This paper explores the non-equilibrium behavior of thermodynamics at the apparent horizon of isotropic and homogeneous universe model in f(G,T) gravity (G and T represent the Gauss-Bonnet invariant and trace of the energy-momentum tensor, respectively). We construct the corresponding field equations and analyze the first as well as generalized second law of thermodynamics in this scenario. It is found that an auxiliary term corresponding to entropy production appears due to the non-equilibrium picture of thermodynamics in first law. The universal condition for the validity of generalized second law of thermodynamics is also obtained. Finally, we check the validity of generalized second law of thermodynamics for the reconstructed f(G,T) models (de Sitter and power-law solutions). We conclude that this law holds for suitable choices of free parameters.

P8: Black Holes – Black holes, Wormholes, etc

Masakatsu Kenmoku (Nara Sci. Academy): "Superradiant Phenomena for Spinor Fields in Rotating Black Hole Geometry"

We derive the results (i) non-existence of zero modes and (ii) the completeness relation for spinor fields in rotating black hole geometry. From these results, we show the superradiant phenomena for spinor fields should be type 2: positive momentum on the horizon (p_H>0) and negative frequency at infinity (\omega<0).

Khalid Saifullah (Quaid-i-Azam U.): "Thermodynamics of rotating non-Kerr black holes"

Some rotating black holes, other than the famous Kerr solution, have attracted a lot of attention in the literature. In this talk the thermodynamical properties of some of these non-Kerr black holes are discussed. First slow rotation and then full rotation has been studied. The relationship between the Hawking temperature and the radius of the black holes is investigated and their physical interpretations discussed. This analysis gives rise to some interesting features of these objects.

Bogeun Gwak (*Invited*, Sejong U.): "Upper Limit of Radiation from Coalescence of Rotating Hayward Black Holes"

We have studied the spin interaction and the gravitational radiation thermally allowed in a head-on collision of two rotating Hayward black holes. The Hayward black hole is a regular black hole in the modified Einstein equation. The potential of the spin interaction can be analytically obtained. For the collision of massive black holes, the gravitational radiation is numerically obtained as the upper bound using the laws of thermodynamics. The effect of the Hayward black hole tends to increase the radiation energy, but we can limit the effect by comparing it with the gravitational waves.

Sung-Won Kim (Ewha Womans U.): "Wormhole cosmological model"

As the McVittie derived the black hole solution in a cosmological model, we derive the cosmological model with a wormhole at the origin of the "cosmical" coordinate system by solving the Einstein's equation with proper matter distribution. We also find the apparent cosmological horizon from the redefined metric and derive the Hawking temperature by using the conventional methods.

Jin Young Kim (Kunsan Nat'l U.): "Wormholes in Einstein-Born-Infeld Gravity"

We study a new approach for the wormhole construction in Einstein-Born-Infeld theory with a cosmological constant, which does not require exotic matters in the Einstein equation. We study the relation of the newly introduced conditions with the usual continuity for the energy-momentum tensor and the gravitational Bianchi identity. We find that there is no violation of energy conditions for the Born-Infeld fields contrary to the usual approaches. The exoticity of energy-momentum tensor is not essential for sustaining the wormholes.

Kimitake Hayasaki (*Invited,* Chungbuk Nat'l U.): "Detection of gravitational wave emission by supermassive black hole binaries through tidal disruption flare"

Galaxy mergers produce binaries of supermassive black holes, which emit gravitational waves prior to their coalescence. We perform the hydrodynamic simulations to study the tidal disruption of stars by such a binary in the final centuries of its life. We find that the gas stream of the stellar debris moves chaotically in the binary potential and forms accretion disks around both black holes. The accretion light curve is modulated over the binary orbital period owing to relativistic beaming. This periodic signal allows to detect the decay of the binary orbit due to gravitational wave emission by observing two tidal disruption events that are separated by more than a decade.

Medeu Abishev (Al Farabi Kazakh Nat'l U.): "Dilatonic dyon black hole solutions in the model with two Abelian gauge fields"

Dilatonic black hole dyon solutions in the gravitational 4d model with a scalar field, two 2-forms, two dilatonic coupling constants λi 6= 0, i = 1, 2, obeying $\lambda 1$ 6= $-\lambda 2$ and sign parameter $\varepsilon = \pm 1$ for scalar field kinetic term are considered. Here $\varepsilon = -1$ corresponds to ghost scalar field. These solutions are defined up to solutions of two master equations for two moduli functions, when $\lambda 2i$ 6= 1/2 for $\varepsilon = -1$. Some physical parameters of the solutions: gravitational mass, scalar charge, Hawking temperature, black hole area entropy and parametrized post-Newtonian (PPN) parameters β and γ are obtained. PPN parameters do not depend on couplings λi and ε. A set of bounds on gravitational mass and scalar charge are found by using a certain conjecture on parameters of solutions, when $1 + 2\lambda\ 2i\varepsilon > 0$, i = 1, 2. The composite dyon solution with $\lambda 1 = \lambda 2$ was considered in ref. [1]. References [1] M.E. Abishev, K.A. Boshkayev, V.D. Dzhunushaliev and V.D. Ivashchuk, Dilatonic dyon black hole solutions, Class. Quantum Grav. 32, No. 16, 165010 (2015).

Seung Hun Oh (Konkuk U.) "Poisson algebra of gauge-invariant quasilocal angular momentum and its asymptotic limit"

In this talk, we will discuss the previously proposed quasilocal angular momentum of gravitational field in the absence of isometries in detail, and find the condition in which it is gauge-invariant. The gauge-invariant angular momentum has the following attractive properties; (i) it follows from one of the Einstein's constraint equations, (ii) it satisfies the Poisson algebra $\{L(\xi), L(\eta) \}{\rm P.B.} = L([\xi, \eta]_{\rm L})$, up to a constant normalizing factor, (iii) its Poisson algebra reduces to the standard $SO(3)$ algebra of angular momentum at null infinity, and (iv) it reproduces the standard value for Kerr spacetime at null infinity. It will be argued that our angular momentum is a quasilocal and canonical generalization of A. Rizzi's geometric definition.

Deniz Olgu Devecioglu (Sogang U.): "Lifshitz black holes in Einstein-Yang-Mills theory"

We find that the four-dimensional cosmological Einstein-Yang-Mills theory with SU(2) gauge group admits Lifshitz spacetime as a base solution for the dynamical exponent $z>1$. Motivated by this, we next demonstrate numerically that the field equations admit black hole solutions which behave regularly on the horizon and at spatial infinity for different horizon topologies. The solutions depend on one parameter, the strength of the gauge field at the horizon, which is fine-tuned to capture the Lifshitz asymptotics at infinity. We also discuss the behavior of solutions and the change in Hawking temperature for black holes that are large or small with respect to the length scale L, which is itself fixed by the value of the cosmological constant.

Lunchakorn Tannukij (Hanyang U.): "Black holes in dRGT Massive Gravity"

Massive gravity is a gravity theory in which the gravitational force is described through a concept of having massive graviton as a force carrier. One of the successful massive gravity theory is known as the dRGT massive gravity in which mass is given to the graviton in a special way so that there is no theoretical pathologies in the theory. By computing its solution, we can obviously see differences between massive gravity and general relativity in an astronomical aspect. In this talk, we present both a sperically symmetric solution and a cylindrically symmetric solution to the dRGT theory. Metric functions of those solutions appear to incorporate a characteristic which can be interpreted as an effective cosmological constant while they also contain a linear term which cannot be obtained from the general relativity. Moreover, some thermodynamical properties like temperatures and heat capacities of those solutions will be mentioned.

Awards for Best Student Presentations

* Best student presentations:

- **Daria Primorac** (ICRANet & Sapienza U. Roma) for *"Analysis of the GRB110731A within the fireshell model"*
- **Zhu Yi** (Huazhong U. Sci. Tech.) for *"Nonminimal coupling and inflationary attractors"*
- **Gihyuk Cho** (Seoul Nat'l U.) for *"Theory of Gravitational Radiation"*
- **Rahim Moradi** (Sapienza U. of Rome and ICRANet) for *"Charged Cosmological Black holes"*
- **Yongwan Gim** (Sogang U.) for *"On the thermodynamic origin of the initial radiation energy density in warm inflation"*

* Selection committee:

Gungwon Kang (KISTI, Chair), Gregory Vereshchagin (ICRANet), Hideki Maeda (Hokkai-Gakuen U.), Muhammad Sharif (U. of Punjab), Jae-Weon Lee (Jungwon U.) and Yungui Gong (Huazhong U. of Sci. Tech.)

The selection was truly based on the excellence of research work and presentation. Five presentations were selected out of 18 in total. A certificate and 100,000KRWs have been awarded to each student.

Group Picture

List of Participants

Name	Affiliation
Abdul WAHAB	KAIST
Aditya Aravind	APCTP
Alex Giacomini	Universidad Austral de Chile
Alexander Yushchenko	Sejong University
Alexey Toporensky	Sternberg Astronomical Institute
Andreas Krut	ICRANet
Anna Pollmann	University of Wuppertal
Atsushi NARUKO	Tohoku University
Bogeun Gwak	Sejong University
Bum-Hoon Lee	Sogang University
Chan Park	KISTI
Chang-Hwan Lee	Pusan National University
Chanju Kim	Ewha Womans University
Chanyong Park	APCTP
Chao Niu	Gwangju Institute of Science and Technology (GIST)
Cheng-Yi Sun	Yukawa Institute for Theoretical Physics, Kyoto University
Chiang-Mei Chen	National Central University
Chulmoon Yoo	Nagoya University
Chunglee Kim	KASI
Daeho Ro	APCTP
Daniel Grumiller	ITP at Vienna University of Technology
Daria Primorac	ICRANet and Dipartimento di Fisica, Sapienza Universia di Roma
Dawoo Park	Seoul National University
Deniz Olgu Devecioglu	Sogang Uni.
Dieter Rudolf Brill	
Dong Woo Kang	Yonsei University
Dong-han Yeom	Leung Center for Cosmology and Particle Astrophysics, National Taiwan University

Dong-Hoon Kim	Seoul National University
Echal Chang	Chungnam national univ.(CNU)
Ehsan Bavarsad	Department of Physics, University of Kashan
Eoin O Colgain	APCTP
Eun ho Lee	Hanyang University
Eunil Won	Korea University
Gansuk Tumurtushaa	IBS-CTPU
Gihyuk Cho	Seoul National University
GiMin Bae	DGIST
Giovanni Battista Pisani	Sapienza University of Rome and ICRANet
Gregory Vereshchagin	ICRANet
Gungwon Kang	KISTI (Korea Inst. of Science and Technology Information)
Hang Bae Kim	Hanyang University
HanGil Choi	Seoul National University
Hee Il Kim	Seoul National University
Heeseung Zoe	DGIST
Hideki Maeda	Hokkai-Gakuen University
Hisaaki Shinkai	Osaka Institute of Technology
Ho Jung Paik	University of Maryland
Hongjun An	Chungbuk National University
Hongsu Kim	Korea astronomy and space science institute(KASI)
Hossein Yavartanoo	ITP/CAS
Hyeong-Chan Kim	Korea National University of Transportation
Hyun Kyu Lee	Hanyang University
Hyun Min Lee	Chung-Ang University
Hyun Seok Yang	Center for Quantum Spacetime, Sogang University
Hyung Mok Lee	Seoul National University
Hyung Won Lee	Inje University
IL HUNG PARK	Sungkyunkwan University (SKKU)
Insik Hahn	Ewha Womans University
Inyong Cho	SeoulTech
Jae-Hyuk Oh	Hanyang University
Jae-Weon Lee	Jungwon University
Jeongcho Kim	Inje University.

Jeong-Hyuck Park	Sogang University, Seoul
Jia-Rui Sun	Sun Yat-Sen University
Jin Young Kim	Kunsan National University
Jingbo Wang	Xinyang Normal University,
jinn-ouk gong	apctp
Jiwon Park	Kunsan National University
John J. Oh	National Institute for Mathematical Sciences
Jong Hyuk Yoon	Konkuk University
Jong-Ping Hsu	University of Massachusetts Dartmouth
Joongoo Lee	Seoul National University
Jorge Armando Rueda Hernandez	ICRANet
Julio David Melon Fuksman	Sapienza University of Rome and ICRANet
Jun Peng	Beijing Normal University
Jun\'ichi Yokoyama	RESCEU, The University of Tokyo
Jungjai Lee	Daejin University
JUNGWON LIM	DGIST
Juyoung Lee	SKKU
Kei-ichi MAEDA	Waseda University
Keun-Young Kim	Gwangju Institute of Science and Technology (GIST)
Khalid Saifullah	Quaid-i-Azam University, Islamabad
Kimitake Hayasaki	Chungbuk National University
Ki-Young Choi	Chonnam National University
Kwanho Park	Seoul National Univ.
Kyung Kiu Kim	Sejong University
Kyungjin Ahn	Chosun University
Laura Marcela Becerra Bayona	Sapienza University of Rome and IcraNet
Lee KwanHun	Hanyang University
Luis Gabriel Gómez	Sapienza University of Rome & ICRANet
Lunchakorn Tannukij	Physics Department, Hanyang University
Masakatsu KENMOKU	Nara Science Academy
Maurice H van Putten	Sejong University
Medeu Abishev	Al Farabi Kazakh National University
Mehbub Khan	University of Massachusetts, Dartmouth
Milos Kovacevic	Sapienza University of Rome and ICRANet

Mohammad Wali Hossain	APCTP
Muhammad Sharif	University of the Punjab
Mu-In Park	Sogang Univ.
Myeong-Gu Park	Kyungpook National University
Myung-Ki Cheoun	Soongsil University
Naser Ahmadiniaz	Institute for Basic Science
O-Kab Kwon	Sungkyunkwan University
Phillial Oh	Sungkyunkwan University
Pil-Jong Jung	GIST
Qing Gao	School of Physical Science and Technology, Southwest University
Rahim Moradi	Sapienza University of Rome and ICRANet
Remo Ruffini	ICRANet
Rong-Gen Cai	Institute of Theoretical Physics, Chinese Academy of Sciences
Run-Qiu Yang	KIAS
Sang Hoon Oh	National Institute for Mathematical Sciences
Sang Pyo Kim	Kunsan National University
Sehrish Iftikhar	Lahore College for Women University
Seodong Shin	Yonsei University
Seokcheon Lee	Gyeongsang National University
Seoktae Koh	Jeju National University
Seung Hun Oh	Konkuk University
Seungho Choe	DGIST
Seungho Lee	Konkuk University
Shinji Miyoki	The University of Tokyo, Institute for Cosmic Ray Research
Sichun Sun	National Taiwan University
Soo Min Choi	Chung-Ang University
Stephen Adiyiah	Addystep Logistics Pty Ltd
Sunghoon Jung	Seoul National University
SungJae Bae	DGIST
Sung-Won Kim	Ewha Womans University
Sungwook E Hong	Korea Astronomy and Space Science Institute (KASI)
Sunly Khimphun	Sogang University
Sunyoung Shin	APCTP
Suzana Bedic	University of Rome La Sapienza, ICRANet

Tae Hoon Lee	Soongsil University
Taejin Lee	Kangwon National University
Taishi Ikeda	Nagoya University
Toshinori Matsui	KIAS
Valeri Frolov	University of Alberta
Wei-Tou Ni	Department of Physics, National Tsing Hua University
Wontae Kim	Sogang Univ.
Wonwoo Lee	CQUeST, Sogang University
Xian Gao	School of Physics and Astronomy, Sun Yat-sen University
Xing-Hui Feng	Beijing Normal University
Yaakov (Yasha) Neiman	Okinawa Institute of Science and Technology (OIST)
Yabo Wu	Liaoning Normal University
Yasuhiro Yamamoto	Yonsei University
Yeongji Kim	Konkuk University
Yerlan K. Aimuratov	Sapienza University of Rome and ICRANet
Yi Zheng	Korea Astronomy and Space Science Institute
Yongmin Cho	Konkuk University
Yong-Seon Song	KASI
Yongsoo Jho	Yonsei University
Yongwan Gim	Sogang University
Yoobin Jeong	POSTECH & CQUeST, Sogang University
Yoo-Jin Kang	Chung Ang University
Yoonbai Kim	Sungkyunkwan University
Youngone Lee	The Korea National University of Transportation is a public
Yungui Gong	Huazhong Univeristy of Science and Technology
Yun-Long Zhang	APCTP (Asia-Pacific Center for Theoretical Physics)
YURI KANG	Ewha Womans University
Zhu Yi	School of Physics, Huazhong University of Science and Technology

EPJ Web of Conferences **168**, 01001 (2018) https://doi.org/10.1051/epjconf/201816801001
Joint International Conference of ICGAC-XIII and IK-15 on Gravitation, Astrophysics and Cosmology

Remarks on non-singular black holes

Valeri P. Frolov[1,⋆]

[1] *Theoretical Physics Institute, Department of Physics*
University of Alberta, Edmonton, Alberta, Canada T6G 2E1

Abstract. We briefly discuss non-singular black hole models, with the main focus on the properties of non-singular evaporating black holes. Such black holes possess an apparent horizon, however the event horizon may be absent. In such a case, the information from the black hole interior may reach the external observer after the complete evaporation of the black hole. This model might be used for the resolution of the information loss puzzle. However, as we demonstrate, in a general case the quantum radiation emitted from the black hole interior, calculated in the given black hole background, is very large. This outburst of the radiation is exponentially large for models with the redshift function $\alpha = 1$. We show that it can be suppressed by including a non-trivial redshift function. However, even this suppression is not enough to guarantee self-consistency of the model. This problem is a manifestation of a general problem, known as the "mass inflation". We briefly comment on possible ways to overcome this problem in the models of non-singular evaporating black holes.

1 Introduction

General Relativity predicts existence of black holes. Hundred years after the creation of this theory this prediction was confirmed by LIGO discovery of the gravitational waves emitted in the black hole coalescence. The general relativity is the most reliable theory of the gravity. However, it has its own problems, connected with inevitable existence of singularities. According to this theory such singularities should exist inside black holes. In the well known exact solutions describing stationary vacuum isolated black holes these singularities are connected with the infinite growth of the curvature invariants, such as the square of the Weyl tensor. This means that anisotropy of the spacetime infinitely grows in the vicinity of such singular points. This behavior is similar to the case of the contracting Kasner universe. In order to escape the black hole singularity one needs to find a mechanism suppressing the infinite growth of the anisotropy.This problem is more severe than the singularity problem in homogeneous isotropic cosmological models, where it is sufficient to suppress the growth of the Ricci tensor.

In order to cure this “disease” several modifications of the gravity theory were proposed. For example, effects of vacuum polarization and particle creation in the black hole interior result in the additional terms in the effective action for gravity. These modifications would contain both, higher derivatives and non-locality. One can also modify the fundamental gravity equations, by considering, for example, a theory with higher derivatives, $f(R)$ and Gauss-Bonnet theories, etc. More recently, it

⋆e-mail: vfrolov@ualberta.ca

was proposed a special non-local modification of the Einstein theory, known as a ghost free gravity. At the more fundamental level, one can treat gravity as an emergent phenomenon (string, loops, etc.). However, in spite of many efforts, till now there is no generally accepted mechanism of black hole singularities removal. In the absence of the adopted theory it is instructive to discuss what kind of modifications one could expect when gravity is UV complete. Such an analysis can be fruitful only if some "natural" assumptions concerning the properties of such "full" theory are imposed.

In this talk I briefly discuss such a 'phenomenological' approach, describe proposed models of a non-singular black hole and their properties. The main new feature of such models is that instead of the event horizon they might have a closed apparent horizon. As a result, the information sent inside the black hole can return to external space after the black hole complete evaporation. This opens an interesting possibility for solving the so-called information paradox. I shall also discuss the quantum effects in the interior of a non-singular black hole and demonstrate that huge out-burst of the quantum radiation from the interior should be expected, unless one specially adapts the metric near the inner horizon. The main original results, presented in this talk, are based on the following publications [1–3].

2 Phenomenological approach and the limiting curvature conjecture

One of the main assumptions in the phenomenological description of non-singular black holes is that there exists a critical energy μ and the corresponding length-scale parameter $\ell = \mu^{-1}$. The metric should be modified when the spacetime curvature $\mathcal{R}$ becomes comparable with ℓ^{-2}. At the same time we assume that one can use the classical metric $g_{\mu\nu}$, which is a solution of the (unknown) modified gravity equations. In other words, the length scale, where quantum gravity effects become important, is much smaller than ℓ. We are looking for black-hole metrics which do not have curvature singularities. The first model of a non-singular black hole was proposed by Bardeen [4], who considered a collapse of a charged matter with a charged matter core inside the black hole instead of its singularity. A model of a non-singular evaporating black hole was first discussed in [5, 6].

Different models of neutral, charged and rotating non-singular black holes were proposed and discussed later [7–18]. A general review of different models of non-singular black holes and additional references on this subject can be found in [1, 19–22].

In a general case, a regular solution besides some critical scale parameter ℓ, which is a parameter of the corresponding UV complete theory, contains also such parameters as mass and charge, which specify a concrete solution. The regularity of the solution means that for a fixed value of these parameters the curvature of the spacetime is finite. However, in a general case there is no guarantee that the maximal value of the curvature would not infinitely grow when, say, the mass (and/or charge) takes special value. It is reasonable to assume that for a viable fundamental theory the absolute maximal value of the curvature is restricted by some fundamental value. In other words, the curvature invariants are uniformly restricted by some universal value $|\mathcal{R}| \leq c\ell^{-2}$, where c is a dimensionless constant which depends only on the type of the curvature invariant. This assumption, called the limiting curvature conjecture, was proposed in [23–25].

3 Non-singular black hole models

A general static spherically symmetric metric in a four-dimensional spacetime can be written in the form

$$ds^2 = -f\alpha^2 dv^2 + 2\alpha dv dr + r^2 d\omega^2 , \tag{1}$$

where $f = f(r)$ and $\alpha = \alpha(r)$ are two arbitrary functions. This metric has the Killing vector $\xi^\mu\partial_\mu = \partial_v$. It is easy to see that

$$f = (\nabla r)^2, \qquad f\alpha^2 = -\xi^2 . \tag{2}$$

We call α a redshift function. In a spacetime with a horizon, $f(r)$ vanishes at the position r_H of the apparent horizon. For a regular static metric such a horizon is at the same time the Killing horizon, so that the redshift function $\alpha(r_H)$ is finite there.

If the metric has a horizon where $f(r_H) = 0$ then

$$\xi^\nu\xi^\mu{}_{;\nu} \overset{\mathrm{H}}{=} \kappa\xi^\mu, \qquad \kappa = \frac{1}{2}(\alpha f')|_{r=r_H} . \tag{3}$$

By definition, κ is the surface gravity. The value of κ depends on the choice of the normalization of the Killing vector. In an asymptotically flat spacetime one usually puts $\xi^2|_{r=\infty} = -1$. A conditions that there is no solid angle deficit implies $f|_\infty = 1$. Hence one also has $\alpha|_\infty = 1$.

Let R be the Ricci scalar, $S_{\mu\nu} = R_{\mu\nu} - \frac{1}{4}g_{\mu\nu}R$, and $C_{\mu\nu\alpha\beta}$ be the Weyl tensor. Let us define the following quadratic in the curvature invariants $\mathcal{S}^2 = S_{\mu\nu}S^{\mu\nu}$ and $C^2 = C_{\mu\nu\alpha\beta}C^{\mu\nu\alpha\beta}$. Then one has

$$R = f'' + \frac{4}{r}f' - 2\frac{f-1}{r^2} + \frac{1}{\alpha}\left(2f\alpha'' + 3f'\alpha' + \frac{4}{r}f\alpha'\right), \tag{4}$$

$$C = \frac{1}{\sqrt{3}}\left[f'' - \frac{2}{r}f' + 2\frac{f-1}{r^2} + \frac{1}{\alpha}\left(2f\alpha'' + 3f'\alpha' - \frac{2}{r}f\alpha'\right)\right] . \tag{5}$$

An expression for $\mathcal{S}$ is similar but quite long and we do not present it here.

We assume that the metric (1) is finite at the origin $r = 0$, so that

$$f = f_0 + f_1 r + f_2 r^2 + O(r^3), \qquad \alpha = \alpha_0 + \alpha_1 r + \alpha_2 r^2 + O(r^3) . \tag{6}$$

In a general case the curvature invariants R and C are singular at the origin and have divergences $\sim r^{-2}$ and $\sim r^{-1}$. Conditions that these divergences are absent and the geometry is regular are $f_0 = 1$ and $f_1 = \alpha_1 = 0$. Under these assumptions one has

$$R = -12\left(f_2 + \frac{\alpha_2}{\alpha_0}\right) + O(r^2), \qquad \mathcal{S} = 2\sqrt{3}\frac{\alpha_2}{\alpha_0} + O(r^2), \ \ C = O(r^2) . \tag{7}$$

4 Hayward model

Let us consider first the case when $\alpha = 1$. We assume also that f is a rational function of r

$$f(r) = \frac{P_n(r)}{\tilde{P}_n(r)} , \tag{8}$$

where P_n and $\tilde{P}_n$ are polynomials of the order $n > 1$. One an check that for $n \le 2$ these metrics cannot be both asymptotically flat at infinity and regular at the center (see [1]). However, for $n = 3$ such metrics exist. An example is the following a non-singular black hole metric was proposed by [26] and discussed in [27]

$$f = 1 - \frac{2Mr^2}{r^3 + 2M\ell^2} . \tag{9}$$

This function has the following asymptotics

$$f = 1 - \frac{2M}{r} + O(r^{-4}), \qquad f = 1 - \frac{r^2}{\ell^2} + O(r^5). \tag{10}$$

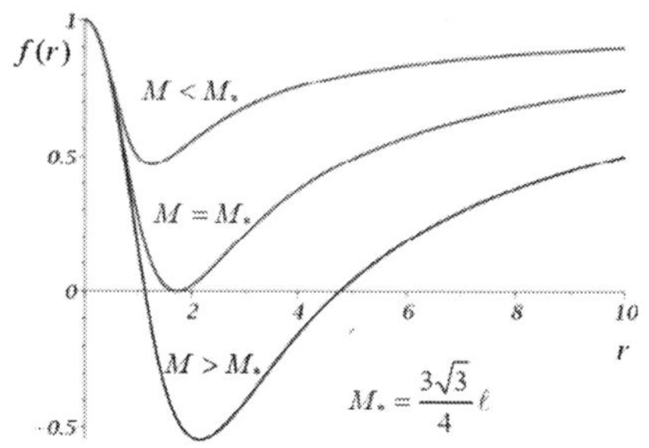

Figure 1. Metric function $f(r)$ for different values of the mass M. Its takes value 1 at $r = 0$ and $r = \infty$. For $M < M_* = \frac{3\sqrt{3}}{4}\ell$ this function is always positive and the metric does not have an apparent horizon. For $M > M_*$ it has two roots, corresponding to outer and inner apparent horizon.

Let us denote $r = (2M\ell^2)^{1/3}z$ and $B = (2M/\ell)^{2/3}$. Then one has

$$f = 1 - B\frac{z^2}{z^3+1}, \quad R = -\frac{6}{\ell^2}\frac{z^3-2}{(z^3+1)^3}, \quad \mathcal{S} = \frac{9}{\ell^2}\frac{z^3}{(z^3+1)^3}, \quad \mathcal{C} = \frac{\sqrt{12}}{\ell^2}\frac{z^3(z^3-2)}{(z^3+1)^3}. \tag{11}$$

All the curvature invariants, besides a common scale factor ℓ^{-2}, contain rational functions of the variable z. The denominators of these functions are positive, while the functions themselves vanish at infinity and are either finite or vanish at $z = 0$. Thus, their absolute value is restricted by some constant, independent of the parameter M. Hence, the metric (9) obeys the limiting curvature condition.

5 Non-singular evaporating black holes

5.1 Sandwich model

In order to describe an evaporating non-singular black hole one can put the mass parameter M in the metric (9) to be time-dependent, $M = M(v)$. This function should be chosen so that it correctly reproduces the black hole mass decrease in the process of the Hawking evaporation. We consider this case later. It is instructive to discuss first a simplified version of the metric, describing a so-called sandwich black hole. Namely, we assume that a black hole is formed as a result of a spherical collapse of a null shell of mass M, it exists for time Δv, and after disappears, as a result of the collapse of another null shell of mass $-M$. During the time interval Δv the metric coincides with (9) [2]. Certainly, this model is quite different from an expected behavior of the evaporating black hole. However, they have the following common property: a finite time of the 'black hole" existence.

The metric function (9) between the shells contains 2 parameters, the mass M and the "fundamental length" ℓ. In what follows we assume that the mass M is large enough, so that the function f has two positive zeros $r_{1,2}$. In such a case it is possible to use $r_{1,2}$ as two new parameters, instead of M and ℓ. We also use the following notation $p = r_1/r_2$. Then the metric function $f(r)$ takes the form

$$f(r) = \frac{(r-r_1)(r-r_2)(r-r_0)}{r^3 - r_1 r_2 r_0}, \qquad r_1 = p, \qquad r_2 = 1, \qquad r_0 = -\frac{p}{p+1}. \tag{12}$$

The parameter p is a position of the outer horizon in the dimensionless units, while the inner horizon is located at $r = 1$. The quantity r_0 is negative. The metric (12) is uniquely specified by two quantities, the mass parameter p and the duration of the black hole existence, $q = \Delta v/r_2$. Because of its rather simple form, many results concerning the properties of such a spacetime can be obtained in an analytical form. The dimensionless surface gravities, calculated for each of these horizons, are

$$\kappa_1 = \frac{(p-1)(p+2)}{2p(p^2+p+1)}, \quad \kappa_2 = -\frac{(p-1)(2p+1)}{2(p^2+p+1)}, \quad \kappa_0 = \frac{(p+1)(p+2)(2p+1)}{2p(p^2+p+1)}. \tag{13}$$

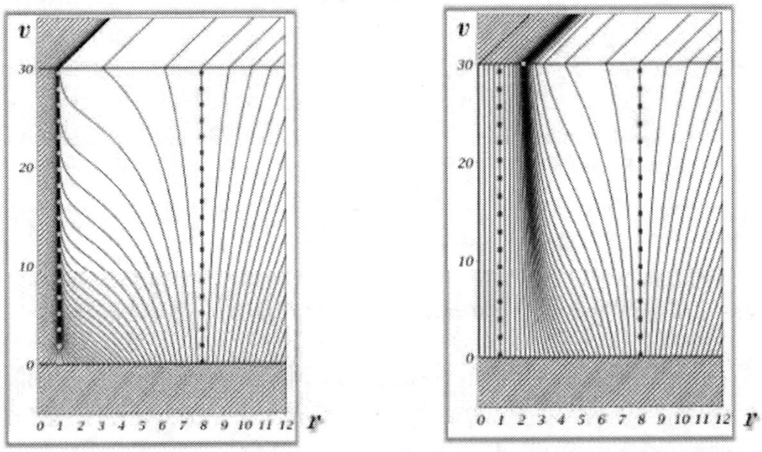

Figure 2. Null rays propagation in the spacetime of a sandwich black hole. The plots are constructed for $p = 8$ and $q = 30$. The left plot illustrates the null rays in the metric with $\alpha = 1$, while the right plot is for $\alpha = \frac{r^6+1}{r^6+1+p^4}$.

The motion of the incoming radial null rays in this geometry is rather simple. They are described by the equation v =const. Such rays pass through the center $r = 0$ and become outgoing. Out-going null rays in the sandwich black-hole geometry are shown at Figure 2. The left plot in the Figure 2 shows the null rays in the sandwich geometry with $\alpha = 1$. The rays that cross the first null shell with $r < 1$ propagate near the origin, being accumulated near the inner horizon from its inner side. The rays that cross the first null shell between $r = 1$ and $r = p$ are also attracted to the inner horizon and are accumulated near it from its outer side. Those rays that cross the first shell at $r > p$ are propagating away from the outer horizon. All incoming rays with the origin at $0 < v < q$, after passing the inner horizon cross the center $r = 0$ and returns to inner horizon from inside. They are also accumulated in the narrow domain in its vicinity. In other words, the inner horizon plays the role of an attractor for the outgoing rays, while the outer horizon repulses the rays. After the collapse of the second shell with a negative mass, all the outgoing null rays freely propagate to the future null infinity $\mathcal{I}^+$. For $q \gg p$ the beam of out-going null rays propagating near the inner horizon is highly squeezed. The corresponding squeezing factor at time $v = q$ is $\sim \exp(-q\kappa_2) \sim \exp(q)$.

The right plot at the Figure 2 illustrates the propagation of null rays in the sandwich black hole geometry with a non-trivial redshift function α. We choose it in the form $\alpha = (r^6+1)/(r^6+1+p^4)$. For large p this factor near the center and the inner horizon $\alpha_0 \sim p^{-k}$ is small. Since $\kappa_2 \sim -\alpha_0$, the surface gravity of the inner horizon is greatly reduced. As a result, it takes longer time to accumulate the null rays near the inner horizon.

5.2 Smooth mass function

At first, we assume that $\alpha = 1$, and put

$$f = 1 - \frac{2M(v)r^2}{r^3 + 2M(v)\ell^2 + \ell^3}. \tag{14}$$

In the static case this metric function differs from (9) by the presence of an extra ℓ^3 term in the denominator. Such a modification makes the geometry near the point $r = 0$ in the limit $M \to 0$ more smooth. Since now the position of the inner horizon depends on time, it is more convenient to use ℓ as a scale parameter. For this choice $\ell = 1$

$$f = 1 - \frac{\mu(v)r^2}{r^3 + \mu(v) + 1}, \qquad \mu(v) \equiv 2M(v). \tag{15}$$

For sufficiently large μ this metric describes the nonsingular evolving black hole with two apparent horizons: the outer horizon $r_1(v)$ and the inner one $r_2(v)$. Their radii are determined by the condition $f = 0$ and can be written explicitly ($\mu = \mu(v)$)

$$r_1 = \frac{1}{2}\left(\mu + 1 + \sqrt{\mu^2 - 2\mu - 3}\right), \qquad r_2 = \frac{1}{2}\left(\mu + 1 - \sqrt{\mu^2 - 2\mu - 3}\right). \tag{16}$$

The third root of the cubic equation $f = 0$ corresponds to a negative value of radius $r_0 = -1$ and is not relevant for our study. The inner and outer apparent horizons exist only when $\mu(v) > 3$. One can easily calculate the surface gravity on the outer and inner horizons

$$\kappa_{1,2} = \mu \frac{r_{1,2}(r_{1,2}^3 - 2\mu - 2)}{2(r_{1,2}^3 + \mu + 1)^2}. \tag{17}$$

One can check that, provided horizons exist, $\kappa_1 \geq 0$, while $\kappa_2 \leq 0$. For large masses $\mu \gg 3$ we have

$$r_1 = \mu - \frac{1}{\mu} + O(\mu^{-2}), r_2 = 1 + \frac{1}{\mu} + O(\mu^{-2}), \kappa_1 = \frac{1}{2\mu} - \frac{1}{\mu^3} + O(\mu^{-4}), \kappa_2 = -1 + \frac{5}{2\mu} + \frac{1}{\mu^3} + O(\mu^{-4}). \tag{18}$$

For $\mu = 3$ the horizons merge and we obtain an extremal case with $r_1 = r_2 = 2$ and $\kappa_1 = \kappa_2 = 0$. For $\mu < 3$ the metric has no the apparent horizon.

For concrete calculations we chose the function $\mu(v)$ in the form

$$\mu(v) = \begin{cases} \dfrac{2v(q-\tau)(q-v)^{1/3}}{C(3(q-\tau)v - \tau v + \tau^2)} & 0 \leq v \leq q, \\ 0 & v < 0, \text{ and } v > q. \end{cases} \tag{19}$$

It is schematically shown at Figure 3. This function has a maximum at $v = \tau$, where $\mu(\tau) = \mu_0$. During the interval of v from 0 to τ the mass parameter μ grows from 0 to μ_0. This is a phase of the null fluid collapse which results in the formation of a black hole. Starting from $v = \tau$ the mass reduces as a result of the Hawking radiation. At this stage $\dot{\mu} \sim \mu^{-2}$. This phase continues up to the time $v = q$

$$q = (C\mu_0)^3 + \tau. \tag{20}$$

The parameter C depends on the number of distinct polarizations of particles which are emitted. Note that overall picture of quantum particle creation depends mostly on the asymptotic properties of the function $\mu(v)$ rather than on variations of its shape.

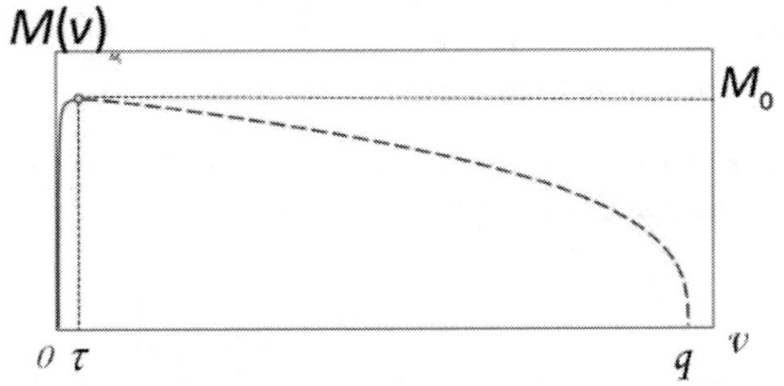

Figure 3. This plot schematically illustrates a typical shape of the mass function $\mu(v) = 2M(v)$. The dashed line describes the stage of the Hawking evaporation of the black hole.

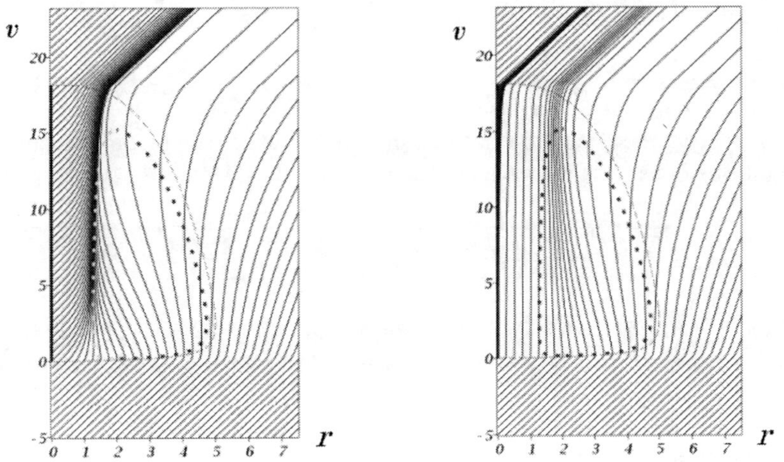

Figure 4. Null rays propagation in the spacetime of a non-singular evaporating black hole. The plots are constructed for the value of the parameters $\mu_0 = 5$, $\tau = 2.5$ and $C = 1/2$. The left plot is for the case $\alpha = 1$, while the right plot is for $\alpha = (1 + r^5)/(1 + r^5 + \mu^3)$.

The left plot of Figure 4 shows the null rays in the geometry of an evaporating non-singular black hole with $\alpha = 1$. The motion of the incoming radial null rays in this geometry is quite simple. They are described by the equation v =const. Such rays pass through the center $r = 0$ and then become outgoing. The incoming null rays with $v < 0$ which cross the collapsing null fluid shell at small radius, $0 < r_- < 1$, are focussed to the inner horizon from its interior. Similar rays for small $r > 1$ are also focussed, but in the exterior of the inner horizon. As a result the corresponding beams of light are squeezed by the factor $\sim \exp(-\kappa_2 q)$, where q is the life-time of the black holes. For the specially chosen α this squeezing is relaxed. The corresponding plot is shown in the right picture at Figure 4. Let us emphasize, that qualitatively the character of the null ray behavior in this geometry is quite similar to the sandwich black hole case.

6 Quantum effects

6.1 Radial null rays and $\mathcal{I}^- \to \mathcal{I}^+$ map

In the chosen coordinates, radial incoming null rays are described by the relation v =const , while the outgoing null rays satisfy the following equation

$$\frac{dr}{dv} = \frac{1}{2}\alpha f . \tag{21}$$

Consider an outgoing radial null ray which reaches the future null infinity $\mathcal{I}^+$ at the retarded time $u = u_+$ and trace it backward in time. We denote by u_- time v when it crosses the center $r = 0$. After this it propagates to the past null infinity $\mathcal{I}^-$ at $v = u_-$. The relation $u_- = u_-(u_+)$ establishes a map between $\mathcal{I}^+$ and $\mathcal{I}^-$. Since the tangent vector to the null ray trajectory is uniquely determined by equation (21), any two outgoing null rays cannot intersect. This means that $u_+(u_-)$ is a monotonically increasing function, so that $du_+/du_- > 0$. The map function contains the information concerning observables at $\mathcal{I}^+$. In particular, it allows one to calculate in 2D approximation the quantum energy flux at $\mathcal{I}^+$ from a non-singular black hole.

6.2 The energy flux

In the 2D approximation, the energy flux of massless particles, created from the initial vacuum state, is given by the following expression

$$\dot{E} = \frac{dE}{du_+} = \frac{1}{48\pi}\left[-2\frac{d^2P}{du_+^2} + \left(\frac{dP}{du_+}\right)^2\right], \qquad P = \ln\left|\frac{du_-}{du_+}\right| . \tag{22}$$

This relation directly follows from a general result obtained by Fulling and Christensen [28] for the quantum average of the stress-energy tensor of a massless scalar field in two dimensions, reconstructed from the conformal anomaly. The same expression can be also obtained by the variation of the Polyakov effective action with respect to the metric [29, 30].

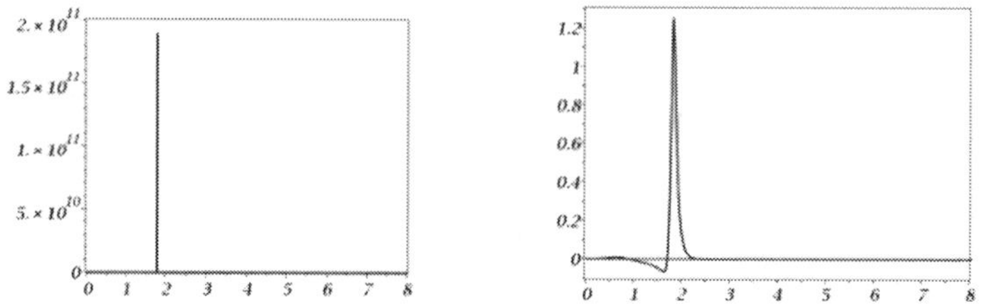

Figure 5. Quantum energy flux dE/du_+ at the future null infinity from the inner horizon of an evaporating black hole. These plots are constructed for $\mu_0 = 3$, $\tau = 6.$ and $C = 1/2$. $\alpha = 1$ for the left plot and $\alpha = (1 + r^5)/(1 + r^5 + \mu^3)$ for the right plot, respectively.

For the sandwich-type black holes the map function can be found analytically (see [2]). For the smooth-type evaporating metric one needs to use numerical calculations. It is quite straightforward to calculate the map function. However, to obtain the energy flux one needs to find its derivatives with respect time u_+ up to the third order. This makes this method not accurate. In order to overcome this problem, a new method, so called "bracket formalism", was developed in [3]. The main idea of this method is to consider a narrow beam of null rays near a fixed one and using the theory of perturbations to derive the equations for rays, propagating in the close vicinity to the chosen "fiducial" ray. It was shown that the obtained set of equations is greatly simplified if instead of usual derivatives to use their "bracket" generalizations, similar to the Schwarzian derivative.

The results of the calculations for sandwich and "realistic" models are quite similar. One can find all the details in [2, 3]. Here we just describe some of the main properties of the quantum energy flux. An external observer first registers rays emitted during the phase of creation of the black hole. For long enough life-time of the black hole, he/she will observe the Hawking radiation. This constant in time radiation is reproduced in our numerical calculations with very high accuracy. This stage is over when the rays from the black hole interior reach the observer. This part of the radiation is sensitive to details of the inner structure of the black hole, but the flux is relatively weak. Next what is seen by the observer is the outburst of radiation from the inner horizon. It intensity depends on the exponent of the factor

$$-[\text{surface gravity of inner horizon}] \times [\text{black hole life time}]\,. \tag{23}$$

These fluxes for sandwich and "realistic" black hole models are similar. For illustration, we show this flux at Figure 5 for "realistic models" with $\alpha = 1$ and a specially chosen non-trivial α.

7 Discussion

The outburst of the quantum energy flux from the inner horizon is related to a so-called mass inflation effect [31, 32] (see also [33]). The inner horizon has a negative surface gravity and it works as an attractor of the outgoing null rays. For a (quasi-) static case, the advanced time v is the Killing time parameter and it differs from the affine parameter λ along the outgoing null ray. Near the inner horizon they are related as follows $\lambda \sim \exp(-\kappa v)$. The momentum $p^\mu = dx^\mu/d\lambda$ of outgoing photons exponentially grows, while their energy, $E \sim -p_v$, remains constant because of the gravitational redshift. However, this compensation mechanism is broken when the position of the inner horizon moves. As a result, quanta leaving a black hole interior possess such exponentially large non-compensated blueshift.

For a special choice of the redshift function the exponential outburst of the radiation emitted from the inner horizon can be suppressed. This happens when the surface gravity of the inner horizon is reduced. As a result, the time of the black hole evaporation is not sufficient for generation the exponential regime, and the energy flux decreases. The energy flux for the model with $\alpha \neq 1$ it still large. Partly this is connected with large contribution of the α-anomaly. This contribution describes the particle creation in the incoming modes, which becomes large for fast change of the redshift function α during the formation of the black hole. As a result, the model also violates a self-consistency requirement (see [34]). The obtained results indicate that the backreaction of the particles created in the black hole interior should be properly taken into account. An interesting problem is a search for dynamical self-consistent non-singular black hole models.

Acknowledgments

The authors thank the Natural Sciences and Engineering Research Council of Canada and the Killam Trust for their financial support.

References

[1] V.P. Frolov, Phys. Rev. **D94**, 104056 (2016), 1609.01758

[2] V.P. Frolov, A. Zelnikov (2016), 1612.05319

[3] V.P. Frolov, A. Zelnikov, Phys. Rev. **D95**, 124028 (2017), 1704.03043

[4] J.M. Bardeen, *Non-singular general-relativistic gravitational collapse*, in *Proceedings of of International Conference GR5,Tbilisi, USSR* (1968), p. 174

[5] V.P. Frolov, G.A. Vilkovisky, , in *The Second Marcel Grossmann Meeting on the Recent Developments of General Relativity (In Honor of Albert Einstein) Trieste, Italy, July 5-11, 1979* (1979), p. 0455

[6] V.P. Frolov, G.A. Vilkovisky, Phys. Lett. **106B**, 307 (1981)

[7] I. Dymnikova, Gen. Rel. Grav. **24**, 235 (1992)

[8] A. Borde, Phys. Rev. **D55**, 7615 (1997), gr-qc/9612057

[9] E. Ayon-Beato, A. Garcia, Phys. Rev. Lett. **80**, 5056 (1998), gr-qc/9911046

[10] J.P.S. Lemos, V.T. Zanchin, Phys. Rev. **D83**, 124005 (2011), 1104.4790

[11] N. Uchikata, S. Yoshida, T. Futamase, Phys. Rev. **D86**, 084025 (2012), 1209.3567

[12] A. Flachi, J.P.S. Lemos, Phys. Rev. **D87**, 024034 (2013), 1211.6212

[13] T. De Lorenzo, C. Pacilio, C. Rovelli, S. Speziale, Gen. Rel. Grav. **47**, 41 (2015), 1412.6015

[14] L. Balart, E.C. Vagenas, Phys. Rev. **D90**, 124045 (2014), 1408.0306

[15] S.G. Ghosh, S.D. Maharaj, Eur. Phys. J. **C75**, 7 (2015), 1410.4043

[16] T. De Lorenzo, A. Giusti, S. Speziale, Gen. Rel. Grav. **48**, 31 (2016), [Erratum: Gen. Rel. Grav.48,no.8,111(2016)], 1510.08828

[17] H.M. Haggard, C. Rovelli, Phys. Rev. **D92**, 104020 (2015), 1407.0989

[18] J.P.S. Lemos, V.T. Zanchin, Phys. Rev. **D93**, 124012 (2016), 1603.07359

[19] S. Ansoldi, ArXiv e-prints (2008), 0802.0330

[20] J.M. Bardeen, ArXiv e-prints (2014), 1406.4098

[21] J.M. Bardeen, ArXiv e-prints (2017), 1706.09204

[22] S. Chinaglia, S. Zerbini, Gen. Rel. Grav. **49**, 75 (2017), 1704.08516

[23] M. Markov, JETP Letters **36**, 265 (1982)

[24] M. Markov, Annals Phys. **155**, 333 (1984)

[25] J. Polchinski, Nucl.Phys. **B325**, 619 (1989)

[26] S.A. Hayward, Phys.Rev.Lett. **96**, 031103 (2006), gr-qc/0506126

[27] V.P. Frolov, JHEP **05**, 049 (2014), 1402.5446

[28] S.M. Christensen, S.A. Fulling, Phys. Rev. **D15**, 2088 (1977)

[29] V.P. Frolov, G.A. Vilkovisky, *Spherically Symmetric Collapse in Quantum Gravity*, in *Quantum Gravity* (Springer US, 1984), pp. 267–290

[30] G.A. Vilkovisky, *The gospel according to DeWitt*, in *Quantum theory of gravity* (Adam Hilger Limited, Bristol UK, 1985), pp. 169–209

[31] E. Poisson, W. Israel, Phys. Rev. Lett. **63**, 1663 (1989)

[32] E. Poisson, W. Israel, Phys. Rev. **D41**, 1796 (1990)

[33] E.G. Brown, R.B. Mann, L. Modesto, Phys. Rev. **D84**, 104041 (2011), 1104.3126

[34] P.A. Bolashenko, V.P. Frolov, in *Physical Effects in the Gravitational Fields of Black Holes , Proceedings of the Lebedev Physics Institute of the Academy of Sciences of the USSR, v.169*, edited by M.A. Markov (Commack, N.Y., Nova Science, 1986), pp. 159–168

EPJ Web of Conferences **168**, 01002 (2018) https://doi.org/10.1051/epjconf/201816801002
Joint International Conference of ICGAC-XIII and IK-15 on Gravitation, Astrophysics and Cosmology

Cosmological Implications of Electroweak Monopole

Y. M. Cho[1,2,*]

[1]*Administration Building 310-4, Konkuk University, Seoul 143-701, Korea*

[2]*School of Physics and Astronomy, Seoul National University, Seoul 151-742, Korea*

In this talk we review the basic features of the electroweak monopole, and estimate the remnant electroweak monopole density of the standard model in the present universe. We show that, although the electroweak phase transition is of the first order, the monopole production comes from the thermal fluctuations of the Higgs field after the phase transition, not the vacuum bubble collisions during the phase transition. Moreover, most of the monopoles produced initially are annihilated as soon as created, and this annihilation continues very long time, longer than the muon pair annihilation time. As the result the remnant monopole density at present universe becomes very small, of 10^{-11} of the critical density, too small to be the dark matter. We discuss the physical implications of our results on the ongoing monopole detection experiments.

PACS numbers: 14.80.Hv, 12.15.-y, 04.20.-q

Keywords: electroweak monopole, mass of electroweak monopole, cosmological production of electroweak monopole, remnant density of electroweak monopole, seed of large scale structure.

I. INTRODUCTION

Ever since Dirac has proposed the Dirac monopole generalizing the Maxwell's theory, the monopole has become an obsession in physics [1]. After the Dirac monopole we have had the Wu-Yang monopole [2], the 't Hooft-Polyakov monopole [3], the grand unification (Dokos-Tomaras) monopole [4], and the electroweak (Cho-Maison) monopole [5, 6]. But only the electroweak monopole become realistic enough to be discovered.

This has made the experimental confirmation of the electroweak monopole one of the most urgent issues in the standard model after the discovery of the Higgs particle, and the newest MoEDAL ("the magnificent seventh") detector at LHC is actively searching for the monopole [7, 8]. On the other hand, the 13 TeV LHC may have no chance to produce the monopole if the monopole mass becomes larger than 6.5 TeV. In this case we must try to detect the remnant monopole at present universe produced in the early universe. so it is fortunate that IceCube, Antares, Auger, and similar experiments are searching for them [9–11]

To detect the remnant monopoles, however, we need to know how many of them are left over in the present universe. There have been discussions on the cosmological production of monopoles, but most of them have been on the grand unification monopoles [12–15]. The general consensus is that the grand unification monopoles would have overclosed the universe without the inflation, but the inflation might have completely diluted them in such a way that they could have no visible impact on the present universe [16].

*Electronic address: `ymcho7@konkuk.ac.kr`

For the electroweak monopole, however, the situation was not so clear [17–19]. There have been claims that the electroweak monopoles could also overclose the universe [18], but we need a more accurate discussion on this issue. *The purpose of this talk is to discuss the cosmological production and successive evolution of the electroweak monopole to predict the remnant monopole density of the standard model at present universe, and discuss the cosmological implications of the electroweak monopole.*

Before we estimate the electroweak monopole density at present universe, we emphasize the basic facts on the electroweak monopole. First, it is the electroweak generalization of the Dirac monopole. But unlike the Dirac monopole which is optional, the electroweak monopole must exist because the electromagnetic U(1) in the standard model becomes non-trivial [5, 6]. This makes the discovery of the monopole, not the Higgs particle, the final (and topological) test of the standard model.

Second, there is a fundamental difference between the Dirac monopole and the electroweak monopole. It carries the magnetic charge twice bigger than the Dirac monopole. This is because, in the course of the electroweak unification, the period of electromagnetic U(1) becomes 4π, not 2π [5, 6]. So we could tell if it is the Dirac monopole or the electroweak monopole, when we discover it.

Third, unlike the Dirac monopole, we can estimate the mass of the electroweak monopole to be of the order of several TeV, roughly $1/\alpha$ times the W-boson mass [20–22]. This is because the monopole mass essentially comes from the same Higgs mechanism which generates the W-boson mass, except that the monopole potential couples to the Higgs multiplet magnetically.

Unfortunately there have been confusions and misunderstandings on the electroweak monopole in the literature. A most serious one is the assertion that the standard model has mo monopole because the theory has no monopole topology [23]. This "non-existence theorem", however, is a false assertion based on wrong topology. To clarify these confusions and misunderstandings, we briefly review the electroweak monopole in the standard model first.

II. ELECTROWEAK MONOPOLE: A REVIEW

Consider the Weinberg-Salam model,

$$\mathcal{L} = -|\mathcal{D}_\mu \phi|^2 - \frac{\lambda}{2}\Big(\phi^\dagger \phi - \frac{\mu^2}{\lambda}\Big)^2 - \frac{1}{4}\vec{F}_{\mu\nu}^2 - \frac{1}{4}G_{\mu\nu}^2,$$
$$\mathcal{D}_\mu \phi = \big(\partial_\mu - i\frac{g}{2}\vec{\tau}\cdot\vec{A}_\mu - i\frac{g'}{2}B_\mu\big)\phi, \qquad (1)$$

where ϕ is the Higgs doublet, $\vec{A}_\mu$ and B_μ are the SU(2) and U(1) gauge potentials. Introducing the Higgs field ρ and the CP^1 field ξ by

$$\phi = \frac{1}{\sqrt{2}}\rho\, \xi, \quad (\xi^\dagger \xi = 1), \qquad (2)$$

we have

$$\mathcal{L} = -\frac{1}{2}(\partial_\mu \rho)^2 - \frac{\rho^2}{2}|\mathcal{D}_\mu \xi|^2 - \frac{\lambda}{8}\big(\rho^2 - \rho_0^2\big)^2$$
$$-\frac{1}{4}\vec{F}_{\mu\nu}^2 - \frac{1}{4}G_{\mu\nu}^2, \quad (\rho_0^2 = \frac{2\mu^2}{\lambda}). \qquad (3)$$

With the ansatz

$$\rho = \rho(r), \quad \xi = i\begin{pmatrix} \sin(\theta/2)\; e^{-i\varphi} \\ -\cos(\theta/2) \end{pmatrix},$$
$$\vec{A}_\mu = \frac{1}{g}A(r)\partial_\mu t\; \hat{r} + \frac{1}{g}(f(r) - 1)\; \hat{r}\times\partial_\mu \hat{r},$$
$$B_\mu = \frac{1}{g'}B(r)\partial_\mu t - \frac{1}{g'}(1 - \cos\theta)\partial_\mu \varphi, \qquad (4)$$

we have the dyon solution of the standard model dressed by the W-boson, Z-boson, and Higgs field, which becomes the monopole solution with $A = B = 0$. The singular monopole solution is shown in Fig. 1 [5, 6, 20].

The most serious misunderstanding on the electroweak monopole is the non-existence. It has been asserted that the vacuum of the Weinberg-Salam model does not allow the monopole topology, and thus has no monopole [23]. Obviously the existence of the monopole solution assures that the standard model has the monopole [5, 6, 20]. If so, the remaining question is if this monopole is optional or not. It is not. The electroweak monopole must exist, if the standard model is correct.

To see this notice that in the ansatz (4) the hypercharge U(1) is made to be non-trivial, which was why the electroweak monopole could exist. So we have to find the

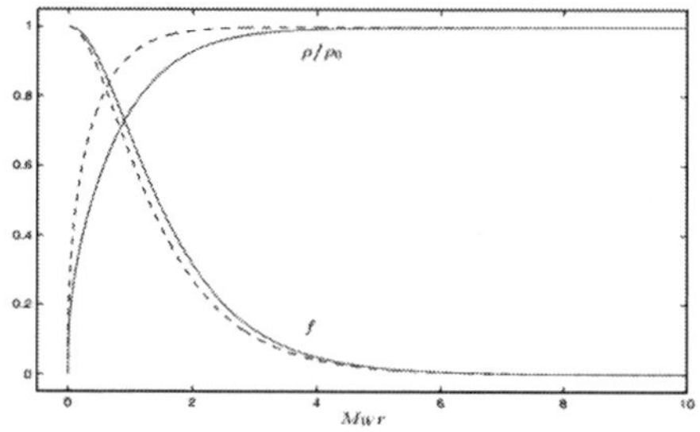

FIG. 1: The finite energy electroweak monopole solution. The solid line (red) represents the regularized monopole and the dotted (blue) line represents the singular electroweak monopole.

reason why the hypercharge U(1) in the standard model must be non-trivial. This follows from two facts. First, the electromagnetic U(1) in the standard model is given by the linear combination of the U(1) subgroup of SU(2) and the hypercharge U(1). Second, the U(1) subgroup of SU(2) is non-trivial. In this case the mathematical consistency requires the hypercharge U(1), and consequently the electromagnetic U(1), to be non-trivial.

But, of course, this has to be confirmed by experiment. This makes the discovery of the monopole, not the Higgs particle, the final (and topological) test of the standard model. This is why the electroweak monopole is so important.

Another important question is whether there is any characteristic feature of the electroweak monopole which is different from the Dirac monopole. The answer is yes. The electroweak monopole carries the magnetic charge twice bigger than the Dirac monopole [5, 20]. The reason is simple. It is well known that the magnetic charge of Dirac monopole must be a multiple of $2\pi/e$. This is because the period of the electromagnetic U(1) is 2π. However, in the course of the electroweak unification the period of electromagnetic U(1) becomes 4π, because the period of the U(1) subgroup of SU(2) becomes 4π. Consequently the magnetic charge g_m of the electroweak monopole must be a multiple of $4\pi/e$.

The third assertion is that we can not estimate the mass of the electroweak monopole, because the electroweak monopole has the point singularity at the origin which makes the energy infinite. Obviously the monopole mass is the most important piece of information from the experimental point of view. There was no way to predict the mass of the Dirac monopole theoretically, and this has made the monopole detection a blind search in the dark room without any theoretical lead.

But this assertion is not true, either. In fact we could easily guess the mass to be of the order of 10 TeV, roughly $4\pi/e^2$ times the W-boson mass. This is because the monopole mass essentially comes from the same Higgs mechanism which generates the W-boson mass, except that the monopole potential couples to the Higgs multiplet magnetically, with the strength $4\pi/e$ [5, 20].

Another way to estimate the mass is to regularize the monopole to make the energy finite [20]. To do this let us modify (3) to the following effective Lagrangian with a non-trivial permittivity ϵ for the hypercharge U(1) gauge field

$$\mathcal{L}_{eff} = -\frac{1}{2}(\partial_\mu\rho)^2 - \frac{\rho^2}{2}|\mathcal{D}_\mu\xi|^2 - \frac{\lambda}{8}(\rho^2-\rho_0^2)^2 \\ -\frac{1}{4}\vec{F}_{\mu\nu}^2 - \frac{\epsilon(\rho)}{4}G_{\mu\nu}^2. \quad (5)$$

The effective Lagrangian still retains the SU(2) $\times$ U(1) gauge symmetry. Moreover, when $\epsilon = 1$, the Lagrangian reproduces the standard model. So this modification affects only the short distance behavior, if ϵ approaches to the unit asymptotically.

To see how ϵ mimics the quantum correction and regularize the monopole, notice that the rescaling of B_μ to B_μ/g' changes g' to $g'/\sqrt{\epsilon}$. So ϵ effectively changes the hypercharge U(1) gauge coupling g' to the "running" coupling $\bar{g}' = g'/\sqrt{\epsilon}$. This means that, by making $\bar{g}'$ infinite (requiring ϵ vanishing) at the origin, we can remove the singularity of the monopole. Moreover, with $\epsilon(\infty) = 1$, we can reproduce the singular monopole asymptotically. In fact, with $\epsilon = (\rho/\rho_0)^8$, we have the regularized monopole solution which has the energy around 7 TeV [20, 22]. This is shown in Fig. 1. Notice that asymptotically the regularized monopole looks almost identical to the singular monopole.

The regularized monopole energy, of course, depends on the functional form of ϵ, so that we could change the energy changing ϵ. But this also affects other things, for example the Higgs to two photon decay rate. And recently Ellis and collaborators noticed that, choosing a more realistic ϵ which can reproduce the experimental value of the Higgs to two photon decay rate, they could reduce the monopole energy less than 5.5 TeV.

III. GRAVITATIONALLY COUPLED ELECTROWEAK MONOPOLE

Now we discuss the gravitational modification of the monopole. Intuitively, the gravitational modification is expected to be negligible. But there is the possibility that the gravitational attraction could change the monopole to a black hole and make the monopole mass arbitrary [24]. We show that this happens only when the Higgs vacuum value approaches to the Planck scale.

Consider the effective Lagrangian (5) minimally coupled to Einstein's theory. For the sake of simplicity, we focus on the static spherically symmetric solutions and assume the space-time metric to be

$$ds^2 = -N^2(r)A(r)dt^2 + \frac{dr^2}{A(r)} + r^2(d^2\theta + \sin^2\theta d\varphi^2), \\ A(r) = 1 - \frac{2Gm(r)}{r}. \quad (6)$$

With the ansatz (4) the reduced Einstein-Weinberg-Salam action becomes

$$S = \int \Big[\frac{1}{4\pi}\dot{m} - AK - U\Big]Ndr, \\ K = \frac{\dot{f}^2}{g^2} + \frac{r^2}{2}\dot{\rho}^2, \\ U = \frac{(1-f^2)^2}{2g^2r^2} + \frac{\lambda}{8}r^2(\rho^2-\rho_0^2)^2 + \frac{\epsilon(\rho)}{2g'^2r^2} + \frac{1}{4}f^2\rho^2, \quad (7)$$

where the dot represents d/dr.

From this we have the following equations of motion

$$\frac{\dot{N}}{N} = 8\pi G\frac{K}{r}, \quad \dot{m} = 4\pi(AK+U), \\ A\ddot{f} + \Big(\dot{A} + A\frac{\dot{N}}{N}\Big)\dot{f} + \frac{1-f^2}{r^2}f - \frac{1}{4}g^2\rho^2 f = 0, \\ A\ddot{\rho} + \Big(\frac{2A}{r} + \dot{A} + A\frac{\dot{N}}{N}\Big)\dot{\rho} - \frac{f}{2r^2}\rho \\ -\frac{\lambda}{2}(\rho^2-\rho_0^2)\rho - \frac{1}{2g'^2r^4}\frac{d\epsilon(\rho)}{d\rho} = 0. \quad (8)$$

This has two limiting solutions. First, when gravitational field decouples (i.e., when $G \to 0$) we have the non-gravitating electroweak monopole solution [20]. Second, when $f = 0$, $\rho = \rho_0$, and $\epsilon = 1$, we have the magnetically charged Reissner-Nordstrom black hole solution with the magnetic charge [24]. But in general the solution depends on three parameters

$$\alpha = \sqrt{G}\rho_0 = \rho_0/M_p, \quad \beta = M_H/M_W, \quad (9)$$

and the Weinberg angle θ_W.

Depending on the boundary conditions the entire solutions of (8) can be classified into two categories: the globally regular solutions and the black holes. Asymptotic flatness of the space-time requires that both $N(r)$ and $m(r)$ become constants at spatial infinity, which requires the following boundary conditions on the W-boson and Higgs field,

$$f(\infty) = 0, \quad \rho(\infty) = \rho_0. \quad (10)$$

We fix the scale of the time coordinate adopting

$$N(\infty) = 1. \quad (11)$$

Notice that $m(\infty)$ which determines the total mass of the monopole is not constrained.

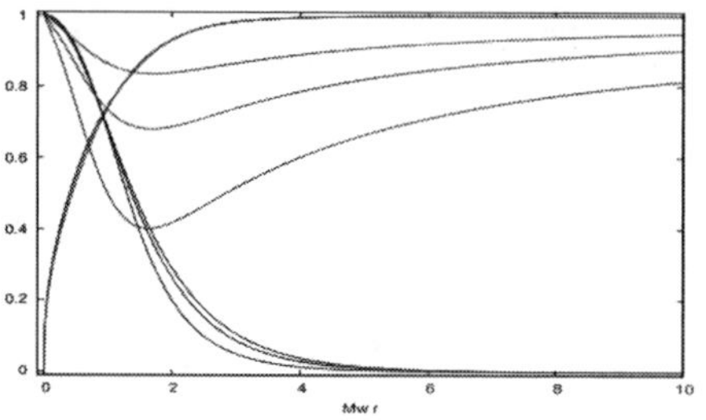

FIG. 2: The W-boson f (blue), Higgs field ρ (red), and the metric function $A(r) = 1 - 2Gm/r$ (green) profiles of the gravitating monopole solutions in the standard model. The black curves represent the finite energy solution in flat spacetime, and the other three are obtained with $\alpha = \rho_0/M_P = 0.1$, 0.2, and 0.38.

For the regular monopole solutions we require the regularity at the origin,

$$f(0) = 1, \quad \rho(0) = 0, \quad m(0) = 0. \tag{12}$$

Now, choosing $\epsilon = (\rho/\rho_0)^8$ for simplicity, we find that the solutions have the following expansions near $r = 0$,

$$f(r) = 1 - f_1 x^2 + ...,$$
$$\rho(r) = h_1 \rho_0 x^{\delta_1} + ..., \quad (\delta_1 = \frac{\sqrt{3}-1}{2})$$
$$m(r) = \frac{2\pi\alpha^2 h_1^2 \delta_1^2}{G M_W \delta_2} x^{\delta_2} + ..., \quad (\delta_2 = \sqrt{3}), \tag{13}$$

where $x = M_W r$. From this we can obtain the solutions by the standard shooting method with f_1 and h_1 as the shooting parameters. The result shows that when $\alpha = \rho_0/M_P$ is less than 0.39, we have a completely regular magnetically charged solution which can describe the monopole star. But when α approaches to 0.39, the solution turns to a magnetic blackhole. This is schematically shown in Fig. 2. Notice that, except for the metric, the gravitating monopole looks very much like the non-gravitating monopole. In particular, the size of the monopole remains roughly about $1/M_W$.

To find the mass of the monopole, first note that with (8) we have

$$m(r) = 4\pi e^{P(r)} \int_0^r (K+U) e^{-P(r')} dr', \tag{14}$$
$$P(r) = 8\pi G \int_r^\infty \frac{K}{r'} dr'. \tag{15}$$

So, with (10) we have the total mass given by

$$\mathcal{M} = m(\infty) = 4\pi \int_0^\infty (K+U) e^{-P(r)} dr, \tag{16}$$

which assures the positivity of the total mass. Moreover, this confirms that the gravitation reduces the monopole mass, which is expected from the attractive nature of the gravitational interaction. In reality, of course, the gravitational modification of the monopole mass becomes negligible because α is very small (of 10^{-16}).

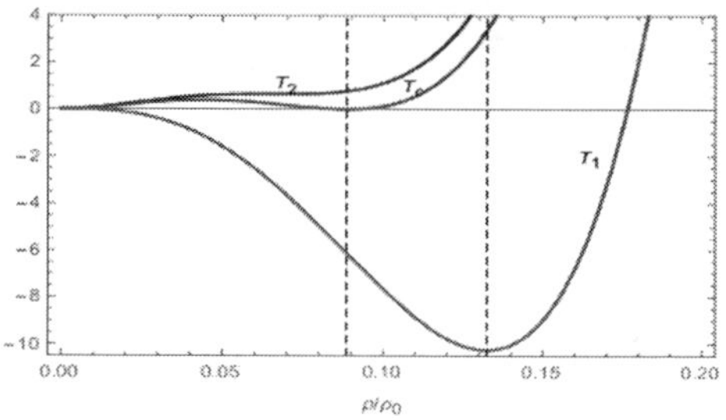

FIG. 3: The amplified effective thermal potential for $T = T_2, T_c$, and T_1. Here the vertical axis represents $V_{eff} - V_0$ in the unit $10^{-5} \times V_0$, and the vertical dashed lines indicate the positions of the local minima of the potentials.

IV. ELECTROWEAK PHASE TRANSITION

Now we discuss the electroweak phase transition. To do this, we have to find the temperature-dependent effective potential of the Higgs field. Fortunately this has been known [17, 19],

$$V_T(\rho) = V_0(\rho) - \frac{C_1}{12\pi}\rho^3\, T + \frac{C_2}{2}\rho^2\, T^2 - \frac{\pi^2}{90} N T^4 + \delta V_T,$$
$$V_0(\rho) = \frac{\lambda}{8}(\rho^2 - \rho_0^2)^2,$$
$$C_1 = \frac{6M_W^3 + 3M_Z^3}{\rho_0^3} \simeq 0.36,$$
$$C_2 = \frac{4M_W^2 + 2M_Z^2 + M_H^2 + 4m_t^2}{8\rho_0^2} \simeq 0.36, \tag{17}$$

where V_0 is the zero-temperature potential, N is the total number of distinct helicity states of the particles with mass smaller than T (counting fermions with the factor 7/8), C_1 and C_2 are the contributions from the ρ dependent part of the masses of gauge bosons, Higgs field, and fermions, M_W, M_Z, M_H, and m_t are the W-boson, Z-boson, Higgs boson, and the top quark masses, and δV_T is the slow-varying logarithmic corrections and the lighter quark contributions which we will neglect from now on.

The effective potential (with $\delta V_T = 0$) is schematically drawn in Fig. 3. It has three local extrema at

$$\rho_s = 0,$$
$$\rho_\pm(T) = \frac{C_1}{4\pi\lambda}\, T \pm \sqrt{\Big(\frac{C_1}{4\pi\lambda}\Big)^2\, T^2 + \rho_0^2 - \frac{2C_2}{\lambda}\, T^2}. \tag{18}$$

The first extremum $\rho_s = 0$ represents the Higgs vacuum of the symmetric (unbroken) phase, and the second extremum $\rho_-(T)$ represents the local maximum, and the third extremum $\rho_+(T)$ represent the local minimum Higgs vacuum of the broken phase. But notice that these two extrema $\rho_\pm$ appear only when T becomes smaller than T_2

$$T_2 = \frac{T_1}{\sqrt{1 - C_1^2/16\pi^2\lambda C_2}} \simeq 146.7 \text{ GeV},$$
$$T_1 = \sqrt{\frac{\lambda}{2C_2}}\,\rho_0 \simeq 146.4 \text{ GeV}, \qquad (19)$$

So above this temperature only $\rho_s = 0$ becomes the true vacuum of the effective potential, and the electroweak symmetry remains unbroken.

At $T = T_2$ we have

$$\rho_- = \rho_+ = (C_1/4\pi\lambda)\, T_2 \simeq 16.3 \text{ GeV}, \qquad (20)$$

but as temperature cools down below T_2 we have two local minima at ρ_s and ρ_+ with $V_T(0) < V_T(\rho_+)$, until T reaches the critical temperature T_c where $V_T(0)$ becomes equal to $V_T(\rho_+)$,

$$T_c = \frac{T_1}{\sqrt{1 - C_1^2/18\pi^2\lambda C_2}} \simeq 146.6 \text{ GeV},$$
$$\rho_+(T_c) = \frac{C_1}{3\pi\lambda}\, T_c \simeq 21.8 \text{ GeV}. \qquad (21)$$

So $\rho_s = 0$ remains the minimum of the effective potential for $T > T_c$. Notice that $T_c/\rho_0 \simeq 0.6$ but $\rho_+/\rho_0 \simeq 0.09$.

Below this critical temperature ρ_+ becomes the true minimum of the effective potential, but $\rho_s = 0$ remains a local (unstable) minimum till the temperature reaches T_1. But at $T = T_c$ the new vacuum bubbles start to nucleate at $\rho = \rho_+$, which takes over the unstable vacuum $\rho_s = 0$ completely at $T = T_1$ where $\rho_+(T_1)$ becomes around 32.6 GeV. From this point ρ_+ becomes the only (true) minimum, which transforms to the well-known Higgs vacuum ρ_0 at zero temperature.

This tells that two local minima ρ_s and ρ_+ are separated by an energy barrier, which means that the electroweak phase transition is of the first order. On the other hand, the energy barrier is very small. Moreover, the barrier lasts only for short period since the temperature difference from T_c to T_1 is very small, $\delta = (T_c - T_1)/T_c \simeq 0.001$. This strongly implies that the phase transition is very mildly the first order, almost the second order.

From (17) we can clculate the temperature-dependent Higgs mass $\bar{M}_H$,

$$\bar{M}_H^2 = \frac{d^2V_{eff}}{d\rho^2}\Big|_{\rho_{min}}$$
$$= \begin{cases} \left[(T/T_1)^2 - 1\right] M_H^2/2, & T \geq T_c, \\ \left[(\rho_+/\rho_0)^2 + 1 - (T/T_1)^2\right] M_H^2/2, & T < T_c. \end{cases} \qquad (22)$$

So $\bar{M}_H$ acquires its minimum value 5.53 GeV at $T = T_c$ and becomes 11.7 GeV at $T = T_1$, and approaches to the zero temperature value 125.7 GeV as the universe cools down.

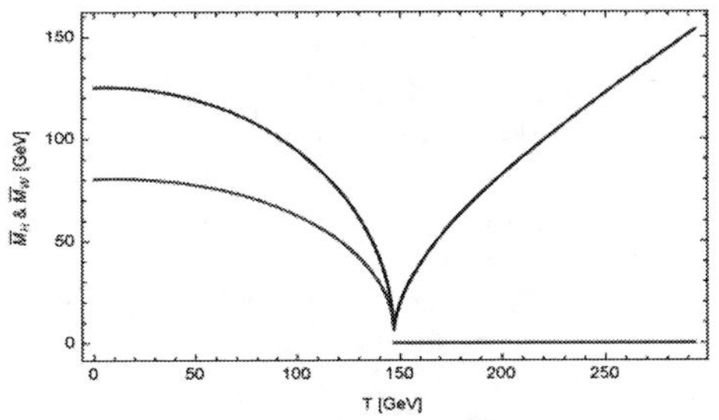

FIG. 4: The temperature-dependent Higgs and W-boson masses. The blue and red curves represent the Higgs and W-boson masses.

Similarly, the W-boson which was massless at high temperature becomes massive at $T = T_c$ when the Higgs field acquires the non-vanishing vacuum expectation value. So, as the Higgs vacuum $\rho = 0$ starts to tunnel to $\rho = \rho_+$ at T_c, the W-boson starts to become massive toward the value $g\rho_+(T_c)/2 \simeq 7.0$ GeV. And it acquires the mass 10.5 GeV at $T = T_1$, which approaches the well-known zero-temperature value at $T = 0$. The temperature-dependent Higgs and W-boson masses are shown in Fig. 4. They play an important role because they determine the correlation length $\xi = \bar{M}_H^{-1}$ and the monopole mass $M_m \simeq (1/\alpha) \times \bar{M}_W$.

The cosmological production of monopole in the second order phase transition was estimated by Kibble and later improved by Zurek [12, 15]. Kibble observed that the fundamental scale which determines the initial monopole density in the second order phase transition is the horizon distance at the time of the transition, and set a upper bound for the initial monopole density. Zurek improved this bound taking into account the fact that the phase transition does not take place instantaneously but continuously.

But this picture does not apply to the electroweak monopole production, because the electroweak phase transition is the first order. In this case, the correlation length becomes finite because the Higgs mass becomes non-zero at the critical temperature. Moreover, what is important in the monopole production is the time of monopole formation. To produce the monopoles we need the change of topology induced by the zero points of the Higgs field, which comes from the thermal fluctuations. And obviously these zero points, the seeds of the monopoles, do not appear instantaneously at the critical temperature, but come to exist some time after the phase transition. And it is well known that the temperature which governs this thermal fluctuation is the Ginzburg temperature, not the critical temperature. This means that even in the second order phase transition, the relevant scale in the monopole production is not the horizon distance at the critical temperature but the correlation length around the Ginzburg temperature.

Of course, in the strongly first order phase transition, the symmetric vacuum becomes meta-stable below the critical temperature, and the vacuum bubbles nucleate during the phase transition. So the vacuum bubble collisions during the phase transition becomes the monopole production mechanism [14]. However, the electroweak phase transition is very mildly first order, so that the bubble formation becomes unimportant. So the monopole formation takes place by the thermal fluctuation of the Higgs field after the phase transition, not by the bubble collisions during the phase transition.

Just below the critical temperature the Higgs field is still subject to large fluctuations which bring $\langle\rho\rangle$ back to zero. This is possible so long as

$$\xi^3 \Delta F \leq T, \tag{23}$$

where $\xi(T)$ is the correlation length and $\Delta F(T) = V(\rho_s) - V(\rho_+)$ is the difference in free energy density between two phases. This large fluctuation disappears when the equality holds, at the Ginzburg temperature. So we have to find the Ginzburg temperature first.

Unfortunately the Ginzburg temperature T_G of the potential (17) can not be expressed in a simple form, but we can find T_G numerically directly from (17) and (22),

$$T_G \simeq 57.6 \text{ GeV}, \qquad \xi_G \simeq 1.7 \times 10^{-16} \text{ cm}. \tag{24}$$

With this we may assume that the period of the monopole formation is between T_1 and T_G, and identify the temperature and correlation length of the initial monopole production to be the mean value of T_1 and T_G,

$$T_i = \frac{T_1 + T_G}{2} \simeq 102.0 \text{ GeV},$$
$$\xi_i = \frac{\xi(T_1) + \xi(T_G)}{2} \simeq 9.3 \times 10^{-16} \text{ cm}. \tag{25}$$

From this we can estimate the initial density of the monopoles

$$\Big(\frac{n_m}{T^3}\Big)_i \simeq \frac{g_P}{\xi_i^3 T_i^3} \simeq 9.1 \times 10^{-4}, \tag{26}$$

where g_P is the probability that one monopole is produced in one correlation volume, which we assume to be around 10 %.

To see how much energy we need to produce the electroweak monopoles, we calculate the energy density of the monopoles from (26)

$$\varrho_{mo}(T_i) = M_m (n_m)_i \simeq 9.0 \times 10^{-3}\; T_i^4 \Big(\frac{M_m}{1 \text{ TeV}}\Big),$$
$$\frac{\varrho_{mo}(T_i)}{\varrho(T_i)} \simeq 2.5 \times 10^{-4} \Big(\frac{M_m}{1 \text{ TeV}}\Big). \tag{27}$$

So, assuming the initial monopole mass to be around 1 TeV, we may conclude that the universe need to consume only a tiny fraction (about 0.025 %) of the total energy to produce the monopoles. This assures that the electroweak monopole production does not alter the standard cosmology.

In fact, we can say that even (26) is an overestimation. The reason is that, as we will see soon, the monopole-antimonopole capture radius becomes much bigger the correlation length ξ_i, so that most of the monopoles annihilate with the anti-monopoles as soon as they are produced.

V. COSMIC EVOLUTION OF ELECTROWEAK MONOPOLE

Although the monopoles are topological and absolutely stable, there are two factors which change the initial monopole density, the Hubble expansion and the annihilation of monopole-antimonopole pairs. So the change of monopole density n_m is determined by the Boltzmann equation [12, 13]

$$\frac{dn_m}{dt} + 3Hn_m = -\sigma n_m^2. \tag{28}$$

where H and σ are the Hubble parameter and the monopole annihilation cross section.

There are two things which affect the annihilation of the monopoles, the thermal Brownian motion (random walk) of the monopole in hot plasma of charged particles and the attraction between monopoles and antimonopoles. After the creation the monopoles diffuse in hot plasma by the Brownian motion with the mean free path l_{free} given by

$$l_{\text{free}} = v_t t_{\text{free}} \simeq \frac{1}{BT}\sqrt{\frac{M_m}{T}}, \quad B = \frac{1}{T}\sum_i n_i \sigma_i, \tag{29}$$

where $v_t \simeq \sqrt{T/M_m}$ and t_{free} are the thermal velocity and the mean free time of the monopoles, n_i and σ_i are the number density and the cross section of the i-th relativistic charged particles and the sum is the sum over all spin states [13].

With

$$n_i \simeq \frac{3\zeta(3)}{4\pi^2} T^3, \quad \sigma_i \simeq \Big(\frac{g_m q_i}{4\pi}\Big)^2 \frac{1}{T^2} = \Big(\frac{q_i}{e}\Big)^2 \frac{1}{T^2}, \tag{30}$$

we have

$$B \simeq \frac{3\zeta(3)}{4\pi^2} \sum_i \Big(\frac{q_i}{e}\Big)^2 \simeq 0.09 \times \sum_i \Big(\frac{q_i}{e}\Big)^2, \tag{31}$$

where q_i is the electric charge of the i-th particle and $\zeta(3) = 1.202...$ is the Riemann zeta function. Since

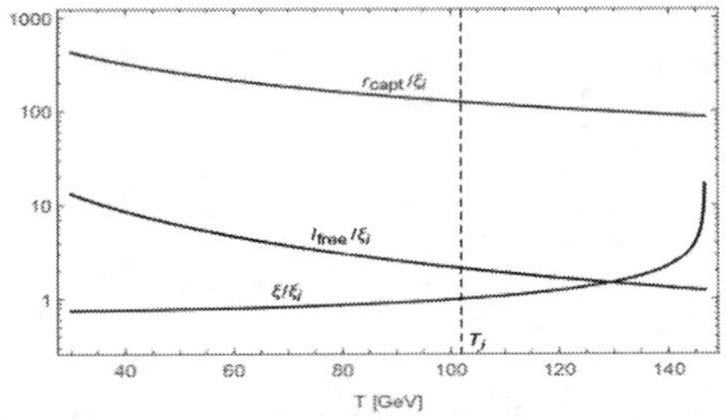

FIG. 5: The relevant scales, ξ in purple, l_{free} in blue, and r_{capt} in red, against T. They are normalized by the correlation length ξ_i at T_i. Here we set $M_m = 5$ TeV.

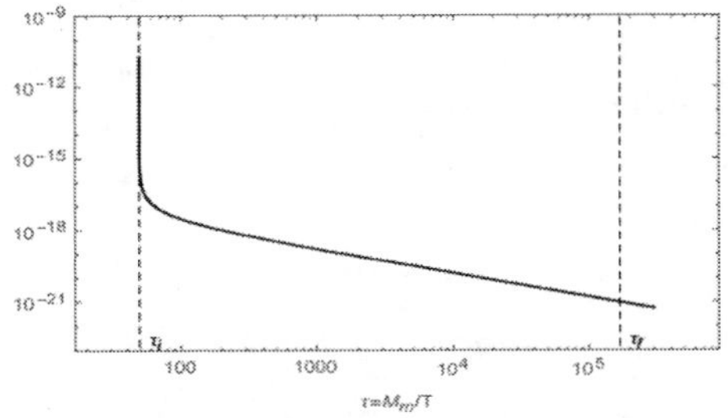

FIG. 6: The evolution of the monopole density n_M/T^3 against $\tau = M_m/T$. The final value of the monopole density is independent of the initial value.

the charged particles in the plasma are the leptons and quarks, we may put $B \simeq 3$. From this we have $v_t \simeq 0.31\times (1\ \text{TeV}/M_m)^{1/2}\ c$ and $l_{\text{free}} \simeq 2.0 \times (M_m/1\ \text{TeV})^{1/2} \times 10^{-16}$ cm around T_i.

Now, against the thermal random walk of the monopoles, we have the attractive Coulomb force between monopoles and anti-monopoles which makes them drift towards each other. The monopole drift velocity v_d at a distance r from the anti-monopole is given by

$$v_d \simeq \frac{1}{\alpha} \times \frac{1}{BT^2r^2} = \frac{\alpha_m}{BT^2r^2}, \tag{32}$$

where α_m is the monopole fine structure constant defined by $\alpha_m = g_m^2/4\pi = 1/\alpha$. Notice that $v_d(r = \xi_i) \simeq 2\ c$. This, of course, is unrealistic but tells that $v_t \ll v_d$. This shows that the Coulomb attraction between the monopoles and anti-monopoles is much stronger than the thermal diffusion.

Moreover, the monopole-antimonopole capture radius given by $r_{\text{capt}} \simeq \alpha_m/T$ is much bigger than the mean free path l_{free}. This strongly implies that the monopole-antimonopole annihilation is much more important around T_i, so that as soon as they are created, they annihilate each other. In Fig. 5 we plot the relevant scales ξ, l_{free}, and r_{capt} against T for comparison. Notice that the capture radius is much bigger than the mean free length and correlation length. This assures that (26) is indeed an overestimation.

Now, if we let the mean distance between the monopole and anti-monopole r be $r \simeq n_m^{-1/3}$, the capture time is given by

$$t_{\text{capt}} \simeq \frac{r}{v_d} \simeq \alpha \times \frac{BT^2}{n_m}. \tag{33}$$

From this we have the monopole-antimonopole annihilation cross section

$$\sigma \simeq \frac{1}{t_{\text{capt}} n_m} \simeq \frac{\alpha_m}{BT^2}. \tag{34}$$

With this we can solve the Boltzmann equation (28),

$$\frac{n_M}{T^3} = \frac{1}{A(\tau - \tau_i) + B},$$
$$A = \alpha_m \times \frac{Cm_p}{BM_m}, \quad B = \Big(\frac{n_M}{T^3}\Big)_i^{-1}. \tag{35}$$

Here M_m is treated as a constant. But in reality it depends on time, so that it should be understood as a mean value.

The evolution of the monopole density n_m/T^3 against M_m/T is shown in Fig. 6, where we have put $M_m = 5$ TeV. Notice that, regardless of the initial condition the monopole density approaches to a constant as the universe is cooled down.

The diffusive capture process is effective only when $\ell_{\text{free}} < r_{\text{capt}}$, which determines the temperature T_f below which the monopole-antimonopole annihilation ceases,

$$T_f \simeq \alpha^2 \times \frac{M_m}{B^2} \simeq 5.9 \times \Big(\frac{M_m}{1\ \text{TeV}}\Big)\ \text{MeV}, \tag{36}$$

so that

$$\tau_f = \frac{M_m}{T_f} \simeq (\alpha_m B)^2 \simeq 1.7 \times 10^5. \tag{37}$$

With $M_m = 5$ TeV, we have $T_f \simeq 29.5$ MeV, which is below the muon decoupling temperature. Actually, around this temperature B becomes very small so that the capture radius becomes smaller. This is because the only charged particles remaining in the plasma are electrons and positrons. But the above analysis clearly shows that the annihilation continues very long time.

The monopole density after annihilation becomes

$$\Big(\frac{n_m}{T^3}\Big)_f \simeq \alpha^3 \times \frac{M_m}{BCm_p} \simeq 1.8 \times 10^{-22}\Big(\frac{M_m}{1\ \text{TeV}}\Big). \tag{38}$$

Obviously this value is much lower than the initial density given by (26). The number of monopole within the comoving volume is conserved thereafter. But they still interact with the electron pairs in the hot plasma before decouple around $T_d \simeq 0.5$ MeV, when the electron pairs disappear and the interaction rate becomes less than the Hubble expansion rate.

Assuming that the expansion is adiabatic, the current number density and the energy density of the monopole is given by

$$\Big(\frac{n_m}{T^3}\Big)_0 = \Big(\frac{g_{s,0}}{g_{s,f}}\Big)\Big(\frac{n_m}{T^3}\Big)_f,$$
$$\rho_{mo,0} = M_m n_{m,0} = M_m \Big(\frac{g_{s,0}}{g_{s,f}}\Big)\Big(\frac{n_m}{T^3}\Big)_f T_0^3, \qquad (39)$$

where T_f is the temperature of universe when the monopole annihilation process effectively ceases and $T_0 = 2.73$ K $= 2.35\times10^{-13}$ GeV is the temperature of the universe today. Here g_s is the effective number of degrees of freedom in entropy.

So the current density parameter of monopole can be written

$$\Omega_{mo}h^2 = \frac{\rho_{mo,0}}{\rho_{c,0}h^{-2}} \simeq 1.2\times10^{-12}\times\Big(\frac{M_m}{1\ \text{TeV}}\Big)^2, \qquad (40)$$

where $\rho_{c,0} = 3H_0^2/8\pi G$ is the critical density of present universe and $h \simeq 0.678$ is the scaled Hubble constant in the unit $H_0/(100\ \text{km s}^{-1}\ \text{Mpc}^{-1})$. With $h \simeq 0.678$ and $M_m \simeq 5$ TeV, we have $\Omega_{mo} \simeq 6.53\times10^{-11}$. This is about 1.31×10^{-9} of the baryon density, which assures that the electroweak monopole cannot be a dark matter candidate.

In terms of the number density, this translates to about 6.1×10^{-20} cm^{-3}*, or about* 2.3×10^{-13} n_b*, where* $n_b \simeq 2.5\times10^{-7}$ cm^{-3} *is the number density of the baryons. Intuitively, this means that there are roughly* 6.6×10^7 *monopoles per unit volume of the earth.* This is a small but significant number which suggests that there are enough electroweak monopoles left over in the universe that we could detect.

VI. PARKER BOUND

Obviously the monopoles are accelerated in magnetic fields, in particular the intergalactic magnetic field. It is well known that the average strength of the galactic magnetic field B is about $3\mu G \simeq 1.2\times10^{-9}$ T. The energy gained by the monopole of charge $g_m = 4\pi/e$ passing across the magnetic field B of scale L is

$$\Delta E = \frac{4\pi}{e}LB \simeq 1.2\times10^{11}\times\Big(\frac{L}{10^{21}\ \text{cm}}\Big)\ \text{GeV}, \qquad (41)$$

where L is normalized to the typical coherence length of galactic magnetic field $L_0 = 300$ pc $\simeq 10^{21}$ cm. Traveling through the distance L_0, the monopole drains energy $\Delta E \simeq 10^{11}$ GeV from the magnetic field and becomes ultra-relativistic [25]. So, although the monopoles when decoupled around 29.5 MeV were non-relativistic, the remnant monopoles at present universe should be treated as relativistic.

Requiring that the rate of this energy loss in the galaxy is small compared to the time scale on which the galactic magnetic field can be regenerated, we can obtain the upper bound on the flux of the monopoles (with mass $M_m \leq 10^{17}$ GeV),

$$F \leq 10^{-15}\ \text{cm}^{-2}\text{sec}^{-1}\text{sr}^{-1}. \qquad (42)$$

This is the Parker bound [26].

The Parker bound sets a limit on the monopole density in the universe. The monopoles with velocity v_m uniformly distributed throughout the universe generate the monopole flux

$$F \simeq \frac{v_m}{4\pi}n_{m,0} = \frac{v_m}{4\pi}\Big(\frac{\rho_{c,0}}{h^2M_m}\Big)\Omega_{mo}h^2$$
$$\simeq 2.3\times10^{-2}\ \Omega_{mo}h^2\ \Big(\frac{v_m}{10^{-3}c}\Big)$$
$$\times\Big(\frac{1\ \text{TeV}}{M_m}\Big)\ \text{cm}^{-2}\text{sec}^{-1}\text{sr}^{-1}, \qquad (43)$$

where v_m is expressed in terms of the average virial velocity of the galaxy $10^{-3}c$. This, with (42), requires

$$\Omega_{mo}h^2 \lesssim 4.3\times10^{-14}\Big(\frac{10^{-3}c}{v_m}\Big)\times\Big(\frac{M_m}{1\ \text{TeV}}\Big). \qquad (44)$$

This set a stringent limit on the density for the relativistic ($v_m \simeq c = 1$) electroweak monopoles. This causes us a serious trouble, because our estimate of the density parameter (40) is roughly 10^5 times bigger than this limit.

We could think of possible ways to circumvent this trouble. First of all, we might suppose that the limit (44) is not trustable. Obviously this is an approximation. For example, the monopoles in the galactic magnetic field, when accelerated near the velocity of light, will make the Chrenkov radiation and will slow down to a limiting velocity considerably less than the velocity of light.

But a more realistic explanation could be that most of the electroweak monopoles in the universe are actually buried inside the galactic cores. This is quite possible, because the electroweak monopoles become the natural source for the premodial black holes and the structure formation. Certainly, as the heaviest stable particles, they could easily cause the density pertubation and become the seeds for the large scale structures in the universe.

Another possibility is that many of the electroweak monopoles are captured inside stellar objects when they hit (large) stellar objects, because they have a large capture cross section due to the magnetic interaction. In fact a relativistic electroweak monopole with mass 5 TeV is expected to travel less than 10 m in Aluminum before they are trapped [27]. So the electroweak monopoles coming to the earth loose most of the energy passing through the earth atmospheric sphere and become non-relativistic, and could easily be trapped near the earth surface. In fact we may conjecture that these trapped monopoles (and buried in large scale structures) could have been the very source of the intergalactic magnetic field.

More importantly, this strongly suggests that the stellar objects could have filtered out and diluted the density of the monopoles in the universe greatly, so that the remnant monopole density at present universe might have become much smaller than (26). So at this moment it is not clear whether our result is in contradiction with the Parker bound.

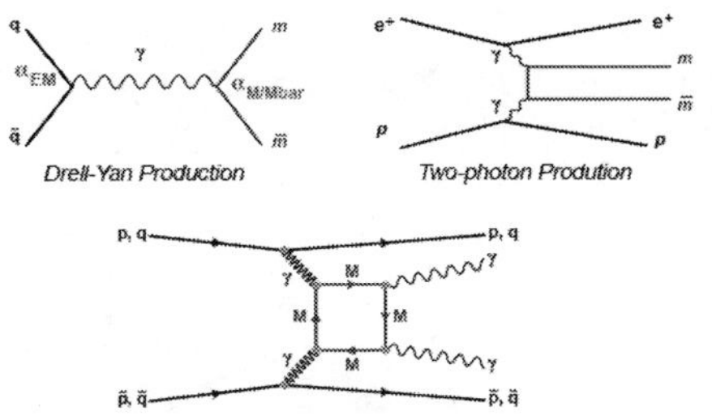

FIG. 7: The conventional electroweak monopole production mechanism. This should be compared with our topologcal production of the electroweak monopole.

VII. DISCUSSIONS

Our analysis confirms that, although the electroweak phase transition is of the first order, it is very mildly first order. So the monopole production mechanism is not the vacuum bubble collisions during the phase transition but the thermal fluctuations of the Higgs field after the phase transition, which plant the seed of the monopoles.

Our result shows that the electroweak monopoles are produced when the temperature of the universe was around 100 GeV (or about 10^{-11} sec after the big bang). And initially the monopoles are produced copiously, perhaps a bit too copiously to be realistic. But most of them are annihilated as soon as created, and this annihilation continues for a long time until the universe cools down to around 39.5 MeV, even after the muon pair annihilation. And eventually they decouple from the other matters at around 0.5 MeV, when the electron pairs annihilate and the monopole interaction rate becomes less than the Hubble expansion rate.

Because of this the electroweak monopole density become very small, $\Omega_{mo} \simeq 6.3 \times 10^{-11}$, in the present universe. This tells that, unlike the grand unification monopole, the electroweak monopole can not overclose the universe. As importantly, this assures that the production of the electroweak monopole does not alter the standard cosmology in any significant way, and exclude the possibility the electroweak monopole to be a candidate of the dark matter.

On the other hand, this means that there are enough monopoles left over, roughly 6.6×10^7 monopoles per every unit volume of the earth in the present universe. This strongly indicates that experimentally there are enough remnant monopoles that we should be able to detect without much difficulty.

However, we should to keep in mind the possibility that the actual density of the remnant monopoles could be much less than this. This is because many of them could have been buried in the the large scale structures of the universe and/or filtered out by stellar objects. In fact, the actual monopole density could be smaller by the factor 10^{-5}, or about 10^3 per every unit volume of the earth, as the Parker bound indicates.

A major concern at LHC is if it can produce the monopole. It has generally been believed that LHC could produce the monopole pair by Drell-Yan process ($p\bar{p} \to \gamma \to M\bar{M}$) and/or photon fusion process ($p\bar{p} \to p\bar{p} + \gamma\gamma \to p\bar{p} + M\bar{M}$). This is shown in Fig. 7, which should be compared with our topological production mechanism. This popular view would be acceptable if we treat the monopole as a point particle. But since the monopole is a topological particle, it is not clear if this is correct.

Obviously for LHC to produce the monopole the monopole mass must be less than 6.5 TeV. In our monopole production mechanism, however, LHC should satisfy two more conditions to produce the monopole. First, it should generate the hot plasma of the beam core of the size at least as big as the initial correlation length $\xi_i \simeq 10^{-15}$ cm. Second, this beam core should last at least ten times longer than 10^{-26} sec to allow enough fluctuations for the Higgs vacuum to plant the monopole seed.

Fortunately, the size of the beam core at LHC has been measured to be about 0.03 cm and lasts for about 10^{-12} sec, long enough for the Higgs field to allow more than enough thermal flucuations. This strongly implies that there is a good chance that LHC could actually produce one monopole-antimonopole pair, which MoEDAL could detect.

For the other experiments searching for the remnant monopoles in the universe, in particular for IceCube, ANTARES, and Auger, an important thing is to know the characteristic features of the remnant monopoles [9–11]. Although the electroweak monopoles were non-relativistic when they decoupled, the intergalactic magnetic field makes them highly relativistic at present universe. On the other hand, they loose most of the energy passing through the earth atmospheric sphere, so that near the earth surface they become non-relativistic again. So on earth these experiments should look for the non-relativistic monopole coming from the sky, which has mass around 4 to 10 TeV and magnetic charge $4\pi/e$.

In principle these experiments have the ability to detect the monopoles, but there is one catch here. These detectors (except Auger) are located underground. This could cause a problem, because the non-relativistic remnant monopoles may be trapped before they reach the detector. For example, IceCube is located under the 2 km thick iceberg, which the remnant monopoles may not be able to penetrate. If so, it would be very difficult for IceCube to detect the monopole. In fact this could be a main reason why IceCube could not find it. So these experiments have to find a way to circumbent this problem to detect the remnant monopoles.

Although the remnant electroweak monopoles in the present universe are insignificant, they have important physical implications. As we have speculated, they could have been the seed of the large scale structures in the universe. Indeed the electroweak monopoles with mass about 10^4 times heavier than the proton, could easily generate the density perturbation and become an excellent candidate for the seed of the large scale structures in the universe. Moreover, as the heaviest relativistic magnetically charged particles in the universe, they become the source of ultra-high energy cosmic rays. Furthermore, they could play an important role in the electroweak baryogenesis.

But the most important point of the electroweak monopole is that it must exist. This makes the detection of the electroweak monopole the final (topological) test of the standard model. In spite of huge efforts, however, the monopole detection so far has not been successful. There could be two reasons for this. First, many of these experiments were the blind seraches in the dark room, with few theoretical leads. Second, many were looking for different type of monopoles, in particular the grand unification monopole.

So, aiming at the electroweak monopole which has the unique characteristic features, we could enhance the probability to detect the monopole greatly. We hope that our analysis in this paper could help confirm the existence of the electroweak monopole.

ACKNOWLEDGMENT

The work is supported in part by the National Research Foundation of Korea funded by the Ministry of Education (Grants 2015-R1D1A1A0-1057578) and by Konkuk University.

[1] P.A.M. Dirac, Proc. Roy. Soc. London, **A133**, 60 (1931); Phys. Rev. **74**, 817 (1948).
[2] T.T. Wu and C.N. Yang, Phys. Rev. **D12**, 3845 (1975); Y.M. Cho, Phys. Rev. Lett. **44**, 1115 (1980); Phys. Lett. **B115**, 125 (1982).
[3] G. 't Hooft, Nucl. Phys. **B79**, 276 (1974); A.M. Polyakov, JETP Lett. **20**, 194 (1974).
[4] C. Dokos and T. Tomaras, Phys. Rev. **D21**, 2940 (1980).
[5] Y.M. Cho and D. Maison, Phys. Lett. **B391**, 360 (1997).
[6] Yisong Yang, Proc. Roy. Soc. London, **A454**, 155 (1998); Yisong Yang, *Solitons in Field Theory and Nonlinear Analysis* (Springer Monographs in Mathematics), p. 322 (Springer-Verlag) 2001.
[7] B. Acharya et al. (MoEDAL Collaboration), Phys. Rev. Lett. **118**, 061801 (2017).
[8] B. Acharya et al. (MoEDAL Collaboration), Int. J. Mod. Phys. **A29**, 1430050 (2014); JHEP 1608, 067 (2016); Y.M. Cho and J. Pinfold, Snowmass White paper, arXiv: [hep-ph] 1307.8390.
[9] R. Abbasi et al. (IceCube Collaboration) Phys. Rev. **D87**, 022001 (2013); M. Aartsen et al. (IceCube Collaboration), Eur. Phys. J. **C74**, 2938 (2014).
[10] S. Adrián-Martinez et al. (ANTARES Collaboration), Astropart. Phys. **35**, 634 (2012).
[11] A. Aab et al. (Pierre Auger Collaboration) Phys. Rev. **D94**, 082002 (2016).
[12] T.W.B. Kibble, J. Phys. **A9** 1387 (1976).
[13] J. P. Preskill, Phys. Rev. Lett. **43**, 1365 (1979).
[14] A. H. Guth and E. J. Weinberg, Nucl. Phys. **B212**, 321 (1983).
[15] W. H. Zurek, Phys. Rep. **276** 177 (1996).
[16] A. H. Guth, Phys. Rev. **D23**, 347 (1981); A. D. Linde, Phys. Lett. **B108** 389 (1982).
[17] D.A. Krizhnits and A.D. Linde, Phys. Lett. **B42**, 471 (1972); C. Bernard, Phys. Rev. **D9**, 3312 (1974); L. Dolan and R. Jakiw, Phys. Rev. **D9**, 3320 (1974); S. Weinberg, Phys. Rev. **D9**, 3357 (1974).
[18] Y. B. Zel'dovich and M.Yu. Khlopov, Phys. Lett. **79B**, 239 (1978); P.J. Adams,V. Canuto, and H.Y. Chiu, Phys. Lett. **61B**, 397 (1976).
[19] G. Anderson and L. Hall, Phys. Rev. **D45**, 2685 (1992); M. Dine, R. Leigh, P Huet, A. Linde, and D. Linde, Phys. Rev. **D46**, 550 (1992).
[20] Kyoungtae Kimm, J.H. Yoon, and Y.M. Cho, Eur. Phys. J. **C75**, 67 (2015); Kyoungtae Kimm, J.H. Yoon, S.H. Oh, and Y.M. Cho, Mod. Phys. Lett. **A31**, 1650053 (2016).
[21] J. Ellis, N.E. Mavromatos, and T. You, Phys. Lett **B756**, 29, (2016).
[22] Y.M. Cho, Kyoungtae Kimm, J.H. Yoon, Phys. Lett. **B761**, 203 (2016).
[23] S. Coleman, *Aspects of Symmetry* (Cambridge Univ. Press, 1985); T. Vachaspati and M. Barriola, Phys. Rev. Lett. **69**, 1867 (1992).
[24] F.A. Bais and R.J. Russell, Phys. Rev. **D11**, 2692 (1975); Y.M. Cho and P.G.O. Freund, Phys. Rev. **D12**, 1588 (1975).
[25] S. Burdin, M. Fairbaim, P. Mermod, D. Milstead, J. Pinfold, T. Sloan, and W. Taylor, Phys. Rep. **582**, 1 (2015).
[26] E.N. Parker, Astrophys. J. 160, 383 (1970).
[27] S. Cecchini, L. Patrizii, Z. Sahnoun, G. Sirri, and V. Togo, arXiv: hep-ph/1606.01220.

Gravitational Wave Polarizations in $f(R)$ Gravity and Scalar-Tensor Theory

Yungui Gong[1,⋆] and *Shaoqi* Hou[1,⋆⋆]

[1] *School of Physics, Huazhong University of Science and Technology, Wuhan, Hubei 430074, China*

Abstract. The detection of gravitational waves by the Laser Interferometer Gravitational-Wave Observatory opens a new era to use gravitational waves to test alternative theories of gravity. We investigate the polarizations of gravitational waves in $f(R)$ gravity and Horndeski theory, both containing scalar modes. These theories predict that in addition to the familiar + and × polarizations, there are transverse breathing and longitudinal polarizations excited by the massive scalar mode and the new polarization is a single mixed state. It would be very difficult to detect the longitudinal polarization by interferometers, while pulsar timing array may be the better tool to detect the longitudinal polarization.

1 Introduction

In September 14th, 2015, the Laser Interferometer Gravitational-Wave Observatory (LIGO) Scientific Collaboration and Virgo Collaboration announced the first detection of gravitational waves (GWs), namely GW150914 [1], nearly 100 years after Einstein's prediction of GWs based on his theory of general relativity (GR). Soon after, two more GW events, GW151226 [2], later that year, and GW170104 [3] early this year were observed. A new era is thus already open, and GW is a tool for testing GR and alternative theories of gravity, for GWs in alterative theories of gravity are predicted to have different polarization contents. The polarizations of null GWS with the propagation speed c were classified by the Newman-Penrose variables [4] in [5]. However, the application of the classification by the Newman-Penrose variables to the polarizations of GWs with the propagation speed different from c leads to confusing results [6]. Although LIGO presently cannot tell the polarizations of the observed GWs, it will become possible in the future when more interferometers are in operation. Other detectors, such as pulsar timing arrays, are also capable of distinguishing the polarization content of GWs [7].

Since the birth of GR, various alterative theories of gravity have been proposed for different motivations. One of the motivations is to cure the nonrenormalizability of GR. For instance, the inclusion of quadratic terms in Riemann tensor $R_{\mu\nu\rho\sigma}$ in the action makes gravity renormalizable [8, 9], and one simple realization of this idea is $f(R)$ gravity [10]. Another motivation is to explain the present accelerating expansion of the universe [11, 12]. The addition of new gravitational degrees of freedom might do the work, and the scalar-tensor theory is the simplest alternative metric theory of gravity

⋆e-mail: yggong@hust.edu.cn
⋆⋆e-mail: shou1397@hust.edu.cn

which contains a scalar field ϕ besides the metric tensor $g_{\mu\nu}$ to describe gravity. In 1974, Horndeski constructed the most general scalar-tensor theory of gravity whose action has higher derivatives of ϕ and $g_{\mu\nu}$, but yields at most second order differential equations of motion [13]. Therefore, there is no Ostrogradsky instability [14].

In this talk, we will discuss the polarizations of GWs predicted by $f(R)$ gravity [6] and Horndeski theory [15]. In Section 2, we focus on $f(R)$ gravity. We will first resolve the debate on how many physical degrees of freedom are contained in $f(R)$ gravity, recently raised by Refs. [16–19]. Next, we determine the polarizations of $f(R)$ gravity using geodesic deviation equations. In Section 3, we consider the polarizations of GWs in the scalar-tensor theory, in particular, Horndeski theory. We first determine the polarizations with similar method used for $f(R)$ gravity, then we consider the possibility to detect the polarizations using the pulsar timing arrays and the interferometers. Finally, there is a brief conclusion.

2 Gravitational Wave Polarizations in $f(R)$ Gravity

Utiyama and DeWitt showed that the addition of the quadratic terms $R_{\mu\nu\rho\sigma}R^{\mu\nu\rho\sigma}$ and R^2 to the Einstein-Hilbert action makes gravity renormalizable at one-loop level [8], and then, Stelle proved the renormalizability at all loop levels [9]. In four dimensions, the volume integration of the Gauss-Bonnet term $\mathcal{G} = R_{\mu\nu\rho\sigma}R^{\mu\nu\rho\sigma} - 4R_{\mu\nu}R^{\mu\nu} + R^2$ vanishes for a simply connected spacetime, as usually considered. So the general quadratic terms can be given by

$$\frac{1}{2\kappa}\int d^4x\sqrt{-g}(\alpha' R_{\mu\nu}R^{\mu\nu} + \alpha R^2), \tag{1}$$

where $\kappa = 8\pi G_{\mathrm{N}}$ with G_{N} Newton's constant, g is the determinant of $g_{\mu\nu}$ and α', α are constants. Setting $\alpha' = 0$, one obtains a model which was first proposed by Starobinsky as an inflationary model [20], and is consistent with the observations of *Planck* [21]. One may thus generalize this action by considering a generic function f of R [10],

$$S = \frac{1}{2\kappa}\int d^4x\sqrt{-g}f(R). \tag{2}$$

A special model with $f(R) = R + \beta R^{-1}$ was applied to explain the late time cosmic acceleration [22–25], but the solar system tests have ruled it out [26, 27]. So more viable $f(R)$ models were proposed recently [28–34].

In fact, $f(R)$ gravity is equivalent to a scalar-tensor theory [35, 36], as the action can be rewritten as

$$S = \frac{1}{2\kappa}\int d^4x\sqrt{-g}[f(\varphi) + (R-\varphi)f'(\varphi)], \tag{3}$$

where $f'(\varphi) = df(\varphi)/d\varphi$. It can be easily shown that $\varphi = R$ on-shell. The polarization content of GWs in $f(R)$ gravity and the detection have been studied in Refs. [16, 17, 37, 38]. Authors of Refs. [16, 17] found out that the massive scalar mode induces the longitudinal polarization, and they claimed that there are four degrees of freedom using the Newman-Penrose (NP) formalism [4, 5]. Kausar *et. al.* supported this claim by arguing that the traceless condition cannot be implemented [18]. However, Myung's work shows that there is no issue with implementing the transverse traceless condition [17], and there are only three degrees of freedom in $f(R)$ gravity [19].

In Ref. [6], we investigated the polarizations of GWs in $f(R)$ gravity and attempted to resolve the debate on how many degrees of freedom propagating in this theory. As it will become clear soon, there are three physical propagating degrees of freedom in $f(R)$ gravity. Therefore, there are the familiar + and × polarizations as in GR, and the transverse and longitudinal polarizations excited by the massive scalar mode $\varphi = R$. We also pointed out that the original NP formalism devised in Ref. [5] for identifying the polarizations of GWs in a generic metric theory of gravity cannot be simply applied to massive GWs. In order to reveal the polarizations, one simply calculates the geodesic deviations caused by the GW, provided that test particles follow geodesics, as usually assumed.

2.1 Equations of Motion

The field equations can be obtained by the variational principle,

$$f'(R)R_{\mu\nu} - \frac{1}{2}f(R)g_{\mu\nu} - \nabla_\mu\nabla_\nu f'(R) + g_{\mu\nu}\Box f'(R) = 0, \tag{4}$$

where $\Box = g^{\mu\nu}\nabla_\mu\nabla_\nu$. Taking the trace of Eq. (4), we get

$$f'(R)R + 3\Box f'(R) - 2f(R) = 0. \tag{5}$$

For the particular model $f(R) = R + \alpha R^2$, Eq. (4) becomes

$$R_{\mu\nu} - \frac{1}{2}\eta_{\mu\nu}R - 2\alpha\left(\partial_\mu\partial_\nu R - \eta_{\mu\nu}\Box R\right) = 0, \tag{6}$$

Take the trace of Eq. (6) or using Eq. (5), we have

$$(\Box - m^2)R = 0, \tag{7}$$

where $m^2 = 1/(6\alpha)$ with $\alpha > 0$. The graviton mass m has been bounded from above by GW170104 as $m < m_b = 7.7 \times 10^{-23}$ eV/c^2 [3], and the observation of the dynamics of the galaxy cluster puts a more stringent limit, $m < 2 \times 10^{-29}$ eV/c^2 [39].

To obtain the GW solutions in the flat spacetime background, perturb the metric around the Minkowski metric $g_{\mu\nu} = \eta_{\mu\nu} + h_{\mu\nu}$ to the first order of $h_{\mu\nu}$, and introduce an auxiliary metric tensor

$$\bar{h}_{\mu\nu} = h_{\mu\nu} - \frac{1}{2}\eta_{\mu\nu}h - 2\alpha\eta_{\mu\nu}R, \tag{8}$$

which transforms in an infinitesimal coordinate transformation $x^\mu \to x'^\mu = x^\mu + \epsilon^\mu$ in the following way,

$$\bar{h}'_{\mu\nu} = \bar{h}_{\mu\nu} - \partial_\mu\epsilon_\nu - \partial_\nu\epsilon_\mu + \eta_{\mu\nu}\partial_\rho\epsilon^\rho, \tag{9}$$

where to the first order of perturbation, the index was raised or lowered by the Minkowski metric $\eta_{\mu\nu}$, i.e., $\epsilon_\mu = \eta_{\mu\nu}\epsilon^\nu$. If we choose ϵ_μ so that it satisfies $\Box\epsilon_\nu = \partial^\mu\bar{h}_{\mu\nu}$ with $\Box = \partial^\mu\partial_\mu$ from now on, then we get the Lorenz gauge condition $\partial^\mu\bar{h}'_{\mu\nu} = 0$. Note that there is still some residual gauge freedom, i.e., $x^{\mu\prime} = x^\mu + \xi^\mu$ with $\Box\xi^\mu = 0$. If ξ^μ also satisfies the relation $\partial_\mu\xi^\mu = -\bar{h}/2$, then $\bar{h}' = 0$ holds. Therefore, it is always possible to choose the transverse traceless gauge condition

$$\partial^\mu\bar{h}_{\mu\nu} = 0, \quad \bar{h} = \eta^{\mu\nu}\bar{h}_{\mu\nu} = 0. \tag{10}$$

With this gauge condition, some algebraic manipulations lead to

$$\Box\bar{h}_{\mu\nu} = 0. \tag{11}$$

Therefore, the equations of motion are Eqs. (7) and (11).

The plane wave solution can be obtained immediately,

$$\bar{h}_{\mu\nu} = e_{\mu\nu}\exp(iq_\mu x^\mu) + c.c., \tag{12}$$

$$R = \phi_1 \exp(ip_\mu x^\mu) + c.c., \tag{13}$$

where $c.c.$ stands for the complex conjugation, $e_{\mu\nu}$ and ϕ_1 are the amplitudes with $q^\nu e_{\mu\nu} = 0$ and $\eta^{\mu\nu} e_{\mu\nu} = 0$, and q_μ and p_μ are the wave numbers satisfying

$$\eta^{\mu\nu} q_\mu q_\nu = 0, \quad \eta^{\mu\nu} p_\mu p_\nu = -m^2. \tag{14}$$

2.2 Physical Degrees of Freedom

In this subsection, we will find the number of physical degrees of freedom in $f(R)$ gravity, using two different methods: examining the energy current carried by GWs and carrying out the Hamiltonian analysis.

2.2.1 Energy Current of GWs

The first method to determine the number of physical degrees of freedom is to calculate the energy current carried by the GW propagating, for example, in the $+z$ direction. Let us consider the null GWs with $q^\mu = \omega(1, 0, 0, 1)$, then the energy current is given by [40],

$$\begin{aligned} t_{0z} =& \frac{1}{\kappa}\langle G^{(1)}_{0z} - G_{0z}\rangle \\ =& \frac{1}{2\kappa}\left\langle \omega^2\left[\left(\frac{e_{xx} - e_{yy}}{2}\right)^2 + e^2_{xy}\right] + 48\alpha^2 R_{,0} R_{,z}\right\rangle, \end{aligned} \tag{15}$$

where $G^{(1)}_{\mu\nu}$ is the first order Einstein tensor. Note that to obtain the above result, the solution (12) is used, but the traceless condition is dropped. Eq. (15) makes it clear that a null GW for which $e_{xx} - e_{yy}$ and e_{xy} are both zero does not transport energy. So if a null wave has a nonvanishing trace $\bar{h}$ such that $e_{xx} + e_{yy} \neq 0$, it does not carry energy, which implies that the trace $\bar{h}$ is not a dynamical degree of freedom. The null GW $\bar{h}_{\mu\nu}$ is physically transverse and traceless, so has two degrees of freedom as in GR. In addition, the Ricci scalar R is the third degree of freedom. So totally, there are three degrees of freedom.

2.2.2 Hamiltonian Analysis

The Hamiltonian analysis of $f(R)$ gravity has been done in Refs. [41–47]. In our work, we did the Hamiltonian analysis with the action (3) for simplicity. With the Arnowitt-Deser-Misner (ADM) foliation [48, 49], the metric takes the standard form

$$ds^2 = -N^2 dt^2 + h_{jk}(dx^j + N^j dt)(dx^k + N^k dt), \tag{16}$$

where N, N^j, h_{jk} are the lapse function, the shift function and the induced metric on the constant t slice Σ_t, respectively. Let $n_\mu = -N\nabla_\mu t$ be the unit normal to Σ_t, and the exterior curvature is $K_{\mu\nu} = \nabla_\mu n_\nu + n_\mu n^\rho \nabla_\rho n_\nu$. In terms of ADM variables and setting $\kappa = 1$, the action (3) is

$$S = \int d^4x N\sqrt{h}\Big[\frac{1}{2}f'(\mathscr{R} - \varphi) + \frac{1}{2}f + \frac{1}{2}f'(K_{jl}K^{jl} - K^2) + \frac{K}{N}(N_j D^j f' - f''\dot{\varphi}) + D_j f' D^j \ln N\Big], \tag{17}$$

where $\mathscr{R}$ is the Ricci scalar for h_{jk} and $K = h^{jk}K_{jk}$. In this action, there are 11 dynamical variables: N, N_j, h_{jk} and φ. Four primary constraints are immediately recognized, i.e., the conjugate momenta for N and N_j vanish weakly,

$$\pi^N = \frac{\delta S}{\delta \dot{N}} \approx 0, \quad \pi^j = \frac{\delta S}{\delta \dot{N}_j} \approx 0. \tag{18}$$

The conjugate momenta for h_{jk} and φ can also be obtained, and the Legendre transformation results in the following Hamiltonian,

$$H = \int_{\Sigma_t} d^3x \sqrt{h}(NC + N_j C^j), \tag{19}$$

where we dropped the boundary terms. Thus, the consistence conditions yield four secondary constraints, i.e., $C \approx 0$ and $C^j \approx 0$ [1], and it can be shown that there are no further secondary constraints. It can also be checked that all the constraints are of the first class, so the number of physical degrees of freedom of $f(R)$ gravity is

$$n = \frac{22 - 8 \times 2}{2} = 3, \tag{20}$$

as expected.

2.3 Polarization Content

To reveal the polarization content of GWs in $f(R)$ gravity, let us calculate the geodesic deviation equations caused by the GW propagating in the $+z$ direction with the wave vectors given by

$$q^\mu = \omega(1, 0, 0, 1), \quad p^\mu = (\Omega, 0, 0, \sqrt{\Omega^2 - m^2}). \tag{21}$$

Inverting Eq. (8), one obtains the metric perturbation,

$$h_{\mu\nu} = \bar{h}_{\mu\nu}(t - z) - 2\alpha\eta_{\mu\nu}R(vt - z), \tag{22}$$

where $v = \sqrt{\Omega^2 - m^2}/\Omega$. It is expected that $\bar{h}_{\mu\nu}$ induces the + and × polarizations. So let us investigate the polarization state caused by the massive scalar field by setting $\bar{h}_{\mu\nu} = 0$. The geodesic deviation equations are

$$\ddot{x} = \alpha\ddot{R}x, \quad \ddot{y} = \alpha\ddot{R}y, \quad \ddot{z} = -\alpha m^2 Rz = -\frac{1}{6}Rz. \tag{23}$$

Therefore, the massive scalar field induces a mix of the pure longitudinal and the breathing modes.

The NP formalism [5] cannot be applied to infer the polarization content of $f(R)$ gravity because the NP formalism was formulated for null GWs. In fact, the calculation shows that Ψ_2 is zero. According to the NP formalism, $\Psi_2 = 0$ means the absence of the longitudinal polarization. From Eq. (23), we see the existence of the longitudinal polarization. Since the six polarization states are completely determined by the electric part of the Riemann tensor R_{itjt}, we can still use the six polarizations classified by the NP formalism as the base states. In terms of these polarization base states, the polarization state caused by the massive scalar field is a mix of the longitudinal and the breathing modes. Since there is no massless limit in $f(R)$ gravity, so we consider more general massive scalar-tensor theory of gravity.

[1] For details, please refer to Ref. [6]

3 Gravitational Wave Polarizations in Scalar-Tensor Theory

As stated before, $f(R)$ gravity is equivalent to a scalar-tensor gravity. We extended our work [6] to the scalar-tensor theory, and study the polarization content of GWs in Horndeski theory [15]. The action is given by[13],

$$S = \int d^4x\sqrt{-g}(L_2 + L_3 + L_4 + L_5), \tag{24}$$

where

$$L_2 = K(\phi, X), \quad L_3 = -G_3(\phi, X)\Box\phi, \quad L_4 = G_4(\phi, X)R + G_{4,X}\left[(\Box\phi)^2 - (\nabla_\mu\nabla_\nu\phi)(\nabla^\mu\nabla^\nu\phi)\right],$$
$$L_5 = G_5(\phi, X)G_{\mu\nu}\nabla^\mu\nabla^\nu\phi - \frac{1}{6}G_{5,X}\left[(\Box\phi)^3 - 3(\Box\phi)(\nabla_\mu\nabla_\nu\phi)(\nabla^\mu\nabla^\nu\phi) + 2(\nabla^\mu\nabla_\alpha\phi)(\nabla^\alpha\nabla_\beta\phi)(\nabla^\beta\nabla_\mu\phi)\right].$$

Here, $X = -\nabla_\mu\phi\nabla^\mu\phi/2$, $\Box\phi = \nabla_\mu\nabla^\mu\phi$, the functions K, G_3, G_4 and G_5 are arbitrary functions of ϕ and X, and $G_{j,X}(\phi, X) = \partial G_j(\phi, X)/\partial X$ with $j = 4, 5$. Horndeski theory includes several interesting theories as its subclasses. For example, to reproduce $f(R)$ gravity, one can set $G_3 = G_5 = 0$, $K = f(\phi) - \phi f'(\phi)$ and $G_4 = f'(\phi)$ with $f'(\phi) = df(\phi)/d\phi$.

3.1 Linearized Equations of Motion

The equations of motion can be derived with the variational principle. Interested readers should be referred to Ref. [50]. We are interested in the GW solutions in the flat spacetime background for which $g_{\mu\nu} = \eta_{\mu\nu}$ and $\phi = \phi_0$ with a constant ϕ_0. This requires that $K(\phi_0, 0) = 0$ and $K_{,\phi_0} = \partial K(\phi, X)/\partial\phi|_{\phi=\phi_0, X=0} = 0$. Now the fields are expanded around the background, $g_{\mu\nu} = \eta_{\mu\nu} + h_{\mu\nu}$ and $\phi = \phi_0 + \varphi$. To the first order of approximation, the equations of motion are

$$(\Box - m^2)\varphi = 0, \quad G^{(1)}_{\mu\nu} - \frac{G_{4,\phi_0}}{G_4(0)}(\partial_\mu\partial_\nu\varphi - \eta_{\mu\nu}\Box\varphi) = 0, \tag{25}$$

where $G_4(0) = G_4(\phi_0, 0)$, $K_{,X_0} = \partial K(\phi, X)/\partial X|_{\phi=\phi_0, X=0}$ and the mass squared of the scalar field is

$$m^2 = -\frac{K_{,\phi_0\phi_0}}{K_{,X_0} - 2G_{3,\phi_0} + 3G^2_{4,\phi_0}/G_4(0)}. \tag{26}$$

Similar to Eq. (8), define a field $\tilde{h}_{\mu\nu}$,

$$\tilde{h}_{\mu\nu} = h_{\mu\nu} - \frac{1}{2}\eta_{\mu\nu}\eta^{\alpha\beta}h_{\alpha\beta} - \frac{G_{4,\phi_0}}{G_4(0)}\eta_{\mu\nu}\varphi, \tag{27}$$

and choose the transverse traceless gauge $\partial_\mu\tilde{h}^{\mu\nu} = 0$, $\eta^{\mu\nu}\tilde{h}_{\mu\nu} = 0$ by using the freedom of coordinate transformation, then the linearized equations (25) become two wave equations,

$$(\Box - m^2)\varphi = 0, \tag{28}$$
$$\Box\tilde{h}_{\mu\nu} = 0. \tag{29}$$

3.2 Polarization Content

A inspection of Eq. (29) makes it clear that the field $\tilde{h}_{\mu\nu}$ denotes the usual massless gravitons and it has two polarization states, the + and × modes. The plane waves (12) and (13) are also the solutions to Eqs. (29) and (28). For GWs propagating in the $+z$ direction with waves vectors given by Eq. (21), one gets the following nonvanishing NP variables

$$\Psi_4 = -\omega^2(\tilde{h}_{xx} - i\tilde{h}_{xy}), \quad \Phi_{22} = \frac{(\Omega + \sqrt{\Omega^2 - m^2})^2}{4}\sigma\varphi, \tag{30}$$

$$\Phi_{00} = \frac{4(\Omega - \sqrt{\Omega^2 - m^2})^2}{(\Omega + \sqrt{\Omega^2 - m^2})^2}\Phi_{22}, \quad \Phi_{11} = -\Lambda = \frac{4m^2}{(\Omega + \sqrt{\Omega^2 - m^2})^2}\Phi_{22}, \tag{31}$$

where $\sigma = G_{4,\phi_0}/G_4(0)$. This result shows that the massive scalar field makes Φ_{22}, Φ_{00}, Φ_{11} and Λ nonzero and they are all proportional to φ. However, for the NP variables associated with the Weyl tensor, only $\Psi_4 \neq 0$ and in particular, $\Psi_2 = 0$. In fact, Ψ_4 denotes the usual + and × modes for the massless gravitons. If the scalar field is massless, only Φ_{22} is nonvanishing and it represents the transverse breathing mode [5].

Next, let us focus on the GW induced by the scalar field φ. Setting $\tilde{h}_{\mu\nu} = 0$ and inverting Eq. (27) to obtain $h_{\mu\nu}$ in terms of φ, the geodesic deviation equations are

$$\ddot{x} = -\frac{1}{2}\Omega^2\sigma\varphi x, \quad \ddot{y} = -\frac{1}{2}\Omega^2\sigma\varphi y, \quad \ddot{z} = -\frac{1}{2}m^2\sigma\varphi z. \tag{32}$$

These expressions take essentially the same form as Eq. (23). Therefore, the massive scalar field excites the longitudinal and transverse breathing polarizations, while the massless scalar field excites only the transverse breathing polarization. As discussed in the previous section, $\Psi_2 = 0$ means the absence of the longitudinal polarization, however, there exists the longitudinal mode even though $\Psi_2 = 0$, so the NP formalism derived for null GWs cannot be directly applied to the massive GWs, and again the polarization state caused by the massive scalar field is a mix of the longitudinal and the breathing modes.

3.3 Experimental Tests

In the interferometer, a photon is emanated from the beam splitter, bounced back by the mirror and returns to the beam splitter. The propagation time when there is no GW does not equal the one when there is. Following [16, 51], we calculate the interferometer response functions of the transverse and longitudinal polarizations in the frequency domain. To calculate the response functions, we assume the beam splitter is placed at the origin of the coordinate system. Figure 1 shows the absolute value of the longitudinal and transverse response functions for aLIGO to a scalar GW with the masses 1.2×10^{-22} eV/c^2 [1] and 7.7×10^{-23} eV/c^2 [3]. From this graph, it is clear that the response functions for the longitudinal polarization are smaller than that of the transverse breathing modes by several orders of magnitude at high frequency regime, so it is difficult to test the existence of longitudinal polarization by interferometer such as aLIGO.

It is possible to tell what the polarization content of GWs is by analyzing the data of the pulsar timing arrays [7, 52–58]. The stochastic gravitational wave background causes the pulse time-of-arrival (TOA) residuals $\tilde{R}(t)$ of pulsars which can be measured [7]. The TOA residuals of any pair of pulsars (named a and b) are correlated. The cross-correlation function is defined to be $C(\theta) = \langle\tilde{R}_a(t)\tilde{R}_b(t)\rangle$ where θ is the angular separation between a and b, and the brackets indicate the ensemble average over the stochastic background. Figure 2 shows the normalized correlation function $\zeta(\theta) = C(\theta)/C(0)$ induced by the massless field $\tilde{h}_{\mu\nu}$ (the left panel) and the scalar field φ (the right panel). The curve on the left panel is actually the same as the one in GR. On the right panel, we calculated $\zeta(\theta)$ for the massless (labeled by Breathing) and the massive (3 different masses in units of m_b) cases.

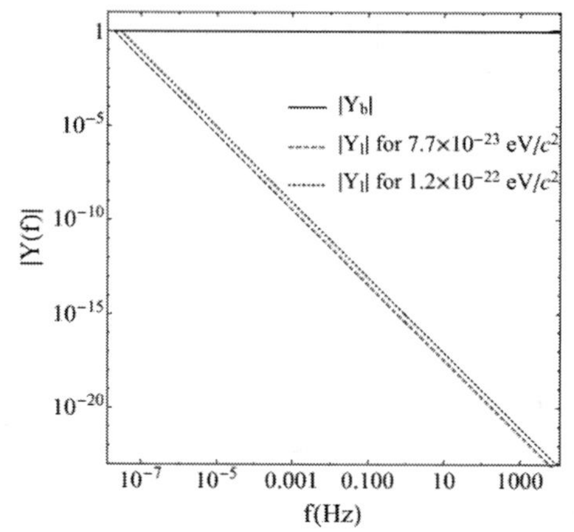

Figure 1. The absolute values of the longitudinal and transverse response functions $|Y_l(f)|$ and $|Y_b(f)|$ as functions of f for aLIGO to a scalar gravitational wave with the masses $1.2 \times 10^{-22}\,\mathrm{eV}/c^2$ [1] (dotted blue curve) and $7.7 \times 10^{-23}\,\mathrm{eV}/c^2$ [3] (dashed red curve). The solid curve denotes $|Y_b(f)|$.

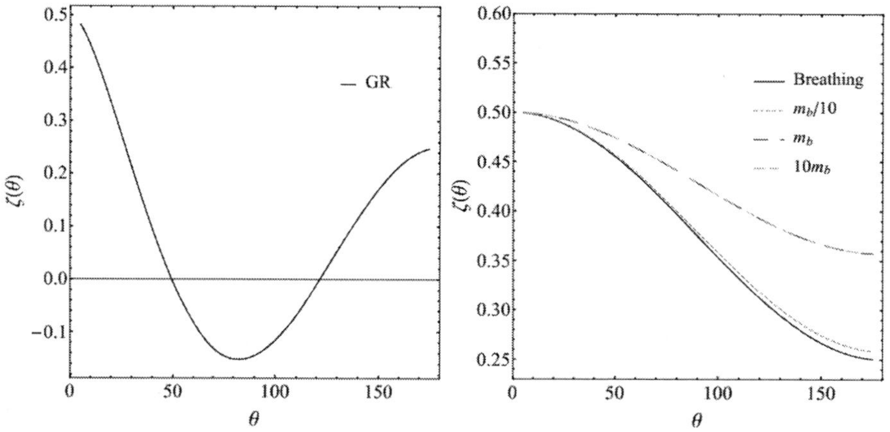

Figure 2. The normalized correlation function $\zeta(\theta) = C(\theta)/C(0)$ as a function of θ for the massless field $\tilde{h}_{\mu\nu}$ (the left panel) and the scalar field φ (the right panel).

Therefore, $\zeta(\theta)$ has rather different behaviors for $\tilde{h}_{\mu\nu}$ and φ. In addition, $\zeta(\theta)$ induced by φ is quite sensitive to small masses with $m \lesssim m_b$, while for larger masses, $\zeta(\theta)$ barely changes. So this provides the possibility to determine the polarization content of GWs. In Ref. [55], Lee also calculated the cross-correlation functions and his results (the right two panels in his Figure 1) are different from those on the right panel in Figure 2, because in his treatment, the longitudinal and the transverse polarizations were assumed to be two independent modes. In our approach, it is not permissible to calculate the cross-correlation function separately for the longitudinal and the transverse polarizations, as they are both excited by the same field φ and the polarization state is a single mode.

4 Conclusion

We first studied the physical degrees of freedom in $f(R)$ gravity with two different approaches. Both of them give the same result: three physical propagating degrees of freedom. We then solved the linearized equations of motion for $f(R)$ gravity and obtained the plane wave solution. The geodesic deviation equations are computed to reveal the polarizations, and there are longitudinal and transverse breathing polarizations, in addition to the + and × polarizations. We also extended the work in Ref. [6] to Horndeski theory [15]. The analysis shows that the scalar field excites both the longitudinal and the transverse breathing polarizations if it is massive, while it excites only the transverse breathing polarization if it is massless. We find that Ψ_2 is zero in both $f(R)$ gravity and Horndeski theory, and that the longitudinal polarization exists even though $\Psi_2 = 0$. The results show the failure of NP formalism [5] in classifying the polarizations of non-null GWs. Since the six polarization states are completely determined by the electric part of the Riemann tensor R_{itjt}, we can still use the six polarizations classified by the NP formalism as the base states. In terms of these polarization base states, the polarization state caused by the massive scalar field is a mix of the longitudinal and the breathing modes. The interferometer responses functions were then computed and found out that it is difficult for interferometers to detect the longitudinal polarization. We also predicted the cross-correlation functions. It implies the possibility of using pulsar timing arrays to detect the polarizations caused by the scalar field.

Acknowledgements: We would like to thank Ke-Jia Lee for useful discussions. This research was supported in part by the Major Program of the National Natural Science Foundation of China under Grant No. 11690021 and the National Natural Science Foundation of China under Grant No. 11475065.

References

[1] B.P. Abbott et al. (LIGO Scientific Collaboration and Virgo Collaboration), Phys. Rev. Lett. **116**, 061102 (2016)
[2] B.P. Abbott et al. (LIGO Scientific Collaboration and Virgo Collaboration), Phys. Rev. Lett. **116**, 241103 (2016)
[3] B.P. Abbott et al. (LIGO Scientific Collaboration and Virgo Collaboration), Phys. Rev. Lett. **118**, 221101 (2017)
[4] E. Newman, R. Penrose, J. Math. Phys. **3**, 566 (1962)
[5] D.M. Eardley, D.L. Lee, A.P. Lightman, Phys. Rev. D **8**, 3308 (1973)
[6] D. Liang, Y. Gong, S. Hou, Y. Liu, Phys. Rev. D **95**, 104034 (2017)
[7] R.W. Hellings, G.S. Downs, Astrophys. J. **265**, L39 (1983)
[8] R. Utiyama, B.S. DeWitt, J. Math. Phys. **3**, 608 (1962)
[9] K.S. Stelle, Phys. Rev. D **16**, 953 (1977)
[10] H.A. Buchdahl, Mon. Not. Roy. Astron. Soc. **150**, 1 (1970)
[11] G.R. Adam et al., Astron. J. **116**, 1009 (1998)
[12] S. Perlmutter et al., Astrophys. J. **517**, 565 (1999)
[13] G.W. Horndeski, Int. J. of Theor. Phys. **10**, 363 (1974)
[14] M. Ostrogradsky, Mem. Acad. St Petersbourg **6**, 385 (1850)
[15] S. Hou, Y. Gong, Y. Liu (2017), 1704.01899
[16] C. Corda, JCAP **2007**, 009 (2007)
[17] C. Corda, Int. J. Mod. Phys. A **23**, 1521 (2008)
[18] H.R. Kausar, L. Philippoz, P. Jetzer, Phys. Rev. D **93**, 124071 (2016)

[19] Y.S. Myung, Adv. High Energy Phys. **2016**, 3901734 (2016)
[20] A.A. Starobinsky, Phys. Lett. B **91**, 99 (1980)
[21] P.A.R. Ade et al. (Planck Collaboration), Astron. Astrophys. **594**, A20 (2016)
[22] D.N. Vollick, Phys. Rev. D **68**, 063510 (2003)
[23] S. Nojiri, S.D. Odintsov, Phys. Rev. D **68**, 123512 (2003)
[24] S.M. Carroll, V. Duvvuri, M. Trodden, M.S. Turner, Phys. Rev. D **70**, 043528 (2004)
[25] E.E. Flanagan, Phys. Rev. Lett. **92**, 071101 (2004)
[26] T. Chiba, Phys. Lett. B **575**, 1 (2003)
[27] A.L. Erickcek, T.L. Smith, M. Kamionkowski, Phys. Rev. D **74**, 121501 (2006)
[28] W. Hu, I. Sawicki, Phys. Rev. D **76**, 064004 (2007)
[29] A.A. Starobinsky, JETP Letters **86**, 157 (2007)
[30] G. Cognola, E. Elizalde, S. Nojiri, S.D. Odintsov, L. Sebastiani, S. Zerbini, Phys. Rev. D **77**, 046009 (2008)
[31] S. Nojiri, S.D. Odintsov, Phys. Rev. D **77**, 026007 (2008)
[32] S. Capozziello, M. De Laurentis, S. Nojiri, S.D. Odintsov, Gen. Relativ. Gravit. **41**, 2313 (2009)
[33] R. Myrzakulov, L. Sebastiani, S. Vagnozzi, Eur. Phys. J. C **75**, 444 (2015)
[34] Z. Yi, Y. Gong, Phys. Rev. D **94**, 103527 (2016)
[35] J. O'Hanlon, Phys. Rev. Lett. **29**, 137 (1972)
[36] P. Teyssandier, P. Tourrenc, J. Math. Phys. **24**, 2793 (1983)
[37] S. Capozziello, C. Corda, M.F. De Laurentis, Phys. Lett. B **669**, 255 (2008)
[38] S. Capozziello, R. Cianci, M. De Laurentis, S. Vignolo, Eur. Phys. J. C **70**, 341 (2010)
[39] A.S. Goldhaber, M.M. Nieto, Phys. Rev. D **9**, 1119 (1974)
[40] C.P.L. Berry, J.R. Gair, Phys. Rev. D **83**, 104022 (2011), [Erratum: Phys. Rev. D 85,089906 (2012)]
[41] E. Yasuo, K. Masahiro, K. Masahiko, S. Jiro, Y. Tadashi, Class. Quantum Grav. **16**, 1127 (1999)
[42] Y. Ezawa, H. Iwasaki, Y. Ohkuwa, S. Watanabe, N. Yamada, T. Yano, Class. Quantum Grav. **23**, 3205 (2006)
[43] N. Deruelle, Y. Sendouda, A. Youssef, Phys. Rev. D **80**, 084032 (2009)
[44] N. Deruelle, M. Sasaki, Y. Sendouda, D. Yamauchi, Prog. Theor. Phys. **123**, 169 (2010)
[45] Y. Sendouda, N. Deruelle, M. Sasaki, D. Yamauchi, Int. J. Mod. Phys.: Conference Series **01**, 297 (2011)
[46] G.J. Olmo, H. Sanchis-Alepuz, Phys. Rev. D **83**, 104036 (2011)
[47] Y. Ohkuwa, Y. Ezawa, Eur. Phys. J. Plus **130**, 77 (2015)
[48] R. Arnowitt, S. Deser, C.W. Misner, *Gravitation: An Introduction to Current Research* (John Wiley and Sons Ltd., New York, 1962), pp. 227–265
[49] R. Arnowitt, S. Deser, C.W. Misner, Gen. Relativ. Gravit. **40**, 1997 (2008)
[50] T. Kobayashi, M. Yamaguchi, J. Yokoyama, Prog. Theor. Phys. **126**, 511 (2011)
[51] M. Rakhmanov, Phys. Rev. D **71**, 084003 (2005)
[52] K.J. Lee, F.A. Jenet, H.P. Richard, Astrophys. J. **685**, 1304 (2008)
[53] K. Lee, F.A. Jenet, R.H. Price, N. Wex, M. Kramer, Astrophys. J. **722**, 1589 (2010)
[54] S.J. Chamberlin, X. Siemens, Phys. Rev. D **85**, 082001 (2012)
[55] K.J. Lee, Class. Quantum Grav. **30**, 224016 (2013)
[56] N. Yunes, X. Siemens, Living Rev. Relativity **16**, 9 (2013)
[57] J. Gair, J.D. Romano, S. Taylor, C.M.F. Mingarelli, Phys. Rev. D **90**, 082001 (2014)

[58] J.R. Gair, J.D. Romano, S.R. Taylor, Phys. Rev. D **92**, 102003 (2015)

Gravitational Wave (GW) Classification, Space GW Detection Sensitivities and AMIGO (Astrodynamical Middle-frequency Interferometric GW Observatory)

Wei-Tou Ni[1]

Center for Gravitation and Cosmology (CGC), Department of Physics, National Tsing Hua University, No. 101, Kuang Fu II Rd., Hsinchu, Taiwan, 300 ROC

Abstract. After first reviewing the gravitational wave (GW) spectral classification. we discuss the sensitivities of GW detection in space aimed at low frequency band (100 nHz–100 mHz) and middle frequency band (100 mHz–10 Hz). The science goals are to detect GWs from (i) Supermassive Black Holes; (ii) Extreme-Mass-Ratio Black Hole Inspirals; (iii) Intermediate-Mass Black Holes; (iv) Galactic Compact Binaries; (v) Stellar-Size Black Hole Binaries; and (vi) Relic GW Background. The detector proposals have arm length ranging from 100 km to 1.35×10^9 km (9 AU) including (a) Solar orbiting detectors and (b) Earth orbiting detectors. We discuss especially the sensitivities in the frequency band 0.1-10 μHz and the middle frequency band (0.1 Hz–10 Hz). We propose and discuss AMIGO as an Astrodynamical Middle-frequency Interferometric GW Observatory.

1 Classification and Spectral Sensitivities

With LIGO's direct detection of the binary black hole merger events [1, 2, 3], we have been fully ushered into the age of Gravitational Wave (GW) astronomy. Detection efforts over all GW frequency bands from cosmological frequency band (1 aHz–10 fHz) to ultra-high frequency band (over 1 THz) have been vigorously exerted (See, e.g. [4]). In Table 1, we list the band ranges and the detection methods [4-6]. We have also plotted the GW detector sensitivities and GW source strengths on single diagrams with ordinates showing characteristic strain, strain power spectral density (psd) amplitude and normalized GW spectral energy density respectively in 2015 [4]. Currently we are updating these diagrams [7]. Fig. 1 shows the strain psd amplitude vs. frequency for various detectors and sources adapting from the corresponding figures of Refs. [4] and [7]. For detailed explanation of the plot, see [4, 7].

Presently, most GW detection efforts are spent in 4 bands – the high frequency band (10 Hz–100 kHz), the upper part (10 μHz–0.1 Hz) of the low frequency band, the very low frequency band (nano-Hz band, 300 pHz –100 nHz) and the extremely low (Hubble)

---

[1] weitou@gmail.com

frequency band (cosmological band or CMB band; 1 aHz–10 fHz). As can be seen in Fi; 1, there are 3 regions which are poor in the near-future projected sensitivities adjacent t these 4 bands: (i) the middle frequency band, (ii) the lower part (100 nHz–10 μHz) of th low frequency band and (iii) the ultralow frequency band (10 fHz–300 pHz). To possibl increase the sensitivity in the frequency band 0.1-10 μHz, Super-ASTROD with arr length of 9 AU has been proposed [8]. To have significant sensitivity in the frequenc band 0.1–10 Hz and yet to be a first-generation candidate for space GW missions, w propose a middle-frequency GW mission AMIGO (Astrodynamical Middle-frequenc Interferometric GW Observatory) with arm length 10,000 km and discuss the concept i section 3 after a review on space GW detection sensitivities in section 2.

Table 1. Frequency classification of gravitational waves and their detection method [4-6]

Frequency band	Detection method
Ultra high frequency band: above 1 THz	Terahertz resonators, optical resonators, an magnetic conversion detectors
Very high frequency band: 100 kHz–1 THz	Microwave resonator/wave guide detectors, lase interferometers and Gaussian beam detectors
High frequency band (audio band)*: 10 Hz–100 kHz	Low-temperature resonators and ground-base laser-interferometric detectors
Middle frequency band: 0.1 Hz–10 Hz	Space laser-interferometric detectors of arr length 100 km – 60,000 km, atom and molecul interferometry, optical clock detectors
Low frequency band (milli-Hz band)†: 100 nHz–0.1 Hz	Radio Doppler tracking of spacecraft, spac laser-interferometric detectors of arm lengt longer than 60,000 km, optical clock detectors
Very low frequency band (nano-Hz band): 300 pHz – 100 nHz	Pulsar timing arrays (PTAs)
Ultralow frequency band: 10 fHz–300 pHz	Astrometry of quasars and their proper motions
Extremely low (Hubble) frequency band (cosmological band): 1 aHz–10 fHz	Cosmic microwave background experiments
Beyond Hubble-frequency band: below 1 aHz	Through the verifications c inflationary/primordial cosmological models

*The range of audio band (also called LIGO band) normally goes only to 10 kHz.
†The range of milli-Hz band is 0.1 mHz to 100 mHz.

2 SPACE GW detection sensitivities

GW detection in space aimed at low frequency band (100 nHz – 100 mHz) and middl frequency band (100 mHz – 10 Hz). Its scientific goals are to detect GWs from (i Supermassive Black Hole Coalescences; (ii) Extreme-Mass-Ratio Black Hole Inspirals (iii) Intermediate-Mass Black Hole Coalescences; (iv) Galactic Compact Binaries, (v Stellar-size Black Hole Binary Inspirals, and (vi) Relic GW Background.

The main technological requirements of GW detection in space are (i) drag-fre requirement; and (ii) requirement of measuring relative distance variation or relativ velocity variation. LISA Pathfinder (LPF) launched on 3 December 2015, has achieve not only the drag-free requirement goal of this technology demonstration mission, bu also has completely met the more stringent LISA drag-free demand [9-11]. In shor LISA Pathfinder has successfully demonstrated the first generation drag-free technolog requirement for space detection of GWs.

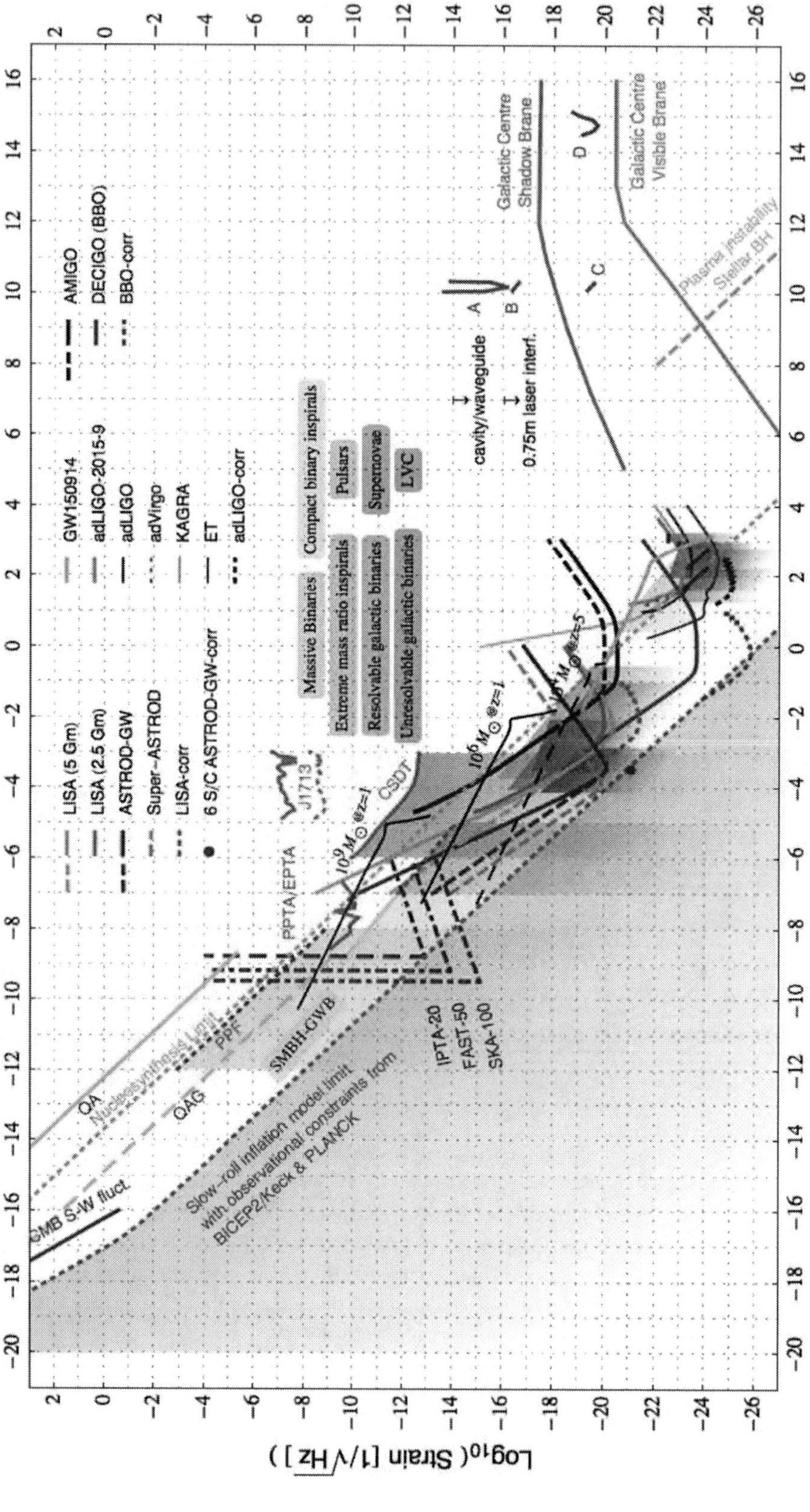

Fig. 1. Strain psd amplitude vs. frequency for various GW detectors and sources. [QA: Quasar Astrometry; QAG: Quasar Astrometry Goal; PPF: Pulsar Parameter Fitting; LVC: LIGO-Virgo Constraints; CSDT: Cassini Spacecraft Doppler Tracking; SMBH-GWB: Supermassive Black Hole-GW Background.]

The requirement of measuring relative distance variation or relative velocity variation is in terms of spectral strain sensitivity. For space GW detection, the first-generation requirement is around 10^{-20} $Hz^{-1/2}$ sensitivity for measurement of strain psd amplitude. For measurement using unequal-arm laser interferometry, the requirement on laser stabilization is similar. However, the present laser stabilization has not reached this kind of stability. One needs to match the two optical paths using Time Delay Interferometry (TDI) to lessen the stability requirement. For TDI configurations and their numerical simulations for various missions, see Tinto and Dhurandhar [12], Wang and Ni [13] and references therein. Experimental demonstration of TDI in laboratory for LISA worked out in 2010-2012 (Vine et al. [14], Mirtyk et al. [15]).

In space, Michelson type interferometry invariably involve large distances. The laser power received at the far end of the optical link is weak. To continue the optical path as required by TDIs, one needs to amplify it. The way of amplification is to track the optical phase of the incoming weak light with the local laser oscillator by optical phase-locking. At National Tsing Hua University, 2 pW weak-light homodyne phase-locking with 0.2 mW local oscillator has been demonstrated (Liao et al. [16, 17]). In JPL (Jet Propulsion Laboratory), Dick et al. [18] have achieved offset phase locking of local oscillator to 40 fW incoming laser light. More recently, Gerberding et al. [19] and Francis et al. [20] have phase-locked and tracked a 3.5 pW weak light signal and a 30 fW weak light signal respectively at reduced cycle slipping rate. For LISA, 85 pW weak-light phase locking is required. For ASTROD-GW, 100 fW weak-light phase locking is required. Hence, the weak level of these weak-light power requirements has achieved. In the future, the frequency-tracking, modulation-demodulation and coding-decoding needs development to make it a mature technology. This is also important for deep space CW (Continuous Wave) optical communication.

As shown in Fig. 1, typical frequency sensitivity spectrum of strain psd amplitude for space GW detection consists of three regions, the acceleration/local gravity gradient/vibration noise dominated region, the shot noise (flat for current space detector projects like LISA in strain psd) dominated region, if any, and the antenna response restricted region. The detector sensitivity in the lower frequency region is constrained by vibration, acceleration noise or gravity-gradient noise. The detector sensitivity of the higher frequency part is constrained by antenna response (or storage time). In a power-limited design, sometimes there is a middle flat region in which the sensitivity is limited by the photon shot noise. [21-24]

The shot noise sensitivity in the strain for GW detection is inversely proportional to $P^{1/2}L$ with P the received power and L the distance or arm length. Since P is inversely proportional to L^2 and $P^{1/2}L$ is constant, this sensitivity limit is independent of the distance. For 1-2 W emitting power, the limit is around 10^{-20}–10^{-21} $Hz^{-1/2}$ (depending on telescope diameter/laser beam divergence). As noted in the LISA study [21], making the arms longer shifts the time-integrated sensitivity curve to lower frequencies while leaving the bottom of the curve at the same level. Hence, ASTROD-GW with longer arm length has better sensitivity at lower frequency. e-LISA, ALIA, TAIJI, and GW interferometers in Earth orbit have shorter arms and therefore have better sensitivities at higher frequency.

In Fig. 1, we plot the sensitivity curves for LISA (5 Gm), LISA (2.5 Gm, i.e. the new LISA), ASTROD-GW, BBO, DECIGO, AMIGO and Super-ASTROD. The sensitivity curves for LISA (5 Gm), ASTROD-GW, BBO and DECIGO are taken from Fig. 3 of [4] and references therein. Others are from [7] and references therein. Fig. 2 is a blowup of Fig. 1 restricted to the frequency range 10^{-10}–10^{4} Hz to show the space frequency regions and the 2 neighboring regions. Section 2.1 discusses new LISA (2.5 Gm) and gives its sensitivity equation. Section 2.2 does it for Super-ASTROD.

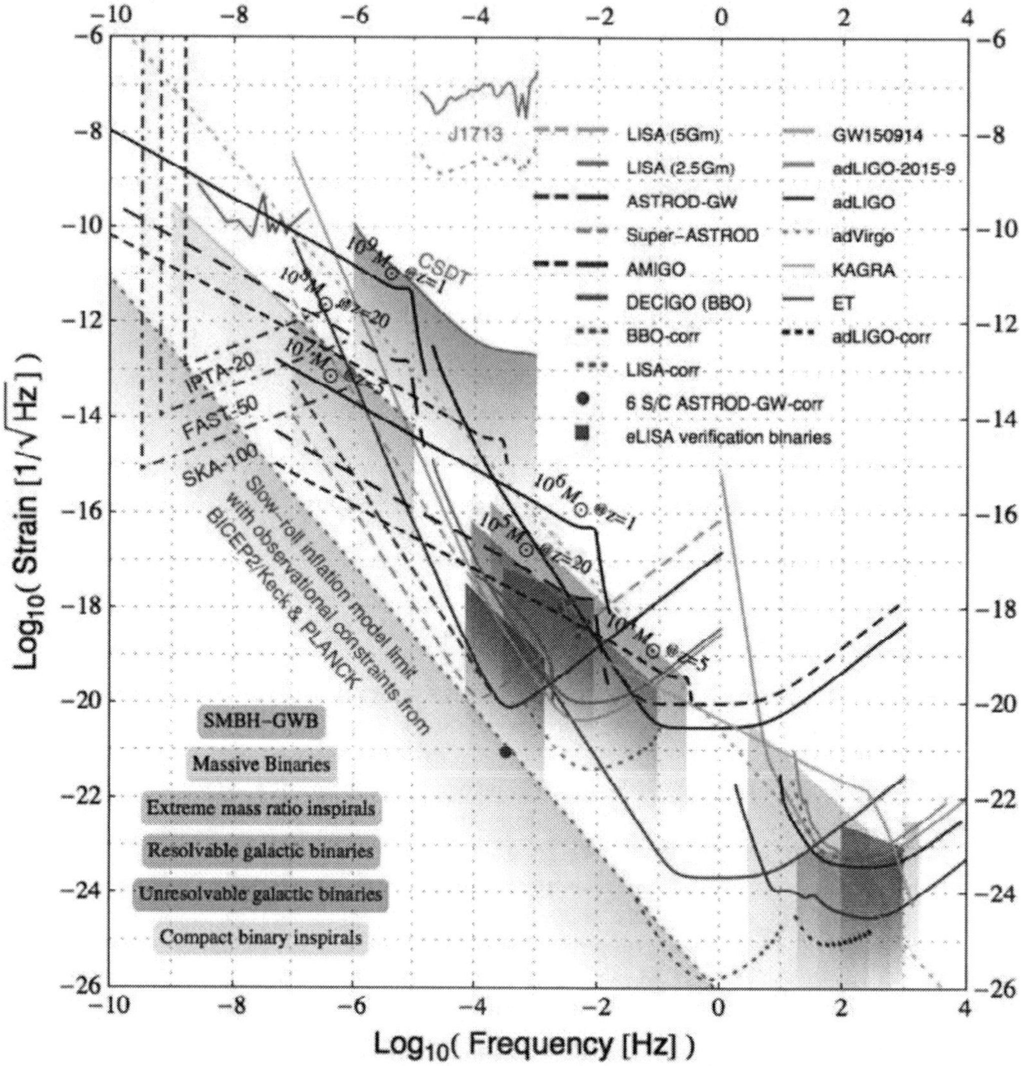

Fig. 2. Strain power spectral density (psd) amplitude vs. frequency for various GW detectors and GW sources. The black lines show the inspiral, coalescence and oscillation phases of GW emission from various equal-mass black-hole binary mergers in circular orbits at various redshift: solid line, z = 1; dashed line, z = 5; long-dashed line z = 20. See text of [24] for more explanation. The strain psd amplitude of GW150914 is calculated from its characteristic amplitude in Figure 1 of [25] using standard formula. The AMIGO design sensitivity is in solid blue while AMIGO baseline sensitivity is in dashed blue. The two curves merge together at lower frequency in the figure. [CSDT: Cassini Spacecraft Doppler Tracking; SMBH-GWB: Supermassive Black Hole-GW Background.]

2.1 LISA (2.5 Gm)

A new LISA proposal (Amaro-Seoane et al. [11]) was submitted to ESA on January 13th in response to the call for missions for the L3 slot in the Cosmic Vision Programme. On 20 June 2017, ESA announced the news that "The LISA trio of satellites to detect gravitational waves from space has been selected as the third large-class mission in ESA's Science programme (ESA 2017)." The basic concept is the same as the original LISA, but with arm length down-scaled to 2.5 Gm from 5 Gm. To distinguish this selected mission proposal from the original one or the NGO/eLISA, we call it LISA (2.5 Gm or new LISA in case of possible ambiguity. Quoting from the proposal [11]:

"The observatory will be based on three arms with six active laser links, between three identical spacecraft in a triangular formation separated by 2.5 million km. Continuously operating heterodyne laser interferometers measure with pm $Hz^{-1/2}$ sensitivity in both directions along each arm, using well-stabilized lasers at 1064 nm delivering 2 W of power to the optical system. Using technology proven in LISA Pathfinder, the Interferometry Measurement System is using optical benches in each spacecraft constructed from an ultra-low expansion glass-ceramic to minimize optical path length changes due to temperature fluctuations. 30 cm telescopes transmit and receive the laser light to and from the other spacecraft. Three independent interferometric combinations of the light travel time between the test masses are possible, allowing, in data processing on the ground, the synthesis of two virtual Michelson interferometers plus a third null-stream, or "Sagnac" configuration." These two virtual Michelson interferometers are two TDIs. They could be two out of three TDI configurations X, Y and Z if they satisfy the noise requirement.

The new LISA design sensitivity is in [10, 11]. A simple analytical approximation of the design sensitivity is in Petiteau et al. [10] and used by Cornish and Robson [26]:

$$S_{Ln}^{1/2}(f) = (20/3)^{1/2}\,(1/L_L) \times [(1 + (f/(1.29f_L))^2)]^{1/2} \times [(S_{Lp} + 4S_a/(2\pi f)^4)]^{1/2}\ \mathrm{Hz}^{-1/2}, \tag{1}$$

over the frequency range 20 μHz $< f <$ 1 Hz. Here L_L = 2.5 Gm is the LISA arm length, $f_L = c/(2\pi L_L)$ is the LISA arm transfer frequency, $S_{Lp} = 8.9 \times 10^{-23}$ m^2 Hz^{-1} is the white position noise, and

$$S_a(f) = 9 \times 10^{-30}\ [1 + (10^{-4}\ \mathrm{Hz}/f)^2 + 16\,(2 \times 10^{-5}\ \mathrm{Hz}/f)^{10}]\ \mathrm{m^2\ s^{-4}\ Hz^{-1}}, \tag{2}$$

is the colored acceleration noise level. This new LISA design sensitivity curve shows in both Fig. 1 and Fig. 2.

2.2 Super-ASTROD

Super-ASTROD (Super Astrodynamical Space Test of Relativity using Optical Devices) is a mission concept with four spacecraft in 5.2 AU orbits together with an Earth–Sun L1/L2 spacecraft ranging optically with one another to probe GWs with frequencies 0.1 μHz–1 mHz, to test fundamental laws of spacetime and to map the outer-solar-system mass distribution and dynamics [8]. In this paper, we address mainly the GW sensitivity of Super-ASTROD for enhancing the GW detection in the sensitivity gap 100 nHz–10 μHz. With three spacecraft in Jupiter-like solar orbits of radius 5.2 AU and period of about 12 year, the desired baseline mission lifetime would be 25 years or more. The orbits of these 3 spacecraft are near the Sun-Jupiter L3, L4, L5 points respectively to form an ASTROD-GW-like configuration [23, 24]. The Super-ASTROD spacecraft configuration will be inclined to the Jupiter orbit plane by 1-3°. The angular precession of the spacecraft configuration will be twice the Jupiter orbiting angular velocity around the Sun, i.e. about 6 years in period. In 6 years, the angular position of quasi-monochromatic GW sources will be resolved.

Due to its large extension, Super-ASTROD is a second-generation GW mission concept. To set its sensitivity goal, we review the GW sensitivity of ASTROD-GW [23, 24] in Fig. 1.

For ASTROD-GW, our goal on the instrumental strain psd amplitude is

$$S_{An}^{1/2}(f) = (1/L_A) \times \{[(1 + 0.5\,(f/f_A)^2)] \times S_{Ap} + [4S_a/(2\pi f)^4]\}^{1/2}\ \mathrm{Hz}^{-1/2}, \tag{3}$$

over the frequency range of 100 nHz $< f <$ 1 Hz. Here $L_A = 260 \times 10^9$ m is the ASTROD-GW arm length, $f_A = c / (2\pi L_A)$ is the ASTROD-GW arm transfer frequency, $S_a = 9 \times 10^{-30}$ m^2 s^{-4} Hz^{-1} is the white acceleration noise level (the same as that for original LISA [21]), and $S_{Ap} = 10816 \times 10^{-22}$ m^2 Hz^{-1} is the (white) position noise level due to laser shot noise which is 2704 (=52^2) times that for original LISA [23]. The corresponding noise curve for the ASTROD-GW instrumental noise psd amplitude $^{(MLDC)}S_{An}{}^{1/2}(f)$ with the same "reddening" factor as specified in MLDC program is

$$^{(MLDC)}S_{An}{}^{1/2}(f) = (1/L_A) \times \{[(1 + 0.5\,(f/f_A)^2\,)] \times S_{Ap} + [1 + (10^{-4}/f)^2]\,(4S_a/(2\pi f)^4)\}^{1/2}\ \mathrm{Hz}^{-1/2}, \quad (4)$$

over the frequency range of 100 nHz $< f <$ 1 Hz. The one without reddening factor are shown with dashed line in the lower frequency part.

For Super-ASTROD, our goal on the instrumental strain noise psd amplitude is

$$S_{Sn}{}^{1/2}(f) = (1/L_S) \times \{[(1 + 0.5\,(f\,/\,f_S)^2\,)] \times S_{Sp} + [4S_a/(2\pi f)^4]\}^{1/2}\ \mathrm{Hz}^{-1/2}, \quad (5)$$

over the frequency range of 100 nHz $< f <$ 1 Hz. Here $L_S = 1350 \times 10^9$ m (9 AU) is the nominal Super-ASTROD arm length, $f_S = c\,/\,(2\pi L_S)$ is the Super-ASTROD arm transfer frequency, $S_a = 9 \times 10^{-30}$ m^2 s^{-4} Hz^{-1} is the white acceleration noise level (the same as that for original LISA and ASTROD-GW), and $S_{Sp} = 291600 \times 10^{-22}$ m^2 Hz^{-1} is the (white) position noise level due to laser shot noise which is 72900 (=270^2) times that for original LISA.

3 AMIGO

A discussion of ground-based GW detector concepts to extend the present ground-based interferometers detection spectral range, i.e., the high-frequency GW band 10 Hz–100 Hz to middle-frequency band 0.1–10 Hz together with the plethora of potential astrophysical sources in this band is given in Harms et al. [27]. Harms et al. examine the potential sensitivity of three detection concepts (atom interferometers, torsion bar antennas and Michelson interferometers), estimate for their event rates and thereby, the sensitivity requirements for these detectors. They find that the scientific payoff from measuring astrophysical gravitational waves in this frequency band is great. However, although they find no fundamental limits to the detector sensitivity in this band, the remaining technical limits will be extremely challenging to overcome. In this conference, Ho Jung presents a middle-frequency GW detector concept SOGRO (Superconducting Omni-directional Gravitational Radiation Observatory) on Earth [28, 29]. The Newtonian-noise cancellation from infrasound and seismic surface fields is very challenging [30].

In this paper, we propose a *first-generation* middle-frequency mission concept AMIGO: Astrodynamical Middle-frequency Interferometric GW Observatory with the following specification:

Arm length: 10,000 km (or a few times of this)

Laser power: 2 – 10 W

Acceleration noise: same as LPF has already achieved

Orbits and formation: 4 options (all LISA-like formations):

(i) Earth-like solar orbits (3-20 degrees behind the Earth orbit)

(ii) 600,000 km high orbit formation around the Earth

(iii) 100,000 km-250,000 high orbit formation around the Earth

(iv) near Earth-Moon L4 (or L5) halo orbit formation

The Scientific Goals of AMIGO are: to bridge the spectra gap between first-generation high-frequency and low-frequency GW sensitivities; Detecting intermediate mass BH coalescence; Detecting inspiral phase and predict time of binary black hole coalescence together with neutron star coalescence for ground interferometers; Detecting compact binary inspirals for studying stellar evolution and galactic population.

For AMIGO, our baseline on the noise psd amplitude assuming 2 W laser power, 30 cm telescopes and same acceleration noise as new LISA is:

$$S_{\rm AMIGOn}{}^{1/2}(f)=(20/3)^{1/2}(1/L_{\rm AMIGO})\times[(1+(f/(1.29f_{\rm AMIGO}))^2)]^{1/2}\times[(S_{\rm AMIGOp}+4S_a/(2\pi f)^4)]^{1/2}{\rm Hz}^{-1/2},(6)$$

over the frequency range of 20 μHz $< f < 1$ kHz. Here $L_{\rm AMIGO} = 0.01 \times 10^9$ m is the AMIGO arm length, $f_{\rm AMIGO} = c/(2\pi L_{\rm AMIGO})$ is the AMIGO arm transfer frequency, $S_{\rm AMIGOp} = 1.424 \times 10^{-28}$ m^2 Hz^{-1} is the (white) position noise level due to laser shot noise which is 16×10^{-6} ($=0.004^2$) times that for new LISA. $S_a(f)$ is the same colored acceleration noise level in (2). The AMIGO baseline sensitivity (6) is plotted as AMIGO dashed curve in both Fig. 1 and Fig. 2.

Since power and lower shot noise is crucial in reach better sensitivity in middle part of the sensitivity curve, we use 10 W laser power and 36 cm ϕ as our design values of the AMIGO mission concept to gain a factor of 10 [$\approx (10/2)\times(36/30)^4$] for shot noise design sensitivity. The AMIGO design sensitivity (6) is plotted as AMIGO solid curve in both Fig. 1 and Fig. 2 by using $S_{\rm AMIGOp} = 0.1424 \times 10^{-28}$ m^2 Hz^{-1}.

In Fig. 1 and Fig. 2, the strain psd amplitude of GW150914 is calculated from its characteristic amplitude in Figure 1 of [25] using standard formula of conversion. AMIGO with either baseline sensitivity or design sensitivity would detect the inspiral phase of GW150914 and predict the coalescence time for the benefit of doing multi-messenger astronomy. However, the design sensitivity has better coverage in detecting the inspiral phase of neutron star coalescence events.

The numerical TDIs for AMIGO would be easier to design compared to new LISA due to AMIGO's shorter arm length. X, Y, Z TDI configurations are well suited for AMIGO. However, experimental requirement on TDI is more stringent and needs developments. More studies on the mission concept and various choices of orbit design are under preparation [31].

4 Discussion and outlook

LIGO's measurement of GWs from black hole coalescence fully ushered us into the era of GW astronomy. LISA space GW mission will explore the GW sources of large part of our Universe with large S/N ratio in the low-frequency part. On the even lower frequency side, PTAs are actively looking for GWs. Between the frequency bands sensitive to PTAs and to first generation space GW detector, Super-ASTROD is discussed as a second-generation space detector concept to have better sensitivity in the interim frequency band to explore GW sources.

Currently, a number of detection methods are proposed and under active research to bridge the middle-frequency band gap between Earth-based and space-borne GW observations with important science goals. In this band, technical limits will be extremely challenging to overcome for Earth-based due to Newtonian noises. In this paper, we propose a first-generation middle-frequency mission concept AMIGO with 10,000 km arm length. The technical readiness level is high. The sensitivity is good to reach science goals considered in the last section.

If a pathfinder mission is desired with 2-spacecraft demonstration of ranging in the solar-system for a LISA-like mission, the case with 2-5 degrees lagging behind the Earth of the first orbit choice in the last section could be considered. Just take one arm of this AMIGO case, it would be good to test many things in the solar system: deployment, both radio and laser communications, noise budget, and drag-free system together with a concentrated effort on distance metrology. It might be simpler than go to L1 or L2 Sun-Earth Lagrange point.

I would like to thank Ling-Wei Luo for his help in drawing Figure 1 and Figure 2.

References

1. B. P. Abbott, et al. (LIGO Scientific and Virgo Collaborations), Observation of gravitational waves from a binary black hole merger, Phys. Rev. Lett. **116**, 061102 (2016)
2. B. P. Abbott et al. (LIGO Scientific and Virgo Collaborations), GW151226: Observation of gravitational waves from a 22-solar-mass binary black hole coalescence, Phys. Rev. Lett. **116**, 241103 (2016)
3. B. P. Abbott et al. (LIGO Scientific and Virgo Collaborations), GW170104: Observation of a 50-solar-mass binary black hole coalescence at redshift 0.2 Phys. Rev. Lett. **118**, 221101 (2017)
4. K. Kuroda, W.-T. Ni and W.-P. Pan, Gravitational waves: Classification, methods of detection, sensitivities, and sources, Int. J. Mod. Phys. D **24,** 1530031 (2015); also in *One Hundred Years of General Relativity: From Genesis and Empirical Foundations to Gravitational Waves, Cosmology and Quantum Gravity*, Chap. 10, ed. W.-T. Ni (World Scientific, Singapore, 2017)
5. W.-T. Ni, Gravitational wave, dark energy and inflation, *Mod. Phys. Lett. A* **25,** 922 (2010); arXiv:1003.3899; Classification of Gravitational Waves, http://astrod.wikispaces.com/file/view/GW-classification.pdf
6. D. Blair, et al., Gravitational wave astronomy: the current status, *Sci. China-Phys. Mech. Astron.* **58,** 120402 (2015)
7. S. Kuroyanagi, L.-W. Luo and W.-T. Ni, Gravitational wave sensitivities over all frequency bands, paper in preparation
8. W.-T. Ni, Super-ASTROD: Probing primordial gravitational waves and mapping the outer solar system, *Class. Quantum Grav.* **26,** 075021 (2009)
9. M. Armano, H. Audley, G. Auger et al., Sub-Femto-g Free Fall for Space-Based Gravitational Wave Observatories: LISA Pathfinder Results, Phys. Rev. Lett. **116**, 231101 (2016)
10. A. Petiteau, M. Hewitson, G. Heinzel, E. Fitzsimons and H. Halloin, LISA noise budget, Tech. rep. LISA Consortium lISA-CST-TN-0001 (2016)
11. P. Amaro-Seoane, H. Audley, S. Babak et al., Laser Interferometer Space Antenna, submitted to ESA on January 13th in response to the call for missions for the L3 slot in the Cosmic Vision Programme, arXiv:1702.00786 [astro-ph.IM]
12. M. Tinto and S. V. Dhurandhar, Time-delay interferometry, Liv. Rev. Rel. **17**, 6 (2014); and references therein
13. G. Wang and W.-T. Ni, Numerical simulation of time delay interferometry for new LISA, TAIJI and other LISA-like missions, arXiv:1707.09127 [astro-ph.IM]
14. G. de Vine, B. Ware, K. McKenzie, R. E. Spero, W. M. Klipstein, and D. A. Shaddock, Experimental Demonstration of Time-Delay Interferometry for the Laser Interferometer Space Antenna, Phys. Rev. Lett. **104**, 211103 (2010)

15. S. J. Mitryk, G. Mueller and J. Sanjuan, Hardware-based demonstration of time-delay interferometry and TDI-ranging with spacecraft motion effects, Phys. Rev. D **86**, 122006 (2012)
16. A.-C. Liao, W.-T. Ni and J.-T. Shy, On the study of weak-light phase locking for laser astrodynamical missions (in Chinese) Publications of the Yunnan Observatory **3**, 88-100 (2002)
17. A.-C. Liao, W.-T. Ni and J.-T. Shy, Pico-watt and femto-watt weak-light phase locking Int. J. Mod. Phys. D **11**, 1075-1085 (2002)
18. G. J. Dick, M. D. Strekalov, K. Birnbaum et al., Optimal phase lock at femtowatt power levels for coherent optical deep-space transponder, IPN Progress Report **42** 1752008 (2008)
19. O. Gerberding, B. Sheard, I. Bykov et al., Phasemeter core for intersatellite laser heterodyne interferometry: modelling, simulations and experiments, Classical Quantum Gravity **30**, 235029 (2013)
20. S. P. Francis, T. T.-Y. Lam, K. McKenzie et al., Weak-light phase tracking with a low cycle slip rate, Optics Letters **39**, 5251-5154 (2014)
21. LISA Study Team, *LISA (Laser Interferometer Space Antenna) — A Cornerstone Mission for the Observation of Gravitational Waves*, ESA System and Technology Study Report, ESA-SCI 11 (2000)
22. W.-T. Ni, ASTROD and gravitational waves, pp. 117-129 in Gravitational Wave Detection, edited by K. Tsubono, M.-K. Fujimoto and K. Kuroda (Universal Academy Press, Tokyo, Japan, 1997)
23. W.-T. Ni, ASTROD-GW: Overview and progress, Int. J. Mod. Phys. D **22**, 1431004 (2013); and references therein
24. W.-T. Ni, Gravitational Wave Detection in Space, Int. J. Mod. Phys. D **25**, 1630001 (2016); also in *One Hundred Years of General Relativity: From Genesis and Empirical Foundations to Gravitational Waves, Cosmology and Quantum Gravity*, Chap. 12, ed. W.-T. Ni (World Scientific, Singapore, 2017); and references therein
25. A. Sesana, Prospects for multiband gravitational-wave astronomy after GW150914, Phys. Rev. Lett. 116, 231102 (2016)
26. N. Cornish and T. Robson, Galactic binary science with the new LISA design, J. Phys.: Conf. Ser. **840,** 012024 (2017)
27. J. Harms, B. J. J. Slagmolen, R. X. Adhikari et al, Low-frequency terrestrial gravitational-wave detectors, Phys. Rev. D **88**, 122003 (2013)
28. H. J. Paik, SOGRO (Superconducting Omni-directional Gravitational Radiation Observatory), Plenary talk given in Joint Meeting of 13th International Conference on Gravitation, Astrophysics, and Cosmology and 15th Italian-Korean Symposium on Relativistic Astrophysics, Ewha Womans University, Seoul, Korea, July 3-7, 2017
29. H. J. Paik, C. E Griggs, M. Moody et al., Low-frequency terrestrial tensor gravitational-wave detector, Class. Quantum Grav. **33,** 075003 (19pp) (2016)
30. J. Harms and H. J. Paik, Newtonian-noise cancellation in full-tensor gravitational-wave detectors, Phys. Rev. D **92**, 022001 (2015)
31. W.-T. Ni, G. Wang and A.-M. Wu, Astrodynamical Middle-frequency Interferometric GW Observatory AMIGO: mission concept and orbit design, paper in preparation

EPJ Web of Conferences **168**, 01005 (2018) https://doi.org/10.1051/epjconf/201816801005
Joint International Conference of ICGAC-XIII and IK-15 on Gravitation, Astrophysics and Cosmology

SOGRO (Superconducting Omni-directional Gravitational Radiation Observatory)

Ho Jung Paik[1,*]

[1]Department of Physics, University of Maryland, College Park, MD 20742, U.S.A.

Abstract. Detection of gravitational waves (GWs) from merging binary black holes (BHs) by Advanced LIGO has ushered in the new era of GW astronomy. Many conceivable sources such as intermediate-mass BH binaries and white dwarf binaries, as well as stellar-mass BH inspirals, would emit GWs below 10 Hz. It is highly desirable to open a new window for GW astronomy in the infrasound frequency band. A low-frequency tensor detector could be constructed by combining six magnetically levitated superconducting test masses. Such a detector would be equally sensitive to GWs coming from anywhere in the sky, and would be capable of resolving the source direction and wave polarization. I will present the design concept of a new terrestrial GW detector, named SOGRO, which could reach a strain sensitivity of 10^{-19}-10^{-21} $Hz^{-1/2}$ at 0.1-10 Hz. Seismic and Newtonian gravity noises are serious obstacles in constructing terrestrial GW detectors at frequencies below 10 Hz. I will explain how these noises are rejected in SOGRO. I will also report the progress made in designing the platform and modelling its thermal noise.

1 Introduction

Detection of gravitational waves (GWs) from merging binary black holes (BHs) by the two Advanced LIGO (aLIGO) detectors has ushered in the new era of GW astronomy [1-3]. Now, the Advanced Virgo (AdV) detector has joined the search for GWs. These laser interferometers cover a frequency band from ~10 Hz to several kilohertz. The LISA mission, selected by ESA for launch in 2034 [4], would observe GWs over a low-frequency band from 0.1 mHz to 0.1 Hz. This would still leave the frequency band f = 0.1-10 Hz uncovered, where interesting astrophysical sources such as intermediate-mass black hole (IMBH) and white dwarf (WD) binaries could be observed.

Ando et al. [5] proposed a space interferometer, DECIGO, to cover the missing mid-frequency band. Harms et al. [6] looked into the possibility of constructing a mid-frequency terrestrial detector by using an atom-laser hybrid interferometer, a differential torsion-bar detector (TOBA), and a conventional laser interferometer. Paik et al. [7] proposed a terrestrial *tensor* GW detector, SOGRO (Superconducting Omni-directional Gravitational Radiation Observatory), consisting of six levitated superconducting test masses, to cover the same frequency band. Such a detector would be equally sensitive to GWs coming from

* Corresponding author: hpaik@umd.edu

anywhere in the sky, and would be capable of resolving the *source direction* and *wave polarization*. In this paper, I discuss the design concept of SOGRO and report the recent progress made in designing the platform and modelling its thermal noise.

Unlike the other detectors, SOGRO requires a large-scale cryogenic operation since the test masses must be levitated from a rigid platform and the entire platform needs to be cooled to 4.2 K or below to reduce its thermal noise. Therefore, our initial research effort was directed to conceptual design of the platform and modelling the platform noise.

2 Design Concept and Operating Principle

The potential of using two widely separated test masses coupled by an electrical circuit to detect GWs was realized in the 1970s by Wagoner et al. [8]. SOGRO is a 3D extension of such a long-baseline resonant-mass detector. Figure 1 shows a perspective view of SOGRO. Six superconducting test masses, each with three linear degrees of freedom, are levitated over three orthogonal mounting tubes. The test masses are made of niobium (Nb) in the shape of a rectangular shell. Superconducting levitation/alignment coils and sensing capacitors (not shown) are located in the gap between the test masses and the mounting tubes, as well as on the outer surfaces of the test masses. The along-axis motions of the two test masses on each coordinate axis are differenced to measure a diagonal component of the wave while the cross-axis (rotational) motions of the four test masses on each coordinate plane are differenced to measure an off-diagonal component of the wave:

$$h_{ii}(t)=\frac{2}{L}\left[x_{+ii}(t)-x_{-ii}(t)\right], \tag{1}$$

$$h_{ij}(t)=\frac{1}{L}\left\{\left[x_{+ij}(t)-x_{-ij}(t)\right]-\left[x_{-ji}(t)-x_{+ji}(t)\right]\right\},\ i\neq j, \tag{2}$$

where $x_{\pm ij}$(t) is displacement of the test mass on the $\pm i$ axis along the j axis and L is the separation between the test masses on each axis.

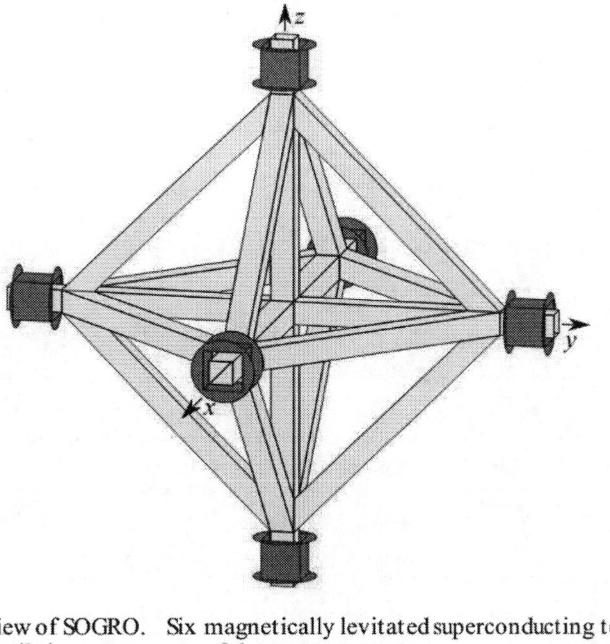

Fig. 1. Perspective view of SOGRO. Six magnetically levitated superconducting test masses are combined to measure all six components of the curvature tensor.

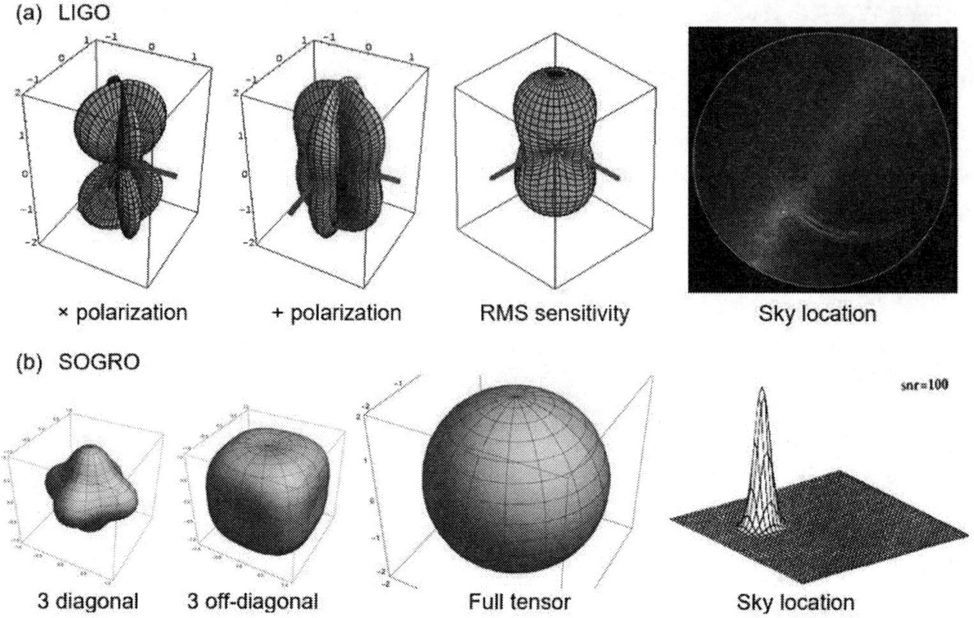

Fig. 2. Antenna pattern of SOGRO compared with that of LIGO. Being a full-tensor detector, SOGRO has full-sky coverage with equal sensitivity to waves coming from any direction with any polarization and a single SOGRO can locate the source in the sky.

Figure 2 compares the antenna pattern of SOGRO with that of LIGO. Being a full-tensor detector, SOGRO has equal sensitivity to waves coming from any direction with any polarization. A single SOGRO can locate the source in the sky, modulo angle π, without necessitating triangulation.

The operating principle of SOGRO follows closely that of the tensor superconducting gravity gradiometer (SGG) under development at the University of Maryland [9]. Since test mass motion is measured with respect to the sensing circuit elements mounted on the platform, SOGRO requires a rigid platform with its mode frequencies above the signal bandwidth. To alleviate excessive demand on cryogenics, the platform must not be too heavy yet sufficiently rigid.

Each test mass weighs 5 tons and has a large enough surface area to be levitated against earth's gravity with a magnetic field of 0.10 T, well below 0.14 T, the first critical field (H_{c1}) of high-purity Nb at 4.2 K. Magnetic levitation allows the test masses to have horizontal differential-mode (DM) resonance frequencies of ~0.01 Hz. In the vertical direction, the DM frequencies become ~1 Hz due to the stiffness arising from nonlinearities in the levitation coils. This implies that the h_{xz}, h_{yz} and h_{zz} channels, which utilize vertical accelerometers, will be noisier. Fortunately, there is a way to detect all the tensor components by combining only the horizontal motions of the test masses. Since the GW tensor is *traceless*, h_{zz} can be obtained from $h_{zz} = -(h_{xx} + h_{yy})$ without necessitating measurement of vertical accelerations. According to Eq. (2), h_{xz} is obtained by combining the relative z displacement of the two test masses on the x axis with the relative x displacement of the two test masses on the z axis:

$$h_{xz} = \frac{1}{L}\left[(z_{x2} - z_{x1}) + (x_{z2} - x_{z1})\right]. \tag{3}$$

Since the platform is completely isolated from the ground tilt (see Sec. 5), $z_{x2} - z_{x1}$ and $x_{z2} - x_{z1}$ are not contaminated by any common-mode (CM) angular acceleration. Further, h_{xz}

drives the two arms symmetrically: $z_{x2} - z_{x1} = x_{z2} - x_{z1}$. The same holds for h_{yz}. Therefore, these tensor components can also be obtained from horizontal displacements only.

$$h_{xz} = \frac{2}{L}(x_{z2} - x_{z1}), \quad h_{yz} = \frac{2}{L}(y_{z2} - y_{z1}). \tag{4}$$

The tiny mechanical energy that a GW deposits into the antenna must be converted into an electrical signal and coupled to a nearly quantum-limited SQUID (Superconducting Quantum Interference Device). Unfortunately, nearly quantum-limited dc SQUIDs have a 1/f noise below several kHz. We need to use an active transducer to upconvert the signal frequency to above the 1/f noise corner frequency. For SOGRO, we propose to employ a superconducting capacitance bridge transducer coupled to a dc SQUID. A similar transducer was developed by Cinquegrana et al. [10] for resonant-mass GW detectors.

Figure 3 shows the transducer for a diagonal-component channel. $C_{ij}(t)$ is a Nb capacitor plate located near the j-th face of the i-th test mass. The bridge output is coupled to a nearly quantum-limited dc SQUID through an impedance-matching superconducting transformer. The sensing capacitances are modulated by GWs at ω and the bridge is driven at $\omega_p >> \omega$, well above the 1/f noise corner frequency of the SQUID. An active transducer introduces a new noise source, the oscillator noise. Although a bridge configuration is employed to null the carrier, the sideband components contained in the carrier can become a dominant noise source, if the bridge balance is not sufficiently precise. In our design, the carrier signal is precisely balanced by applying feedback forces to the test masses, eliminating the oscillator noise at the output.

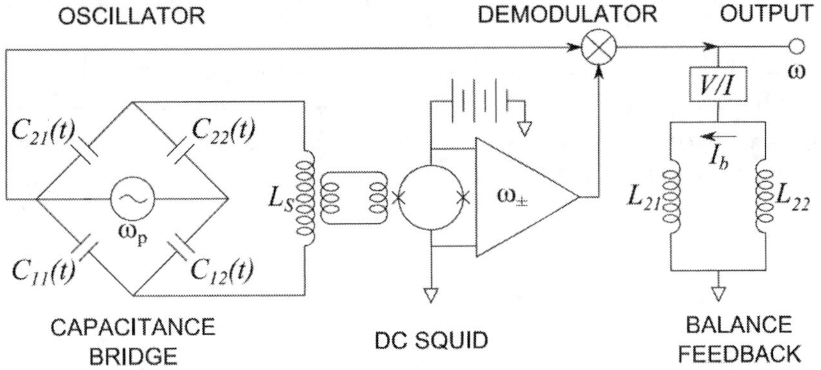

Fig. 3. Capacitance-bridge transducer coupled to a nearly quantum-limited dc SQUID. The bridge is driven at the resonance frequency of the capacitance bridge tuned with the SQUID input coil inductance. Feedback is applied to keep the bridge precisely balanced.

3 Platform Design and Noise Analysis

An efficient design of the platform and the cryostat is of utmost importance for SOGRO. We have carried out FEM analysis for various platform design options with $L = 30$, 50 and 100 m. Our ANSYS analysis shows that, by building the 3D cross with aluminum (Al) tubes triangulated by Al tube struts and fortified by additional braces at the points where these tubes meet, a 50-m platform can be constructed with all the mode frequencies above 10 Hz and a total mass of ~220 tons (see Fig. 4). To increase their internal mode frequencies, the test masses are made compact by using folded Nb shells and are levitated inside the orthogonal Al tubes near the six vertices of the platform. This gives an additional advantage of being able to shrink the size of the cryostat.

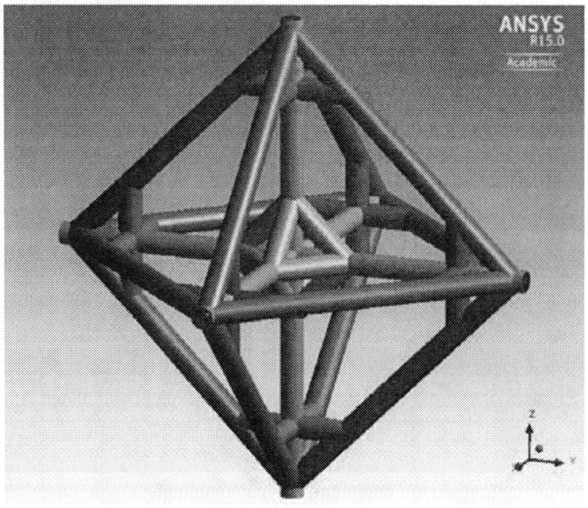

Fig. 4. 50-m platform constructed with 2.5-m diameter Al tube with 1-cm thick wall, triangulated with diagonal Al tube struts and fortified by additional braces.

For the platform noise, we assume an internal damping model, which modifies Hooke's law as $F = -k(1 + i\phi)x$, where ϕ is the loss angle. The thermal noise displacement power spectral density (PSD) is given [11] by

$$S_\xi(f) = \frac{4k_B T\omega_{pl}^2}{M_{pl}\omega Q_{pl}\left[(\omega_{pl}^2 - \omega^2)^2 + (\omega_{pl}^2/Q_{pl})^2\right]}, \tag{5}$$

where ω_{pl} is the resonance frequency of the mode under consideration, M_{pl} is the effective mass of the platform, which depends on where the displacement ξ is measured, and Q_{pl} is the quality factor of the mode. To compute $S_\xi(f)$, a pre-stressed modal analysis was performed using ANSYS finite element software.

Figure 5 shows the platform thermal noise coming from dominant xx longitudinal platform modes in (a), and dominant xy scissor modes of the platform in (b), at $T = 4.2$ K assuming $Q_{pl} = 10^7$. All the mode frequencies are above 10 Hz except for one scissor mode. The noise summed over six dominant modes gives a level lower than or comparable to the intrinsic detector noise of Advanced SOGRO (aSOGRO) for both h_{xx} and h_{xy} channels (see Fig. 6). (Other modes make negligible contributions.) For initial SOGRO, $Q_{pl} = 10^6$ would be sufficient to keep the platform thermal noise below the intrinsic detector noise level.

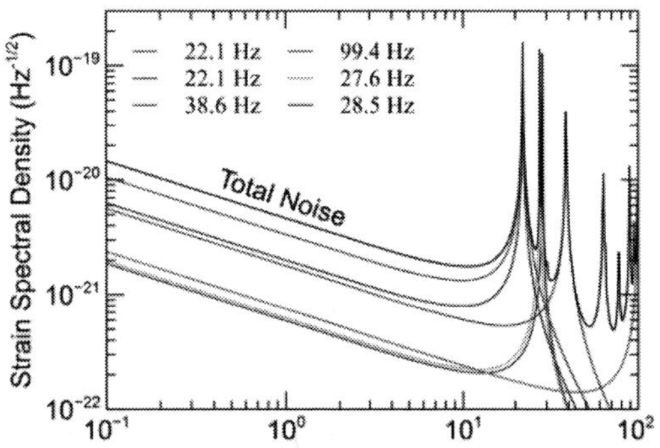

(a) Longitudinal modes

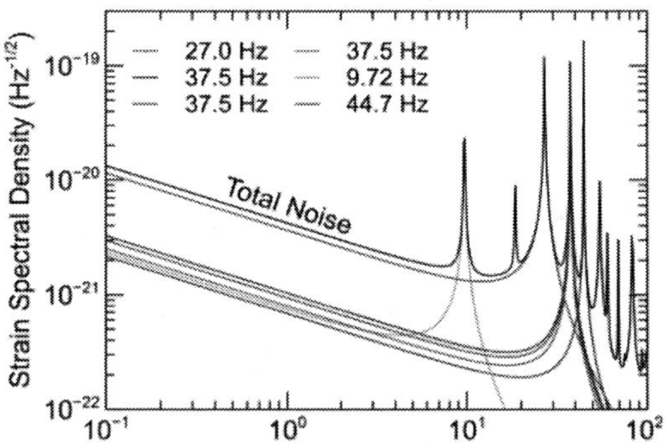

(b) Scissor modes

Fig. 5. Platform thermal noise coming from (a) dominant longitudinal modes and (b) dominant scissor modes at $T = 4.2$ K assuming $Q_{pl} = 10^7$.

Obtaining $Q_{pl} = 10^7$ in a composite structure will be extremely challenging even at cryogenic temperatures. However, we expect that the Q requirement can be relaxed by at least an order of magnitude as we optimize the design of the platform. M_{pl} for a particular mode becomes infinity at its nodes, resulting in $S_\xi(f) = 0$. With the additional degrees of freedom introduced by the corner braces, one could adjust the shapes and positions of these braces to move the nodal points of dominant modes nearer to the test mass locations. An in-depth engineering study is required to optimize the platform design.

4 Detector Sensitivity

The detector sensitivity is limited by the intrinsic detector noise (antenna thermal plus amplifier noise) and the platform thermal noise. Seismic and Newtonian gravity noises (NN) are important external noise sources which can also limit the detector resolution. In this section, I will discuss the intrinsic detector noise and the platform noise. Seismic noise isolation and NN mitigation will be described in Sec. 5.

The detector noise PSD of an off-diagonal-component GW detector with an active transducer can be shown [7] to be

$$S_h(f) = \frac{16}{ML^2\omega^4}\left[\frac{k_B T\omega_D}{Q_D} + \frac{|\omega^2 - \omega_D^2|}{\omega_p}\left(1+\frac{1}{\beta^2}\right)^{1/2} k_B T_N\right], \quad (6)$$

where M and L are the mass of each test mass and the arm-length of the detector; T is the temperature; ω_D and Q_D are the DM (angular) resonance frequency and Q; and β and T_N are the energy coupling constant of the transducer and the noise temperature of the SQUID, respectively. One obvious way of increasing the sensitivity is enlarging M and L. However, it will require cooling a larger mass. In this paper, we consider a detector with $M = 5 \times 10^3$ kg and $L = 50$ m, which brings the total mass including the platform to ~250 tons. In an active transducer, it is possible to achieve $\beta > 1$. For our tuned capacitance-bridge transducer, we have

$$\beta = \frac{2CE_p^2 Q_p}{M|\omega^2 - \omega_D^2|}\frac{1}{\sqrt{1+(2Q_p\omega/\omega_p)^2}}, \quad (7)$$

where C is the equilibrium capacitance of each sensing capacitor, E_p is the amplitude of the driving electric field at ω_p, and Q_p is the electrical Q.

Table 1 shows the proposed detector parameter values and detector noise for SOGRO and aSOGRO. Both detectors have an arm-length of 50 m and test masses weighing 5 tons each. The amplifier noise number is defined by $n \equiv k_B T_N/\hbar\omega_p$. SOGRO is cooled to 4.2 K by liquid helium and aSOGRO to 0.1 K by using He^3/He^4 dilution refrigerators. White noise levels of $10\hbar$ have been demonstrated at 0.1 K by using a two-stage dc SQUID, in which a commercial sensor was used as the preamplifier for the first SQUID [12]. We set $20\hbar$ and $5\hbar$ as the target SQUID noise for SOGRO and aSOGRO, respectively. SOGRO has very stringent Q requirement. Although no fundamental limit is known for the Q of levitated superconducting masses, parasitic forces due to trapped magnetic flux and patch fields will cause damping in the transducer. To minimize such parasitic forces, the test masses will be constructed with high-purity Nb and with the surface highly polished.

In Fig. 6, we plot the expected strain sensitivities of SOGRO and aSOGRO, as well as those of aLIGO, AdV and eLISA. The NN due to the Rayleigh waves and infrasound waves before mitigation are also plotted [6, 13]. The SOGRO detector noise is dominated by the

Table 1. Proposed detector parameters and detector noise of SOGRO.

Parameter	**SOGRO**	**aSOGRO**	**Method employed (/aSOGRO)**
Each test mass M	5 ton	5 ton	Nb shell
Arm-length L	50 m	50 m	Over "rigid" platform
Antenna temp T	4.2 K	0.1 K	LHe /He^3-He^4 dilution refrigerator
Platform temp T_{pl}	4.2 K	4.2 K	Large-scale cryogenics
Platform Q factor Q_{pl}	10^5	10^6	Al tube construction
DM frequency f_D	0.01 Hz	0.01 Hz	Magnetic levitation (horizontal)
DM quality factor Q_D	10^7	10^8	Surface polished pure Nb
Pump frequency f_p	50 kHz	50 kHz	Tuned capacitor bridge transducer
Amplifier noise no. n	20	5	2-stage dc SQUID cooled to 0.1 K
Detector noise $S_h^{1/2}(f)$	2.9×10^{-20} Hz$^{-1/2}$	5.6×10^{-21} Hz$^{-1/2}$	Computed at 1 Hz

antenna thermal noise below 0.2 Hz, by the amplifier noise at 0.2-1 Hz, and by the platform thermal noise above 1 Hz. aSOGRO has potential sensitivity $\leq 10^{-20}$ Hz$^{-1/2}$ at 0.5-10 Hz. To reach the detector noise limit, the NN must be mitigated by up to 70 dB. The sensitivity of aSOGRO surpasses those of terrestrial and space laser interferometers between 0.1 and 10 Hz and fills the frequency gap between them. In addition to detecting the IMBH and WD binaries, aSOGRO will be sensitive enough to be able to detect stellar mass BH binaries like GW150914 and warn the laser interferometers days before merger.

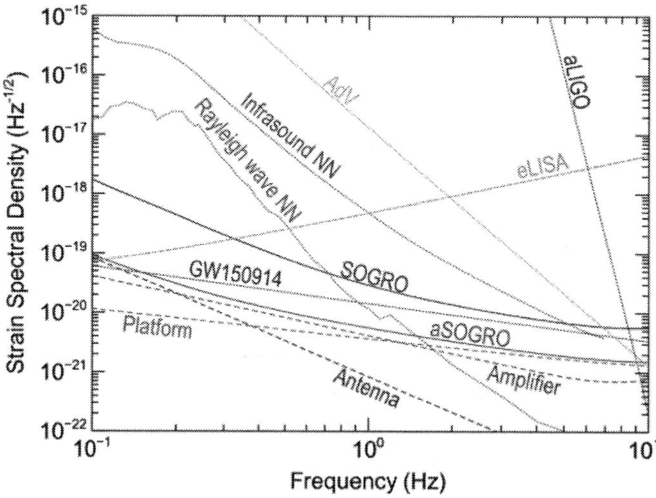

Fig. 6. Expected strain sensitivities of SOGRO and aSOGRO as well as those of aLIGO, AdV and eLISA. aSOGRO will be sensitive enough to be able to detect stellar mass BH binaries like GW150914 and to warn the laser interferometers days before merger.

5 Seismic and Newtonian Noise

We propose to construct SOGRO underground at a depth ≥ 500 m. The seismic noise level of an underground lab at 0.5-10 Hz is $\leq 5\times10^{-8}$ m s^{-2} Hz$^{-1/2}$ [14], 10 orders of magnitude above the target sensitivity of SOGRO. Vibration isolation at such low frequencies is extremely difficult. The CM rejection techniques demonstrated with SGGs [15], which produced a stable CM rejection level of three parts in 10^8, will be applied to SOGRO and improved to one part in 10^9. The highly polished, high-purity Nb surfaces will allow for exceptional stability of the balance and error rejection. This will reduce the vibration isolation requirement to 20 dB.

The entire platform will be suspended as a pendulum from outside the cryostat with the suspension cable attached to just above its center of mass. The resulting 25-m pendulum will have its pendulum frequency at 0.1 Hz and a torsional mode frequency ≤ 1 mHz. This pendulum suspension will completely decouple the platform from the ground tilt, and provide passive isolation of 40 dB to horizontal accelerations and 120 dB to angular accelerations at 1 Hz, and much more at 10 Hz. Combined with the overall CM rejection of 180 dB, the seismic noise could be reduced to below the intrinsic detector noise level in all degrees of freedom except for the vertical. For the vertical direction, 20-dB isolation must be provided by combining passive and active isolation.

The NN generated by moving local masses poses a formidable challenge to approaching the detector noise limit at $f \leq 1$ Hz. At low frequencies, the NN is dominated by Rayleigh waves and infrasound waves. Both the Rayleigh and infrasound waves are exponentially suppressed with depth. For the Rayleigh waves (c_R = 3.5 km s^{-1}), there is hardly any attenuation at a 500-m depth. However, for the infrasound (c_{IS} = 330 m s^{-1}), the NN is attenuated by up to a factor 6 at 0.2 Hz and to < 10^{-8} at 1 Hz at the same depth. At 0.2 Hz, where microseisms peak, the NN must be rejected by another three orders of magnitude to reach the intrinsic noise limit. SOGRO is able to meet this challenge by utilizing the full tensor measurement and simultaneous measurement of environmental density fluctuations. A detailed analysis has been published elsewhere [16]. I summarize our findings here.

Consider a GW coming from (θ, ϕ) direction in the presence of Rayleigh and infrasound waves, where θ and ϕ are the polar (with respect to zenith) and the azimuthal angle, respectively. The full strain tensor in the GW coordinates has the form:

$$h'(\omega) = \begin{pmatrix} h_+(\omega) + h'_{N11}(\omega) & h_\times(\omega) + h'_{N12}(\omega) & h'_{N13}(\omega) \\ h_\times(\omega) + h'_{N12}(\omega) & -h_+(\omega) + h'_{N22}(\omega) & h'_{N23}(\omega) \\ h'_{N13}(\omega) & h'_{N23}(\omega) & h'_{N33}(\omega) \end{pmatrix}, \tag{7}$$

where $h_+(\omega)$ and $h_\times(\omega)$ are the two polarization components of the GW, and $h'_{Nij}(\omega)$ are the NN components. Due to the *transverse nature* of the GW, h'_{13}, h'_{23}, h'_{33}, and $h'_{11} + h'_{22}$ contain only the NN components. Since a Rayleigh wave has three unknowns (amplitude, phase, incident angle ϕ) and an infrasound wave has four (amplitude, phase, incident angles θ, ϕ) while eight NN variables (amplitude and phase of h'_{N13}, h'_{N23}, h'_{N33}, $h'_{N11} + h'_{N22}$) are measured, SOGRO can determine up to a combination of *any two* waves.

In the general case where more than two waves are incident simultaneously, SOGRO must be supplemented with external seismometers and microphones and the NN must be removed by using a correlation method. Figure 7(a) shows the residual NN achieved for Rayleigh waves *in the absence of* infrasound waves by using h'_{13}, h'_{23}, h'_{33} and a_z (CM), plus seven seismometers with SNR of 10^3 at the radius of 5 km as the input of the Wiener filter. The NN has been removed to ~10^{-3} with environmental sensors (seismometers) alone. The local channels of SOGRO improve the noise significantly only near $\theta=\pi/2$.

Figure 7(b) is the residual NN achieved for infrasound waves *in the absence of* Rayleigh waves by using h'_{13}, h'_{23}, h'_{33} and 15 microphones of SNR of 10^4, one at the detector, seven each at radii of 600 m and 1 km around the detector. With the environmental sensors (microphones) alone, the NN *cannot* be mitigated except at θ = 0, π /2, and π. This is because the infrasound waves come from a half-space and the microphones deployed over a surface are insufficient to measure the effect of 3D density variations of the atmosphere. The vertical strain component h'_{33} of SOGRO largely makes up for this deficiency. With the aid of the local strain channels, the NN has been rejected to 10^{-3} for *all* θ.

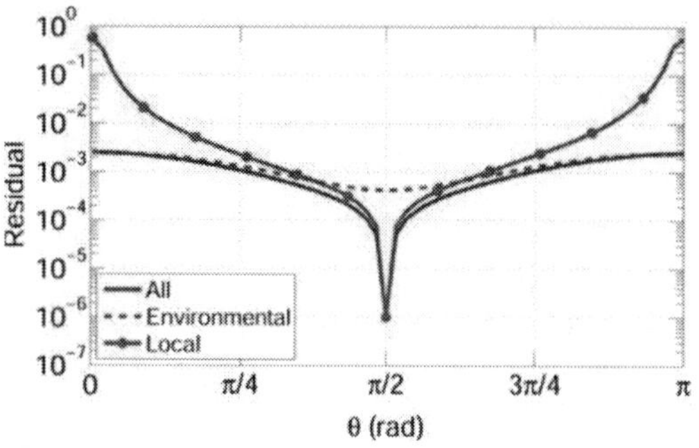

(a) Residual Rayleigh wave NN

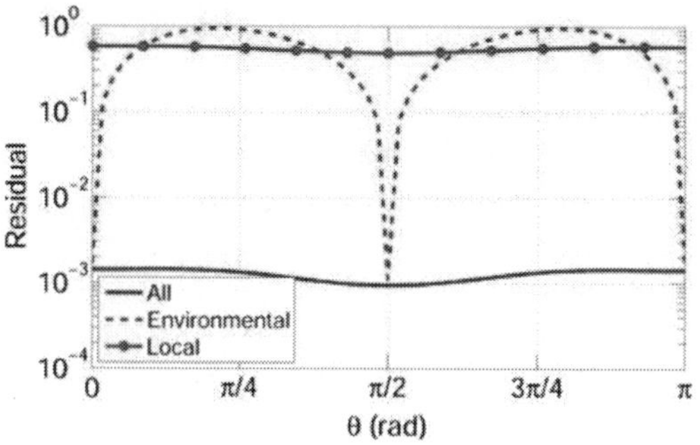

(b) Residual infrasound wave NN

Fig. 7. (a) Residual NN achieved for Rayleigh waves by using h'_{13}, h'_{23}, h'_{33} and a_z, plus seven seismometers with SNR = 10^3 at R = 5 km. (b) Residual NN achieved for infrasound by using h'_{13}, h'_{23}, h'_{33} and 15 microphones of SNR = 10^4, one at the detector, seven each at R = 600 m and 1 km

With *both waves present*, the resulting mixed NN *cannot* be removed by using these methods. We propose to use an array of external seismometers to first remove the Rayleigh waves independently of the infrasound to one part in 10^3, and then combine the *cleaned-up* h'_{13}, h'_{23}, h'_{33} and microphone signals to remove the remaining NN due to the infrasound to one part in 10^3. This will satisfy the rejection requirement for both types of NN for SOGRO as well as aSOGRO.

5 Conclusions

A wideband terrestrial tensor GW detector with a strain sensitivity of 10^{-19}-10^{-21} Hz$^{-1/2}$ for the frequency band of 0.1-10 Hz could be constructed by using six widely separated, magnetically levitated superconducting test masses. Being a full-tensor detector, a single SOGRO antenna would be capable of determining the source direction and wave polarization for signals with sufficient SNR. The tensor outputs could be combined to better reject the NN caused by the ground motion and atmospheric density fluctuations. This way, NN mitigation can be achieved at the same level as in conventional GW detectors *independently of the GW direction of propagation* [16].

A major technical challenge is the construction of a large (50 m, 200+ ton), rigid enough (≥ 10 Hz), 3D platform with high Q (≥ 10^6) that can be cooled to the liquid helium temperature. Detailed engineering studies need to be performed to find a platform design which reduces the weight while providing sufficient rigidity. Another challenge is obtaining the required high Q (~10^8) in the levitated superconducting test masses. A dedicated experimental research effort will be required to achieve such a high Q.

Although very challenging, SOGRO has a number of attractive features and deserves a serious consideration as an option for future low-frequency terrestrial GW detectors. SOGRO will be the first tensor detector that can perform genuine *full-sky search* for GW events. With SOGRO, one could set a new limit for GW flux at f = 0.1-10 Hz. Binaries composed of two IMBHs of 10^4 $M_\odot$ at the distance of a few billion light years or binaries composed of two WDs within the Local Group could be detected.

A sub-Hz high-precision GW detector has other useful applications. A recent study [17] shows that a gravity strain sensor with 10^{-20} Hz$^{-1/2}$ sensitivity at 0.1 Hz would detect an early gravity perturbation induced by an earthquake of magnitude $M > 6$ with SNR > 5 at epicentral distances up to 500 km. The prompt detection of earthquakes would enable an early warning to be issued before destructive seismic waves reach populated areas. Another potential application of SOGRO is mitigation of the NN for future third-generation GW detectors such as post advanced LIGO interferometers [18] and ET [19]. A set of shorter arm-length (~5 m) SOGROs located in close proximity to the interferometer test masses could detect the NN affecting the interferometer test masses directly [20]. The mini-SOGRO signals could then be used to clean out the NN from the interferometer.

I acknowledge useful discussions with Peter Shawhan, Rai Weiss, Rana Adhikari, Yanbei Chen, Brian Lantz, Hyung Mok Lee, Ettore Majorana, Runqiu Liu, Gungwon Kang, John Oh, Chunglee Kim and Yongho Lee. I have benefited from collaborations with Jan Harms, Vol Moody, Cornelius Griggs, Xing Bian and Krishna Venkateswara, and am indebted to Ron Norton for laboriously carrying out FEM analysis for numerous platform designs and performing many other calculations. This work was partly supported by NSF grant PHY1105030 and NASA grant NNX14AI43G.

References

1. B. P. Abbott et al. (LIGO Scientific Collaboration and Virgo Collaboration), Phys. Rev. Lett. **116**, 061102 (2016)
2. B. P. Abbott et al. (LIGO Scientific Collaboration and Virgo Collaboration), Phys. Rev. Lett. **116**, 241103 (2016)
3. B. P. Abbott et al. (LIGO Scientific and Virgo Collaboration), Phys. Rev. Lett. **118**, 221101 (2017)
4. http://www.aei.mpg.de/2068798/lisaselection
5. M. Ando et al., Class. Quantum Grav. **27**, 084010 (2010)
6. J. Harms et al., Phys. Rev. D **88**, 122003 (2013)
7. H. J. Paik et al., Class. Quantum Grav. **33**, 075003 (2016)
8. R. V. Wagoner, C. M. Will and H. J. Paik, Phys. Rev. D **19**, 2325 (1979)
9. C. E. Griggs et al., Proc. 46th Lunar and Planetary Conference (Houston, Texas, 2015), pp. 1735-1736 http://www.hou.usra.edu/meetings/lpsc2015/pdf/1735.pdf
10. C. Cinquegrana et al., Phys. Rev. D **48**, 448 (1993)
11. P. R. Saulson, Phys. Rev. D **42**, 2437 (1990)
12. P. Falferi et al., Appl. Phys. Lett. **93**, 172506 (2008)
13. J. R. Bowman, G. E. Baker and M. Bahavar, Geophys. Res. Lett. **32**, L09803 (2005)
14. M. G. Beker et al., J. Phys.: Conf. Ser. **363**, 012004 (2012)
15. M. V. Moody, H. J. Paik and E. R. Canavan, Rev. Sci. Instrum. **73**, 3957 (2002)
16. J. Harms and H. J. Paik, Phys. Rev. D **92**, 022001 (2015)
17. J. Harms et al., Geophys. J. Int. **201**, 1416 (2015)
18. LIGO Science Collaboration, Instrument science white paper LIGO DCC T1400316-v4 (2015) https://dcc.ligo.org/public/0113/T1400316/004/T1400316-v5.pdf
19. ET Science Team, Einstein gravitational wave telescope conceptual design study (2011) https://tds.ego-gw.it/itf/tds/cod ET-0106C-10
20. H. J. Paik and J. Harms, Proc. 11th Edoardo Amaldi Conf. on Gravitational Waves (Gwangju, Korea, 2015)

The binary progenitors of short and long GRBs and their gravitational-wave emission

J. A. Rueda[1,2,3,⋆], *R.* Ruffini[1,2,3,4], *J. F.* Rodriguez[1,2], *M.* Muccino[1,2], *Y.* Aimuratov[1,2], *U.* Barres de Almeida[3], *L.* Becerra[1,2], *C. L.* Bianco[1,2], *C.* Cherubini[5,6], *S.* Filippi[5,6], *M.* Kovacevic[1,2], *R.* Moradi[1,2], *G. B.* Pisani[1,2], and *Y.* Wang[1,2]

[1]*Dipartimento di Fisica and ICRA, Sapienza Università di Roma, P.le Aldo Moro 5, I–00185 Rome, Italy*
[2]*ICRANet, P.zza della Repubblica 10, I–65122 Pescara, Italy*
[3]*ICRANet-Rio, Centro Brasileiro de Pesquisas Físicas, Rua Dr. Xavier Sigaud 150, 22290–180 Rio de Janeiro, Brazil*
[4]*Université de Nice Sophia Antipolis, CEDEX 2, Grand Château Parc Valrose, Nice,*
[5]*Unit of Nonlinear Physics and Mathematical Modeling, Università Campus Bio-Medico di Roma, Via A. del Portillo 21, I–00128 Rome, Italy*
[6]*ICRA, Università Campus Bio-Medico di Roma, Via A. del Portillo 21, I–00128 Rome, Italy*

Abstract. We have sub-classified short and long-duration gamma-ray bursts (GRBs) into seven families according to the binary nature of their progenitors. Short GRBs are produced in mergers of neutron-star binaries (NS-NS) or neutron star-black hole binaries (NS-BH). Long GRBs are produced via the induced gravitational collapse (IGC) scenario occurring in a tight binary system composed of a carbon-oxygen core (CO_{core}) and a NS companion. The CO_{core} explodes as type Ic supernova (SN) leading to a hypercritical accretion process onto the NS: if the accretion is sufficiently high the NS reaches the critical mass and collapses forming a BH, otherwise a massive NS is formed. Therefore long GRBs can lead either to NS-BH or to NS-NS binaries depending on the entity of the accretion. We discuss for the above compact-object binaries: 1) the role of the NS structure and the nuclear equation of state; 2) the occurrence rates obtained from X and gamma-rays observations; 3) the predicted annual number of detections by the Advanced LIGO interferometer of their gravitational-wave emission.

1 Introduction

There has been a traditional phenomenological classification of gamma-ray bursts (GRBs) based on the observed prompt duration, T_{90}: long GRBs for $T_{90} > 2$ s and short GRBs for $T_{90} < 2$ s[1–5]. Progress has been made in the meantime in the understanding of the nature of both long and short GRBs leading to a physical, instead of empirical, classification of GRBs based on the progenitor systems [6–8].

⋆e-mail: jorge.rueda@icra.it

1.1 Long GRBs

In the case of long GRBs we stand on the induced gravitational collapse (IGC) scenario that introduces as their progenitors short-period binaries composed of a carbon-oxygen core (CO_{core}) with a NS companion [9–15]. The core-collapse of the CO_{core}, which forms a NS as central remnant (hereafter νNS), leads also to a SN explosion that triggers a massive, hypercritical accretion process onto the NS companion. The parameters of the *in-state*, i.e. of the CO_{core}-NS binary, lead to two sub-classes of long GRBs with corresponding *out-states* [6]:

- *X-ray flashes (XRFs).* Long bursts with $E_{iso} \lesssim 10^{52}$ erg are produced by CO_{core}-NS binaries with relatively large binary separations ($a \gtrsim 10^{11}$ cm). The accretion rate of the SN ejecta onto the NS in these systems is not high enough to bring the NS mass to the critical value M_{crit}, hence no BH is formed. The out-state of this GRB sub-class can be either a νNS-NS binary if the system keeps bound after the SN explosion, or two runaway NSs if the binary system is disrupted.
- *Binary driven hypernovae (BdHNe).* Long bursts with $E_{iso} \gtrsim 10^{52}$ erg are instead produced by more compact CO_{core}-NS binaries ($a \lesssim 10^{11}$ cm, see e.g., Refs. [13, 15]). In this case the SN triggers a larger accretion rate onto the NS companion, e.g. $\gtrsim 10^{-2}$–10^{-1} $M_\odot$ s^{-1}, bringing the NS to its critical mass M_{crit},[11–13] namely to the point of gravitational collapse with consequent formation of a BH. Remarkably, in Ref. [14] it was recently shown that the large majority of BdHNe leads naturally to NS-BH binaries owing to the high compactness of the binary that avoids the disruption of it even in cases of very high mass loss exceeding 50% of the total mass of the initial CO_{core}-NS binary.

In addition, it exists the possibility of *BH-SNe*.[6] Long burst with $E_{iso} \gtrsim 10^{54}$ erg occurring in close CO_{core}-BH binaries in which the hypercritical accretion produces, as *out-states*, a more massive BH and a νNS. These systems have been considered in Ref. [6] as a subset of the BdHNe but no specific example have been yet observationally identified.

1.2 Short GRBs

There is the consensus within the GRB community that the progenitors of short GRBs are mergers of NS-NS and/or NS-BH binaries (see, e.g., Refs. [16–19], and Ref. [20], for a recent review). Similarly to the case of long GRBs, in Ref. [6] short GRBs have been split into different sub-classes:

- *Short gamma-ray flashes (S-GRFs).* Short bursts with energies $E_{iso} \lesssim 10^{52}$ erg, produced when the post-merger core do not surpass the NS critical mass M_{crit}, hence there is no BH formation. Thus these systems left as byproduct a massive NS and possibly, due to the energy and angular momentum conservation, orbiting material in a disk-like structure or a low-mass binary companion.
- *Authentic short gamma-ray bursts (S-GRBs).* Short bursts with $E_{iso} \gtrsim 10^{52}$ erg, produced when the post-merger core reaches or overcome M_{crit}, hence forming a Kerr or Kerr-Newman BH,[8] and also in this case possibly orbiting material.
- *Ultra-short GRBs (U-GRBs).* A new sub-class of short bursts originating from νNS-BH merging binaries. They can originate from BdHNe (see Ref. [14]) or from BH-SNe.

In addition, it exists the possibility of *gamma-ray flashes (GRFs).* These are bursts with hybrid properties between short and long, they have $10^{51} \lesssim E_{iso} \lesssim 10^{52}$ erg. This sub-class of sources originates in NS-WD mergers.[6]

We focus here on the physical properties of the above progenitors, as well as on the main properties of NSs that play a relevant role in the dynamics of these systems and that lead to the above different GRB sub-classes. We shall discuss as well recent estimates of the rates of occurrence on all the above subclasses based on X and gamma-ray observations, and also elaborate on the possibility of detecting the gravitational wave (GW) emission originated in these systems.

2 IGC, Hypercritical Accretion, and Long GRBs

We turn now to the details of the accretion process within the IGC scenario. Realistic simulations of the IGC process were performed in Ref. [12], including: 1) detailed SN explosions of the CO_{core}; 2) the hydrodynamic details of the hypercritical accretion process; 3) the evolution of the SN ejecta material entering the Bondi-Hoyle region all the way up to its incorporation into the NS. Here the concept of hypercritical accretion refers to the fact the accretion rates are highly super-Eddington. The accretion process in the IGC scenario is allowed to exceed the Eddington limit mainly for two reasons: i) the photons are trapped within the infalling material impeding them to transfer momentum; ii) the accreting material creates a very hot NS atmosphere ($T \sim 10^{10}$ K) that triggers a very efficient neutrino emission which become the main energy sink of these systems unlike photons.

The hypercritical accretion process in the above simulations was computed within a spherically symmetric approximation. A further step was given in Ref. [13] by estimating the angular momentum that the SN ejecta carries and transfer to the NS via accretion, and how it affects the evolution and fate of the system. The calculations are as follows: first the accretion rate onto the NS is computed adopting an homologous expansion of the SN ejecta and introducing the pre-SN density profile of the CO_{core} envelope from numerical simulations. Then, it is estimated the angular momentum that the SN material might transfer to the NS: it comes out that the ejecta have enough angular momentum to circularize for a short time and form a disc around the NS. Finally, the evolution of the NS central density and rotation angular velocity (spin-up) is followed computing the equilibrium configurations from the numerical solution of the axisymmetric Einstein equations in full rotation, until the critical point of collapse of the NS to a BH taking into due account the equilibrium limits given by mass-shedding and the secular axisymmetric instability.

Now we enter into the details of each of the above steps. The accretion rate of the SN ejecta onto the NS can be estimated via the Bondi-Hoyle-Lyttleton accretion formula:

$$\dot{M}_B(t) \;=\; \pi \rho_{\rm ej} R_{\rm cap}^2 \sqrt{v_{\rm rel}^2 + c_{\rm s,ej}^2}, \qquad R_{\rm cap}(t) \;=\; \frac{2GM_{\rm NS}(t)}{v_{\rm rel}^2 + c_{\rm s,ej}^2}, \tag{1}$$

where G is the gravitational constant, $\rho_{\rm ej}$ and $c_{\rm s,ej}$ are the density and sound speed of the SN ejecta, $R_{\rm cap}$ is the NS gravitational capture radius (Bondi-Hoyle radius), $M_{\rm NS}$, the NS mass, and $v_{\rm rel}$ the ejecta velocity relative to the NS: $\vec{v}_{\rm rel} = \vec{v}_{\rm orb} - \vec{v}_{\rm ej}$, with $|\vec{v}_{\rm orb}| = \sqrt{G(M_{\rm core} + M_{\rm NS})/a}$, the module of the NS orbital velocity around the CO_{core}, and $\vec{v}_{\rm ej}$ the velocity of the supernova ejecta (see Fig. 1).

Extrapolating the results for the accretion process from stellar wind accretion in binary systems, the angular momentum per unit time that crosses the NS capture region can be approximated by:$\dot{L}_{\rm cap} = (\pi/2)\left(\epsilon_\rho/2 - 3\epsilon_v\right)\rho_{\rm ej}(a,t)v_{\rm rel}^2(a,t)R_{\rm cap}^4(a,t)$, where ϵ_ρ and ϵ_v are parameters measuring the inhomogeneity of the flow (see Ref. [13] for details).

In order to simulate the hypercritical accretion it is adopted an homologous expansion of the SN ejecta, i.e. the ejecta velocity evolves as $v_{\rm ej}(r,t) = nr/t$, where r is the position of every ejecta layer from the SN center and n is called expansion parameter. The ejecta density is given by $\rho_{\rm ej}(r,t) = \rho_{\rm ej}^0(r/R_{\rm star}(t), t_0)\frac{M_{\rm env}(t)}{M_{\rm env}(0)}\left(\frac{R_{\rm star}(0)}{R_{\rm star}(t)}\right)^3$, where $M_{\rm env}(t)$ the mass of the CO_{core} envelope, namely the mass of the ejected material in the SN explosion and available to be accreted by the NS, $R_{\rm star}(t)$ is the position of the outermost layer of the ejected material, and $\rho_{\rm ej}^0$ is the pre-SN density profile. The latter can be approximated with a power law: $\rho_{\rm ej}(r,t_0) = \rho_{\rm core}(R_{\rm core}/r)^m$, where $\rho_{\rm core}$, $R_{\rm core}$ and m are the profile parameters which are fixed by fitting the pre-SN profiles obtained from numerical simulations.

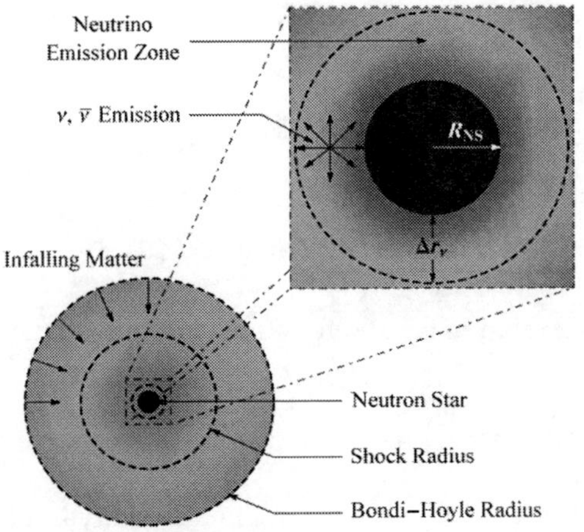

Figure 1. Scheme of the IGC scenario: the CO_{core} undergoes SN explosion, the NS accretes part of the SN ejecta and then reaches the critical mass for gravitational collapse to a BH, with consequent emission of a GRB. The SN ejecta reach the NS Bondi-Hoyle radius and fall toward the NS surface. The material shocks and decelerates as it piles over the NS surface. At the neutrino emission zone, neutrinos take away most of the infalling matter gravitational energy gain. The neutrinos are emitted above the NS surface in a region of thickness Δr_ν about half the NS radius that allow the material to reduce its entropy to be finally incorporated to the NS. The image is not to scale. For further details and numerical simulations of the above process see Refs. [12, 13, 15].

For the typical parameters of pre-SN CO_{core} and assuming a velocity of the outermost SN layer $v_{sn}(R_{star}, t_0) \sim 10^9$ cm s^{-1} and a free expansion $n = 1$ (for details of typical initial conditions of the binary system see Refs. [12] and [13]), Eq. (1) gives accretion rates around the order of $10^{-4} - 10^{-2} M_\odot$ s^{-1}, and an angular momentum per unit time crossing the capture region $\dot{L}_{cap} \sim 10^{46}$–$10^{49}$ g cm^2 s^{-2}.

We consider the NS companion of the CO_{core} initially as non-rotating, thus at the beginning the NS exterior spacetime is described by the Schwarzschild metric. The SN ejecta approach the NS with specific angular momentum, $l_{acc} = \dot{L}_{cap}/\dot{M}_B$, thus they will circularize at a radius r_{st} if they have enough angular momentum. What does the word "enough" means here? The last stable circular orbit (LSO) around a non-rotating NS is located at a distance $r_{lso} = 6GM_{NS}/c^2$ and has an angular momentum per unit mass $l_{lso} = 2\sqrt{3}GM_{NS}/c$. The radius r_{lso} is larger than the NS radius for masses larger than 1.67 $M_\odot$, 1.71 $M_\odot$, and 1.78 $M_\odot$ for the GM1, TM1, and NL3 nuclear equation of state (EOS).[13] If $l_{acc} \geq l_{lso}$ the material circularizes around the NS at locations $r_{st} \geq r_{lso}$. For the values of the IGC systems under discussion here, $r_{st}/r_{lso} \sim 10 - 10^3$, thus the SN ejecta have enough angular momentum to form a sort of disc around the NS. Even in this case, the viscous forces and other angular momentum losses that act on the disk will allow the matter in the disk to reach the inner boundary at $r_{in} \sim r_{lso}$, to then be accreted by the NS.

Within this picture, the NS accretes the material from r_{in} and the NS mass and angular angular momentum evolve as:

$$\dot{M}_{NS} = \left(\frac{\partial M_{NS}}{\partial M_b}\right)_{J_{NS}} \dot{M}_b + \left(\frac{\partial M_{NS}}{\partial J_{NS}}\right)_{M_b} \dot{J}_{NS}, \qquad \dot{J}_{NS} = \xi\, l(r_{in}) \dot{M}_B, \tag{2}$$

where M_b is the NS baryonic mass, $l(r_{in})$ is the specific angular momentum of the accreted material at r_{in}, which corresponds to the angular momentum of the LSO, and $\xi \leq 1$ is a parameter that measures the efficiency of angular momentum transfer. We assume in our simulations $\dot{M}_b = \dot{M}_B$. The baryonic and gravitational mass are related by [21]:

$$\frac{M_b}{M_\odot} = \frac{M_{NS}}{M_\odot} + \frac{13}{200}\left(\frac{M_{NS}}{M_\odot}\right)^2 \left(1 - \frac{1}{137} j_{NS}^{1.7}\right), \tag{3}$$

where $j_{NS} \equiv cJ_{NS}/(GM_\odot^2)$. In addition, since the NS will spin up with accretion, we need information of the dependence of the specific angular momentum of the LSO as a function of both the NS mass

Table 1. Critical NS mass in the non-rotating case and constants k and p needed to compute the NS critical mass in the non-rotating case given by Eq. (5). The values are given for the NL3, GM1 and TM1 EOS.

EOS	$M_{crit}^{J=0}$ $(M_\odot)$	p	k
NL3	2.81	1.68	0.006
GM1	2.39	1.69	0.011
TM1	2.20	1.61	0.017

and angular momentum. For corotating orbits the following relation is valid for all the aforementioned EOS:[13]

$$l_{\rm lso} = \frac{GM_{\rm NS}}{c}\left[2\sqrt{3} - 0.37\left(\frac{j_{\rm NS}}{M_{\rm NS}/M_\odot}\right)^{0.85}\right]. \tag{4}$$

The NS accretes mass until it reaches a region of instability. There are two main instability limits for rotating NSs: mass-shedding or Keplerian limit and the secular axisymmetric instability. The critical NS mass along the secular instability line is approximately given by:[21]

$$M_{\rm NS}^{\rm crit} = M_{\rm NS}^{J=0}(1 + kj_{\rm NS}^{p}), \tag{5}$$

where the parameters k and p depends of the nuclear EOS (see Table 1). These formulas fit the numerical results with a maximum error of 0.45%.

2.1 Most recent simulations of the IGC process

Additional details and improvements of the hypercritical accretion process leading to XRFs and BdHNe have been recently presented in Ref. [15]. In particular:

1. It was there improved the accretion rate estimate including the density profile finite size/thickness and additional CO_{core} progenitors leading to different SN ejecta masses were also considered.

2. It was shown in Ref. [13] the existence of a maximum orbital period, P_{max}, over which the accretion onto NS companion is not high enough to bring it to the critical mass for gravitational collapse to a BH. Therefore, CO_{core}-NS binaries with $P > P_{max}$ lead to XRFs while the ones with $P \lesssim P_{max}$ lead to BdHNe. In Ref. [15] the determination of P_{max} was extended to all the possible initial values of the mass of the NS companion and the angular momentum transfer efficiency parameter was also allowed to vary.

3. It was computed the expected luminosity during the hypercritical accretion process for a wide range of binary periods covering XRFs and BdHNe.

4. It was there shown that the presence of the NS companion originates large asymmetries (see, e.g., simulation in Fig. 2) in the SN ejecta leading to observable signatures in the X-rays.

Fig. 2 shows a simulation of an IGC process presented in Ref. [15]. We considered the effects of the gravitational field of the NS on the SN ejecta including the orbital motion as well as the changes in the NS gravitational mass owing to the accretion process via the Bondi formalism. The supernova matter was described as formed by point-like particles whose trajectory was computed by solving the Newtonian equation of motion. The initial conditions of the SN ejecta are computed assuming an homologous velocity distribution in free expansion. The initial power-law density profile of the CO

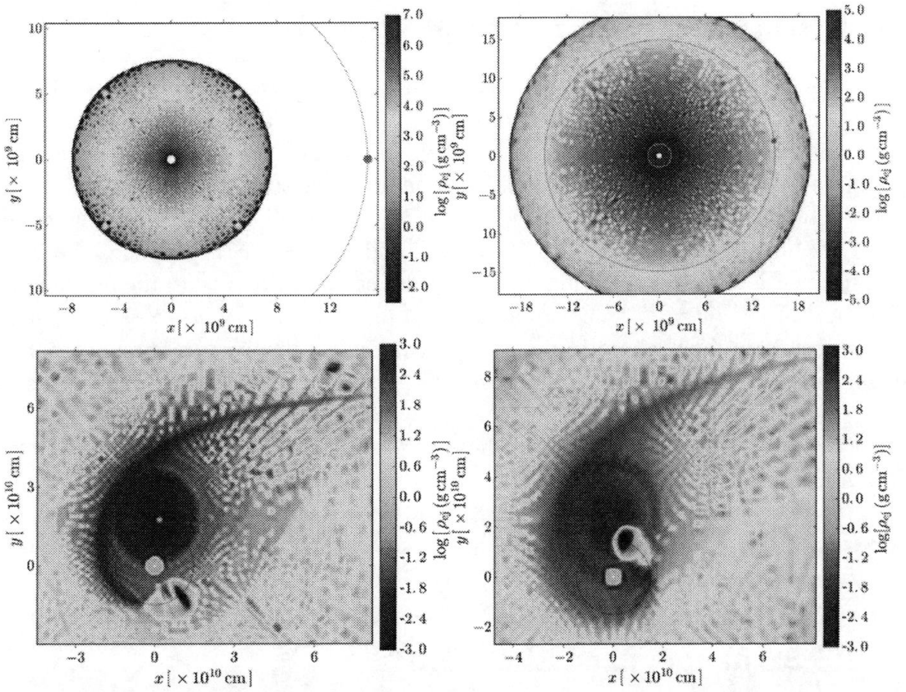

Figure 2. Hypercritical accretion process in the IGC binary system at selected evolution times. In this example the CO_{core} has a total mass of 9.44 $M_{\odot}$ divided in an ejecta mass of 7.94 $M_{\odot}$ and a νNS of 1.5 $M_{\odot}$ formed by the collapsed high density core. The supernova ejecta evolve homologously with outermost layer velocity $v_{0,star} = 2 \times 10^9$ cm s^{-1}. The NS binary companion has an initial mass of 2.0 $M_{\odot}$. The binary period is $P \approx 5$ min, which corresponds to a binary separation $a \approx 1.5 \times 10^{10}$ cm. The system of coordinates is centered on the νNS represented by the white-filled circle at $(0,0)$. The NS binary companion, represented by the gray-filled circle, orbits counterclockwise following the thin-dashed circular trajectory. The colorbar indicates values of ejecta density in logarithmic scale. *Left upper panel:* initial time of the process. The supernova ejecta expand radially outward and the NS binary companion is at $(a, 0)$. *Right upper panel:* the accretion process starts when the first supernova layers reach the Bondi-Hoyle region. This happens at $t = t_{acc,0} \approx a/v_{0,star} \approx 7.7$ s. *Left lower panel:* the NS binary companion reaches the critical mass by accreting matter from the SN with consequent collapse to a BH. This happens at $t = t_{coll} \approx 254$ s $\approx 0.85\,P$. The newly-formed BH of mass $M_{BH} = M_{crit} \approx 3\,M_{\odot}$ is represented by the black-filled circle. It is here evident the asymmetry of the supernova ejecta induced by the presence of the accreting NS companion at close distance. *Right lower panel:* $t = t_{coll} + 100$ s $= 354$ s $\approx 1.2P$, namely 100 s after the BH formation. It appears here the new binary system composed of the νNS and the newly-formed BH.

envelope is simulated by populating the inner layers with more particles. For the $M_{ZAMS} = 30\,M_{\odot}$ progenitor which gives a CO_{core} with envelope profile $\rho^0_{ej} \approx 3.1 \times 10^8 (8.3 \times 10^7/r)^{2.8}$ g cm^{-3}, we adopt for the simulation a total number of $N = 10^6$ particles. We assume that particles crossing the Bondi-Hoyle radius are captured and accreted by the NS so we removed them from the system as they reach that region. We removed these particles according to the results obtained from the numerical integration explained above. Fig. 2 shows the orbital plane of an IGC binary at selected times of its evolution. The NS has an initial mass of 2.0 $M_{\odot}$; the CO_{core} leads to a total ejecta mass 7.94 $M_{\odot}$ and a νNS of 1.5 $M_{\odot}$. The orbital period of the binary is $P \approx 5$ min, i.e. a binary separation $a \approx 1.5 \times 10^{10}$ cm. For these parameters the NS reaches the critical mass and collapses to form a BH.

2.2 Post-Explosion Orbits and Formation of NS-BH Binaries

We have seen how in BdHNe the accretion process can lead to BH formation in a time-interval as short as the orbital period. We here deepen this analysis to study the effect of the SN explosion in such a scenario following Ref. [14]. As the ejecta timescale becomes a fraction of the orbital one, the fate of the post-explosion binary is altered. For these models, we assumed tight binaries in circular orbit with an initial orbital separation of 7×10^9 cm. With CO_{core} radii of 1–4×10^9 cm, such a separation is small, but achievable. The binary consists of a CO_{core} and a 2.0 $M_\odot$ NS companion. When the CO_{core} collapses, it forms a 1.5 $M_\odot$ NS, ejecting the rest of the core. We then vary the ejecta mass and time required for most of the ejected matter to move out of the binary. Ref. [14] showed that even if 70% of the mass is lost from the system (in the 8 $M_\odot$ ejecta case), the system remains bound as long as the explosion time is just above the orbital time ($T_{\rm orbit}$ = 180 s) with semi-major axes of less than 10^{11} cm.

The tight compact binaries produced in these explosions will emit GWs driving the system to merge. For typical massive star binaries, the merger time is many Myr. For BdHNe, the merger time is typically 10^4 yr, or less [14]. Since the merger should occur within the radius swept clean by the BdHN we expect a small baryonic contamination around the merger site which might lead to a new family of events which we term ultrashort GRBs, U-GRBs, to this new family of events.

3 NS-NS/NS-BH mergers and Short GRBs

Let us turn to short GRBs. We first proceed to estimate the mass and the angular momentum of the post-merger core via baryonic mass and angular momentum conservation of the system. We adopt for simplicity that non-rotating binary components. We first compute the total baryonic mass of the NS-NS binary $M_b = M_{b_1} + M_{b_2}$ using the relation between the gravitational mass M_i and the baryonic mass M_{b_i} ($i = 1,2$) recently obtained in Ref. [21] and given in Eq. (3) assuming $j_{\rm NS} = cJ_{\rm NS}/(GM_\odot^2) = 0$. The post-merger core will have approximately the entire baryonic mass of the initial binary, i.e. $M_{b,\rm core} \approx M_b$, since little mass is expected to be ejected during the coalescence process. However, the gravitational mass of the post-merger core cannot be estimated using again the above formula since, even assuming non-rotating binary components, the post-merger core will necessarily acquire a fraction $\eta \leq 1$ of the binary angular momentum at the merger point. One expects a value of η smaller than unity since, during the coalesce, angular momentum is loss e.g. by gravitational wave emission and it can be also redistributed e.g. into a surrounding disk.

To obtain the gravitational mass of the post-merger core, we can use again Eq. (3) relating the baryonic mass $M_{b,\rm NS}$ and the gravitational mass $M_{\rm NS}$ in this case with $j_{\rm NS} \neq 0$. The mass and angular momentum of the post-merger core, respectively $M_{\rm core}$ and $J_{\rm core}$, are therefore obtained from baryon mass and angular momentum conservation, i.e.: $M_{\rm core} = M_{\rm NS}$, $M_{b,\rm core} = M_{b,\rm NS} = M_{b_1} + M_{b_2}$, $J_{\rm core} = J_{\rm NS} = \eta J_{\rm merger}$, where $J_{\rm merger}$ is the system angular momentum at the merger point. The value of $J_{\rm merger}$ is approximately given by $J_{\rm merger} = \mu r_{\rm merger}^2 \Omega_{\rm merger}$, where $\mu = M_1 M_2/M$ is the binary reduced mass, $M = M_1 + M_2$ is the total binary mass, and $r_{\rm merger}$ and $\Omega_{\rm merger}$ are the binary separation and angular velocity at the merger point. If we adopt the merger point where the two stars enter into contact we have $r_{\rm merger} = R_1 + R_2$, where R_i is the radius (which depend on the EOS) of the i-component of the binary.

Given the parameters of the merging binary, the above equations lead to the merged core properties. For the sake of exemplifying, let us assume a mass-symmetric binary, $M_1 = M_2 = M/2$. In this case the above equations lead to the angular momentum of the merged core $J_{\rm core} = (\eta/4)(GM^2/c)C^{-1/2}$, where $C \equiv GM_1/(c^2R_1) = GM_2/(c^2R_2)$ is the compactness of the merging binary components. Therefore, if we adopt $M_1 = 1.4\ M_\odot$ and $C = 0.15$, $M_{\rm core} = (2.61, 2.65)\ M_\odot$ for

$\eta = (0, 1)$, i.e. for a dimensionless angular momentum of the merged core $j_{\rm core} = (0, 5.06)$. Whether or not these pairs $(M_{\rm core}, j_{\rm core})$ correspond to stable NSs depend on the nuclear EOS. A similar analysis can be done for any other pair of binary masses.

4 Detectability of GWs produced by the GRB progenitors

Having established the nature of the progenitors of each GRB sub-class, we turn now to briefly discuss the detectability of their associated GW emission. The minimum GW frequency detectable by the broadband aLIGO interferometer is $f_{\rm min}^{\rm aLIGO} \approx 10$ Hz.[35] Since during the binary inspiral the GW frequency is twice the orbital one, this implies that a binary enters the aLIGO band for orbital periods $P_{\rm orb} \lesssim 0.2$ s. Thus, $\rm CO_{core}$-NS binaries, *in-states* of XRFs and BdHNe, and $\rm CO_{core}$-BH binaries, *in-states* of BH-SN, are not detectable by aLIGO since they have orbital periods $P_{\rm orb} \gtrsim 5$ min $\gg 0.2$ s. Concerning their *out-states* after the corresponding hypercritical accretion processes, namely νNS-NS, *out-states* of XRFs, and νNS-BH, *out-states* of BdHNe and BH-SNe, they are not detectable by aLIGO at their birth but only when approaching the merger. Clearly, the analysis of the νNS-NS mergers is included in the analysis of the S-GRFs and S-GRBs and, likewise, the merger of νNS-BH binaries is included in the analysis of U-GRBs. In the case of NS-WD binaries the WD is tidally disrupted by the NS making their GW emission hard to be detected (see, e.g., Ref. [36]).

A coalescing binary evolves first through the *inspiral regime* to then pass over a *merger regime*, the latter composed by the plunge leading to the merger itself and by the ringdown (oscillations) of the newly formed object. During the inspiral regime the system evolves through quasi-circular orbits and is well described by the traditional point-like quadrupole approximation.[37–39] The GW frequency is twice the orbital frequency ($f_s = 2f_{\rm orb}$) and grows monotonically. The energy spectrum during the inspiral regime is: $dE/df_s = (1/3)(\pi G)^{2/3} M_c^{5/3} f_s^{-1/3}$, where $M_c = \mu^{3/5} M^{2/5} = \nu^{3/5} M$ is the so-called *chirp mass* and $\nu \equiv \mu/M$ is the symmetric mass-ratio parameter. A symmetric binary ($m_1 = m_2$) corresponds to $\nu = 1/4$ and the test-particle limit is $\nu \to 0$. The GW spectrum of the merger regime is characterized by a GW burst.[40] Thus, one can estimate the contribution of this regime to the signal-to-noise ratio with the knowledge of the location of the GW burst in the frequency domain and of the energy content. The frequency range spanned by the GW burst is $\Delta f = f_{\rm qnm} - f_{\rm merger}$, where $f_{\rm merger}$ is the frequency at which the merger starts and $f_{\rm qnm}$ is the frequency of the ringing modes of the newly formed object after the merger, and the energy emitted is $\Delta E_{\rm merger}$. With these quantities defined, one can estimate the typical value of the merger regime spectrum as: $dE/df_s \approx \Delta E_{\rm merger}/\Delta f$. Unfortunately, the frequencies and energy content of the merger regime of the above merging binaries are such that it is undetectable by LIGO.[41].

Since the GW signal is deep inside the detector noise, the signal-to-noise ratio (ρ) is usually estimated using the matched filter technique.[42] The exact position of the binary relative to the detector and the orientation of the binary rotation plane are usually unknown, thus it is a common practice to average over all the possible locations and orientations, i.e.:[42] $\langle \rho^2 \rangle = 4 \int_0^\infty \langle |\tilde{h}(f)|^2 \rangle / S_n(f) df = 4 \int_0^\infty h_c^2(f)/[f^2 S_n(f)] df$, where f is the GW frequency in the detector frame, $\tilde{h}(f)$ is the Fourier transform of $h(t)$, and $\sqrt{S_n(f)}$ is the one-sided amplitude spectral density of the detector noise, and $h_c(f)$ is the characteristic strain, $h_c = (1+z)/(\pi d_l) \sqrt{(1/10)(G/c^3)(dE/df_s)}$. We recall that in the detector frame the GW frequency is redshifted by a factor $1+z$ with respect to the one in the source frame, f_s, i.e. $f = f_s/(1+z)$ and d_l is the luminosity distance to the source. We adopt a ΛCDM cosmology with $H_0 = 71$ km s^{-1} Mpc^{-1}, $\Omega_M = 0.27$ and $\Omega_\Lambda = 0.73$.[43]

A threshold $\rho_0 = 8$ in a single detector is adopted by LIGO.[44] This minimum ρ_0 defines a maximum detection distance or GW horizon distance, say $d_{\rm GW}$, that corresponds to the most optimistic case when the binary is just above the detector and the binary plane is parallel to the detector plane.

In order to give an estimate the annual number of merging binaries associated with the above GRB sub-classes detectable by aLIGO we can use the lower and upper values of the aLIGO *search volume* defined by $\mathcal{V}_s = V_{\rm max}^{\rm GW}\mathcal{T}$, where $V_{\rm max}^{\rm GW} = (4\pi/3)\mathcal{R}^3$, where $\mathcal{T}$ is the observing time and $\mathcal{R}$ is the so-called *detector range* defined by $\mathcal{R} = \mathcal{F} d_{\rm GW}$, with $\mathcal{F}^{-1} = 2.2627$ (see, Ref. [44, 45], for details). For a (1.4+1.4) $M_\odot$ NS binary and the three following different observational campaigns we have:[44] 2015/2016 (O1; $\mathcal{T} = 3$ months) $\mathcal{V}_S = (0.5\text{–}4) \times 10^5$ Mpc^3 yr, 2017/2018 (O3; $\mathcal{T} = 9$ months) $\mathcal{V}_S = (3\text{–}10)\times 10^6$ Mpc^3 yr, and the entire network including LIGO-India at design sensitivity (2022+; $\mathcal{T} = 1$ yr) $\mathcal{V}_S = 2 \times 10^7$ Mpc^3 yr. The maximum possible sensitivity reachable in 2022+ leads to $d_{\rm GW} \approx 0.2$ Gpc, hence $V_{\rm max}^{\rm GW} \approx 0.033$ Gpc^3, for such a binary. One can use this information for other binaries with different masses taking advantage of the fact that $d_{\rm GW}$ scales with the binary chirp mass as $M_c^{5/6}$. The expected GW detection rate by aLIGO can be thus estimated as: $\dot{N}_{\rm GW} \equiv \rho_{\rm GRB} V_{\rm max}^{\rm GRB}$, where $\rho_{\rm GRB}$ is the inferred occurrence rate of GRBs computed in Ref. [6]. Bearing the above in mind it is easy to check that there is a low probability for aLIGO to detect the GW signals associated with the GRB binary progenitors: indeed in the best case of the 2022+ observing rung one obtains, respectively, ~ 1 detection every 3 and 5 yr for U-GRBs and S-GRFs.

References

[1] E.P. Mazets, S.V. Golenetskii, V.N. Ilinskii, V.N. Panov, R.L. Aptekar, I.A. Gurian, M.P. Proskura, I.A. Sokolov, Z.I. Sokolova, T.V. Kharitonova, Ap&SS**80**, 3 (1981)

[2] R.W. Klebesadel, *The durations of gamma-ray bursts*, in *Gamma-Ray Bursts - Observations, Analyses and Theories*, edited by C. Ho, R.I. Epstein, E.E. Fenimore (Cambridge University Press, 1992), pp. 161–168

[3] J.P. Dezalay, C. Barat, R. Talon, R. Syunyaev, O. Terekhov, A. Kuznetsov, *Short cosmic events - A subset of classical GRBs?*, in *American Institute of Physics Conference Series*, edited by W. S. Paciesas & G. J. Fishman (1992), Vol. 265 of *American Institute of Physics Conference Series*, pp. 304–309

[4] C. Kouveliotou, C.A. Meegan, G.J. Fishman, N.P. Bhat, M.S. Briggs, T.M. Koshut, W.S. Paciesas, G.N. Pendleton, ApJ**413**, L101 (1993)

[5] M. Tavani, ApJ**497**, L21 (1998), `arXiv:astro-ph/9802192`

[6] R. Ruffini, J.A. Rueda, M. Muccino, Y. Aimuratov, L.M. Becerra, C.L. Bianco, M. Kovacevic, R. Moradi, F.G. Oliveira, G.B. Pisani et al., ApJ**832**, 136 (2016), `1602.02732`

[7] R. Ruffini, Y. Wang, M. Enderli, M. Muccino, M. Kovacevic, C.L. Bianco, A.V. Penacchioni, G.B. Pisani, J.A. Rueda, ApJ**798**, 10 (2015), `1405.5723`

[8] R. Ruffini, M. Muccino, M. Kovacevic, F.G. Oliveira, J.A. Rueda, C.L. Bianco, M. Enderli, A.V. Penacchioni, G.B. Pisani, Y. Wang et al., ApJ**808**, 190 (2015), `1412.1018`

[9] R. Ruffini, M.G. Bernardini, C.L. Bianco, L. Vitagliano, S.S. Xue, P. Chardonnet, F. Fraschetti, V. Gurzadyan, *Black Hole Physics and Astrophysics: The GRB-Supernova Connection and URCA-1 - URCA-2*, in *The Tenth Marcel Grossmann Meeting*, edited by M. Novello, S. Perez Bergliaffa, R. Ruffini (Singapore: World Scientific, 2006), p. 369, `astro-ph/0503475`

[10] L. Izzo, J.A. Rueda, R. Ruffini, A&A**548**, L5 (2012), `1206.2887`

[11] J.A. Rueda, R. Ruffini, ApJ**758**, L7 (2012), `1206.1684`

[12] C.L. Fryer, J.A. Rueda, R. Ruffini, ApJ**793**, L36 (2014), `1409.1473`

[13] L. Becerra, F. Cipolletta, C.L. Fryer, J.A. Rueda, R. Ruffini, ApJ**812**, 100 (2015), `1505.07580`

[14] C.L. Fryer, F.G. Oliveira, J.A. Rueda, R. Ruffini, Physical Review Letters **115**, 231102 (2015), `1505.02809`

[15] L. Becerra, C.L. Bianco, C.L. Fryer, J.A. Rueda, R. Ruffini, ApJ**833**, 107 (2016), `1606.02523`

[16] J. Goodman, ApJ**308**, L47 (1986)
[17] B. Paczynski, ApJ**308**, L43 (1986)
[18] D. Eichler, M. Livio, T. Piran, D.N. Schramm, Nature**340**, 126 (1989)
[19] R. Narayan, T. Piran, A. Shemi, ApJ**379**, L17 (1991)
[20] E. Berger, ARAA**52**, 43 (2014), 1311.2603
[21] F. Cipolletta, C. Cherubini, S. Filippi, J.A. Rueda, R. Ruffini, Phys. Rev. D**92**, 023007 (2015), 1506.05926
[22] C.L. Fryer, W. Benz, M. Herant, ApJ**460**, 801 (1996), astro-ph/9509144
[23] R.A. Chevalier, ApJ**346**, 847 (1989)
[24] Y.B. Zel'dovich, L.N. Ivanova, D.K. Nadezhin, Soviet Astronomy **16**, 209 (1972)
[25] R. Ruffini, J. Wilson, Physical Review Letters **31**, 1362 (1973)
[26] C.L. Fryer, F. Herwig, A. Hungerford, F.X. Timmes, ApJ**646**, L131 (2006), astro-ph/0606450
[27] C.L. Fryer, ApJ**699**, 409 (2009), 0711.0551
[28] L. Izzo, R. Ruffini, A.V. Penacchioni, C.L. Bianco, L. Caito, S.K. Chakrabarti, J.A. Rueda, A. Nandi, B. Patricelli, A&A**543**, A10 (2012)
[29] N.R. Tanvir, A.J. Levan, A.S. Fruchter, J. Hjorth, R.A. Hounsell, K. Wiersema, R.L. Tunnicliffe, Nature**500**, 547 (2013), 1306.4971
[30] N.R. Sibgatullin, R.A. Sunyaev, Astronomy Letters **26**, 772 (2000), astro-ph/0011344
[31] C.L. Fryer, S.E. Woosley, D.H. Hartmann, ApJ**526**, 152 (1999), astro-ph/9904122
[32] M. Dominik, K. Belczynski, C. Fryer, D.E. Holz, E. Berti, T. Bulik, I. Mandel, R. O'Shaughnessy, ApJ**759**, 52 (2012), 1202.4901
[33] K.A. Postnov, L.R. Yungelson, Living Reviews in Relativity **17**, 3 (2014), 1403.4754
[34] J.G. Hills, ApJ**267**, 322 (1983)
[35] LIGO Scientific Collaboration, J. Aasi, B.P. Abbott, R. Abbott, T. Abbott, M.R. Abernathy, K. Ackley, C. Adams, T. Adams, P. Addesso et al., Classical and Quantum Gravity **32**, 074001 (2015), 1411.4547
[36] V. Paschalidis, M. MacLeod, T.W. Baumgarte, S.L. Shapiro, Phys. Rev. D**80**, 024006 (2009)
[37] P.C. Peters, J. Mathews, Phys. Rev. **131**, 435 (1963)
[38] P.C. Peters, Phys. Rev. **136**, 1224 (1964)
[39] M. Rees, R. Ruffini, J.A. Wheeler, *Black holes, gravitational waves and cosmology* (Gordon and Breach Science Publishers Inc., New York, 1974)
[40] M. Davis, R. Ruffini, W.H. Press, R.H. Price, Phys. Rev. Lett. **27**, 1466 (1971)
[41] R. Ruffini, J. Rodriguez, M. Muccino, J.A. Rueda, Y. Aimuratov, U. Barres de Almeida, L. Becerra, C.L. Bianco, C. Cherubini, S. Filippi et al., ArXiv:1602.03545 (2016), 1602.03545
[42] É.É. Flanagan, S.A. Hughes, Phys. Rev. D**57**, 4535 (1998), gr-qc/9701039
[43] M. Rigault, G. Aldering, M. Kowalski, Y. Copin, P. Antilogus, C. Aragon, S. Bailey, C. Baltay, D. Baugh, S. Bongard et al., ApJ**802**, 20 (2015), 1412.6501
[44] B.P. Abbott, R. Abbott, T.D. Abbott, M.R. Abernathy, F. Acernese, K. Ackley, C. Adams, T. Adams, P. Addesso, R.X. Adhikari et al., Living Reviews in Relativity **19** (2016), 1304.0670
[45] L.S. Finn, D.F. Chernoff, Phys. Rev. D**47**, 2198 (1993)

EPJ Web of Conferences **168**, 01007 (2018) https://doi.org/10.1051/epjconf/201816801007
Joint International Conference of ICGAC-XIII and IK-15 on Gravitation, Astrophysics and Cosmology

Towards causal patch physics in dS/CFT

Yasha Neiman[1,★]

[1]*Okinawa Institute of Science and Technology, 1919-1 Tancha, Onna-son, Okinawa 904-0495, Japan*

Abstract. This contribution is a status report on a research program aimed at obtaining quantum-gravitational physics inside a cosmological horizon through dS/CFT, i.e. through a holographic description at past/future infinity of de Sitter space. The program aims to bring together two main elements. The first is the observation by Anninos, Hartman and Strominger that Vasiliev's higher-spin gravity provides a working model for dS/CFT in 3+1 dimensions. The second is the proposal by Parikh, Savonije and Verlinde that dS/CFT may prove more tractable if one works in so-called "elliptic" de Sitter space – a folded-in-half version of global de Sitter where antipodal points have been identified. We review some relevant progress concerning quantum field theory on elliptic de Sitter space, higher-spin gravity and its holographic duality with a free vector model. We present our reasons for optimism that the approach outlined here will lead to a full holographic description of quantum (higher-spin) gravity in the causal patch of a de Sitter observer.

1 Introduction to higher-spin dS/CFT

For decades, the great conceptual problem in theoretical physics has been that of quantum gravity. In the real world, this problem remains unsolved – we don't know how nature reconciles the two simple facts $\hbar \neq 0$ and $G \neq 0$. However, one can also frame quantum gravity as a less demanding question: how can these two facts be reconciled even in principle, in any fictitious but mathematically consistent universe? This version of the question is now essentially solved. The first fully convincing solution (by the author's standards) was the discovery of AdS/CFT in the late 1990's [1–3]. There, a non-gravitating conformal field theory (CFT) provides a complete description of quantum-gravitational observables at spatial infinity, in a universe with negative cosmological constant $\Lambda < 0$. Remarkably, at around the same time that AdS/CFT was discovered by theorists, astronomical observations [4, 5] have falsified its core premise: it appears that our Universe has a cosmological constant of the "wrong sign", i.e. $\Lambda > 0$. Thus, despite some wide-spread impressions, quantum gravity is both theoretically tractable and observationally falsifiable: an entire class of theoretical models has been found, and has been falsified by observation! Naively, this is very strange: one would expect that any observation relevant to quantum gravity should involve the tiny Planck scale. Instead, due to the way in which holography interchanges UV and IR, we find a crucial dependence of the theory on Λ, i.e. on the behavior of spacetime at the *largest* distances.

The sign of Λ may at first seem like a technical detail, but in fact it has a crucial implication for our place in the Universe. In asymptotically anti-de Sitter (AdS) spacetime, with $\Lambda < 0$, *spatial infinity is*

★e-mail: yashula@gmail.com

a physical place. Lightlike signals can go there and back again in finite time, while massive probes and observers can get arbitrarily close. Thus, the observables calculated by AdS/CFT make operational physical sense, and it's easy to accept that only this kind of observables that can be precisely defined. After all, the bulk spacetime's geometry is experiencing quantum fluctuations, or perhaps is fundamentally ill-defined altogether. It is only at infinity, where the Planck length becomes negligible, that one can have a fixed classical geometry, and therefore pin down precise notions for what is being prepared and measured. All of this is quite satisfactory and consistent, except it is not applicable to the real world. In asymptotically de Sitter (dS) spacetime, with $\Lambda > 0$, spatial infinity is completely irrelevant to observable physics, even though it may still exist from a "God's eye" perspective. Instead, every observer is trapped inside a cosmological horizon of finite area. In effect, when passing from AdS to dS, the roles of space and time get interchanged. In dS, instead of taking finite time to cross an infinite distance, causal signals take infinite time to cross a finite distance – the distance to the horizon. While AdS has a conformal boundary at spatial infinity, dS has boundaries at past and future infinity. While the AdS boundary is a physically meaningful *place*, the dS boundary is instead a pair of *times*. Moreover, for every observer, most of the dS boundary is inaccessible: at infinity, the horizon radius appears pointlike, so the observer can only access the initial and final endpoints of her wordline.

Thus, we now have on our hands an updated version of the problem of quantum gravity: to reconcile the *three* simple facts $\hbar \neq 0$, $G \neq 0$ and $\Lambda > 0$. The implication is *to understand quantum gravity in finite regions of space*, or at least in the largest region that is physically relevant – the causal patch of an observer in asymptotically dS spacetime. Given the success of AdS/CFT, it is tempting to try a similar approach, i.e. to construct dS/CFT [6] – holography in de Sitter space. As already covered by the above discussion, the move from AdS to dS introduces a host of conceptual issues: how to treat the two boundaries at past & future infinity? How can time emerge holographically from a spacelike boundary? Since the boundary is inaccessible to any single observer, how do we translate the CFT's output into physics inside an observer's horizon? Is the Hilbert space of an observer's causal patch finite-dimensional, as implied by Bekenstein's entropy formula? Is this Hilbert space the same for all observers? What are the implications for the measurement procedure in quantum mechanics, which relies on a measurement apparatus with an infinite-dimensional state space?

For a number of years, these issues were widely discussed – see e.g. [7–14]. However, active interest has gradually died down, due to the absence of a reliable working model on which to test the various ideas. Indeed, while AdS/CFT models abound, the vast majority of them turn pathological if we try to flip the sign of Λ. There are two simple reasons for this. First, string theory in general and AdS/CFT in particular are well-understood only in the presence of unbroken supersymmetry, which is incompatible with the dS spacetime symmetry structure. Second, in addition to the graviton and other massless or light fields, string theory generically contains a tower of massive excited modes. When infinity is a time instead of a place, these massive fields oscillate – in analogy with the e^{-imt} behavior in Minkowski space – which implies complex conformal weights on the boundary. Both of these problems are difficult, though perhaps potentially solvable. For the time being, if one wants a working model of dS/CFT, one had better work with a theory that does not require supersymmetry, and does not contain any massive fields. In, 2011, a model of just this type has been put forward [15].

The model of [15] is a duality between type-A higher-spin gravity in dS_4 and a free vector model on the 3d boundary at future infinity. It results from flipping the sign of Λ in the corresponding AdS/CFT model of [16]. Higher-spin gravity [17, 18] is an interacting theory of an infinite tower of massless fields with increasing spin; in the minimal version, there is one field for every even integer spin ($s = 0, 2, 4, \dots$). The infinite tower of massless fields corresponds to an infinite-dimensional gauge symmetry – higher-spin symmetry – which extends the ordinary spacetime symmetry of General Relativity (GR). In the holographic model of [15, 16], a global version of the infinite-dimensional higher-spin symmetry is unbroken, which can be realized in the boundary CFT if and only if the latter is free [19]. One can think of higher-spin gravity as the "opposite of supergravity": it extends the spacetime gauge symmetry in a bosonic direction rather than a fermionic one. Unlike supergravity, though, higher-spin theory is very different from GR coupled to matter, since the spin-2 graviton is now joined by fields of spin $s > 2$. Moreover, unlike in string theory, where most of the fields are heavy and decouple at low energy, here all of the higher-spin fields are massless, and they all interact at once and at all orders in derivatives. As a result, the interactions of higher-spin theory are *nonlocal at the cosmological scale*, and are nothing like those of GR. Thus, this model is highly unrealistic. However, it permits us to do holography with $\Lambda > 0$, and, as an added bonus, does not require the extra dimensions of string theory: the bulk is really (A)dS$_4$, without the additional compact factor that is always present in stringy AdS/CFT.

By the time the model of [15] was proposed, work on the conceptual underpinnings of dS/CFT has largely died down. The main conceptual framework that was left standing, and was used in [15], is that of [11]. There, the Lorentzian dS space is treated as a natural analytic continuation of Euclidean AdS, sharing the same symmetries and the same boundary. Using the well-understood Euclidean AdS/CFT dictionary, this leads one to recognize the CFT partition function as the *Hartle-Hawking wavefunction [20]* of the Lorentzian dS at future infinity. While this interpretation is very solid, it fails to address the issue of cosmological horizons: the wavefunction at future infinity is not accessible to any observer, and a dictionary is still needed between that and the physics inside an observable patch.

Our goal, then, is to extract the physics within causal patches for the higher-spin dS/CFT model of [15]. In other words, we wish to revisit the old conceptual questions of dS/CFT, this time within the context of a working model. Specifically, we will pick up the suggestion of [10], where it was proposed that dS/CFT should become more tractable on a folded-in-half version of de Sitter space, where every observer can access all of space (though not all of spacetime). We contend that, while this idea has some well-known problems when applied to GR, it is a match made in heaven with the higher-spin model. If our marriage between the two proves successful, we will have the first-ever working model of quantum (sort-of-)gravity inside a cosmological horizon. Given the great differences between real-world GR and higher-spin theory, such a model may not address all the outstanding questions about $\Lambda > 0$ quantum gravity. However, some questions, such as the dimension of the causal patch's Hilbert space, should be general enough that they can be usefully addressed even within this toy system.

2 Folding de Sitter space in half

Global de Sitter space is the hyperboloid of constant spacelike radius in the ambient flat spacetime $\mathbb{R}^{1,4}$:

$$dS_4 = \left\{ x^\mu \in \mathbb{R}^{1,4} \,|\, x_\mu x^\mu = 1 \right\} . \quad (1)$$

The asymptotes to this hyperboloid are the past-pointing and future-pointing null directions in $\mathbb{R}^{1,4}$. These make up the boundaries of dS_4 at past and future infinity:

$$\mathcal{I}^\pm = \left\{ \ell^\mu \in \mathbb{R}^{1,4} \,|\, \ell_\mu \ell^\mu = 0 \,;\; \ell^0 \gtrless 0 \,;\; \ell^\mu \cong \lambda \ell^\mu \;\; \forall \lambda > 0 \right\} . \quad (2)$$

An observer in dS_4 is specified by her worldline's asymptotic endpoints $p_i \in \mathcal{I}^-$ and $p_f \in \mathcal{I}^+$. We identify the observable patch as the region inside both past and future horizons; for a detailed argument on this point, see [24].

In several ways, the spacetime (1) is "twice too big": it has two boundaries (2) rather than one; only half the spacetime (the Poincare patch) is relevant to cosmology; and, most crucially for us,

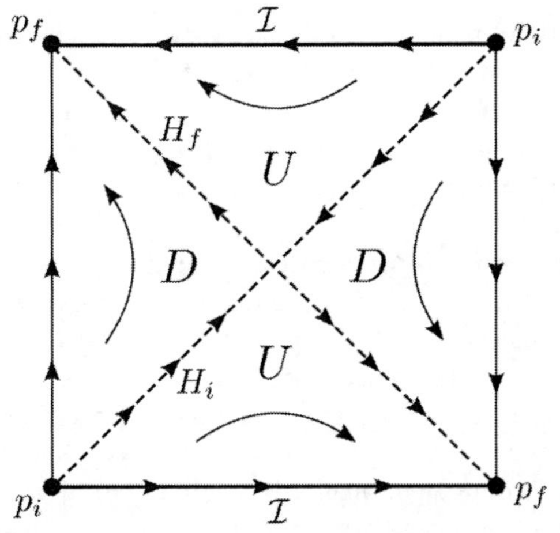

Figure 1. A Penrose diagram of $dS_4/\mathbb{Z}_2$. Opposite points in the diagram are identified. Past and future infinity are identified into a single boundary $\mathcal{I}$. The observer's worldline endpoints $p_i, p_f \in \mathcal{I}$ define horizons H_i, H_f, which divide the spacetime into two independent "quadrants": the causal patch D (identified with its antipode) and its complement U, which contains $\mathcal{I}$. The arrows denote the direction of the Killing vector ξ^μ that generates time translations in the causal patch. The observer induces a time orientation in regions where ξ^μ is causal, i.e. in the causal patch and on the horizons.

every observer sees only half of space. Thus, before we can construct a dictionary from $\mathcal{I}^\pm$ into the observable patch, we must somehow disentangle the relevant information from that pertaining to the antipodal patch. A way out was suggested by [10], in a modern echo of an old proposal by Schrodinger. The idea is to fold de Sitter space in half, in the unique way that preserves its full symmetry group $O(1,4)$. In the $\mathbb{R}^{1,4}$ language, this is accomplished by identifying every point x^μ with its antipodal point $-x^\mu$. The spacetime and its boundary thus become:

$$dS_4/\mathbb{Z}_2 = \left\{x^\mu \in \mathbb{R}^{1,4} \,|\, x_\mu x^\mu = 1 \,;\; x^\mu \cong -x^\mu\right\} \;; \quad \mathcal{I} = \left\{\ell^\mu \in \mathbb{R}^{1,4} \,|\, \ell_\mu \ell^\mu = 0 \,;\; \ell^\mu \cong \lambda \ell^\mu \;\; \forall \lambda\right\} . \quad (3)$$

The folded-in-half spacetime $dS_4/\mathbb{Z}_2$ is known as elliptic de Sitter space; the word "elliptic" refers to the spacelike separation between the identified pairs of points $\pm x^\mu$. The spacetime (3) has only one boundary: past and future infinity are identified with each other. Crucially, every observer's causal patch now covers all of space, though not all of spacetime. See figure 1 for the causal structure of $dS/\mathbb{Z}_2$ with a chosen observer.

As we discuss in the next section, $dS/\mathbb{Z}_2$ has a peculiar but consistent causal structure. The main reason why the $dS/\mathbb{Z}_2$ approach of [10] was abandoned is that this subtle consistency is not preserved by gravitational perturbations. However, we have already made the compromise of committing to the higher-spin model of [15], where the bulk theory is *very different from GR*. As we discuss below, this may allow us to evade the conflict between the causal structure and bulk perturbations.

3 Observer-dependent quantum mechanics

The folding in half of de Sitter space carries one obvious complication: since we're identifying opposite points in both space and time, global time orientability is lost. In $dS/\mathbb{Z}_2$, there is no consistent distinction between past and future lightcones. Nevertheless, there is never an *observable* violation of causality: there are no closed timelike curves, and every observer's causal patch is time-orientable. Since $\mathcal{I}^-$ and $\mathcal{I}^+$ have been identified, an observer in $dS/\mathbb{Z}_2$ is defined by a pair of points p_i, p_f on the single boundary $\mathcal{I}$. The time orientation in the causal patch is then induced by the ordering of p_i and p_f, i.e. by the orientation of the observer's worldline. Different observers induce time orientations on different causal patches. Where their causal patches intersect, the time orientations of two different observers may or may not agree.

This peculiar status of the arrow of time carries crucial implications for quantum mechanics in $dS/\mathbb{Z}_2$. Indeed, the operator algebra $[q, p] = i$ for e.g. a particle depends on our convention for the sign of the velocity, and therefore on the signs of time derivatives. As a result, while quantum mechanics is consistent in observable patches of $dS/\mathbb{Z}_2$, there is no global, observer-independent quantum mechanics of the entire Universe. This situation is reminiscent of the idea of horizon complementarity [9, 22]. Horizon complementarity sets out to solve the black hole information paradox by positing that it depends on a consistent quantum description of the entire Universe, which need not exist. If every observer can only access a partial region of spacetime due to horizons, then a consistent quantum description of every such region should suffice. Complementarity then goes on to posit that these descriptions in the different patches are in fact equivalent – that the same Hilbert space is being encoded in different ways.

As we can see, apart from its broader significance for holography, field theory on elliptic de Sitter space forms a unique theoretical laboratory for the idea of complementarity. Indeed, complementarity is essentially "forced" by the two sides of every horizon being identified. In addition, at the classical level, every observer indeed sees a different encoding of the same information: while the observers see different causal patches, each of them covers a complete time slice. Finally, unlike in most discussions of complementarity, the system is well-defined and under complete control.

In [23, 24], we carried out a detailed study of quantum field theory (QFT) in $dS/\mathbb{Z}_2$ and its observer dependence. The key idea is to introduce ordinary dS as a double cover of $dS/\mathbb{Z}_2$, and to use QFT on dS as the observer-independent global structure. The subtlety lies in the different mappings $dS/\mathbb{Z}_2 \to dS$ that are induced by different observers. As a result, $dS/\mathbb{Z}_2$ inherits the local field operators from dS, but the operator algebra becomes observer-dependent. In addition, mixed states in $dS/\mathbb{Z}_2$ naturally arise from pure antipodally symmetric states in dS.

Within this framework, we were able to show that observers agree on all the expectation values in the intersections of their causal patches, provided that they consistently agree (or consistently disagree) on the time orientation. However, when two observers agree on the arrow of time in one region and disagree in another, they will generally disagree on the entanglement between these two regions. In more physical terms, they will disagree on the entanglement across one of their horizons. In particular, one observer's pure state can appear as mixed to another. As far as we can tell, this picture, while peculiar, is perfectly consistent. It therefore demonstrates that horizon complementarity may have been conceived too narrowly. In the $dS/\mathbb{Z}_2$ universe, we see a *generalized* version of complementarity: the quantum world-pictures of different observers are *not* fully equivalent, but they are mutually consistent where it is required.

4 Massless holography in $dS_4/\mathbb{Z}_2$

Having understood the structure of QFT on elliptic de Sitter space, one can proceed towards the main task: reconstructing the Lorentzian quantum physics inside a causal patch from the Euclidean partition function of the boundary CFT. As we've seen in the previous section, this reconstructed quantum physics (e.g. the operator algebra) will necessarily depend on the choice of observer. In [24], we accomplished this task in the limit of non-interacting (but quantum) bulk fields. Already at this level, it is important that we're working specifically with the higher-spin model, because our construction depends crucially on the fact that the bulk fields are *massless*.

The specific problem solved in [24] can be phrased as follows, using a spin-0 field for simplicity. A choice of observer in $dS_4/\mathbb{Z}_2$ breaks the spacetime symmetry group $O(1,4)$ down to the subgroup $\mathbb{R}\times O(3)$ of time translations and rotations. This suggests a basis $a_{lm}(\omega), a^*_{lm}(\omega)$ for the field's modes in the causal patch, where the frequency ω and the angular momentum numbers (l,m) are conserved. Since the field is free, each of these modes is a harmonic oscillator, with ω already specified as its frequency. On the other hand, a similar decomposition into modes $c_{lm}(\omega), c^*_{lm}(\omega)$ can be performed for the boundary data at past/future infinity $\mathcal{I}$ (note that the generator of time translations is spacelike outside the causal patch and tangential to $\mathcal{I}$). Since the causal patch covers all of space, there must be a one-to-one mapping between the oscillators a, a^* and the boundary modes c, c^*. Using the $\mathbb{R}\times O(3)$ symmetry and the linearity of the field equations, we see that this mapping must take the form:

$$a_{lm}(\omega) = N_l(\omega)c_{lm}(\omega)\ , \tag{4}$$

where $N_l(\omega)$ are some c-number coefficients. Taking the oscillators a, a^* to be normalized, we can read off the *operator algebra and Hamiltonian* that are induced on the Euclidean boundary data at $\mathcal{I}$ by our choice of observer:

$$\left[\hat{c}_{lm}(\omega), \hat{c}^\dagger_{l'm'}(\omega')\right] = \frac{1}{|N_l(\omega)|^2}\cdot 2\pi\delta(\omega-\omega')\delta_{ll'}\delta_{mm'}\ ; \tag{5}$$

$$\hat{H} = \int_0^\infty \frac{d\omega}{2\pi}\sum_{lm}\omega\,|N_l(\omega)|^2\,\hat{c}^\dagger_{lm}(\omega)\,\hat{c}_{lm}(\omega)\ . \tag{6}$$

Thus, our task is to derive the coefficients $|N_l(\omega)|^2$. Also, since we are constructing a holographic dictionary, this must be done *strictly from the CFT partition function, with no bulk calculations*. This turns out to be possible, thanks to a peculiar property of *free massless bulk fields*: their two types of boundary data (Dirichlet vs. Neumann, electric vs. magnetic) correspond directly to the two possible signs of antipodal symmetry, i.e. to the two possible kinds of fields in the $dS_4/\mathbb{Z}_2$ bulk [25–27]. We refer to [24] for the full construction, simply citing here the result:

$$|N_l(\omega)|^2 = S_l(\omega)\cdot\frac{1}{2}\left(\coth\frac{\pi\omega}{2}\right)^{\eta(-1)^l}\ , \tag{7}$$

where $S_l(\omega)$ are the coefficients in the Gaussian approximation to the CFT partition function:

$$Z_{\mathrm{CFT}}[c_{lm}(\omega), c^*_{lm}(\omega)] \sim \exp\left(-\int_0^\infty\frac{d\omega}{2\pi}\sum_{lm}S_l(\omega)\,|c_{lm}(\omega)|^2\right)\ . \tag{8}$$

The significance of the result (7) is that it shows how, as a matter of principle, one can derive a quantum operator algebra and a Lorentzian time evolution out of the CFT on the Euclidean boundary. However, since we've used so extensively the special features of free massless fields on $dS_4/\mathbb{Z}_2$, one may worry that a generalization to the interacting level will prove too difficult. If there is hope, it must lie in the fact that the interactions of higher-spin gravity are themselves very special. In the remaining sections, we will discuss some recent work in higher-spin theory that gives us cause for optimism.

5 Higher-spin gravity on a fixed geometry

The free-field result of the previous section relied on the $\mathbb{R}\times O(3)$ symmetry of a causal patch in $dS_4/\mathbb{Z}_2$. Even putting aside worries about causality, what are we to do once the geometry is gravitationally altered and the symmetry destroyed? At first, this problem seems even worse in higher-spin gravity than it does in GR. Indeed, in its usual formulation, higher-spin theory is "even more" generally covariant: the metric is not just dynamical, it is now relegated to a gauge-dependent component of the infinitely larger higher-spin connection.

In [28], we uncovered a feature of higher-spin gravity that offers a way around this problem. Through a slight (and locally equivalent) reformulation, higher-spin theory can change from a "more-than-gravitational" theory on a featureless manifold into a "non-gravitational" theory on a fixed (A)dS$_4$ geometry! Crucially, one can do this *without deforming the higher-spin gauge symmetry*, i.e. without substantially complicating the field equations. The reason why this is possible is the different relationship in GR vs. higher-spin theory between diffeomorphisms and the dynamical spin-2 field. In GR, the dynamical spin-2 field is quite crucially the "gauge field" for local translations, i.e. for diffeomorphisms, i.e. it must be identified with the spacetime metric. In higher-spin theory, one instead has, as a component of the higher-spin gauge group, a local "translation" symmetry that is completely divorced from diffeomorphisms, i.e. that only acts on fields at a given point. The reason for this is that the field equations of higher-spin gravity are written in an "unfolded" formulation, in which both the ordinary fields and their full tower of spacetime derivatives are packaged together into a single "master field". Thus, a translation can be performed by accessing the fields' derivatives that are already encoded at the given point, without the need to go to an adjacent point, i.e. without a diffeomorphism.

With the gauge symmetry thus divorced from diffeomorphisms, there is no longer any need to identify the dynamical spin-2 field as the spacetime metric. Instead, it can live, alongside the fields of all other spins, on a fixed external geometry. We must note, though, that this external geometry must then be pure (A)dS$_4$ (or a topological modification thereof, such as $dS_4/\mathbb{Z}_2$). While this result arises naturally in [28] through a mechanism peculiar to higher-spin gravity, it can be understood already at the level of free massless fields: for spin 2 and higher, the very consistency of free massless field equations on a fixed background severely constrains the allowed background geometries [29].

To sum up, on one hand, higher-spin gravity is similar to GR, and therefore interesting: it contains a spin-2 massless field, and has a diff-invariant formulation. On the other hand, it is different from GR, and therefore tractable, by virtue of having a formulation on a fixed background geometry. For our present purpose, the implication is that we can continue working on a pure $dS_4/\mathbb{Z}_2$ geometry, with its causal structure and symmetries, even at the interacting level.

6 Bulk and CFT from twistor space

Even if we are allowed to keep the $dS_4/\mathbb{Z}_2$ geometry at the interacting level, the interactions of higher-spin gravity must still be contended with. In the present section, we outline an angle of attack on this problem. The key intuition is that in higher-spin holography with unbroken global higher-spin symmetry, there is a sense in which *the free theory is enough*. To see one way in which this is true, consider the partition function, i.e. the effective bulk "on-shell action", as a functional of boundary data. This can be expressed equivalently as a non-linear functional of the *linearized bulk solution* corresponding to this boundary data. Now, for ordinary local theories, the linearized bulk solution is a very *awkward* parameterization of the boundary data, so this observation is only of philosophical interest. In higher-spin gravity, on the other hand, thanks to the unfolded formulation, the entire bulk solution is conveniently encoded within the master field *at any single point*. Thus, expressing the partition function via the linearized solution becomes practical. Moreover, the answer is almost completely fixed by higher-spin symmetry, up to an overall coefficient at every order in the perturbative expansion [30, 31].

In a recent work [32], we were able to go further, and obtain the complete expression for the partition function in terms of the linearized bulk solution. To accomplish this, we made use of the fact that even with the bulk interactions taken into account, the boundary CFT remains free. In other words, all the effects of bulk interactions are contained within the correlators of a free boundary theory. However, this can be put to practical use only if we express the bulk and boundary theories in some common language. This is the main achievement of [32].

The common language that we used to unify bulk and boundary in the higher-spin/free-CFT model is Penrose's twistor theory [33, 34]. Twistors are spinors of the spacetime symmetry group – in our case, $O(1,4)$. They correspond geometrically to totally-null planes (in the complexified 4d bulk) or to null geodesics (on the complexified 3d boundary). These objects are ideally suited for describing massless fields, and indeed play a central role in the formulation of higher-spin gravity. In particular, the Penrose transform relates functions in twistor space to solutions of the free massless field equations in the 4d bulk. In [32], we rewrite the Penrose transform within the geometric framework of [28], i.e. in the context of higher-spin gravity on a fixed (A)dS$_4$ geometry. We then present a "holographic dual" of the Penrose transform, which relates the same twistor functions to the boundary CFT's sources and operators. Finally, we express the CFT partition function in the twistor language, which is directly related via the Penrose transform to the linearized bulk solution.

Without going into any technical detail, we present here some of the key formulas. The CFT action can be written as follows, with bilocal sources $\Pi(\ell',\ell)$ coupled to the bilocal single-trace operators $\phi^I(\ell)\bar{\phi}_I(\ell')$, where ℓ,ℓ' are boundary points:

$$S_{\mathrm{CFT}}[\Pi(\ell',\ell)] = -\int d^3\ell\,\bar{\phi}_I\Box\phi^I - \int d^3\ell' d^3\ell\,\bar{\phi}_I(\ell')\Pi(\ell',\ell)\phi^I(\ell)\,. \tag{9}$$

The source $\Pi(\ell',\ell)$ can be packaged into a twistor function $f(Y)$ via:

$$F(Y) = -\frac{1}{4\pi}\int d^3\ell\,d^3\ell'\,\Pi(\ell',\ell)\,\sqrt{-2\ell\cdot\ell'}\,\delta_\ell(Y)\star\delta_{\ell'}(Y)\,, \tag{10}$$

where $\delta_\ell(Y)$ is a certain delta function depending on the boundary point ℓ, and $\star$ is the star product of higher-spin algebra. The partition function can now be written in terms of $F(Y)$ as:

$$Z_{\mathrm{CFT}}[F(Y)] \sim \exp\left(-\frac{N}{4}\operatorname{tr}_\star \ln_\star[1+F(Y)]\right) = (\det{}_\star[1+F(Y)])^{-N/4}\,, \tag{11}$$

where $\operatorname{tr}_\star$ is the higher-spin-invariant trace operation, and $\det_\star$ is the corresponding determinant. On the bulk side, the twistor function $F(Y)$ is related to a linearized bulk solution via the Penrose transform:

$$F(Y) = C(x;Y)\star i\delta_x(Y)\,, \tag{12}$$

where the master field $C(x;Y)$ serves as a generating function for fields and their derivatives at the bulk point x, while $\delta_x(Y)$ is a bulk version of $\delta_\ell(Y)$.

7 Outlook and summary

In this note, we motivated higher-spin holography in "elliptic" de Sitter space $dS_4/\mathbb{Z}_2$ as a promising approach for developing the first-ever working model of quantum (higher-spin) gravity inside a cosmological horizon. We described the successful solution [24] of this problem at the level of free bulk fields, and presented some results to indicate that the full interacting theory may prove tractable as well. The next step in this program should be to rewrite the free-field result of [24] in the twistor language of [32], described in the previous section. Then, if we are lucky, the full interacting result will readily present itself, similarly to the partition function (11) in [32].

In the work described here, we are making heavy use of the very special and wildly unrealistic properties of higher-spin gravity. As a result, even if we succeed at extracting quantum physics inside the horizon, many of the original questions of $\Lambda > 0$ quantum gravity may not even make sense within our treatment. However, this shouldn't be the case for *every* interesting question. In particular, we hope to find out whether the Hilbert space inside the horizon is finite-dimensional, and whether its dimension is captured by the Bekenstein-Hawking black hole entropy formula.

Finally, we must note that higher-spin gravity is of interest for reasons other than its applicability to $\Lambda > 0$. The developments described in the previous two sections are of general relevance to the study of this theory. A successful formulation of the quantum physics inside a causal patch of $dS_4/\mathbb{Z}_2$ should be of similarly broad relevance, as it will constitute a more direct quantization of higher-spin gravity than what is currently available.

Acknowledgements

I thank the organizers of the ICGAC-13/IK-15 conference at Ewha Womans University for their hospitality. Parts of the work reported here were performed in collaboration with Lucas Hackl and Illan Halpern. YN is supported by the Quantum Gravity Unit of the Okinawa Institute of Science and Technology Graduate University (OIST). During earlier stages of the work, YN was employed at Perimeter Institute and Penn State University, where he was supported by the Government of Canada through Industry Canada, by the Province of Ontario through the Ministry of Research & Innovation, by NSERC Discovery grants, as well as by the NSF grant PHY-1205388 and the Eberly Research Funds of Penn State.

References

[1] J. M. Maldacena, Int. J. Theor. Phys. **38**, 1113 (1999) [Adv. Theor. Math. Phys. **2**, 231 (1998)] doi:10.1023/A:1026654312961 [hep-th/9711200].

[2] E. Witten, Adv. Theor. Math. Phys. **2**, 253 (1998) [hep-th/9802150].

[3] O. Aharony, S. S. Gubser, J. M. Maldacena, H. Ooguri and Y. Oz, Phys. Rept. **323**, 183 (2000) doi:10.1016/S0370-1573(99)00083-6 [hep-th/9905111].

[4] A. G. Riess *et al.* [Supernova Search Team], Astron. J. **116**, 1009 (1998) doi:10.1086/300499 [astro-ph/9805201].

[5] S. Perlmutter *et al.* [Supernova Cosmology Project Collaboration], Astrophys. J. **517**, 565 (1999) doi:10.1086/307221 [astro-ph/9812133].

[6] A. Strominger, JHEP **0110**, 034 (2001) doi:10.1088/1126-6708/2001/10/034 [hep-th/0106113].

[7] E. Witten, hep-th/0106109.

[8] V. Balasubramanian, J. de Boer and D. Minic, Class. Quant. Grav. **19**, 5655 (2002) [Annals Phys. **303**, 59 (2003)] doi:10.1016/S0003-4916(02)00020-9 [hep-th/0207245].

[9] L. Dyson, M. Kleban and L. Susskind, JHEP **0210**, 011 (2002) [hep-th/0208013].

[10] M. K. Parikh, I. Savonije and E. P. Verlinde, Phys. Rev. D **67**, 064005 (2003) [hep-th/0209120].

[11] J. M. Maldacena, JHEP **0305**, 013 (2003) [astro-ph/0210603].

[12] T. Banks, W. Fischler and S. Paban, JHEP **0212**, 062 (2002) doi:10.1088/1126-6708/2002/12/062 [hep-th/0210160].

[13] D. Klemm and L. Vanzo, JCAP **0411**, 006 (2004) doi:10.1088/1475-7516/2004/11/006 [hep-th/0407255].

[14] N. Arkani-Hamed, S. Dubovsky, A. Nicolis, E. Trincherini and G. Villadoro, JHEP **0705**, 055 (2007) doi:10.1088/1126-6708/2007/05/055 [arXiv:0704.1814 [hep-th]].

[15] D. Anninos, T. Hartman and A. Strominger, arXiv:1108.5735 [hep-th].
[16] I. R. Klebanov and A. M. Polyakov, Phys. Lett. B **550**, 213 (2002) [hep-th/0210114].
[17] M. A. Vasiliev, Int. J. Mod. Phys. D **5**, 763 (1996) [hep-th/9611024].
[18] M. A. Vasiliev, In *Shifman, M.A. (ed.): The many faces of the superworld* 533-610 [hep-th/9910096].
[19] J. Maldacena and A. Zhiboedov, J. Phys. A **46**, 214011 (2013) [arXiv:1112.1016 [hep-th]].
[20] J. B. Hartle and S. W. Hawking, Phys. Rev. D **28**, 2960 (1983).
[21] I. F. Halpern and Y. Neiman, JHEP **1512**, 057 (2015) doi:10.1007/JHEP12(2015)057 [arXiv:1509.05890 [hep-th]].
[22] L. Susskind, L. Thorlacius and J. Uglum, Phys. Rev. D **48**, 3743 (1993) [hep-th/9306069].
[23] L. Hackl and Y. Neiman, Phys. Rev. D **91**, no. 4, 044016 (2015) [arXiv:1409.6753 [hep-th]].
[24] I. F. Halpern and Y. Neiman, JHEP **1512**, 057 (2015) doi:10.1007/JHEP12(2015)057 [arXiv:1509.05890 [hep-th]].
[25] M. A. Vasiliev, J. Phys. A **46**, 214013 (2013) [arXiv:1203.5554 [hep-th]].
[26] G. S. Ng and A. Strominger, Class. Quant. Grav. **30**, 104002 (2013) [arXiv:1204.1057 [hep-th]].
[27] Y. Neiman, JHEP **1410**, 153 (2014) [arXiv:1406.3291 [hep-th]].
[28] Y. Neiman, JHEP **1504**, 144 (2015) doi:10.1007/JHEP04(2015)144 [arXiv:1502.06685 [hep-th]].
[29] R. Penrose and W. Rindler, "Spinors And Space-time. Vol. 1: Two-Spinor Calculus and Relativistic Fields," section 5.8, Cambridge, Uk: Univ. Pr.
[30] N. Colombo and P. Sundell, arXiv:1208.3880 [hep-th].
[31] V. E. Didenko and E. D. Skvortsov, JHEP **1304**, 158 (2013) [arXiv:1210.7963 [hep-th]].
[32] Y. Neiman, "The Holographic Dual of the Penrose Transform," to appear.
[33] R. Penrose and W. Rindler, "Spinors And Space-time. Vol. 2: Spinor And Twistor Methods In Space-time Geometry," Cambridge, Uk: Univ. Pr. (1986) 501p
[34] R. S. Ward and R. O. Wells, "Twistor geometry and field theory," Cambridge, UK: Univ. Pr. (1990) 520p

Standard sirens and dark sector with Gaussian process *

Rong-Gen Cai[1,2,**], Tao Yang[3,***]

[1] *CAS Key Laboratory of Theoretical Physics, Institute of Theoretical Physics, Chinese Academy of Sciences, P.O. Box 2735, Beijing 100190, China*
[2] *School of Physical Sciences, University of Chinese Academy of Sciences, No. 19A Yuquan Road, Beijing 100049, China*
[3] *Department of Astronomy, Beijing Normal University, Beijing, 100875, China*

Abstract. The gravitational waves from compact binary systems are viewed as a standard siren to probe the evolution of the universe. This paper summarizes the potential and ability to use the gravitational waves to constrain the cosmological parameters and the dark sector interaction in the Gaussian process methodology. After briefly introducing the method to reconstruct the dark sector interaction by the Gaussian process, the concept of standard sirens and the analysis of reconstructing the dark sector interaction with LISA are outlined. Furthermore, we estimate the constraint ability of the gravitational waves on cosmological parameters with ET. The numerical methods we use are Gaussian process and the Markov-Chain Monte-Carlo. Finally, we also forecast the improvements of the abilities to constrain the cosmological parameters with ET and LISA combined with the *Planck*.

1 Introduction

More than 100 years, Einstein himself have predicated the existence of the gravitational waves (GWs) by his general relativity of gravitational theory. On 11 February 2016, the Laser Interferometer Gravitational Wave Observatory (LIGO) collaboration reported the first direct detection of the gravitational wave source GW150914 [1], which means new era of cosmology and astronomy with GW is coming. So far another two gravitational wave signals have also been detected [2, 3]. These three signals are all from the merger of two black holes. The information gathered from these GW sources will lead to advances not only in fundamental physics and astrophysics, but also in cosmology [4]. In this paper, we summarize some of our recent works in which we use GWs as the standard sirens and explore its ability in constraining the cosmological parameters [5–7]. Our works are inspired by Schutz's idea which was proposed in 1986 [8]. Schutz showed that it is possible to determine the Hubble constant from GW observations, by using the fact that GWs from the binary systems encode the absolute distance information. We can assume other techniques are available to obtain the redshift of a GW event, for example, we can measure the redshift through the identification of an accompanying electromagnetic (EM) signal which is called EM counterpart. Thus we can get the $d_L - z$ relation and use it

*The invited talk is given by the first author of this paper.
**e-mail: cairg@itp.ac.cn
***e-mail: yangtao2017@bnu.edu.cn

to explore the cosmic expansion. We use the Einstein Telescope [1] (ET, the third-generation ground-based detector) and the Laser Interferometric Space Antenna [2] (LISA, the space based detector) to simulate the GWs data. ET is mainly envisaged to detect the GWs produced by the binary systems consisting of the neutron stars (NSs) and black holes (BHs). The frequency range is about $1 - 10^4$ Hz. We use it to forecast the GW's constraints on cosmological parameters such as the dark matter density Ω_m, Hubble constant H_0, and the equation of state w of dar energy, while, the LISA mission aims at measuring gravitational waves (GWs) in the frequency band around the mHz. Among the richness of astrophysical sources expected to produce a detectable GW signal at those frequencies, the signal produced by the massive black hole binaries (MBHBs) from 10^4 to 10^7 solar masses is expected to the loudest [9]. For the MBHB standard siren datasets constructed with LISA contain events that could be detected up to $z \sim 10$ which is much larger than that of the type Ia supernova (SNIa), we expect that GW information can extend the reconstructed dark interaction to higher redshifts. In what follows, after briefly reviewing how the dark sector interaction can be reconstructed by GP [5], the concept of standard sirens and the analysis of reconstructing the dark sector with LISA from [6] are introduced. Then we show the results of using the ET to constrain the cosmological parameters in [7]. Finally, we give the conclusions and future prospects.

2 Dark sector interaction with Gaussian process

The cosmological constant together with the cold dark matter (CDM) (called the ΛCDM model) turned out to be the standard model which fits the current observational data sets consistently. However, it is faced with the fine-tuning problem and the coincidence problem. Especially, to alleviate the coincident problem,an interaction between the dark energy and dark matter is introduced. Usually, to study the interaction of the dark energy and dark matter, the interaction form has to be assumed (for a review, see [10]). Such an assumed interaction form between dark energy and dark matter may introduce some bias. Therefore a model independent investigation is called for. In [11], we investigated the possible interaction in a way independent of specific interacting form by dividing the whole range of redshift into a few bins and setting the interacting term to be a constant in each redshift, it was found that the interaction is likely to cross the non-interacting line and has an oscillation behavior. Some years later, Salvatelli *et al* [12] showed that the null interaction is excluded at 99% confidence level (C.L.) when they added the redshift-space distortions (RSD) data to the *Planck* data for the decaying vacuum energy model.

2.1 The model

In [5], we present a nonparametric approach to reconstruct the interaction term between dark energy and dark matter directly from the observational data using the Gaussian process. In a flat universe with an interaction between dark energy and dark matter, we can change the conservation equation to be

$$\dot{\rho}_m + 3H\rho_m = -Q\,, \tag{1}$$

$$\dot{\rho}_{DE} + 3H(1+w)\rho_{DE} = Q\,. \tag{2}$$

[1] http://www.et-gw.eu/
[2] https://www.elisascience.org/

Here we don't parameterize the interaction form Q to be any form. Combing the Friedmann equation and after some calculations, one can finally get

$$-wq = 2\left(\frac{3D''^2}{D'^5} - \frac{D'''}{D'^4} + \frac{w'D''}{wD'^4}\right)(1+z)^2 + \left[2(5+3w)\frac{D''}{D'^4} + \frac{3w'}{wD'^3}\right](1+z) + \frac{9(1+w)}{D'^3}, \tag{3}$$

where, $Q = qH_0^3$ and $D(z) = (H_0/c)(1+z)^{-1}d_L(z)$. From this, we can see that once the equation of state w of dark energy is given, we can use the observed distance-redshift relationship to reconstruct the interaction.

2.2 The Gaussian process

The Gaussian process [3] is designed to use the supervised learning to build a model (or function) from the training data and then use the model to forecast the new samples. In our works, it can reconstruct a function from data without assuming a parametrization for it. The GP method has been used in many works [13–15], We use the GaPP code developed in [16] to derive our results. The distribution over functions provided by GP is suitable to describe the observed data. At each point z, the reconstructed function $f(z)$ is also a Gaussian distribution with a mean value and Gaussian error. The functions at different points z and $\tilde{z}$ are related by a covariance function $k(z,\tilde{z})$, which only depends on a set of hyperparameters ℓ and σ_f. Here ℓ gives a measure of the coherence length of the correlation in x-direction and σ_f denotes the overall amplitude of the correlation in the y-direction. Both of them will be optimized by GP with the observed data set. In contrast to actual parameters, GP does not specify the form of the reconstructed function. Instead it characterizes the typical changes of the function. According to [17], we choose the kernel function as the matérn ($\nu = 9/2$) form

$$k(z,\tilde{z}) = \sigma_f{}^2 \exp\left(-\frac{3\,|z-\tilde{z}|}{\ell}\right)\times\left[1 + \frac{3\,|z-\tilde{z}|}{\ell} + \frac{27(z-\tilde{z})^2}{7\ell^2} + \frac{18|z-\tilde{z}|^3}{7\ell^3} + \frac{27(z-\tilde{z})^4}{35\ell^4}\right]. \tag{4}$$

For detailed description of the Gaussian process, one can refer to [16], see also [5, 18, 19].

2.3 Reconstruct the dark sector interaction

To see the ability of the GP method to distinguish different models and recover the correct behaviors of the models, we create mock data sets of future SNIa according to Dark Energy Survey (DES) [20] for two fiducial models: the LCDM model and a decaying vacuum energy model with $\rho_{DE} = 3\alpha H$ with $w = -1$. Here we set $\Omega_{m0} = 0.3$ for both. The mock data results are shown in figure 1 (left and middle). We can see the GP method can recover and distinguish both of these two fiducial models. Now we can apply our GP method to the real data set Union2.1. We firstly test the vacuum dark energy with the dark matter. The result is shown in figure 1 (right). It tells us that the ΛCDM model (no interaction) fits well with the SNIa Union 2.1 data sets. More results for the wCDM model and CPL parametrization can be found in [5].

Here we have reviewed how to reconstruct the dark sector interaction by Gaussian process. In the following, we will introduce our works on using the GWs as the standard sirens to reconstruct the dark sector interaction [6] and constrain the cosmological parameters [7].

[3] http://www.gaussianprocess.org/gpml/

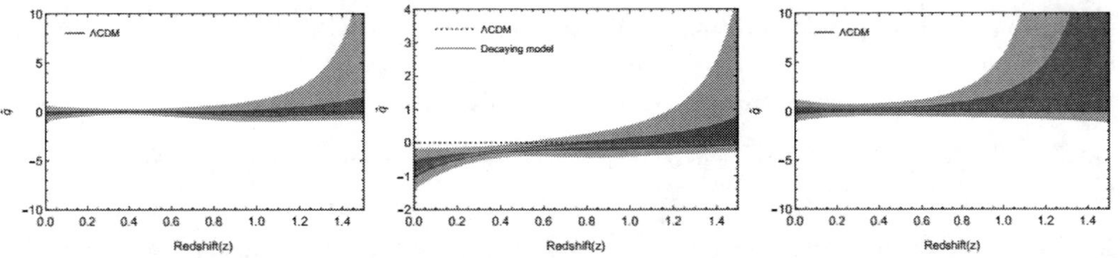

Figure 1. The results from [5] for the reconstruction of the interaction term. The left is from the mock data of DES for the ΛCDM model, the middle is from the mock data of DES for the decaying model, and the right is from the real data sets Union2.1 for the vacuum dark energy and dark matter. Here $\tilde{q}(z) = q/(1+z)^6$.

3 Reconstruct the dark sector with LISA

In this section, we summarize the results from the paper [6] in which we use the MBHBs as the standard sirens to reconstruct the dark sector. In the most recent configuration for LISA, conceived to answer the call from ESA for its L3 mission, the observatory will be based on three arms with six active laser links, between three identical spacecraft in a triangular formation separated by 2.5 million km. The proposed nominal mission duration in science mode is of 4 years, but the mission should be designed with consumables and orbital stability to facilitate a total mission duration of up to 10 years. The simulated datasets of MBHB standard sirens are taken from [21], where realistic catalogues of GW events detected by LISA and for which an EM counterpart can be identified by future telescopes were constructed. The MBHB standard siren datasets constructed with LISA contain events that could be detected up to $z \sim 10$. It is thus expected that GW information can extend the reconstructed dark interaction to a much higher redshift.

3.1 Massive black hole binary mergers as standard sirens for LISA

Compact binary inspirals are excellent examples of standard sirens, irrespective of the masses of the inspiralling objects. In fact the waveform h produced by these systems during the inspiral phase, is theoretically well described by the analytical solution

$$h_\times = \frac{4}{d_L(z)}\left(\frac{G\mathcal{M}_c(z)}{c^2}\right)^{5/3}\left(\frac{\pi f}{c}\right)^{2/3}\cos\iota\,\sin[\Phi(f)]\,, \tag{5}$$

which is valid at the lowest (Newtonian) order for the "cross" GW polarization (the same expression holds for the "plus" polarization with a difference dependence on the orientation of the binary's orbital plane ι). Here $\mathcal{M}_c(z)$ is the (redshifted) chirp mass, f the GW frequency at the observer and $\Phi(f)$ its phase. For our purposes the luminosity distance $d_L(z)$ is the most important parameter entering the waveform. Parameter estimation over the observed GW signal directly yields the value of the luminosity distance of the GW source, together with an uncertainty due to the detector noise. This implies that once the signal is detected and characterised, the luminosity distance of the source can be extracted, without the need of cross calibrating with other distance indicators or unwanted systematics, as in the case of SNIa. If also the redshift z of the GW source (or the one of its hosting galaxy) is measured, one obtains a data point in the distant-redshift diagram which can be used to constrain the distance-redshift relation.

The point now is to predict how many MBHB standard sirens will be detected by LISA and what kind of cosmological data they will provide. Based on [22], Three different competing scenarios for the initial conditions of the massive BH population at high redshift have been taken into account:

- **popIII**: a "light-seeds" scenario where the massive BHs form from the remnants of population III stars;
- **Q3d**: a "heavy-seeds" scenario where massive BHs form from the collapse of protogalactic disks, but a time delay between the merger of the hosting galaxies and the BHs is present;
- **Q3nod**: another "heavy-seeds" scenario with no time delay between the merger of hosting galaxies and BHs.

There are 118 catalogues of 5 years produced for each MBHB formation model to reduce the statistical uncertainty due to the low number of events in each catalogue. We select a representative catalogue among all 118 of them by the figure of merit. This will allow us to use the Gaussian process using the representative catalogue, which provides a fairly realistic realisation of MBHB standard sirens data detected by LISA. We will also repeat the analysis considering a LISA mission duration of 10 years, for which the number of standard sirens events on average doubles with respect to a 5 years mission.

3.2 The results

We use the same methods developed in [5] for reconstructing the dark sector interaction with LISA. Our results are summarised by three different "elements": three initial conditions for the massive BH, two mission years, and two different data combinations (LISA alone or LISA + DES). All results from [6] are shown in figures 2-5. We can see that using MBHB standard siren alone, the simulated LISA data can reconstruct the interaction well from about $z \sim 1$ to $z \sim 3$ (for a 5 years mission) and $z \sim 4$ or even $z \sim 5$ (for a 10 years mission). However the reconstruction is not efficient at redshift lower than $z \sim 1$ as MBHB events are much rare at late times. When combined with the simulated DES datasets, Gaussian processes can reconstruct the interaction well in the whole redshift region from $z \sim 0$ to $z \sim 3$ for 5 years LISA and from $z \sim 0$ to $z \sim 5$ for 10 years LISA, respectively. Thus the addition of DES data points, which cover the $0 < z < 1.5$ range, to the LISA MBHB standard siren catalogues, covering the range $1 < z < 10$, effectively reconstruct the dark interaction at all accessible redshift values, nicely integrating with each other. We can also note that in all our results, the MBHB scenarios of Q3d and Q3nod are better than popIII at reconstructing the dark sector interaction, in agreement with the results of [23].

In order to understand how MBHB cosmological data will improve the constraints provided by traditional EM signal such as SNIa (standard candles), we can compare the reconstruction uncertainty provided by [5], i.e. using DES only data, with the ones obtained by DES+LISA data. The results are shown in figure 6. We see that when the contribution of LISA MBHB standard siren is added, the errors remain small up to ~ 3 (for popIII) or $z \sim 5$ (for Q3d), where they reach levels comparable to the value at $z \sim 1.5$ given by DES only. Thus we can conclude that using only LISA data, the dark sector interaction can be well reconstructed from $z \sim 1$ to $z \sim 3$ (for a 5 years mission) and from $z \sim 1$ to $z \sim 5$ (for a 10 years mission). While, with DES datasets, the interaction is well reconstructed in the whole redshift range from $z = 0$ to $z \sim 3$ (5 years mission) and from $z = 0$ to $z \sim 5$ (10 years mission). Thus the massive BH binary can be used to constrain the dark sector interaction at redshift ranges not reachable by the usual supernoave datasets which probe only the $z < 1.5$ range. So, LISA will extend the ability of SNIa data to constrain the dark sector interaction in a model independent manner up to higher redshift.

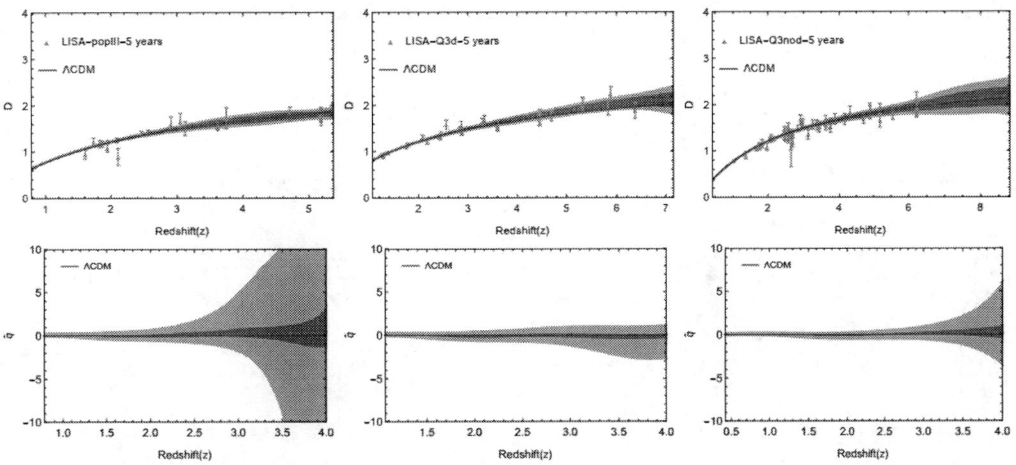

Figure 2. Reconstructions of the interaction using LISA alone for a 5 years mission. From left to right each column reports the results for popIII, Q3d, Q3nod.

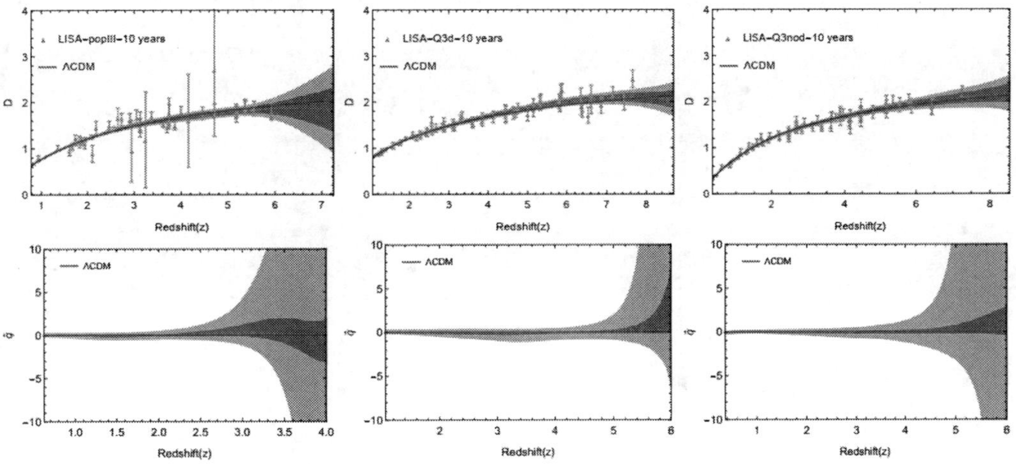

Figure 3. Reconstructions of the interaction using LISA alone for a 10 years mission. From left to right each column reports the results for popIII, Q3d, Q3nod.

4 Constraint ability of GW on cosmological parameters with ET

This section briefly summarizes the results from the paper [7] in which we use GWs as the standard sirens and check their abilities on constraining the cosmological parameters with ET. For ET, the GWs we mainly focus on are emitted by the binaries of NS-NS and NS-BH [24]. The binary merger of a NS with either a NS or BH is hypothesized to be the progenitor of a short and intense burst of γ-rays (SGRB) [25]. An EM counterpart like SGRB can provide the redshift information if the host galaxy of the event can be pinpointed. The expected rates of BNS and BHNS detections per year for the ET are about the order $10^3 - 10^7$ (ET project). However, only a small fraction ($\sim 10^{-3}$) is expected to have the observation of SGRB. In this work we take the detection rate in the middle range (10^5), thus 10^2

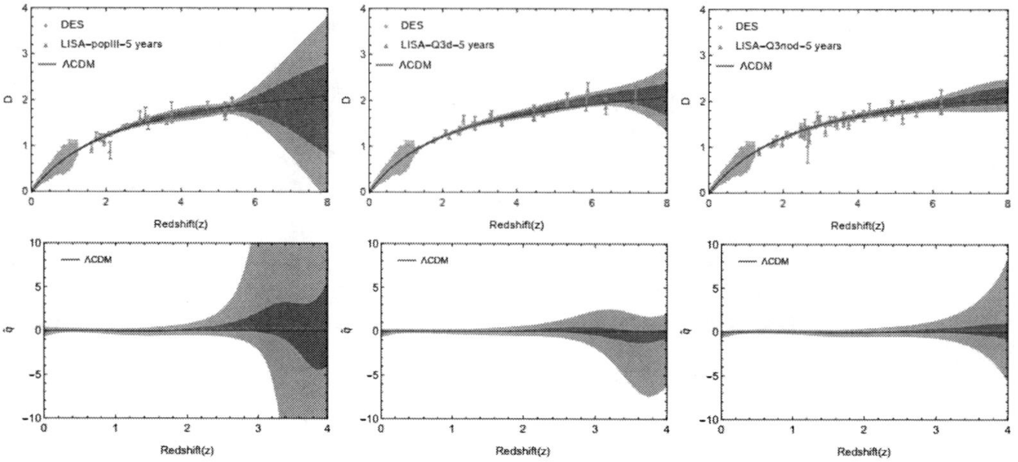

Figure 4. Reconstruction of the interaction using DES+LISA (5 years) data. From left to right each column reports the results for popIII, Q3d, Q3nod.

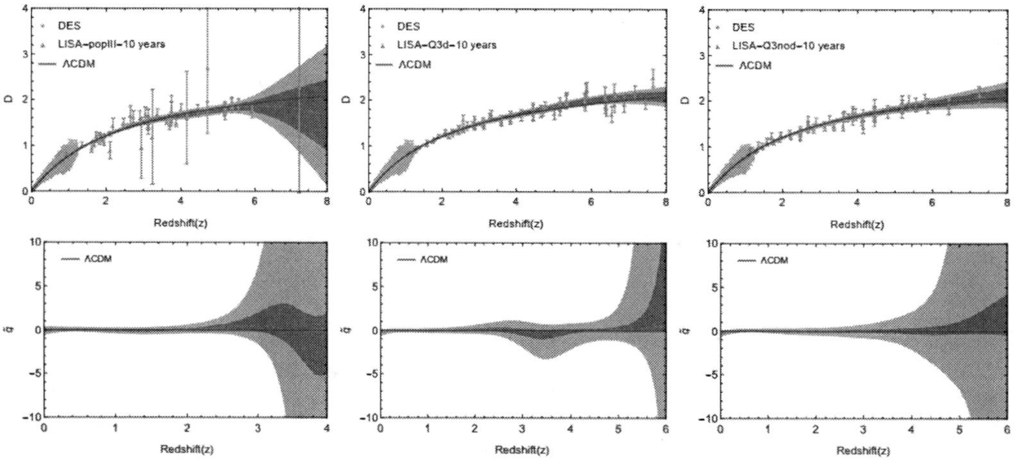

Figure 5. Reconstruction of the interaction using DES+LISA (10 years) data. From left to right each column reports the results for popIII, Q3d, Q3nod.

events with SGRB per year. With this postulate, we can simulate 100 to 1000 numbers of GW events to see that with how many events we can constrain the cosmological parameters as precisely as the current *Planck* results. We adopt the MCMC method and the results from [7] are shown in figure 7. We find that with about 500-600 GW events we can constrain the Hubble constant with an accuracy comparable to *Planck* temperature data and *Planck* lensing combined results. As for the dark matter density parameter, the GW data alone seem not able to provide a constraint as good as for the Hubble constant, the sensitivity of 1000 GW events is a little lower than that of *Planck* data. It should require more than 1000 events to match the *Planck* sensitivity.

For constraining the equation of state $w(z)$, we adopt the nonparametric method Gaussian process to reconstruct the $w(z)$ function. We find 700 GW events can give the same constraint accuracy to

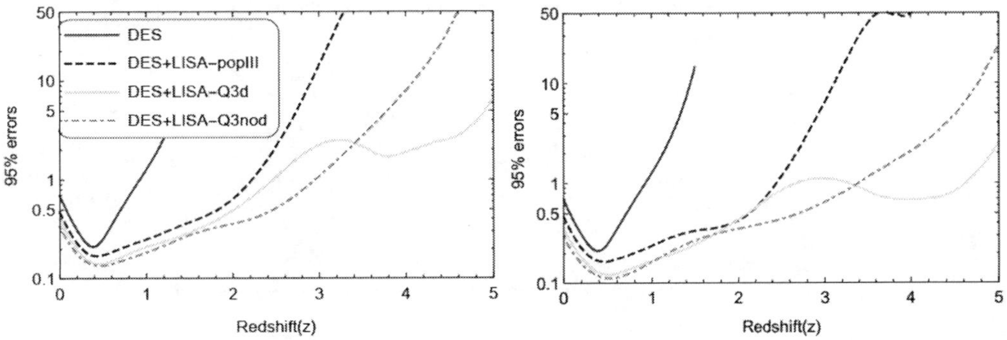

Figure 6. 95% confidence errors on the reconstructed interaction term [6].

$w(z)$ as *Planck* (the constant w) in the low redshift region as shown in figure 8. With 1000 GW events, we can constrain $\Delta h_0 \sim 5 \times 10^{-3}$ and $\Delta\Omega_m \sim 0.02$. The equation of state can be constrained to $\Delta w(z) \sim 0.03$ in the low redshift region with Gaussian process.

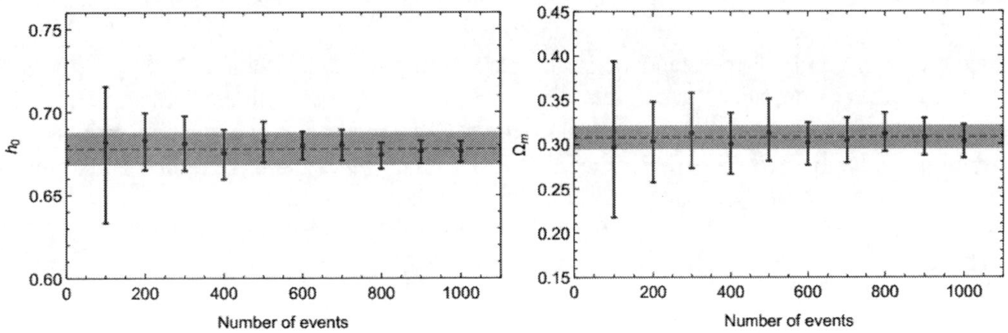

Figure 7. Sixty-eight percent confidence level (C.L.) (red line) and the best fit (red dot) for H_0 (left) and Ω_m (right) for a variable number of GW events with EM counterpart. The fiducial model is shown as the dashed line. For a comparison, the blue shaded area is the 68% C.L. constrained by the *Planck* temperature data combined with *Planck* lensing in the current *Planck* 2015 results.

5 Conclusions and future prospects

GW standard sirens will in general offer a new independent way, complementary to ordinary EM observations, to probe the cosmic expansion. Even if rare, GW standard sirens will complement, and increase confidence in other cosmic distance indicators, in particular SNIa standard candles. ET and LISA as the future GW detectors will play important roles in the GW astronomy. The different sources and the reshift regions supplied by the EM counterpart will enrich the information of the traditional EM signals. With the model-independent numerical method such as the Gaussian process, one can easily processes the GW's data sets and apply them to our researches.

The machine learning as a big data analysis technique will become more and more important in the future. As one of the machine learning technique, Gaussian process can provide us a new and

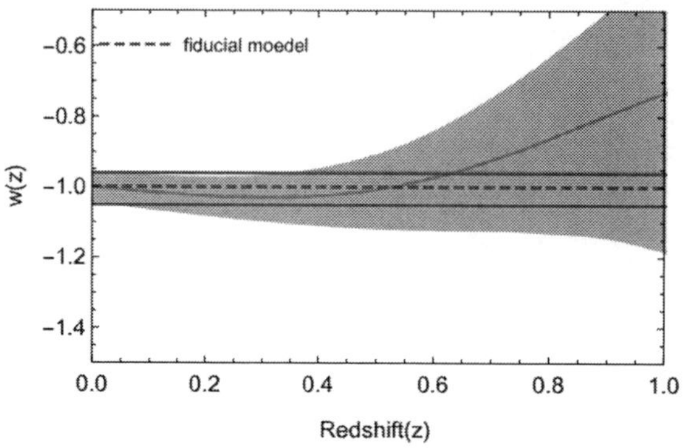

Figure 8. Reconstructions of $w(z)$ from the simulated data sets with 700 GW events. The shaded blue regions are the 68% C.L. of the constraint. The fiducial value $w = -1$ is shown as the red dashed line. For a comparison, we add the two red lines that bound the 68% C.L. of the constraint of the constant equation of state w from the *Planck* data combined with the type-Ia supernovae. This figure is from [7].

model-independent method to analyse the huge and complex data sets. It can help us to find the physical meaning and forecast the tendency and property which are implied in these data sets.

Also, we can combine the standard sirens data with the *Planck* data and expect giving better constraints on the cosmological parameters. For example, we can use both the LISA and ET simulated data sets combined with *Planck* data, then adopt the MCMC method to give the constraints of a series of cosmological parameters. By adding the GWs data to the *Planck* data, the tighter constraints of the parameters such as the H_0, Ω_m, w_0, and w_a are anticipated.

6 Acknowledgements

We thank Z.K. Guo and Nicola Tamanini for helpful collaborations in those works discussed in this paper. This work was supported in part by the National Natural Science Foundation of China Grants No.11690022, No.11375247, No.11435006, and No.11647601, and by the Strategic Priority Research Program of CAS Grant No.XDB23030100 and by the Key Research Program of Frontier Sciences of CAS.

References

[1] B. P. Abbott *et al.* [LIGO Scientific and Virgo Collaborations], Phys. Rev. Lett. **116**, no. 6, 061102 (2016) doi:10.1103/PhysRevLett.116.061102 [arXiv:1602.03837 [gr-qc]].

[2] B. P. Abbott *et al.* [LIGO Scientific and Virgo Collaborations], Phys. Rev. Lett. **116**, no. 24, 241103 (2016) doi:10.1103/PhysRevLett.116.241103 [arXiv:1606.04855 [gr-qc]].

[3] B. P. Abbott *et al.* [LIGO Scientific and VIRGO Collaborations], Phys. Rev. Lett. **118**, no. 22, 221101 (2017) doi:10.1103/PhysRevLett.118.221101 [arXiv:1706.01812 [gr-qc]].

[4] R. G. Cai, Z. Cao, Z. K. Guo, S. J. Wang and T. Yang, doi:10.1093/nsr/nwx029 arXiv:1703.00187 [gr-qc].

[5] T. Yang, Z. K. Guo and R. G. Cai, Phys. Rev. D **91**, no. 12, 123533 (2015) doi:10.1103/PhysRevD.91.123533 [arXiv:1505.04443 [astro-ph.CO]].

[6] R. G. Cai, N. Tamanini and T. Yang, JCAP **1705**, no. 05, 031 (2017) doi:10.1088/1475-7516/2017/05/031 [arXiv:1703.07323 [astro-ph.CO]].

[7] R. G. Cai and T. Yang, Phys. Rev. D **95**, no. 4, 044024 (2017) doi:10.1103/PhysRevD.95.044024 [arXiv:1608.08008 [astro-ph.CO]].

[8] B. F. Schutz, Nature **323**, 310 (1986). doi:10.1038/323310a0

[9] N. Tamanini, J. Phys. Conf. Ser. **840**, no. 1, 012029 (2017) doi:10.1088/1742-6596/840/1/012029 [arXiv:1612.02634 [astro-ph.CO]].

[10] B. Wang, E. Abdalla, F. Atrio-Barandela and D. Pavon, Rept. Prog. Phys. **79**, no. 9, 096901 (2016) doi:10.1088/0034-4885/79/9/096901 [arXiv:1603.08299 [astro-ph.CO]].

[11] R. G. Cai and Q. Su, Phys. Rev. D **81**, 103514 (2010) doi:10.1103/PhysRevD.81.103514 [arXiv:0912.1943 [astro-ph.CO]].

[12] V. Salvatelli, N. Said, M. Bruni, A. Melchiorri and D. Wands, Phys. Rev. Lett. **113**, no. 18, 181301 (2014) doi:10.1103/PhysRevLett.113.181301 [arXiv:1406.7297 [astro-ph.CO]].

[13] T. Holsclaw, U. Alam, B. Sanso, H. Lee, K. Heitmann, S. Habib and D. Higdon, Phys. Rev. D **82**, 103502 (2010) doi:10.1103/PhysRevD.82.103502 [arXiv:1009.5443 [astro-ph.CO]].

[14] T. Holsclaw, U. Alam, B. Sanso, H. Lee, K. Heitmann, S. Habib and D. Higdon, Phys. Rev. Lett. **105**, 241302 (2010) doi:10.1103/PhysRevLett.105.241302 [arXiv:1011.3079 [astro-ph.CO]].

[15] T. Holsclaw, U. Alam, B. Sanso, H. Lee, K. Heitmann, S. Habib and D. Higdon, Phys. Rev. D **84**, 083501 (2011) doi:10.1103/PhysRevD.84.083501 [arXiv:1104.2041 [astro-ph.CO]].

[16] M. Seikel, C. Clarkson and M. Smith, JCAP **1206**, 036 (2012) doi:10.1088/1475-7516/2012/06/036 [arXiv:1204.2832 [astro-ph.CO]].

[17] M. Seikel and C. Clarkson, arXiv:1311.6678 [astro-ph.CO].

[18] R. G. Cai, Z. K. Guo and T. Yang, JCAP **1608**, no. 08, 016 (2016) doi:10.1088/1475-7516/2016/08/016 [arXiv:1601.05497 [astro-ph.CO]].

[19] R. G. Cai, Z. K. Guo and T. Yang, Phys. Rev. D **93**, no. 4, 043517 (2016) doi:10.1103/PhysRevD.93.043517 [arXiv:1509.06283 [astro-ph.CO]].

[20] J. P. Bernstein *et al.*, Astrophys. J. **753**, 152 (2012) doi:10.1088/0004-637X/753/2/152 [arXiv:1111.1969 [astro-ph.CO]].

[21] N. Tamanini, C. Caprini, E. Barausse, A. Sesana, A. Klein and A. Petiteau, JCAP **1604**, no. 04, 002 (2016) doi:10.1088/1475-7516/2016/04/002 [arXiv:1601.07112 [astro-ph.CO]].

[22] E. Barausse, Mon. Not. Roy. Astron. Soc. **423**, 2533 (2012) doi:10.1111/j.1365-2966.2012.21057.x [arXiv:1201.5888 [astro-ph.CO]].

[23] C. Caprini and N. Tamanini, JCAP **1610**, no. 10, 006 (2016) doi:10.1088/1475-7516/2016/10/006 [arXiv:1607.08755 [astro-ph.CO]].

[24] W. Zhao, C. Van Den Broeck, D. Baskaran and T. G. F. Li, Phys. Rev. D **83**, 023005 (2011) doi:10.1103/PhysRevD.83.023005 [arXiv:1009.0206 [astro-ph.CO]].

[25] E. Nakar, Phys. Rept. **442**, 166 (2007) doi:10.1016/j.physrep.2007.02.005 [astro-ph/0701748 [ASTRO-PH]].

EPJ Web of Conferences **168**, 01009 (2018) https://doi.org/10.1051/epjconf/201816801009
Joint International Conference of ICGAC-XIII and IK-15 on Gravitation, Astrophysics and Cosmology

New entropy formula for Kerr black holes

Hernán A. González[1,⋆], *Daniel* Grumiller[1,⋆⋆], *Wout* Merbis[1,⋆⋆⋆], and *Raphaela* Wutte[1,⋆⋆⋆⋆]

[1] *Institute for Theoretical Physics, TU Wien, Wiedner Hauptstr. 8-10/136, A-1040 Vienna, Austria*

Abstract. We introduce a new entropy formula for Kerr black holes inspired by recent results for 3-dimensional black holes and cosmologies with soft Heisenberg hair. We show that also Kerr–Taub–NUT black holes obey the same formula.

1 Introduction

Recently a new horizon entropy formula has emerged [1]

$$S = 2\pi\,(J_0^+ + J_0^-) \tag{1}$$

that is more universal than the Bekenstein–Hawking [2, 3] or Wald's [4, 5] entropy formulas, albeit only in three dimensions. The quantities $J_0^\pm$ are zero-mode charges of $\mathfrak{u}(1)$ current algebras that appear to arise generically in near horizon descriptions.

Formula (1) was first derived for black holes in 3-dimensional Anti-de Sitter space within Einstein gravity [1] inspired by related earlier discussions of near horizon boundary conditions [6, 7] and the concept of soft hair [8]. During the past 18 months the entropy formula (1) was shown to apply also to flat space cosmologies in Einstein gravity [9], to black holes (flat space cosmologies) in higher spin theories [10] ([11]) where Bekenstein–Hawking fails, and to higher derivative theories with gravitational Chern–Simons term [12] where Wald's formula fails.

Besides the entropy itself also Cardyology simplifies [1, 9, 13] including log-corrections [14], while semi-classical considerations allow for Hardyology [15–18], by which we mean a counting of explicitly constructed semi-classical black hole microstates that combinatorially reduces to partitions of large integers. Nearly all results so far apply only to three spacetime dimensions, with the exception of [18] that applies Hardyology to extremal Kerr black holes. The phenomenologically most interesting non-extremal Kerr black hole [19] was not discussed in detail so far.

The main purpose of this proceedings contribution is to make modest progress towards lifting the exciting results in three dimensions to four dimensions by providing an analog of the entropy formula (1) for non-extremal Kerr black holes. Our main result is a new entropy formula

$$S_{\text{Kerr}} = 4\pi\, J_0^+ J_0^- \tag{2}$$

⋆e-mail: hgonzale@hep.itp.tuwien.ac.at
⋆⋆e-mail: grumil@hep.itp.tuwien.ac.at
⋆⋆⋆e-mail: merbis@hep.itp.tuwien.ac.at
⋆⋆⋆⋆e-mail: rwutte@hep.itp.tuwien.ac.at

that is equivalent to the Bekenstein–Hawking law or Wald's entropy, but expressed in terms of zero mode charges of u(1) current algebras that are expected to appear in a suitable near horizon description of non-extremal Kerr.

This work is organized as follows. In section 2 we summarize briefly salient aspects of the black hole entropy (1) in three dimensions. In section 3 we present our new results for the entropy of non-extremal Kerr black holes. In section 4 we elaborate on inner horizon entropy, limits of Kerr and the generalization to Kerr–Taub–NUT. In section 5 we provide a brief outlook.

2 Black hole entropy in three dimensions

In this section we review material that appeared in [1, 9], whose conventions and notations we use. We start with the near horizon expansion of non-extremal black holes (or cosmologies).

The vicinity of non-extremal horizons allows an expansion around Rindler space

$$\mathrm{d}s^2 = -\alpha^2 r^2\, \mathrm{d}t^2 + \mathrm{d}r^2 + \gamma^2\, \mathrm{d}\varphi^2 + \cdots \tag{3}$$

where the ellipsis denotes terms that vanish as the radial coordinate tends to zero, $r \to 0$. The angular coordinate is periodic, $\varphi \sim \varphi + 2\pi$. The exact solutions of Einstein's equations with negative cosmological constant $\Lambda = -1/\ell^2$ that asymptote to the near horizon metric (3) are given by

$$\mathrm{d}s^2 = -\alpha^2\ell^2 \sinh^2(r/\ell)\, \mathrm{d}t^2 + \mathrm{d}r^2 + 2\alpha\omega\ell^2 \sinh^2(r/\ell)\, \mathrm{d}t\, \mathrm{d}\varphi + (\gamma^2 \cosh^2(r/\ell) - \omega^2\ell^2 \sinh^2(r/\ell))\, \mathrm{d}\varphi^2\,. \tag{4}$$

As usual, for defining the theory it is crucial to declare which variations of the metric are allowed, i.e., to impose some consistent set of boundary conditions.

The near horizon boundary conditions of [1, 9] allow for arbitrary variations of the horizon radius function γ and the rotation function ω but no variations of Rindler acceleration α, i.e., $\delta\gamma \neq 0 \neq \delta\omega$ and $\delta\alpha = 0$. For constant Rindler acceleration the equations of motion imply conservation of these functions in time, $\partial_t\gamma = 0 = \partial_t\omega$, which are near horizon analogs of the holographic Ward identities $\partial_\mp T^{\pm\pm} = 0$ for the usual Brown–Henneaux boundary conditions [20].

The canonical boundary charges associated with these choices are given by the Fourier modes

$$J_n^\pm = \frac{1}{16\pi G} \oint \mathrm{d}\varphi\, e^{\pm in\varphi}\, (\gamma \pm \omega\ell) \tag{5}$$

where G is the 3-dimensional Newton constant. Their algebra consists of two u(1) current algebras

$$[J_n^\pm,\, J_m^\pm] = \frac{\ell}{8G}\, n\, \delta_{n+m,0} \qquad [J_n^+,\, J_m^-] = 0\,. \tag{6}$$

The zero mode charges $J_0^\pm$ commute with all generators $J_n^\pm$, which encodes algebraically the "soft hair" property. Since the algebra (6) can be mapped to infinite copies of the Heisenberg algebra frequently the name "soft Heisenberg hair" is used to distinguish from other soft hair constructions.

The Bekenstein–Hawking entropy is given by

$$S_{\text{BH}} = \frac{2\pi \oint \gamma}{4G} = 2\pi\, (J_0^+ + J_0^-) \tag{7}$$

which coincides with the result announced in (1). While in Einstein gravity the result (7) is perhaps not too surprising, its simplicity is still remarkable and should be contrasted with the usual form of the entropy expressed in terms of Virasoro zero modes $L_0^\pm$ [21–23],

$$S_{\text{Cardy}} = 2\pi \left(\sqrt{\frac{cL_0^+}{6}} + \sqrt{\frac{cL_0^-}{6}} \right) \tag{8}$$

with the Brown–Henneaux central charge

$$c = \frac{3\ell}{2G}\,. \tag{9}$$

More suprisingly, formula (1) applies also to flat space, where the usual Cardy-like formula is given by [24, 25] (L_0, M_0 are BMS$_3$ zero modes and c_M is the Barnich–Compère central charge [26])

$$S_{\text{BMS}_3} = 2\pi\, L_0\, \sqrt{\frac{c_M}{2M_0}} \tag{10}$$

and to situations where Bekenstein–Hawking fails, such as higher spin [27, 28] black holes [29, 30], where even for the simplest case of spin-3 black holes the entropy looks fairly complicated [31] ($L_0^\pm$ and c are the same as in the spin-2 case; $W_0^\pm$ denotes the spin-3 zero mode charges with suitable normalization; for $W_0^\pm = 0$ the Cardy formula (8) is recovered)

$$S_{\text{spin 3}} = 2\pi\bigg(\sqrt{\frac{cL_0^+}{6}}\cos\Big(\frac{1}{3}\,\arcsin\big(W_0^+(L_0^+)^{-3/2}\big)\Big) + \sqrt{\frac{cL_0^-}{6}}\cos\Big(\frac{1}{3}\,\arcsin\big(W_0^-(L_0^-)^{-3/2}\big)\Big)\bigg). \tag{11}$$

Similar remarks and results apply to higher spin flat space [32, 33] cosmologies [34–36].

Let us note for later purposes that in the flat space case the algebra (6) appears naturally in an off-diagonal basis, $J_n = J_n^+ + J_{-n}^-$ and $K_n = \frac{1}{\ell}\,(J_n^+ - J_{-n}^-)$, such that

$$[J_n,\, K_m] = \frac{1}{4G}\, n\, \delta_{n+m,0} \tag{12}$$

and the entropy (1) reads

$$S = 2\pi\, J_0\,. \tag{13}$$

The algebraic reasons behind these surprising results are Sugawara-type of relations between the near horizon generators $J_n^\pm$ and the usual asymptotic variables, like Virasoro generators, BMS$_3$ generators or higher spin generators. For example, the Virasoro generators $L_n^\pm$ emerge uniquely from a Miura-type of transformation (spelled out in [1]) that leads to a Sugawara construction

$$L_n^\pm = \frac{4G}{\ell}\sum_p J_{n-p}^\pm J_p^\pm + inJ_n^\pm \tag{14}$$

with a precise twist-term that allows to recover the Brown–Henneaux central charge (9). Inserting this central charge together with the vacuum expectation value of the zero-mode generator

$$L_0^\pm = \frac{4G}{\ell}\,(J_0^\pm)^2 \tag{15}$$

into the Cardy formula (8) recovers formula (1). In the spin-3 case the Sugawara-type construction involves tri-linears in the spin-2 and spin-3 generators $J_{(2,3)\,n}^\pm$ [10] which in a similar way allows to recover formula (1) from the complicated expression (11) and vice versa. A summary of some of these algebraic constructions is provided in [9].

Besides these unexpected simplifications of the results for entropy, at least in three dimensions one can go further and exploit the near horizon theory for a microstate counting (including log corrections), see [14] and references therein. Moreover, based on the $\mathfrak{u}(1)$ current algebras (6) a specific proposal for all microstates of semi-classical BTZ black holes was provided in [15, 17].

While it is not yet clear how these more advanced considerations generalize to higher dimensions, the more basic ones reviewed here show at least that there is a simple and fairly universal entropy law (1) in three dimensions that often considerably simplifies the results and paves the way for a deeper analysis of black hole entropy. We show in the next section that something similar can be achieved for non-extremal Kerr black holes.

3 Kerr black hole entropy

3.1 Bekenstein–Hawking entropy

Kerr black holes are characterized uniquely by their mass M and angular momentum $J = aM$, in terms of which their Bekenstein–Hawking entropy reads (we set Newton's constant to unity)

$$S_{\text{Kerr}} = 2\pi M^2\Big(1 + \sqrt{1 - a^2/M^2}\Big)\,. \tag{16}$$

3.2 Near horizon symmetry algebra by Donnay et al.

To continue with our agenda we need some analog of the near horizon boundary conditions reviewed in section 2, which so far do not exist. Thus, we use instead the near horizon boundary conditions introduced in [6] (see also [37]), which lead to a symmetry algebra whose non-vanishing commutators are displayed below.

$$[L_n^\pm,\, L_m^\pm] = (n-m)\, L_{n+m}^\pm \tag{17a}$$

$$[L_n^+,\, T_{(m,k)}] = -m\, T_{(n+m,\,k)} \tag{17b}$$

$$[L_n^-,\, T_{(m,k)}] = -k\, T_{(m,\,n+k)} \tag{17c}$$

The Witt generators $L_n^\pm$ produce "superrotations" and the remaining generators $T_{(n,m)}$ "supertranslations". Note, however, that the algebra (17) differs from the Barnich–Troessaert generalization [38, 39] of BMS [40, 41].

3.3 Sugawara deconstruction

There are three key observations for our purposes. The first one, made in [6], relates the supertranslation double-zero-mode to the Kerr black hole entropy (16),[1]

$$S_{\text{Kerr}} = 4\pi\, T_{(0,0)} \tag{18}$$

The second one (made in the same paper) is that for the Kerr solution the Witt zero mode generators, up to an imaginary factor, are given by the Kerr angular momentum.

$$L_0^\pm = \pm\frac{i}{2}\, J \tag{19}$$

The third one, made in [15], provides a "Sugawara-deconstruction" of the algebra (17) in terms of four $\mathfrak{u}(1)$ current algebras (commutators not displayed vanish)

$$[\mathcal{J}_n^\pm,\, \mathcal{J}_m^\pm] = -[\mathcal{K}_n^\pm,\, \mathcal{K}_m^\pm] = \frac{n}{2}\,\delta_{n+m,\,0} \tag{20}$$

namely

$$T_{(n,m)} = (\mathcal{J}_n^+ + \mathcal{K}_n^+)(\mathcal{J}_m^- + \mathcal{K}_m^-) \qquad\qquad L_n^\pm = \sum_p (\mathcal{J}_{n-p}^\pm + \mathcal{K}_{n-p}^\pm)(\mathcal{J}_p^\pm - \mathcal{K}_p^\pm)\,. \tag{21}$$

The "deconstruction" above is of course not unique and we can change to a more convenient basis, which we do now. Defining the generators

$$J_n^\pm := \mathcal{J}_n^\pm + \mathcal{K}_n^\pm \qquad\qquad K_n^\pm := \mathcal{J}_n^\pm - \mathcal{K}_n^\pm \tag{22}$$

[1] For convenience our normalization of $T_{(0,0)}$ differs from the one in [6] by a factor of surface gravity and a factor 2.

yields an algebra consisting of two copies of the 3-dimensional flat space near horizon symmetry algebra (12) (again all commutators not displayed vanish)

$$[J_n^\pm,\, K_m^\pm] = n\,\delta_{n+m,0}\,. \tag{23}$$

The generators (21) are given by the bilinears

$$T_{(n,m)} = J_n^+ J_m^- \qquad\qquad L_n^\pm = \sum_p J_{n-p}^\pm K_p^\pm\,. \tag{24}$$

3.4 Determination of zero modes

Comparing with [6] for the non-extremal Kerr solution the four zero modes of the new generators have to obey three algebraic constraints (determined by their relations to $L_0^\pm$ and $T_{(0,0)}$). To solve uniquely for these zero modes we need a fourth condition. We impose

$$J_0^+ + J_0^- + K_0^+ + K_0^- \stackrel{!}{=} 2M \tag{25}$$

based on the rationale that this chirally symmetric sum of zero modes should not see the angular momentum, while dimensional analysis shows it must be linear in mass. The least obvious aspect of the postulate (25) is the factor 2 on the right hand side. This factor is fixed uniquely by demanding that in the Schwarzschild limit of vanishing angular momentum, $J \to 0$, the zero modes agree with each other pairwise, $J_0^+ = J_0^-$ (the other relation, $K_0^+ = K_0^-$, holds automatically in this limit since both quantities tend to zero). Solving the algebraic equations (25), $T_{(0,0)} = J_0^+ J_0^-$ and $L_0^\pm = J_0^\pm K_0^\pm$ establishes uniquely results for the zero modes of the new generators.

$$J_0^\pm = \frac{1}{2}\big(M + \sqrt{M^2 - a^2} \pm ia\big) \qquad\qquad K_0^\pm = \frac{1}{2}\big(M - \sqrt{M^2 - a^2} \pm ia\big) \tag{26}$$

3.5 Results for non-extremal Kerr in terms of $\mathfrak{u}(1)$ zero modes

We express now standard Kerr parameters in terms of the zero mode charges (26). The black hole and Cauchy horizon radii

$$r_+ = J_0^+ + J_0^- \qquad\qquad r_- = K_0^+ + K_0^- \tag{27}$$

as well as the black hole mass

$$M = J_0^+ + K_0^- = J_0^- + K_0^+ \tag{28}$$

are linear in the zero mode charges, while angular momentum

$$J = -2i\,J_0^+ K_0^+ = 2i\,J_0^- K_0^- \tag{29}$$

and, equivalently, the Virasoro zero modes (19)

$$L_0^\pm = J_0^\pm K_0^\pm \tag{30}$$

are bilinear in them.

The Bekenstein–Hawking entropy (18) is also bilinear in the zero modes.

$$\boxed{S_{\text{Kerr}} = 4\pi\, J_0^+ J_0^-} \tag{31}$$

The new entropy formula (31) is the main result of this work. The non-trivial aspect is not just that the entropy can be written as in (31), but rather that the quantities therein, $J_0^\pm$, are (linear combinations of) zero modes of $\mathfrak{u}(1)$ generators that are related in a Sugawara-like way to the near horizon charges (24) discovered in [6]. As compared to the 3-dimensional flat space result (13) the entropy now contains a product of two zero-mode charges instead of a single one. The factor 4π in (31) is naturally interpreted as volume of the round unit 2-sphere.

4 Elaborations

In this section we discuss some elaborations. We address inner horizon entropy in section 4.1, mention Schwarzschild and extremal limits in section 4.2, and present the generalization to Kerr–Taub–NUT in section 4.3. We comment on some further interesting generalizations in the concluding section 5.

4.1 Inner horizon entropy

The "inner horizon entropy" defined and used in [42–46] is bilinear in the other pair of zero modes [the ones not appearing in the new entropy formula (31)].

$$S_{\text{inner}} = 2\pi\, K_0^+ K_0^- = 2\pi\, M^2\left(1 - \sqrt{1-a^2/M^2}\right) \tag{32}$$

The result above suggests that the generators $J_n^\pm$ are associated with the outer (black hole) horizon, while the generators $K_n^\pm$ are associated with the inner (Cauchy) horizon. (Note that one can define generators $\tilde{T}_{(n,m)} = K_n^+ K_m^-$ that also satisfy (17), but with T replaced by $\tilde{T}$, which can be interpreted as generating supertranslations on the inner horizon.) This suggestion is confirmed by the expressions for outer and inner horizon radii (27). As may have been anticipated, angular momentum (29) is a quantity that involves information about both horizons. The same is true for the mass (28).

4.2 Schwarzschild and extremal Kerr

For Schwarzschild the inner horizon charges vanish, $K_0^\pm = 0$. Other than that all the results of section 3 apply.

By contrast, in the extremal limit the outer and inner horizon charges are identical, $J_0^\pm = K_0^\pm$, which physically makes sense as both horizons then coalesce to a single one. In that case the new entropy formula (31) by virtue of the angular momentum formulas (29) can be rewritten as

$$S_{\text{EK}} = 4\pi\,\sqrt{J_0^+ K_0^+ J_0^- K_0^-} = 2\pi\,|J| \tag{33}$$

which coincides with the near horizon extremal Kerr result [47, 48]. Note, however, that the boundary conditions of [6] on which our construction is based solely apply to non-extremal Kerr black holes, so one has to be careful with this limit. See [18] for more on extremal Kerr entropy in a near horizon context, including a proposal for its microstates.

4.3 Kerr–Taub–NUT

We consider now a generalization to Kerr–Taub–NUT [49, 50] whose metric in Boyer–Lindquist coordinates is given by

$$\mathrm{d}s^2 = -\frac{\Delta}{\Sigma}\left(\mathrm{d}t - (a\sin^2\theta - 2n\cos\theta)\,\mathrm{d}\varphi\right)^2 + \frac{\Sigma}{\Delta}\,\mathrm{d}r^2 + \Sigma\,\mathrm{d}\theta^2 + \frac{\sin^2\theta}{\Sigma}\left(a\,\mathrm{d}t^2 - (r^2+a^2+n^2)\,\mathrm{d}\varphi\right)^2 \tag{34}$$

with Kerr parameter $a = J/M$, nut charge n and the two functions

$$\Delta = r^2 - 2Mr + a^2 - n^2 \qquad \text{and} \qquad \Sigma = r^2 + (n + a\cos\theta)^2\,. \tag{35}$$

Rather than repeating in detail the analysis we did for Kerr we merely state the result for the zero mode charges.

$$J_0^\pm = \frac{1}{2}\left(M + \sqrt{M^2-a^2+n^2} \pm ia\right) \qquad K_0^\pm = \frac{1}{2}\left(M - \sqrt{M^2-a^2+n^2} \pm ia\right) \tag{36}$$

Besides mass M and angular momentum J the zero mode charges (36) depend also on the nut charge n.

Repeating the analysis of section 3.5 yields expressions for the horizon radii

$$r_+ = J_0^+ + J_0^- = M + \sqrt{M^2 - a^2 + n^2} \qquad r_- = K_0^+ + K_0^- = M - \sqrt{M^2 - a^2 + n^2} \tag{37}$$

and the mass.

$$M = J_0^+ + K_0^- = J_0^- + K_0^+ \tag{38}$$

The main change as compared to Kerr is that the left and right Witt zero mode charges $L_0^\pm = J_0^\pm K_0^\pm$ do not add up to zero, $L_0^+ + L_0^- \neq 0$, but instead yield the square of the nut charge.

$$-\frac{n^2}{2} = J_0^+ K_0^+ + J_0^- K_0^- \tag{39}$$

The angular momentum $J = aM$ is determined from the difference of the Witt zero mode charges.

$$iJ = J_0^+ K_0^+ - J_0^- K_0^- \tag{40}$$

The Bekenstein–Hawking entropy is again compatible with our main result (31).

$$S_{\text{Kerr-Taub-NUT}} = 4\pi J_0^+ J_0^- = 2\pi\left(M^2 + n^2 + M\sqrt{M^2 - a^2 + n^2}\right) \tag{41}$$

For completeness we also state the corresponding result for inner horizon entropy.

$$S_{\text{inner}} = 4\pi K_0^+ K_0^- = 2\pi\left(M^2 + n^2 - M\sqrt{M^2 - a^2 + n^2}\right) \tag{42}$$

4.4 Summary of relations between $\mathfrak{u}(1)$ zero mode charges and black hole parameters

Table 4.4 summarizes various linear and bilinear relations between the $\mathfrak{u}(1)$ zero mode charges $J_0^\pm$, $K_0^\pm$ and black hole quantities.

zero mode combination	black hole quantity	physical interpretation
$J_0^+ + J_0^-$	r_+	black hole radius
$K_0^+ + K_0^-$	r_-	Cauchy horizon radius
$J_0^+ + K_0^+$	M	black hole mass
$J_0^+ - K_0^+$	$\sqrt{M^2 - a^2 + n^2}$	extremality parameter
$J_0^+ - J_0^-$	$ia = iJ/M$	Kerr parameter
$\boldsymbol{J_0^+ J_0^-}$	$\boldsymbol{S/(4\pi)}$	**black hole entropy**
$K_0^+ K_0^-$	$S_{\text{inner}}/(4\pi)$	"inner horizon entropy"
$J_0^+ K_0^+ - J_0^- K_0^-$	iJ	angular momentum
$J_0^+ K_0^+ + J_0^- K_0^-$	$-n^2/2$	nut charge (squared)

Table 1. List of linear and bilinear relations between $\mathfrak{u}(1)$ zero mode charges $J_0^\pm$, $K_0^\pm$ and black hole quantities

5 Outlook

Our main result (31) is a reformulation of the usual Kerr black hole entropy. As indicated, we view this as the first modest step in a more ambitious program, namely to provide near horizon boundary conditions, discuss associated symplectic and/or asymptotic symmetries, derive Cardyology and propose Hardyology for non-extremal Kerr. The endgame of this research avenue is to provide an explicit set of all microstates for semi-classical non-extremal Kerr black holes and to discuss consequences for black hole evaporation, (avoidance of) information loss and related semi-classical black hole puzzles.

In four and five spacetime dimensions Cvetič and collaborators have found that the product of black hole and "inner horizon" entropies is some integer multiple of $4\pi^2$, see [44] and references therein as well as papers by Ansorg and collaborators, e.g. [51]. Our results (31) and (32) actually suggest that already the individual entropies are some integer multiples of 2π, provided the zero mode charges $J_0^\pm$ and $K_0^\pm$ are quantized in the integers. Results for three-dimensional black holes suggest that this is the case [15, 17], but clearly it would be desirable to verify this in four dimensions.

Another interesting research direction is to relate our results to the hidden conformal symmetry for Kerr black holes, see [52] and references therein. While we do not see direct evidence for conformal symmetry, our $\mathfrak{u}(1)$ currents $J_n^\pm$ and $K_n^\pm$ can be Sugawara-combined to Virasoro generators, which then indirectly generate conformal symmetry, see [17] for a summary of the relevant algebraic statements.

The new entropy formula for Kerr (31) and some of the directions alluded to above should be generalizable to other black holes in four dimensions with cosmological constant, electromagnetic sources and acceleration (e.g. Carter–Kottler a.k.a. Kerr-(anti-)de Sitter [53, 54], Kerr–Newman [55, 56], accelerated black holes [57], Plebański–Demiański [58] or its cosmological generalization [59], and black holes with toroidal or higher genus horizons [60]) as well as to black holes (and possibly other black objects such as black rings [61]) in more than four spacetime dimensions. Studying these generalization would also allow to verify if the simple entropy formula (31) is again as universal as its 3-dimensional pendant (1).

Besides the four- and higher-dimensional applications mentioned above there is still some interesting work to be done in three spacetime dimensions. Namely, so far formula (1) was verified solely for locally maximally symmetric metrics, such as Bañados–Teitelboim–Zanelli black holes [62, 63] or flat space cosmologies [64, 65]. However, the near horizon approximation (3) should be valid for any horizon, even if the solution is not maximally symmetric. It would be interesting to verify whether our general conclusions still apply to such scenarios, for instance to warped black holes, see e.g. [66]. Further open issues in three spacetime dimensions are listed in the concluding section of [9].

Whether all these research avenues will turn out to be fruitful is unclear. However, in our opinion it is exciting that there is now a new approach available — based on near horizon considerations and soft Heisenberg hair — to address macro- and microscopic aspects of black hole entropy.

Acknowledgments

DG thanks Soo-Jong Rey and the organizers of the 13th International Conference on Gravitation, Astrophysics, and Cosmology and the 15th Korea-Italy Relativistic Astrophysics Symposium for the invitation and the participants for enjoyable discussions. We thank Hamid Afshar, Martin Ammon, Stephâne Detournay, Laura Donnay, Gaston Giribet, Alfredo Perez, Miguel Pino, Stefan Prohazka, Max Riegler, Shahin Sheikh-Jabbari, David Tempo, Ricardo Troncoso and Hossein Yavartanoo for collaboration on various aspects of near horizon symmetries, soft Heisenberg hair, black hole entropy and semi-classical black hole microstates. Additionally, we thank Glenn Barnich, Steve Carlip, Geoffrey Compère, Jakob Salzer, Friedrich Schöller and Andy Strominger for discussions.

This work was supported by the Austrian Science Fund (FWF), projects P 27182-N27 and P 28751-N27.

References

[1] H. Afshar, S. Detournay, D. Grumiller, W. Merbis, A. Perez, D. Tempo, R. Troncoso, Phys. Rev. **D93**, 101503 (2016), 1603.04824
[2] J.D. Bekenstein, Phys. Rev. **D7**, 2333 (1973)
[3] S.W. Hawking, Commun. Math. Phys. **43**, 199 (1975)
[4] R.M. Wald, Phys. Rev. **D48**, 3427 (1993), gr-qc/9307038
[5] V. Iyer, R.M. Wald, Phys. Rev. **D50**, 846 (1994), gr-qc/9403028
[6] L. Donnay, G. Giribet, H.A. Gonzalez, M. Pino, Phys. Rev. Lett. **116**, 091101 (2016), 1511.08687
[7] H. Afshar, S. Detournay, D. Grumiller, B. Oblak, JHEP **03**, 187 (2016), 1512.08233
[8] S.W. Hawking, M.J. Perry, A. Strominger, Phys. Rev. Lett. **116**, 231301 (2016), 1601.00921
[9] H. Afshar, D. Grumiller, W. Merbis, A. Perez, D. Tempo, R. Troncoso, Phys. Rev. **D95**, 106005 (2017), 1611.09783
[10] D. Grumiller, A. Perez, S. Prohazka, D. Tempo, R. Troncoso, JHEP **10**, 119 (2016), 1607.05360
[11] M. Ammon, D. Grumiller, S. Prohazka, M. Riegler, R. Wutte, JHEP **05**, 031 (2017), 1703.02594
[12] M.R. Setare, H. Adami, Nucl. Phys. **B914**, 220 (2017), 1606.05260
[13] A. Pérez, D. Tempo, R. Troncoso, JHEP **06**, 103 (2016), 1605.04490
[14] D. Grumiller, A. Perez, D. Tempo, R. Troncoso, JHEP **08**, 107 (2017), 1705.10605
[15] H. Afshar, D. Grumiller, M.M. Sheikh-Jabbari (2016), 1607.00009
[16] M.M. Sheikh-Jabbari, H. Yavartanoo, Phys. Rev. **D95**, 044007 (2017), 1608.01293
[17] H. Afshar, D. Grumiller, M.M. Sheikh-Jabbari, H. Yavartanoo, JHEP **08**, 087 (2017), 1705.06257
[18] K. Hajian, M.M. Sheikh-Jabbari, H. Yavartanoo (2017), 1708.06378
[19] R.P. Kerr, Phys. Rev. Lett. **11**, 237 (1963)
[20] J.D. Brown, M. Henneaux, Commun. Math. Phys. **104**, 207 (1986)
[21] J.L. Cardy, Nucl. Phys. **B270**, 186 (1986)
[22] A. Strominger, JHEP **02**, 009 (1998), hep-th/9712251
[23] S. Carlip, Class. Quant. Grav. **15**, 3609 (1998), hep-th/9806026
[24] A. Bagchi, S. Detournay, R. Fareghbal, J. Simon, Phys. Rev. Lett. **110**, 141302 (2013), 1208.4372
[25] G. Barnich, JHEP **1210**, 095 (2012), 1208.4371
[26] G. Barnich, G. Compère, Class.Quant.Grav. **24**, F15 (2007), gr-qc/0610130
[27] M. Henneaux, S.J. Rey, JHEP **1012**, 007 (2010), 1008.4579
[28] A. Campoleoni, S. Fredenhagen, S. Pfenninger, S. Theisen, JHEP **1011**, 007 (2010), 1008.4744
[29] M. Gutperle, P. Kraus, JHEP **1105**, 022 (2011), 1103.4304
[30] M. Ammon, M. Gutperle, P. Kraus, E. Perlmutter, J.Phys. **A46**, 214001 (2013), 1208.5182
[31] C. Bunster, M. Henneaux, A. Perez, D. Tempo, R. Troncoso, JHEP **1405**, 031 (2014), 1404.3305
[32] D. Grumiller, M. Irakleidou, I. Lovrekovic, R. McNees, Phys.Rev.Lett. **112**, 111102 (2014), 1310.0819

[33] H.A. Gonzalez, J. Matulich, M. Pino, R. Troncoso, JHEP **1309**, 016 (2013), 1307.5651
[34] M. Gary, D. Grumiller, S. Prohazka, S.J. Rey, JHEP **1408**, 001 (2014), 1406.1468
[35] J. Matulich, A. Perez, D. Tempo, R. Troncoso, JHEP **05**, 025 (2015), 1412.1464
[36] M. Riegler, Phys.Rev. **D91**, 024044 (2015), 1408.6931
[37] L. Donnay, G. Giribet, H.A. González, M. Pino, JHEP **09**, 100 (2016), 1607.05703
[38] G. Barnich, C. Troessaert, Phys. Rev. Lett. **105**, 111103 (2010), 0909.2617
[39] G. Barnich, C. Troessaert, JHEP **1005**, 062 (2010), 1001.1541
[40] H. Bondi, M. van der Burg, A. Metzner, Proc. Roy. Soc. London **A269**, 21 (1962)
[41] R. Sachs, Phys. Rev. **128**, 2851 (1962)
[42] F. Larsen, Phys.Rev. **D56**, 1005 (1997), hep-th/9702153
[43] M. Cvetic, F. Larsen, Phys.Rev. **D56**, 4994 (1997), hep-th/9705192
[44] M. Cvetic, G.W. Gibbons, C.N. Pope, Phys. Rev. Lett. **106**, 121301 (2011), 1011.0008
[45] A. Castro, M.J. Rodriguez, Phys.Rev. **D86**, 024008 (2012), 1204.1284
[46] S. Detournay, Phys.Rev.Lett. **109**, 031101 (2012), 1204.6088
[47] M. Guica, T. Hartman, W. Song, A. Strominger (2008), 0809.4266
[48] G. Compere, Living Rev.Rel. **15**, 11 (2012), 1203.3561
[49] A.H. Taub, Annals Math. **53**, 472 (1951)
[50] E. Newman, L. Tamubrino, T. Unti, J. Math. Phys. **4**, 915 (1963)
[51] M. Ansorg, J. Hennig, Phys. Rev. Lett. **102**, 221102 (2009), 0903.5405
[52] A. Castro, A. Maloney, A. Strominger, Phys.Rev. **D82**, 024008 (2010), 1004.0996
[53] F. Kottler, Ann. Phys. **361**, 401 (1918)
[54] B. Carter, in *Black Holes*, edited by C. DeWitt, B. DeWitt (Gordon and Breach, 1973)
[55] E.T. Newman, R. Couch, K. Chinnapared, A. Exton, A. Prakash, R. Torrence, J. Math. Phys. **6**, 918 (1965)
[56] E.T. Newman, A.I. Janis, J. Math. Phys. **6**, 915 (1965)
[57] M. Appels, R. Gregory, D. Kubiznak, Phys. Rev. Lett. **117**, 131303 (2016), 1604.08812
[58] J.F. Plebanski, M. Demianski, Annals Phys. **98**, 98 (1976)
[59] J.B. Griffiths, J. Podolsky, Int. J. Mod. Phys. **D15**, 335 (2006), gr-qc/0511091
[60] D.R. Brill, J. Louko, P. Peldan, Phys. Rev. **D56**, 3600 (1997), gr-qc/9705012
[61] R. Emparan, H.S. Reall, Phys. Rev. Lett. **88**, 101101 (2002), hep-th/0110260
[62] M. Bañados, C. Teitelboim, J. Zanelli, Phys. Rev. Lett. **69**, 1849 (1992), hep-th/9204099
[63] M. Bañados, M. Henneaux, C. Teitelboim, J. Zanelli, Phys. Rev. **D48**, 1506 (1993), gr-qc/9302012
[64] L. Cornalba, M.S. Costa, Phys.Rev. **D66**, 066001 (2002), hep-th/0203031
[65] L. Cornalba, M.S. Costa, Fortsch.Phys. **52**, 145 (2004), hep-th/0310099
[66] D. Anninos, W. Li, M. Padi, W. Song, A. Strominger, JHEP **03**, 130 (2009), 0807.3040

EPJ Web of Conferences **168**, 01010 (2018) https://doi.org/10.1051/epjconf/201816801010
Joint International Conference of ICGAC-XIII and IK-15 on Gravitation, Astrophysics and Cosmology

Stringy Gravity: Solving the Dark Problems at 'short' distance

Jeong-Hyuck Park*

[1] *Department of Physics, Sogang University, 35 Baekbeom-ro, Mapo-gu, Seoul 04107, Korea*
[2] *Center for Theoretical Physics of the Universe, Institute for Basic Science (IBS), Seoul 08826, Korea*

Abstract. Dictated by Symmetry Principle, string theory predicts *not* General Relativity *but* its own gravity which assumes the entire closed string massless sector to be geometric and thus gravitational. In terms of $R/(MG)$, *i.e.* the dimensionless radial variable normalized by mass, Stringy Gravity agrees with General Relativity toward infinity, but modifies it at short distance. At far short distance, gravitational force can be even repulsive. These may solve the dark matter and energy problems, as they arise essentially from small $R/(MG)$ observations: long distance divided by much heavier mass. We address the pertinent differential geometry for Stringy Gravity, stringy Equivalence Principle, stringy geodesics and the minimal coupling to the Standard Model. We highlight the notion of 'doubled-yet-gauged' coordinate system, in which a gauge orbit corresponds to a single physical point and proper distance is defined between two gauge orbits by a path integral.

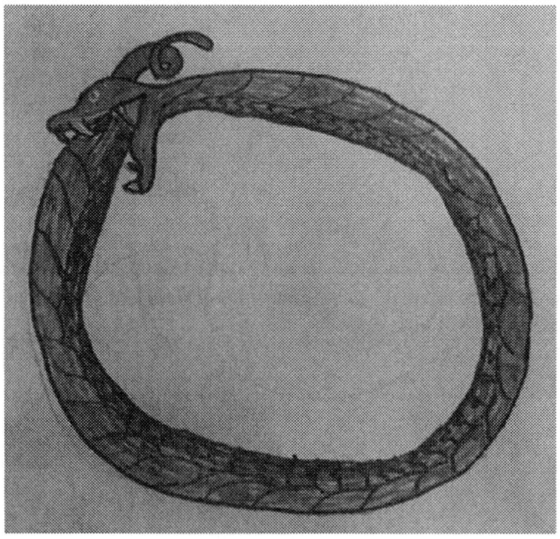

Uroboros : An ancient Egyptian symbol for a serpent which eats its own tail

*e-mail: park@sogang.ac.kr

1 Introduction

Ever since Einstein formulated his theory of gravity, *i.e.* General Relativity (GR), the metric, $g_{\mu\nu}$, has been privileged to be the only geometric and hence gravitational field, on account of the adopted Riemannian geometry. All other fields or particles are meant to be 'extra' matters. In particular, the Riemannian metric sets a proper length which every free particle would minimize, following the geodesic,

$$\ddot{x}^{\lambda} + \left\{ {}^{\,\lambda}_{\mu\,\nu} \right\} \dot{x}^{\mu}\dot{x}^{\nu} = 0\,. \tag{1}$$

The gravitational coupling to matters, or to the Standard Model, is 'minimally' fixed by the metric and covariant derivatives which ensure both diffeomorphisms and local Lorentz symmetry. In the words of Cheng-Ning Yang, *symmetry dictates interaction.*

On the other hand, string theory suggests us to put a two-form gauge potential, $B_{\mu\nu}$, and a scalar dilaton, ϕ, on an equal footing along with the metric. Forming the massless sector of closed strings, they are ubiquitous in all string theories, with the conventional (Riemannian) description given by

$$\int \mathrm{d}^{D}x\ \sqrt{-g}e^{-2\phi}\left(R_{g} + 4\partial_{\mu}\phi\partial^{\mu}\phi - \tfrac{1}{12}H_{\lambda\mu\nu}H^{\lambda\mu\nu}\right) \qquad \text{where} \qquad H = \mathrm{d}B\,. \tag{2}$$

Furthermore, a genuine stringy symmetry, called T-duality, can mix three of them [1, 2]. These may well hint at the existence of *Stringy Gravity* which should take the entire closed string massless sector to be geometric and gravitational. For this, a novel differential geometry beyond Riemann is required, in particular to shed light on the $\mathbf{O}(D,D)$ T-duality symmetry hidden in the above action.

After series of pioneering works on 'doubled sigma models' [3–8] and 'double field theory' (DFT) [9–13] (*c.f.* [14–16]), such an idea of Stringy Gravity has materialized. The word 'double' refers to the fact that doubled $(D+D)$-dimensional coordinates are used for the description of D-dimensional physical spacetime. While such a usage was historically first made in the case of a torus background – by introducing a dual coordinate conjugate to the string winding momentum – the doubled coordinates turned out to be far more general: they can be applied to any compact or non-compact spacetime, as well as to not only string but also to particle physics.

2 Stringy Gravity: Uroboros solution to Dark Problems

2.1 Stringy differential geometry

Stringy Gravity of our interest adopts the *doubled-yet-gauged coordinate system* [17] which meets two properties. Firstly, an $\mathbf{O}(D,D)$ group is *a priori* postulated, having the constant invariant "metric",

$$\mathcal{J}_{AB} = \begin{pmatrix} \mathbf{0} & \mathbf{1} \\ \mathbf{1} & \mathbf{0} \end{pmatrix}. \tag{3}$$

With the inverse, $\mathcal{J}^{AB}$, it can freely raise and lower the $\mathbf{O}(D,D)$ vector indices (capital letters, $A, B, \cdots$). Secondly, the doubled coordinates are gauged by an equivalence relation,

$$x^{A} \sim x^{A} + \Delta^{A}(x)\,, \tag{4}$$

where Δ^{A} is an arbitrary 'derivative-index-valued' vector. This means that its superscript index must be identifiable as that of derivative, $\partial^{A} = \mathcal{J}^{AB}\partial_{B}$. For example, with arbitrary functions, Φ_i, belonging to the theory, $\Delta^{A} = \sum_{i,j}\Phi_{i}\partial^{A}\Phi_{j}$. The equivalence relation (4) can be realized by requiring that every field or function in Stringy Gravity – such as Φ_i itself, physical fields, local symmetry parameters and their arbitrary derivatives – should be invariant under the coordinate gauge symmetry shift,[1]

$$\Phi_k(x+\Delta) = \Phi_k(x) \quad \Longleftrightarrow \quad \Delta^A\partial_A = 0\,. \tag{5}$$

In this way, it becomes not a point but a gauge orbit in the doubled coordinate system that corresponds to a single physical point. The above coordinate gauge symmetry invariance is equivalent to so called the 'section condition' in DFT,

$$\partial_A\partial^A = 0\,. \tag{6}$$

With respect to the off block-diagonal form of the $\mathbf{O}(D,D)$ metric (3), the doubled coordinates split into two parts: $x^A = (\tilde{x}_\mu, x^\nu)$, and hence $\partial_A\partial^A = 2\tilde{\partial}^\mu\partial_\mu$. The general solution to the section condition is then given by $\tilde{\partial}^\mu \equiv 0$, up to $\mathbf{O}(D,D)$ rotations [10, 11].

General covariance in doubled-yet-gauged spacetime, or doubled diffeomorphisms, are given by

$$\delta x^A = V^A\,, \qquad \delta\partial_A = -\partial_A V^B\partial_B = (\partial^B V_A - \partial_A V^B)\partial_B\,, \tag{7}$$

and for a covariant tensor (or tensor density with weight ω),[2]

$$\delta T_{A_1\cdots A_n} = -\omega\partial_B V^B T_{A_1\cdots A_n} + \textstyle\sum_{i=1}^n (\partial_B V_{A_i} - \partial_{A_i}V_B)T_{A_1\cdots A_{i-1}}{}^B{}_{A_{i+1}\cdots A_n}\,. \tag{8}$$

The whole massless sector of closed strings, or *stringy gravitons*, consist of a unit-weighted scalar density given by an exponentiation of DFT-dilaton, e^{-2d}, and a symmetric projector,

$$P_{AB} = P_{BA}\,, \qquad P_A{}^B P_B{}^C = P_A{}^C\,, \tag{9}$$

which accompanies the complementary, orthogonal projector, $\bar{P}_{AB} = \mathcal{J}_{AB} - P_{AB}$. It follows that the difference of the two projectors sets a symmetric $\mathbf{O}(D,D)$ element, *i.e.* DFT-metric, $P_{AB} - \bar{P}_{AB} = \mathcal{H}_{AB}$, satisfying $\mathcal{H}_{AB} = \mathcal{H}_{BA}$ and $\mathcal{H}_A{}^C\mathcal{H}_B{}^D\mathcal{J}_{CD} = \mathcal{J}_{AB}$.

The above $\mathbf{O}(D,D)$ covariant defining properties of the stringy gravitons can be conventionally solved by $\{g_{\mu\nu}, B_{\mu\nu}, \phi\}$. However, not all the solutions can be parametrized by these GR variables. Stringy Gravity is more general than GR: it encompasses 'non-Riemannian' geometries where the Riemannian metric, $g_{\mu\nu}$, cannot be defined, even locally (see [18–20] for examples and [21] for the complete classification and relations to Newton-Cartan/Carroll gravities, chiral strings *etc.*).

Requiring compatibility with all the stringy gravitons, *i.e.* $\nabla_A d = 0$, $\nabla_A P_{BC} = 0$, $\nabla_A\bar{P}_{BC} = 0$, and imposing some torsionless conditions, one can uniquely determine the *connection* for a diffeomorphism covariant derivative, $\nabla_A = \partial_A + \Gamma_A$ [22],

$$\begin{array}{ll}\Gamma_{CAB} = & 2\left(P\partial_C P\bar{P}\right)_{[AB]} + 2\left(\bar{P}_{[A}{}^D\bar{P}_{B]}{}^E - P_{[A}{}^D P_{B]}{}^E\right)\partial_D P_{EC} \\ & -4\left(\frac{1}{P_M{}^M-1}P_{C[A}P_{B]}{}^D + \frac{1}{\bar{P}_M{}^M-1}\bar{P}_{C[A}\bar{P}_{B]}{}^D\right)\left(\partial_D d + (P\partial^E P\bar{P})_{[ED]}\right)\,.\end{array} \tag{10}$$

This stringy analogue of the Christoffel symbol constitutes subsequently a scalar curvature, $S_{(0)}$, and a two-indexed "Ricci" curvature, $P_A{}^C\bar{P}_B{}^D S_{CD}$, which further leads to the stringy counterpart of the conserved Einstein tensor: $G_{AB} = 4P_{[A}{}^C\bar{P}_{B]}{}^D S_{CD} - \frac{1}{2}\mathcal{J}_{AB}S_{(0)}$ satisfying $\nabla_A G^{AB} = 0$ [23].

The scalar curvature multiplied by the unit-weighted integral measure, $e^{-2d}S_{(0)}$, sets then the action for *pure* Stringy Gravity, which agrees with (2) upon Riemannian parametrization and up to total derivatives. Its equations of motion amount to the vanishing of the stringy Einstein tensor, $G_{AB} = 0$, or equivalently $S_{(0)} = 0$ and $P_A{}^C\bar{P}_B{}^D S_{CD} = 0$. In particular, constant background solutions include not only the flat Minkowskian spacetime but also non-Riemannian ones, *e.g.* $\mathcal{H}_{AB} = \mathcal{J}_{AB}$, *etc.* [21].

[1] See also Eq.(13) for the cases where the doubled coordinates are dynamical fields.

[2] Eq.(8) corresponds to the passive counterpart of "generalized Lie derivative" *à la* Siegel [10].

2.2 Minimal coupling to the Standard Model

To couple to fermions, it is necessary to introduce *a pair of vielbeins*, V_{Ap}, $\bar{V}_{A\bar{p}}$, which form the mutually orthogonal projectors, $P_{AB} = V_{Ap}V_B{}^p$, $\bar{P}_{AB} = \bar{V}_{A\bar{p}}\bar{V}_B{}^{\bar{p}}$, and satisfy their own defining properties:

$$V_{Ap}V^A{}_q = \eta_{pq}\,, \qquad \bar{V}_{A\bar{p}}\bar{V}^A{}_{\bar{q}} = \bar{\eta}_{\bar{p}\bar{q}}\,, \qquad V_{Ap}\bar{V}^{A\bar{p}} = 0\,, \qquad V_{Ap}V_B{}^p + \bar{V}_{A\bar{p}}\bar{V}_B{}^{\bar{p}} = \mathcal{J}_{AB}\,. \quad (11)$$

The unbarred and barred lowercase letters are the vector indices of twofold local Lorentz symmetries, $\mathbf{Spin}(1, D{-}1)$ and $\mathbf{Spin}(D{-}1, 1)$, subject to Minkowskian metrics, η_{pq} (mostly plus) and $\bar{\eta}_{\bar{p}\bar{q}}$ (mostly minus), respectively. The vielbeins diagonalize $\mathcal{J}_{AB}$ and $\mathcal{H}_{AB}$ simultaneously, with the eigenvalues, $(\eta, +\bar{\eta})$ and $(\eta, -\bar{\eta})$. Physically, the doubling of the spin groups implies a pair of locally inertial frames which exist separately for the left and right closed string modes [24].[3] Now, to take care of all the local symmetries in Stringy Gravity, the covariant derivative generalizes to $\nabla_A = \partial_A + \Gamma_A + \Phi_A + \bar{\Phi}_A$, where the last two are $\mathbf{Spin}(1, D{-}1)$ and $\mathbf{Spin}(D{-}1, 1)$ connections. These are fixed by the compatibility with the vielbeins, $\nabla_A V_{Bp} = 0$, $\nabla_A \bar{V}_{B\bar{p}} = 0$, and thus related to the stringy Christoffel symbol (10) [26].

With the basics of the stringy differential geometry, $\{d, V_{Ap}, \bar{V}_{A\bar{p}}, \nabla_A, S_{(0)}\}$, the Standard Model can be straightforwardly, and minimally, coupled to Stringy Gravity [27]. In particular, fermionic kinetic terms read

$$e^{-2d}\,\bar{\psi}\gamma^A\nabla_A\psi = e^{-2d}\,\bar{\psi}V^A{}_p\gamma^p(\partial_A\psi + \tfrac{1}{4}\Phi_{Apq}\gamma^{pq}\psi) \equiv \sqrt{-g}\,\bar{\chi}\gamma^\mu(\partial_\mu\chi + \tfrac{1}{4}\omega_{\mu pq}\gamma^{pq}\chi + \tfrac{1}{24}H_{\mu pq}\gamma^{pq}\chi)\,. \quad (12)$$

The second equality, or '$\equiv$', holds up to the Riemannian reduction of $\tilde{\partial}^\mu \equiv 0$, and a field redefinition of the fermion, $\chi \equiv 2^{-\frac{1}{4}}e^{-\phi}\psi$, which couples to the Riemannian vierbein and B-field, as if fundamental string. That is to say, dilaton ϕ is *dark* to the fermion, χ.[4] Although quantum field theory is supposed to describe "point-like" particles, the Dirac spinor χ seems to remember its stringy origin.

From Stringy Gravity point of view, H-flux is part of the torsionless connection, and the stringy Christoffel symbol (10) cannot be transformed point-wise to vanish by doubled diffeomorphisms. That is to say, locally inertial (normal) frames do not exist in Stringy Gravity: the Equivalence Principle is broken for a string. This should not be a surprise, since string is an extended object, subject to "tidal force", or H-flux. It is also consistent with the absence of a four-indexed "Riemann" curvature in Stringy Gravity [22, 28]. Eq.(12) then implies the violation of the Equivalence Principle for fermions (*c.f* [29] or "fifth force"). We continue to discuss the recovery of the principle in section 2.4.

Given the twofold local Lorentz symmetries, one needs to decide the spin group for each fermion in the Standard Model after preparing two sets of gamma matrices. An open question, which could be answered experimentally, is whether the spin groups of quarks and leptons are different (or not) [27].

2.3 Uroboros solution to the Dark Matter and the Dark Energy problems

Motivated by the coupling of the Standard Model to the closed string massless sector, such as (12), the most general, asymptotically flat, spherically symmetric *vacuum* solution to $D=4$ Stringy Gravity was studied in [30] (see also [31]). The solution lets the stringy Einstein tensor vanish identically, $G_{AB} = 0$, and therefore corresponds to the stringy analogue of the Schwarzschild solution to GR. Yet, it possesses three free parameters: not only mass, M (measurable at infinity), but also, say H-flux and dilaton charge. All of them ought to reflect the intrinsic properties of the implicitly assumed

[3] Ensuring the twofold spin groups in the maximally supersymmetric DFT [25] and the doubled-yet-gauged Green-Schwarz superstring action [20], the conventional IIA and IIB theories are unified into a single unique theory that is chiral with respect to both spin groups. The distinction of IIA and IIB then refers to two different types of (Riemannian) 'solutions' rather than 'theories'.

[4] On the other hand, gauge bosons couple to $g_{\mu\nu}$ and ϕ only: H-flux is *dark* to the gauge bosons.

matter localized at the center, which deserve, combined with (12), further explorations, possibly in connection with 'energy condition' in GR.

The generic spherical solution would be regarded to feature a naked singularity, if it were interpreted from GR point of view, *c.f.* (2). But, strictly within the framework of Stringy Gravity, it may be non-singular: all the stringy curvature tensors are trivial. Inevitably, the notion of singularity depends on the differential geometry in use. Investigation into the geodesic completeness is of interest.[5]

Analysis of the circular geodesic around the center in 'string frame' – to be justified in section 2.4 below – shows that the rotation curve generically features a maximum and thus non-Keplerian over a finite range, while becoming asymptotically Keplerian toward infinity [30]. Specifically, with appropriately tuned H-flux and dilaton charge, in terms of $R/(MG)$, *i.e.* the dimensionless radial variable normalized by mass, Stringy Gravity agrees with GR becoming Keplerian for large $R/(MG) \gtrsim 10^8$, but modifies it at shorter distance, $R/(MG) \lesssim 2 \times 10^6$ (*c.f.* MOND [32]). At far shorter distance, gravitational force can be even repulsive. These may solve the dark matter and energy problems which are essentially based on small $R/(MG)$ observations: long distance divided by much heavier mass.

Namely, *the self-interaction of the closed string massless sector makes Stringy Gravity modify GR at 'short' distance in terms of $R/(MG)$*, such that it may explain the observed galaxy rotation curves, hence solve the dark matter problem. Further, *the repulsive force at far shorter distance can be responsible for the accelerating expansion of the Universe,* and thus solve the dark energy problem.

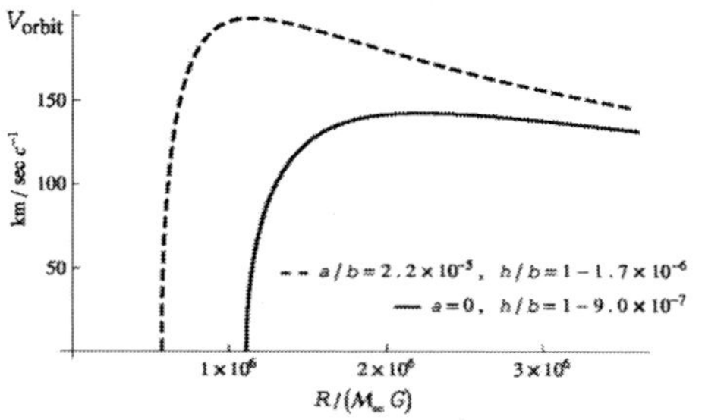

Rotation curves in Stringy Gravity [30]

Dimensionless rotation curves, $V_{\rm orbit}$ *vs.* $R/(MG)$, around the spherical vacuum in Stringy Gravity feature generically a maximum. By tuning the free parameters, one can make the maximum velocity comparable to the observations of galaxy rotations. Intriguingly as unintended consequence, the horizontal scale, $R/(MG)$, also appears consistent with the observatory data of visible matters, *c.f.* 'Uroboros' spectrum below. For sufficient small $R/(MG)$, the gravitational force can be repulsive and the orbital velocity is not defined. Changing the horizontal axis from $R/(MG)$ to the physical radius, R, the rotation curves can oscillate for nontrivial radial mass distribution functions, $M(R)$.

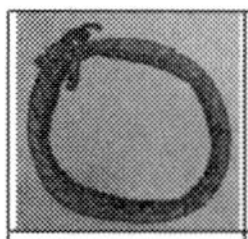	Electron ($R \simeq 0$)	Proton	Hydrogen Atom	Billiard Ball	Earth	Solar System ($1\mathrm{AU}/M_{\odot}G$)	Milky Way (visible)	Galaxy Cluster	Universe ($M \propto R^3$)
$R/(MG)$	0^+	7.1×10^{38}	2.0×10^{43}	2.4×10^{26}	1.4×10^{9}	1.0×10^{8}	1.5×10^{6}	$\sim 10^{5}$	0^+

'Uroboros' spectrum of the dimensionless radial variable normalized by mass in natural units. Observing stars and galaxies far away, we can probe deep into the short distance natrue of GRAVITY.

[5] *c.f.* both (1) and (17), as well as cosmic censorship *à la* Penrose.

2.4 Proper distance in doubled-yet-gauged spacetime: Justification for String Frame

Shifting gears, we now discuss how to define the *proper distance* in doubled-yet-gauged spacetime. First of all, we note that the usual infinitesimal one-form, $\mathrm{d}x^A$, is neither doubled diffeomorphism covariant, $\delta(\mathrm{d}x^A) = \mathrm{d}x^B\partial_B V^A \neq \mathrm{d}x^B(\partial_B V^A - \partial^A V_B)$, (7), (8), nor coordinate gauge symmetry invariant, $\mathrm{d}\Delta^A = \mathrm{d}x^B\partial_B\Delta^A \neq 0$, (5). Thus, the naive contraction with the DFT-metric, $\mathrm{d}x^A\mathrm{d}x^B\mathcal{H}_{AB}$, cannot give any sensible definition of the proper length. To cure the problem, one needs to gauge $\mathrm{d}x^A$ explicitly, introducing a connection, $\mathcal{A}^A$, which should satisfy the same property as the coordinate gauge symmetry generator, Δ^A (5),

$$\mathrm{D}x^A := \mathrm{d}x^A - \mathcal{A}^A\,, \qquad\qquad \mathcal{A}^A\partial_A = 0\,. \tag{13}$$

Provided the connection transforms appropriately, $\mathrm{D}x^A$ becomes a well-behaved vector of stringy geometry, meaning doubled diffeomorphism covariant and coordinate gauge symmetry invariant [18]. We propose then to define the proper distance in doubled-yet-gauged spacetime by a *path integral* [33],

$$\|\, x_1\,,\, x_2\,\| := -\ln\left[\int \mathcal{D}\mathcal{A}\,\exp\left(-\int_1^2 \sqrt{\mathrm{D}x^A\mathrm{D}x^B\mathcal{H}_{AB}}\right)\right]. \tag{14}$$

By letting $\tilde{\partial}^\mu \equiv 0$ and $\mathcal{A}^A \equiv (A_\mu, 0)$, we get $\mathrm{D}x^A \equiv (\mathrm{d}\tilde{x}_\mu - A_\mu, \mathrm{d}x^\nu)$. That is to say, only the half of the doubled coordinates, *i.e.* $\tilde{x}_\mu$ directions, are gauged. Furthermore, with the Riemannian parametrization of the DFT-metric, we get

$$\mathrm{D}x^A\mathrm{D}x^B\mathcal{H}_{AB} \equiv \mathrm{d}x^\mu\mathrm{d}x^\nu g_{\mu\nu} + \left(\mathrm{d}\tilde{x}_\mu - A_\mu + \mathrm{d}x^\rho B_{\rho\mu}\right)(\mathrm{d}\tilde{x}_\nu - A_\nu + \mathrm{d}x^\sigma B_{\sigma\nu})\,g^{\mu\nu}\,. \tag{15}$$

Thus, after integrating out the auxiliary connection, A_μ, our proposal (14) reduces, at least classically, to the conventional proper distance in Riemannian geometry, $\int_1^2 \sqrt{\mathrm{d}x^\mu\mathrm{d}x^\nu g_{\mu\nu}}$, which is certainly independent of the gauged $\tilde{x}_\mu$ coordinates:

$$\left\|\, x_1^A\,,\, x_2^A\,\right\| = \left\|\, x_1^\mu\,,\, x_2^\mu\,\right\|\,. \tag{16}$$

In this way, indeed the formula (14) measures the distance between two 'gauge orbits'.

The exponent in (14) sets immediately the action for a point particle propagating in the doubled-yet-gauged spacetime, or its square root free einbein formulation [30, 34]. As a consequence, after the auxiliary connection being integrated out, the point particle couples to $g_{\mu\nu}$ only. That is to say, both ϕ and $B_{\mu\nu}$ are *dark* to particles: they are hard to detect by any experiment based on particle concept. Further, the point particle follows the geodesic defined *not* in the Einstein frame *but* in the string frame. Namely, *the Equivalence Principle is restored in the string frame for a point particle.* This preferred choice of the frame is all due to the fundamental symmetries of Stringy Gravity.

The above path integral definition of the proper length can be easily generalized to (Nambu-Goto type) area and volume, which in turn gives the doubled-yet-gauged actions for string or Green-Schwarz superstring [18, 20]. In particular, with auxiliary worldsheet metric, h_{ij}, and the stringy Christoffel symbol (10), the Euler-Lagrange equations of the string action reduce, upon Riemannian backgrounds (*c.f.* non-Riemannian [21]), to

$$\tfrac{1}{\sqrt{-h}}\partial_i\left(\sqrt{-h}\mathcal{H}_{AB}D^i x^B\right) + \Gamma_{ABC}\left(\bar{P}^B{}_D D_i x^D\right)\left(P^C{}_E D^i x^E\right) = 0\,, \tag{17}$$

which can be viewed as the stringy analogue of the point particle geodesic equation (1).

Acknowledgements

This note is based on series of works by the author in collaborations with Imtak Jeon, Kanghoon Lee, Soo-Jong Rey, Yuho Sakatani, Yoonji Suh, Wonyoung Cho, Jose Fernández-Melgarejo, Woohyun Rim, Sung Moon Ko, Charles Melby-Thompson, Rene Meyér, Minwoo Suh, Kang-Sin Choi, Kevin Morand and Xavier Bekaert, as well as discussions with Sang-Hyeon Ahn, David Berman, Chris Blair, Robert Brandenberger, Martin Cederwall, Jaewon Lee, Chan Park, Hyun Seok Yang and Sang Heon Yi. The author wishes to thank them all. This work was supported by the National Research Foundation of Korea through the Grant NRF-2016R1D1A1B01015196.

References

[1] T. H. Buscher, "A Symmetry of the String Background Field Equations," Phys. Lett. B **194** (1987) 59.
[2] T. H. Buscher, "Path Integral Derivation of Quantum Duality in Nonlinear Sigma Models," Phys. Lett. B **201** (1988) 466.
[3] M. J. Duff, "Duality Rotations in String Theory," Nucl. Phys. B **335** (1990) 610.
[4] A. A. Tseytlin, "Duality Symmetric Formulation of String World Sheet Dynamics," Phys. Lett. B **242** (1990) 163.
[5] A. A. Tseytlin, "Duality symmetric closed string theory and interacting chiral scalars," Nucl. Phys. B **350** (1991) 395.
[6] C. M. Hull, "A Geometry for non-geometric string backgrounds," JHEP **0510** (2005) 065.
[7] C. M. Hull, "Global aspects of T-duality, gauged sigma models and T-folds," JHEP **0710** (2007) 057.
[8] C. M. Hull, "Doubled Geometry and T-Folds," JHEP **0707** (2007) 080 [hep-th/0605149].
[9] W. Siegel, "Two vierbein formalism for string inspired axionic gravity," Phys. Rev. D **47** (1993) 5453 [hep-th/9302036].
[10] W. Siegel, "Superspace duality in low-energy superstrings," Phys. Rev. D **48** (1993) 2826 [hep-th/9305073].
[11] C. Hull and B. Zwiebach, "Double Field Theory," JHEP **0909** (2009) 099 [arXiv:0904.4664 [hep-th]].
[12] C. Hull and B. Zwiebach, "The Gauge algebra of double field theory and Courant brackets," JHEP **0909** (2009) 090 [arXiv:0908.1792 [hep-th]].
[13] O. Hohm, C. Hull and B. Zwiebach, "Generalized metric formulation of double field theory," JHEP **1008** (2010) 008 [arXiv:1006.4823 [hep-th]].
[14] G. Aldazabal, D. Marques and C. Nunez, "Double Field Theory: A Pedagogical Review," Class. Quant. Grav. **30** (2013) 163001 doi:10.1088/0264-9381/30/16/163001 [arXiv:1305.1907 [hep-th]].
[15] D. S. Berman and D. C. Thompson, "Duality Symmetric String and M-Theory," Phys. Rept. **566** (2014) 1 doi:10.1016/j.physrep.2014.11.007 [arXiv:1306.2643 [hep-th]].
[16] O. Hohm, D. Lüst and B. Zwiebach, "The Spacetime of Double Field Theory: Review, Remarks, and Outlook," Fortsch. Phys. **61** (2013) 926 doi:10.1002/prop.201300024 [arXiv:1309.2977].
[17] J. H. Park, "Comments on double field theory and diffeomorphisms," JHEP **1306** (2013) 098 [arXiv:1304.5946 [hep-th]].
[18] K. Lee and J. H. Park, "Covariant action for a string in doubled-yet-gauged spacetime," Nucl. Phys. B **880** (2014) 134 [arXiv:1307.8377 [hep-th]].
[19] S. M. Ko, C. Melby-Thompson, R. Meyer and J. H. Park, "Dynamics of Perturbations in Double Field Theory & Non-Relativistic String Theory," JHEP **1512** (2015) 144 [arXiv:1508.01121 [hep-th]].

[20] J. H. Park, "Green-Schwarz superstring on doubled-yet-gauged spacetime," JHEP **1611** (2016) 005 [arXiv:1609.04265 [hep-th]].

[21] K. Morand and J. H. Park, "Classification of non-Riemannian doubled-yet-gauged spacetime," arXiv:1707.03713 [hep-th].

[22] I. Jeon, K. Lee and J. H. Park, "Stringy differential geometry, beyond Riemann," Phys. Rev. D **84** (2011) 044022 [arXiv:1105.6294 [hep-th]].

[23] J. H. Park, S. J. Rey, W. Rim and Y. Sakatani, "O(D, D) covariant Noether currents and global charges in double field theory," JHEP **1511** (2015) 131 [arXiv:1507.07545 [hep-th]].

[24] M. J. Duff, "Hidden String Symmetries?," Phys. Lett. B **173** (1986) 289.

[25] I. Jeon, K. Lee, J. H. Park and Y. Suh, "Stringy Unification of Type IIA and IIB Supergravities under $N = 2D = 10$ Supersymmetric Double Field Theory," Phys. Lett. B **723** (2013) 245 [arXiv:1210.5078 [hep-th]].

[26] I. Jeon, K. Lee and J. H. Park, "Incorporation of fermions into double field theory," JHEP **1111** (2011) 025 [arXiv:1109.2035 [hep-th]].

[27] K. S. Choi and J. H. Park, "Standard Model as a Double Field Theory," Phys. Rev. Lett. **115** (2015) no.17, 171603 [arXiv:1506.05277 [hep-th]].

[28] O. Hohm and B. Zwiebach, "On the Riemann Tensor in Double Field Theory," JHEP **1205** (2012) 126.

[29] T. D. Lee and C. N. Yang, "Conservation of Heavy Particles and Generalized Gauge Transformations," Phys. Rev. **98** (1955) 1501.

[30] S. M. Ko, J. H. Park and M. Suh, "The rotation curve of a point particle in stringy gravity," arXiv:1606.09307 [hep-th]. To appear in JCAP.

[31] C. P. Burgess, R. C. Myers and F. Quevedo, "On spherically symmetric string solutions in four-dimensions," Nucl. Phys. B **442** (1995) 75 [hep-th/9410142].

[32] M. Milgrom, "A Modification of the Newtonian dynamics as a possible alternative to the hidden mass hypothesis," Astrophys. J. **270** (1983) 365.

[33] J. H. Park, "Green-Schwarz superstring and Stringy Gravity in doubled-yet-gauged spacetime," talk at Banff International Research Station, http://www.birs.ca/events/2017/5-day-workshops/17w5018/videos/watch/201701231429-Park.html.

[34] C. D. A. Blair, "Particle actions and brane tensions from double and exceptional geometry," arXiv:1707.07572 [hep-th].

Scalar Dyon Production In Near Extremal Kerr-Newman Black Holes

Chiang-Mei Chen[1,⋆], *Sang Pyo* Kim[2,⋆⋆], *Jia-Rui* Sun[3,⋆⋆⋆], and *Fu-Yi* Tang[1,⋆⋆⋆⋆]

[1] *Department of Physics, National Central University, Chungli 32001, Taiwan*
[2] *Department of Physics, Kunsan National University, Kunsan 54150, Korea*
[3] *School of Physics and Astronomy, Sun Yat-Sen University, Guangzhou 510275, China*

Abstract. The pair production of charged scalar dyons is analytically studied in near-extremal Kerr-Newman (KN) dyonic black holes. The pair production rate and its thermal interpretation are given. Moreover, the absorption cross section ratio has been compared with the two-point function of the conformal field theories (CFTs) holographically dual to the near horizon geometry, namely warped AdS_3, of the near extremal Kerr-Newman black holes to verify the threefold dyonic KN/CFTs correspondence.

1 Introduction

Two independent processes, the Schwinger mechanism [1] and the Hawking radiation [2] are mixed for the spontaneous pair production in charged black holes. The leading contribution is expected to happen in the near horizon region. In particular, the results of production can be analytically obtained for the near extremal black holes [3–5]. Thus we investigate the pair production of scalar dyons in the spacetime of the near-horizon region of the near extremal dyonic Kerr-Newman (KN) black holes [6] without considering the back-reactions. The spacetime has the geometric structure of a warped AdS_3 plus an electromagnetic field. In such a background we can analytically solve the Klein-Gorden (KG) equation for the probe scalar dyons, and obtain the exact solutions in terms of the well-known hypergeometric functions. Further, by imposing the particle viewpoint of boundary condition on the solutions, the physical quantities such as the vacuum persistence amplitude, the mean number of produced pairs and the absorption cross section ratio are obtained. The corresponding thermal interpretation [7–9] has been discussed. In addition, the physical quantities calculated from the gravity side are shown to match well with those results of the scalar dyon operator in the dual boundary conformal field theories (CFTs) side, based on the KN/CFTs correspondence [10, 11].

⋆e-mail: cmchen@phy.ncu.edu.tw
⋆⋆e-mail: sangkim@kunsan.ac.kr
⋆⋆⋆e-mail: sunjiarui@sysu.edu.cn
⋆⋆⋆⋆e-mail: foue.tang@gmail.com

2 The near horizon near extreme dyonic KN black holes

The geometry of the near horizon of a near extreme dyonic KN black hole has the structure of a warped AdS_3 as [6]

$$ds^2 = \Gamma(\theta)\left[-(\rho^2 - B^2)d\tau^2 + \frac{d\rho^2}{\rho^2 - B^2} + d\theta^2\right] + \gamma(\theta)(d\varphi + b\rho d\tau)^2, \quad (1)$$

$$A_{[1]} = -\frac{Q(r_0^2 - a^2\cos^2\theta) - 2Pr_0 a\cos\theta}{\Gamma(\theta)}\rho d\tau - \frac{Qr_0 a\sin^2\theta - P(r_0^2 + a^2)\cos\theta \pm P\Gamma(\theta)}{\Gamma(\theta)}d\varphi, \quad (2)$$

$$\tilde{A}_{[1]} = -\frac{P(r_0^2 - a^2\cos^2\theta) + 2Qr_0 a\cos\theta}{\Gamma(\theta)}\rho d\tau - \frac{Pr_0 a\sin^2\theta + Q(r_0^2 + a^2)\cos\theta \mp Q\Gamma(\theta)}{\Gamma(\theta)}d\varphi, \quad (3)$$

where

$$\Gamma(\theta) = r_0^2 + a^2\cos^2\theta, \quad \gamma(\theta) = \frac{(r_0^2 + a^2)^2\sin^2\theta}{r_0^2 + a^2\cos^2\theta}, \quad b = \frac{2ar_0}{r_0^2 + a^2}, \quad r_0 = \sqrt{Q^2 + P^2 + a^2}. \quad (4)$$

The ρ is the radial coordinate of the warped AdS_3 section, B labels a deviation from the extreme limit and acts as the new horizon radius in this geometry, a is the angular parameter and Q, P are the electric and magnetic charges of the original dyonic KN black holes. The magnetic monopole induces a string-like singularity causing different choices of the gauge potential: the upper sign is regular in the northern hemisphere ($0 \le \theta < \pi/2$), and the lower sign in the southern hemisphere ($\pi/2 < \theta \le \pi$). Moreover, the dual gauge potential, $\tilde{A}_{[1]}$, is defined as ($\star$ denoting the Hodge dual)

$$\tilde{F}_{[2]} = d\tilde{A}_{[1]} \quad \text{such that} \quad \tilde{F}_{[2]} = {}^\star F_{[2]}. \quad (5)$$

The black hole's thermodynamical quantities, such as the Hawking temperature, entropy, angular velocity and chemical potentials in the new coordinates reduce to

$$T_H = \frac{B}{2\pi}, \quad S_{BH} = \pi(r_0^2 + a^2 + 2Br_0), \quad \Omega_H = -\frac{2ar_0 B}{r_0^2 + a^2},$$
$$\Phi_H = \frac{Q(Q^2 + P^2)B}{r_0^2 + a^2}, \quad \tilde{\Phi}_H = \frac{P(Q^2 + P^2)B}{r_0^2 + a^2}. \quad (6)$$

3 Particle creation

3.1 Boundary conditions

There is a potential barrier due to the electromagnetic and gravitational forces in the near horizon region, against which the pair production becomes a tunneling process by considering the fluxes in the scattering matrix theory. Two boundary conditions can be imposed (see the Fig. in [12]):

The first one is the outer boundary condition (particle viewpoint), which restricts no incoming fluxes at the asymptotic outer boundary in the new coordinates (1). In the Stückelberg-Feynman picture, the outgoing (transmitted) flux at the asymptotic region represents the spontaneous produced *particle*, the outgoing (incident) flux at the horizon represents the total particles created by vacuum fluctuations, and the incoming (reflected) flux represents the re-annihilated pairs.

The second one is the inner boundary condition (antiparticle viewpoint) that restricts no outgoing fluxes at the inner boundary, which means that the incoming (transmitted) flux at the horizon represents the spontaneous produced *antiparticle*, the incoming (incident) and the outgoing (reflected) fluxes at the asymptotic region represents the total created antiparticles and the re-annihilated pairs. Since particles and anti-particles are always produced in pairs due to the charge conservation and/or the energy-momentum conservation, these two boundary conditions are actually equivalent [3].

3.2 Physical quantities

For bosonic particles, the flux conservation [13]

$$|D_{\rm incident}| = |D_{\rm reflected}| + |D_{\rm transmitted}|, \tag{7}$$

is related to the Bogoliubov relation

$$|\mathcal{A}|^2 - |\mathcal{B}|^2 = 1, \tag{8}$$

where the vacuum persistence amplitude $|\mathcal{A}|^2$ and the mean number of produced pairs $|\mathcal{B}|^2$ are given by the ratio of the flux components in the Coulomb gauge

$$|\mathcal{A}|^2 \equiv \frac{D_{\rm incident}}{D_{\rm reflected}}, \qquad |\mathcal{B}|^2 \equiv \frac{D_{\rm transmitted}}{D_{\rm reflected}}. \tag{9}$$

Moreover, from the viewpoint of scattering of an incident flux from the asymptotic boundary, we can define the absorption cross section ratio as

$$\sigma_{\rm abs} \equiv \frac{D_{\rm transmitted}}{D_{\rm incident}} = \frac{|\mathcal{B}|^2}{|\mathcal{A}|^2}. \tag{10}$$

4 Scalar dyons production

The EoM of the probe scalar field Φ with mass m, electric and magnetic charges q, p propagating in the near horizon geometry (1) is

$$D_\alpha D^\alpha \Phi - m^2 \Phi = 0, \tag{11}$$

where $D_\alpha \equiv \nabla_\alpha - iqA_\alpha - ip\tilde{A}_\alpha$ with ∇_α being the usual covariant derivative. The corresponding radial flux is

$$D = i\sqrt{-g}g^{\rho\rho}(\Phi D_\rho \Phi^* - \Phi^* D_\rho \Phi). \tag{12}$$

Using the following ansatz

$$\Phi(\tau, \rho, \theta, \varphi) = \mathrm{e}^{-i\omega\tau + i[n \mp (qP - pQ)]\varphi} R(\rho) S(\theta), \tag{13}$$

the KG equation can be completely separated into the angular part, see [6], and the radial part

$$\partial_\rho\left[(\rho^2 - B^2)\partial_\rho R\right] + \left(\frac{\left[\omega(r_0^2 + a^2) - (qQ + pP)(Q^2 + P^2)\rho + 2nar_0\rho\right]^2}{(r_0^2 + a^2)^2(\rho^2 - B^2)} - m^2(r_0^2 + a^2) - \lambda_l\right)R = 0, \tag{14}$$

where λ_l is the separation constant. The radial equation resembles the EoM of a scalar field with an effective mass

$$m_{\rm eff}^2 = m^2 - \frac{[2nar_0 - (qQ + pP)(Q^2 + P^2)]^2}{(r_0^2 + a^2)^3} + \frac{\lambda_l}{r_0^2 + a^2}, \tag{15}$$

propagating in the AdS_2 space with the radius $L_{\rm AdS} = r_0^2 + a^2$. The instability in the AdS_2 space occurs if the Breitenlohner-Freedman (BF) bound [14, 15] is violated, i.e., $m_{\rm eff}^2 < -\frac{1}{4L_{\rm AdS}^2}$, whose relation becomes

$$m_{\rm eff}^2 < -\frac{1}{4(r_0^2 + a^2)} \quad \Rightarrow \quad \frac{[2nar_0 - (qQ + pP)(Q^2 + P^2)]^2}{(r_0^2 + a^2)^2} - m^2(r_0^2 + a^2) - \left(\lambda_l + \frac{1}{4}\right) > 0. \tag{16}$$

Now Eq. (14) can be exactly solved and leads to the following solution

$$\begin{aligned} R(\rho) &= c_1(\rho - B)^{-\frac{i}{2}(\tilde{\kappa}-\kappa)}(\rho + B)^{\frac{i}{2}(\tilde{\kappa}+\kappa)} F\left(\frac{1}{2} + i\tilde{\kappa} + i\mu, \frac{1}{2} + i\tilde{\kappa} - i\mu; 1 + i\tilde{\kappa} - i\kappa; \frac{1}{2} - \frac{\rho}{2B}\right) \\ &+ c_2(\rho - B)^{-\frac{i}{2}(\tilde{\kappa}-\kappa)}(\rho + B)^{\frac{i}{2}(\tilde{\kappa}+\kappa)} F\left(\frac{1}{2} + i\kappa + i\mu, \frac{1}{2} + i\kappa - i\mu; 1 + i\kappa - i\tilde{\kappa}; \frac{1}{2} - \frac{\rho}{2B}\right), \end{aligned} \tag{17}$$

with three essential parameters

$$\tilde{\kappa} = \frac{\omega}{B}, \qquad \kappa = \frac{(qQ + pP)(Q^2 + P^2) - 2nar_0}{r_0^2 + a^2}, \qquad \mu = \sqrt{\kappa^2 - m^2(r_0^2 + a^2) - \lambda_l - \frac{1}{4}}, \tag{18}$$

in which μ^2 is positive due to the BF bound violation in Eq. (16). Then the Bogoliubov coefficients and the absorption cross section ratio in the particle viewpoint can be obtained as [6]

$$|\mathcal{A}|^2 = \frac{\cosh(\pi\kappa - \pi\mu)\cosh(\pi\tilde{\kappa} + \pi\mu)}{\cosh(\pi\kappa + \pi\mu)\cosh(\pi\tilde{\kappa} - \pi\mu)}, \tag{19}$$

$$|\mathcal{B}|^2 = \frac{\sinh(2\pi\mu)\sinh(\pi\tilde{\kappa} - \pi\kappa)}{\cosh(\pi\kappa + \pi\mu)\cosh(\pi\tilde{\kappa} - \pi\mu)}, \tag{20}$$

$$\sigma_{\rm abs} = \frac{|\mathcal{B}|^2}{|\mathcal{A}|^2} = \frac{\sinh(2\pi\mu)\sinh(\pi\tilde{\kappa} - \pi\kappa)}{\cosh(\pi\kappa - \pi\mu)\cosh(\pi\tilde{\kappa} + \pi\mu)}. \tag{21}$$

5 Thermal Interpretation

Following our previous studies [3, 5], the mean number of produced pairs (20) can be reexpressed as

$$\mathcal{N} = |\mathcal{B}|^2 = \left(\frac{e^{-2\pi\kappa+2\pi\mu} - e^{-2\pi\kappa-2\pi\mu}}{1 + e^{-2\pi\kappa-2\pi\mu}}\right)\left(\frac{1 - e^{-2\pi\tilde{\kappa}+2\pi\kappa}}{1 + e^{-2\pi\tilde{\kappa}+2\pi\mu}}\right). \tag{22}$$

Note that the mean number (22) has a similar form except for different quantum numbers as that of charged scalars in a near-extremal RN [3] and KN [5] since the near-horizon geometry is an $AdS_2 \times S^2$ for the near-extremal RN black hole while it is a warped AdS_3 for the near-extremal KN black hole. Following Refs. [7, 8], we introduce an effective temperature and its associated counterpart

$$T_{\rm KN} = \frac{\bar{m}}{2\pi\kappa - 2\pi\mu} = T_U + \sqrt{T_U^2 + \frac{\mathcal{R}}{8\pi^2}}, \qquad \bar{T}_{\rm KN} = \frac{\bar{m}}{2\pi\kappa + 2\pi\mu} = T_U - \sqrt{T_U^2 + \frac{\mathcal{R}}{8\pi^2}}, \tag{23}$$

where the effective mass $\bar{m}$ is

$$\bar{m} = \sqrt{m^2 - \frac{\lambda + 1/4}{2}\mathcal{R}}, \tag{24}$$

and the corresponding Unruh temperature T_U and AdS curvature $\mathcal{R}$ are

$$T_U = \frac{\kappa}{2\pi\bar{m}(r_0^2 + a^2)} = \frac{(qQ + pP)(Q^2 + P^2) - 2nar_0}{2\pi\bar{m}(r_0^2 + a^2)^2}, \qquad \mathcal{R} = -\frac{2}{r_0^2 + a^2}. \tag{25}$$

Therefore, we have

$$2\pi\kappa - 2\pi\mu = \frac{\bar{m}}{T_{\rm KN}}, \qquad 2\pi\kappa + 2\pi\mu = \frac{\bar{m}}{\bar{T}_{\rm KN}}, \qquad 2\pi\tilde{\kappa} = \frac{\omega}{T_{\rm H}}, \qquad 2\pi\kappa = \frac{q\Phi_{\rm H} + p\bar{\Phi}_{\rm H} + n\Omega_H}{T_{\rm H}}, \tag{26}$$

and the mean number (22) can be expressed as

$$\mathcal{N} = e^{\frac{\bar{m}}{T_{\rm KN}}} \times \left(\frac{e^{-\frac{\bar{m}}{T_{\rm KN}}} - e^{-\frac{\tilde{m}}{T_{\rm KN}}}}{1 + e^{-\frac{\tilde{m}}{T_{\rm KN}}}} \right) \times \left\{ \frac{e^{-\frac{\bar{m}}{T_{\rm KN}}} \left(1 - e^{-\frac{\omega - q\Phi_{\rm H} - p\bar{\Phi}_{\rm H} - n\Omega_{\rm H}}{T_{\rm H}}} \right)}{1 + e^{-\frac{\omega - q\Phi_{\rm H} - p\bar{\Phi}_{\rm H} - n\Omega_{\rm H}}{T_{\rm H}}} e^{-\frac{\bar{m}}{T_{\rm KN}}}} \right\}. \tag{27}$$

Now the physical meaning of each term in Eq. (27) becomes clear; we interpret the first parenthesis as the Schwinger effect with the effective temperature $T_{\rm KN}$ in AdS_2 [16] and the second parenthesis as the Schwinger effect in the Rindler space [17], in which the Unruh temperature is given by the Hawking temperature and the charges have the chemical potentials of $\Phi_{\rm H}, \bar{\Phi}_{\rm H}$, and $\Omega_{\rm H}$, while the effective temperature for the Schwinger effect due to the electric field on the horizon is still determined by $T_{\rm KN}$.

The mean number of produced pairs above and the absorption cross section ratio in the previous section have been obtained using the exact solution in the near horizon geometry of an extremal or near extremal KN black hole. In addition, by applying the phase-integral formula, we can derive the instanton actions from the Hamilton-Jacobi action for the field equation, which lead to the mean number, for details see [6]. This method allows one to understand the physical origin of each term as a consequence of simple poles in the complex plane of space [18] and further connects the interpretation to other physical systems involving the Schwinger effect in curved spacetimes [19].

6 Holographic CFT description

According to the KN/CFTs duality [10, 11], the absorption cross section ratio of scalar field in Eq. (21) corresponds to that of its dual operator in the dual two-dimensional CFT with left- and right-hand sectors

$$\sigma_{\rm abs} \sim T_{\rm L}^{2h_{\rm L}-1} T_{\rm R}^{2h_{\rm R}-1} \sinh\left(\frac{\tilde{\omega}_{\rm L}}{2T_{\rm L}} + \frac{\tilde{\omega}_{\rm R}}{2T_{\rm R}}\right) \left|\Gamma\left(h_{\rm L} + i\frac{\tilde{\omega}_{\rm L}}{2\pi T_{\rm L}}\right)\right|^2 \left|\Gamma\left(h_{\rm R} + i\frac{\tilde{\omega}_{\rm R}}{2\pi T_{\rm R}}\right)\right|^2, \tag{28}$$

where $T_{\rm L}, T_{\rm R}$ are the temperatures, $h_{\rm L}, h_{\rm R}$ are the conformal dimensions of the dual operator, $\tilde{\omega}_{\rm L} = \omega_{\rm L} - q_{\rm L}\Phi_{\rm L}$ and $\tilde{\omega}_{\rm R} = \omega_{\rm R} - q_{\rm R}\Phi_{\rm R}$ are the total excited energy in which $(q_{\rm L}, q_{\rm R})$ and $(\Phi_{\rm L}, \Phi_{\rm R})$ are respectively the charges and chemical potentials (both including the electric and the magnetic contributions for the dyonic KN black hole case) of the dual left and right-hand operators.

From the field/operator duality in the AdS/CFT correspondence, the complex conformal dimensions $(h_{\rm L}, h_{\rm R})$ of the dual operator in the two-dimensional CFT can be determined from the asymptotic expansion of the bulk dyonic charged scalar field (17) at the AdS boundary [6]

$$h_{\rm L} = h_{\rm R} = \frac{1}{2} \pm i\mu. \tag{29}$$

Next, we will analyze the dual CFTs descriptions of the pair production rate and the absorption cross section ratio in the near extremal dyonic KN black hole in the J- and Q- as well as the P-pictures, respectively.

6.1 J-picture

In the J-picture, the left- and right-hand central charges of the dual CFT are determined by the angular momentum [10, 11]

$$c_{\rm L}^J = c_{\rm R}^J = 12J, \tag{30}$$

and the associated left- and right-hand temperatures for the near extremal dyonic KN black hole are

$$T_{\rm L}^{J} = \frac{r_0^2 + a^2}{4\pi a r_0}, \qquad T_{\rm R}^{J} = \frac{B}{2\pi a}. \tag{31}$$

The CFT microscopic entropy is calculated from the Cardy formula

$$S_{\rm CFT} = \frac{\pi^2}{3}(c_{\rm L}^J T_{\rm L}^J + c_{\rm R}^J T_{\rm R}^J) = \pi(r_0^2 + a^2 + 2r_0 B), \tag{32}$$

which agrees with the macroscopic entropy (6) of the near extremal KN black hole.

Besides, by matching the first law of black hole thermodynamics with that of the dual CFT, i.e., $\delta S_{\rm BH} = \delta S_{\rm CFT}$, the following relation holds

$$\frac{\delta M - \Omega_{\rm H}\delta J - \Phi_{\rm H}\delta Q - \bar{\Phi}_{\rm H}\delta P}{T_{\rm H}} = \frac{\tilde{\omega}_{\rm L}}{T_{\rm L}} + \frac{\tilde{\omega}_{\rm R}}{T_{\rm R}}, \tag{33}$$

where the angular velocity and chemical potentials at $\rho = B$ are given in Eq. (6). To probe the rotation we need to turn off the charges of the probe scalar field and set $T_{\rm L} = T_{\rm L}^J$ and $T_{\rm R} = T_{\rm R}^J$, then for the dyonic KN black hole $\delta M = \omega,\ \delta J = n,\ \delta Q = 0,\ \delta P = 0$. Thus, we have

$$\tilde{\omega}_{\rm L}^J = n \quad \text{and} \quad \tilde{\omega}_{\rm R}^J = \frac{\omega}{a} \quad \Rightarrow \quad \frac{\tilde{\omega}_{\rm L}^J}{2T_{\rm L}^J} = -\pi\kappa \quad \text{and} \quad \frac{\tilde{\omega}_{\rm R}^J}{2T_{\rm R}^J} = \pi\tilde{\kappa}, \tag{34}$$

where q, p are set to zero. Consequently, the agreement between the absorption cross section ratio (21) of the scalar field (with $q = p = 0$) in the near extremal dyonic KN black hole and that of its dual scalar operator in Eq. (28) is found in the J-picture.

6.2 Q-picture

In the Q-picture, the left- and right-hand central charges of the dual CFT are [10, 11]

$$c_{\rm L}^Q = c_{\rm R}^Q = \frac{6Q(Q^2 + P^2)}{\ell}, \tag{35}$$

where the parameter ℓ is the measure of the $U(1)$ bundle formed by the background Maxwell field, which can be interpreted as the radius of the embedded extra circle in the fifth dimension. The associated temperatures of the left- and right-hand sectors for near horizon geometry of the near extremal dyonic KN black hole are

$$T_{\rm L}^Q = \frac{(r_0^2 + a^2)\ell}{2\pi Q(Q^2 + P^2)}, \qquad T_{\rm R}^Q = \frac{r_0 B\ell}{\pi Q(Q^2 + P^2)}. \tag{36}$$

Then, the microscopic entropy of the dual CFT obtained from the Cardy formula

$$S_{\rm CFT} = \frac{\pi^2}{3}(c_{\rm L}^Q T_{\rm L}^Q + c_{\rm R}^Q T_{\rm R}^Q) = \pi(r_0^2 + a^2 + 2r_0 B), \tag{37}$$

again reproduces the area law for the entropy of the near extremal dyonic KN black hole.

In the Q-picture, the modes n and p, which characterize the rotation and magnetic charge of the probe scalar field, should be turned off in order that the electrically charged probe only detects the electric charge of the black hole, i.e., $\delta M = \omega,\ \delta J = 0,\ \delta Q = q,\ \delta P = 0$. Then, using Eq. (33), with $T_{\rm L} = T^Q_{\rm L}$ and $T_{\rm R} = T^Q_{\rm R}$, we have

$$\tilde{\omega}^Q_{\rm L} = -q\ell \quad \text{and} \quad \tilde{\omega}^Q_{\rm R} = \frac{2\omega r_0 \ell}{Q(Q^2+P^2)}, \tag{38}$$

and $\frac{\tilde{\omega}^Q_{\rm L}}{2T^Q_{\rm L}} = -\pi\kappa$ and $\frac{\tilde{\omega}^Q_{\rm R}}{2T^Q_{\rm R}} = \pi\tilde{\kappa}$ for the dual CFT. Therefore, the absorption cross section ratio of the charged scalar field (with $n = p = 0$) in Eq. (21) also agrees with that of its dual scalar operator in Eq. (28) in the Q-picture.

6.3 P-picture

The left- and right-hand central charges of the P-picture CFT can be obtained from the electromagnetic duality, which are

$$c^P_{\rm L} = c^P_{\rm R} = \frac{6P(Q^2+P^2)}{\ell}, \tag{39}$$

and the associated left- and right-hand temperatures are

$$T^P_{\rm L} = \frac{(r_0^2+a^2)\ell}{2\pi P(Q^2+P^2)}, \qquad T^P_{\rm R} = \frac{r_0 B\ell}{\pi P(Q^2+P^2)}. \tag{40}$$

Then, the microscopic and macroscopic entropies are the same:

$$S_{\rm CFT} = \frac{\pi^2}{3}(c^P_{\rm L}T^P_{\rm L} + c^P_{\rm R}T^P_{\rm R}) = \pi(r_0^2 + a^2 + 2r_0 B). \tag{41}$$

Along the same line as the Q-picture, we need to turn off the modes of n and q in the P-picture, i.e. $\delta M = \omega,\ \delta J = 0,\ \delta Q = 0,\ \delta P = p$. Then, applying Eq. (33), with $T_{\rm L} = T^P_{\rm L}$ and $T_{\rm R} = T^P_{\rm R}$, we obtain

$$\tilde{\omega}^P_{\rm L} = -p\ell \quad \text{and} \quad \tilde{\omega}^P_{\rm R} = \frac{2\omega r_0 \ell}{P(Q^2+P^2)}, \tag{42}$$

and $\frac{\tilde{\omega}^P_{\rm L}}{2T^P_{\rm L}} = -\pi\kappa$ and $\frac{\tilde{\omega}^P_{\rm R}}{2T^P_{\rm R}} = \pi\tilde{\kappa}$. Therefore, in the P-picture of the KN/CFTs duality, we find the agreement between the absorption cross section ratio of the charged scalar field (with $n = q = 0$) in Eq. (21) and that of its dual scalar operator in Eq. (28).

7 Conclusions

We study the scalar dyon production for the near extremal dyonic KN black hole without back-reaction. The KG equation is eactly solved in the near horizon region where the geometry includes a warped AdS_3 structure. The near horizon region contains a causal horizon and a dominating electric field which capture both contributions: the Hawking radiation and the Schwinger mechanism. The exact solutions were obtained in terms of the hypergeometric functions. By imposing the particle viewpoint of boundary condition, the physical quantities associated to the pair production can be derived by the ratio of boundary fluxes. In particular, the expressions of the vacuum persistence amplitude, the mean number of pairs and the absorption section ratio were obtained. The existence condition for the pair production corresponds to the instability of the probe fields in the AdS_2, i.e. the violation of the BF bound.

The thermal interpretation of the pair production rate for the dyonic KN black holes includes two contributions: one is the Schwinger effect in the AdS_2 space caused mainly by the electromagnetic field of black holes and the other is the Schwinger effect in the Rindler space due to the Hawking temperature, i.e. the deviation from the extremality. The leading term of the production rate is determined by the effective temperature T_{KN}, which is a kind of Unruh temperature originated from the acceleration due to the electromagnetic force and the AdS_2 curvature. There is duality between the electric charges and magnetic charges of the black hole and produced pairs. The effect of the AdS_2 curvature is to bind pairs and thereby decrease the pair production while the electromagnetic field increases the Unruh temperature for the effective temperature and thus enhances the pair production. The angular momentum, electric and magnetic charges of the produced dyonic particles are expected to couple with the corresponding thermal quantities, i.e. the angular velocity, electric chemical potential, and magnetic chemical potential, of the dyonic KN black hole as the subleading terms to the leading Boltzmann factor.

The twofold dual CFTs descriptions (i.e., J-picture and Q-picture) of the Schwinger pair production of the near extremal KN black hole [5] are generalized into the threefold dual CFTs pictures for the dyonic KN black hole which includes an additional magnetic charge. The presence of the P-picture is associated with the dual gauge potential which can be regarded as a new "magnetic hair" of the dyonic KN black hole and provides the $U(1)$ fiber on the base manifold. It can only be probed by the magnetic charge of the scalar field. Furthermore, based on the threefold dyonic KN/CFTs duality, the dual CFTs descriptions of the absorption cross section ratios and the pair production rate of the dyonic charged scalar field are found in the J-, Q-, and P-pictures, respectively.

Acknowledgments

The work of C. M. C. was were supported by the Ministry of Science and Technology of the R.O.C. under the grants MOST 105-2112-M-008-018 and 106-2112-M-008-010. The work of S. P. K. was supported by the Basic Science Research Program through the National Research Foundation of Korea (NRF) funded by the Ministry of Education (NRF-2015R1D1A1A01060626). The work of J. R. S. was supported by the NSFC under Grant No. 11675272 and the Open Project Program of State Key Laboratory of Theoretical Physics, Institute of Theoretical Physics, Chinese Academy of Sciences, China (No. Y5KF161CJ1) and the Fundamental Research Funds for the Central Universities.

References

[1] J. S. Schwinger, "On Gauge Invariance and Vacuum Polarization," Phys. Rev. **82**, 664 (1951).

[2] M. K. Parikh and F. Wilczek, "Hawking Radiation as Tunneling," Phys. Rev. Lett. **85**, 5042 (2000).

[3] C.-M. Chen, S. P. Kim, I.-C. Lin, J.-R. Sun and M.-F. Wu, "Spontaneous Pair Production in Reissner-Nordstrom Black Holes," Phys. Rev. D **85**, 124041 (2012) [arXiv:1202.3224 [hep-th]].

[4] C.-M. Chen, J.-R. Sun, F.-Y. Tang and P.-Y. Tsai, "Spinor Particle Creation in Near Extremal Reissner-Nordstrom Black Holes," Class. Quant. Grav. **32**, 195003 (2015) [arXiv:1412.6876 [hep-th]].

[5] C.-M. Chen, S. P. Kim, J.-R. Sun and F.-Y. Tang, "Pair Production in Near Extremal Kerr-Newman Black Holes," Phys. Rev. D **95**, 044043 (2017) [arXiv:1607.02610 [hep-th]].

[6] C.-M. Chen, S. P. Kim, J.-R. Sun and F.-Y. Tang, "Pair Production of Scalar Dyons in Kerr-Newman Black Holes," arXiv:1705.10629 [hep-th].

[7] S. P. Kim, H. K. Lee and Y. Yoon, "Thermal Interpretation of Schwinger Effect in Near-Extremal RN Black Hole," arXiv:1503.00218 [hep-th].
[8] S. P. Kim, "Schwinger Effect, Hawking Radiation and Gauge-Gravity Relation," Int. J. Mod. Phys. A **30**, 1545017 (2015) [arXiv:1506.03990 [hep-th]].
[9] S. P. Kim, "Schwinger Effect in (A)dS and Charged Black Hole," arXiv:1509.05532 [hep-th].
[10] C.-M. Chen, Y.-M. Huang, J.-R. Sun, M.-F. Wu and S.-J. Zou, "Twofold Hidden Conformal Symmetries of the Kerr-Newman Black Hole," Phys. Rev. D **82**, 066004 (2010) [arXiv:1006.4097 [hep-th]].
[11] C.-M. Chen and J.-R. Sun, "The Kerr-Newman/CFTs Correspondence," Int. J. Mod. Phys. Conf. Ser. **07**, 227 (2012) [arXiv:1201.4040 [hep-th]].
[12] C.-M. Chen, J.-R. Sun and F.-Y. Tang, "Pair Production In Near Extremal Charged Black Holes," arXiv:1510.00141 [hep-th].
[13] S. P. Kim and D. N. Page, "Schwinger Pair Production in Electric and Magnetic Fields," Phys. Rev. D **73**, 065020 (2006) [hep-th/0301132].
[14] P. Breitenlohner and D. Z. Freedman, "Stability in Gauged Extended Supergravity," Annals Phys. **144**, 249 (1982).
[15] P. Breitenlohner and D. Z. Freedman, "Positive Energy in anti-De Sitter Backgrounds and Gauged Extended Supergravity," Phys. Lett. B **115**, 197 (1982).
[16] R. G. Cai and S. P. Kim, "One-Loop Effective Action and Schwinger Effect in (Anti-) de Sitter Space," JHEP **1409**, 072 (2014) [arXiv:1407.4569 [hep-th]].
[17] Cl. Gabriel and Ph. Spindel, "Quantum charged fields in (1+1) Rindler space," Annals Phys. **284**, 263 (2000) [gr-qc/9912016].
[18] S. P. Kim and D. N. Page, "Improved Approximations for Fermion Pair Production in Inhomogeneous Electric Fields," Phys. Rev. D **75**, 045013 (2007) [hep-th/0701047].
[19] S. P. Kim, "Geometric Origin of Stokes Phenomenon for de Sitter Radiation," Phys. Rev. D **88**, 044027 (2013) [arXiv:1307.0590 [hep-th]].

EPJ Web of Conferences **168**, 01012 (2018) https://doi.org/10.1051/epjconf/201816801012
Joint International Conference of ICGAC-XIII and IK-15 on Gravitation, Astrophysics and Cosmology

Experiments on the CMB Spectrum, Big Jets Model and Their Implications for the Missing Half of the Universe

Leonardo Hsu[1,*] and *Jong-Ping* Hsu[2,**]

[1] *College of Education and Human Development, University of Minnesota, Minneapolis, MN 55455, USA*
[2] *Department of Physics, University of Massachusetts Dartmouth, North Dartmouth, MA 02747, USA*

Abstract. Based on the limiting continuation of Lorentz-Poincaré invariance, we propose an alternative formulation of the generalized Planck distribution for inertial and non-inertial frames. The Lorentz invariant Planck distribution law leads to a new physical interpretation of the dipole anisotropy of the Cosmic Microwave Background. The Big Jets model predicts a distant 'antimatter blackbody,' whose radiations could make 50% of the sky very slightly warmer than the isotropic CMB temperature T_{CMB} with a cosine function. The other 50% of the sky has the same isotropic temperature T_{CMB}. Thus, we could have a pseudo-dipole anisotropy because the microwaves emitted from the anti-matter blackbody are totally absorbed by our matter blackbody. We suggest that accurate data of satellite experiments might be used to search for the pseudo-dipole anisotropy and the missing half of the antimatter universe.

1 Introduction

We discuss the concept of Lorentz-Poincaré invariance and its limiting continuation, which could be considered as a minimum generalization of the principle of relativity from inertial to non-inertial frames[1, 2]. Such a limiting Lorentz-Poincaré invariance enables us to obtain physical laws for a 'general frame' (inertial and non-inertial). As a result, we propose an alternative formulation of the generalized invariant Planck distribution that leads to a new physical interpretation of the dipole anisotropy of the Cosmic Microwave Background (CMB)[3, 4]. We stress that satellite measurements of the CMB might provide the data necessary for testing different formulations of the Planck distribution law, which is associated with different transformations of the temperature. Such a test is significant because the alternative invariant Planck distribution implies that, if there is a temperature difference (or asymmetry) between two opposite directions in sky with the pattern, $T(\theta) = T_{CMB} + T_e cos\theta$, for $cos\theta > 0$, and $T(\theta) = T_{CMB}$, for $cos\theta < 0$.[1] Big Jets model suggests that a pseudo-dipole anisotropy could be caused by the presence of an extremely distant antimatter-blackbody.[5]

We note that satellite data from the COBE, WMAP, and Planck satellite experiments all show a six-month variation corresponding to the Earth's orbital motion. However, this effect is normally considered an undesirable source of noise in the signal so that it is typically removed from the data without

*e-mail: lhsu@umn.edu
**e-mail: jhsu@umassd.edu

[1] The pattern $T_e cos\theta$ and zero for two opposite hemispheres shows a pseudo-dipole anisotropy and is the characteristics of the total absorption of microwaves emitted from the antimatter blackbody by our matter-blackbody.

much comment. It appears that there is no published data on the magnitude of this variation.[3] The magnitude of this variation is certainly detectable by satellite experiments. It can also be used to test the law of blackbody radiation in a rotational frame and the alternative 'general-frame' (GF) covariant blackbody distribution.[6] However, one must be more careful. As indicated by the analysis of the Wilson experiment[7], it is not correct to use the Lorentz transformations to analyze situations involving non-inertial circular motion, even locally (irrespective of the fact that it sometimes gives the correct answer, as in the case of lifetime dilations of unstable particles in a circular storage ring). Instead, we must first correctly generalize the blackbody distribution law to a rotational frame.

2 Lorentz invariant blackbody distribution law

To accomplish this, we first generalized the original Planck distribution law $P(\omega, T) = 1/[exp(\omega/k_B T) - 1]$ for the rest frame of the blackbody to the Lorentz invariant blackbody distribution law for all inertial frames. We propose that the quantity $\omega = k_0$ should be replaced by the product of the four-velocity of the thermo-system $U^\mu = dx^\mu/ds$ with the wave four-vector $k^\mu = (k_0, k_x, k_y, k_z)$. In the rest frame of a thermo-system, its four-velocity is $U^\mu = (1, 0, 0, 0)$, so $k_\mu U^\mu = k_0 = \omega$, where $c = \hbar = 1$. Thus, we insure that this new formulation remains consistent with all experimental results, as well as remains consistent with the framework of special relativity. Because Planck's law was devised before the advent of special relativity, one does not necessarily expect its original formulation to be relativistically consistent, and such a replacement is natural. We have also assumed that the quantity $k_B T$ is Lorentz-invariant. Although the universality of the Boltzmann constant k_B has never been established in a relativistic framework, such an assumption is consistent with many previous discussions of the transformation properties of heat and temperature in the literature[8–13].

Thus, in this alternative formulation, the generalized Planck distribution in an inertial frame F is

$$B(k_\mu U^\mu, T) = \frac{1}{[exp(k_\mu U^\mu)/(k_B T) - 1]}, \quad k_\mu U^\mu = \eta_{\mu\nu} k^\mu U^\nu, \quad T = T_{inv}, \tag{1}$$

$$U^\mu = \frac{dx^\mu}{ds}, \quad \eta_{\mu\nu} U^\mu U^\nu = 1, \quad \eta_{\mu\nu} = (1, -1, -1, -1), \quad c = \hbar = 1,$$

where T in $B(k_\mu U^\mu, T)$ is understood as a scalar, and U^μ is the 4-velocity of the thermo-system as measured in F. We have used $B(k_\mu U^\mu, T)$ instead of $P(\omega, T)$ to distinguish this version of the Planck distribution law from its conventional formulation[3, 4].

To see the transformation properties of $B(k_\mu U^\mu, T)$ in (1), consider a blackbody at rest and located at the origin of F. In another inertial frame F' moving with a relative velocity V along the $+x$ axis relative to F, the 4-velocity U'^μ of the blackbody is,

$$U'^\mu = \left(\frac{1}{\sqrt{1 - V^2}}, \frac{-V}{\sqrt{1 - V^2}}, 0, 0 \right), \tag{2}$$

so that

$$k'_\mu U'^\mu = k'_0 U'_0 - k'_x U'_x = k'_0 \frac{[1 + V cos\theta']}{\sqrt{1 - V^2}} = k_\mu U^\mu = k_0, \tag{3}$$

where $cos\theta' = k'_x/k'_0$. The relation (3) between k_0 and k'_0 is the well-known relativistic Doppler shift. Writing the invariant Planck law (1) in terms of the non-invariant radiation energy $\omega' = k'_0$ and the 4-velocity (2) in F', we have

$$B'(k'_\mu U'^\mu, T) = \frac{1}{exp[\omega'(1 + V cos\theta')/(k_B T \sqrt{1 - V^2})] - 1}. \tag{4}$$

In contrast, in the conventional formulation[3, 4], one accepts the original form of the Planck distribution, $P(\omega, T) = 1/[exp(\omega/k_B T) - 1]$, for a blackbody at rest in an inertial frame $F(t, x, y, z)$, which involves the quantity $\omega/k_B T = k_0/k_B T$. One reasons that since the zeroth component of the wave 4-vector $k^\mu = (\omega, k_x, k_y, k_z)$ transforms as $\omega = \omega'[1 + V cos\theta']/\sqrt{1 - V^2}$, $cos\theta' = k'_x/k'_0$ if the temperature transforms as

$$T = T'(\theta')(1 + V cos\theta')/\sqrt{1 - V^2}, \tag{5}$$

then one has $\omega/k_B T = \omega'/k_B T'$. Therefore, the Planck's distribution $P(\omega, T) = 1/(exp(\omega/k_B T) - 1)$ has the same form in both F and F' frames: $P(\omega, T) = P'(\omega', T')$.

However, in this case, the Planck distribution law $P(\omega, T)$ is incompatible with the principle of relativity in that the laws of physics now do NOT have the same form in all inertial frames. This can be seen explicitly if one considers three frames F, F' and F'', the transformations of $P(\omega, T)$ and the ratio $\omega/k_B T$ in them. One can see that there is a preferred frame in which the Planck law $P(\omega, T)$ takes the simplest form, if the angle-dependent temperature (5) alone is used.

Furthermore, this conventional formulation with the angle dependent temperature (5) has two subtle conceptual difficulty. First, in special relativity, since ω is the zeroth component of a 4-vector, the ratio $\omega/k_B T = \omega'/k_B T'$ is invariant only if $k_b T$ is also the zeroth component of a 4-vector parallel to the k^μ 4-vector, and what this 4-vector that involves the temperature might be is never fully specified in the conventional formulation. Instead, one simply introduces[3] equation (5) for the transformation of temperature rather than a temperature 4-vector, begging the question, just what kind of quantity is temperature? Since it is not invariant, it cannot be a scalar. Second, none of the Lorentz transformations of the temperature of a thermo-system, including those developed by Einstein[8], Planck[9], Ott[10], Arzelies[11], Landsberg[12], van Kampen[13] and others, have ever considered an angle-dependent transformation of temperature. As such, the angle-dependent temperature transformation in (5) is simply an ad hoc assumption and neither complete nor completely satisfactory.

Comparing the result (4) with the conventional formulation,

$$P(\omega, T) = P'(\omega', T') = 1/[exp(\omega'(1 + V cos\theta')/(k_B T \sqrt{1 - V^2})) - 1],$$

with the angle-dependent temperature (5), one can see that the two are quantitatively identical. However, they have an important conceptual and physical difference.

First, there was no need to construct a new temperature four-vector with an unusual angle-dependent temperature transformation.

Second, because $k_\mu U^\mu$ and $k_B T$ are now separately Lorentz invariant and $k_B T$ is a Lorentz scalar, if the invariant blackbody distribution $B(k_\mu U^\mu, T)$ and the CMB temperature are isotropic in one frame F, then they will both be isotropic in every other inertial frame as well.

3 Generalized Planck distribution for non-inertial frames

Actually, we can go further towards the non-inertial frame. In the literature, there exists a fundamental space-time symmetry framework called limiting four-dimensional symmetry (or limiting Lorentz-Poincaré invariance) that can help us generalize physical laws in inertial frames to non-inertial frames of reference.[1, 2] In particular, one can use this framework to derive a set of coordinate transformations between an inertial frame F' and a rotating frame F_r that (a) are exact, (b) simplify to the Lorentz transformations in the appropriate limit (i.e., as the orbital radius goes to infinity while the product of the orbital radius and angular velocity remains finite[15]), (c) are consistent with the results of experiments such as the Davies-Jennison experiment, Thim's experiment involving radio sources in circular motion[2, 15], and high energy experiments involving unstable particles in a circular storage ring, and (d) support the analysis of the Wilson experiment.[7] These transformations can also give us the metric tensors $P_{\mu\nu}(x)$ necessary for generalizing the blackbody distribution to rotating and constant-linear-acceleration frames.

To investigate the effect of the Earth's orbital motion on the blackbody distribution (1), for simplicity, let us ignore the relative linear motion between the solar system and the (possible) 'CMB rest frame' and approximate the Earth's orbital motion as a circular orbit. Consider an inertial frame $F'(t', x', y', z')$ and a rotational frame $F_r(t_r, x_r, y_r, z_r)$, where the origin of F_r orbits the origin of F' at a constant distance R with a constant angular velocity $\Omega \equiv \Omega_r$ (as measured in the rotating frame). A Cartesian coordinate system is used in both frames, set up in such a way that the positive portion of the y_r-axis of the frame F_r always extends through the origin of F'. Let the Earth be at rest relative to the rotating frame and located at $x_r = \rho,\ y_r = 0,\ z_r = 0$. The exact rotational coordinate transformations are then [15]

$$t' = \gamma(t_r + \boldsymbol{\rho} \cdot \boldsymbol{\beta}), \quad x' = \gamma[x_r cos(\Omega t_r) - (y_r - R)sin(\Omega t_r)],$$

$$y' = \gamma[x_r sin(\Omega t_r) + (y_r - R)cos(\Omega t_r)], \quad z' = z_r, \tag{6}$$

$$\beta = |\boldsymbol{\Omega} \times \mathbf{S}| = \Omega\sqrt{x_r^2 + (y_r - R)^2} = \Omega S < 1, \quad \boldsymbol{\rho} \cdot \boldsymbol{\beta} = x_r R\Omega, \quad \gamma = \frac{1}{\sqrt{1-\beta^2}}, \tag{7}$$

$$\boldsymbol{\rho} = (x_r, y_r, 0), \quad \mathbf{S} = (x_r, y_r - R, 0), \quad \boldsymbol{\beta} = \boldsymbol{\Omega} \times \mathbf{S}, \quad \boldsymbol{\Omega} = (0, 0, \Omega).$$

The times t' and t_r are measured in units of length.[2] The rotational transformation (6) form a pseudo-group (in the sense of Whitehead and Veblen)[17]. In the limit $R \to \infty$ and $\Omega \to 0$ such that the product $R\Omega = \beta_o$ is a finite non-zero constant velocity, the transformations (6) with (7) reduce to the exact Lorentz transformations.

For simplicity of discussions below, it suffices to set $R = 0$ in (6). To the first order in $\rho\Omega$, the non-vanishing components of the covariant metric tensors $P_{r\mu\nu}(x)$ of the rotating frame F_r are then [14],

$$P_{r00} = -P_{r11} = -P_{r22} = -P_{r33} = 1, \quad P_{r01} = \Omega y_r, \quad P_{r02} = -\Omega x_r. \tag{8}$$

In this approximation, the contravariant metric tensors $P_r^{\mu\nu}$ are the same, e.g., $P_r^{01} = \Omega y_r,\ P_r^{02} = -\Omega x_r$, etc.

The covariant wave vectors for the blackbody radiation in the inertial and rotating frames are denoted by $k'_\mu = (k'_o, k'_1, k'_2, k'_3)$ and $k_{r\mu} = (k_{r0}, k_{r1}, k_{r2}, k_{r3})$, respectively. They are related by the rotational wave-vector transformations,[15]

$$k'_0 \approx k_{r0} - \Omega\rho k_{r2}, \quad k'_1 \approx k_{r1} cos(\Omega t_r) - k_{r2} sin(\Omega t_r), \tag{9}$$

$$k'_2 \approx k_{r1} sin(\Omega t_r) + k_{r2} cos(\Omega t_r), \quad k'_3 = k_{r3}.$$

For purposes of making predictions in F' and F_r frames, we shall concentrate on a simple case, where $k'_2 = k'_3 = 0$. The quantity $k_{r\mu}$ denotes the corresponding wave vector as measured in F_r at the location of the detector, $(x_r, y_r, x_r) = (\rho, 0, 0)$. In this case, (9) and $k'_0 = k'_1 > 0$ lead to the relations

$$\frac{k_{r2}}{k_{r0}} \approx \frac{-sin(\Omega t_r)}{1 - \Omega\rho sin(\Omega t_r)}, \quad k_{r0} \approx \frac{k'_0}{1 + \Omega\rho\ sin(\Omega t_r)}. \tag{10}$$

While k'_0 and k'_1 are time-independent, k_{r0}, k_{r1} and k_{r2} measured in F_r are time-dependent.

[2]The operational meaning of space-time coordinates in $F'(t', x', y', z')$ is well-known. The operational meaning of coordinates in a non-inertial frame such as $F_r(t_r, x_r, y_r, z_r)$ can be realized by a grid of (computerized) 'space-time clocks'[16] located in F_r and satisfies (6).

Now consider a thermo-system at rest in F' with $U'^{\mu} = (1, 0, 0, 0)$. To first order in $\rho\Omega$, its velocity in F_r is $U_r^{\mu} = dx_r^{\mu}/ds \approx (1, \Omega y_r, -\Omega x_r, 0)$. Thus, we have the relationship,

$$P_r^{\mu\nu} k_{r\mu} U_{r\nu} \approx k_{r0}[1 + \rho\Omega \; sin(\Omega t_r)], \tag{11}$$

for the rotational frame F_r, where $\rho\Omega$ is the Earth's orbital velocity, and we have used $U_r^{\mu} = P_r^{\mu\nu} U_{r\nu}$ and (10).

In general, one can use the metric tensor $P^{\mu\nu}$ in place of the Minkowski metric tensor $\eta^{\mu\nu}$ in order to generalize the invariant blackbody distribution (1) from an inertial frame to a non-inertial frame, such as a constant-linear-acceleration (CLA) frame F_{CLA} or a constant rotational frame F_r.

(A) Constant-linear-acceleration frame F_{CLA}:

$$B_{CLA}(k_{\lambda} U^{\lambda}, T) = \frac{1}{[exp(P^{\mu\nu} k_{\mu} U_{\nu})/(k_B T) - 1]}, \tag{12}$$

where $P_{00} = \gamma^2(\gamma_o^2 - \alpha_o x)$, $P_{11} = P_{11} = P_{22} = -1$;[16] and

(B) Rotational frame F_r with constant angular velocity:

$$B_r(k_{r\lambda} U_r^{\lambda}, T) = \frac{1}{[exp(P_{r\mu\nu} k_r^{\mu} U_r^{\nu})/(k_B T) - 1]}, \tag{13}$$

where $P_{r\mu\nu}$ is given by (8).

To the first order in $\rho\Omega$, we have the generalized blackbody distribution for a rotational frame,

$$B_r(k_{r\lambda} U_r^{\lambda}, T) \approx \frac{1}{[exp(k_0'/k_B T) - 1]}, \qquad T = T_{inv}, \tag{14}$$

where $k_{r\lambda} U_r^{\lambda} = P_r^{\lambda\nu} k_{r\lambda} U_{r\nu} \approx k_{r0}[1 + \rho\Omega \; sin(\Omega t_r)]/k_B T$, and we have used equations (10) and (11). As expected, the GF covariant blackbody distribution law (14) does not depend on time.

Therefore, based on (1), (10) and (14), when the GF covariant distribution law (14) is measured in a rotational frame F_r, we obtain the time-independent blackbody distribution law,

$$B_r(k_{r\lambda} U_r^{\lambda}, T) = B'(\eta^{\mu\nu} k_{\mu}' U_{\nu}', T), \tag{15}$$

where we have used the relation $P_r^{\alpha\beta} k_{r\alpha} U_{r\beta} = \eta^{\alpha\beta} k_{\alpha}' U_{\beta}'$.

4 Non-invariant spectral radiance and six-month variation

However, we stress that satellite experiments do not measure the GF covariant distribution law $B_r(P_r^{\mu\nu} k_{r\mu} U_{r\nu}, T)$. Instead, they measure the non-invariant spectral radiance in frequency (or power per unit area) W_r, which is the power emitted per unit projected area of a blackbody at temperature T, into a unit solid angle, in the interval k_r to $k_r + dk_r$, $k_r = |\mathbf{k}_r| > 0$. In the $F_r(t_r, x_r, y_r, z_r)$ frame with curvilinear coordinates, the spectral radiance W_r is related to the covariant differential of the photon energy-momentum tensor $DT_{r\mu\nu}$,

$$DT_{r\mu\nu} = \frac{2}{\pi^2} k_{r\mu} k_{r\nu} B_r(k_{r\lambda} U_r^{\lambda}, T) \delta(P_r^{\alpha\beta} k_{r\alpha} k_{r\beta}) \sqrt{-det P_r^{\lambda\sigma}} d^4 k_r, \tag{16}$$

$$W_r = DT_{r00}/(dk_r d\overline{\Omega}_r), \quad k_r = |\mathbf{k}_r|, \tag{17}$$

where the integration over $k_{r0} > 0$ in (16) is understood and $\overline{\Omega}_r$ is the solid angle. To carry out this integration in the general case, $\delta(f(k_{r0}, \mathbf{k}_r))dk_{r0}$, one may use

$$f(k_{r0}, \mathbf{k}_r) = P_r^{\alpha\beta} k_{r\alpha} k_{r\beta} = \eta^{\alpha\beta} k'_\alpha k'_\beta$$

and (9) to obtain

$$f(k_{r0}, \mathbf{k}_r) = (k_{r0} - \Omega\rho k_{r2})^2 - k_r^2$$

to the first order in $\rho\Omega$. Thus, we have

$$DT_{r00} = \left[\frac{2}{\pi^2} k_0 k_0 B_r(k_{r\lambda} U_r^\lambda, T) \frac{d^3 k_r}{2k_r} \right]_{k_{r0}=k_r+\rho\Omega k_{r2}} \tag{18}$$

To see explicitly the six-month variation in the measured W_r, it suffices to consider the simple case $k'_\mu = (k'_0, k'_1, 0, 0)$ at the location of the detector $(x_r, y_r, z_r) = (\rho, 0, 0)$ using relation (12). Based on (16), (17) and the corresponding relation $dT'_{00} = W' dk' d\overline{\Omega}'$ for the inertial frame F', we obtain the following relationship for spectral radiances measured in the two frames,

$$\frac{W_r}{W'} \approx \left(\frac{\omega_r}{\omega'}\right)^3 [1 + \Omega\rho\ sin(\Omega t_r)] \approx 1 - 2\Omega\rho\ sin(\Omega t_r), \tag{19}$$

where we have used the rotational frequency shift in (9) and the invariance of the blackbody distributions (14) and (4). For Earth's orbital motion, we may consider the Earth to be at the location of the detector, $(x_r, y_r, x_r) = (\rho, 0, 0)$ in the rotational frame F_r. Thus we have

$$\Omega \approx 0.2 \times 10^{-6}/sec \qquad and \qquad \Omega\rho \approx 9.9 \times 10^{-5}$$

(in natural units) for (19).

Therefore, we predict that the blackbody spectral radiance W_r measured from the orbiting Earth will indeed display a six-month variation (19), consistent with the qualitative results of satellite experiments. What remains now is to test the quantitative predictions of (19) against satellite experiments measuring the CMB.

5 Big Jets model and the missing half of the antimatter universe

According to the Big Jets model[16, 18], the presence of an extremely distant 'antimatter-galaxies blackbody' could contribute to the observed dipole anisotropy in CMB. The Big Jets model suggests that the universe began with two Big Jets rather than one Big Bang, similar to the particle collision phenomena one might encounter in high-energy laboratories. The model suggests that the universe originated with the formation of two diametrically opposed jets, composed of baryons and anti-baryons, etc. in each jet. We may picture these two jets as two big fire balls moving away from each other. The processes of their annihilations and decays eventually lead to a baryon-electron dominated fireball and an antibaryon-positron dominated fireball. Then, from the vantage point of an observer in either fireball, the evolution of that 'observer's fireball' would be similar to the general features of a hot Big Bang. Now the two gigantic fire balls cool down to become one baryon-electron dominated blackbody (called 'Yang blackbody') and a antibaryon-positron dominated blackbody (called 'Yin blackbody') with a temperature of roughly 3K, which are separated by an extremely large distance D_{YY}. As a result, lights emitted from anti-supernovae and anti-galaxies are too faint to be detected. Only the microwaves emitted from the Yin blackbody as a whole may be detected by our apparatus.

To search the missing half of the universe suggested by the Big Jets model, let us approximate, for simplicity, blackbodies Yin and Yang as two gigantic spheres of radius R_Y. Microwaves emitted from the antimatter Yin blackbody toward our Yang blackbody are completely absorbed, causing the Yang hemisphere facing Yin to be slightly warmer than the opposite hemisphere and giving the Yang blackbody temperature a 'pseudo-dipole distribution' roughly. Suppose the Yang blackbody receiving microwaves from the Yin blackbody so that its hemisphere facing Yin is warmer by ΔT, depending on angle. One may try to estimate ΔT by using the Stefan-Boltzmann law and the condition of equilibrium. But it will involve some unknown parameters. Thus, we just use ΔT as a parameter to be determined by experiments. Suppose $D_{YY} >> D_Y$, the Big Jets model implies that, approximately, the observed temperature T_{ob} in our Yang blackbody has a simple angle dependence,

$$T_{ob} \approx T_{CMB} + Y(\theta)\Delta T, \tag{20}$$

$$Y(\theta) = cos\theta, \quad for\ the\ surface\ of\ the\ hemisphere\ toward\ Yin\ blackbody,$$

$$Y(\theta) = 0, \quad for\ the\ surface\ of\ the\ hemisphere\ away\ from\ Yin\ blackbody,$$

where $\theta = 0$ corresponds to the direction from the center of Yang blackbody to that of Yin blackbody.

The reason for $Y(\theta) = 0$ in (20) is that the (parallel) microwaves from the Yin blackbody are totally absorbed by the Yang blackbody, so that the temperature on the Yang hemisphere away from the Yin blackbody remains uniform, $T_{ob} = T_{CMB}$, independent of angles.

In the presence of the Yin blackbody, the spectral radiance W'_B of the blackbody radiation measured in the inertial frame F' involves the Lorentz invariant temperature T and the invariant blackbody distribution $B'(k'_\lambda U'^\lambda, T_{ob})$. The non-invariant spectral radiance $W'_B(k'_\lambda U'^\lambda, T_{ob})$ can be approximated by the Taylor series expansion in the inertial frame F'. We have

$$\begin{aligned} W'_B(k'_\lambda U'^\lambda, T_{ob}) &= \frac{1}{4\pi^3}(k'_0)^3 B'(k'_\lambda U'^\lambda, T_{ob}) \\ &\approx W'_B(k'_\lambda U'^\lambda, T_{CMB}) + W'_{BJ}, \end{aligned} \tag{21}$$

where the second term W'_{BJ} in (21) determines the CMB dipole spectrum due to the blackbody Yin in the Big Jets model. We obtain the pseudo-dipole spectrum due to the antimatter blackbody,

$$\begin{aligned} W'_{BJ} &\approx \Delta T cos\theta \left(\frac{\partial W'_B(k'_\lambda U'^\lambda, T_{ob})}{\partial T_{ob}} \right)_{T_{ob}=T_{CMB}} \\ &\approx \Delta T cos\theta \left(\frac{\omega'^4}{4\pi^3 k_B T^2_{CMB}} \frac{(exp(\omega'/k_B T_{CMB})}{[exp(\omega'/k_B T_{CMB}) - 1]^2} \right), \end{aligned} \tag{22}$$

where we have taken the non-relativistic limit of the 4-velocity U'^λ, i.e., $U'^\lambda = (1, 0, 0, 0)$, and $k'_\sigma U'^\sigma = k'_0 = \omega'$.

For the sake of discussion, suppose that anti-matter blackbody happened to be in the direction of the constellation Leo, then the prediction (22) by the Big-Jets model turns out to be the same shape as the frequency spectrum of the conventional theory[3, 19] and, hence, both are consistent with the experiment of Fixsen et al.[19] On the other hand, when one measures the spectral radiance $W'_B(k'_\lambda U'^\lambda, T))$ in (21) in the direction away from Leo, the second term W'_{BJ} in (21) does not contribute because $Y(\theta^*) = 0$ in (21) due to the complete absorption of the microwaves by our blackbody Yang. Thus, in the direction away from the antimatter blackbody Yin, (20) and (21) lead to the usual CMB spectral radiance,

$$W'_B(k'_\lambda U'^\lambda, T) = W'_B(k'_\lambda U'^\lambda, T_{CMB}), \tag{23}$$

which is independent of the angle θ. The result (23) hold for the surface of the hemisphere of the Yang blackbody away from the Yin blackbody and suggests a new method to search for the missing half of the antimatter universe, as we shall discuss below.

In contrast to (22) and (23), the conventional theory with angle-dependent temperature (5) leads to the spectrum

$$B'_\nu(T') \approx B'_\nu(T_{CMB}) + VT_{CMB}\, cos\theta' \left(\frac{\partial B'_\nu(T')}{\partial T'}\right)_{T'=T_{CMB}}, \tag{24}$$

$$T' \approx T_{CMB}[1 + Vcos\theta'],$$

where θ' is the angle between the direction of the motion of the solar system and the line of sight. The velocity $V \approx 370km/s$ (in MKS units) is interpreted as the motion of the solar system (F' frame) in the CMB. The spectrum (or the Planck function) $B_\nu(T)$ takes the simplest form $\omega^3/[exp(\omega/(k_B T)) - 1]$, $\omega = 2\pi\nu$ in the preferred frame F[20, 21].

6 Discussions

The conclusion based on making the blackbody distribution laws (1) and (4) explicitly invariant does not imply that the conventional theory with (24) and the angle-dependent temperature (5) is wrong because they use different definitions (or 'interpretations') of the temperature. Nevertheless, from the viewpoint of Lorentz-Poincaré invariance, the dipole anisotropy in CMB temperature in the conventional theory with (24) is due to an unnatural interpretation of the invariant distribution law $B'(k'_\lambda U'^\lambda, T)$ given by (4), in terms of the non-invariant variable ω' and an ad hoc angle-dependent temperature (5), i.e. $1/T'(\theta') = (1 + Vcos\theta')/(T\sqrt{1-V^2})$. This can be seen from the fact that the Lorentz invariant law $B'(k'_\lambda U'^\lambda, T)$ in (4) turns out to be quantitatively identical to $P'(\omega', T')$,

$$B'(k'_\lambda U'^\lambda, T) = P'(\omega', T') = 1/[exp(\omega'(1 + Vcos\theta')/(k_B T\sqrt{1-V^2})) - 1]. \tag{25}$$

Thus, the same set of satellite data can be viewed from two conceptually different viewpoints. We believe that the Lorentz invariant view of the distribution law $B'(k'_\lambda U'^\lambda, T)$ in terms of the scalar function $k'_\lambda U'^\lambda = k_\lambda U^\lambda$ and the invariant temperature $T = T_{inv}$ is more natural. The reasons are that the blackbody distribution law $B'(k'_\lambda U'^\lambda, T)$ satisfies the principle of relativity and that the invariant temperature (or $K_B T$) is consistent with those developed by Einstein et al for relativistic thermodynamics. [8–13]. On the contrary, if one considers (24) and the angle-dependent temperature (5) in three inertial frames explicitly, one sees that they are not consistent with the principle of relativity in that the ad hoc temperature (5) itself does not have proper transformation properties within the 4-dimensional symmetry framework.

Based on the general-frame covariant blackbody distributions (1), (4), (12)-(15), it is unnatural and inconsistent with the principle of relativity to attribute the 'dipole anisotropy of the CMB' to the motion of the solar system relative to the CMB. The general-frame covariant result (15) implies that if the Lorentz invariant distribution law $B(k'_\mu U'^\mu, T)$ and the CMB temperature are isotropic in one inertial frame F', then they will both be isotropic in every inertial and rotational frames as well, according to the principle of limiting Lorentz-Poincaré invariance.

For experimental tests and search for the missing half of the universe, the Big Jets model predicts that the physical universe is made of two gigantic blackbodies with the following properties:

(i) Within our 'matter blackbody', there are no 'antimatter stars' or 'anti-matter galaxies' due to annihilation processes in the early epoch of evolution. This has been supported by observations and experiments in our 'matter blackbody.'

(ii) It is remarkable that the dipole spectrum (22) of the antimatter blackbody has the same shape as that of the usual 'hot dipole' in the direction of the constellation Leo. Thus, it is interesting to see whether evidence for the missing half of the universe may be hidden in the data of the hot dipole anisotropy. Their spectra differ only in the 'cold dipole', as shown in (23). Therefore, the Big Jets model suggests that, in general, one should examine whether there is a small asymmetry of the spectrum in two opposite directions in the sky, towards and away from the constellation Leo. Such a small asymmetry could be tested by, say, the experiment of Fixsen et al[19], which measured the specific spectrum of the dipole for approximately 60 % of the sky (including the 'hot pole', $cos\theta > 0$).[3]

The work was supported in part by the Jing Shin Research Fund of the UMass Dartmouth Foundation.

References

[1] J. P. Hsu and L. Hsu, Phys. Letters, A **196**, 1 (1994). Such a broad formulation of space-time transformations based solely on the first postulate is essential for generalization to non-inertial frames, especially the rotational frame for our discussion here.

[2] J. P. Hsu and L. Hsu, Eur. Phys. J. Plus, **128**, 74 (2013), arXiv: 1307.0662.

[3] D. Scott and G. F. Smoot, "27. COSMIC MICROWAVE BACKGROUND", section 27.2.2, for The Review of Particle Physics, K.A. Olive et al. (Particle Data Group), Chinese Phys. C **38**, 09000, (2014); (http://pdg.lbl.gov). See also ref. 21.

[4] S. Weinberg, *Gravitation and Cosmology, Principles and Applications of the General Theory of Relativity* (John Wiley and Sons, 1972) pp. 520-522.

[5] J. P. Hsu and L. Hsu, in the *Proceedings of the Twelfth International Conference on Gravitation, Astrophysics and Cosmology*, Moscow, 2015 (Ed. V. Melnikov and J. P. Hsu, World Scientific, 2016) p. 3; and reference therein.

[6] L. Hsu and J. P. Hsu, 'Experiments on Invariance of Planck's Law and Anisotropy of Cosmic Microwave Background' (UMassD preprint, 2017).

[7] G. N. Pellegrini and A. R. Swift, Am. J. Phys. **63**, 694 (1995) and references therein.

[8] A. Einstein, Jahrb. Rad. u. Elektr, **4**,411 (1907).

[9] M. Planck, Ann. d. Phys. **26**, 792 (1908).

[10] H. Ott, Zeitschr. d. Phys. **174**, 70 (1963).

[11] H. Arzelies, Nuovo Cimento **35**, 792 (1965).

[12] P.T. Landsberg, Nature **212**, 571 (1966).

[13] N.G. van Kampen, Phys. Rev. **173**, 295 (1968).

[14] L. Landau and E. Lifshitz, *The Classical Theory of Fields*, (Addison-Wesley, 1951), p. 31.

[15] L. Hsu and J. P. Hsu, Eur. Phys. J. Plus, **129**, 16 (2014).

[16] J. P. Hsu and L. Hsu, *Space-Time Symmetry and Quantum Yang-Mills Gravity* (Singapore, World Scientific) pp. 26-28; p. 225, note 20.

[17] O. Veblen and J. H. C. Whitehead, *The Foundation of Differential Geometry* (Cambridge Press, 1953), pp. 37-38.

[3] At present, the dipole spectrum has been fit almost to the entire sky (with the exception of about 1% of the sky missing due to the lack of FIRAS data there and another $(10-20)\%$ of the sky being ignored because the galactic plane is too bright and difficult to normalize.[22]). Fixsen has performed a quick fit separately to the hot pole (where $cos\theta > 0$) and the cold pole (where $cos\theta < 0$), then looked for differences between the two spectra. There is no difference to a sensitivity of a few $\times\ 10^{-4}$ relative to the peak of the CMB. One may use a simple property of a maximum related to the energy density of radiation to roughly estimate the corresponding temperature difference between two opposite directions in the sky as $\Delta T < 3 \times 10^{-4}$ K, due to the possible presence of a distant antimatter blackbody.

[18] J. P. Hsu, Chin. J. Phys. bf 41, 015101 (2017).

[19] D. J. Fixsen, et al, Astrophys. J. **420**, 445 (1994).

[20] G. F. Smoot, Nobel Lecture, (https://www.nobelprize.org/nobel_prizes/physics/laureates/2006/smoot-lecture.html) pp.123-124.

[21] V. S. Weisskoph, Am. Sci. **71** 473 (1983).

[22] D. J. Fixsen, private correspondence, 2017.

EPJ Web of Conferences **168**, 01013 (2018) https://doi.org/10.1051/epjconf/201816801013
Joint International Conference of ICGAC-XIII and IK-15 on Gravitation, Astrophysics and Cosmology

The revival of General Relativity at Princeton: Daring Conservatism

Dieter Brill[1,*], *Alexander* Blum[2,**]

[1]Department of Physics, University of Maryland, College Park, MD 20742, USA

[2]Max Planck Institute for the History of Science, Boltzmannstraβe 22, 14195 Berlin, Germany

Abstract. After General Relativity was established in essentially its present form in 1915 it was celebrated as a great success of mathematical physics. But the initial hopes for this theory as a basis for all of physics began to fade in the next several decades, as General Relativity was relegated to the margins of theoretical physics. Its fate began to rise in the 1950's in a revival of interest and research that over time made gravitational physics one of the hottest research topics it is today. One center of this renaissance was Princeton, where two relative newcomers explored new and different approaches to gravitational physics. Robert Dicke showed that gravity is not as inaccessible to experiment as was thought, and John Wheeler propelled it into the mainstream by proposing highly original and imaginative consequences of Einstein's theory. We will concentrate on these ideas that, in his characteristically intriguing style, Wheeler called "Daring Conservatism" -- a term well known to his associates, but one he never mentioned in print. With the aid of unpublished manuscripts and notes we will explore Daring Conservatism's origin and motivation, its successes and failures, and the legacy it left behind.

After Einstein's founding of General Relativity in 1915, and its initial spectacular success, there were several decades when it was relegated to the margins of theoretical physics. The reason, we feel, was two-fold. There were no further fundamental experiments to test the theory, and on the theory side there was no program that could bring general relativity back into contact with the part of physics that was making the great advances in such areas as quantum, nuclear, particle, light and solid state physics. The 1950's marked the beginning of a renaissance in both experimental and general relativity, which propelled it to the hot research topic it is today. An important center of this re-awakening was Princeton, where new experiments and fresh ideas were being developed, primarily by Robert Dicke and John Wheeler. Figure 1 shows them at this most creative time.

Robert Dicke was able to test the basis of a fundamental theory like General Relativity by measurements of unprecedented accuracy that was made possible by the new electronics available after the war. To assess the significance of high precision experiments he

* Speaker and Corresponding author: brill@physics.umd.edu
** e-mail: ablum@mpiwg-berlin.mpg.de

considered alternative theories patterned after Mach's Principle, and formalized in Brans-Dicke theory. The history of his contributions has been expertly told by Peebles [1]. Wheeler's path to General Relativity is less familiar (except through his autobiography [2]), and worth telling today.

Dicke Bleakney Wheeler

Figure 1. Robert Dicke, Walter Bleakney and John Wheeler, detail from a faculty photograph, Department of Physics, Princeton University, about 1950

1 Particles

When his position at Princeton after the war allowed Wheeler to think about fundamental physics, it was not clear which of the new theoretical ideas being developed would be lasting and fundamental. Instead of engaging in such free inventions, as he called them, he preferred the conservative approach and asked, can one exclude the presence, in the physics we already know, of concepts more fundamental than the rest, which promise to explain all of physics? The spirit of the explanation itself, however, was to use fertile, "anything goes" imagination. As Feynman (apocryphally?) remarked

Some people think Wheeler's gotten crazy in his later years, but he's always been crazy.

A master of coining catchy phrases, Wheeler later gave his grand vision a more polite name, Daring Conservatism. This phrase is well-known to his associates from that time, but has never appeared in print. (In fact, some remember it as radical conservatism or dynamic conservatism.) Wheeler himself attributes Daring Conservatism to Bohr:

His judgment and courage, his daring conservatism, carried him to wonderful conclusions of a kind that free invention has not strength to reach, nor conviction to maintain.

At first the fundamental concept was that "everything is particles," specifically electrons. He showed that electrons and positrons can bind together into "polyelectrons" [3], and suggested this, rather than Yukawa field theory, as the explanation of mesons. Due to the discrete, quantized nature of particles, their continuous fields are required to be replaced by direct action between particles. First instances are the papers with Feynman [4] in which action at a distance between charges is shown as a complementary description to the field theory of classical electrodynamics. The radiative forces of this theory gave Wheeler the hope to explain nuclei, as expressed in a question to Bohr [5]:

Is it possible to exclude a picture of elementary particle constitution based entirely on positive and negative electrons?

Bohr may have affirmed the exclusion, there is no record of further work on this idea.

For the same reason of particle atomicity, in mass as well as in charge, Wheeler felt that it should in principle be possible to dispense with the concepts of continuous space and time in Einstein's theory of the gravitational field. He proposed the elementary concepts of world line and light cone as the basis of the description of nature [6]. Points on a world line are to be specified by a parameter (proper time), and light rays provide the connection, called a liaison (later [7]: association), between these points and those on any other world line. Liaisons can lead to future as well as past points on the other world line, similar to half advanced, half retarded potentials of Wheeler-Feynman electrodynamics. This makes them a candidate for implementation of Mach's principle, which Wheeler considered an essential basis for a theory of gravity. For a number of years Wheeler searched for an action written only in terms of liaisons that would yield equations of motion for the particle world lines and would reproduce the Einstein-Hilbert action in the limit of a large number of world lines [8]. The last notebook entry is from 1953, without any clear conclusion.

2 Teaching and Transition

No event was more important in Wheeler's search than his 1952 assignment to teach the graduate course on relativity at Princeton, for he practiced his often-quoted credo that one can only learn by teaching – and he prepared "to give the best possible course", started a new notebook [9], planned to publish his course as a book, and involved his students with reports and research-level problems. His new and daring ideas pervaded the course, and made him confront the unfinished nature of creations like liaison theory and Mach's principle, as well as the nitty-gritty details of differential geometry calculations vis-à-vis their physical meaning. Firmly believing in Mach's principle he did not favor the physically unexplained asymptotic Minkowskian region of the Schwarzschild solution, for "Geometry cannot be part of physics in some regions and not part of physics in others." Instead, Mach was taken to demand that the universe should be closed. Here is how he explained this to his class on March 31, 1952.

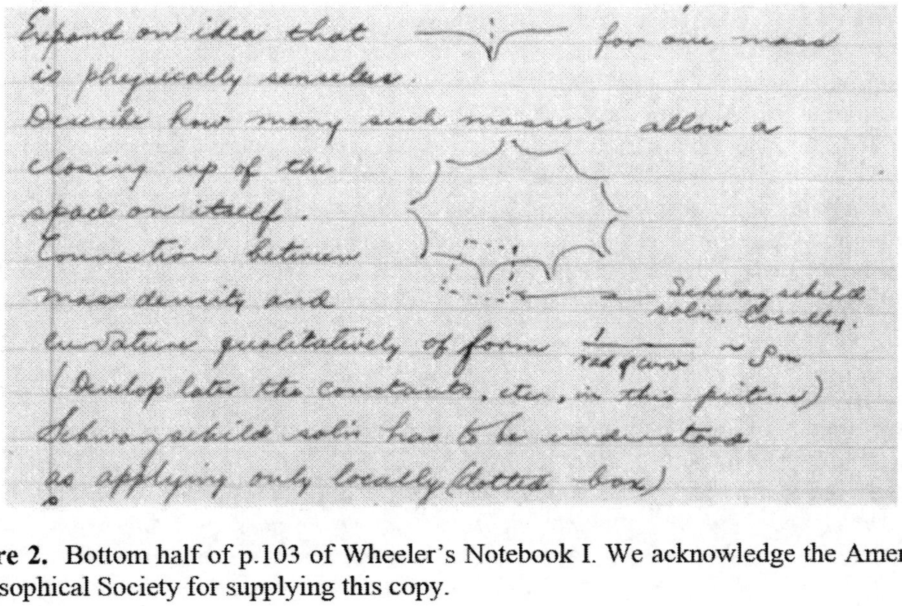

Figure 2. Bottom half of p.103 of Wheeler's Notebook I. We acknowledge the American Philosophical Society for supplying this copy.

A passage from the next notebook page shows his strong feeling about space closure:

> Question raised in class whether mass density enough to permit open or closed universe, in view of expansion rate.
>
> Answer: Conclude from above that closure comes first; density knowledge too poor to permit proof of contradiction; closure so fundamental to whole Mach idea that in present state of knowledge think of density value having to yield precedence to Mach principle.

During the years of the course it became apparent that fields may be a more fundamental basis for physics than particles. To Wheeler it was not an abrupt break with the previous particle picture, but part of a *Study (or Critique) of Classical Field Theory*. This was the name of a series of papers [4, 11, 12] that starts with electromagnetic action at a distance as a tool to analyse the field concept, continues with the Wheeler-Feynman papers that account for EM interaction without involving the EM field, and ends with structures built purely from fields or geometry. These six papers may be taken to define Wheelerian Daring Conservatism.

3 Daringly Conservative about Fields and Geometry

In 1954 Wheeler was ready to announce this program to the world of physics in the Twelfth Richtmyer Memorial Lecture [10]:

I propose that we should raise at the head of our enterprise the flag of daring conservatism. Let us be conservative in the sense of staying on the solid ground of what is firm and harmonious and well established; ... conservative in the sense that we repel free invention of new theories; daring in the sense that we follow out the consequences of our already existing ideas to the uttermost extremes; daring in the sense that we use what we already know to enter into quite new fields of thought.

He discusses five most fundamental concepts that conservatism would allow for building new structures (particles): electromagnetism, gravitation theory, quantization, the neutrino, and electron theory. He mentions meson theory, the cosmological constant, and unified field theory as examples of free invention that should be rejected. He singles out gravitation as a field with physical consequences that have been little investigated, such as laws of test particle motion that follow from the field equation. Comparing this to the motion of hurricanes as a consequence of hydrodynamics he asks
How different is a particle from a kind of hurricane in the fields of zero rest mass?
and outlines such a structure as an electromagnetic disturbance held together by gravitational attraction. Arthur Wightman suggested the name Kugelblitz, but it was later called a geon. For small geons quantum effects are expected to lead to a spectrum of masses, therefore
In what points of principle, if any, do the disturbances of smallest mass differ from elementary particles?
This field point of view leads him to question electron theory:
Is the theory incomplete because it regards the electron as a primitive entity, endowed with a rest mass of its own? If instead the electron is a special kind of disturbance in the electromagnetic field and in other fields of zero rest mass, will there not be many phenomena ... in which it is not legitimate to think of primordial mass as done in the Dirac theory?

The incomplete transcript of the Richtmyer lecture stops abruptly at this point, before accounting for the electron charge in terms of sourceless electromagnetic fields.

A way to account for electric charge appeared less than 8 months after the Richtmyer lecture in the famous geon paper [11]. Here Wheeler shows detailed calculations for spherical geons, and introduces graphically a doubly connected space, later called a wormhole (originally, in Notebook II: woodchuck hole). He also gives estimates for transformations of, and interactions between, geons. The geon calculations involved a careful analysis of high frequency and high angular momentum EM waves, proof that a sufficiently spherically symmetric distribution is possible, and a reduction of the Einstein-Maxwell equations to three coupled differential equations of the eigenvalue type. The solution required numerical integration, one of the first uses of electronic computers in General Relativity. The start of the geon calculation, according to Notebook II, was on January 25, 1954, and he received the computer results on March 24 together with a bill of $195 for 6½ hours of computer time.

The mathematics of the model of electric charge as wormholes with trapped flux was published with Misner as the final Part VI of the Critique series [12]. For those working in General Relativity as "index pushers" this paper was the first introduction to the modern mathematics of differentiable manifolds and differential forms as applied to physical problems. It also contained a new point of view of the well-established physics of Maxwell and Einstein, for it showed a way of stating the content of Maxwell's equations purely in terms of the curvature that Einstein's equations prescribe for a spacetime containing Maxwell fields. This procedure yields purely geometrical equations, which both unify and keep unchanged gravitation and electromagnetism – an "already unified theory".

This final paper of the Critique series summarizes the essential results of Daring Conservatism in a paradoxical form, which Wheeler had already used in talks and lectures to promote his ideas and insights: Empty Curved Space can provide

- Gravitation without gravitation
- Electromagnetism without electromagnetism
- Charge without charge
- Mass without mass

but not (yet)

- Spin without spin
- Elementary particles without elementary particles.

To these he would add

- Equations of motion without equations of motion
- Law without Law.

The already unified theory took some ribbing

- A theory without a theory.

In fact, the already unified theory did not work in all situations. Asked about its status (at a Stevens meeting), Wheeler recounted being at a restaurant, worrying about the problems of this unification, when the fortune cookies came around. His read "If it became true, it wasn't much of a dream" – and he accepted. Still, Wheeler persisted in the view that the ultimate nature of physics is geometry as described by some quantum version of the sourceless Einstein equations. Complementary to seeing a solution of the classical equations as a fixed spacetime, he promoted the view of it as a developing sequence of spacelike surfaces. Mathematicians had shown that the equations form a dynamical system of initial geometry and its time development. Wheeler gave it a name: Geometrodynamics.

4 Collapse

No paradoxical byword was associated with the most influential topic Wheeler proposed in the Richtmyer lecture for investigation by Daring Conservatism; it would be black holes in today's terms. At the time the question was whether an actual star would follow the same collapse scenario as Oppenheimer and Snyder's model of collapsing dust [12]. The answer to Wheeler's quest for the endstate of gravitational collapse was made possible by the 1960 publication (at Wheeler's urging) of Kruskal's paper [13] on the Maximal Extension of the Schwarzschild Metric. Wheeler analyzed and popularized in detail, from conservative Einstein theory without free invention, that the collapse produces a Black Hole. By 1963 Daring Conservatism had done its job in attracting attention to General Relativity, as the first "Texas Conference" [14] was devoted to theory and observation of this new physics.

The Texas Conference marked, in a sense, the end of strict Daring Conservatism with Wheeler, for he briefly violated one of its rules: no free invention. Interactions and transformations of black holes and ordinary physics suggested further properties. For example, Wheeler's question, what happens to the entropy of a system that falls into a Schwarzschild black hole, led to the idea that black holes themselves carry entropy (related to their mass). The similar question, what happens to (dimensionless) nucleon number conservation when a number of nucleons is dropped into a black hole, cannot be answered in the same way, for black holes have no "hair" other than (dimensionful) mass, charge, and angular momentum [15]. Wheeler therefore proposed that nucleon number is carried away by a new radiation he called delta rays [16]. He must have quickly dropped this free invention[1] for the only mention in print is one sentence in Hong-Yee Chu's introductory Report on the Texas Symposium in the conference proceedings [14].

5 Conclusion

The ideas from Daring Conservatism as paradoxically stated above all had an influence on the direction of General Relativity research but with varying longevity. Gravitation as geometrodynamics persists, with the latest confirmation of gravitational waves, but the already unified theory of electromagnetism fell by the wayside. Charge has become only one of many applications of topology in physics. The geon idea still appears in various connections [18] but the original geon, although providing stable orbits for each constituent photon, turned out unstable to collective collapse or dispersal of all its photons [19].

Wheeler tried many ideas for a geometrical model of spin but none were ultimately successful. Sum-over-histories quantization of geometry is also mentioned already in the Richtmyer lecture. Still a viable approach to quantum spacetime, it led Wheeler to the idea of its foam-like structure. Elementary particles were found to consist of quarks and gluons, not the fundamental massless fields Wheeler had in mind, but Wheeler rejoined [20]
... it would still seem reasonable to expect that one must have some perspective on what happens at 10^{-33} cm [scale of the foam] *before one can find the rationale of quarks and particles.*

Daring conservatism was a breath of fresh air in a field where many regarded exact solutions as the only sure way to explore the richness of Einstein's theory. In this search

[1] His later view was that such conservation laws can be transcended [17].

physical meaning was often secondary, as when it was not clear whether a newly-discovered solution of the field equations described the same physics as a well-known one in new coordinates. Working with the same, well-established theories Wheeler produced constructions that were maybe weird or crazy, or only good approximations, but they were objects of physics, their appeal was undeniable, failure to recognize them was not an option. Most of all he gave his students and the whole GR community new hope and inspiration that there was much – or everything – of physics to be found in Einstein's theory.

References

1. P. J. E. Peebles, Eur. Phys. J. H **42**, 177-259 (2017), arXiv:1603.06474
2. J. A. Wheeler with K. Ford, *Geons, Black Holes, and Quantum Foam* (W.W. Norton, New York, 1998)
3. J. A. Wheeler, Ann. New York Acad. Sci. **XLVIII**, 219-238 (1946)
4. J. A. Wheeler, R. P. Feynman, Rev. Modern Phys. **17**, 157 (1945) – *Part II, study of classical field theory*; J. A. Wheeler, R. P. Feynman, Rev. Modern Phys. **21**,425 (1949) – *Part III, study of classical field theory*
5. J. Wheeler letter to N. Bohr, January 21, 1950
6. J. A. Wheeler, Proc. Am. Philos. Soc. 90, 36 (1946); Notebook *ACTION AT A DISTANCE I* (1950-53) from American Philosophical Society in Philadelphia
7. D. H. Wesley, J. A. Wheeler, in A. Ashtekar et al (eds.), *Revising the Foundations of Relativistic Physics*, 421-436, Kluwer (2003)
8. Notebook RELATIVITY II (May 1953 – April 1954), American Philosophical Society
9. Notebook RELATIVITY I (Sep. 1952 – May 1953), Mss.B.W564_SerV_Vol39, American Philosophical Society
10. J. A. Wheeler, *FIELDS AND PARTICLES*, Adapted from the Twelfth Richtmyer Memorial Lecture, delivered in New York, 29th January 1954 before the joint session of the American Association of Physics Teachers and the American Physical Society (unfinished Typescript)
11. J. A. Wheeler, Phys. Rev. **97**, 511-536 (1955) – *Part V of a study of classical field theory*
12. C. W. Misner, J. A. Wheeler, Ann. Phys. **2**, 525-603 (1957)
13. M. D, Kruskal, Phys. Rev. **119,** 1743 (1960)
14. I. Robinson, A. Schild, E. Schucking, *eds Quasistellar Sources and Gravitaional Collapse* (U. of Chicago P., Chicago, 1965)
15. R. Ruffini, J. A. Wheeler, Physics Today **24**, 30 (1971)

16. D. Brill, Bull. Am. Phys. Soc. **60** (2015) Abstract J5.00004

17. J. A. Wheeler, Quaderno **157**, 133-164 (1971)

18. J. Louko, J. Phys. Conf. Ser. **222**:012038 (2009), arXiv:1001.0124v1 [gr-qc]

19. G. Gundlach, Phys. Rept. **376**, 339 (2003), gr-qc/0210101

20. J.A. Wheeler in R. Ruffini (ed.), *Quantum Cosmology* (World Sci.Pub.Co.1987) p.300

EPJ Web of Conferences **168**, 01014 (2018) https://doi.org/10.1051/epjconf/201816801014
Joint International Conference of ICGAC-XIII and IK-15 on Gravitation, Astrophysics and Cosmology

Status of the GroundBIRD Telescope

J. Choi[1], *R.* Génova-Santos[2], *M.* Hattori[3], *M.* Hazumi[4], *H.* Ishitsuka[4], *F.* Kanno[3], *K.* Karatsu[5], *K.* Kiuchi[6], *R.* Koyano[7], *H.* Kutsuma[3], *K.* Lee[8], *S.* Mima[6], *M.* Minowa[9], *M.* Nagai[10], *T.* Nagasaki[4], *M.* Naruse[7], *S.* Oguri[6], *T.* Okada[3], *C.* Otani[6], *R.* Rebolo[2], *J.* Rubiño-Martín[2], *Y.* Sekimoto[10], *J.* Suzuki[4], *T.* Taino[7], *O.* Tajima[4,11], *N.* Tomita[9], *T.* Uchida[4], and *E.* Won[8,⋆] *M.* Yoshida[4]

for the GroundBIRD Collaboration

[1]*Center for Axion and Precision Physics Research, Institute for Basic Science (IBS), Daejeon 34051*
[2]*Instituto de Astrofísica de Canarias, E38205 - La Laguna, Tenerife, Canary Islands, Spain*
[3]*Tohoku University, Aramaki Aza Aoba 6-3, Aoba-ku, Sendai, 980-8578, Japan*
[4]*Institute of Particle and Nuclear Studies, High Energy Accelerator Research Organization (KEK), Tsukuba, Ibaraki 305-0801, Japan*
[5]*TU Delft, 2600 AA Delft, Netherland*
[6]*RIKEN, 2-1 Hirosawa, Wako, Saitama, 351-0198, Japan*
[7]*Saitama University, Saitama-shi 338-8570, Japan'*
[8]*Korea University, 145 Anam-ro, Seongbuk-gu, Seoul, 02841, Korea*
[9]*The University of Tokyo, 7-3-1, Hongo, Bunkyo-ku, Tokyo 113-0033, Japan*
[10]*Advanced Technology Center, National Astronomical Observatory of Japan, Tokyo 181-8588*
[11]*Department of Physics, Kyoto University, Kyoto 606-8501*

Abstract. Our understanding of physics at very early Universe, as early as 10^{-35} s after the Big Bang, relies on the scenario known as the inflationary cosmology. Inflation predicts a particular polarization pattern in the cosmic microwave background, known as the B-mode yet the strength of such polarization pattern is extremely weak. To search for the B-mode of the polarization in the cosmic microwave background, we are constructing an off-axis rotating telescope to mitigate systematic effects as well as to maximize the sky coverage of the observation. We will discuss the present status of the GroundBIRD telescope.

1 Introduction

The present standard model of Universe [1], or smoothly expanding Universe without any non-smooth feature from the beginning, has at least two known problems. One is the observed isotropic distribution of the cosmic microwave background (CMB) temperature, down to 10 ppm level [2] to its mean value of 2.7 K, yet any CMB photons seen today separated by a couple of degree in the sky today were

⋆e-mail: eunil@hep.korea.ac.kr

out of causal contact at the time of the last scattering before reaching to us as millimeter electromagnetic wave. This is known as the horizon problem in cosmology [3]. Also, the recent measurement on the curvature of the space of Universe indicates that Universe is very flat to sub percent level [4] and it is interpreted that the mass density in early Universe must have been fine-tuned to more than ten digits according to Friedmann equations. This is known as the flatness problem [5].

These two problems[1] are considered as fine-tunning problems or initial-value problems of the standard model of Universe right after the big bang, and in order to solve them, a model of rapid, superluminal expansion of space in very early Universe, commonly called the inflationary cosmology is proposed in 1980's [6]. This exponential expansion in nature allows all photons to be in causal contact to explain the isotropic temperature of the CMB and flatness of the space at present. It has become clear that the ultimate test of the inflationary Universe is to search for the curl mode of the polarization in the CMB [7]. The strength of the B-mode polarization, that is related with the energy scale of the inflation, is unfortunately not predicted by theory precisely and commonly parameterized by the ratio, r of the tensor to scalar perturbation of the power spectra of the CMB anisotropy.

The search for the primordial B-mode polarization of the CMB has become one of major fields in the modern cosmology and recent ground based and satellite set upper limits on r to be order of 0.1 [8, 9]. Our telescope, GroundBIRD (B-mode Imaging Radiation Detector) [10], is being constructed in Japan, aiming at the detection of the B-mode polarization down to r to percent level. We will discuss advantages of our design, the construction status, and our observation plan in the following section.

2 GroundBIRD Telescope

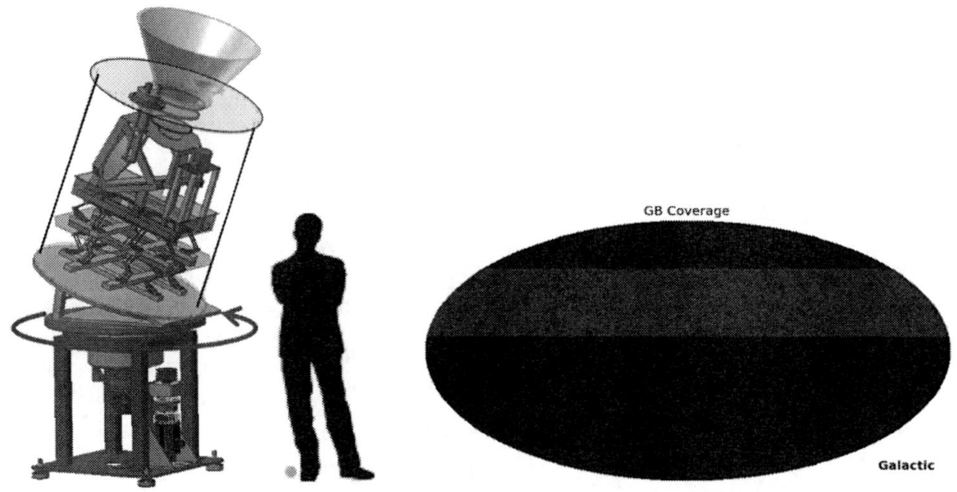

Figure 1. Left figure shows a drawing of the GroundBIRD telescope. To set the scale, a human is also drawn. Right figure show the sky coverage of the GroundBIRD telescope when it is located at the latitude of 28° N, having a zenith angle of 20°.

The GroundBIRD (GB) telescope is under construction at KEK in Japan. The GB telescope adapts a fast rotation scan strategy of 20 rpm, or 120°/s. A drawing of the GB telescope is shown at the left

[1] There is also a magnetic monopole problem but it assumes grand unified theories that are not settled yet and therefore not mentioned here.

side in Fig. 1. As indicated here in Fig. 1, GB rotates with a finite zenith angle at the chosen site. A single rotation with a 20° zenith angle forms a large toroidal coverage in the sky. On top of that, the rotation of Earth itself shifts this toroidal coverage by a constant speed of 15°/h, making the sky coverage of the GB to be approximately 40% of the entire sky that is about 10 times larger than coverage from other ground based telescopes that use constant elevation azimuthal scan method [11]. This large sky coverage is demonstrated at the right side in Fig. 1 for the location at at the latitude of 28° N, having the zenith angle to be 20°. This fast rotation scan strategy also allows to mitigate so called $1/f$-noise effect from the baseline fluctuation of the detector. This is one of the unique features of the GB telescope. Note that rotation of the entire telescope is technically extremely challenging since at the same time, electric power lines and helium gas to the cryogenic system must be provided while the temperature of the focal plane must be stable at 250 mK. The implementation and stable operation of rotating table will be discussed in the following section.

This fast rotation enables GB to detect both the reionization bump of the multipole $\ell < 10$ and the recombination peak at $\ell \approx 100$ of the B-mode power spectrum predicted by inflation scenario of the very early Universe. This can be demonstrated by computing the detector noise power spectrum N_ℓ^{BB} where

$$N_\ell^{BB} = \left(\frac{\pi}{10800} \frac{w_p^{-1/2}}{\mu\text{K arcmin}} \right)^2 \mu\text{K}^2 \text{ str}. \quad (1)$$

Here $w_p^{-1/2}$ is the noise in Stokes parameters Q or U per pixel whose solid angle Ω_{pix}, gives $\sqrt{\Omega_{\text{pix}}} = 1$ arcmin [12]. Using this one can estimate sensitivity and a projected GB sensitivity on the B-mode polarization is shown in Fig. 2. Two bottom curves (green and blue) are B-mode power spectra for r=0.1, and 0.01. A thick (red) curve is the GB sensitivity estimation when $w_p^{-1/2}$ = 3.5 μK arcmin. A curve above (purple) is when $w_p^{-1/2}$ = 10 μK arcmin. The almost straight line (black) except right corner indicates the gravitational lensing effect, and the dashed curve is EE mode of the power spectrum. Depending on the value of r, GB may be able to detect both bumps that are mentioned above.

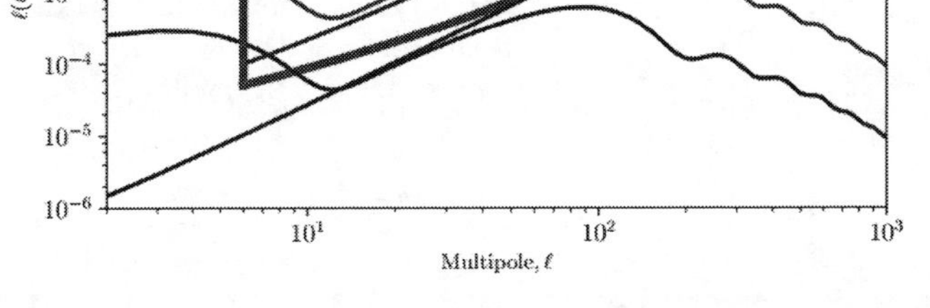

Figure 2. A sensitivity plot of the GB telescope. Two bottom curves (green and blue) are B-mode power spectra for r=0.1, and 0.01. A thick (red) curve is the GB sensitivity estimation when $w_p^{-1/2}$ = 3.5 μK arcmin. A curve above is when $w_p^{-1/2}$ = 10 μK arcmin. The almost straight line (black) except right corner indicates the gravitational lensing effect, and the dashed curve is EE mode of the power spectrum.

The optical system of GB telescope hosts a Mizuguchi-Dragone dual reflecting mirrors [13] with a focal plane. The field of view of the GB is ±10°, and the angular resolution of each pixel is 0.5° (full-width at the half maximum) for the 145 GHz band. Including these, the design parameters of the GB telescope are summarized in Table 1.

Table 1. List of the parameters specified in the GB telescope.

Aperture diameter	300 mm
Optics	Cross Dragone, FOV = ±10°
Detector	lens-antenna-filter coupled KID
Frequency bands	145 and 220 GHz
Resolution	0.5° at 145 GHz and 0.3° at 220 GHz
Scan speed	120°/s or 20 rpm

2.1 Observation site

The proposed observation site is located at del Teide observatory in Tenerife Island that is one of Canary Islands in Spain. The latitude of the site is 28° N and this allows a unique CMB observation from the Northern hemisphere. Note that most of ground-based CMB telescopes are located in South pole or at a desert in Southern hemisphere. Also, the headquarter of the Institute of Astrophysics at Canary islands (Instituto de Astrofísica de Canarias, IAC) is located near to the observation site, allowing a close collaborative local support from IAC. The proposed site is at the elevation of 2400 m above the sea level.

2.2 Cryogenics

The two rotating structure [14] consists of two sets of tables and support structure. At the bottom of the tables, rotary joints for helium gas circulation and for electric power lines are mounted. The rotary joint for helium gas circulation was specially developed but the rotary joint for electricity is a commercial one. On the table, a cryostat with cryocoolers are mounted. Other instruments such as computer, vacuum gauge, and thermometer monitor are also placed. The overall configuration can be seen in Fig. 3.

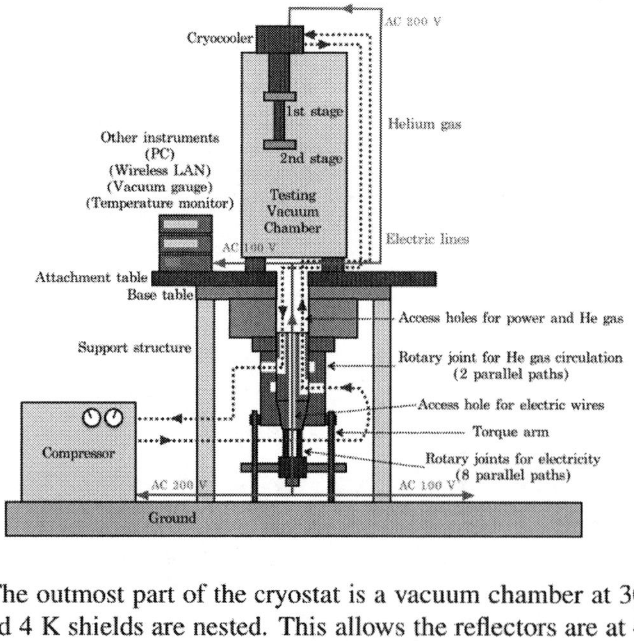

Figure 3. Layout of the system maintaining a dry cooling on a rotating table. A cryostat and other instruments are mounted on the table. A custom-made rotary joint for the helium gas transfer and a commercial rotary joint for the electricity are mounted below the table structure.

The outmost part of the cryostat is a vacuum chamber at 300 K. Inside the vacuum chamber, 40 K and 4 K shields are nested. This allows the reflectors are at 4 K and that mitigates radiation noise from the mirror surface. The fake polarization through non-zero emissivity of mirror surface is also minimized by 1/100 of the intensity compared with ambient temperature reflectors. The cryostat that allows sub-Kelvin refrigeration consists of a pulse tube cryocooler and a helium sorption cooler [15]. The cryostat on the table holds a two-stage pulse tube cryocooler from PT415-RM, Cryomech, Inc., Syracuse, NY. The cooling power for the first stage is 36 W at 45 K and for the second stage 1.35 W at 4.2 K. A series of two rotary joints mentioned above maintains the simultaneous circulation of electricity and helium gas to the coolers. For the further cooling below 4 K, a three-stage helium sorption cooler from CRC_GL10, Chase Research Cryogenics Ltd., Sheffield, UK was used at the bottom of the second stage of the cryocooler. This three stage contain sequentially ^{4}He, ^{3}He, and ^{3}He, and are thermally connected in series. They require cold and warm "switches" for the operation: the cold switch maintains the cryopump to adsorb the helium vapor and the warm switch raises the cryopump temperature. When the cold switch is turned on, each cold head reverts to the cold stage and when the warm switch is on, each tank is refilled with the liquid helium. A stable long-term operation at 234 mK on the rotating table of 20 rpm is achieved and the detailed operation is described in Ref. [16].

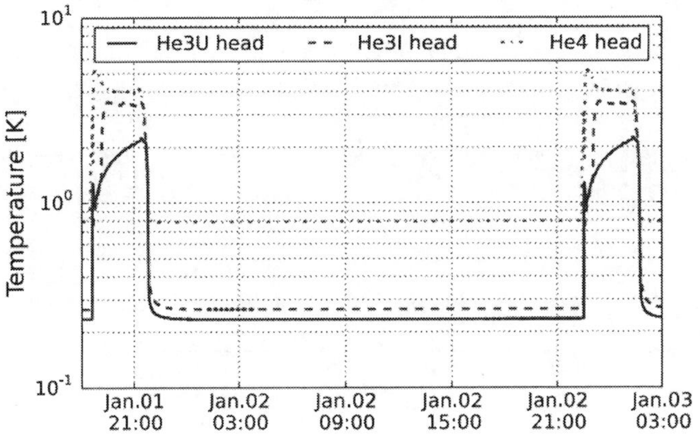

Figure 4. A refrigeration cycle for the sorption cooler on the rotation table at 20 rpm. The variation of temperature at each cold stage as a function of time is shown in this plot. Structure seen at two ends indicate the periods of warm switch that is described in the text.

The cryostat described above allows two mirrors in the Cross Dragone optics system to be at less than 4 K and the focal plane to be at 250 mK and a demonstration of such capability while the cryostat is under rotation of 20 rpm is shown in Fig. 4. The coldest head, labeled as He3U head reaches down to 234 mK when the cryostat is under rotation for long period. This is a clear demonstration of a stable operation of the cryostat under the 20 rpm rotation [16].

2.3 Calibration Strategy

Sparsely arranged straight wires are placed in front of the field of view of the telescope. A combination of these wires and the rotation of the telescope enables us to have an ideal modulation of the polarized calibration signals. This calibration can by construction be continuous during the observation period. In this way, spurious polarization can be as small as less than 0.3% level [10].

2.4 Detector

For the CMB millimeter photon detection, a superconducting thin film array of LC circuits, or kinetic inductance detectors (KIDs) [18] are chosen for the GB telescope. For two millimeter wave bands of GB observation, parameters of KIDs sensors are listed in Table 2. The incoming CMB radiation

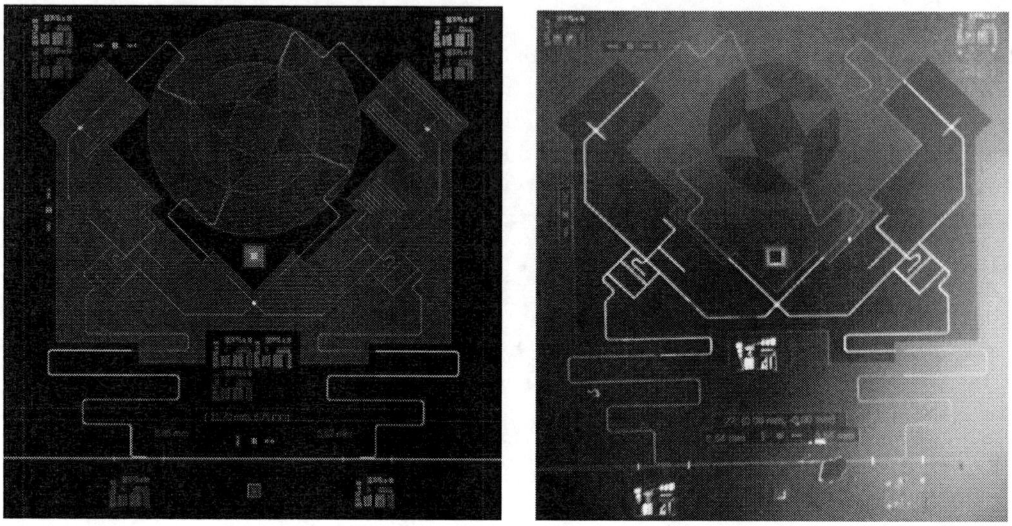

Figure 5. Our KID system mask design for simulation and fabrication (left) and a photograph [17] of the fabricated KID system using the hybrid fabrication process (right). The four triangular arms at the top-center are orthomode transducer based polarization antennas. From there, two 180° hybrid are seen at the left and right sides. The two snake-like structure correspond to two KIDs and the signals from them are read out by a feedline located at the bottom that runs horizontally.

Table 2. List of the parameters specified on KIDs for the GB telescope. Note that the noise temperature in this case is integrated over the time with the photon noise limit from sky temperature.

Band (GHz)	D_{pixel} (3λ) (mm)	NET (μK $\sqrt{\mathrm{s}}$)	Pixel # per wafer	Wafer #	KID #	NET_{array} (μK $\sqrt{\mathrm{s}}$)
145	6.3	310	55	6	660	12
220	4.1	530	112	1	224	35

is guided to the detector system through homemade corrugated horn. Then the polarization of CMB is detected with dual polarization antennas that is a pair of two planar orthomode transducers. Then 180° hybrid selects the signal mode of TE11 and suppress TM01 mode. Since the hybrid has to be of micro strip line while antennas are coplanar ones, we design a Chebyshev transformer to match impedance. After the hybrid, a reverse process has to be made since the KID has coplanar waveguide structure, and for that another Chebyshev transformer is designed. The end-to-end simulation of the entire detector system from antenna to KID front is carried out. The transmittance for our signal TE11 is $\sim$ 90% for GB observation band and the crosstalk of TE11 $\leftrightarrow$ TE11 and TM01 $\rightarrow$ TE11 are found to be negligible.

The actual fabrication of the KID sensor system is a complicated process due to the fact that it contains both coplanar and micro strip line. A hybrid fabrication approach is applied for the fabrication: KID and feedline are fabricated at RIKEN and millimeter-wave circuits are fabricated at National Astronomical Observatory of Japan. A prototype KIDs with and without antenna circuit using this hybrid fabrication approach are made and showed similar performance between with and without antenna circuits. This demonstrates the validity of the hybrid fabrication approach.

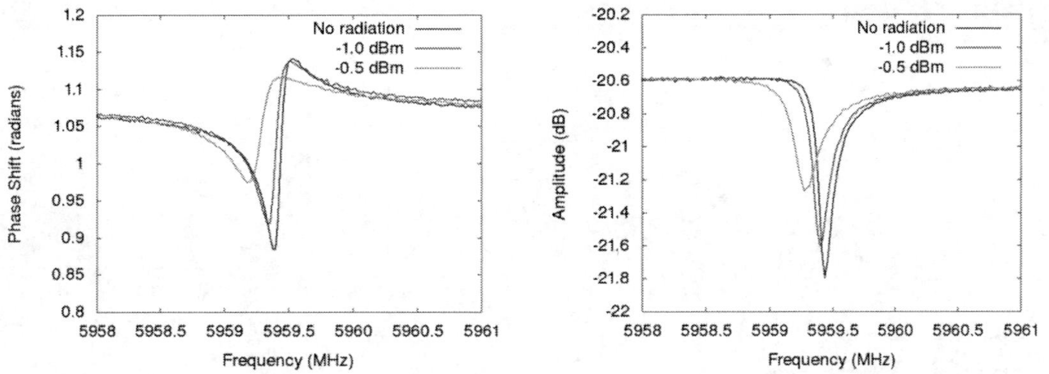

Figure 6. KID response on millimeter wave radiation recorded by our readout system. The left plot shows the phase shift and the right plot shows the amplitude change as a function of the frequency of a KID sensor.

2.5 Readout Electronics

For the readout of the KID signal, we use a commercially available field programmable gate array (FPGA) board. At this moment, we use Xilinx Kintex Ultrascale digital board with a daughter board having dual channel ADC and DAC of both 14 bit and 200 MSps. This daughter board, called RHEA board is a homemade board running at 200 MHz. A digital radio frequency (RF) waves are generated from FPGA and fed into the DAC. Analog RF signals are modulated with a local oscillator and send to KID sensors. The return signals are demodulated and received by ADC.

Figure 6 show an example of real signal from a KID in a cryostat with the described readout system above. With respect to the resonance curve when there is no radiation, there are clear shifts in phase and in amplitude distribution when millimeter wave radiation at different power levels are applied, which are identical observation with a high quality network analyzer, demonstrating that the basic functionality of the readout system that we developed.

3 Summary

The GroundBIRD telescope is aiming at the detection of *B*-mode of CMB polarization down to percent level for *r*. To realize this, a fast rotation of the entire telescope when superconducting millimeter photon sensors are cooling down to 250 mK is implemented. A noble hybrid fabrication method is applied to implement the detector system with both planner and micro strip structure, and its basic principle is demonstrated through comparisons of a prototype with and without antenna circuits. The readout electronics utilize a mixture of commercial FPGA and homemade ADC/DAC boards, and are shown to work as expected. A picture of the assembled GB telescope system is shown in Fig. 7. The current plan is to ship the GB telescope to the observation site in 2018 and commission the GB telescope afterward.

This work is supported by Grants-in-Aid for Scientific Research from The Ministry of Education, Culture, Sports, Science, and Technology, Japan (KAKENHI Grant Nos. 23684017, 25610064, 26105519, 26247045, 14J10972, 15H05448, and 15H05743), the Center for the Promotion of Integrated Sciences (CPIS) of SOKENDAI, the FY 2012 Joint Development Research on an Open Application Basis program of the NAOJ, Research Grants in the Natural Sciences from the Mitsubishi Foundation, and the Basic Science Research Program through the National

Figure 7. A picture of fully assembled GB telescope in a safety cage. A rack on the left contains service electronics to the telescope.

Research Foundation of Korea (NRF) funded by the Ministry of Education, Science and Technology (Grant No. NRF-2017R1A2B3001968). EW acknowledges partially Korea University Future Research Grant.

References

[1] S. Dodelson, *Modern Cosmology*, (Academic Press, Oxford, 2003) 1-20.
[2] The COBE Team: G. F. Smoot *et. al*, Astrophysical Journal Letters, **396**, L1-5 (1992).
[3] W. Rindler, Monthly Notices of the Royal Astronomical Society, **116**, 662-677 (1956).
[4] D. N. Spergel, *et al.*, Astrophysical Journal Supplement Series. **170** (2): 337–408 (2007).
[5] R. Dicke, *Gravitation and the Universe: Jayne Lectures for 1969*, (American Philosophical Society).
[6] A. Guth, Phy. Rev. D **23**, 347 (1981); A. Linde, Phys. Lett. **108B**, 389 (1982); A. Albrecht and P. J. Steinhardt, Phys. Rev. Lett. **48**, 1220 (1982).
[7] U. Seljak, and M. Zaldarriaga, Phys. Rev. Lett. **78**, 2054 (1997).
[8] Planck Collaboration, P. A. R. Ade *et al.*, A&A, **594**, A20 (2016).
[9] BICEP2 Collaboration, P. A. R. Ade *et al.*, Phys. Rev. Lett. **116**, 031302, (2016).
[10] GroundBIRD Group, S. Oguri *et al.*, J. Low Temp. Phys. **176** 691-697 (2014); GroundBIRD Group, S. Oguri *et al.*, J. Low Temp. Phys. **184** 786-792 (2016).
[11] The QUIET Collaboration, Astrophys. J. **741**, 111 (2011).

[12] N. Katayama and E. Komatsu, Astrophys. J. **737**, 78 (2011).
[13] Y. Mizuguchi, M. Akagawa, H. Yokoi, in *Proceedings of IEEE Antennas and Propagation Society, International Symposium*, **14** 2-5 (1976).
[14] S. Oguri, J. Choi, M. Kawai, and O. Tajima, Rev. of Sci. Instr. **84**, 055116 (2013).
[15] R. Bhatia *et al.*, Cryogenics, **40**, 685-691 (2000).
[16] S. Oguri *et. al.*, Rev. of Sci. Instr. **84**, 055116 (2013); S. Oguri *et. al.*, Rev. of Sci. Instr. **85**, 086101 (2014).
[17] The picture on the right courtesy of K. Kiuchi of RIKEN.
[18] P. Day *et. al.*, Nature, **425**, 817 (2003).

What can we learn from GRBs?

Marco Muccino[1,2,⋆], *Remo* Ruffini[1,2,3,4], *Yerlan* Aimuratov[1,3], *Laura M.* Becerra[1,2], *Carlo L.* Bianco[1,2], *Mile* Karlica[1,3], *Milos* Kovacevic[1,3], *Julio D.* Melon Fuksman[1,2], *Rahim* Moradi[1,2], *Ana V.* Penacchioni[5,6], *Giovanni B.* Pisani[1,2], *Daria* Primorac[1,2], *Jorge A.* Rueda[1,2,4], *Soroush* Shakeri[2,7], *Gregory V.* Vereshchagin[1,2], *She-Sheng* Xue[1,2], and *Yu* Wang[1,2]

[1] *Dip. di Fisica, Sapienza Università di Roma, Piazzale Aldo Moro 5, I-00185 Rome, Italy*
[2] *ICRANet-Pescara, Piazza della Repubblica 10, I-65122 Pescara, Italy*
[3] *Université de Nice Sophia-Antipolis, Grand Château Parc Valrose, Nice, CEDEX 2, France*
[4] *ICRANet-Rio, Centro Brasileiro de Pesquisas Fisicas, Rua Dr. Xavier Sigaud 150, Rio de Janeiro, RJ, 22290-180, Brazil*
[5] *University of Siena, Dept. of Physical Sciences, Earth and Environment, Via Roma 56, I-53100 Siena, Italy*
[6] *ASI Science Data Center, via del Politecnico s.n.c., I-00133 Rome Italy*
[7] *Department of Physics, Isfahan University of Technology, 84156-83111, Iran*

Abstract. We review our recent results on the classification of long and short gamma-ray bursts (GRBs) in different subclasses. We provide observational evidences for the binary nature of GRB progenitors. For long bursts the induced gravitational collapse (IGC) paradigm proposes as progenitor a tight binary system composed of a carbon-oxygen core (CO_{core}) and a neutron star (NS) companion; the supernova (SN) explosion of the CO_{core} triggers a hypercritical accretion process onto the companion NS. For short bursts a NS–NS merger is traditionally adopted as the progenitor. We also indicate additional sub-classes originating from different progenitors: (CO_{core})–black hole (BH), BH–NS, and white dwarf–NS binaries. We also show how the outcomes of the further evolution of some of these sub-classes may become the progenitor systems of other sub-classes.

1 Introduction

Gamma-ray bursts (GRBs) are traditionally classified based on their duration [1–5]: short GRBs last $\lesssim 2$ s, while long GRBs last $\gtrsim 2$ s.

Thanks to the extensive data collected by γ-ray telescopes, such as AGILE, BATSE, BeppoSAX, *Fermi*, *HETE-II*, *INTEGRAL*, Konus/WIND and *Swift*, to more sofisticated time-resolved spectral analyses, and to the theoretical treatment of the *fireshell* model [6–8] it has become evident that both long and short bursts originate from binary progenitors and that they can be further subdivided into a variety of sub-classes, depending on the evolution of these binary systems [9–11].

Short bursts are associated to NS-NS or BH-NS mergers [12–22]: their host galaxies are of both early- and late-type, their localization with respect to the host galaxy often indicates a large offset [23–29] or a location of minimal star-forming activity with typical circumburst medium (CBM) densities of $\sim 10^{-5}$–10^{-4} cm^{-3}, and no supernovae (SNe) have ever been associated to them.

Long bursts have been traditionally associated to the death of single massive stars [30]. The large majority of long bursts is related to SNe and are spatially correlated with bright star-forming regions in their host galaxies [31, 32] with a typical CBM density of ~ 1 cm^{-3} [33, 34]. However, the above single progenitor model contrasts with the fact that most massive stars are found in binary systems [35], that most type Ib/c SNe occur in binary systems [36] and that SNe associated to long GRBs are indeed of type Ib/c [37]. Indeed, recently we have found evindence for multiple components in long GRB emissions evidencing the presence of a precise sequence of different astrophysical processes [33, 34], which led to the formulation of the Induced Gravitational Collapse (IGC) paradigm [6, 38–40] expliciting the role of binary systems as progenitors of the long GRBs. The IGC paradigm explains the GRB-SN connection by proposing as progenitors (or *in-state*) a tight binary system composed of a carbon-oxygen core (CO_{core}) undergoing a SN explosion and a companion neutron star (NS) [39–41]. The SN explosion triggers hypercritical accretion onto the companion NS [39, 40, 42].

Recent observations of a prolonged 0.1–100 GeV high energy emission by the *Fermi*-LAT instrument evidenced its correlation with both long [9] and short bursts [10] with isotropic energy $E_{iso} \gtrsim 10^{52}$ erg. On the basis of this correlation in such systems with different progenitors, we have identified the onset of the GeV emission with the moment of the formation of a black hole (BH) [9, 10]. This implies that systems with energy $E_{iso} \lesssim 10^{52}$ erg, which do not exhibit GeV emission, do not form BHs.

Indeed, we proposed the following classification scheme. Long GRBs, according to the IGC paradigm [9], are classified into two sub-classes [42]:

⋆e-mail: marco.muccino@icranet.it

- X-ray flashes (XRFs) with $E_{iso} \lesssim 10^{52}$ erg and rest-frame spectral peak energy $E_{p,i} \lesssim 200$ keV. These systems, already pioneered in a different context [43–45], originate in widely separated CO_{core}–NS binaries with an orbital separation $a > 10^{11}$ cm [41], therefore the hypercritical accretion onto the NS companion is insufficient to induce gravitational collapse to a BH [39–41] and, therefore, as expected in our theory no GeV emission is observed. The outcomes (or *out-states*) of XRFs are binaries composed of a new NS (νNS) produced from the SN explosion, and a massive NS (MNS) which accreted matter from the SN ejecta. Their occurrence rate is $\rho_{\rm XRF} = 100^{+45}_{-34}$ Gpc^{-3}yr^{-1} [42] (see figure 1).
- Binary-driven hypernovae (BdHNe) with $E_{iso} \gtrsim 10^{52}$ erg and $E_{p,i} \gtrsim 200$ keV. BdHNe occur in tighter binaries ($a < 10^{11}$ cm), where the hypercritical accretion onto the companion NS leads to the formation of a BH [41] and, therefore, to the emission of the associated prolonged GeV emission (observable when inside the LAT field of view). Specific constant power-law behaviors are observed in their high energy GeV and X-rays luminosity light curves [9, 46, 47]. The out-states of BdHNe are νNS-BH binaries [41, 42, 48, 49]. The BdHN occurrence rate is $\rho_{\rm BdHN} = 0.77^{+0.09}_{-0.08}$ Gpc^{-3}yr^{-1} [42] (see figure 1).

For progenitor system composed of a CO_{core} in binary with an already formed BH [8, 42], leading to bursts with $E_{iso} \gtrsim 10^{54}$ erg and $E_{p,i} \gtrsim 2$ MeV, the observational identification is still pending. In these systems, which correspond to the late evolutionary stages of X-ray binaries as Cyg X-1 and Cyg X-3 [50], or microquasars [51], the hypercritical accretion produces, as *out-states*, a more massive BH and a νNS. Their occurrence rate can contribute to that of BdHNe, being CO_{core}–BH a particular case of BdHN progenitors (see figure 1).

In total analogy, the formation of a BH can occur in short bursts, depending whether or not the mass of the merged core of the binary system exceeds the NS critical mass. For NS–NS binaries, which are the outcomes of XRFs, we have [10, 42, 48]:

- Short gamma-ray flashes (S-GRFs), with $E_{iso} \lesssim 10^{52}$ erg and $E_{p,i} \lesssim 2$ MeV. They occur when the merged core does not exceed the NS critical mass and does not collapse into a BH, but still remains as a MNS with some additional orbiting material to guarantee the angular momentum conservation. As a consequence, no GeV emission is expected from these systems and, indeed, is not observed. The S-GRF occurrence rate is $\rho_{\rm S\text{-}GRF} = 3.6^{+1.4}_{-1.0}$ Gpc^{-3} y^{-1} [42] (see figure 1).
- Authentic short GRBs (S-GRBs) with $E_{iso} \gtrsim 10^{52}$ erg and $E_{p,i} \gtrsim 2$ MeV. They occur when a BH is formed in the NS–NS merger. Thus, these systems exhibit GeV emission. Their occurrence rate is $\rho_{\rm S\text{-}GRB} = \left(1.9^{+1.8}_{-1.1}\right) \times 10^{-3}$ Gpc^{-3} y^{-1} [42] (see figure 1).

We have recently proposed the existence of ultra-short GRBs (U-GRBs), a new hybrid sub-class of (yet unobserved) short bursts originating from the BdHNe out-states (νNS–BH binaries), which remain bound nearly in 100% of the cases [48]. These systems represent a yet unaccounted family of merging NS-BH binaries in the current standard population synthesis analyses [48], therefore, including other possible channels of formation for NS-BH binaries, the lower limit of the U-GRB occurrence rate can be assumed equal to the BdHN rate, i.e., $\rho_{\rm U\text{-}GRB} \gtrsim 0.77^{+0.09}_{-0.08}$ Gpc^{-3}yr^{-1} [42] (see figure 1).

Finally, we proposed another sub-class of sources originating in NS–WD mergers: gamma-ray flashes (GRFs), a class of long GRBs occurring in a CBM environment with low density, e.g., $\sim 10^{-3}$ cm^{-3}, typical of the halos of the GRB host galaxies [42, 52, 53], not associated with SNe [54], and characterized by the presence of a macronova emission in the optical afterglow [55]. GRFs have $10^{51} \lesssim E_{iso} \lesssim 10^{52}$ erg and $0.2 \lesssim E_{p,i} \lesssim 2$ MeV and, therefore, the NS–WD merger forms a MNS and not a BH [42]. NS–WD binaries, are notoriously very common astrophysical systems [56] and their possible formation channels have been studied [57, 58]; as proposed in Ref. [42], another less likely but yet possible channel of formation is the merger of a NS–WD binary produced from an S-GRF. The GRF rate of occurrence is $\rho_{\rm GRF} = 1.02^{+0.71}_{-0.46}$ Gpc^{-3} y^{-1} [42] (see figure 1).

In all the above systems, the 10^{52} erg limit is clearly a function of the yet unknown precise value of the NS critical mass. As already pointed out in Ref. [42] the direct observation of the separatrix energy between S-GRFs and S-GRBs, and also between BdHNe and XRFs, gives fundamental informations for the determination of the actual value of maximum NS mass and for the minimum mass of the newly-formed BH.

In this paper we review the latest observational and theoretical results which led to above burst classification scheme. In Section 2, we briefly summarize the fireshell model. In Section 3, we describe the observational properties of XRFs and BdHNe and their interpretation within the IGC paradism. In Section 4, we focus on the S-GRBs and specially on the theoretical interpretation of their GeV emission. In Section 4, we summarize our Conclusions.

2 The fireshell model

Before going into details in the observational and theoretical description of XRFs, BdHNe, S-GRFs and S-GRBs, we briefly summarize the fireshell model which is at the basis for the above classification of all bursts.

In the fireshell model [6–8], the GRB emission originate from an optically thick e^+e^- plasma of total energy $E^{\rm tot}_{e^+e^-}$ – the fireshell. Its expansion and self-acceleration is due to the gradual e^+e^- annihilation [59]. Even after engulfing the baryonic mass M_B left over by the progenitor system, quantified by the baryon load $B = M_B c^2/E^{\rm tot}_{e^+e^-}$ [60], the fireshell remains still optically thick and continues its self-acceleration up to ultrarelativistic velocities [61, 62]. When the fireshell reaches the transparency condition, a first flash of radiation, the P-GRB, is emitted [7, 59, 60]. The spectrum of the P-GRB is determined by the geometry of the pair-creation region: in the case of the spherically symmetric dyadosphere, the P-GRB spectrum is generally described by a single thermal component

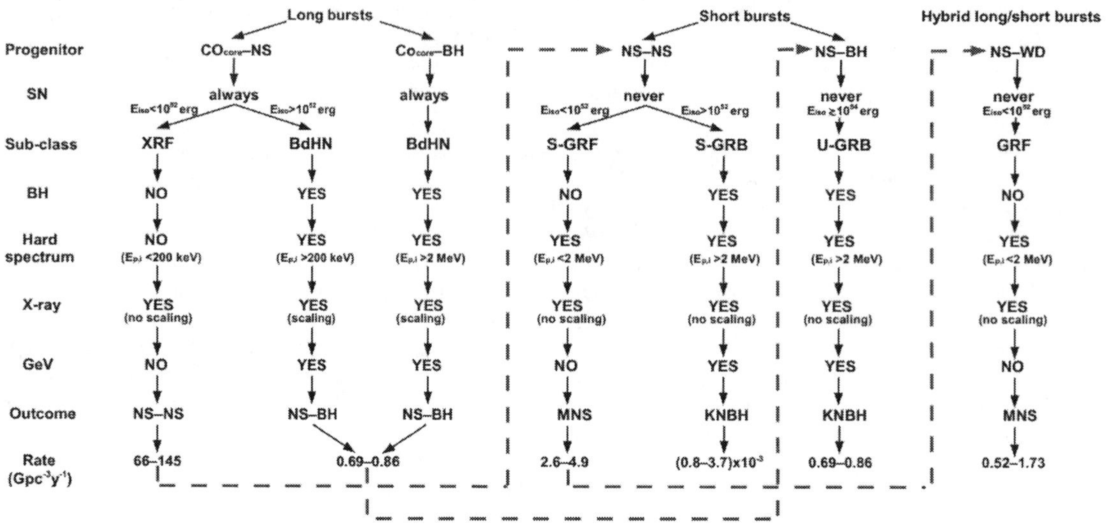

Figure 1. Summary of the properties of the burst sub-classes discussed in the Introduction. The red dashed lines indicate the evolutionary tracks linking out-states and in-states of some of the sub-classes. Additional details can be found in Ref. [42].

[10, 63]; in the case of an axially symmetric dyadotorus, the resulting P-GRB spectrum is a convolution of thermal spectra of different temperatures which resembles more a power-law spectral energy distribution with an exponential cutoff [64, 65].

After transparency, the accelerated baryons propagate through and interact in fully radiative regime with the CBM, giving rise to the prompt emission [7]. The structures observed in the prompt emission of a GRB depend on the CBM density n_{CBM} and its inhomogeneities [66–68]. In both long and short bursts the CBM clouds have similar masses (10^{22}–10^{24} g), sizes ($\approx 10^{15}$–10^{16} cm), and typical distances from the BH ($\approx 10^{16}$–10^{17} cm) [10, 33]. The observed prompt emission spectrum results from the convolution of a large number of comoving spectra with decreasing temperatures and Lorentz and Doppler factors, due to each collision with the CBM, over the surfaces of constant arrival time for photons at the detector [69, 70] over the entire observation time.

To conclude, the evolution of an optically thick baryon-loaded pair plasma, is generally described in terms of $E^{\rm tot}_{e^+e^-}$ and B and it is independent of the way the pair plasma is created. This general formalism can also be applied to any optically thick e^+e^- plasma, like the one created via $\nu\bar{\nu} \leftrightarrow e^+e^-$ mechanism in a NS merger as described in [17, 71, 72].

3 XRFs and BdHNe in the IGC paradigm

We here focus on the comparison between XRF and BdHN sub-classes within the IGC paradigm, giving a special attention to the latest theoretical results on the BdHNe.

In the IGC scenario, both XRFs and BdHNe originate in the hypercritical accretion process of the SN ejecta onto the NS binary companion. In this phenomenon photons are trapped in the accreting material and the accretion energy is lost through a large associated neutrino emission [39, 40, 73, 74]. In the XRFs, the $CO_{\rm core}$-NS binary is widely separated ($a \gtrsim 10^{11}$ cm), thus the accretion rate $< 10^{-2}\, M_\odot\, {\rm s}^{-1}$ can only push the binary companion NS to become a MNS. The resulting emission, dubbed *Episode 1*, lasts $\sim 10^2$–10^4 s. Its spectrum is characterized by: 1) a thermal component spectrum with temperatures in the range of 0.1–2 keV and corresponding radii of 10^{10}–10^{12} cm (see figure 2, left-panel), possibly originating from the outflow within the NS atmosphere driven out by Rayleigh-Taylor convection instabilities [40], and 2) a power-law component, possibly related to the excess of angular momentum of the system which necessarily leads to a jetted emission [41]. The long lasting X-ray emission does not exhibit any specific common late power-law behavior (see figure 2, right panel) and can be explained by the emission of the SN ejecta shocked by the hypercritical accretion emission of the XRF. This energy injection into the SN ejecta leads to the occurrence of a broad-lined SN Ic [75] with a kinetic energy larger than that of the traditional SNe Ic [42]. The absence of GeV emissions is implicit in the nature of the hypercritical accretion process not leading to a BH. Of course, all XRFs at redshift $z \lesssim 1$ exhibit an optical SN with a luminosity similar to the one of SN 1998bw [76], which occurs after 10–15 days in the source cosmological rest-frame.

In the IGC paradigm, the shorter the $CO_{\rm core}$-NS binary period, the larger the accretion rate and the values of $E_{\rm iso}$ and $E_{\rm p,i}$, and correspondingly the shorter the prompt emission duration [41]. Indeed, in BdHNe the $CO_{\rm core}$-NS binary is more tightly bound ($a \lesssim 10^{11}$ cm) and the accretion rates of the SN ejecta can be $\gtrsim 10^{-2}$–$10^{-1}\, M_\odot\, {\rm s}^{-1}$, leading the companion NS to collapse to a BH [40, 41]. For

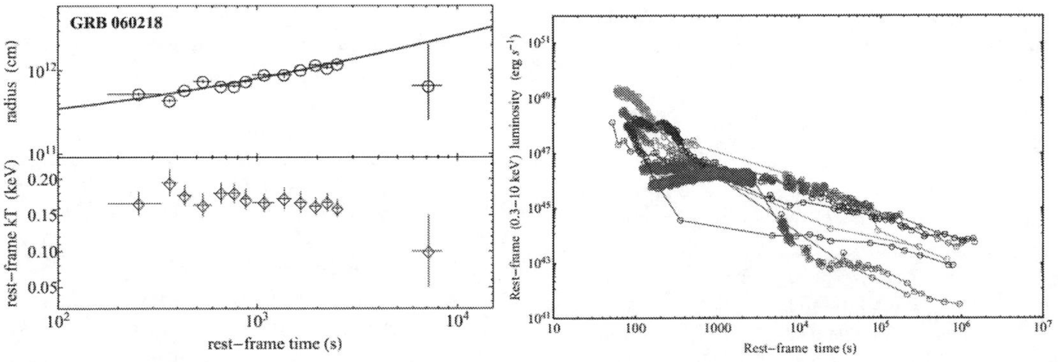

Figure 2. Left panel: the evolution of the radius of the thermal component detected in GRB 060218 (black circles) and its linear fit (solid red curve) and of the corresponding rest-frame temperature (blue diamonds). Reproduced from Ref. [77]. Right panel: rest-frame X-ray 0.3–10 keV luminosity light curves of selected XRFs: 050416A (red), 060218 (dark green), 070419A (orange), 081007 (magenta), 100316D (brown), 101219B (purple), and 130831A (green).

this reason BdHNe exhibit a more complex structure than XRFs composed of distinct Episodes.

- *Episode 1* in BdHNe, like in the case of XRFs, also originates in the hypercritical accretion process. The corresponding spectrum again exhibits: 1) an expanding thermal component with a decreasing temperature, typical radii of 10^9–10^{10} cm and an average expansion speed of $\sim 10^8$–10^9 cm s^{-1} (see figure 3 (a)), and b) a power-law function [33, 34, 78].
- *Episode 2* corresponds to the γ-ray prompt emission of an authentic long GRB (see figure 3 (b)), stemming from the collapse of the companion NS to a BH and leading to the vacuum polarization process and the creation of an e^+e^- plasma. The analysis of the P-GRB emission indicates that BdHNe have a baryon load of $B \approx 10^{-4}$–10^{-2} and at transparency they reach a Lorentz factor of $\Gamma = 10^2$–10^3. The prompt emission is produced by the interaction of the fireshell with the CBM clouds located at $\sim 10^{16}$–10^{17} cm from the burst site with average number density of ~ 1 cm^{-3} [33, 78].
- *Episode 3* occurs after the prompt emission in the X-rays. It composed of a steep decay characterized by the presence of an early X-ray flare, a plateau and a late power-law decay which we refer as to the afterglow. These three components are dubbed flare-plateau-afterglow (FPA) phase [79]. During the early X-ray flare phase (typically at a rest-frame time of $\sim 10^2$ s) an expanding thermal component has been observed in its spectrum [9, 47, 79]. The inferred radii are typically $\sim 10^{12}$–10^{13} cm and they expand at mildly relativistic speed with $\Gamma \lesssim 4$ [9, 47, 78, 79]. The size of the corresponding emitting region is clearly incompatible with the radii inferred from Episodes 1 and 2. When computed in the source cosmological rest-frame, the plateau and the late power-law decay exhibit new features (see figure 3 (c)): 1) the overlapping of the afterglow phases, which have typical slopes of $-1.7 \lesssim \alpha \lesssim -1.3$ and show a characteristic common power-law behavior [46]; the nested property, which shows that the duration (the luminosity) of the plateau phase is inversely (directly) proportional to the energy of the GRB emission, i.e., the more energetic the source, the smaller (higher) the duration (the luminosity) of the plateau [47]. The use of the overlapping of the afterglows as a distance indicator has been explored by inferring the redshifts of GRB 101023 [34], and has been applied to predict the occurrence of the SN associated to GRB 130427A before its discovery [80], later confirmed by the observations [81–84]. In the IGC scenario, the FPA originates from the SN ejecta [9, 79]. In BdHNe the SN ejecta experiences an energy injection from GRB emission leading to the occurrence of a broad-lined SN Ic [75] with a kinetic energy larger than that of the traditional SNe Ic. This energy injection results in an isotropic energy emission of 10^{51}–10^{52} erg for the FPA phase. In particular, the X-ray flare can be modeled by considering the impact of the GRB on the SN ejecta and the propagation of the optically thick e^+e^- plasma into a medium largely baryon-contaminated ($B \approx 10$–10^2). A numerical integration starting at 10^{10} cm all the way to 10^{12} cm, where the transparency is reached, gives a perfect agreement between the radius of the emitter at transparency and the observations, as well a coincidence of the observed time of the peak emission of the flare [79]. The plateau and the afterglow phases are still under study (M. Karlica et al., in preparation).
- *Episode 4* corresponds to the optical SN emission observable in all BdHNe at $z \lesssim 1$ after ≈ 10–15 days in the cosmological rest-frame. All these SNe have a standard luminosity similar to the one of SN 1998bw [76].
- *Episode 5* is identified with the distinctive long-lived GeV emission, observed in the majority of BdHNe when within the LAT field of view. Though this emission follows a precise power-law behavior with index ≈ -1.2 [42, 85] (see figure 3 (d)), this emission is conceptually distinct in its underlying physical process from that of Episode 3: it originate, in facts, in the further accretion of matter onto the newly-formed BH and it is ob-

servable only after the transparency emission, i.e., the P-GRB [42].

4 The S-GRBs in the NS–NS merger paradigm

In Section 1 we discussed the rates of S-GRFs and S-GRBs, showing that S-GRFs are the most frequent events among the short bursts. This result is also in good agreement with the NS–NS binaries observed within our Galaxy: only a subset of them has a total mass larger than the NS critical mass $M_{\rm crit}^{\rm NS}$ and can form a BH in their merging process [10] if we assume that $M_{\rm crit}^{\rm NS} = 2.67 M_\odot$ for a non-rotating, globally neutral NS within the NL3 nuclear model [87]. In this light S-GRBs are very important for inferring constraints on $M_{\rm crit}^{\rm NS}$, on the NS equation of state, and on the minimum mass of the newly-formed BH.

To date, within the fireshell model we have analized five authentic S-GRBs: 090227B [63], 140619B [10], GRB 090510 [64], 081024B and 140402A [65]. The analyses of the P-GRB emission and the correlation between the spikes of the prompt emission and CBM inhomogeneities gave the most successful test for the fireshell model. S-GRBs share some remarkable analogies but also some differences with BdHNe in view of the simplicity of the underlying physical system of S-GRBs, which unlike the BdHNe, do not exhibit any of the extremely complex activities related to the SN (see Section 3).

- *Episode 1* here is related to the NS–NS merger activity before the gravitational collapse into a BH and possibly corresponds to faint precursors observed in some short bursts [64, 88]. Because of the compactness of the systems this process at times is not observable.
- *Episode 2* corresponds to the GRB emission from the NS-NS merger. Within the fireshell model it is composed of the P-GRB, which occur before the onset of the GeV emission, and the prompt emission (see figure 4, left panel). From the analysis of their P-GRB emission, all S-GRBs have a standard values of the baryon load ($B \approx 5 \times 10^{-5}$), which is consistent with the crustal masses of NS-NS mergers [10, 89], and of the Lorentz factors at the transparency $\Gamma \approx 10^4$ [10, 63–65]. From the fit of the prompt emission (see figure 4, right panel), it came out that S-GRBs occur in a standard CBM with average density $\langle n_{\rm CBM} \rangle \approx 10^{-5}$ cm^{-3} [10, 63–65], which is typical of galactic halos where NS–NS mergers migrate, owing to natal kicks imparted to the binaries at birth [22].
- *Episode 3*, which corresponds to the traditional X-ray afterglow, differs from that of BdHNe which results from the interaction between the GRB and the SN ejecta. Work on this topic is still ongoing.
- *Episode 4*, identified with the optical emission of a SN, is here missing.
- *Episode 5* coincides with the GeV emission turning on soon after the P-GRB and being coeval with the prompt emission. With the exception of GRB 090227B, which was outside the nominal *Fermi*-LAT field of view [86], all S-GRBs consistently exhibit this emission, which appears to be strictly correlated to that observed in the BdHNe. Since the presence of a BH is the only commonality between BdHNe and S-GRBs, by analogy we assume that the GeV emission originate from the activity of the newly-born BH produced in the merger [10]. The rest-frame 0.1–100 GeV luminosity light curves of all S-GRBs with LAT data follow a common power-law behavior with the rest-frame time which goes as $t^{-1.29\pm0.06}$ (see dashed black line in figure 5).

Table 1 lists the redshift, $E_{p,i}$, E_{iso} (in the rest-frame energy band 1–10000 keV), and the GeV isotropic emission energy E_{LAT} (in the rest-frame energy band 0.1–100 GeV) of all S-GRBs. The values of E_{LAT} represent lower limits to the actual GeV isotropic emission energies, since at late times the observations of GeV emission could be prevented due to instrumental threshold of the LAT. Using the maximum GeV photon observed energy $E_{\rm GeV}^{\rm max}$ listed in table 1, we derive a lower limit on the Lorentz factor of the GeV emission $\Gamma_{\rm GeV}^{\rm min}$ by requiring the optically thin condition to the high energy photons [90]. For each S-GRB we estimate lower limits in each time interval of the GeV luminosity light curves in figure 5. Then, $\Gamma_{\rm GeV}^{\rm min}$ for each S-GRB has been then determined as the largest among the inferred lower limits. It follows that the GeV emission is produced by an ultrarelativistic outflows with $\Gamma_{\rm GeV}^{\rm min} \gtrsim 300$ (see table 1).

We propose that the GeV emission in S-GRBs is produced by accretion onto the new-born BH of a certain amount of mass that remains bound to it because of the conservation of energy and angular momentum from the merger moment to the BH birth [64]. Lower limits on the amount of accreted mass can be attained by considering the accretion process onto a maximally rotating Kerr BH. Depending whether the infalling material is in co- or counter-rotating orbit with the spinning BH, the maximum efficiency of the conversion of gravitational energy into radiation is $\eta_+ = 42.3\%$ or $\eta_- = 3.8\%$, respectively [91] and, therefore, $E_{\rm LAT}$ can be expressed as

$$E_{\rm LAT} = f_{\rm b}^{-1} \eta_\pm M_{\rm acc}^{\eta_\pm} c^2 \,, \qquad (1)$$

where $f_{\rm b}$ is the beaming factor which depends on the geometry of the GeV emission, and $M_{\rm acc}^{\eta_\pm}$ is the amount of accreted mass depending on the choice of the efficiency. The observational evidence that the totality of S-GRBs exhibit GeV emission and that its absence is due instrumental absence of alignment between the LAT and the source at the time of the GRB emission suggest no significant beaming. Therefore, in the following we set $f_{\rm b} \equiv 1$. The corresponding estimates of $M_{\rm acc}^{\eta_\pm}$ in our sample of S-GRBs are listed in table 1.

5 Conclusions

Remarkable progresses in the understanding of GRBs have been made possible thanks to the great amount X- and γ-rays and high energy data and to a deeper theoretical understanding of WD [92], NS [87, 89] and BH [93], leading to a new paradigm purporting the role of binary systems as progenitors of GRBs: $\rm CO_{core}$–NS binaries for long

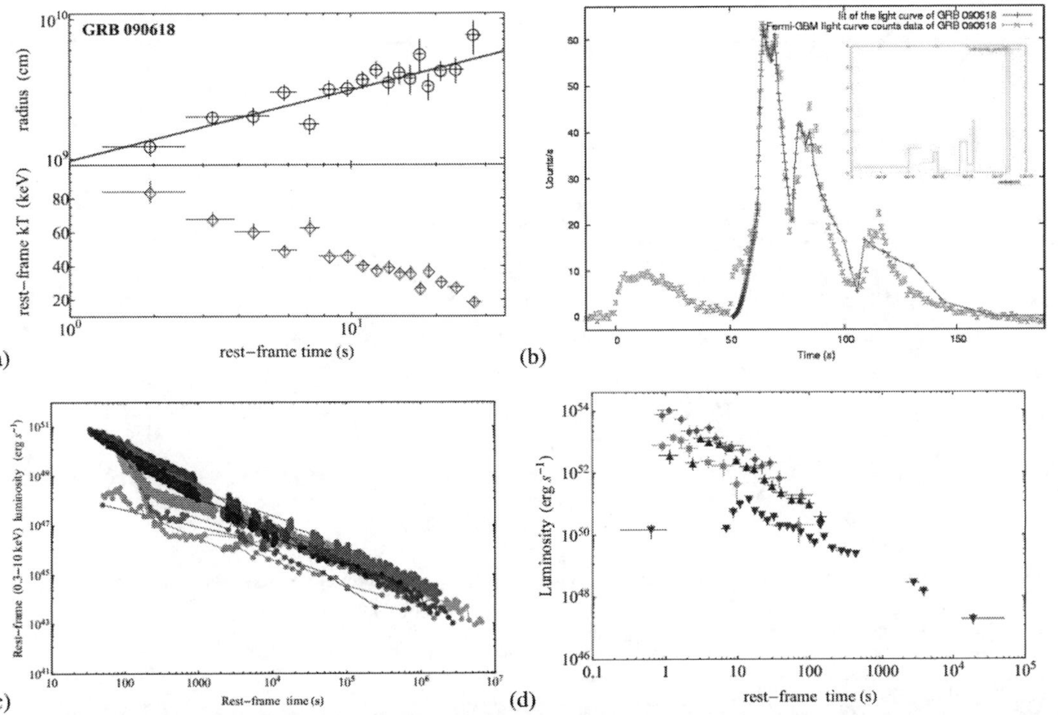

Figure 3. (a) The evolution of the radius of the thermal component detected in the Episode 1 of GRB 090618 (black circles) and its linear fit (solid red curve), and the decay of the corresponding rest-frame temperature (blue diamonds). (b) The fireshell simulation (red line) of the light curve of Episode 2 of the prototype GRB 090618 (green data). The small inset reproduces the CBM profile required for the simulation. Reproduced from Ref. [33]. (c) The rest-frame 0.3–10 keV luminosity light curves of selected BdHNe: 050525 (brown), 060729 (pink), 061007 (black), 080319B (blue), 090618 (green), 091127 (red), 100816A (orange), 111228A (light blue), and 130427A (purple) [79]. (d) The rest-frame 0.1–100 GeV luminosity light curves [86] of selected BdHNe: GRB 080916C (magenta circles), GRB 090902B (purple triangles), GRB 110731A (orange squares), GRB 130427A (blue reversed triangles).

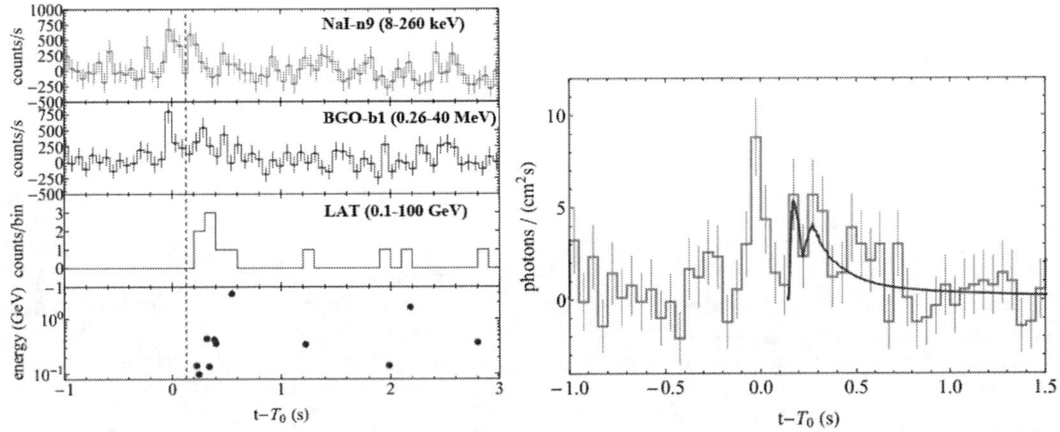

Figure 4. Left panel: background subtracted 50 ms binned data from the NaI-n9 (8 – 260 keV, top panel) and BGO-b1 (0.26 – 40 MeV, second panel) detectors, the 100 ms binned high energy data (0.1 – 100 GeV, third panel, without error bars), and the high energy photons detected by the of the *Fermi*-LAT (bottom panel) for the S-GRB 081024B; the vertical dashed line marks the on-set of the LAT light curve. Right panel: the NaI-n9 light curve of the prompt emission of the S-GRB 081024B (green data) and the simulation within the fireshell model (red curve). All plots are reproduced from Ref. [65].

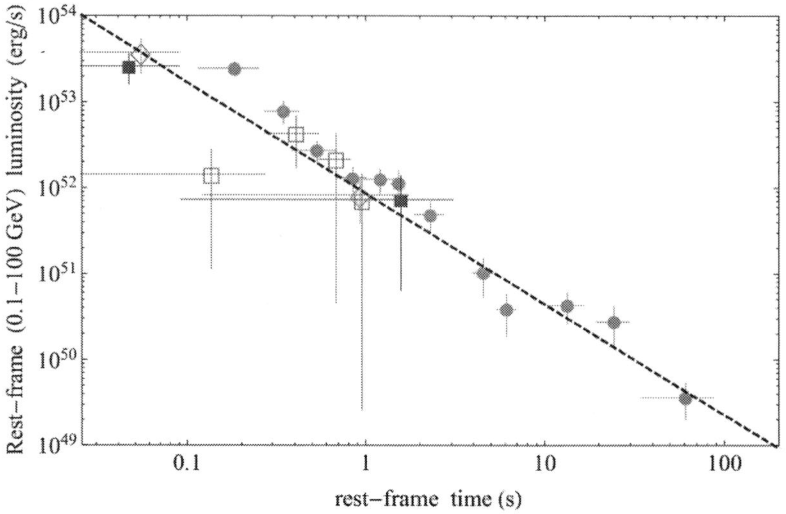

Figure 5. The rest-frame 0.1–100 GeV isotropic luminosities of the S-GRBs: 081024B (orange empty diamonds), 090510 (gray filled circles), 140402A (red filled squares), and 140619B (green empty squares). All the light curves start after the P-GRB emission. The dashed black line marks the common behavior of all the S-GRB light curves which goes as $t^{-1.29\pm0.06}$. Reproduced from Ref. [65].

GRB	z	$E_{p,i}$ (MeV)	E_{iso} (10^{52} erg)	E_{GeV}^{max} (GeV)	Γ_{GeV}^{min}	E_{LAT} (10^{52} erg)	$M_{acc}^{\eta_+}$ ($M_\odot$)	$M_{acc}^{\eta_-}$ ($M_\odot$)
081024B	3.12 ± 1.82	9.56 ± 4.94	2.64 ± 1.00	3	$\gtrsim 779$	$\gtrsim 2.79 \pm 0.98$	$\gtrsim 0.04$	$\gtrsim 0.41$
090227B	1.61 ± 0.14	5.89 ± 0.30	28.3 ± 1.5	–	–	–	–	–
090510	0.903 ± 0.003	7.89 ± 0.76	3.95 ± 0.21	31	$\gtrsim 551$	$\gtrsim 5.78 \pm 0.60$	$\gtrsim 0.08$	$\gtrsim 0.86$
140402A	5.52 ± 0.93	6.1 ± 1.6	4.7 ± 1.1	3.7	$\gtrsim 354$	$\gtrsim 4.5 \pm 2.2$	$\gtrsim 0.06$	$\gtrsim 0.66$
140619B	2.67 ± 0.37	5.34 ± 0.79	6.03 ± 0.79	24	$\gtrsim 471$	$\gtrsim 2.34 \pm 0.91$	$\gtrsim 0.03$	$\gtrsim 0.35$

Table 1. S-GRB properties: z, $E_{p,i}$, the maximum GeV photon observed energy E_{GeV}^{max}, the minimum Lorentz factor of the GeV emission Γ_{GeV}^{min}, E_{iso}, E_{LAT}, and the amount of infalling accreting mass co-rotating (counter-rotating) with the BH $M_{acc}^{\eta_+}$ ($M_{acc}^{\eta_-}$), needed to explaing E_{LAT}.

bursts within the IGC paradigm [6, 38–40], and NS–NS (or NS-BH) binaries for short bursts, as widely accepted and confirmed by strong observational and theoretical evidences [12–22]. These paradigms have led to the classification of GRBs in seven different sub-classes (see figure 1). We here focus our attention on the sub-classes of XRFS, BdHNe, S-GRFs and S-GRBs.

In Section 2, we review the fireshell model for GRBs [6–8] and its essential role in order to disentangle the various emission episodes characterizing each of the above sub-classes.

In Section 3, we summarize the commonalities and the differences between the observational properties of XRFs and BdHNe and provide their theoretical interpretation within the IGC paradigm, namely, whether or not the hypercritical accretion process leads to the formation of a BH.

In Section 4, we outline the properties S-GRFs and S-GRBs originating in NS–NS mergers leading to a MNS and the formation of a BH, respectively. Then, we focus on S-GRBs and on the key role of the P-GRB identification for their description, as well as the analysis of the GeV emission. We finally discuss in details the GeV emission uniquely observed in both BdHNe and S-GRBs, when within the Fermi-LAT FoV. In both cases it starts after the P-GRB emission and it is coeval with the occurrence of the prompt emission [42]. Moreover, the rest-frame 0.1–100 GeV luminosities in BdHNe and S-GRBs share a common luminosity pattern, a precise power-law behavior with time $\propto t^{-1.2}$ [9, 10, 42, 65, 85]. These commonalities, in such different systems, as well as their energy requirements are naturally explained if we assume that the GeV emission originates by accretion processes in the newly-born BH [9, 10]. In all the identified S-GRBs, within the Fermi-LAT FoV, GeV photons are always observed [42, 64]. This implies that no intrinsic beaming is necessary to explain the S-GRB GeV emission. Within the hypothesis of isotropic emission, in the case of S-GRBs we point out how the total energy of the GeV emission can attained from the gravitational binding energy of mass accretion of $M \gtrsim 0.03$–$0.08 M_\odot$ or $M \gtrsim 0.35$–$0.86 M_\odot$ for co- or counter-rotating orbits with a maximally rotating BH, respectively (see table 1). A lower limit on the Lorentz factor of the GeV emission of $\Gamma_{GeV}^{min} \gtrsim 300$ can be obtained by requiring the optically thin condition to the high energy photons [90].

From the above consideration and the proposed classification scheme some considerations follow.

- The knowledge of the separatrix energy of 10^{52} erg, which discriminates between systems forming or not BHs and on which our classification scheme of GRBs is based, represents an observational constraints on the value of the NS critical mass $M_{\rm crit}$, certainly in the range of 2.2–2.7 $M_\odot$ for a non-rotating NS depending on the equations of state [87], and the minimum mass of a BH. This value is consistent and can be derived in BdHNe by considering the hypercritical accretion process onto a NS leading to an energy release in form of neutrinos and photons, given by the gain of gravitational potential energy of the matter accreted in the NS. This includes the change of binding energy of the NS while accreting both matter and angular momentum [42].
- Most noteworthy, the existence of a precise common power-law behavior in the rest-frame 0.1–100 GeV luminosities of S-GRBs (see figure 5), following the BH formation, points to a commonality in the mass and spin of the newly-formed BH. This result is explainable with the expected mass of the merging NSs, each one of mass $M \approx 1.3$–$1.5 M_\odot$ [94], and the above expected range of the non-rotating NS critical mass, leading to a standard value of the BH mass and of its Kerr parameter [10].
- Finally, we discuss the gravitational wave (GW) detectability by advanced LIGO (aLIGO) from S-GRBs. We have already shown that binaries in which each NS has a mass $M_{\rm NS} = 1.34\ M_\odot = 0.5 M_{\rm crit}^{\rm NS}$ produce GW signals which are well below the signal to noise ratio S/N= 8 needed for a positive detection by aLIGO: a positive GW detection may occur only for sources located at $z \lesssim 0.14$ for the aLIGO 2022+ run, a redshift well above that of GRB 090510, to date the closest S-GRB located at $z = 0.903$ [10, 95, 96].

M. M. and J. A. R. acknowledge the partial support of the project N 3101/GF4 IPC-11, and the target program F.0679 0073-6/PTsF of the Ministry of Education and Science of the Republic of Kazakhstan.

References

[1] E.P. Mazets, S.V. Golenetskii, V.N. Ilinskii, V.N. Panov, R.L. Aptekar, I.A. Gurian, M.P. Proskura, I.A. Sokolov, Z.I. Sokolova, T.V. Kharitonova, Astrophysics and Space Science **80**, 3 (1981)

[2] J.P. Dezalay, C. Barat, R. Talon, R. Syunyaev, O. Terekhov, A. Kuznetsov, *Short cosmic events - A subset of classical GRBs?*, in *American Institute of Physics Conference Series*, edited by W. S. Paciesas & G. J. Fishman (1992), Vol. 265 of *American Institute of Physics Conference Series*, pp. 304–309

[3] R.W. Klebesadel, *The durations of gamma-ray bursts*, in *Gamma-Ray Bursts - Observations, Analyses and Theories*, edited by C. Ho, R.I. Epstein, E.E. Fenimore (Cambridge University Press, 1992), pp. 161–168

[4] C. Kouveliotou, C.A. Meegan, G.J. Fishman, N.P. Bhat, M.S. Briggs, T.M. Koshut, W.S. Paciesas, G.N. Pendleton, ApJ Letters **413**, L101 (1993)

[5] M. Tavani, ApJ Letters **497**, L21 (1998), arXiv:astro-ph/9802192

[6] R. Ruffini, C.L. Bianco, F. Fraschetti, S.S. Xue, P. Chardonnet, ApJ Letters **555**, L117 (2001), arXiv:astro-ph/0106534

[7] R. Ruffini, C.L. Bianco, F. Fraschetti, S.S. Xue, P. Chardonnet, ApJ Letters **555**, L113 (2001), arXiv:astro-ph/0106532

[8] R. Ruffini, C.L. Bianco, F. Fraschetti, S.S. Xue, P. Chardonnet, ApJ Letters **555**, L107 (2001), arXiv:astro-ph/0106531

[9] R. Ruffini, Y. Wang, M. Enderli, M. Muccino, M. Kovacevic, C.L. Bianco, A.V. Penacchioni, G.B. Pisani, J.A. Rueda, ApJ **798**, 10 (2015), 1405.5723

[10] R. Ruffini, M. Muccino, M. Kovacevic, F.G. Oliveira, J.A. Rueda, C.L. Bianco, M. Enderli, A.V. Penacchioni, G.B. Pisani, Y. Wang et al., ApJ **808**, 190 (2015), 1412.1018

[11] R. Ruffini, J.A. Rueda, M. Muccino, G.B. Pisani, Y. Wang, L.M. Becerra, M. Kovacevic, F.G. Oliveira, Y. Aimuratov, C.L. Bianco et al., ArXiv e-prints (2016), 1602.02732

[12] J. Goodman, ApJ Letters **308**, L47 (1986)

[13] B. Paczynski, ApJ Letters **308**, L43 (1986)

[14] D. Eichler, M. Livio, T. Piran, D.N. Schramm, Nature **340**, 126 (1989)

[15] R. Narayan, T. Piran, A. Shemi, ApJ Letters **379**, L17 (1991)

[16] P. Meszaros, M.J. Rees, ApJ Letters **482**, L29 (1997), astro-ph/9609065

[17] S. Rosswog, E. Ramirez-Ruiz, M.B. Davies, MNRAS **345**, 1077 (2003), astro-ph/0110180

[18] W.H. Lee, E. Ramirez-Ruiz, D. Page, ApJ Letters **608**, L5 (2004), astro-ph/0404566

[19] E. Nakar, Physics Reports **442**, 166 (2007), astro-ph/0701748

[20] A. Endrizzi, R. Ciolfi, B. Giacomazzo, W. Kastaun, T. Kawamura, ArXiv e-prints (2016), 1604.03445

[21] M. Ruiz, R.N. Lang, V. Paschalidis, S.L. Shapiro, ApJ Letters **824**, L6 (2016), 1604.02455

[22] E. Berger, Annual Review of A&A **52**, 43 (2014), 1311.2603

[23] K.C. Sahu, M. Livio, L. Petro, F.D. Macchetto, J. van Paradijs, C. Kouveliotou, G.J. Fishman, C.A. Meegan, P.J. Groot, T. Galama, Nature **387**, 476 (1997), arXiv:astro-ph/9705184

[24] J. van Paradijs, P.J. Groot, T. Galama, C. Kouveliotou, R.G. Strom, J. Telting, R.G.M. Rutten, G.J. Fishman, C.A. Meegan, M. Pettini et al., Nature **386**, 686 (1997)

[25] J.S. Bloom, J.X. Prochaska, D. Pooley, C.H. Blake, R.J. Foley, S. Jha, E. Ramirez-Ruiz, J. Granot, A.V. Filippenko, S. Sigurdsson et al., ApJ **638**, 354 (2006), arXiv:astro-ph/0505480

[26] E. Troja, A.R. King, P.T. O'Brien, N. Lyons, G. Cusumano, MNRAS **385**, L10 (2008), 0711.3034

[27] W. Fong, E. Berger, D.B. Fox, ApJ **708**, 9 (2010), 0909.1804

[28] E. Berger, New Astron. Rev. **55**, 1 (2011), 1005.1068

[29] D. Kopač, P. D'Avanzo, A. Melandri, S. Campana, A. Gomboc, J. Japelj, M.G. Bernardini, S. Covino, S.D. Vergani, R. Salvaterra et al., MNRAS **424**, 2392 (2012), 1203.1864

[30] S.E. Woosley, J.S. Bloom, Annual Review of A&A **44**, 507 (2006), astro-ph/0609142

[31] A.S. Fruchter, A.J. Levan, L. Strolger, P.M. Vreeswijk, S.E. Thorsett, D. Bersier, I. Burud, J.M. Castro Cerón, A.J. Castro-Tirado, C. Conselice et al., Nature **441**, 463 (2006), arXiv:astro-ph/0603537

[32] K.M. Svensson, A.J. Levan, N.R. Tanvir, A.S. Fruchter, L.G. Strolger, MNRAS **405**, 57 (2010), 1001.5042

[33] L. Izzo, R. Ruffini, A.V. Penacchioni, C.L. Bianco, L. Caito, S.K. Chakrabarti, J.A. Rueda, A. Nandi, B. Patricelli, A&A **543**, A10 (2012), 1202.4374

[34] A.V. Penacchioni, R. Ruffini, L. Izzo, M. Muccino, C.L. Bianco, L. Caito, B. Patricelli, L. Amati, A&A **538**, A58 (2012), 1112.2970

[35] N. Smith, Annual Review of A&A **52**, 487 (2014), 1402.1237

[36] N. Smith, W. Li, A.V. Filippenko, R. Chornock, MNRAS **412**, 1522 (2011), 1006.3899

[37] M. Della Valle, International Journal of Modern Physics D **20**, 1745 (2011)

[38] R. Ruffini, M.G. Bernardini, C.L. Bianco, L. Caito, P. Chardonnet, M.G. Dainotti, R. Fraschetti, R. Guida, G. Vereshchagin, S. Xue, *The Role of GRB 031203 in Clarifying the Astrophysical GRB Scenario*, in *ESA Special Publication* (2007), Vol. 622 of *ESA Special Publication*, p. 561, 0705.2456

[39] J.A. Rueda, R. Ruffini, ApJ **758**, L7 (2012)

[40] C.L. Fryer, J.A. Rueda, R. Ruffini, ApJ Letters **793**, L36 (2014), 1409.1473

[41] L. Becerra, F. Cipolletta, C.L. Fryer, J.A. Rueda, R. Ruffini, ApJ **812**, 100 (2015), 1505.07580

[42] R. Ruffini, J.A. Rueda, M. Muccino, Y. Aimuratov, L.M. Becerra, C.L. Bianco, M. Kovacevic, R. Moradi, F.G. Oliveira, G.B. Pisani et al., ApJ **832**, 136 (2016), 1602.02732

[43] J. Heise, *X-Ray Flashes and X-Ray Counterparts of Gamm-Ray Bursts*, in *Gamma-Ray Burst and Afterglow Astronomy 2001: A Workshop Celebrating the First Year of the HETE Mission*, edited by G.R. Ricker, R.K. Vanderspek (2003), Vol. 662 of *American Institute of Physics Conference Series*, pp. 229–236, astro-ph/0111246

[44] L. Amati, F. Frontera, J.J.M. in't Zand, M. Capalbi, R. Landi, P. Soffitta, L. Vetere, L.A. Antonelli, E. Costa, S. Del Sordo et al., A&A **426**, 415 (2004), astro-ph/0407166

[45] A.M. Soderberg, S.R. Kulkarni, E. Nakar, E. Berger, P.B. Cameron, D.B. Fox, D. Frail, A. Gal-Yam, R. Sari, S.B. Cenko et al., Nature **442**, 1014 (2006), astro-ph/0604389

[46] G.B. Pisani, L. Izzo, R. Ruffini, C.L. Bianco, M. Muccino, A.V. Penacchioni, J.A. Rueda, Y. Wang, A&A **552**, L5 (2013), 1304.1764

[47] R. Ruffini, M. Muccino, C.L. Bianco, M. Enderli, L. Izzo, M. Kovacevic, A.V. Penacchioni, G.B. Pisani, J.A. Rueda, Y. Wang, A&A **565**, L10 (2014), 1404.3946

[48] C.L. Fryer, F.G. Oliveira, J.A. Rueda, R. Ruffini, Physical Review Letters **115**, 231102 (2015), 1505.02809

[49] L. Becerra, C.L. Bianco, C.L. Fryer, J.A. Rueda, R. Ruffini, ApJ **833**, 107 (2016), 1606.02523

[50] R. Giacconi, R. Ruffini, *Physics and astrophysics of neutron stars and black holes* (1978)

[51] I.F. Mirabel, L.F. Rodríguez, Nature **392**, 673 (1998)

[52] L. Caito, M.G. Bernardini, C.L. Bianco, M.G. Dainotti, R. Guida, R. Ruffini, A&A **498**, 501 (2009), 0810.4855

[53] L. Caito, L. Amati, M.G. Bernardini, C.L. Bianco, G. de Barros, L. Izzo, B. Patricelli, R. Ruffini, A&A **521**, A80+ (2010), 1006.4842

[54] M. Della Valle, G. Chincarini, N. Panagia, G. Tagliaferri, D. Malesani, V. Testa, D. Fugazza, S. Campana, S. Covino, V. Mangano et al., Nature **444**, 1050 (2006), arXiv:astro-ph/0608322

[55] Z.P. Jin, X. Li, Z. Cano, S. Covino, Y.Z. Fan, D.M. Wei, ApJ Letters **811**, L22 (2015), 1507.07206

[56] M. Cadelano, C. Pallanca, F.R. Ferraro, M. Salaris, E. Dalessandro, B. Lanzoni, P.C.C. Freire, ApJ **812**, 63 (2015), 1509.01397

[57] T.M. Tauris, E.P.J. van den Heuvel, G.J. Savonije, ApJ Letters **530**, L93 (2000), astro-ph/0001013

[58] P. Lazarus, T.M. Tauris, B. Knispel, P.C.C. Freire, J.S. Deneva, V.M. Kaspi, B. Allen, S. Bogdanov, S. Chatterjee, I.H. Stairs et al., MNRAS **437**, 1485 (2014), 1310.5857

[59] R. Ruffini, J.D. Salmonson, J.R. Wilson, S.S. Xue, A&A **350**, 334 (1999), arXiv:astro-ph/9907030

[60] R. Ruffini, J.D. Salmonson, J.R. Wilson, S. Xue, A&A **359**, 855 (2000), arXiv:astro-ph/0004257

[61] A.G. Aksenov, R. Ruffini, G.V. Vereshchagin, Physical Review Letters **99**, 125003 (2007), 0707.3250

[62] A.G. Aksenov, R. Ruffini, G.V. Vereshchagin, Physical Review D **79**, 043008 (2009), 0901.4837

[63] M. Muccino, R. Ruffini, C.L. Bianco, L. Izzo, A.V. Penacchioni, ApJ **763**, 125 (2013), 1205.6600

[64] R. Ruffini, M. Muccino, Y. Aimuratov, C.L. Bianco, C. Cherubini, M. Enderli, M. Kovacevic, R. Moradi, A.V. Penacchioni, G.B. Pisani et al., ApJ **831**, 178 (2016), 1607.02400

[65] Y. Aimuratov, R. Ruffini, M. Muccino, C.L. Bianco, A.V. Penacchioni, G.B. Pisani, D. Primorac, J.A. Rueda, Y. Wang, ApJ **844**, 83 (2017), 1704.08179

[66] R. Ruffini, C.L. Bianco, P. Chardonnet, F. Fraschetti, S. Xue, ApJ Letters **581**, L19 (2002), arXiv:astro-ph/0210648

[67] R. Ruffini, C.L. Bianco, S.S. Xue, P. Chardonnet, F. Fraschetti, V. Gurzadyan, International Journal of Modern Physics D **13**, 843 (2004), arXiv:astro-ph/0405284

[68] R. Ruffini, M.G. Bernardini, C.L. Bianco, P. Chardonnet, F. Fraschetti, V. Gurzadyan, L. Vitagliano, S.S. Xue, *The Blackholic energy: long and short Gamma-Ray Bursts (New perspectives in physics and astrophysics from the theoretical understanding of Gamma-Ray Bursts, II)*, in *XIth Brazilian School of Cosmology and Gravitation*, edited by M. Novello & S. E. Perez Bergliaffa (2005), Vol. 782 of *American Institute of Physics Conference Series*, p. 42, arXiv:astro-ph/0503476

[69] C.L. Bianco, R. Ruffini, ApJ Letters **633**, L13 (2005), arXiv:astro-ph/0509621

[70] C.L. Bianco, R. Ruffini, ApJ Letters **620**, L23 (2005), arXiv:astro-ph/0501390

[71] R. Narayan, B. Paczynski, T. Piran, ApJ Letters **395**, L83 (1992), arXiv:astro-ph/9204001

[72] J.D. Salmonson, J.R. Wilson, ApJ **578**, 310 (2002), astro-ph/0203349

[73] Y.B. Zel'dovich, L.N. Ivanova, D.K. Nadezhin, Soviet Astronomy **16**, 209 (1972)

[74] R. Ruffini, J. Wilson, Physical Review Letters **31**, 1362 (1973)

[75] K. Maeda, K. Nomoto, ApJ **598**, 1163 (2003), astro-ph/0304172

[76] T.J. Galama, P.M. Vreeswijk, J. van Paradijs et al., Nature **395**, 670 (1998)

[77] S. Campana, V. Mangano, A.J. Blustin, P. Brown, D.N. Burrows, G. Chincarini, J.R. Cummings, G. Cusumano, M. Della Valle, D. Malesani et al., Nature **442**, 1008 (2006), arXiv:astro-ph/0603279

[78] R. Ruffini, *Black Holes, Supernovae and Gamma Ray Bursts*, in *Thirteenth Marcel Grossmann Meeting: On Recent Developments in Theoretical and Experimental General Relativity, Astrophysics and Relativistic Field Theories*, edited by K. Rosquist (2015), pp. 242–314

[79] R. Ruffini, Y. Wang, Y. Aimuratov, L. Becerra, C.L. Bianco, M. Karlica, M. Kovacevic, L. Li, J.D. Melon Fuksman, R. Moradi et al., ArXiv e-prints (2017), 1704.03821

[80] R. Ruffini, C.L. Bianco, M. Enderli, M. Muccino, A.V. Penacchioni, G.B. Pisani, J.A. Rueda, N. Sahakyan, Y. Wang, L. Izzo, GRB Coordinates Network **14526**, 1 (2013)

[81] A. de Ugarte Postigo, D. Xu, G. Leloudas, T. Kruehler, D. Malesani, J. Gorosabel, C.C. Thoene, R. Sanchez-Ramirez, S. Schulze, J.P. Fynbo et al., GRB Coordinates Network **14646**, 1 (2013)

[82] A.J. Levan, A.S. Fruchter, J. Graham, N.R. Tanvir, J. Hjorth, J. Fynbo, D. Perley, S.B. Cenko, E. Pian, Z. Cano et al., GRB Coordinates Network **14686**, 1 (2013)

[83] A.M. Watson, N. Butler, A. Kutyrev, W.H. Lee, M.G. Richer, C. Klein, O. Fox, J.X. Prochaska, J. Bloom, A. Cucchiara et al., GRB Coordinates Network **14666**, 1 (2013)

[84] D. Xu, A. de Ugarte Postigo, T. Kruehler, D. Malesani, G. Leloudas, J.P.U. Fynbo, J. Hjorth, S. Schulze, P. Jakobsson, Z. Cano et al., GRB Coordinates Network **14597**, 1 (2013)

[85] L. Nava, G. Vianello, N. Omodei, G. Ghisellini, G. Ghirlanda, A. Celotti, F. Longo, R. Desiante, R. Barniol Duran, MNRAS **443**, 3578 (2014), 1406.6693

[86] M. Ackermann, M. Ajello, K. Asano, M. Axelsson, L. Baldini, J. Ballet, G. Barbiellini, D. Bastieri, K. Bechtol, R. Bellazzini et al., ApJ Supplement **209**, 11 (2013), 1303.2908

[87] F. Cipolletta, C. Cherubini, S. Filippi, J.A. Rueda, R. Ruffini, Physical Review D **92**, 023007 (2015), 1506.05926

[88] E. Troja, S. Rosswog, N. Gehrels, ApJ **723**, 1711 (2010), 1009.1385

[89] R. Belvedere, J.A. Rueda, R. Ruffini, Journal of Korean Physical Society **65**, 897 (2014)

[90] Y. Lithwick, R. Sari, ApJ **555**, 540 (2001)

[91] L.D. Landau, E.M. Lifshitz, *The classical theory of fields* (Course of theoretical physics - Pergamon International Library of Science, Technology, Engineering and Social Studies, Oxford: Pergamon Press, 4th rev.engl.ed., 1975)

[92] K. Boshkayev, J.A. Rueda, R. Ruffini, I. Siutsou, ApJ **762**, 117 (2013), 1204.2070

[93] R. Ruffini, G. Vereshchagin, S. Xue, Physics Reports **487**, 1 (2010), 0910.0974

[94] F. Özel, P. Freire, Annual Review of A&A **54**, 401 (2016), 1603.02698

[95] F.G. Oliveira, J.A. Rueda, R. Ruffini, ApJ **787**, 150 (2014)

[96] R. Ruffini, J. Rodriguez, M. Muccino, J.A. Rueda, Y. Aimuratov, U. Barres de Almeida, L. Becerra, C.L. Bianco, C. Cherubini, S. Filippi et al., ArXiv e-prints (2016), 1602.03545

EPJ Web of Conferences **168**, 02001 (2018) https://doi.org/10.1051/epjconf/201816802001
Joint International Conference of ICGAC-XIII and IK-15 on Gravitation, Astrophysics and Cosmology

Hamiltonian reduction of Einstein's equations without isometries

Jong Hyuk Yoon[1,⋆]

[1] *Department of Physics, Konkuk University, 120 Neungdong-ro, Gwangjin-gu, Seoul 05029, Korea*

Abstract. I apply the Hamiltonian reduction procedure to 4-dimensional spacetimes without isometries and find privileged spacetime coordinates in which the physical Hamiltonian is expressed in terms of the conformal two metric and its conjugate momentum. Physical time is the area element of the cross section of null hypersurface, and the physical radial coordinate is defined by *equipotential* surfaces on a given spacelike hypersurface of constant physical time. The physical Hamiltonian is local and positive in the privileged coordinates. Einstein's equations in the privileged coordinates are presented as Hamilton's equations of motions obtained from the physical Hamiltonian.

1 Introduction

It is well-known that the true gravitational degrees of freedom of general relativity reside in the conformal two metric of the spatial cross section of null hypersurfaces[1, 2]. Eliminating unphysical degrees of freedom by identifying arbitrarily specifiable spacetime coordinates with certain functions in phase space and thereby presenting the theory in terms of physical degrees of freedom in the privileged coordinates, free from constraints, is known as the ADM Hamiltonian reduction[3–5]. Prof. Kuchař and others applied this procedure to spacetimes that admit two commuting Killing vector fields, known as midi-superspace[6–8], and showed that Einstein's theory was equivalent to cylindrical massless scalar field theory propagating in the 1+1 dimensional Minkowski spacetime.

In this paper, we apply the ADM Hamiltonian reduction to general spacetimes of 4 dimensions under no symmetry assumption[9–13]. The area element of the spatial cross section of a null hypersurface emerges as the physical time, and the physical radial coordinate is defined by *equipotential* surfaces on a given spacelike hypersurface of constant physical time. We present the fully reduced physical Hamiltonian in these privileged coordinates[14], which turns out to be local and positive. The momentum constraints turn out to be simply the *defining* equations of the physical linear and angular momentum densities in term of the conformal two metric and its conjugate momentum[6]; hence, there remain no constraints to solve, and the theory becomes constraint-free. Moreover, we find that our Hamiltonian reduction is self-consistent because Hamilton's equations of motion obtained through this Hamiltonian reduction are *identical* to the Ricci-flat equations in the privileged coordinates. As a by-product of this Hamiltonian reduction, we found an independent proof of *topological censorship*[15–20], which followed directly from one of the Einstein's equations in these coordinates.

⋆e-mail: yoonjh@konkuk.ac.kr

2 The action in the (2+2) formalism

Let us recall that the metric in the (2+2) decomposition[1, 2, 9–13] of 4 dimensional spacetimes can be written as

$$ds^2 = 2dudv - 2hdu^2 + \tau\rho_{ab}\left(dy^a + A_+{}^a du + A_-{}^a dv\right) \times\left(dy^b + A_+{}^b du + A_-{}^b dv\right). \quad (1)$$

The vector fields $\hat{\partial}_\pm$ defined as

$$\hat{\partial}_\pm := \partial_\pm - A_\pm{}^a \partial_a, \quad (2)$$

where

$$\partial_+ = \partial/\partial u, \quad \partial_- = \partial/\partial v, \quad \partial_a = \partial/\partial y^a \quad (a = 2, 3), \quad (3)$$

are horizontal vector fields orthogonal to the two-dimensional spacelike surface N_2 generated by ∂_a. The inner products of the horizontal vector fields are given by

$$<\hat{\partial}_+, \hat{\partial}_+> = -2h, \quad <\hat{\partial}_+, \hat{\partial}_-> = 1, \quad <\hat{\partial}_-, \hat{\partial}_-> = 0, \quad (4)$$

which tells us that $\hat{\partial}_-$ is a null vector field, and that $\hat{\partial}_+$ has a norm $-2h$, which can be either positive, negative, or zero, depending on the sign of h. In this paper, we choose the sign $-2h > 0$ so that $v =$ constant is a spacelike hypersurface. The metric on N_2 is $\tau\rho_{ab}$, where τ is the area element of N_2 and ρ_{ab} is the conformal two metric with $\det\rho_{ab} = 1$.

As was shown in Ref.[13], the Einstein's equations can be obtained from the variational principle of the following action integral:

$$S = \int dvdud^2y\{\pi_\tau\dot{\tau} + \pi_h\dot{h} + \pi_a\dot{A}_+{}^a + \pi^{ab}\dot{\rho}_{ab} - \text{"1"}\cdot C_- - \text{"0"}\cdot C_+ - A_-{}^a C_a\}, \quad (5)$$

where the overdot $\dot{}$ $= \partial_-$, and "1", "0", and $A_-^{\ a}$ are Lagrange multipliers that enforce the constraints $C_- = 0$, $C_+ = 0$, and $C_a = 0$, which are given by

$$\begin{aligned}(\text{i})\ C_- :=& \frac{1}{2}\pi_h\pi_\tau - \frac{h}{4\tau}\pi_h^2 - \frac{1}{2\tau}\pi_h D_+\tau + \frac{1}{2\tau^2}\rho^{ab}\pi_a\pi_b \\ &-\frac{\tau}{8h}\rho^{ab}\rho^{cd}(D_+\rho_{ac})(D_+\rho_{bd}) - \frac{1}{2h\tau}\rho_{ab}\rho_{cd}\pi^{ac}\pi^{bd} \\ &-\frac{1}{2h}\pi^{ac}D_+\rho_{ac} - \tau \mathrm{R}_{(2)} + D_+\pi_h - \partial_a(\tau^{-1}\rho^{ab}\pi_b) \\ &= 0, \end{aligned} \tag{6}$$

$$\begin{aligned}(\text{ii})\ C_+ :=& \pi_\tau D_+\tau + \pi_h D_+ h + \pi^{ab}D_+\rho_{ab} \\ &-2D_+(h\pi_h + D_+\tau) + 2\partial_a(h\tau^{-1}\rho^{ab}\pi_b + \rho^{ab}\partial_b h) \\ &= 0, \end{aligned} \tag{7}$$

$$\begin{aligned}(\text{iii})\ C_a :=& \pi_\tau\partial_a\tau + \pi_h\partial_a h + \pi^{bc}\partial_a\rho_{bc} - 2\partial_b(\rho_{ac}\pi^{bc}) \\ &-D_+\pi_a - \partial_a(\tau\pi_\tau) = 0, \end{aligned} \tag{8}$$

respectively. Here, $\mathrm{R}_{(2)}$ is the scalar curvature of N_2, and the diffN_2-covariant derivative[13] of a tensor density q_{ab} with weight s is defined as

$$\begin{aligned} D_+q_{ab} :=& \partial_+q_{ab} - [A_+,\ q]_{\mathrm{L}ab} = \partial_+q_{ab} - A_+^{\ c}\partial_c q_{ab} \\ &-q_{cb}\partial_a A_+^{\ c} - q_{ac}\partial_b A_+^{\ c} - s(\partial_c A_+^{\ c})q_{ab}, \end{aligned} \tag{9}$$

where $[A_+,\ q]_{\mathrm{L}ab}$ is the Lie derivative of q_{ab} along $A_+ := A_+^{\ a}\partial_a$. For instance, the diff$N_2$-covariant derivatives of the area element τ and the conformal metric ρ_{ab}, which are a scalar density and a tensor density with weight 1 and -1 with respect to the diffN_2 transformations, respectively, are given by

$$D_\pm\tau = \partial_\pm\tau - A_\pm^{\ a}\partial_a\tau - (\partial_a A_\pm^{\ a})\tau, \tag{10}$$

$$\begin{aligned} D_\pm\rho_{ab} =& \partial_\pm\rho_{ab} - A_\pm^{\ c}\partial_c\rho_{ab} - \rho_{cb}\partial_a A_\pm^{\ c} - \rho_{ac}\partial_b A_\pm^{\ c} \\ &+(\partial_c A_\pm^{\ c})\rho_{ab}. \end{aligned} \tag{11}$$

On the other hand, h is a scalar under the diffN_2 transformations, whose diffN_2-covariant derivative is given by

$$D_\pm h = \partial_\pm h - A_\pm^{\ a}\partial_a h, \tag{12}$$

and the diffN_2-covariant field strength $F_{+-}^{\ \ a}$ is defined as

$$\begin{aligned} F_{+-}^{\ \ a} :=& \partial_+ A_-^{\ a} - \partial_- A_+^{\ a} - [A_+,\ A_-]^a \\ =& \partial_+ A_-^{\ a} - \partial_- A_+^{\ a} - A_+^{\ b}\partial_b A_-^{\ a} + A_-^{\ b}\partial_b A_+^{\ a}. \end{aligned} \tag{13}$$

The diffN_2-covariant derivatives of the conjugate momenta π_τ, π_h, π_a, and π^{ab}, which are tensor densities of weights 0, 1, 1 and 2, respectively, are given by

$$D_\pm\pi_\tau = \partial_\pm\pi_\tau - A_\pm^{\ a}\partial_a\pi_\tau, \tag{14}$$

$$D_\pm\pi_h = \partial_\pm\pi_h - A_\pm^{\ c}\partial_c\pi_h - (\partial_c A_\pm^{\ c})\pi_h, \tag{15}$$

$$D_\pm\pi_a = \partial_\pm\pi_a - A_\pm^{\ c}\partial_c\pi_a - \pi_c\partial_a A_\pm^{\ c} - (\partial_c A_\pm^{\ c})\pi_a, \tag{16}$$

$$\begin{aligned} D_\pm\pi^{ab} =& \partial_\pm\pi^{ab} - A_\pm^{\ c}\partial_c\pi^{ab} + \pi^{cb}\partial_c A_\pm^{\ a} + \pi^{ac}\partial_c A_\pm^{\ b} \\ &-2(\partial_c A_\pm^{\ c})\pi^{ab}. \end{aligned} \tag{17}$$

We note that because ρ_{ab} has a unit determinant, its derivatives satisfy the conditions

$$\rho^{bc}\partial_\pm\rho_{bc} = \rho^{bc}\partial_a\rho_{bc} = \rho^{bc}D_\pm\rho_{bc} = 0, \tag{18}$$

and the conjugate momentum π^{ab} is traceless,

$$\rho_{ab}\pi^{ab} = 0. \tag{19}$$

3 Hamiltonian reduction I: τ as physical time

Let us define a *potential* function R and its conjugate momentum π_R as

$$\partial_+ R := -h\pi_h, \tag{20}$$

$$\pi_R = -\partial_+\ln(-h), \tag{21}$$

respectively. The transformation from (h, π_h) to (R, π_R) is a *canonical* transformation, as it changes the action integral by total derivatives only. If we impose the constraints $C_+ = 0$ and $C_a = 0$ and choose the Lagrange multiplier $A_-^{\ a} = 0$, then the action in Eq. (5) becomes

$$\begin{aligned} S =& \int dv du d^2y\{\pi_\tau\dot{\tau} + \pi_R\dot{R} + \pi_a\dot{A}_+^{\ a} + \pi^{ab}\dot{\rho}_{ab} - C_-\} \\ &+ \text{total derivatives}, \end{aligned} \tag{22}$$

where the constraint C_- in the new variables is given by

$$\begin{aligned} C_- =& -\frac{1}{2h}\pi_\tau\partial_+ R - \frac{1}{4h\tau}(\partial_+R)^2 + \frac{1}{2h\tau}(D_+\tau)(\partial_+R) \\ &-\frac{1}{h}D_+(\partial_+R) + \frac{1}{h}(\partial_+R)\partial_+\ln(-h) \\ &-\frac{1}{h}(\partial_+R)A_+^{\ a}\partial_a\ln(-h) \\ &-\tau R^{(2)} + \frac{1}{2\tau^2}\rho^{ab}\pi_a\pi_b - \partial_a(\tau^{-1}\rho^{ab}\pi_b) \\ &-\frac{1}{2h\tau}\rho_{ab}\rho_{cd}\pi^{ac}\pi^{bd} - \frac{\tau}{8h}\rho^{ab}\rho^{cd}(D_+\rho_{ac})(D_+\rho_{bd}) \\ &-\frac{1}{2h}\pi^{ac}D_+\rho_{ac} = 0. \end{aligned} \tag{23}$$

The function h still appears in Eq. (23), but as h is related to R and π_R by Eqs. (20) and (21), the constraint function C_- given by Eq. (23) may be viewed as a function of the new variables $(\tau, Q^I; \pi_\tau, \Pi_I)$, where $Q^I = (R, A_+^{\ a}, \rho_{ab})$ and $\Pi_I = (\pi_R, \pi_a, \pi^{ab})$. Notice that the first term in Eq. (23) is linear in π_τ, and that all the remaining terms are independent of π_τ. Thus, the equation of motion for τ is given by

$$\dot{\tau} = \int du d^2y\frac{\delta C_-}{\delta\pi_\tau} = -\frac{1}{2h}\partial_+R. \tag{24}$$

Now, recall that $\tau = \tau(v, u, y^a)$. If we solve this equation for v, then v may be viewed as a function of (τ, u, y^a) and consequently, $(R,\ A_+^{\ a},\ \rho_{ab})$ may be regarded as functions of (τ, u, y^a). Therefore,

$$\begin{aligned} &\dot{R} = \dot{\tau}\partial_\tau R,\ \dot{A}_+^{\ a} = \dot{\tau}\partial_\tau A_+^{\ a},\ \dot{\rho}_{ab} = \dot{\tau}\partial_\tau\rho_{ab}, \\ &C_- = -(\frac{2h}{\partial_+R})\dot{\tau}C_-, \end{aligned} \tag{25}$$

because $\partial u/\partial v = \partial y^a/\partial v = 0$. Then, the action in Eq. (22) becomes

$$\begin{aligned} S =& \int dv du d^2y\dot{\tau}\{\pi_\tau + \pi_R\partial_\tau R + \pi_a\partial_\tau A_+^{\ a} \\ &+\pi^{ab}\partial_\tau\rho_{ab} + (\frac{2h}{\partial_+R})C_-\} \\ =& \int d\tau du d^2y\{\pi_R\partial_\tau R + \pi_a\partial_\tau A_+^{\ a} + \pi^{ab}\partial_\tau\rho_{ab} - C_{(1)}\} \\ =& \int d\tau du d^2y\{\sum_I\Pi_I\partial_\tau Q^I - C_{(1)}\}, \end{aligned} \tag{26}$$

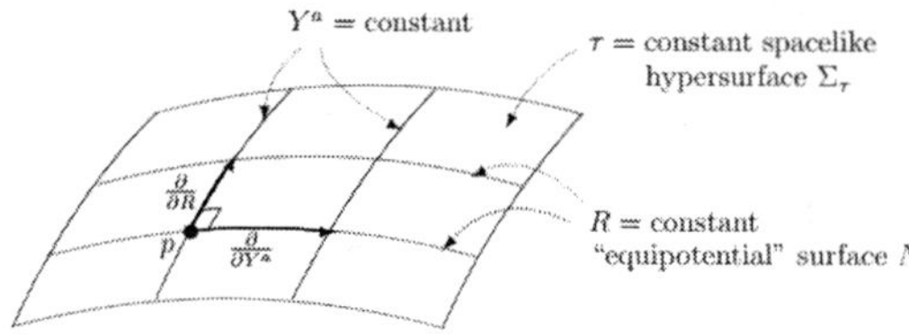

Figure 1. On R = constant surface N_2 on Σ_τ, Y^a = constant is chosen to be normal to N_2 at each point p on Σ_τ. In this case the "shift" vector $A_+{}^a$ is zero at p.

where we replaced $dv\dot{\tau}$ by $d\tau$ in the second line, and $C_{(1)}$ is defined as

$$C_{(1)} = -(\frac{2h}{\partial_+ R})C_- - \pi_\tau. \tag{27}$$

Notice that if we impose the constraint $C_- = 0$, the function $C_{(1)}$ becomes

$$C_{(1)} = -\pi_\tau. \tag{28}$$

4 Hamiltonian reduction II: Fixation of $u = R$ and $y^a = Y^a$ such that $A_+{}^a = 0$

The second step in the Hamiltonian reduction consists of *identifying* arbitrarily specifiable coordinates u and y^a as

$$u = R, \tag{29}$$

$$y^a = Y^a \tag{30}$$

such that the "shift" vector $A_+{}^a$ is zero, i.e.,

$$A_+{}^a = 0. \tag{31}$$

For the class of spacetimes whose spatial topology is either $N_2 \times \mathbb{R}$ or $N_2 \times S^1$, where N_2 is a compact two-dimensional space with the genus g, the condition in Eq. (31) can always be met by relabeling N_2 such that Y^a = constant is normal to u = constant at each point on Σ_τ. If we continue to label N_2 this way, then the two dimensional "shift" A_+^a can always be made zero for the class of spacetimes under consideration (see Fig. 1). Thus, we will work in the coordinates $X^A := (\tau, R, Y^a)$, which satisfy the coordinate conditions

$$\frac{\partial X^A}{\partial X^B} = \delta^A_B. \tag{32}$$

Then, it follows from Eq. (24) that

$$\dot{\tau} = -\frac{1}{2h} > 0, \tag{33}$$

which means that τ increases monotonically along the outgoing null vector field.

Hamilton's equations of motion follow from the variational principle of the action integral in Eq. (26):

$$\partial_\tau Q^I = \int_{\Sigma_\tau} du d^2 y \frac{\delta C_{(1)}}{\delta \Pi_I}|_{u=R, y^a=Y^a}, \tag{34}$$

$$\partial_\tau \Pi_I = -\int_{\Sigma_\tau} du d^2 y \frac{\delta C_{(1)}}{\delta Q^I}|_{u=R, y^a=Y^a}, \tag{35}$$

where $Q^I = (R, A_+{}^a, \rho_{ab})$, $\Pi_I = (\pi_R, \pi_a, \pi^{ab})$, and Σ_τ is a spacelike hypersurface defined by τ = constant.

5 Main results

In the following, we present the main results of this paper.
Einstein's evolution equations I

1. $\frac{\partial R}{\partial \tau} = 0 \Rightarrow$

$$\tau \mathrm{R}_{(2)} = \frac{1}{2}\tau^{-2}\rho^{ab}\pi_a\pi_b - \partial_a(\tau^{-1}\rho^{ab}\pi_b) \tag{36}$$

(topological censorship)

2. $\frac{\partial A_+{}^a}{\partial \tau} = 0 \Rightarrow$

$$\tau^{-1}\pi_a = -\partial_a \ln(-h) \text{ (superpotential)} \tag{37}$$

3. $$\frac{\partial \tau}{\partial \tau} = 1 \text{ (trivial)} \tag{38}$$

4. $$\frac{\partial \ln(-h)}{\partial \tau} = H - \frac{1}{\tau} \text{ (superpotential)} \tag{39}$$

5. $$\frac{\partial \pi_a}{\partial \tau} = 2\tau^{-1}\pi_a + (\pi^{bc} + \frac{\tau}{2}\rho^{bd}\rho^{ce}\partial_R\rho_{de})\partial_a\rho_{bc} \\ -\partial_b(2\pi^{bc}\rho_{ac} + \tau\rho^{bc}\partial_R\rho_{ac}) \tag{40}$$

(evolution equation of π_a)

6. $$\frac{\partial \pi_\tau}{\partial \tau} = \frac{1}{2}\tau^{-2} + \tau^{-2}\rho_{ab}\rho_{cd}\pi^{ac}\pi^{bd} \\ -\frac{1}{4}\rho^{ab}\rho^{cd}(\partial_R\rho_{ac})(\partial_R\rho_{bd}) - 2\tau^{-2}\partial_a(h\rho^{ab}\pi_b) \tag{41}$$

(evolution equation of π_τ)

Einstein's constraint equations

7. $$C_- = 0 \Rightarrow \pi_\tau = -H + 2\partial_R \ln(-h) \\ -2h\{\tau \mathrm{R}^{(2)} - \frac{1}{2\tau^2}\rho^{ab}\pi_a\pi_b + \frac{\partial}{\partial Y^a}(\tau^{-1}\rho^{ab}\pi_b)\} \tag{42}$$

(definition of physical Hamiltonian)

8. $$C_+ = 0 \Rightarrow \pi_R = -\pi^{ab}\partial_R\rho_{ab} \tag{43}$$

(definition of physical linear momentum)

9. $$C_a = 0 \Rightarrow \tau^{-1}\pi_a = -\pi^{bc}\partial_a\rho_{bc} + 2\partial_b(\pi^{bc}\rho_{ac}) \\ -\tau\partial_a(H + \pi_R) \tag{44}$$

(definition of physical angular momentum)

Superpotential $\ln(-h)$

10. $$\partial_\tau \ln(-h) = H - \tau^{-1} \tag{45}$$

11. $$-\partial_R \ln(-h) = \pi_R \tag{46}$$

12. $$-\partial_a \ln(-h) = \tau^{-1}\pi_a \tag{47}$$

Integrability conditions

13. $$\partial_R(\tau^{-1}\pi_a) = \partial_a\pi_R \tag{48}$$

14. $$\partial_\tau\pi_R = -\partial_R H \tag{49}$$

15. $$\partial_\tau(\tau^{-1}\pi_a) = -\partial_a H. \tag{50}$$

In the above equations, H is defined by

$$H = \frac{1}{\tau}\rho_{ab}\rho_{cd}\pi^{ac}\pi^{bd} + \frac{1}{4}\tau\rho^{ab}\rho^{cd}(\partial_R\rho_{ac})(\partial_R\rho_{bd}) \\ +\pi^{ac}\partial_R\rho_{ac} + \frac{1}{2\tau} \geq \frac{1}{2\tau}. \tag{51}$$

The evolution equations of ρ_{ab} and π^{ab} can be found from the reduced action principle

$$S^* = \int_{\Sigma_\tau} dR d^2 Y \{\pi^{ab}\partial_\tau\rho_{ab} - C^*_{(1)}\}, \tag{52}$$

where $C^*_{(1)}$ is the restriction of $C_{(1)}$ to the coordinates $u = R$ and $y^a = Y^a$:

$$C^*_{(1)} := C_{(1)}|_{u=R, y^a=Y^a} = -\pi_\tau = H - 2\partial_R \ln(-h). \quad (53)$$

Einstein's evolution equations II

$$16. \; \frac{\partial \rho_{ab}}{\partial \tau} = \int_{\Sigma_\tau} dR d^2Y \frac{\delta C^*_{(1)}}{\delta \pi^{ab}} \quad (54)$$

$$17. \; \frac{\partial \pi^{ab}}{\partial \tau} = -\int_{\Sigma_\tau} dR d^2Y \frac{\delta C^*_{(1)}}{\delta \rho_{ab}}. \quad (55)$$

The spacetime metric in these privileged coordinates becomes

$$ds^2 = -4hdRd\tau - 2hdR^2 + \tau\rho_{ab}dY^a dY^b. \quad (56)$$

6 discussion

Let us discuss some key features of the Hamiltonian reduction discussed in this article.

(i) First of all, let us mention that the whole set of equations summarized in Section 5 is *identical to the vacuum Einstein's equations*

$$\mathrm{R}_{AB} = 0 \quad (57)$$

for spacetimes whose metric is given by Eq. (56). Thus, the whole procedure of the Hamiltonian reduction respects general covariance, as it must, even though the final theory is written in the privileged coordinates.

(ii) The integral of Eq. (36) over a closed two surface N_2 becomes

$$\int_{N_2} d^2Y \tau^{-2} \rho^{ab} \pi_a \pi_b = 16\pi(1-g) \geq 0, \quad (58)$$

where g is the genus of N_2. This identity states that, as long as the out-going null hypersurface forms a congruence of null geodesics which admits a cross section, the spatial topology of that null hypersurface is either a two sphere or a torus. This is a remarkably simple proof of *topological censorship*, as it does not rely on assumptions such as global hyperbolicity, asymptotic conditions, energy conditions, and so on, either inside or outside the null hypersurface, which are normally assumed in the literature[15–20], although a condition that *no caustics exist on the null hypersurface* is necessary for the existence of well-defined cross sections of the null congruences.

(iii) The spatial integral of $C^*_{(1)}$ defined in Eq. (53) is the sought-for *physical* Hamiltonian K of vacuum spacetimes, haha

$$K := \int_{\Sigma_\tau} dR d^2Y H \geq 0, \quad (59)$$

which is positive-definite in the privileged coordinates.

(iv) The logarithm of the conformal factor in the (τ, R) subspace is the *superpotential* $\ln(-h)$, whose gradients yield $(H - \tau^{-1}, -\pi_R, -\tau^{-1}\pi_a)$ through Eqs. (45), (46), and (47), respectively. The superpotential[6] is a local function of $X^A = (\tau, R, Y^a)$, and is determined by the line integral

$$\ln\frac{h(X)}{h(X_0)} = \int_{X_0}^{X} (H - \tau^{-1})d\tau - \pi_R dR - \tau^{-1}\pi_a dY^a\} \quad (60)$$

along *any* contour from X_0 to X in a given spacetime.

(v) The integrability conditions in Eqs. (48), (49), and (50) are the consistency conditions, which follow from the definition of the superpotential $\ln(-h)$.

(vi) We would like to emphasize that *all* the constraint equations, $C_\pm = 0$ and $C_a = 0$ in the privileged coordinates, are solved in such a way that the constraints simply turn out to be the defining equations of the physical Hamiltonian density π_τ, the linear momentum density π_R, and the angular momentum density $\tau^{-1}\pi_a$, as is summarized in Eqs. (42), (43), and (44), respectively. Namely, π_R and H given by Eqs. (43) and (51) are functions of $(\tau; \rho_{ab}, \pi^{ab})$, and the superpotential $\ln(-h)$ given by Eq. (60) is a function of H, π_R, and $\tau^{-1}\pi_a$, which is again determined by $(\tau; \rho_{ab}, \pi^{ab})$, up to an overall constant factor. Thus, all the terms on the right-hand sides of Eqs. (42), (43), and (44) are functions of $(\tau; \rho_{ab}, \pi^{ab})$. This means that the constraints are completely solved for π_τ, π_R, and $\tau^{-1}\pi_a$ in terms of the *free, unconstrained* data (ρ_{ab}, π^{ab}).

(vii) If we define the total linear momentum Π_R and total angular momentum Π_a as

$$\Pi_R := \int_{\Sigma_\tau} dR d^2Y \pi_R, \quad \Pi_a := \int_{\Sigma_\tau} dR d^2Y \tau^{-1}\pi_a, \quad (61)$$

then we find, by integrating Eqs. (43) and (44),

$$\Pi_R = -\int_{\Sigma_\tau} dR d^2Y \pi^{bc} \partial_R \rho_{bc}, \quad (62)$$

$$\Pi_a = -\int_{\Sigma_\tau} dR d^2Y \pi^{bc} \frac{\partial}{\partial Y^a} \rho_{bc}. \quad (63)$$

Eqs. (62) and (63) show that Π_R and Π_a are the generating functions of translations of ρ_{ab} and π^{ab} along ∂_R and $\partial/\partial Y^a$. This justifies our interpretations of Π_R and Π_a as the total linear and angular momentum carried by the conformal two metric, respectively. Moreover, if no spatial boundaries exist, then by virtue of the integrability conditions in Eqs. (49) and (50), Π_R and Π_a are *conserved* in the time τ.

References

[1] R. Sachs, Proc. Roy. Soc. A **270**, 103 (1962).
[2] R. A. d'Inverno and J. Stachel, J. Math. Phys. **19** (12), 2447 (1978).
[3] R. Arnowitt, S. Deser, and C. W. Misner, Phys. Rev. **117**, 1595 (1960).
[4] C. Rovelli and L. Smolin, Phys. Rev. Lett. **72**, 446 (1994).
[5] J. D. Brown and K. V. Kuchař, Phys. Rev. D **51**, 5600 (1995).
[6] K. V. Kuchař, Phys. Rev. D **4**, 955 (1971).
[7] V. Husain, Phys. Rev. D **50**, 6207 (1994).

[8] J. D. Romano and C. G. Torre, Phys. Rev. D **53**, 5634 (1996).

[9] Y. M. Cho, Q. H. Park, K. S. Soh and J. H. Yoon, Phys. Lett. B **286**, 251 (1992).

[10] J. H. Yoon, Phys. Lett. B **308**, 240 (1993).

[11] J. H. Yoon, Phys. Lett. B **451**, 296 (1999).

[12] J. H. Yoon, J. Korean Phys. Soc. **34**, 108 (1999).

[13] J. H. Yoon, Phys. Rev. D **70**, 084037 (2004).

[14] V. Husain and T. Pawłowski, Phys. Rev. Lett. **108**, 141301 (2012).

[15] S. W. Hawking, Comm. Math. Phys. **25**, 152 (1972).

[16] J. L. Friedman, K. Schleich, and D. M. Witt, Phys. Rev. Lett. **71**, 1486 (1993).

[17] P. T. Chruściel and R. M. Wald, Class. Quantum Grav. **11**, L147 (1994).

[18] T. Jacobson, and S. Venkataramani, Class. Quant. Grav. **12**, 1055 (1995).

[19] G. J. Galloway, Class. Quant. Grav. **13**, 1471 (1996).

[20] G. J. Galloway, K. Schleich, D. M. Witt, and E. Woolgar, Phys. Rev. D **60**, 104039 (1999).

Gravitational Collapse of Massless Fields in an Expanding Universe

Chul-Moon Yoo[1,⋆]

[1] *Division of Particle and Astrophysical Science, Graduate School of Science, Nagoya University, Nagoya 464-8602, Japan*

Abstract. Gravitational collapse of a massless scalar field with the periodic boundary condition in a cubic box is reported. This system can be regarded as a lattice universe model. The initial data is constructed for a Gaussian like profile of the scalar field taking the integrability condition associated with the periodic boundary condition into account. For a large initial amplitude, a black hole is formed after a certain period of time. While the scalar field spreads out in the whole region for a small initial amplitude. The difference of the late time expansion law of the lattice universe depending on the final fate of the gravitational collapse is discussed.

1 Introduction

When we consider gravitational collapse, the most common boundary condition is the asymptotically flat boundary condition. Obviously, the asymptotically flat boundary condition is relevant for stellar collapse. Nonetheless, when we consider collapse of a cosmological object, we may not neglect the effect of the expansion of the universe. For instance, for simulation of a primordial black hole, whose horizon scale may be comparable to the Hubble horizon scale. Here, we report a numerical experiment of gravitational collapse in an expanding background.

In this paper, we consider a massless scalar field in a cubic box with the periodic boundary condition for each pair of opposite boundary surfaces. This system can be regarded as a lattice universe model in which the same cubic region is spatially repeated. Black hole systems with the same boundary condition is studied in Refs. [1–3]. In practice, we simulate the only octant region of the cubic box imposing reflecting boundary condition on each boundary surface. The initial scalar field configuration is set to be a Gaussian like profile. Then, we report how the dynamics of the system changes depending on the initial amplitude of the scalar field profile focusing on black hole formation and expansion law of the lattice universe.

In Sec. 2, settings and initial data construction are given. In Sec. 3, we report the time evolution for large amplitude and small amplitude cases. We use the geometrized units in which both the speed of light and Newton's gravitational constant are one.

⋆e-mail: yoo@gravity.phys.nagoya-u.ac.jp

2 Settings and initial data construction

2.1 scalar field

We consider the massless scalar field, whose equations of motion are given by $\nabla^\mu \nabla_\mu \phi = 0$. Introducing the momentum $\Pi := -n^\mu \nabla_\mu \phi$, we can rewrite the field equations in the form of the 3+1 decomposition as follows:

$$\left(\partial_t - \beta^i \partial_i\right) \phi = -\alpha \Pi, \tag{1}$$
$$\left(\partial_t - \beta^i \partial_i\right) \Pi = -\alpha \Delta \phi - \gamma^{\mu\nu} \partial_\mu \alpha \partial_\nu \phi + \alpha K \Pi, \tag{2}$$

where α, β^i, $\gamma^{\mu\nu}$ and K are the lapse function, shift vector, spatial metric and the trace of the extrinsic curvature K_{ij}.

The stress energy tensor $T_{\mu\nu} = \nabla_\mu \phi \nabla_\nu \phi - g_{\mu\nu}\left(\frac{1}{2}\nabla^\rho \phi \nabla_\rho \phi\right)$ can be also expressed in the form of the 3+1 decomposition as follows:

$$\rho \ := \ n_\mu n_\nu T^{\mu\nu} = \frac{1}{2}\Pi^2 + \frac{1}{2}\mathrm{e}^{-4\psi}\tilde{\gamma}^{ij}\partial_i \phi \partial_j \phi, \tag{3}$$
$$j_i \ := \ -\gamma_{i\mu} n_\nu T^{\mu\nu} = -\Pi \partial_i \phi, \tag{4}$$
$$S_{ij} \ := \ \gamma_{i\mu}\gamma_{j\nu} T^{\mu\nu} = \partial_i \phi \partial_j \phi - \frac{1}{2}\tilde{\gamma}_{ij}\tilde{\gamma}^{kl}\partial_k \phi \partial_l \phi + \frac{1}{2}\mathrm{e}^{4\psi}\tilde{\gamma}_{ij}\Pi^2, \tag{5}$$

where $\gamma_{ij} = \mathrm{e}^{4\psi}\tilde{\gamma}_{ij}$.

2.2 Initial data

Constraint equations are given by the followings:

$$\mathcal{R} + K^2 - K_{ij}K^{ij} \ = 16\pi\rho, \tag{6}$$
$$D_j K^{ji} - D^i K \ = 8\pi j_i. \tag{7}$$

We adopt the conventional decomposition given by $K_{ij} = \mathrm{e}^{4\psi}\tilde{A}_{ij} + \frac{1}{3}K\gamma_{ij}$. The momentum constraints are trivially satisfied under the following assumptions: $\Pi = 0$, $K = \text{const.}$, $\tilde{\gamma}_{ij} = \delta_{ij}$, $\tilde{A}_{ij} = 0$. The Hamiltonian constraint is given by

$$\Delta\psi + \delta^{ij}\partial_i \psi \partial_j \psi - \frac{1}{12}K^2 \mathrm{e}^{4\psi} = -2\pi\rho \mathrm{e}^{4\psi} = -\pi\delta^{ij}\partial_i \phi \partial_j \phi. \tag{8}$$

We consider the following Gaussian like initial scalar field profile:

$$\phi_0 = A \exp\left[-\frac{r^2}{2\sigma^2}\right] W(r), \tag{9}$$

where L is the length of the edge of the cubic box and $W(r)$ is a function to suppress the Gaussian tail and make it compatible to the periodic boundary condition.

In order to solve the equation (8), we need to impose the following integrability condition:

$$0 \ = \ \int_{\text{box}} \mathrm{d}^3x \left(\Delta\psi - \frac{1}{12}K^2 \mathrm{e}^{4\psi} + (\partial\psi)^2 + \pi(\partial\phi)^2\right) \Leftrightarrow K^2 = 12\frac{\int_{\text{box}} \mathrm{d}^3x \left((\partial\psi)^2 + \pi(\partial\phi)^2\right)}{\int_{\text{box}} \mathrm{d}^3x \mathrm{e}^{4\psi}}, \tag{10}$$

where the integral of the Laplacian part vanishes because of the periodic boundary condition. We numerically solve the elliptic differential equation (8) by using iterative method imposing the integrability condition at each step of the iteration. The length scale of the coordinates are fixed by imposing $\psi(L, L, L) = 0$. We numerically solve the Hamiltonian constraint equation imposing the integrability condition.

We set $\sigma = 0.3L$ and consider the following two cases:

- Case1:$A \simeq 2.35$,
- Case2:$A \simeq 0.235$.

The initial profile of ϕ and ψ are shown in Figs. 1 and 2 for the case 1 and 2, respectively. The initial

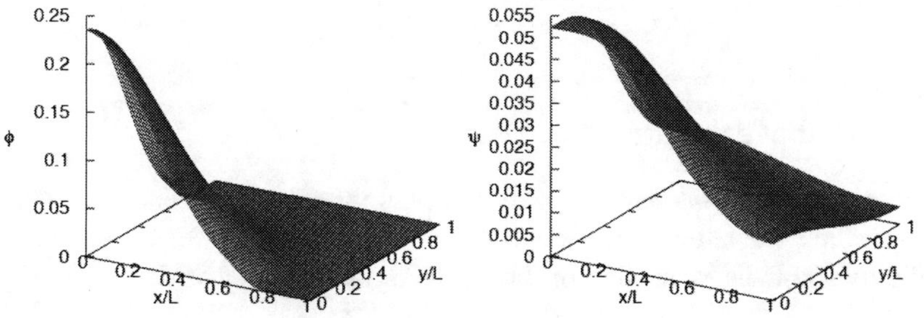

Figure 1. Initial profiles of ϕ and ψ for the case 1.

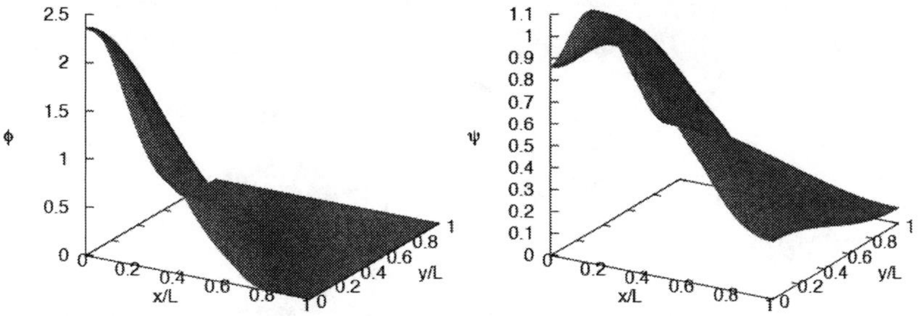

Figure 2. Initial profiles of ϕ and ψ for the case 2.

values of $1/H := 3/|K|$ are given by $3.62L$ for the case1 and $0.843L$ for the case2.

3 Time evolution

For the simulation of the time evolution, we use the 4th order Runge-Kutta method with the so-called BSSN formalism [4, 5]. We use the same gauge condition as in Ref. [2]. The grid spacing is uni-grid with Cartesian coordinates. Excision of a black hole interior region[6] is performed when a black hole is formed.

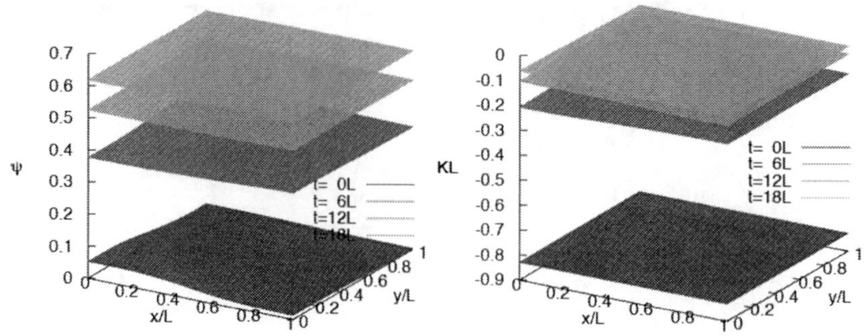

Figure 3. Evolution of ψ(left) and K(right) for the case 1.

3.1 Case 1:small initial amplitude

We show the time evolution of ψ and K on the x-y plane in Fig. 3. It can be seen that the value of ψ increases with time and the value of the effective Hubble parameter $-K/3$ decreases with time. These behaviors explicitly show the expansion of the lattice universe.

3.2 Case 2:large initial amplitude

For the case 2, we observe black hole formation. Here, we show the time evolution of the shape of the horizon in the coordinate space in Fig. 4. The horizon is shrinking in the coordinate space as is also reported in Refs. [2, 3].

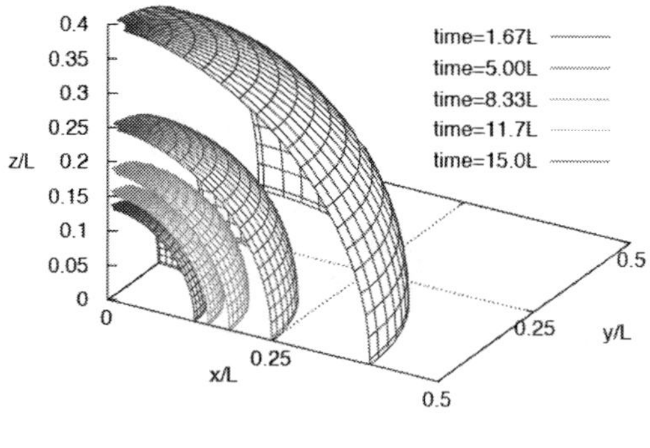

Figure 4. Evolution of the horizon shape in the coordinate space.

3.3 Time evolution of the effective scale factor

Finally, we show the dependence of the expansion law on the initial amplitude of the scalar field profile. We consider a set of observers each of which stays at a point of fixed spatial coordinates. Then, we calculate the area of a boundary surface on the constant proper time slice for the set of observers. Fig. 5 shows the time dependence of the square root of the area of a boundary surface. For the case 1, the dependence is approximated by the radiation dominated universe($\propto \tau^{1/2}$) in a late time. While, for the case 2, the numerical result looks approaching that of the matter dominated universe($\propto \tau^{2/3}$).

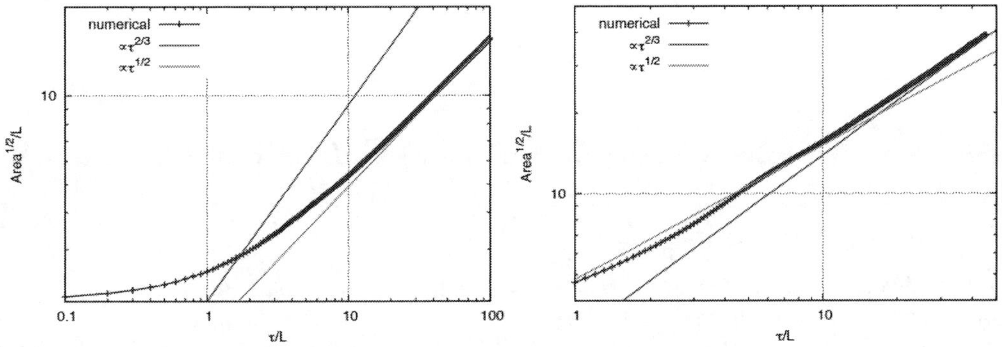

Figure 5. Square root of the area of a boundary surface is depicted as a function of the proper time measured by a set of observers with fixed spatial coordinates.

4 summary

We have considered gravitational collapse of a scalar field with the periodic boundary condition in a cubic box. This system can be regarded as a lattice universe model. We have shown that, when the initial amplitude of the scalar field is so small that the scalar field can finally spread in the whole region, the expansion law of the lattice universe model can be approximated by that of the radiation dominated universe in a late time. While, when the initial amplitude is so large that a black hole can be formed, the expansion law approaches that of the matter dominated universe. The all results are preliminary and we will report more concrete results elsewhere.

Acknowledgements

This work was supported by JSPS KAKENHI Grant Numbers JP16K17688, JP16H01097 (CY),

References

[1] C.M. Yoo, H. Abe, K.i. Nakao, Y. Takamori, Phys.Rev. **D86**, 044027 (2012), `1204.2411`
[2] C.M. Yoo, H. Okawa, K.i. Nakao, Phys.Rev.Lett. **111**, 161102 (2013), `1306.1389`
[3] C.M. Yoo, H. Okawa, Phys. Rev. **D89**, 123502 (2014), `1404.1435`
[4] M. Shibata, T. Nakamura, Phys.Rev. **D52**, 5428 (1995)
[5] T.W. Baumgarte, S.L. Shapiro, Phys.Rev. **D59**, 024007 (1999), `gr-qc/9810065`
[6] M. Alcubierre, B. Bruegmann, Phys. Rev. **D63**, 104006 (2001), `gr-qc/0008067`

EPJ Web of Conferences **168**, 02003 (2018) https://doi.org/10.1051/epjconf/201816802003
Joint International Conference of ICGAC-XIII and IK-15 on Gravitation, Astrophysics and Cosmology

A viable compactification scenario in Gauss-Bonnet gravity

Alexey Toporensky[1,2,⋆] and *Sergey* Pavluchenko[3,⋆⋆]

[1] *Sternberg Astronomical Institute, Moscow State University, Moscow 119991 Russia*
[2] *Kazan Federal University, Kazan 420008 Russia*
[3] *Programa de Pós-Graduação em Física, Universidade Federal do Maranhão (UFMA), 65085-580, São Luís, Maranhão, Brazil*

Abstract. In this talk we address two important issues which could affect reaching the exponential and Kasner asymptotes in Einstein-Gauss-Bonnet cosmologies – spatial curvature and anisotropy in both three- and extra-dimensional subspaces. In the first part we consider cosmological evolution of spaces being the product of two isotropic and spatially curved subspaces. We consider all possible number of spatial dimensions and provide description of the curvature effects in these dimensions. It is demonstrated that the dynamics in $D = 2$ (the number of extra dimensions) and $D \geqslant 3$ is different. In particular, the regime with the "stabilization" of extra dimensions could be reached only if $D \geqslant 3$. In the second part we study the influence of initial anisotropy. Our study of reveals that transition from Gauss-Bonnet Kasner regime to anisotropic exponential expansion (with expanding three and contracting extra dimensions) is stable with respect to breaking the symmetry within both three- and extra-dimensional subspaces in any number of extra dimensions. This allows us to construct a scenario where isotropisation of outer and inner subspaces is reached dynamically from rather general anisotropic initial conditions.

In our study we deal with the cosmological dynamics in Lovelock gravity [1]. It has the following structure: its Lagrangian is constructed from terms

$$L_n = \frac{1}{2^n}\delta^{i_1 i_2 \dots i_{2n}}_{j_1 j_2 \dots j_{2n}} R^{j_1 j_2}_{i_1 i_2} \dots R^{j_{2n-1} j_{2n}}_{i_{2n-1} i_{2n}}, \qquad (1)$$

where $\delta^{i_1 i_2 \dots i_{2n}}_{j_1 j_2 \dots j_{2n}}$ is the generalized Kronecker delta of the order $2n$. One can verify that L_n is Euler invariant in $D < 2n$ spatial dimensions and so it would not give nontrivial contribution into the equations of motion. So that the Lagrangian density for any given D spatial dimensions is sum of all Lovelock invariants (1) upto $n = \left[\frac{D}{2}\right]$ which give nontrivial contributions into equations of motion (see e.g. [2]:

$$\mathcal{L} = \sqrt{-g} \sum_n c_n L_n, \qquad (2)$$

where g is the determinant of metric tensor, c_n is a coupling constant of the order of Planck length in $2n$ dimensions and summation over all n in consideration is assumed.

The most general spatially-flat metric *Ansatz* has the form

⋆e-mail: atopor@rambler.ru
⋆⋆e-mail: sergey.pavluchenko@gmail.com

$$g_{\mu\nu} = \mathrm{diag}\{-1, a_1^2(t), a_2^2(t), \ldots, a_n^2(t)\}; \tag{3}$$

substituting it into the Lagrangian and varying with respect to the metric gives us the equations of motion:

$$2\left[\sum_{j\neq i}(\dot{H}_j + H_j^2) + \sum_{\substack{\{k>l\}\\ \neq i}} H_k H_l\right] + 8\alpha\left[\sum_{j\neq i}(\dot{H}_j + H_j^2)\sum_{\substack{\{k>l\}\\ \neq\{i,j\}}} H_k H_l + 3\sum_{\substack{\{k>l>\\ m>n\}\neq i}} H_k H_l H_m H_n\right] - \Lambda = 0 \tag{4}$$

as the ith dynamical equation. The first Lovelock term—the Einstein-Hilbert contribution—is in the first set of brackets and the second term—Gauss-Bonnet—is in the second set; α is the coupling constant for the Gauss-Bonnet contribution and we put the corresponding constant for Einstein-Hilbert contribution to unity. Also, since we a consider spatially flat cosmological model, scale factors do not hold much in the physical sense and the equations are rewritten in terms of the Hubble parameters $H_i = \dot{a}_i(t)/a_i(t)$. Apart from the dynamical equations, we write down a constraint equation

$$2\sum_{i>j} H_i H_j + 24\alpha \sum_{i>j>k>l} H_i H_j H_k H_l = \Lambda. \tag{5}$$

Prior to this study, we completed investigation of the most simple (but the most important as well) cases. The spatial part of these cases is the product of three- and extra-dimensional subspaces which are spatially flat and isotropic [3–5]. So that the obvious next step is consideration of these subspaces being non-flat and anisotropic, and that is what we have done in current investigation. Non-flatness is addressed by assuming that both subspaces have constant curvature while anisotropy – by breaking the symmetry between the spatial directions. The results of the curvature study suggest that the only viable regimes are those from the flat case with $\gamma_{(\mathbf{D})} = 0$ requirement. Additionally, in the Λ-term case there is "geometric frustration" regime, described in [6, 7] and further investigated in [8] with $\gamma_{(\mathbf{D})} = -1$ requirement.

First we investigate the impact of the spatial curvature on the cosmological regimes. As a "background" we use the results obtained in [3–5] – exact regimes for $\gamma_{(3)} = \gamma_{(\mathbf{D})} \equiv 0$ for both vacuum and Λ-term cases. As we use them as a "background" solutions, it is worth to quickly describe them all. All solutions found for both vacuum and Λ-term cases could be splitted into two groups – those with "standard" regimes as both past and future asymptotes and those with nonstandard singularity as one (or both) of the asymptotes. By the "standard" regimes we mean generalized Kasner solutions(when dependence of scale factor upon time is power-law) and exponential solutions (when scale factors depend exponentially on time, so that corresponding Hubble parameters are constant). In our study me encounter two different Kasner regimes – "classical" GR Kasner regime (with $\sum p_i = \sum p_i^2 = 1$ where p_i is Kasner exponent from the definition of power-law behavior $a_i(t) = t^{p_i}$), which we denote as K_1 (as $\sum p_i = 1$) and it is low-energy regime; and GB Kasner regime (with $\sum p_i = 3$), which we denote as K_3 and it is high-energy regime. For realistic cosmology we should have high-energy regime as past asymptote and low-energy as future, but our investigation demonstrates that potentially both K_1 and K_3 could play a role as past and future asymptotes [3]. Also we should note that K_1 exist only in the vacuum regime, while K_3 as past asymptotes we encounter in both vacuum and Λ-term regimes (see [4] for details). The exponential regimes [9–11] could be seen in both vacuum and Λ-term regimes and there are two of them – isotropic and anisotropic ones. The former of them corresponds to the case where all the directions are isotropized and, since we work in the multidimensional case, it does not fit the observations. On contrary, the latter of them have different Hubble parameters for three- and extra-dimensional subspaces. For realistic compactification we demand expansion of the three- and contraction of the extra-dimensional spaces. The exponential solutions are denoted as E_{iso} for isotropic and E_{3+D} for anisotropic, where D is the number of extra dimensions (so that, say, in $D = 2$ the anisotropic exponential solution is denoted as E_{3+2}).

The second large group are the regimes which have nonstandard singularity as either of the asymptotes or even both of them. The nonstandard singularity is the situation which arises in nonlinear theories and in our particular case it corresponds to the point of the evolution where $\dot{H}$ (the derivative of the Hubble parameter) diverges at the final H; we denote it as nS. Our previous research reveals that nonstandard singularity is a wide-spread phenomena in EGB cosmology, for instance, in $(4 + 1)$-dimensional Bianchi-I vacuum case all the trajectories have nS as either past or future asymptote [12]. Since a nonstandard singularity means the beginning or end of dynamical evolution, either higher or lower values of H do not reached and so the entire evolution from high to low energies cannot be restored; for this reason we disregard the trajectories with nS in the present study.

So that the viable (or realistic) regimes are limited to $K_3 \to K_1$ and $K_3 \to E_{3+D}$ for vacuum case and $K_3 \to E_{3+D}$ for Λ-term; these regimes we further investigate in the presence of curvature.

The results of the investigation suggest that all three considered cases have the original regime unchanged as long as $\gamma_D = 0$. This means that the curvature of three-dimensional world alone cannot change the future asymptote. For nonzero curvature of the extra dimensions, the situation is different in all three cases: in vacuum $K_3 \to E_{3+D}$ case all trajectories with $\gamma_D \neq 0$ are singular; in vacuum $K_3 \to K_1$ we have two new regimes but both of them are nonviable; finally, in Λ-term $K_3 \to E_{3+D}$ case if $\gamma_D > 0$ the future asymptote is singular while for $\gamma_D < 0$ there could be viable regime with stabilization of extra dimensions, but this regime occurs only when isotropic exponential solution cannot exist and in $D \geqslant 3$.

Our study also demonstrate the difference between $D = 2$ (the lowest dimension where anisotropic exponential solution could exist [4]) and $D \geqslant 3$ case: in the former of them, the regime with stabilization cannot be reached while in the latter it is reachable.

To conclude, it seems that the only important player in this case is the curvature of extra dimensions. And this is clear why is it so – from requirements of viability we demand that three-dimensional subspace should expand while extra dimensions should contract. The expansion of the three dimensions cannot be stopped by neither curvature alone. On the other hand, extra dimensions are contracting, so both signs of extra-dimensional curvature affect it – positive usually leads to singularity (standard or not) while negative could turn it to expansion. The latter could even change the dynamics in three-dimensional sector, what we also see in K^{iso}_{3+D} regime.

The anisotropy was studied numerically – we found an exact $(3 + D)$ anisotropic exponential solution (according to [3–5]) and "anizotropize" metric around it (consider totally anisotropic cosmology with $H_1 \sim H_2 \sim H_3 \sim H$ and $H_4 \sim \cdots \sim H_{D+3} \sim h$ where H and h are Hubble parameters for exact solution E_{3+D}). Then we fix all but two Hubble parameters and vary the remaining two to see how the evolution changes with variation of the initial conditions.

The results of anisotropy study reveal that the viable $K_3 \to K_1$ (in vacuum) and $K_3 \to E_{3+D}$ (in both vacuum and Λ-term) regimes are always stable with respect to breaking the isotropy in both subspaces, meaning that within some vicinity of exact transition, all initial conditions still lead to this regime (see Fig. 1). The difference is in the area of the basin of attraction – for $K_3 \to K_1$ it is quite small (see Fig. 1(a)) while for $K_3 \to E_{3+D}$ (the particular example is $K_3 \to E_{3+2}$) it is much larger and unbounded (see Fig. 1(b)). As a way of example, we also provided the typical evolution curves for $K_3 \to E_{3+D}$ in Fig. 2(a) and $K_3 \to E_{iso}$ in Fig. 2(b).

Our study clearly demonstrates that the dynamics of the non-flat cosmologies could be different from flat cases and even some new regimes could emerge. Here we covered only the simplest case with constant-curvature subspaces leaving the most complicated cases aside – we are going to investigate some of them deeper in near future.

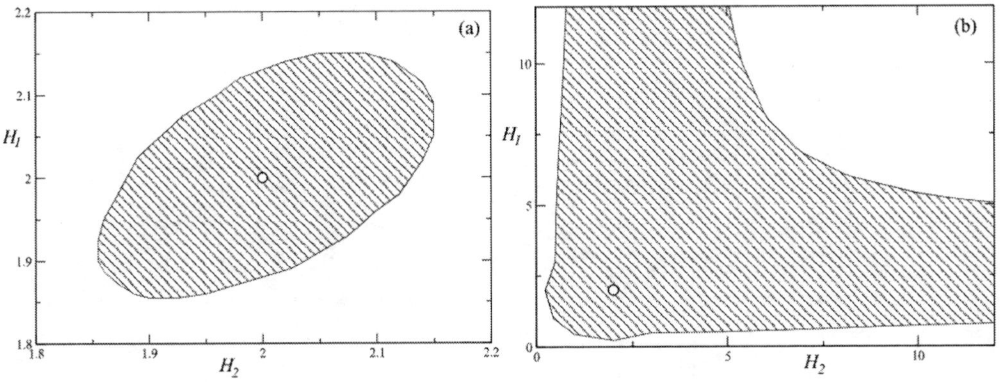

Figure 1. Typical stability areas for vacuum $K_3 \to K_1$ regime on (a) panel; vacuum and Λ-term $K_3 \to E_{3+D}$ regime on (b) panel (see the text for more details).

On the other hand, anisotropy of the subspaces can decrease the initial conditions leading to exact solution and through this diminish the chances for successful compactification. Our thorough investigation of totally anisotropic (Bianchi-I-type) five-dimensional EGB model [12] is a good example – the only future asymptote in the vacuum case is nonstandard singularity. And since the isotropy (even within the subspace) could be treated as some sort of fine-tuning, independence of the final asymptote on the isotropy highly improve the chances for the model to be accepted, so that this direction of study is also important.

The previously described results allow us to construct a scenario of compactification which satisfy two important requirements:

- the evolution starts from a rather general anisotropic initial conditions,
- the evolution ends in a state with three isotropic big expanding dimensions and stabilized isotropic extra dimensions.

The first part of the scenario in question uses the results of anisotropy investigation. We have seen there that while starting from a quite wide area of the initial conditions, the flat anisotropic Universe tends to the exponential solution with three equal expanding dimensions.

Suppose also, that a negative spatial curvature is small enough at the beginning and starts to be important only after this transition to exponential solution (which is established in the present investigation only for a flat Universe) already occured. This condition allows us to glue the second part of the scenario which requires negative spatial curvature of the inner space. When describing the effects of curvature we saw that the exponential solution turn to the solution with stabilized extra dimensions in this case. As a result of these two stages a Universe starting from initially anisotropic both outer expanding three dimensional space and contracting extra dimensional inner space evolve naturally to the final stage with isotropic three big dimensions and isotropic and stabilized inner extra dimensional space. The only additional condition for this scenario to realize is that spatial curvature should become dynamically important only after the transition to exponential solution occurs.

This finalize our talk. The presented analysis suggests that more in-depth investigation of both curvature and anisotropy effects are required – we have investigated and described the most simple but still very important cases – constant-curvature and flat anisotropic (Bianchi-I-type) geometries; in the near future we are going to consider more complicated topologies.

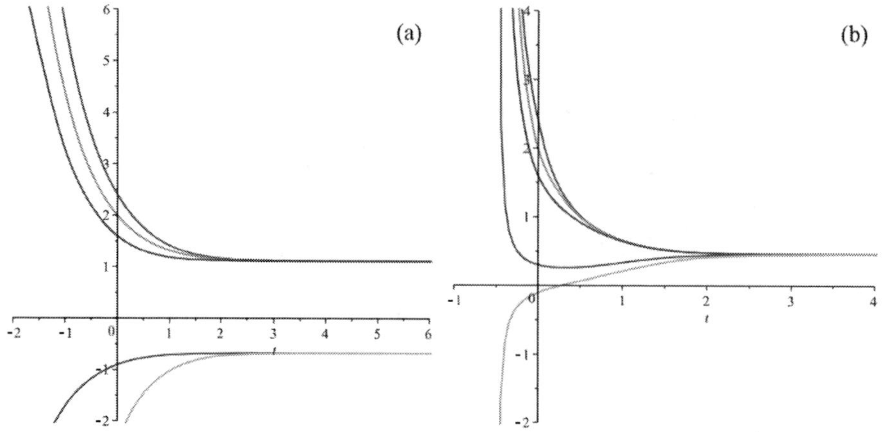

Figure 2. Typical evolution curve for stable anisotropic (a) and isotropic (b) exponential solutions with broken symmetry in both subspaces (see the text for more details).

Acknowledgments S.A.P. was supported by FAPEMA under project BPV-00038/16. The work of A.T. is supported by RFBR grant 17-02-01008 and by the Russian Government Program of Competitive Growth of Kazan Federal University. Authors are grateful to Alex Giacomini (ICFM-UACh, Valdivia, Chile) for discussions.

References

[1] D. Lovelock, J. Math. Phys. (N.Y.) **12**, 498 (1971).
[2] S.A. Pavluchenko, Phys. Rev. D **80**, 107501 (2009).
[3] S.A. Pavluchenko, Phys. Rev. D **94**, 024046 (2016).
[4] S.A. Pavluchenko, Phys. Rev. D **94**, 084019 (2016).
[5] S.A. Pavluchenko, Eur. Phys. J. C **77**, 503 (2017).
[6] F. Canfora, A. Giacomini and S. A. Pavluchenko, Phys. Rev. D **88**, 064044 (2013).
[7] F. Canfora, A. Giacomini and S. A. Pavluchenko, Gen. Rel. Grav. **46**, 1805 (2014).
[8] F. Canfora, A. Giacomini, S. A. Pavluchenko and A. Toporensky, *Friedmann dynamics recovered from compactified Einstein-Gauss-Bonnet cosmology*, Grav. Cosmol. *(accepted)*, arXiv:1605.00041
[9] D. Chirkov, S. Pavluchenko, A. Toporensky, Mod. Phys. Lett. **A29**, 1450093 (2014).
[10] D. Chirkov, S. Pavluchenko, A. Toporensky, Gen. Rel. Grav. **46**, 1799 (2014).
[11] D. Chirkov, S. Pavluchenko, A. Toporensky, Gen. Rel. Grav. **47**, 137 (2015).
[12] S.A. Pavluchenko, Phys. Rev. D **82**, 104021 (2010).
[13] S.A. Pavluchenko, Phys. Rev. D **92**, 104017 (2015).
[14] V. D. Ivashchuk, Eur. Phys. J. C **76**, 431 (2016).

EPJ Web of Conferences **168**, 02004 (2018) https://doi.org/10.1051/epjconf/201816802004
Joint International Conference of ICGAC-XIII and IK-15 on Gravitation, Astrophysics and Cosmology

Experiments on Frequency Dependence of the Deflection of Light in Yang-Mills Gravity

Yun Hao[1,⋆], *Yiyi* Zhu[1,⋆⋆], and *Jong-Ping* Hsu[1,⋆⋆⋆]

[1] *Department of Physics, University of Massachusetts Dartmouth, North Dartmouth, MA 02747, USA*

Abstract. In Yang-Mills gravity based on flat space-time, the eikonal equation for a light ray is derived from the modified Maxwell's wave equations in the geometric-optics limit. One obtains a Hamilton-Jacobi type equation, $G_L^{\mu\nu}\partial_\mu\Psi\partial_\nu\Psi = 0$ with an effective Riemannian metric tensor $G_L^{\mu\nu}$. According to Yang-Mills gravity, light rays (and macroscopic objects) move as if they were in an effective curved space-time with a metric tensor. The deflection angle of a light ray by the sun is about 1.53″ for experiments with optical frequencies $\approx 10^{14}$Hz. It is roughly 12% smaller than the usual value 1.75″. However, the experimental data in the past 100 years for the deflection of light by the sun in optical frequencies have uncertainties of (10-20)% due to large systematic errors. If one does not take the geometric-optics limit, one has the equation, $G_L^{\mu\nu}[\partial_\mu\Psi\partial_\nu\Psi cos\Psi + (\partial_\mu\partial_\nu\Psi)sin\Psi] = 0$, which suggests that the deflection angle could be frequency-dependent, according to Yang-Mills gravity. Nowadays, one has very accurate data in the radio frequencies $\approx 10^9$Hz with uncertainties less than 0.1%. Thus, one can test this suggestion by using frequencies $\approx 10^{12}$ Hz, which could have a small uncertainty 0.1% due to the absence of systematic errors in the very long baseline interferometry.

1 Introduction

Yang-Mills gravity is a gauge theory based on the 4-dimensional translational group T_4 in flat space-time. Such a theory can be quantized to obtain Feynman-Dyson rules for quantum Yang-Mills gravity and is consistent with known experiments.[1–4] A new interesting property is that the classical objects such as light rays and macroscopic objects move as if they were in an effective curved space-time. The reason is that their equations of motion are derived from the quantum wave equations in flat space-time, which involves T_4 gauge covariant derivative $\Delta_\mu = \partial_\mu - ig\phi_{\mu\nu}p^\nu$ where $p_\mu = i\partial_\mu$ is the generator of the T_4 group, and $\phi_{\mu\nu}$ is the symmetric tensor gauge field associated with space-time translational gauge symmetry. In the geometric-optics (i.e., classical) limit, quantum wave equations reduce to the form of relativistic Hamilton-Jacobi type equations with an effective Riemannian metric tensor for classical objects such as light rays, $G_L^{\mu\nu}\partial_\mu\Psi\partial_\nu\Psi = 0$. The emergence of an effective metric tensor in the classical limit of quantum equation is a new and key property of the Yang-Mills gravity for its consistence with experiments.[3, 4] Therefore, Yang-Mills gravity reveals that the apparent curvature

⋆e-mail: yhao@umassd.edu
⋆⋆e-mail: yzhu3@umassd.edu
⋆⋆⋆e-mail: jhsu@umassd.edu

of space-time appears to be simply a manifestation of the flat space-time translational gauge symmetry for the motion of quantum particles in the classical limit.

2 Modified Maxwell's equations and eikonal equations

Let us consider Yang-Mills gravity in an inertial frame $F(w, x, y, z)$ with a broad 4-dimensional symmetry.[5] The Lagrangian for the electromagnetic potential field A_μ in the presence of gravity is obtained by replacing the partial space-time derivatives ∂_μ in the Lagrangian by the T_4 gauge covariant derivative Δ_μ. Thus, we have[1]

$$L_{em} = -\frac{1}{4}F_{\mu\nu}F^{\mu\nu}, \qquad F_{\mu\nu} = \Delta_\mu A_\nu - \Delta_\nu A_\mu, \tag{1}$$

$$\Delta_\mu = \partial_\mu - ig\phi_{\mu\nu}p^\nu \equiv J_{\mu\nu}\partial^\nu, \qquad J_{\mu\nu} = (\eta_{\mu\nu} + g\phi_{\mu\nu}),$$

$$\eta_{\mu\nu} = (1, -1, -1, -1), \qquad g = \sqrt{8\pi G_N}\,, \qquad c = \hbar = 1.$$

Since we are interested in the classical light rays, it is not necessary to consider the gauge-fixing terms in the Lagrangian (1). One may carry out variation of the Lagrangian (1) to arrive at modified Maxwell's equations in the presence of gravity. The modified wave equations are

$$\partial_\mu(J^\mu_\alpha F^{\alpha\nu}) = 0. \tag{2}$$

To derive the equation of a light ray, we use the usual limiting expression [6]

$$A_\mu(x) = a_\mu e^{i\Psi(x)}, \qquad a_\mu = \epsilon_\mu(k, \lambda). \tag{3}$$

where the eikonal Ψ and the wave 4-vector $\partial_\mu\Psi = k_\mu$ are large in the geometric-optics limit.

The modified electromagnetic wave equation (2) with the limiting expression (3) lead to

$$\Delta_\mu(\Delta^\mu A^\lambda - \Delta^\lambda A^\mu) + (\partial_\alpha J^\alpha_\mu)(\Delta^\mu A^\lambda - \Delta^\lambda A^\mu) = 0. \tag{4}$$

Using the gauge condition $\partial_\mu A^\mu = 0$, the second term $-\Delta_\mu\Delta^\lambda A^\mu$ can be simplified,

$$-\Delta_\mu\Delta^\lambda A^\mu = -J^\sigma_\mu\partial_\sigma(J^{\lambda\alpha}\partial_\alpha A^\mu) = -(\delta^\sigma_\mu + g\phi^\sigma_\mu)\partial_\sigma(J^{\lambda\alpha}\partial_\alpha A^\mu)$$

$$\approx -g\phi^\sigma_\mu J^{\lambda\alpha}\partial_\alpha\partial_\mu A^\mu - g\phi^\sigma_\mu J^{\lambda\alpha}\partial_\sigma\partial_\alpha A^\mu = -g\phi^\sigma_\mu J^{\lambda\alpha}\partial_\sigma\partial_\alpha A^\mu. \tag{5}$$

Note that the terms involving $(\partial_\alpha J^\alpha_\mu)$ are small and the terms with $\partial_\alpha\partial_\sigma A^\mu$ and $\partial_\alpha\Psi\partial_\sigma\Psi = k_\alpha k_\sigma$ are large. The eikeonal equation (4) can be written as

$$\delta^\lambda_\mu J^\alpha_\beta J^{\beta\sigma}\partial_\alpha\partial_\sigma A^\mu - gJ^{\lambda\sigma}\phi^\alpha_\mu\partial_\alpha\partial_\sigma A^\mu = Z^\lambda_\mu\epsilon^\mu = 0, \tag{6}$$

where

$$Z^\lambda_\mu = (\delta^\lambda_\mu G^{\alpha\sigma}\partial_\alpha\Psi\partial_\sigma\Psi - gJ^{\lambda\sigma}\phi^\alpha_\mu\partial_\alpha\Psi\partial_\sigma\Psi), \qquad G^{\alpha\sigma} = J^\alpha_\beta J^{\beta\sigma},$$

Since we are interested in the law for the propagation of light rays and the further simplification of (4), we use the relation for the polarization vector $\sum_\lambda \epsilon^\mu(k, \lambda)\epsilon^\nu(k, \lambda) \to -\eta^{\mu\nu}$.[7] Multiplying $Z^\lambda_\mu\epsilon^\mu$ in (4) by $\epsilon^\nu\eta_{\nu\lambda}$ and summing over all polarizations, we obtain

$$\sum_\lambda Z^\lambda_\mu\epsilon^\mu\epsilon^\nu\eta_{\nu\lambda} = -\delta^\mu_\lambda Z^\lambda_\mu = 0. \tag{7}$$

After some straightforward calculations, we obtain a new eikonal equation in Yang-Mills gravity that involves an effective metric tensor $G_L^{\mu\nu}$,

$$G_L^{\mu\nu}\partial_\mu\Psi\partial_\nu\Psi = 0, \tag{8}$$

$$G_L^{\mu\nu} = J^\mu_\lambda J^{\lambda\nu} - \frac{g}{4}\phi^\mu_\lambda J^{\lambda\nu},$$

in the geometric-optics (i.e., classical) limit. It leads to the deflection angle of a light ray by the sun to be about 1.53″.[3, 4] It is stressed that we have the effective Riemannian metric tensor $G_L^{\mu\nu}$ and eikonal equation (8) only in this classical limit.

In order to see the frequency dependence of the eikonal equation in Yang-Mills gravity, let us consider the eikonal equation involving both dominate term $\partial_\alpha\Psi\partial_\beta\Psi$ and non-dominate term $\partial_\alpha\partial_\beta\Psi$. We calculate the real part of (2),

$$Re[\partial_\mu(J^\mu_\alpha F^{\alpha\nu})] = (\epsilon^\nu J^\alpha_\mu J^{\mu\sigma} - \epsilon^\mu g\phi^\alpha_\mu J^{\nu\sigma})[\partial_\alpha\Psi(\partial_\sigma\Psi)cos\Psi + (\partial_\alpha\partial\sigma\Psi)sin\Psi] \tag{9}$$

$$-g\partial_\alpha\phi^\alpha_\mu[-\epsilon^\nu J^{\mu\sigma}\partial_\sigma\Psi + \epsilon^\mu J^{\nu\sigma}\partial_\sigma\Psi]sin\Psi,$$

where the imaginary part is ignored because the electromagnetic wave is a real function. The last term in (9) is small and ignored because the gravitational strength on the surface of the sun is approximately $g\phi^{00} \approx 10^{-6}$. Eq. (9) can be simplified,

$$G_L^{\alpha\sigma}[\partial_\alpha\Psi(\partial_\sigma\Psi)cos\Psi + (\partial_\alpha\partial_\sigma\Psi)sin\Psi] \approx 0, \tag{10}$$

$$G_L^{\alpha\sigma} = (J^\alpha_\mu J^{\mu\sigma} - \frac{g}{4}\phi^\alpha_\mu J^{\mu\sigma}) \tag{11}$$

where we have used $\sum_\lambda \epsilon_\alpha(k,\lambda)\epsilon^\nu(k,\lambda) = -\delta^\nu_\alpha$.

3 Discussions of frequency dependence and tests

The simple eikonal equation (8) describes the trajectories of light rays with optical frequencies ($\approx 10^{14}$Hz) because it is derived in the limit of geometric-optics. This property is reliable and supported by optical experiments. Thus, the eikonal equation (8) may not be a good approximation for the rays with radio frequencies ($\approx 10^9 Hz$). However, the general ‘eikonal equation’ (10) is obtained without taking the geometric-optics limit, so it should be better in principle for describing the rays with radio frequencies. Although the numerical results are difficult to obtain from the equation (10), it implies that, experimentally, the angle of deflection by the sun depend on frequencies in Yang-Mills gravity. In other words, the eikonal equation takes the simplest form (8) only in the limit of geometric-optics. Suppose one does the deflection of light experiments, Yang-Mills gravity predicts that the angles of deflection for rays with optical frequencies ($\approx 10^{14}$Hz) and for rays with radio frequencies ($\approx 10^9 Hz$) will not be the same because equations (8) and (10) are different.

In Yang-Mills gravity, the deflection angle of a light ray by the sun is found to be about 1.53″ in the geometric-optics limit.[3, 4] This deflection angle is roughly 12% smaller than the usual value 1.75″ in general relativity.[1] The 1919 measurements made during a solar eclipse by Dyson, Eddington and Davidson[8] were very difficult and any repeat would be equally difficult. As a result, the experimental results for the bending of light by the sun with optical frequencies in the past 100 years have large

[1] The situation is quite different for the perihelion shift. In this case, the result for the advance of the perihelion differs from that of general relativity by correction terms which are of the order of 10^{-7} for Mercury.[3] The observational accuracy of the perihelion shift of Mercury is about 1 %.

uncertainties of $(10 - 20)\%$,[9–11] due to large systematic errors. In the experiments with optical frequencies, the systematic errors completely dominate the analysis. For example, when one measures in the daytime looking at the sun, there are extreme temperature variations and optical distortions within the optics. As a result, the precision of image measurement from space station to avoid earth's atmospheric inhomogeneities does not help much in reducing the uncertainty.

All recent accurate measurements of the light deflection angle by the sun have been performed in radio frequencies near 10^9 Hz (with uncertainties less than 0.1%), which are too far away from optical frequencies to test Yang-Mills gravity with the eikeonal equation (8). However, Yang-Mills gravity could be tested by detecting the frequency-dependence of the deflection angle by the sun with a better accuracy, say, with $(0.1 - 1)\%$ uncertainties.

We propose carrying out a new experiment using higher frequencies $\approx 10^{12}$ Hz which would have an uncertainty of $\approx 0.1\%$. Nowadays, this is technically possible due to the absence of systematic errors in the very long baseline interferometry. Such an experiment could test the frequency dependence of the angle of light deflection by the sun, as implied by the difference of eikonal equations (8) and (10) in Yang-Mills gravity. The eikonal equation (10) seems too complicated to be solved numerically (due to the presence of the functions $cos\Psi$ and $sin\Psi$). However, the experimental test of Yang-Mills gravity could be done, independent of the solution to (10). One can simplify concentrate on whether there is an experimental difference between the deflection angle measured with the radio frequencies $\approx 10^9 Hz$ and that measured in the frequencies $\approx 10^{12} Hz$. We believe that such an experimental test is feasible by using the very long baseline interferometry and can test the prediction of frequency-dependence of the deflection angle in Yang-Mills gravity.

The work was supported in part by the Jing Shin Research Fund of the UMassD Foundation. One of us (JP) would like to thank Elliott Horch, Bill van Altena and Titov Oleg for useful discussions on technical aspects of experiments.

References

[1] J. P. Hsu, Int. J. Mod. Phys. A **21**, 5119 (2006).

[2] Eur. Phys. J. Plus, **128**, 31 (2013). arXiv 1302.6956[gr-qc].

[3] J. P. Hsu and L. Hsu, *Space-Time Symmetry and Quantum Yang-Mills Gravity* (World Scientific, 2013) pp. 134-137.

[4] K. O. Cottrell and J. P. Hsu, Eur. Phys. J. Plus **130**, 147 (2015).

[5] J. P. Hsu and L. Hsu, Phys. Letters, A **196**, 1 (1994).

[6] L. Landau and E. Lifshitz, *The Classical Theory of Fields*, (Addison-Wesley, 1951), p. 136.

[7] J. D. Bjorken and S. D. Drell, *Relativistic Quantum Fields* (McGraw-Hill, New York, 1965) p. 386.

[8] F. W. Dyson, A. S. Eddington and C. Davidson, Phil. Trans. Roy. Soc., **220A**, 291 (1920).

[9] S. Weinberg, *Gravitation and Cosmology, Principles and Applications of the General Theory of Relativity* (John Wiley and Sons, 1972) pp. 190-194.

[10] See, for example, E. B. Fomalont and S. M. Kopeikin, Astrophys. J. **598**, 704 (2003), for a recent light deflection experiment,

[11] Wei-Tou Ni, in *100 Years of Gravity and Accelerated Frames, The Deepest Insights of Einstein and Yang-Mills* (Ed. J. P. Hsu and D. Fine, World Scientific, 2005) p. 478 and p. 481.

On the Induced Gravitational Collapse

Laura M. Becerra[1,2,⋆], *Carlo* Bianco[1,2], *Chris* Fryer[4], *Jorge* Rueda[1,2,3], and *Remo* Ruffini[1,2,3]

[1] *Dipartimento di Fisica and ICRA, Sapienza Università di Roma, P.le Aldo Moro 5, I–00185 Rome, Italy*
[2] *ICRANet, P.zza della Repubblica 10, I–65122 Pescara, Italy*
[3] *ICRANet-Rio, Centro Brasileiro de Pesquisas Físicas, Rua Dr. Xavier Sigaud 150, 22290–180 Rio de Janeiro, Brazil*
[4] *CCS-2, Los Alamos National Laboratory, Los Alamos, NM 87545*

Abstract. The induced gravitational collapse (IGC) paradigm has been applied to explain the long gamma ray burst (GRB) associated with type Ic supernova, and recently the X-ray flashes (XRFs). The progenitor is a binary systems of a carbon-oxygen core (CO) and a neutron star (NS). The CO core collapses and undergoes a supernova explosion which triggers the hypercritical accretion onto the NS companion (up to $10^{-2}\, M_{\odot} s^{-1}$). For the binary driven hypernova (BdHNe), the binary system is enough bound, the NS reach its critical mass, and collapse to a black hole (BH) with a GRB emission characterized by an isotropic energy $E_{iso} > 10^{52}$ erg . Otherwise, for binary systems with larger binary separations, the hypercritical accretion onto the NS is not sufficient to induced its gravitational collapse, a X-ray flash is produced with $E_{\rm iso} < 10^{52}$ erg. We're going to focus in identify the binary parameters that limits the BdHNe systems with the XRFs systems.

1 Introduction

In [1], the progenitors of long gamma-ray bursts (GRBs) have been explained within the Induced Gravitational Collapse (IGC) paradigm. The initial configuration is a compact binary system, formed by a Carbon-Oxygen (CO) core, a star that has lost its helium and hydrogen layers, with a neutron star (NS). The CO-core collapses and produces a supernova explosion that triggers an hypercritical accretion onto the NS companion. If the hypercritical accretion on the NS is not enough to make it to reach the critical mass for gravitational collapse, a first scenario takes place characterized by a x-ray flash (XRF) emission. The final system will be a NS-NS binary, formed by the ν-NS, the remnant from the collapse of the CO-core, and a more massive NS. This kind of events are characterized by isotropic energies of $E_{\rm iso} \lesssim 10^{52}$ erg, peak energies between $4 < E_{\rm p,i} < 200$ keV and a prompt emission duration of about $\sim 10^2 - 10^4$ s. Otherwise if the system is enough compact, the NS companion reaches by accretion the critical mass and collapses to a black hole (BH). These systems have been called binary-driven hypernovae (BdHNe) and has been deeply studied in [1–5, 10, 11] . The BdHNe events are characterized by isotropic energies $E_{\rm iso} \gtrsim 10^{52}$ erg, spectral peak energy in $0.2 < E_{\rm p,i} < 2$ MeV and prompt emission with durations of up to 100 s. The final system is a νNS-BH binary [6].

In this proceedings, we will identify the principal parameters of the initial binary configurations that can leave either to XRFs or BdHNe. In section 2, we estimate the accretion rate on the NS and

⋆e-mail: laura.maarcela@icranet.com

follow its mass and angular momentum evolution in order to establish critical binary initial parameters that determined the final fate of the system after the supernova explosion. In section 3, we model the light-curve of the XRF 060218 with the IGC scenario and finally in section 4 we resume our results.

2 Hypercritical Accretion process

Following the Bondy-Hoyle-Lyttleton formalism [7–9], an estimation of the accretion rate of the ejected material onto the NS is given by:

$$\dot{M}_B = \pi \rho_{\rm ej} R_{\rm cap}^2 \sqrt{v_{\rm ej}^2 + v_{\rm orb}^2 + c_{\rm s,ej}^2}, \quad \text{with} \quad R_{\rm cap} = \frac{2GM_{\rm NS}}{v_{\rm ej}^2 + v_{\rm orb}^2 + c_{\rm s,ej}^2} \tag{1}$$

where $R_{\rm cap}$ is the NS gravitational capture radius, G is the gravitational constant, $\rho_{\rm ej}$ and $\vec{v}_{\rm ej}$ the density and velocity of the supernova ejecta, $M_{\rm NS}$ the NS mass, $v_{\rm orb} = \sqrt{G(M_{\rm NS} + M_{\rm CO})/a}$ is the orbital velocity of the binary system, with a the binary separation, and $M_{\rm CO} = M_{\rm env} + M_{\rm Fe}$ the total mass of the CO core which consist of the envelope mass $M_{\rm env}$ that will be ejected in the SN explosion and the central iron core mass, $M_{\rm Fe}$. In order to integrate equation (1), we have fitted the pre-supernova density profile of the CO envelope with a power-law, $\rho_{\rm ej} \propto r^{-m}$ with $m < 3.0$, and adopted an homologous explosion model for the supernova expansion (see [10, 11] for details). The left panel of figure 1 shows the evolution of the mass accretion rate onto the NS. We have used, as our canonical model, an initial binary system of a $2\,M_\odot$ NS and the CO-core of a $30\,M_{\rm zams}$ progenitor, that ejects $M_{\rm env} = 7.94\,M_\odot$ in the SN explosion and leaves an iron core of $M_{\rm Fe} = 1.5\,M_\odot$. The accretion on the NS can be as high as $\sim\ 10^{-1}\,M_\odot\,s^{-1}$. For these high accretions, the photons would be trapped in the accretion flow and the gain of gravitational energy of the accreted material is mainly radiated via neutrino emission, specifically, by the e^+e^- pair annihilation process [12–14].

2.1 Mass and angular momentum evolution of the NS companion during the hypercritical accretion

If we want to discriminate the binary parameters of the systems in which the NS can reach, by accretion, its critical mass ($M_{\rm crit}$) and consequently collapse to a BH, from the systems in which the

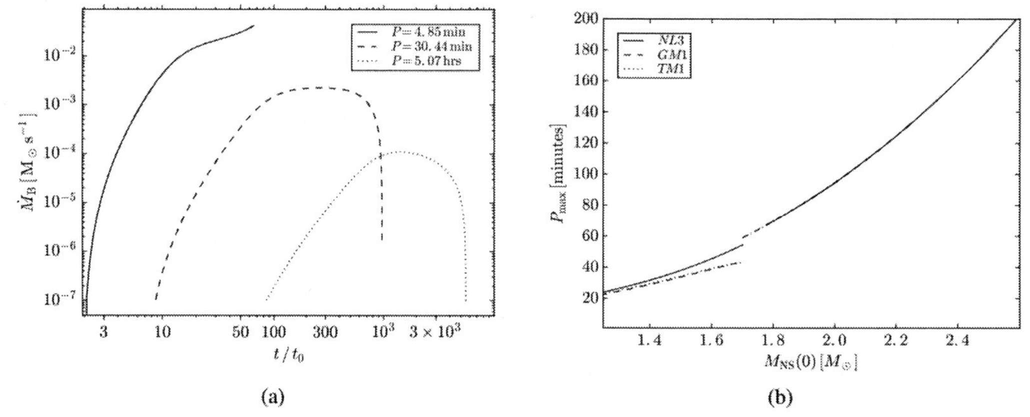

(a) (b)

Figure 1. Left: Evolution of the mass accretion rate on the NS companion calculated with the Bondy-Hoyle-Lyttleton accretion formalism for three selected orbital period. The initial binary system is formed by a $2\,M_\odot$ NS and the CO-core of a $30\,M_{\rm zamns}$ progenitor. Right: Maximum orbital period for which the NS reaches its critical mass by the accretion of material from the SN ejecta and collapse in a BH.

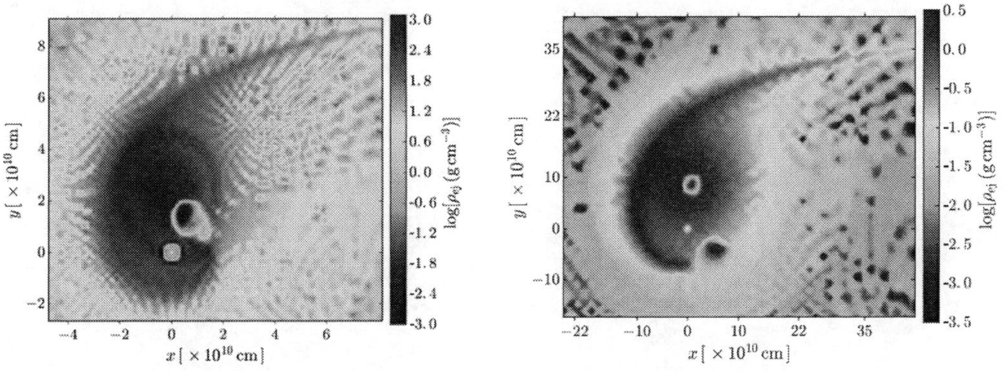

(a) (b)

Figure 2. Snapshots of the supernova ejecta density on the binary equatorial plane in the IGC scenario. We used the binary parameter of our canonical model. The left panel corresponds to a binary initial period of $P_{\rm orb} = 5$ min and the picture is made at 100 s after the collapse of the NS. The right panel corresponds to a period of $P_{\rm orb} = 50$ min. In this case the NS does not collapse and the snapshot is at $t \approx 44$ min.

accretion is not sufficient to induce such a collapse, we need to determine how the NS evolves during the accretion process. In general, the accretion of baryonic mass, M_b, as well as the accretion of angular momentum $J_{\rm NS}$, will modify the NS gravitational mass, $M_{\rm NS}$ in the following way:

$$\dot{M}_{\rm NS}(t) = \frac{\partial M_{\rm NS}}{\partial M_b}\dot{M}_b + \frac{\partial M_{\rm NS}}{\partial J_{\rm NS}}\dot{J}_{\rm NS}, \tag{2}$$

In [10] it was demonstrated that the supernova ejecta material has enough angular momentum to circularize around the NS and form a disk-like structure. Then, the NS angular momentum evolution is dictated by the disk accretion torque: $\dot{J}_{\rm NS} = \xi l(R_{\rm in})\dot{M}_B$, with $l(R_{\rm in})$ the angular momentum per unit mass of the material located at $r = R_{\rm in}$, the disk inner boundary radius, that will be the maximum between the radius of the last stable circular orbit, $r_{\rm lso}$, and the NS radius, $R_{\rm NS}$. We have also introduced a parameter $\xi \leq 1$ that accounts for the efficiency of the angular momentum transfer. Additionally, if we assume that all the mass entering the NS capture region will be accreted by the NS as baryonic mass: $M_b(t) = M_b(t_0) + M_B(t)$, then $\dot{M}_b = \dot{M}_B$, we can integrate equation (2) to follow the evolution of the NS during the accretion process. We have used the fitting formulas obtained in [15, 16] for the relations between the NS gravitational mass with its baryonic mass and angular momentum as well as the angular momentum of the last stable orbit.

Thus, a NS with initial mass $M_{\rm NS}(t_0)$ can reach $M_{\rm crit}$ if it accretes an amount of mass $\Delta M_{\rm acc} = M_{\rm crit} - M_{\rm NS}(t_0)$ from the supernova ejecta. This critical mass will depend, in general, on the EoS consider to model the NS matter and, in the case of a rotating NS, n the star angular momentum [15]. Therefore, given the initial NS mass, the CO core mass, and the supernova ejecta profile and its velocity, the accretion rate increases for shorter orbital periods. Therefore, there exists a maximum orbital period, $P_{\rm max}$, up to which, given $M_{\rm NS}(t_0)$ (and all the other binary parameters), the NS can accreate this precise amount of mass, $\Delta M_{\rm acc}$. The right panel of figure 1 shows $P_{\rm max}$ as a function of the NS initial mass, for a CO core progenitor of 30 $M_{\rm zmans}$.

3 SN-asymmetries induced by the NS companion presence

For supernova explosions occurring in close binaries with compact companions such as NSs or BHs, like the case of the IGC scenario, the supernova ejecta is subjected to a strong gravitational field

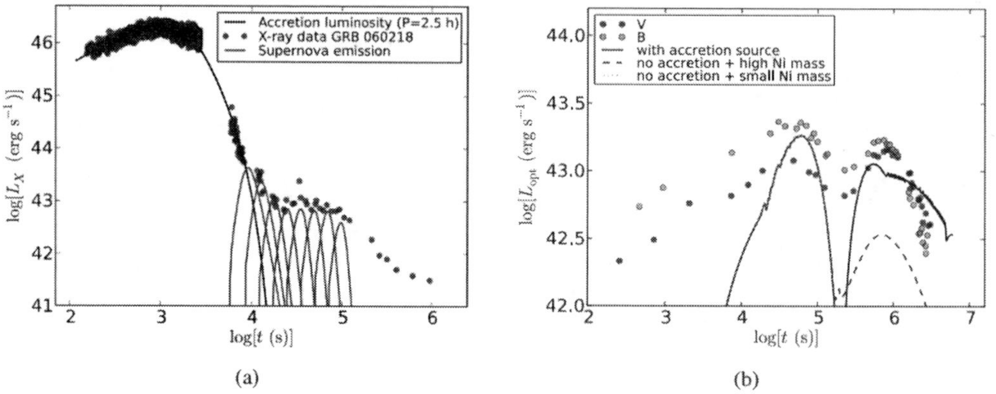

Figure 3. Left: Comparison of the accretion and the supernova luminosity with the observed X-ray luminosity of XRF 060218. The binary system has the following parameters: supernova velocity $v_{\rm star,0} = 2 \times 10^9$ cm s^{-1}, a pre-supernova core obtained from the $M_{\rm ZAMS} = 20\ M_\odot$ evolution, initial NS mass $M_{\rm NS}(t_0) = 1.4\ M_\odot$, and orbital period of 2.5 h. Right: Optical and UV luminosity of XRF 060218 [18]. The red dotted curve shows the supernova optical emission without either ^{56}Ni decay or accretion energy. The blue solid curve includes the energy deposition from the accretion onto the NS. The dashed green curve increases the total ^{56}Ni yield but cannot explain the observational data.

which produces a deformation of the supernova fronts closer to the accreting companion. In order to visualize this, we have simulated the evolution of the supernova layers in the binary system by dividing the SN ejecta in N particles of different mass and following its three-dimensional motion under the action of the gravitational field of the orbiting NS. We have varied the NS gravitational mass with equations (1) and (2) and also, we have removed from the simulation the particles that are crossing the Bondi-Hoyle radius. Figure 2 shows the surface density on the equatorial plane for the canonical initial binary system parameters. The left panel corresponds to an orbital period of about $P \approx 5$ min and its made at 100 s after the collapse of the NS companion and the formation of the BH . It can be seen the increasing asymmetry of the supernova ejecta around the orbital plane. Even for longer binary periods, from which the accretion does not lead to the collapse of the NS, the asymmetries can still be formed, as in shown in the right panel of figure 2. This system corresponds to a orbital period of about $P \approx 50$ min.

In order to validate the IGC scenario, we have to contrast it with the observations. In the left panel of figure 3 we present the observed X-ray luminosity of XRF 060218. The early part of the light-curve ($t \lesssim 10^3$ s) has been fitted with the luminosity expected from the accretion process on the NS companion, that was estimated as: $L_{\rm acc} = (\dot{M}_b - \dot{M}_{\rm NS})c^2$, the amount of gravitational energy gained by the accreted matter by falling to the NS surface and which is not spent in changing the gravitational binding energy of the NS.

For the long-lasting X-ray plateau in the afterglow (at times $t \sim 10^3$–10^6 s) we need to analyze the emission of the supernova at early stages. We have calculated the shock breakout luminosity using the light-curve code described in [17]. To simulate the energy that the hypercritical accretion process onto the NS adds to the ejecta, we injected it as an energy source at the base of the explosion and to mimic the asymmetries in the SN ejecta, caused by the NS companion presence (see figures 2), we have modeled a series of spherical explosions with different densities. For XRF 060218, we assume an initial explosion energy of 2×10^{51} erg, ranging the spherical equivalent-mass from 0.05–4 $M_\odot$. Figure 3 shows light-curves rising quickly at $t \lesssim 10^4$ s for the lowest mass to $\sim 10^5$ s for the 4 $M_\odot$

explosion. The observed emission would come from the sum of this full range of explosions (see [11] for details).

Finally, the right panel of figure 3 shows the V and B band light-curves for XRF 060218 [18]. The light-curve has two peaks: near 50,000 s and at 500,000 s. Using our 1 $M_\odot$ 1D model from our X-ray emission, we simulate the V and B band light-curves. Without either ^{56}Ni decay or accretion energy, the supernova explosion only explains the first peak. However, if we include the energy deposition from the accretion onto the NS (for our energy deposition, we use 4×10^{46} erg s^{-1} over a 2500 s duration), our simulations produce a second peak at roughly 500,000 s (see [11] for details).

4 Conclusions

We have studied the IGC paradigm to explain the nature of BdHNe and XRFs. We have followed the accretion onto the NS and computed the binary period, $P_{\rm max}$, over which the NS does not accrete sufficient matter to reach the critical mass and produce a BH. We have shown that the presence of the NS in very compact orbit produces large asymmetries in the supernova ejecta around the orbital plane due to the accretion and the action of the NS gravitational field on the supernova layers. These asymmetries lead to observable effects in the supernova emission. We have analyzed the light curve of XRF 060218 and explained its prompt emission ($t \lesssim 10^3$ s) with the accretion luminosity. We have shown that the observed long-lasting ($t > 10^3$ s) afterglow X-ray emission can be powered by a sequence of shock breakouts in different directions: the more massive directions produce later shock breakouts. Finally, we have simulated the optical emission of the supernova and compared our theoretical expectation with the optical luminosity of XRF 060218 which shows a peculiar double-peaked shape. We demonstrated that the source of energy given by the hypercritical accretion onto the NS provides a double-peaked light-curve consistent with the observational data.

References

[1] Ruffini, R., Rueda, J. A., Muccino, M., et al., ApJ **832**, 136 (2016)
[2] Ruffini, R., Bernardini, M. G., Bianco, C. L., et al. *Proceedings of the MG10 Meeting* (Singapore: World Scientific Publishing, Rio de Janeiro, Brazil, 2003) 369
[3] Izzo, L., Rueda, J. A., & Ruffini, R., A&A **548**, L5 (2012)
[4] Rueda, J. A., & Ruffini, R., ApJL, **758**, L7 (2012)
[5] Fryer, C. L. and Rueda, J. A. and Ruffini, R., ApJL, **793**, L36 (2014)
[6] Fryer, C. L. and Oliveira, F. G. and Rueda, J. A. and Ruffini, R., PhRvL, **115**, 231102 (2015)
[7] Hoyle, F., & Lyttleton, R. A., Proceedings of the Cambridge Philosophical Society **35**, 405 (1939)
[8] Bondi, H., & Hoyle, F., MNRAS **104**, 273 (1944)
[9] Bondi, H., MNRAS **112**, 195 (1952)
[10] Becerra, L., Cipolletta, F., Fryer, C. L., Rueda, J. A., & Ruffini, R., ApJ **812**, 100 (2015)
[11] Becerra, L., Bianco, C., Fryer, C. L., Rueda, J. A., & Ruffini, R., ApJ **833**, 107 (2016)
[12] Zel'dovich, Y. B., Ivanova, L. N., & Nadezhin, D. K., Soviet Ast **16**, 209 (1972)
[13] Chevalier, R. A., ApJ **346**, 847 (1989)
[14] Fryer, C. L., Benz, W., & Herant, M., ApJ **460**, 801 (1996)
[15] Cipolletta, F., Cherubini, C., Filippi, S., Rueda, J. A., & Ruffini, R., PhRvD **92**, 023007 (2015)
[16] Cipolletta, F. and Cherubini, C. and Filippi, S. and Rueda, J. A., & Ruffini, R., PhRvD **96**, 024046 (2017)
[17] Bayless, A. J., Even, W., Frey, L. H., et al., ApJ **805**, 98 (2015)
[18] Pian, E., Mazzali, P. A., Masetti, N., et al. 2006, Nature **442**, 1011 (2006)

Strong-field gravitational-wave emission in Schwarzschild and Kerr geometries: some general considerations

J. F. Rodríguez[1,2,*], *J. A.* Rueda[1,2,**], and *R.* Ruffini[1,2,***]

[1] *ICRANet, Piazza della Repubblica 10, I–65122 Pescara, Italy*
[2] *Dipartimento di Fisica and ICRA, Sapienza Università di Roma, P.le Aldo Moro 5, I–00185 Rome, Italy*

Abstract. We have used the perturbations of the exact solutions of the Einstein equations to estimate the relativistic wave emission of a test particle orbiting around a black hole. We show how the hamiltonian equations of motion of a test particle augmented with the radiation-reaction force can establish a priori constraints on the possible phenomena occurring in the merger of compact objects. The dynamical evolution consists of a helicoidal sequence of quasi-circular orbits, induced by the radiation-reaction and the background spacetime. Near the innermost stable circular orbit the evolution is followed by a smooth transition and finally plunges *geodesically* into the black hole horizon. This analysis gives physical insight of the merger of two equal masses objects.

1 Introduction

The problem of the merger of two massive objects is too complicated to be directly analyzed by solving the full Einstein equations. One reasonable attempt is to limit to the case of perturbations in the strong-field limit. We are going to neglect all post-Newtonian approaches to this problem. There are idealized processes in the strong-field limit to describe the motion of a test particle plunging into the field of an already formed black hole (BH). One case is purely geodesic where a test particle initially at rest or with a finite kinetic energy plunges from an infinite distance. Broadly speaking, all particles starting from infinite distance give rise to "*a precursor, a main burst and a ringing tail*". A finite energy emission occurs, $\Delta E = \eta m(m/M)$, where m and M are the particle and BH mass, with η ranging from 0.01 to almost 1, as a function of the nature of the BH and of the initial energy and angular momentum of the particle [1–4]. Another case, on which we focus here, it is analyzed by the study of the hamiltonian equations augmented with radiative forces. It consists of a "helicoidal" sequence of quasi-circular orbits, mainly driven by the radiation-reactionin the early stages. It follows a smooth transition near the innermost stable circular orbit (ISCO), and from this point it plunges describing a pure geodesic of the background spacetime. In Sec. 2 we recall the results on the case of circular orbits without taking into account radiation reaction and find the gravitational-wave energy and angular momentum flux at infinity, following the Sasaki-Nakamura method. In Sec. 3 we stufy the evolution of the particle, taking into account the radiation-reaction, obtaining the "helicoidal drifting sequence" passing through the last circular orbit and finally plunging.

*e-mail: jose.rodriguez@icranet.org
**e-mail: jorge.rueda@icra.it
***e-mail: ruffini@icra.it

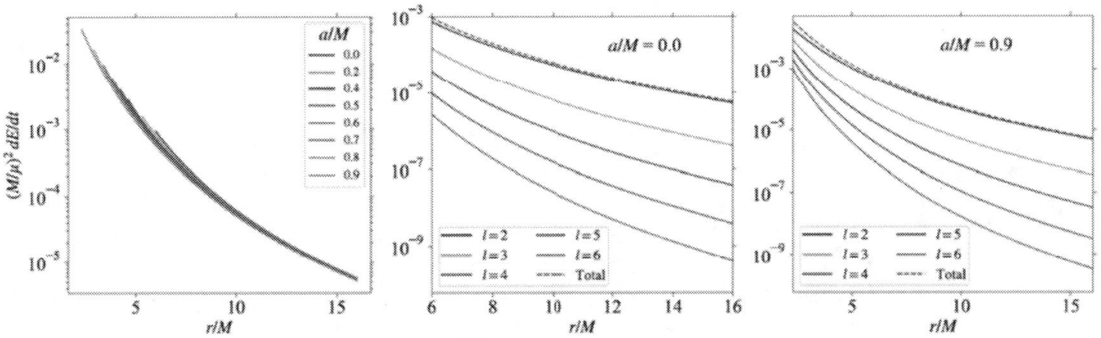

Figure 1. The figure on the left shows the total gravitational-wave energy flux at infinity, $\dot{E} \equiv dE/dt$, emitted by the test particle in circular orbits around the BH for different values of the normalized spin, as a function of r/M. The center and right figures show the contribution of the gravitational-wave modes $l \geq 2$ to the total gravitational-wave energy flux at infinity in the case a Schwarzschild BH and a Kerr black hole with $a/M = 0.9$, correspondingly.

2 Gravitational Radiation Fluxes for Circular Orbits

The calculation of the energy and momentum fluxes due to a particle orbiting a BH can be done by using standard metric perturbation theory, e.g. the work of Regge-Wheeler for perturbations in the Schwarzschild spacetime. Nevertheless, it is convenient to use an alternative approach developed by Teukolsky [5, 6] which deals with curvature perturbations. After performing a variable separation, it is possible to solve the corresponding equations. Unfortunately, the Teukolsky radial equation has a long-range potential and its numerical integration with boundary conditions is difficult. Sasaki and Nakamura [7, 8] found a change of variables that introduces a short-range and well-behaved potential $U(r)$. The Sasaki-Nakamura equation is:

$$X''_{lm\omega} - F(r)X'_{lm\omega} - U(r)X_{lm\omega} = \mathcal{S}_{lm\omega}. \tag{1}$$

Details on the functions F, U and $\mathcal{S}_{lm\omega}$ and their numerical solution can be found in Refs. [8] and [9]). The solution must satisfy the boundary conditions of out-going radiation at ∞ and in-going radiation at the horizon. To solve the radial equation with this suitable boundary conditions we used the Green's function technique. The procedure we used was the following: first, we solved numerically the Eq. (1) with the suitable boundary conditions, and then inverted the transformation to find the original radial function. Since we have used the Green's function technique, there are two solutions of the homogeneous Sasaki-Nakamura equation. The solution which satisfies the boundary condition at the outer horizon is $X^{H}_{lm\omega}$ and the solution that satisfies the boundary condition at ∞ is $X^{\infty}_{lm\omega}$. In the case of circular orbits, the energy and angular momentum fluxes carried by the gravitational waves to infinity are given by

$$\frac{dE}{dt} = \sum_{l\geq 2}^{\infty} \sum_{m=-l}^{l} \frac{|Z^{H}_{lm\omega}|^2}{4\pi\omega_m^2}, \tag{2}$$

$$\frac{dJ}{dt} = \sum_{l\geq 2}^{\infty} \sum_{m=-l}^{l} \frac{m|Z^{H}_{lm\omega}|^2}{4\pi\omega_m^3}, \tag{3}$$

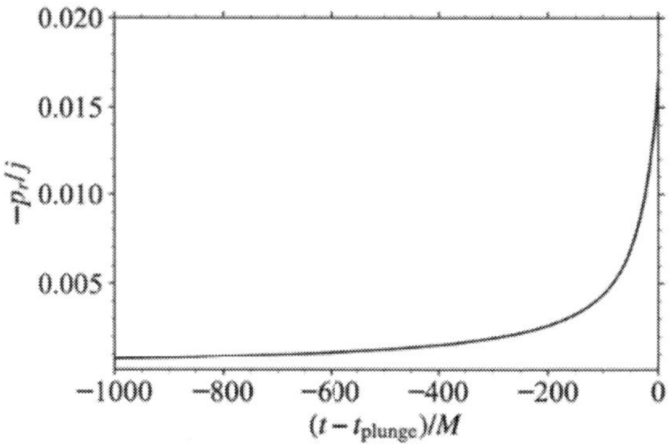

Figure 2. p_r/j: ratio of the radial momentum to the angular momentum per unit mass p_ϕ/μ of a test particle in the helicoidal drifting sequence around a Kerr black hole with $a/M = 0.7$. The mass-ratio is $\mu/M = 1/100$. The plot shows the ratio from $r \approx 5.32M$ up to the location of the last circular orbit at $r \approx 3.39M$. The time t_{plunge} is here defined as the time of the passage of the particle through the ISCO.

where $\omega_m = m\omega$, and $Z^H_{lm\omega}$ is a complex number found from $X^H_{lm\omega}$ (see eg. [9] for the explicit expression). It turns out that $Z^H_{lm\omega}$ is a function of the orbital frequency Ω. The results for different values of the normalized spin of the BH are shown in Fig. 1. We found agreement with the results of Refs. [10, 11].

3 Helicoidal drifting sequence

After calculating the gravitational radiation induced by a particle in a circular orbit, we proceed to find the dynamical effect in the particle's trajectory. In order to calculate this we used the Hamiltonian formalism augmented with a non-conservative force coming from the radiation. The Hamiltonian of the test particle of mass μ in the field of the Kerr BH of mass M is given by (see e.g. [12], and references therein)

$$H = -p_t = -N^i p_i + N\sqrt{\mu^2 + \gamma^{ij} p_i p_j}, \tag{4}$$

where $N = 1/\sqrt{-g^{00}}$, $N^i = -g^{ti}/g^{tt}$ and $\gamma^{ij} = g^{ij} + N^i N^j/N^2 = g^{ij} - g^{ti}g^{tj}/g^{tt}$. Here Latin index stands for the spatial Boyer-Lindquist coordinates (r, θ, ϕ), p_r and p_ϕ are, respectively, the radial and the angular momentum of the particle. The Hamilton canonical equations on the equatorial plane $\theta = \pi/2$ under the action of radial and azimuthal dissipative effects can then be written as (for a review on the subject see e.g. [13] in the extreme mass-ratio limit)

$$\frac{dr}{dt} = \frac{\partial H}{\partial p_r}, \quad \frac{d\phi}{dt} \equiv \Omega = \frac{\partial H}{\partial p_\phi}, \quad \frac{dp_r}{dt} = -\frac{\partial H}{\partial r} + \mathcal{F}^{\text{nc}}_r, \quad \frac{dp_\phi}{dt} = \mathcal{F}^{\text{nc}}_\phi, \tag{5}$$

We adopt for the radial and azimuthal non-conservative radiation-reaction forces:

$$\mathcal{F}^{\text{nc}}_r = 0, \tag{6}$$

$$\mathcal{F}^{\text{nc}}_\phi = -\frac{dJ}{dt}. \tag{7}$$

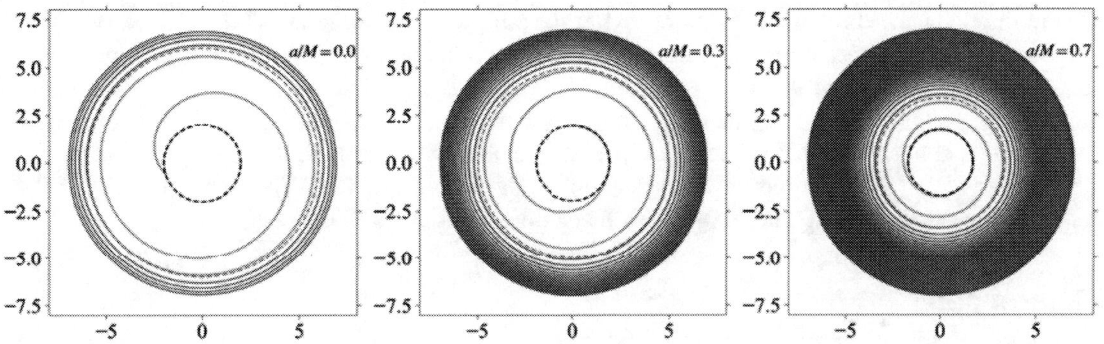

Figure 3. Trajectory of a test particle in the helicoidal drifting sequence around a Schwarzschild black hole, $a/M = 0$ (left panel) and around a Kerr black hole with $a/M = 0.3$ (center panel) and with $a/M = 0.7$ (right panel). The blue part of the trajectory goes from $r = 7M$ up to the location of the corresponding ISCO (gray-dashed circle), while the orange color indicates the plunge regime. The black-hole horizon is indicated with a black-dashed circle. The mass-ratio is $\mu/M = 1/100$.

where J is the angular momentum carried out to infinity by the gravitational waves given by Eq. (3). The assumption (6) is supported from the results of [14] in which it can be seen that the linear momentum carried out by the waves to infinity satisfies

$$\frac{dp_r^{\rm GW}}{dt} \ll \left|\frac{\partial H}{\partial r}\right|. \tag{8}$$

We have checked from our numerical computations that $|1 - \Omega(dJ/dt)/(dE/dt)| \sim 10^{-6}$ during the evolution.

In the traditional literature the evolution of the particle has been treated "adiabatically" . Within this formalism, the test particle that it is initially on a stable circular orbit, evolves to another circular orbit by the emission of gravitational radiation. This approach relies on the assumption that the radial velocity $\dot{r}$ is much less that the azimuthal velocity $r\Omega$. The adiabatic condition breaks down near the ISCO. On the other hand, in our formalism there is an additional effect induced by the radial momentum p_r. During the early stages of the evolution, the two approaches are practically equivalent. This is due to the fact that during this phase the radial momentum is negligible. However, near the ISCO p_r becomes appreciable (see Fig. 2). This fact allows a smooth transition through the ISCO. Since the radiation-reaction was calculating by assuming circular orbits, after the ISCO it can not be used. After this point we turn off the gravitational-reaction force and the test particles follows a pure geodesic down to the horizon. It is worthwhile to mention that when the particle arrives at $r = r_{\rm ISCO}$, the energy and angular momentum are different from the corresponding values of the ISCO. Furthermore, as already mentioned the radial momentum is different from zero as opposed to the adiabtic approach. All this induces the particle to plunge into the black even without radiating. The Fig. 3 shows the numerical results of the formalism for selected values of the normalized spin of the BH.

4 Conclusions

The case of a particle starting from a finite radius in a quasi-circular orbit leads to a novel "helicoidal drifting sequence", leading finally to a smooth merging into the black hole. There is a smooth transition in the Schwarzschild and Kerr cases. After the particle crosses the ISCO the trajectory is purely geodesic. When the radiation reaction is taken into account, as evidenced in Fig. 3, there is a clear appearance of "helicoidal drifting sequence". The more the particle gets closer to the horizon the more is the particle drift due to the increasing contribution of the radial momentum (see Fig. 2). The ISCO in this context does not witness any special role and only assists to the passage of the plunging particle in view of the acquired radial momentum in its previous history. The final result is the one of a test particle smoothly merging in the Kerr black hole without any burst.

References

[1] M. Davis, R. Ruffini, W.H. Press, R.H. Price, Physical Review Letters **27**, 1466 (1971)
[2] S.L. Detweiler, E. Szedenits, Jr., ApJ**231**, 211 (1979)
[3] R. Ruffini, M. Sasaki, Progress of Theoretical Physics **66**, 1627 (1981)
[4] Y. Kojima, T. Nakamura, Progress of Theoretical Physics **71**, 79 (1984)
[5] S.A. Teukolsky, ApJ**185**, 635 (1973)
[6] S.A. Teukolsky, W.H. Press, ApJ**193**, 443 (1974)
[7] M. Sasaki, T. Nakamura, Progress of Theoretical Physics **67**, 1788 (1982)
[8] M. Sasaki, T. Nakamura, Physics Letters A **89**, 68 (1982)
[9] S.A. Hughes, Phys. Rev. D**61**, 084004 (2000), gr-qc/9910091
[10] S.L. Detweiler, ApJ**225**, 687 (1978)
[11] R. Fujita, H. Tagoshi, Progress of Theoretical Physics **112**, 415 (2004), gr-qc/0410018
[12] R.T. Jantzen, P. Carini, D. Bini, Annals of Physics **215**, 1 (1992), gr-qc/0106043
[13] W.B. Han, Z. Cao, Phys. Rev. D**84**, 044014 (2011), 1108.0995
[14] M.J. Fitchett, S. Detweiler, MNRAS**211**, 933 (1984)

Quantum Gravity Effects in Cosmology

Je-An Gu[1], *Sang Pyo* Kim[2,⋆], and *Che-Min* Shen[3]

[1] *Leung Center for Cosmology and Particle Astrophysics, National Taiwan University, Taipei 10617, Taiwan*
[2] *Department of Physics, Kunsan National University, Kunsan 54150, Korea*
[3] *Department of Physics and Leung Center for Cosmology and Particle Astrophysics, National Taiwan University, Taipei 10617, Taiwan*

Abstract. Within the geometrodynamic approach to quantum cosmology, we studied the quantum gravity effects in cosmology. The Gibbons-Hawking temperature is corrected by quantum gravity due to spacetime fluctuations and the power spectrum as well as any probe field will experience the effective temperature, a quantum gravity effect.

1 Introduction

Recent remarkable observations have made cosmology a science of precision. The more precise and accurate cosmological observations are, the more likely the possibility of quantum gravity effects are to be measured. In this talk, we review a quantum gravity effect in cosmological observations. There have been proposed different approaches to quantum gravity (for a review, see [1]). In partciular, the geometrodynamical approach to quantum gravity and cosmology seems to be practically useful in that the transition from quantum gravity to semiclasical gravity and then to classical gravity including quantum corrections for the Einstein and the energy-stress tensors prescribes dynamical laws and correlations in terms of classical cosmology variables and parameters [2].

The Wheeler-DeWitt (WDW) equation is the quantum law for the geometrodynamical approach to quantum gravity, which becomes a relativistic wave equation in the superspace of the scale factor a and an inflaton ϕ for a Friedmann-Robertson-Walker universe. The scattering formulation of the two-component wave was advanced in Refs. [3, 4]. The Born-Oppenheimer interpretation of the WDW equation with respect to the Planck scale for gravity part and the energy scale for the inflaton separates the gravity part from the inflaton part and a further application of de Broglie-Bohm pilot-wave theory results in the classical equation together with quantum corrections of gravitational and field fluctuations [5, 6] (for references, see [2]). The quantum cosmology for an inflationary model with inhomogeneous fluctuations predicted the power spectrum with quantum corrections, which is suppressed at large scales and provides a weaker upper bound on the Hubble constant H than the tensor-to-scalar ratio [7]. Other quantum gravity effects may be found in Ref. [8].

The main purpose of this proccedings' article is to compliment the recent discovery that the quantum cosmology results in an effective Gibbons-Hawking temperature due to quantum fluctuations [9]. To do this, we provide two new methods to find approximately the solution to the WDW equation for

⋆e-mail: sangkim@kunsan.ac.kr

the FRW universe with a cosmological constant: the WKB approximation method and the Liouville-Green function method. From the wave function we can calculate the quantum potential, which gives rise to the quantum corrections to the semiclassical gravity. We then solve the semiclassical equation for the Euclieam geometry to find a periodic solution, whose period is the inverse temperature for the dS space in the sense of Ref. [10]. This approach may be an alternative to the gravitational instanton method [11]. Finally, we show that a probe field experiences this periodicity, which justifies the physical meaning of the temperature for dS space with quantum corrections.

2 de Broglie-Bohm Pilot-Wave Theory for Quantum Cosmology

The quantum cosmology based on the WDW equation leads to the semiclassical cosmology with quantum corrections as will be shown below. The WDW equation for a spatially closed FRW universe with a cosmological constant Λ, i.e, pure dS space, takes the form

$$\Big[-l_p^2\frac{\partial^2}{\partial a^2}+\frac{1}{l_p^2}(a^2-H_\Lambda^2 a^4)\Big]\Psi(a)=0, \tag{1}$$

where $l_p = \sqrt{4\pi G/2}$ is the Planck length. Introducing a dimensionless variable $x = a/l_p$ and a constant $\bar{H}_\Lambda = H_\Lambda l_p$, the WDW equation describes a quantum problem with the zero energy for an inverted Mexican hat potential

$$\Big[-\frac{d^2}{dx^2}+x^2-\bar{H}_\Lambda^2 x^4\Big]\Psi(x)=0. \tag{2}$$

We now apply the de Broglie-Bohm pilot theory to Eq. (2) by expressing the solution in the oscillatory region as

$$\Psi(x)=F(x)e^{iS(x)}, \tag{3}$$

where $F = |\Psi|$. Then, the real part of Eq. (2) becomes

$$S'^2-F''+x^2-\bar{H}_\Lambda^2x^4=0, \tag{4}$$

and so does the imaginary part

$$S''F+2S'F'=0. \tag{5}$$

Substituting the solution $F = 1/|S|^{1/2}$ of Eq. (5) into Eqs. (3) and (4) is equivalent to the exact Wentzel-Kramers-Brillouin (WKB) solution or phase-integral formula. The essence of the de Broglie-Bohm pilot theory is to solve Eqs. (4) and (5) at the same time by regarding F as a functional of S, i.e., $F(S)$. In the nonoscillatory region corresponding to a Euclidean spacetime, Eqs. (3)–(5) may be understood as an analytical continuation from the oscillatory region. In fact, introducing the cosmological time as the directional derivative of x along the peak of the wave packet

$$\frac{\partial}{\partial t}=-\frac{1}{x}\frac{\partial S}{\partial x}\frac{\partial}{\partial x}, \tag{6}$$

Eq. (4) becomes the semiclassical gravity equation [5, 6]

$$\Big(\frac{\dot{x}}{x}\Big)^2+\frac{1}{x^2}=\bar{H}_\Lambda^2-\frac{1}{x^4}V_q(x), \tag{7}$$

where a dot denotes the derivative with respect to t and $V_q = -F''/F$ is the quantum potential.

A passing remark is that Eq. (5) and thereby the quantum potential correspond to the flux (probability) conservation in the de Broglie-Bohm pilot theory of quantum mechanics [12]. On the other hand, the quantum potential provides a quantum correction in quantum cosmology to the classical gravity equation since there is no probability concept due to the single nature as a quantum system. The quantum potential is a consequence of spherical fluctuations of the FRW geometry.

3 Semiclassical Gravity Equation

The quantum gravity effect in a pure gravity contributes the quantum potential to the semiclassical gravity as shown in Eq. (7). We may apply either the WKP approximation or the Liouville-Green (LG) transformation method, to the WDW equation (1) or (2). The LG transformation method, however, reduces to the WKB approximation in the limit of large variable while it also applies to the turning point up to an additional constant term [3].

From now on, we shall employ the LG transformation method, which rewrites Eq. (2) in terms of a new variable as [3]

$$\Big[\frac{\partial^2}{\partial\eta^2} - (\eta^m + \Delta(\eta))\Big]\tilde{\Psi}(\eta) = 0, \tag{8}$$

where

$$\Big(\frac{\partial\eta}{\partial x}\Big)^2 \eta^m = V_g(x), \quad \Delta(\eta) = -\frac{3}{4}\frac{\left(\frac{d^2\eta}{dx^2}\right)^2}{\left(\frac{d\eta}{dx}\right)^4} + \frac{1}{2}\frac{\left(\frac{d^3\eta}{dx^3}\right)^2}{\left(\frac{d\eta}{dx}\right)^3}. \tag{9}$$

and

$$\tilde{\Psi} = \sqrt{\frac{d\eta}{dx}}\Psi(x), \quad V_g(x) = x^2 - \bar{H}_\Lambda^2 x^4. \tag{10}$$

Note that $\Delta(\eta)$ is proportional to the Schwazian derivative as $\Delta(\eta) = (1/2(d\eta/dx)^2)(S\eta)(x)$. The degree m is chosen by the order of roots of $V_g(x) = 0$. The classical motion is allowed in the region outside the simple root $x_H = 1/\bar{H}_\Lambda$, so we choose $m = 1$.

For large x, i.e., the scale factor a bigger than the Hubble radius $1/\bar{H}_\Lambda$ in the Planckian unit, $\Delta(\eta) \approx 1/\eta^2$. In this asymptotic region, we can neglect $\Delta(\eta)$ and solve Eq. (8) in terms of the Hankel functions as

$$\Psi_{\rm exp}(x) = D_1 \frac{\xi^{1/2}}{V_g^{1/4}} H^{(2)}_{1/3}(\xi), \quad \Psi_{\rm con}(x) = D_2 \frac{\xi^{1/2}}{V_g^{1/4}} H^{(1)}_{1/3}(\xi), \tag{11}$$

where D's are integration constants and

$$\xi = i\int^x \sqrt{V_g(x')}dx'. \tag{12}$$

Here, $\Psi_{\rm exp}(x)$ corresponds to an expanding branch of the wave function while $\Psi_{\rm con}(x)$ corresponds to a contracting branch. In the limit of large x, which belongs to the WKB region,

$$\xi = -\int^x \sqrt{\bar{H}_\Lambda^2 x'^4 - x'^2}dx' = -S(x), \tag{13}$$

the asymptotic form of the Hankel function

$$H^{(1)}_{1/3}(\xi) = \sqrt{\frac{2}{\pi\xi}}e^{i(\xi-\frac{5\pi}{12})}, \quad H^{(2)}_{1/3}(\eta) = \sqrt{\frac{2}{\pi\xi}}e^{-i(\xi-\frac{5\pi}{12})} \tag{14}$$

gives the WKB approximation, up to a constant phase,

$$\Psi_{\rm exp}(x) = D_1 \frac{1}{|V_g|^{1/4}} e^{i(S(x)-\frac{5\pi}{12})}, \quad \Psi_{\rm con}(x) = D_2 \frac{1}{|V_g|^{1/4}} e^{-i(S(x)-\frac{5\pi}{12})}. \tag{15}$$

Including correction (9) at the turning point, a straightforward calculation approximately leads to the quantum potential for large and small x as

$$V_q(x) = -\frac{2}{x^2} \quad (x \gg \frac{1}{\bar{H}_\Lambda}), \quad V_q(x) = -\frac{3}{4x^2} \quad (x \le \frac{1}{\bar{H}_\Lambda}). \tag{16}$$

4 Gibbons-Hawking (GH) Temperature corrected by QC

The semiclassical gravity equation (7) now reads

$$\left(\frac{\dot{a}}{a}\right)^2 + \frac{1}{a^2} = H_\Lambda^2 + \frac{c(l_p)^4}{a^6}, \tag{17}$$

which is Eq. (16) of Ref. [9]. Here, $c = 2$ for large a and $c = 3/4$ for small a. Note that a homogeneous massless field in the FRW geometry contributes the same form as the spherical fluctuations of the geometry [13]. The classical gravity has the solution $a(t) = \cosh(H_\Lambda t)/H_\Lambda$, which has the Euclidean solution $a(\tau) = \cos(H_\Lambda \tau)/H_\Lambda$ in the classically forbidden region. The Gibbons-Hawking temperature is the inverse of the periodicity $T_{\mathrm{GH}} = H_\Lambda/(2\pi)$ as shown in Ref. [10]. A probe field in the dS space experiences this periodicity, whose correlation function exhibits the Bose-Einstein or Fermi-Dirac distribution with the Gibbons-Hawking temperature.

We now solve Eq. (17) in the Euclidean time $\tau = it$ in the Euclidean spacetime by the quadrature method, which leads to the periodicity

$$\tau = 2\int_{a_-}^{a_+} \frac{da}{\sqrt{1 - H_\Lambda^2 a^2 - \frac{3(l_p)^4}{4a^4}}} = \frac{2}{H_\Lambda}\int_{u_-}^{u_+} \sqrt{\frac{u}{\frac{u^2}{H_\Lambda^2} - u^3 - \frac{3(l_p)^4}{4H_\Lambda^2}}}\, du, \tag{18}$$

where $u = a^2$ and another factor of two is multiplied in the second line in order to include the motion in the region of $a \leq 0$. Equation (18) is the complete elliptic integral of third kind [14]

$$\tau = \frac{2}{H_\Lambda}\int_{u_-}^{u_+} \sqrt{\frac{u}{(u_+ - u)(u - u_-)(u - u_0)}}\, du, \tag{19}$$

where $u_\pm$ are two positive roots of the cubic equation of u in Eq. (18). We may perturbatively find the roots as

$$u_+ = \frac{1}{H_\Lambda^2} - \frac{3}{4}(H_\Lambda l_p)^2, \quad u_- = \frac{\sqrt{3}}{2} l_p^2 \tag{20}$$

and, by neglecting the term $l_p^4/H_\Lambda^2 = (R_H l_p)^2 l_p^2$ with the Hubble radius $R_H = 1/H_\Lambda$, integrate approximately as

$$\tau_q \approx \frac{2}{H_\Lambda}\Big(\arcsin(1 - \frac{3}{2}(H_\Lambda l_p)^4) + \arcsin(1 - \sqrt{3}(H_\Lambda l_p)^2)\Big). \tag{21}$$

Thus, the temperature from the inverse periodicity is approximately given by

$$T_q = \frac{H_\Lambda}{2\pi}\Big(1 + \frac{(12)^{1/4}}{\pi} H_\Lambda l_p\Big) \tag{22}$$

and confirms the result from Ref. [9]. The quantum correction to the Gibbons-Hawking temperature is the order of $H_\Lambda l_p$.

5 Conclusion

In this paper, we have studied the quantum effect of spherical fluctuations of spacetime to the Gibbons-Hawking temperature. For this purpose, we have employed the semiclassical gravity theory based on the WDW equation and provided the semiclassical gravity for a pure gravity with the LG transformation method to calculate the quantum potential. The LG transformation method reduces to the WKB approximation in the limit of large variable. The LG transformation method approximately gives the wave function itself in terms the Hankel functions, from which the amplitude and thereby the quantum potential can be analytically derived. We noted that both the methods confirm the Gibbons-Hawking temperature corrected by quantum gravity to be consistent [9].

We now provide a physical reasoning why a probe field could measure the effective Gibbons-Hawking temperature in this paper. A homogeneous probe field ϕ of a light mass m does not give a back-reaction enough to change the spacetime itself governed by the semiclassical gravity and obeys the equation of motion in the Cauchy formulation of the first order:

$$\frac{d}{dt}\begin{pmatrix}\phi\\ \dot{\phi}\end{pmatrix} = \begin{pmatrix}0 & 1\\ -m^2 & -3\frac{\dot{a}}{a}\end{pmatrix}\begin{pmatrix}\phi\\ \dot{\phi}\end{pmatrix}. \tag{23}$$

The Floquet theory guarantees the periodicity for the probe field and its correlation function due to the periodicity of the semiclassical gravity (17) for the FRW universe. Thus, the semiclassical gravity goes beyond the classical FRW geometry with a stationary Killing vector but preserves a similar property. The Gibbons-Hawking temperature corrected by quantum gravity is a physical one to be measured by the probe field [15] and seems to be universal. The power spectrum of inflation models is similarly expected to have the quantum gravity effect due to the effective temperature.

6 Acknowledgements

S. P. K. would like to thank Rong-Gen Cai, Jorge Armando Rueda Hernández, Remo Ruffini and Gregory Vereshchagin for useful discussions during ICGAC13-IK15 and Maurice van Putten for the question of radiation for the Gibbons-Hawking temperature. This work was supported by the Basic Science Research Program through the National Research Foundation of Korea (NRF) funded by the Ministry of Education (NRF-2015R1D1A1A01060626).

References

[1] *Approaches to Fundamental Physics*, edited by I-O. Stamatescu and E. Seiler, Lect. Note Phys. **721** (Springer-Verlag, Berlin, 2007).
[2] S. P. Kim and W. Kim, Int. J. Mod. Phys. Conf. Ser. **43**, 1660205 (2016) [arXiv:1603.08087].
[3] S. P. Kim, Phys. Rev. D **46**, 3403 (1992).
[4] S. P. Kim, The Universe **1**, no. 1, 11 (2013) [arXiv:1304.7439].
[5] S. P. Kim, Phys. Rev. D **52**, 3382 (1995) [gr-qc/9511038].
[6] S. P. Kim, Phys. Rev. D **55**, 7511 (1997) [gr-qc/9611040].
[7] C. Kiefer and M. Kraemer, Phys. Rev. Lett. **108**, 021301 (2012) [arXiv:1103.4967].
[8] S. Carlip, D. W. Chiou, W. T. Ni and R. Woodard, Int. J. Mod. Phys. D **24**, 1530028 (2015) [arXiv:1507.08194].
[9] J. A. Gu, S. P. Kim and C. M. Shen, Int. J. Mod. Phys. D **26**, 1750116 (2017).
[10] R. Brout, Found. Phys. **17**, 603 (1987).
[11] S. W.Hawking and I. G. Moss, Phys. Lett. B **110**, 35 (1983).
[12] I. Licata, D. Fiscaletti, *Quantum potential: Physics, geometry and algebra*, (Springer, Heidelberg, 2014).
[13] S. P. Kim, Phys. Lett. A **236**, 11 (1997) [gr-qc/9703065].
[14] I. S. Gradshteyn and I. M. Ryzhik, *Table of Integrals, Series, and Products*, (Academic Press, London, 1994).
[15] B. .Kay and R. M. Wald, Phys. Rep. **207**, 49 (1991).

EPJ Web of Conferences **168**, 03002 (2018) https://doi.org/10.1051/epjconf/201816803002
Joint International Conference of ICGAC-XIII and IK-15 on Gravitation, Astrophysics and Cosmology

Schwinger mechanism in electromagnetic field in de Sitter spacetime

Ehsan Bavarsad[1,⋆], *Sang Pyo* Kim[2,3], *Clément* Stahl[4], and *She-Sheng* Xue[5,6]

[1]*Department of Physics, University of Kashan, 8731753153, Kashan, Iran*
[2]*Department of Physics, Kunsan National University, Kunsan 54150, Korea*
[3]*Center for Relativistic Laser Science, Institute for Basic Science, Gwangju 61005, Korea*
[4]*Instituto de Física, Pontificia Universidad Católica de Valparaíso, Casilla 4950, Valparaíso, Chile*
[5]*ICRANet, Piazzale della Repubblica 10, 65122 Pescara, Italy*
[6]*Dipartimento di Fisica, Universita di Roma "La Sapienza", Piazzale Aldo Moro 5, 00185 Rome, Italy*

Abstract. We investigate Schwinger scalar pair production in a constant electromagnetic field in de Sitter (dS) spacetime. We obtain the pair production rate, which agrees with the Hawking radiation in the limit of zero electric field in dS. The result describes how a cosmic magnetic field affects the pair production rate. In addition, using a numerical method we study the effect of the magnetic field on the induced current. We find that in the strong electromagnetic field the current has a linear response to the electric and magnetic fields, while in the infrared regime, is inversely proportional to the electric field and leads to infrared hyperconductivity.

1 Introduction

The Schwinger effect [1] is a phenomenon of pair production due to a strong electromagnetic field background. Since the early universe scenarios predict the existence of strong electromagnetic fields, it is natural to consider the Schwinger effect in this context. The pioneering works studied this effect in the 1+1 dimensional [2] and 1+3 dimensional [3] de Sitter spacetimes. To describe Schwinger effect in a curved spacetime, the more relevant quantity is the induced current [2, 3]. Indeed, it is not plagued by the notion of particle in the adiabatic future, which allows one to explore a broader parameter space. In [4], the Schwinger mechanism in 1+2 dimensions was explored as an example of odd dimensional field theory in dS.

In this proceedings' paper, we present some results of [5] and further elaborate the numerical results. The object of this work is to study the effect of a constant magnetic field background to the already present dS and electric backgrounds. This is a common generalization in flat spacetime where the analytic results are known for long [1], but it has never been investigated properly in dS. The origin of the observed large scale magnetic field is now an open question in cosmology but two main scenarios are emerging: an origin after recombination or a primordial one; for a recent review see, e.g., [6]. The effect of a magnetic field background on the scalar pair production in the spatially flat Friedmann-Lemaitre-Robertson-Walker type universes has been investigated in [7]. The perturbative amplitude for the fermion pair production in a magnetic field in dS has also been computed in [8].

⋆e-mail: bavarsad@kashanu.ac.ir

2 Schwinger Effect

In order to analyze the Schwinger effect, we assume that the complex scalar field is a test field probing two background fields: the gravitational field described by the dS metric

$$ds^2 = \Omega^2(\tau)\big(d\tau^2 - dx^2 - dy^2 - dz^2\big), \qquad \Omega(\tau) = -\frac{1}{H\tau}, \qquad \tau \in (-\infty, 0), \qquad (x, y, z) \in \mathbb{R}^3, \tag{1}$$

where H is the Hubble constant, and the electromagnetic field

$$A_\mu(x) = -\frac{E}{H^2\tau}\delta^3_\mu + By\delta^1_\mu, \tag{2}$$

where E and B are constants. The vector potential (2) leads to a constant electric and magnetic field parallel to each other in the conformal metric (1). The Schwinger pair production rate is obtained from the Bogoliubov coefficients. For a state with the comoving momentum k and the Landau level n, the Bogoliubov coefficients are given by

$$\alpha = \frac{(2|\gamma|)^{\frac{1}{2}}\Gamma(2\gamma)}{\Gamma(\frac{1}{2} + \kappa + \gamma)} e^{\frac{i\pi}{2}(\kappa-\gamma)}, \qquad \beta = -i\frac{(2|\gamma|)^{\frac{1}{2}}\Gamma(-2\gamma)}{\Gamma(\frac{1}{2} + \kappa - \gamma)} e^{\frac{i\pi}{2}(\kappa+\gamma)}, \tag{3}$$

where the parameters have been defined as

$$k = \sqrt{k_z^2 + (2n+1)eB}, \qquad \mu = \frac{m}{H}, \qquad \lambda = \frac{eE}{H^2},$$
$$\kappa = \frac{i\lambda k_z}{k}, \qquad \gamma = \sqrt{\frac{9}{4} - \lambda^2 - \mu^2}, \qquad \ell = eB\tau^2. \tag{4}$$

where m and e are the mass and charge of the complex scalar field, respectively. In terms of the Bogoliubov coefficient the pair production rate Γ is given by [5]

$$\Gamma = \frac{1}{\Omega^4(\tau) T L_y} \sum_{n=0}^{\infty} \int \frac{dk_z}{(2\pi)} \frac{dk_x}{(2\pi)} \left|\beta(k_z, n)\right|^2, \tag{5}$$

where T is the time interval of the pair production and L_y is the auxiliary parameter determining the normalization volume. Since the Bogoliubov coefficient β is independent of the momentum k_x, the integral gives $eBL_y/(2\pi)$ from the position of the center of the wave packet. Under the semiclassical condition [2, 3]

$$\lambda^2 + \mu^2 \gg 1, \tag{6}$$

the time when most of the particles are created, estimates $\tau k_z \sim -|\gamma|$ [2, 3], hence the k_z integral can be transformed into a τ integral. We then obtain

$$\Gamma = \Big(\frac{eB\Omega^{-2}}{2\pi}\Big)\Big(\frac{H^2|\gamma|}{2\pi}\Big) \sum_{n=0}^{\infty} \Big[\frac{e^{2\pi|\kappa|} - 1}{e^{2\pi|\gamma|} - e^{-2\pi|\gamma|}} + \frac{1}{e^{2\pi|\gamma|} - 1}\Big], \tag{7}$$

where

$$|\kappa| = \frac{\lambda|\gamma|}{\sqrt{|\gamma|^2 + (2n+1)\ell}}. \tag{8}$$

In fig. 1a the pair production rate Γ given by eq. (7) is plotted as a function of the electric field for different intensities of the magnetic field. This figure illustrates that in the strong electric field regime

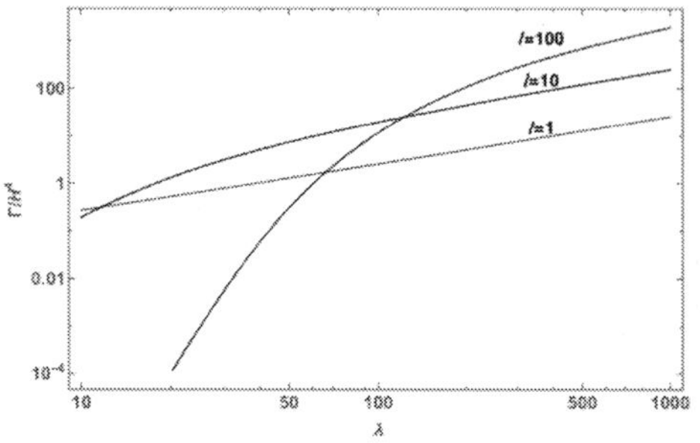

(a) For different values of ℓ, the normalized pair production rate Γ/H^4 is plotted as a function of λ, in the lowest Landau state $n = 0$ with $\mu = 1$.

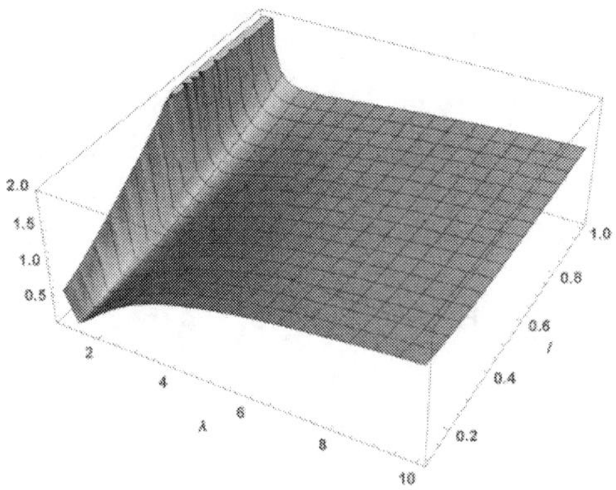

(b) The ratio of the induced current to the semiclassical current $J/J_{\rm sem}$ is plotted as a function of λ and ℓ, in the lowest Landau state $n = 0$ with $\mu = 1$.

Figure 1

$\lambda \gg \max(1, \mu, \ell)$, the pair production rate is significant: $\Gamma \gg 1$. The first term in the square bracket in eq. (7) is the pair production rate from the electromagnetic field while the second term is the dS radiation weighted by the density of states for the electromagnetic field. Since the second term in Γ (see e.q. (7)) is independent of Landau levels, its sum apparently diverges. However, the Riemann zeta function prescription of $\zeta(0) = -1/2$ together with the $n = 0$ term gives a constant factor of $1/2$. Thus, the pair production from the zeta regularization technique leads to a finite result

$$\Gamma = \Big(\frac{eB\Omega^{-2}}{2\pi}\Big)\Big(\frac{H^2|\gamma|}{2\pi}\Big)\Big(\frac{1}{e^{4\pi|\gamma|}-1}\Big)\Big[\frac{1}{2} + \sum_{n=0}^{\infty} e^{2\pi(|\kappa|+|\gamma|)}\Big]. \tag{9}$$

In the limit of zero electric field $E = 0$, the first term in the square bracket of eq. (7) vanishes and the second term is the dS radiation weighted by the density of states of the magnetic field background. Therefore the gravitational pair production rate in dS is enhanced by the magnetic field. One can verify that the pair production expression given by eq. (7) leads to the well known Schwinger formula in the flat spacetime limit ($H = 0$).

3 Induced Current

Once the pair production rate Γ is calculated, one can read the number density of the Schwinger pairs as $\mathcal{N} \sim \Gamma/H$. Then, the semiclassical conductive current $J_{\rm sem}$ of the pairs having a charge e and a velocity v due to the background electric field is given by $J_{\rm sem} = 2e\mathcal{N}v$. In this section we investigate numerically the expectation value of the current operator (which is referred to as the induced current) in the Hadamard in-state. The only nonvanishing component of the induced current, which is parallel to the electric field background, is given by [5]

$$J = \frac{eH^3\ell}{4\pi^2}\sum_{n=0}^{\infty}\int_{-\infty}^{+\infty}\frac{dk_z}{k}(\lambda - k_z\tau)e^{i\pi\kappa}\Big|W_{\kappa,\gamma}(2ik\tau)\Big|^2, \tag{10}$$

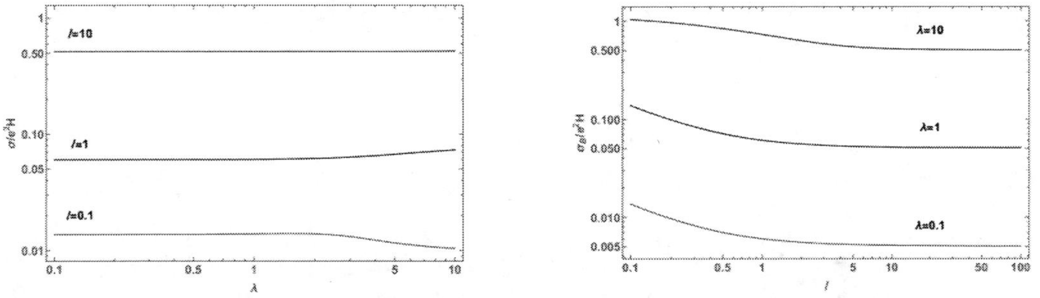

Figure 2: Left (right) panel: for different values of ℓ (λ), the normalized conductivity σ/e^2H $\left(\sigma_B/e^2H\right)$ is plotted as a function of λ (ℓ), in the lowest Landau state $n = 0$ with $\mu = 1$.

where W is the Whittaker function. In the regime of $\lambda \gg 1$ the semiclassical condition (6) is satisfied, and the induced current (10) is comparable to the semiclassical current $J_{\rm sem}$. Considering the ultrarelativistic particles with velocity $v \sim 1$, fig. 1b shows that the induced current J approaches the semiclassical current $J_{\rm sem}$ in the strong electric field regime $\lambda \gg \max(1, \mu, \ell)$. For a better understanding of the physics of the results, it is convenient to define the electric and magnetic conductivity σ and σ_B as

$$\sigma = \frac{J}{E}, \qquad\qquad \sigma_B = \frac{J}{\Omega^{-2}B}. \tag{11}$$

In fig. 2 we plot σ and σ_B as a function of the electric and magnetic fields, respectively. The figure reveals that the induced current of a massive scalar field responds to the strong electromagnetic field as $J \propto B \cdot E$. In the infrared regime where the relations $\ell \ll \mu \ll \lambda \ll 1$ are satisfied, fig. 3 illustrates that the conductivity increases as the electric field decreases. This phenomena was first reported in [2] and dubbed infrared hyperconductivity.

4 Conclusion

We have reported here some results of [5]: the effect of a constant magnetic field on Schwinger mechanism in dS. We have described its effects and investigated some various limiting cases. These considerations are relevant to magnetogenesis scenarios. Using the zeta function regularization, we have obtained a finite expression for the Schwinger pair production rate. We have found that the Schwinger effect is enhanced by the magnetic field background. In the zero electric field limit, our result agrees with the Gibbons-Hawking radiation modulo the density of states for magnetic field. In addition, we have numerically examined the induced current. We have found that in the strong electromagnetic field the induced current is dominated by a term proportional to $E \cdot B$ and is thus approximately proportional to the pseudo-scalar of the Maxwell theory, which corresponds to the chiral magnetic effect for spin-1/2 fermions [9]. In the infrared regime, the current and conductivity are increasing as the electric field decreases, which leads to the infrared hyperconductivity phenomenon.

Acknowledgement

S. P. K. would like to thank Remo Ruffini at ICRANet, where this work was initiated and also W-Y. Pauchy Hwang for the warm hospitality at National Taiwan University. The work of S. P. K. was

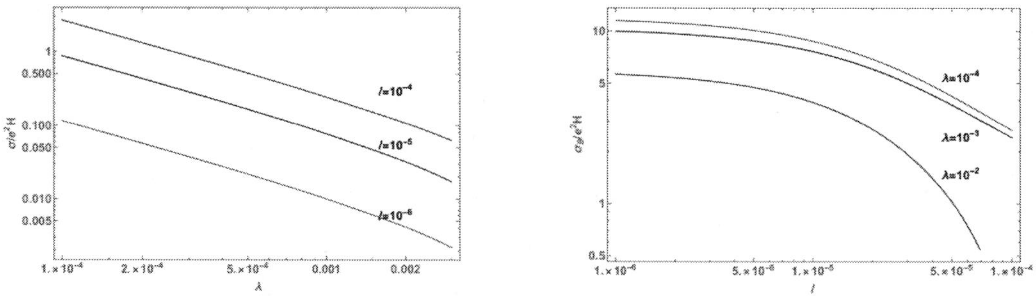

Figure 3: Left (right) panel: in the infrared regime, for different values of ℓ (λ), the normalized conductivity $\sigma/e^2H\left(\sigma_B/e^2H\right)$ is plotted as a function of λ (ℓ), in the lowest Landau state $n = 0$ with $\mu = 10^{-4}$.

supported by IBS (Institute for Basic Science) under IBS-R012-D1 and also by the Basic Science Research Program through the National Research Foundation of Korea (NRF) funded by the Ministry of Education (NRF-2015R1D1A1A01060626). E. B. and S.-S. X. would like to thank S. P. K. for the warm hospitality at Kunsan National University. E. B. is supported by the University of Kashan Grant No. 317203/2.

References

[1] J. S. Schwinger, Phys. Rev. **82**, 664 (1951).

[2] M. B. Fröb, J. Garriga, S. Kanno, M. Sasaki, J. Soda, T. Tanaka and A. Vilenkin, JCAP **1404**, 009 (2014).

[3] T. Kobayashi and N. Afshordi, JHEP **1410**, 166 (2014).

[4] E. Bavarsad, C. Stahl and S. S. Xue, Phys. Rev. D **94**, no. 10, 104011 (2016).

[5] E. Bavarsad, S. P. Kim, C. Stahl and S. S. Xue, [arXiv:1707.03975 [hep-th]].

[6] K. Subramanian, Rept. Prog. Phys. **79**, no. 7, 076901 (2016).

[7] S. Moradi, Mod. Phys. Lett. A **24**, 1129 (2009); S. Haouat and R. Chekireb, Eur. Phys. J. C **72**, 2034 (2012).

[8] C. Crucean and M. A. Băloi, Phys. Rev. D **93**, no. 4, 044070 (2016).

[9] K. Fukushima, D. E. Kharzeev and H. J. Warringa, Phys. Rev. D **78**, 074033 (2008).

EPJ Web of Conferences **168**, 03003 (2018) https://doi.org/10.1051/epjconf/201816803003
Joint International Conference of ICGAC-XIII and IK-15 on Gravitation, Astrophysics and Cosmology

Holographic Entanglement Entropy, SUSY & Calibrations

Eoin Ó Colgáin[1,⋆]

[1] *Asia Pacific Center for Theoretical Physics, Postech, Pohang 37673, Korea*

Abstract. Holographic calculations of entanglement entropy boil down to identifying minimal surfaces in curved spacetimes. This generically entails solving second-order equations. For higher-dimensional AdS geometries, we demonstrate that supersymmetry and calibrations reduce the problem to first-order equations. We note that minimal surfaces corresponding to disks preserve supersymmetry, whereas strips do not.

1 Introduction

Determining entanglement entropy in Quantum Field Theory is a challenging problem. For QFTs with holographic duals, Ryu-Takayanagi (RT) argued that the task is equivalent to calculating the area of codimension-two minimal surfaces in a dual geometry [1, 2]. The results for AdS_3 show perfect agreement with 2D CFTs [3–5], suggesting that this method may have merit in higher dimensions. Regularisation aside, even for time-independent QFTs, this approach is plagued by one noticeable technicality: the task of identifying minimal surfaces involves solving second-order equations.

In this letter, we consider two approaches, one inspired by physics, the second by mathematics, to recast the minimal surface problem so that the resulting equations are first-order. From the physics perspective, as demonstrated for locally AdS_3 geometries [6], e. g. BTZ black holes [7, 8], one can derive the minimal surface from supersymmetry (SUSY) [9]. An immediate corollary is that in 3D the RT minimal surfaces are supersymmetric. Here, we show in higher-dimensional AdS_{d+2} geometries that a disk minimal surface preserves supersymmetry, whereas the strip does not. Working backwards to derive the surface is expected to mirror the 3D analysis [9]. This suggests that supersymmetry may be exploited to identify minimal surfaces in supersymmetric geometries, such as LLM [10].

Our second approach involves importing calibrations [11] from mathematics. At the heart of this method is the fact that a closed differential form is guaranteed to be a minimal surface in its homology class. For locally AdS_3 geometries, it has been shown that the minimal surfaces correspond to special Lagrangian (sLag) submanifolds on a 2D spacelike hypersurface [12]. In this letter, we demonstrate the the calibration conditions for strips and disks in AdS_{d+2} geometries agree with the equations that result from extremising actions [2]. In particular, we will show that second-order equations are implied.

⋆e-mail: ocolgain.eoin@apctp.org

2 Supersymmetric minimal surfaces

We consider two minimal surfaces embedded in unit radius AdS_{d+2} spacetimes, namely

$$\mathrm{d}s^2_{\mathrm{disk}} = \frac{-\mathrm{d}t^2 + \mathrm{d}\eta^2 + \eta^2 \mathrm{d}s^2(S^{d-1}) + \mathrm{d}r^2}{r^2}, \tag{1}$$

and

$$\mathrm{d}s^2_{\mathrm{strip}} = \frac{-\mathrm{d}t^2 + \mathrm{d}x_1^2 + \mathrm{d}x_2^2 + \cdots + \mathrm{d}x_d^2 + \mathrm{d}r^2}{r^2}, \tag{2}$$

which we will refer to as disk and strip geometries, respectively. In each case, we assume the minimal surface is embedded along $\mathbb{R}^d$, but in the former the radial direction only depends on η, $r(\eta)$, while in the latter, it depends on x_1, $r(x_1)$. Symmetry reduces the complexity of the problem to an effective 3D problem. We consider time-independent configurations and set $t = 0$ for simplicity.

In order to discuss supersymmetric configurations, we introduce the supersymmetry condition, or Killing spinor equation (KSE), of AdS_{d+2} spacetime,

$$\nabla_\mu \epsilon = \frac{1}{2}\gamma_\mu \epsilon, \tag{3}$$

where ϵ denotes the Killing spinor, γ_μ are matrices satisfying the Clifford algebra, and $\nabla_\mu \equiv \partial_\mu + \frac{1}{4}\omega_{\mu\rho\sigma}\gamma^{\rho\sigma}$, where ω is the spin connection. All information about the geometry is encoded in the spin connection, and acting twice with ∇, one derives the Einstein equation $R_{\mu\nu} = -(d+1)g_{\mu\nu}$.

For $d = 1$, there is no distinction between strip and disk, and it was shown that RT minimal surfaces for AdS_3 spacetimes preserve *locally* half the supersymmetry [9]. Here we extend the analysis to higher dimensions. For simplicity, we focus on $d = 3$, which is most relevant to AdS/CFT [13].

2.1 Disk

We consider the geometry (1) with a circular disk on the boundary $\eta \leq h$, with h constant. For this surface to be consistent with supersymmetry, we expect to be able to identity a subset of the Killing spinors that "commute" with the bulk minimal surface. More precisely, there must be Killing spinors that are solutions to the following projection condition:

$$\Gamma\epsilon = \frac{i}{\sqrt{1 + \left(\frac{\mathrm{d}r}{\mathrm{d}\eta}\right)^2}}\gamma_{\theta\phi}\left(\gamma_\eta + \frac{\mathrm{d}r}{\mathrm{d}\eta}\gamma_r\right)\epsilon = \epsilon. \tag{4}$$

The projection condition is inspired by supersymmetric branes and embeddings [14], where $\Gamma^2 = 1$, so that this defines a valid projection condition.

The solution to the Killing spinor equation (3), evaluated at $t = 0$, is

$$\epsilon = \left(r^{\frac{1}{2}} + r^{-\frac{1}{2}}\eta\gamma_\eta\right)\tilde{\epsilon}_+ + r^{-\frac{1}{2}}\tilde{\epsilon}_-, \tag{5}$$

where we have defined $\tilde{\epsilon}_\pm = e^{\frac{\theta}{2}\gamma_{\eta\theta}} e^{\frac{\phi}{2}\gamma_{\theta\phi}}\epsilon_\pm$ in terms of the usual two-sphere coordinates and constant spinors satisfy $\gamma_r\epsilon_\pm = \pm\epsilon_\pm$. Evaluating this condition on the known minimal surface [2],

$$\eta^2 + r^2 = h^2, \tag{6}$$

where h is a constant, we find that the projection condition implies the relation,

$$i\gamma_{\theta\phi\eta}\,\epsilon_- = h\,\epsilon_+. \tag{7}$$

As a result, we see that $\epsilon_\pm$ are not independent, which reduces the number of supersymmetry by half. This result closely mirrors earlier 3D analysis [9] and we conclude that the disk minimal surface will be supersymmetric in other dimensions.

2.2 Strip

This time around the Killing spinor is

$$\epsilon = \left(r^{\frac{1}{2}} + r^{-\frac{1}{2}} x_i \gamma_i\right) \epsilon_+ + r^{-\frac{1}{2}} \epsilon_-, \tag{8}$$

where repeated indices $i = 1, 2, 3$ are summed and the constant spinors are subject to the same condition. The projector Γ is analogous:

$$\Gamma = \frac{i}{\sqrt{1 + \left(\frac{dr}{dx_1}\right)^2}} \gamma_{x_2 x_3} \left(\gamma_{x_1} + \frac{dr}{dx_1} \gamma_r\right), \tag{9}$$

which is simply an artifact of the large amount of symmetry. Since the surface is embedded along x_2 and x_3, for supersymmetry to be preserved, these coordinates must drop out of the projection condition. It is easy to convince oneself that this cannot happen, so supersymmetry is not preserved.

3 Calibrations

In [12], where AdS_3 geometries were largely discussed, proposals were given for candidate calibrations in higher dimensions. For geometries (1) and (2), we recall the suggested d-forms [12]:

$$\varphi_{\text{disk}} = e^{i\chi} \left(\frac{\eta}{r}\right)^{d-1} \left(\frac{d\eta}{r} + i\frac{dr}{r}\right) \wedge \text{vol}(S^{d-1}), \tag{10}$$

and

$$\varphi_{\text{strip}} = e^{i\chi} \frac{1}{r^{d-1}} \left(\frac{dx_1}{r} + i\frac{dr}{r}\right) \wedge dx_2 \wedge \cdots \wedge dx_d, \tag{11}$$

where χ is an arbitrary phase that is fixed through the calibration conditions:

$$0 = \text{Im}(\varphi) = d\text{Re}(\varphi). \tag{12}$$

Only in the special case $d = 1$ do the forms correspond to sLag calibrations and otherwise, their identity in mathematics is not immediately obvious. Admittedly, we have pulled φ_{disk}, φ_{strip} and (12) out of a hat. However, we justify the expressions by the fact that they correspond to sLag calibration conditions for $d = 1$ and also lead to reasonable results; in this section we will show that the conditions (12) recover the minimal surface equations identified in [2]. Our analysis here improves on [12] by showing agreement at the level of equations and not for specific solutions. This suggests that φ_{disk} and φ_{strip} are indeed valid calibrations. In contrast to the last section, here we consider general d.

3.1 Disk

We return to the disk geometry (1), where the minimal surface has the action [2],

$$\mathcal{L}_{\text{disk}} = \frac{\eta^{d-1}}{r^d} \sqrt{1 + \left(\frac{dr}{d\eta}\right)^2}. \tag{13}$$

Varying the action, we encounter a second-order equation [2]:

$$\eta r r'' + (d-1) r (r')^3 + (d-1) r r' + d\eta (r')^2 + d\eta = 0, \tag{14}$$

where $r' = \frac{dr}{d\eta}$, etc. Assuming our calibration (10) is correct, we expect that the calibration conditions (12) imply (14). To demonstrate this, let us note that we can rewrite (14) as

$$\frac{\eta r r''}{1+(r')^2} + (d-1)rr' + d\eta = 0. \tag{15}$$

Taking the derivative of the $\mathrm{Im}(\varphi) = 0$ condition, we see that

$$\partial_r \chi = -\tan\chi \partial_\eta \chi. \tag{16}$$

We can use this expression to bring $\mathrm{dRe}(\varphi) = 0$ to the form:

$$0 = -d\eta\cos\chi + r(d-1)\sin\chi + \frac{r\eta\partial_\eta\chi}{\cos\chi}. \tag{17}$$

At this point we substitute,

$$\cos\chi = \frac{1}{\sqrt{1+(r')^2}}, \quad \sin\chi = -\frac{r'}{\sqrt{1+(r')^2}}, \quad \partial_\eta\chi = -\frac{r''}{(1+(r')^2)^2}, \tag{18}$$

where all expressions follow from the first equation in (12). Substituting back into the (17), we get (14). So, we have demonstrated that any solution to the first order equations (12) satisfies the second-order equation (14), thus validating the proposed calibration $\varphi_{\rm disk}$.

Before concluding, it is known that (6) solves (14). It is worth noting that taking the large d limit, (14) simplifies, but still retains (6) as a solution. In fact, we can rewrite (14) as

$$(\eta r r'' - r(r')^3 - rr') + d\left(r(r')^3 + rr' + \eta(r')^2 + \eta\right) = 0. \tag{19}$$

where it can be checked that both bracketed terms vanish on the known solution. The other solutions, which are not common, do not make contact with the AdS boundary at the boundary of the disk. Our observations suggest (6) is the only solution relevant for holographic entanglement entropy.

3.2 Strip

Moving along, we consider the strip geometry, where the area functional is given by the action [2],

$$\mathcal{L}_{\rm strip} = \frac{1}{r^d}\sqrt{1+\left(\frac{dr}{dx_1}\right)^2}, \tag{20}$$

which corresponds to an x_1-independent Hamiltonian. Viewing x_1 as time, we infer,

$$\frac{dr}{dx_1} = \frac{\sqrt{h^{2d} - r^{2d}}}{r^d}, \tag{21}$$

where h is a constant. Note, this first-order equation can be compared directly with (12) for $\varphi_{\rm strip}$.

The calibration conditions that arise from (11) can be solved using the method of characteristics (see, for example [15]) and are equivalent to a systems of ODEs,

$$r\,dr = -\frac{r\sqrt{1-f^2}}{f}dx = \frac{r^2\,df}{p\,f}, \tag{22}$$

where we have defined $f = \cos\chi$. It is easy to find one first integral,

$$c_1 = \frac{f}{r^p}, \tag{23}$$

with constant c_1. Substituting this into the remaining equation and integrating, one can identify a second first integral, but we will not need the expression. It is enough to redefine $c_1 = -1/h^p$ so that we recover (21), which makes it clear that extremisation of the action (20) and the calibration conditions are equivalent.

We have thus demonstrated that $\varphi_{\rm disk}$ and $\varphi_{\rm strip}$ are valid calibrations, despite not being obviously sLag, except when $d = 1$.

Acknowledgements

We are grateful to I. Bakhmatov, N. Deger, J. Gutowski and H. Yavartanoo for collaboration on this topic. We thank M. M. Sheikh-Jabbari for related discussions.

References

[1] S. Ryu and T. Takayanagi, "Holographic derivation of entanglement entropy from AdS/CFT," Phys. Rev. Lett. **96**, 181602 (2006) [hep-th/0603001].

[2] S. Ryu and T. Takayanagi, "Aspects of Holographic Entanglement Entropy," JHEP **0608**, 045 (2006) [hep-th/0605073].

[3] C. Holzhey, F. Larsen and F. Wilczek, "Geometric and renormalized entropy in conformal field theory," Nucl. Phys. B **424**, 443 (1994) [hep-th/9403108].

[4] P. Calabrese and J. L. Cardy, "Entanglement entropy and quantum field theory," J. Stat. Mech. **0406**, P06002 (2004) [hep-th/0405152].

[5] P. Calabrese and J. Cardy, "Entanglement entropy and conformal field theory," J. Phys. A **42**, 504005 (2009) [arXiv:0905.4013 [cond-mat.stat-mech]].

[6] M. Banados, "Three-dimensional quantum geometry and black holes," AIP Conf. Proc. **484**, 147 (1999) [hep-th/9901148].

[7] M. Banados, C. Teitelboim and J. Zanelli, "The Black hole in three-dimensional space-time," Phys. Rev. Lett. **69**, 1849 (1992) [hep-th/9204099].

[8] M. Banados, M. Henneaux, C. Teitelboim and J. Zanelli, "Geometry of the (2+1) black hole," Phys. Rev. D **48**, 1506 (1993) Erratum: [Phys. Rev. D **88**, 069902 (2013)] [gr-qc/9302012].

[9] E. Ó Colgáin and H. Yavartanoo, "Banados and SUSY: on supersymmetry and minimal surfaces of locally AdS_3 geometries," Class. Quant. Grav. **34**, no. 9, 095008 (2017) [arXiv:1610.05638 [hep-th]].

[10] H. Lin, O. Lunin and J. M. Maldacena, "Bubbling AdS space and 1/2 BPS geometries," JHEP **0410**, 025 (2004) [hep-th/0409174].

[11] R. Harvey and H. B. Lawson, Jr., "Calibrated geometries," Acta Math. **148**, 47 (1982).

[12] I. Bakhmatov, N. S. Deger, J. Gutowski, E. Ó Colgáin and H. Yavartanoo, "Calibrated Entanglement Entropy," JHEP **1707**, 117 (2017) [arXiv:1705.08319 [hep-th]].

[13] J. M. Maldacena, Int. J. Theor. Phys. **38**, 1113 (1999) [Adv. Theor. Math. Phys. **2**, 231 (1998)] [hep-th/9711200].

[14] D. P. Sorokin, "Superbranes and superembeddings," Phys. Rept. **329**, 1 (2000) [hep-th/9906142].

[15] R. Courant, D. Hilbert, "Methods of Mathematical Physics", Vol. II, John Wiley (1962)

EPJ Web of Conferences **168**, 03004 (2018) https://doi.org/10.1051/epjconf/201816803004
Joint International Conference of ICGAC-XIII and IK-15 on Gravitation, Astrophysics and Cosmology

Tunneling decay of self-gravitating vortices

Éric Dupuis[1,⋆], *Yan* Gobeil[1,2,⋆⋆], *Bum-Hoon* Lee[3,4,⋆⋆⋆], *Wonwoo* Lee[3,⋆⋆⋆⋆], *Richard* MacKenzie[1,†], *Manu B.* Paranjape[1,‡], *Urjit A.* Yajnik[5,§], and *Dong-han* Yeom[6,7,¶]

[1] *Groupe de Physique des Particules, Département de physique, Université de Montréal, C. P. 6128, Succursale Centre-ville, Montréal, Québec, Canada H3C 3J7*
[2] *Department of Physics, Ernest Rutherford Physics Building, McGill University, 3600 rue University, Montréal, Québec, Canada H3A 2T8*
[3] *Center for Quantum Spacetime, Sogang University, Seoul 04107, Korea*
[4] *Department of Physics, Sogang University, Seoul 04107, Korea*
[5] *Department of Physics, Indian Institute of Technology Bombay, Mumbai 400076, India*
[6] *Leung Center for Cosmology and Particle Astrophysics, National Taiwan University, Taipei 10617, Taiwan*
[7] *Asia Pacific Center for Theoretical Physics, and Department of Physics, Pohang University of Science and Technology, Pohang 37673, Korea*

Abstract. We investigate tunneling decay of false vortices in the presence of gravity, in which vortices are trapped in the false vacuum of a theory of scalar electrodynamics in three dimensions. The core of the vortex contains magnetic flux in the true vacuum, while outside the vortex is the appropriate topologically nontrivial false vacuum. We numerically obtain vortex solutions which are classically stable; however, they could decay *via* tunneling. To show this phenomenon, we construct the proper junction conditions in curved spacetime. We find that the tunneling exponent for the vortices is half that for Coleman-de Luccia bubbles and discuss possible future applications.

Brief summary of the talk

Gauged vortices are topological soliton solutions in the theory of a complex scalar field ϕ with $U(1)$ symmetry in three spacetime dimensions. For standard vortices, the potential is minimized for a nonzero value of $|\phi|$, so the vacuum manifold is a circle, as is spatial infinity. In a finite-energy field configuration, the field ϕ must tend towards a vacuum value at spatial infinity and gauge field is necessary. The phase of the field ϕ changes by $2\pi n$, where the integer n is the winding number of the configuration representing topologically distinct classes. The energy of the vortex solution consists of three parts as follows: The contribution from the magnetic flux stabilizes the configuration against collapse, while the contribution from the potential stabilizes the configuration against expansion. The

⋆e-mail: eric.dupuis.1@umontreal.ca
⋆⋆e-mail: yan.gobeil@mail.mcgill.ca
⋆⋆⋆e-mail: bhl@sogang.ac.kr
⋆⋆⋆⋆e-mail: warrior@sogang.ac.kr✠
†e-mail: richard.mackenzie@umontreal.ca
‡e-mail: paranj@lps.umontreal.ca
§e-mail: yajnik@iitb.ac.in
¶e-mail: innocent.yeom@gmail.com

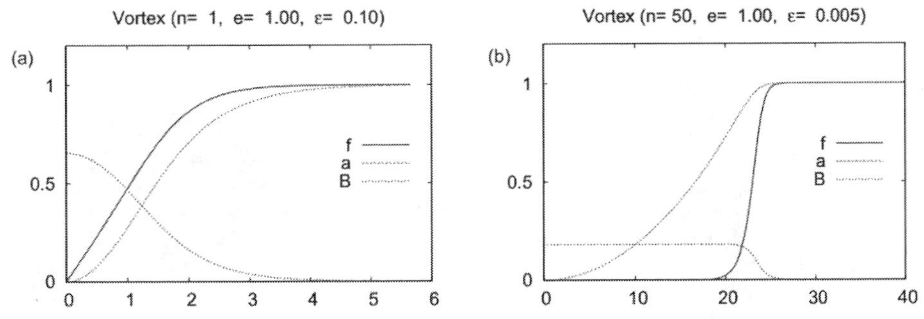

(a) (b)

Figure 1. Vortex profile for (a) thick-wall and (b) thin-wall vortices. Displayed are the functions $f(r)$ and $a(r)$ and the magnetic field $B(r) = na'(r)/er$. (a) Vortex ($n = 1$, $e = 1.00$, $\varepsilon = 0.10$) and (b) Vortex ($n = 50$, $e = 1.00$, $\varepsilon = 0.005$). For more details, see Ref. [7].

covariant derivative of the scalar field also contributes an energy as a linear function of the radius. The profile of the scalar field can sometimes be flat inside and outside a thin transition region between these two values; we call such configurations thin-wall vortices. In this case, an approximate analytical treatment known as the thin-wall approximation can be used.

In such a model in four spacetime dimensions, the soliton becomes an one-dimensional topological defect [1, 2]. The configuration can be infinite or form a closed loop. In the cosmological context they could appear as cosmic strings [3, 4], possibly playing a role in structure formation or having other observable effects [5, 6].

In earlier work [7], we studied the decay of vortices trapped in the symmetry-breaking false vacuum, in which the scalar field interpolates between a true vacuum state at the origin and a false vacuum state at spatial infinity. We chose the vacuum energy in the symmetry-restoring true vacuum to be negative, while that in the false vacuum vanishes. We obtained the numerical solutions representing both thick and thin vortices as shown in Figure 1. Thin-wall vortices require a large winding number and a small difference, ε, between the false and true vacuum energies. For thin-wall vortices, if the false vacuum contains a metastable vortex solution instead of the homogeneous configuration at the initial moment, the vortex was classically stable at a given radius. However, through quantum tunneling, the vortex could become a larger one and expand leaving behind it the symmetric true vacuum. It is possible that conventional vacuum decay proceeds by tunneling and is exponentially suppressed, while vortex tunneling is unsuppressed or is only slightly suppressed. In this case, the presence of vortices in the universe would catalyze vacuum decay. This occurs as one approaches the dissociation limit, in which the tunneling barrier between a classically stable vortex and an unstable, expanding vortex shrinks away. Thus, there is a range of parameter space for which vortices are classically stable but trigger the decay of the false vacuum in an essentially unsuppressed manner.

As extension of the work [8], we studied the decay of cosmic strings that are trapped in the false vacuum in a theory of scalar electrodynamics in four spacetime dimensions. We restricted our analysis to the case of thin-walled cosmic strings which occur when large magnetic flux is trapped inside the string. Thus the string looks like a tube of fixed radius, at which it is classically stable. The core of the string contains magnetic flux in the true vacuum, while outside the string, separated by a thin wall, is the false vacuum. The string decays by tunneling to a configuration which is represented by a bulge as shown in Figure 2, where the region of true vacuum within is ostensibly enlarged. The bulge

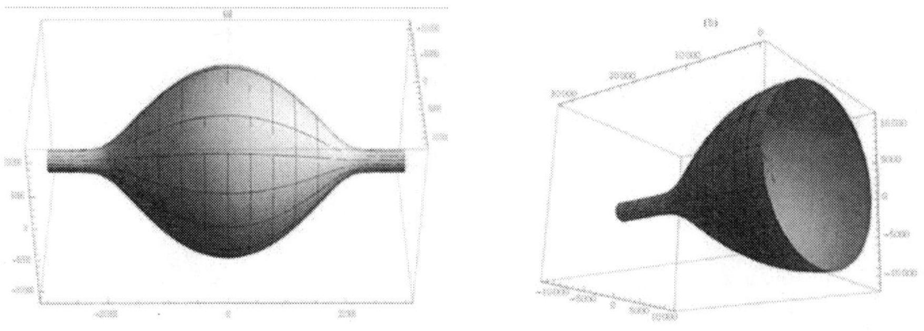

(a) (b)

Figure 2. (a) Cosmic string profile at the bounce point. (b) Cut away of the cosmic string profile at bounce point. For more details, see Ref. [8].

can be described as the meeting of a kink soliton-antisoliton pair along the length of the string. It can be described as a bulge appearing in the initial string, starting from one of small, classically stable radius, expanding to a fat string of large, classically unstable (to expansion) radius and then returning back to one of small radius along its length. This configuration is the bounce point of a corresponding $O(2)$-symmetric instanton, which we can determine numerically. Once the bulge appears it expands rapidly in real time. The paired soliton and antisoliton recede from each other along the length of the string with a velocity that quickly approaches the speed of light, leaving behind a fat tube. At the same time the radius of the fat tube that is being formed expands (transversely) as it is no longer classically stable, converting false vacuum to the true vacuum with ever-diluting magnetic field within. The rate of this expansion is determined by the energy difference between the true vacuum and the false vacuum. Our analysis could be applied to a network of cosmic strings formed in the very early Universe or vortex lines in a superheated superconductor.

In the present work [9], we are studying the effect of vortices on the tunneling decay of a symmetry-breaking false vacuum in three spacetime dimensions with gravity. We derive the equations of motion for the gauge, scalar field, and metric functions and impose the appropriate boundary conditions to get the finite energy configuration. We present numerical solutions for both thick- and thin-wall vortices. The numerical solutions have inside true vacuum state and the magnetic flux with the wall in the presence of gravity. By varying the difference between the false and true vacuum states, we investigate how the existence of the solution can be affected by the strength of $\kappa = 8\pi G$. The existence of static vortex solutions with gravity is established. After that, we investigate more closely the metastability of thin-wall vortices. For this purpose, we employ the Israel junction conditions [10] to understand if the vortex can tunnel through a certain potential barrier and expand dynamically. In the framework of junction conditions, the wall is considered to be a surface layer in the sense that a nonvanishing surface energy density σ exists. This implies a modification of the spacetime. The exterior spacetime with a vortex confined within a finite radius is locally Minkowski with a conical defect [11]. The conical defect is the analog of the Schwarzschild mass parameter that is familiar in the $3+1$ dimensional context.

Figure 3 shows the causal structure of thin-wall vortices, in which the yellow region is inside the wall and the green region is outside the wall. Left: as the vortex tunnels toward a collapsing phase, it becomes a timelike cusp (that behaves like a massive particle) which is located at $r = 0$. Right: as

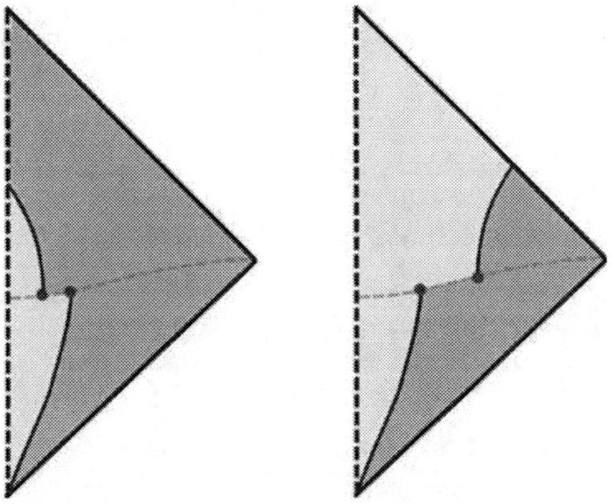

Figure 3. The causal structure of thin-wall vortices.

the vortex tunnels toward an expanding phase, the radius increases to infinity. We are interested in the right case.

If the scalar field is in a metastable vacuum state, the tunneling process from the false vacuum state to the true vacuum state can occur *via* potential barrier penetration, which is the nucleation process of a vacuum bubble [12, 13]. In the semiclassical approximation, the decay rate of the metastable vacuum state per unit time per unit volume is given by

$$\Gamma/V = Ae^{-B}[1 + O(\hbar)] , \tag{1}$$

where the coefficient A comes from the determinant arising in the saddle-point evaluation of the path integral and the exponent B is the difference between the Euclidean action of the bounce solution and the action of the background solution, i.e. $B = S^E_{\rm bounce} - S^E_{\rm bckg}$. The determinant factor must exclude the integration over the zero modes. The Coleman-de Luccia (CdL) bounce has translation invariance in all direction, giving three zero modes. The vortex only has time translation invariance, giving one zero mode. The position of the vortex is fixed once and for all. These modes are removed. Instead, the corresponding degrees of freedom (either the center of the vacuum bounce in spacetime or the center of the vortex bounce in time) are integrated over. We are interested in finding the exponent B. In the presence of a vortex, a similar process exists. As can be deduced in the collective coordinate approximation suggested in [9], quantum fluctuations of the vortex radius can also lead to a phase transition. In this case, the relevant instanton describes the expansion of the wall. To understand the cosmological relevance of the vortex, we compare the false vacuum lifetime in presence and absence of a vortex.

For the decay probability, we can consider a variety of decay channels, i.e. CdL bounce, oscillating bounce, Hawking-Moss instanton, tunneling vortices, etc. The remarkable thing about the decay probability is that the channel for the tunneling vortex can be more dominant than that for the CdL bounce. Computations in the thin-wall limit have shown that the false vortex's contribution to metastability remains dominant, and more so as $\varepsilon \to 0$ limit. Another effect is the suppression of tunneling when the vacuum energy density reaches a limiting value, $\varepsilon \to \varepsilon_c$. This critical value is increased as gravity is turned on, but only very slightly for reasonable values of κ. In this respect, this is another minor way in which gravity stabilizes solutions of this model.

The extension to the decay of the string with gravity could be interesting. We postpone to discuss this issue in future works.

Acknowledgements

We would like to thank Sung-Won Kim, Remo Ruffini, Sang-Pyo Kim, Gungwon Kang, Jonghyuk Yoon, Hyung Won Lee, and Hyeong-Chan Kim for their hospitality at Joint Conference of 13th International Conference on Gravitation, Astrophysics, and Cosmology, and 15th Italian-Korean Symposium on Relativistic Astrophysics in Korea, 03-07 Jul 2017. This work was financially supported in part by the Natural Sciences and Engineering Research Council of Canada and the Fonds de recherche du Québec — Nature et technologies. RM, MP and UY thank the Ministre des Relations internationales et de la Francophonie du Gouvernement du Québec under the Cooperation Québec-Maharashtra for continuing financial support. BHL is supported by Basic Science Research Program through the National Research Foundation of Korea (NRF-2017R1D1A1B03028310). WL is supported by Basic Science Research Program through the National Research Foundation of Korea (2016R1D1A1B01010234). DY is supported by the Leung Center for Cosmology and Particle Astrophysics (LeCosPA) of National Taiwan University (103R4000).

References

[1] A. A. Abrikosov, Sov. Phys. JETP **5**, 1174 (1957) [Zh. Eksp. Teor. Fiz. **32**, 1442 (1957)].

[2] H. B. Nielsen and P. Olesen, Nucl. Phys. B **61**, 45 (1973).

[3] T. W. B. Kibble, J. Phys. A **9**, 1387 (1976).

[4] T. W. B. Kibble, Phys. Rept. **67**, 183 (1980).

[5] R. H. Brandenberger, Nucl. Phys. Proc. Suppl. **246-247**, 45 (2014).

[6] H. Tashiro, T. Sekiguchi and J. Silk, JCAP **1401**, 013 (2014).

[7] B.-H. Lee, W. Lee, R. MacKenzie, M. B. Paranjape, U. A. Yajnik and D.-h. Yeom, Phys. Rev. D **88**, 085031 (2013).

[8] B.-H. Lee, W. Lee, R. MacKenzie, M. B. Paranjape, U. A. Yajnik and D.-h. Yeom, Phys. Rev. D **88**, 105008 (2013).

[9] É. Dupuis, Y. Gobeil, B.-H. Lee, W. Lee, R. MacKenzie, M. B. Paranjape, U. A. Yajnik and D.-h. Yeom, arXiv:1709.03839 [hep-th].

[10] W. Israel, Nuovo Cim. B **44S10**, 1 (1966) [Nuovo Cim. B **44**, 1 (1966)] (erratum: Nuovo Cim. B **48**, 463 (1967)).

[11] S. Deser, R. Jackiw and G. 't Hooft, Annals Phys. **152**, 220 (1984).

[12] S. R. Coleman, Phys. Rev. D **15**, 2929 (1977) Erratum: [Phys. Rev. D **16**, 1248 (1977)].

[13] C. G. Callan, Jr. and S. R. Coleman, Phys. Rev. D **16**, 1762 (1977).

Oscillon in Einstein-scalar system with double well potential and its properties.

Taishi Ikeda[1,⋆], *Chul-Moon* Yoo[1], and *Vitor* Cardoso[2,3]

[1]*Department of Physics, Graduate School of Science, Nagoya University, Nagoya 464-6602, Japan*
[2]*CENTRA, Departamento de Física, Instituto Superior Técnico, Universidade de Lisboa, Avenida Rovisco Pais 1, 1049 Lisboa, Portugal*
[3]*Perimeter Institute for Theoretical Physics, 31 Caroline Street North Waterloo, Ontario N2L 2Y5, Canada*

Abstract. The dynamical evolution of self-interacting scalar field has many nontrivial behaviors, which tell us many lessons in a nonlinear dynamics. On Minkowski spacetime, the scalar field with double well potential has localized, non-singular, time-dependent, long-lived solutions, which are called *oscillons*. The lifetime of the oscillon depends on the initial conditions. Furthermore, when the initial parameter is fine-tuned, oscillons can be infinitely, and type I critical behavior is observed. Here, we investigate the Einstein-scalar system with double well potential. We show that oscillons exist in this system, and discuss the behavior when the initial parameter is fine-tuned. Our results suggests that a new type of critical behavior appears in this theory.

1 Introduction

Nonlinear field equations often appear, and its nonlinearity plays an important role in many situations in physics. However, our understanding of the properties of the nonlinear effects is inadequate. In order to understand them more deeply, it is useful to focus on the universal feature of the nonlinear equations. Nonlinear scalar field equations in some classes have solitonic solutions, and it's formation, time evolution and stability are also important problems.
Here, we consider an Einstein-scalar system with a double well potential of the form

$$V(\Phi) = \frac{\lambda}{4}(\Phi^2 - \sigma^2)^2. \tag{1}$$

This theory has two interesting sources of non-linearity. One is the gravitational interaction, and the other is non-linear potential of the scalar field. When the gravity is turned off, there is localized, longevity, nonsingular solutions, which are called "oscillons", in this theory. On flat space, the oscillons have been studied by Bogolyubsky et al [1], and Copeland et al[2]. They found that the oscillons can be formed after a bubble collapses. The typical energy of the oscillon is about $43/\sigma$, and the lifetime τ depends on the initial bubble radius. Typically, the lifetime is about $10^2 - 10^3(1/\sqrt{\lambda\sigma^2})$. However, when the initial bubble radius R_0 is fine-tuned to some value R_*, the lifetime obeys the scaling law:

$$\tau = -\gamma \ln |R_0 - R_*| + C, \tag{2}$$

⋆e-mail: ikeda@gravity.phys.nagoya-u.ac.jp

where γ and C are constants. This behavior is similar to the type I critical behavior, which is known as universal feature of the gravitational system[3–6]. In this study, we investigate the effects of the gravity on the oscillon in the weak gravity case.

2 Setting and numerical formulation

2.1 System

We consider the Einstein-scalar theory described by the action

$$S = \int d^4x \sqrt{-g}\left(\frac{R}{8\pi G} - g^{\mu\nu}\Phi_{,\mu}\Phi_{,\nu} - 2V(\Phi)\right), \tag{3}$$

where $V(\Phi)$ is the double well potential which is given by eq.(1). This system has a typical length scale $L = 1/\sqrt{\lambda\sigma^2}$. $G\sigma^2$ is a dimensionless parameter which characterizes the strength of the gravitational interaction.

2.2 formulation

We solve the field equations numerically by using the generalized Baumgarte-Shapiro-Shibata-Nakamura (G-BSSN) formulation. G-BSSN formulation is a generalization of the BSSN formulation [7, 8] to the case of curvilinear coordinate.[9] This formulation is useful in spherically symmetric spacetime.

G-BSSN formulation is based on the 3+1 decomposition and conformal decomposition. In the 3+1 decomposition, the line element is given by

$$ds^2 = -\alpha^2 dt^2 + \gamma_{ij}(dx^i + \beta^i dt)(dx^j + \beta^j dt), \tag{4}$$

where α is a lapse function, β^i is a shift vector, and γ_{ij} is a 3-metric. Extrinsic curvature $K_{ij} \equiv (\frac{\partial}{\partial t} - \mathcal{L}_\beta)\gamma_{ij}$ is also important quantity. In the conformal decomposition, the 3 metric and the extrinsic curvature are decomposed to $\phi, \tilde{\gamma}_{ij}, K, \tilde{A}_{ij}$, which are defined by

$$\gamma_{ij} = e^{4\phi}\tilde{\gamma}_{ij}, \tag{5}$$

$$K_{ij} = e^{4\phi}\tilde{A}_{ij} + \frac{1}{3}\gamma_{ij}K. \tag{6}$$

The determinant of $\tilde{\gamma}_{ij}$ is given from the determinant of the reference metric $\bar{\gamma}_{ij}$. Furthermore, for the numerical stability, we introduce the auxiliary field:

$$\tilde{\Lambda}^k \equiv \tilde{\gamma}^{ij}(\tilde{\Gamma}^k_{ij} - \bar{\Gamma}^k_{ij}) = \tilde{\gamma}^{ij}\Delta\tilde{\Gamma}^k_{ij}. \tag{7}$$

In the case of the spherically symmetric spacetime, the nontrivial components of the $\tilde{\gamma}_{ij}, \tilde{A}_{ij}, \beta^i, \tilde{\Lambda}$ are as follows:

$$\tilde{\gamma}_{ij} = \mathrm{diag}(a, br^2, br^2\sin^2\theta) \tag{8}$$

$$\tilde{A}_{ij} = \mathrm{diag}(A, Br^2, Br^2\sin^2\theta), \tag{9}$$

$$\beta^i = (\beta, 0, 0), \tag{10}$$

$$\tilde{\Lambda}^i = (\tilde{\Lambda}, 0, 0). \tag{11}$$

We can get the time evolution equations for $a, b, \phi, A, B, K, \tilde{\Lambda}, \Phi$ and Π which is a momentum conjugate of the scalar field (see [11]).

2.3 gauge condition

We use the harmonic gauge condition:

$$\partial_t \alpha = \mathcal{L}_\beta \alpha - K\alpha^2, \tag{12}$$

and $\beta = 0$.

2.4 numerics

We use the iterative Crank-Nicolson scheme [12] for the integration in time, and a 2nd order finite difference method for spatial derivative. In order that the solutions remain smooth, we add the Kreiss-Oliger dissipation.

2.5 boundary condition

For numerical simulation, it is necessary to impose the boundary conditions. For the regularity at the origin, we must impose the Neumann boundary conditions,

$$\begin{aligned} \alpha'|_{t=0} &= \phi'|_{r=0} = K'|_{r=0} = \tilde{\Lambda}|_{r=0} \\ &= \Phi'|_{r=0} = \Pi'|_{r=0} = 0, \\ a'|_{r=0} &= b'|_{r=0} = A'|_{r=0} = B'|_{r=0} = 0. \end{aligned}$$

For the local flatness at the origin, we must impose the Dirichlet boundary condition for a, b, A, B:

$$a|_{r=0} - 1 = b|_{r=0} - 1 = A|_{r=0} = B|_{r=0} = 0. \tag{13}$$

If this Dirichlet boundary condition is satisfied on the initial data, the evolution equations guarantee this boundary condition at any time. However, for the numerical simulation, this condition can be violated, and the violations is the origin of the numerical instability. In this study, we explicitly impose the Neumann boundary condition for $\alpha, \phi, K, \tilde{\Lambda}, \Phi, \Pi$ and the Dirichlet boundary condition for a, b, A, B. The outer boundary condition is the asymptotically Minkowski spacetime.

2.6 definition of the lifetime

We define the lifetime τ of the oscillon as follows:

$$\frac{M(\tau; r_0)}{M(0; r_0)} = \epsilon \ll 1, \tag{14}$$

where $M()\tau; r_0$ is a Kodama mass inside a sphere of the radius r_0[13, 14].

2.7 Initial data

In this study, we use momentary static Gaussian bubble and a spatially conformally flat initial data:

$$\begin{aligned} \Phi(t = 0, r) &= -\sigma + 2\sigma e^{-r^2/R_0^2}, \\ a(t = 0, r) &= b(t = 0, r) = 1, \end{aligned} \tag{15}$$

where R_0 is a radius of the bubble.

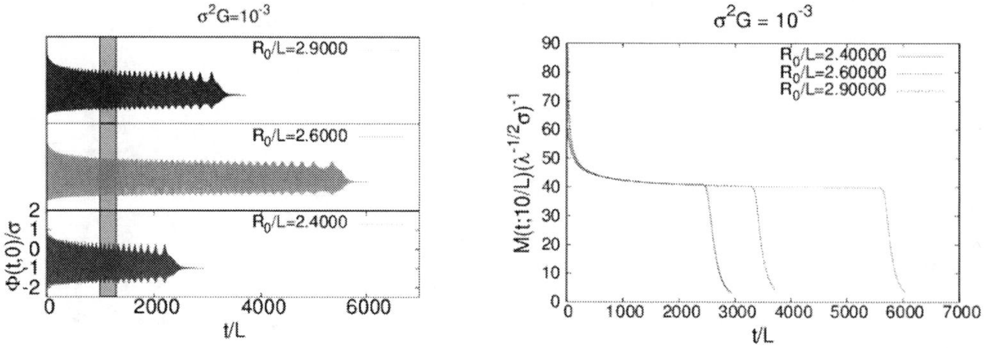

Figure 1. Left panel shows envelope of the time evolution of $\Phi(t, r = 0)$ for $\sigma^2 G = 1.0 \times 10^{-3}$. Right panel show the time evolution of the Kodama mass.

3 Results

This system has one model parameter $\sigma^2 G$ which characterize the strength of the coupling between the scalar field and gravity. In this study, we study the oscillon's formation and it's property for $\sigma^2 G = 1.0 \times 10^{-4}, 5.0 \times 10^{-4}, 1.0 \times 10^{-3}$ and 2.0×10^{-3}.

We find that the self-gravitating oscillon appears after the bubble collapses, and its typical behavior is similar to the case of flat background(see Fig.1) Oscillon's typical energy depends on $\sigma^2 G$ (see Table.1). Furthermore, when the bubble radius is fine-tuned, the type I critical behavior appears for

Table 1. The dependence of $M_{\rm Oscillon}$ on $\sigma^2 G$.

$\sigma^2 G$	$M_{\rm Oscillon}/L$
1.0×10^{-4}	43
5.0×10^{-4}	41
1.0×10^{-3}	40
2.0×10^{-3}	38

$\sigma^2 G = 1.0 \times 10^{-4}, 5.0 \times 10^{-4}, 1.0 \times 10^{-3}$, that is τ obeys the scaling low:

$$\frac{\tau}{L} = -\gamma \ln \left|\frac{R_0 - R_*}{L}\right| + C. \tag{16}$$

For $\sigma^2 G = 2.0 \times 10^{-3}$, τ shows periodic modulations around the scaling (see Fig.2):

$$A \cos(-\log \left|\frac{R_0 - R_*}{L}\right| + \varphi) \equiv \frac{\tau}{L} - (-\gamma \log \left|\frac{R_0 - R_*}{L}\right|) - \delta. \tag{17}$$

This behavior may be new type of the critical behavior.

References

[1] I. L. Bogolyubsky and V. G. Makhankov, JETP Lett. **24**, 12 (1976).

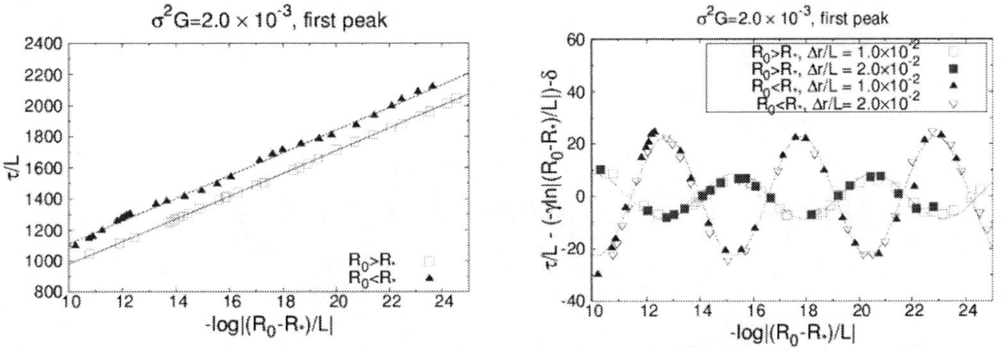

Figure 2. The left panels show the relation between the lifetime τ and the initial bubble radius near the first three peaks for $\sigma^2 G = 2.0 \times 10^{-3}$. The right panel shows the deviation between the lifetime as predicted by scaling law and our actual numerical results. We checked the convergence of our result.

[2] E. J. Copeland, M. Gleiser and H.-R. Muller, Phys. Rev. D **52**, 1920 (1995) doi:10.1103/PhysRevD.52.1920 [hep-ph/9503217].

[3] M. W. Choptuik, Phys. Rev. Lett. **70**, 9 (1993). doi:10.1103/PhysRevLett.70.9

[4] C. Gundlach and J. M. Martin-Garcia, Living Rev. Rel. **10**, 5 (2007) doi:10.12942/lrr-2007-5 [arXiv:0711.4620 [gr-qc]].

[5] K. Clough and E. A. Lim, arXiv:1602.02568 [gr-qc].

[6] T. Ikeda and C. M. Yoo, Phys. Rev. D **94**, no. 12, 124032 (2016) doi:10.1103/PhysRevD.94.124032 [arXiv:1610.07280 [gr-qc]].

[7] M. Shibata and T. Nakamura, Phys. Rev. D **52**, 5428 (1995). doi:10.1103/PhysRevD.52.5428

[8] T. W. Baumgarte and S. L. Shapiro, Phys. Rev. D **59**, 024007 (1999) doi:10.1103/PhysRevD.59.024007 [gr-qc/9810065].

[9] J. D. Brown, Phys. Rev. D **79**, 104029 (2009) doi:10.1103/PhysRevD.79.104029 [arXiv:0902.3652 [gr-qc]].

[10] A. Akbarian and M. W. Choptuik, Phys. Rev. D **92**, no. 8, 084037 (2015) doi:10.1103/PhysRevD.92.084037 [arXiv:1508.01614 [gr-qc]].

[11] T. Ikeda, C. M. Yoo and V. Cardoso, arXiv:1708.01344 [gr-qc].

[12] S. A. Teukolsky, Phys. Rev. D **61**, 087501 (2000) doi:10.1103/PhysRevD.61.087501 [gr-qc/9909026].

[13] H. Kodama, Prog. Theor. Phys. **63**, 1217 (1980). doi:10.1143/PTP.63.1217

[14] T. Harada, C. M. Yoo, T. Nakama and Y. Koga, Phys. Rev. D **91**, no. 8, 084057 (2015) doi:10.1103/PhysRevD.91.084057 [arXiv:1503.03934 [gr-qc]].

EPJ Web of Conferences **168**, 03006 (2018) https://doi.org/10.1051/epjconf/201816803006
Joint International Conference of ICGAC-XIII and IK-15 on Gravitation, Astrophysics and Cosmology

Dark Energy and Dark Matter from Emergent Gravity Picture

Hyun Seok Yang[1,*]

[1] *Center for Quantum Spacetime, Sogang University, Seoul 04107, Korea*

Abstract. We suggest that dark energy and dark matter may be a cosmic uroboros of quantum gravity due to the coherent vacuum structure of spacetime. We apply the emergent gravity to a large N matrix model by considering the vacuum in the noncommutative (NC) Coulomb branch satisfying the Heisenberg algebra. We observe that UV fluctuations in the NC Coulomb branch are always paired with IR fluctuations and these UV/IR fluctuations can be extended to macroscopic scales. We show that space-like fluctuations give rise to the repulsive gravitational force while time-like fluctuations generate the attractive gravitational force. When considering the fact that the fluctuations are random in nature and we are living in the (3+1)-dimensional spacetime, the ratio of the repulsive and attractive components will end in $\frac{3}{4} : \frac{1}{4} = 75 : 25$ and this ratio curiously coincides with the dark composition of our current Universe. If one includes ordinary matters which act as the attractive gravitational force, the emergent gravity may explain the dark sector of our Universe more precisely.

1 Introduction

Dark energy and dark matter are a great mystery of the 20th century physics, which has not been resolved yet. In retrospect, the resolution of a great puzzle requires the upheaval of a radical new physics. Recall how the blackbody radiation and the photoelectric effect had been resolved at the beginning of the 20th century. We know that these problems could not be solved by simply modifying the Newtonian dynamics and the classical electrodynamics. A radical new paradigm, the so-called quantum mechanics, was necessary to solve these problems. If dark energy and dark matter would be such a case, i.e., the 21st century version of the blackbody radiation and the photoelectric effect, they would not be understood by simply modifying the general relativity and the quantum field theory. Another novel paradigm, a.k.a. quantum gravity, may be necessary to understand the nature of dark energy and dark matter.

The concept of emergent gravity and spacetime recently activated by the AdS/CFT correspondence advocates that spacetime is not a fundamental entity existed from the beginning but an emergent property from a primal monad such as matrices. The emergent spacetime is a new fundamental paradigm that allows a background-independent formulation of quantum gravity and opens a new perspective to resolve the notorious problems in theoretical physics such as the cosmological constant problem, hierarchy problem, dark energy, and dark matter. Moreover the emergent spacetime picture admits a background-independent description of the inflationary universe which has a sufficiently elegant and

*e-mail: hsyang@sogang.ac.kr

explanatory power to defend the integrity of physics against the multiverse hypothesis. We emphasize that noncommutative (NC) spacetime necessarily implies emergent spacetime if spacetime at microscopic scales should be viewed as NC [1]. We will elaborate the emergent gravity from a large N matrix model by considering the vacuum in the NC Coulomb branch satisfying the Heisenberg algebra and argue that dark energy and dark matter may arise as a cosmic uroboros of quantum gravity due to the coherent vacuum structure of spacetime.

2 Emergent Spacetime and the Dark Side of Universe

Let us start with a zero-dimensional matrix model with four $N \times N$ Hermitian matrices, $\{\phi_a \in \mathcal{A}_N | a = 1, \cdots, 4\}$. The equations of motion for the matrix model are given by

$$[\phi_b, [\phi_a, \phi_b]] = 0, \tag{1}$$

which must be supplemented with the Jacobi identity

$$[\phi_a, [\phi_b, \phi_c]] + [\phi_b, [\phi_c, \phi_a]] + [\phi_c, [\phi_a, \phi_b]] = 0. \tag{2}$$

If we consider the $N \to \infty$ limit, the large N limit opens a new phase of the Coulomb branch given by

$$[\phi_a, \phi_b]|_{\text{vac}} = -iB_{ab} \qquad \Rightarrow \qquad \langle \phi_a \rangle_{\text{vac}} = p_a \equiv B_{ab} y^b \tag{3}$$

where the vacuum moduli y^a satisfy the Moyal-Heisenberg algebra $[y^a, y^b] = i\theta^{ab}$ with $B_{ab} = (\theta^{-1})_{ab}$. This vacuum will be called the NC Coulomb branch. Note that the NC Coulomb branch saves the NC nature of matrices while the conventional commutative vacuum dismisses the property.

Suppose that the fluctuations around the vacuum (3) take the form

$$\phi_a = p_a + \widehat{A}_a(x, y). \tag{4}$$

We denote the NC $\star$-algebra on $\mathbb{R}^4_\theta$ by $\mathcal{A}_\theta$ and $\widehat{A}_a(y) \in \mathcal{A}_\theta$ are four-dimensional NC $U(1)$ gauge fields. The adjoint scalar fields in Eq. (4) now obey the deformed algebra given by

$$[\phi_a, \phi_b] = -i(B_{ab} - \widehat{F}_{ab}) \in \mathcal{A}_\theta, \tag{5}$$

where $\widehat{F}_{ab} = \partial_a \widehat{A}_b - \partial_b \widehat{A}_a - i[\widehat{A}_a, \widehat{A}_b]$ with the definition $\partial_a \equiv \text{ad}_{p_a} = -i[p_a, \cdot]$. Note that $[\phi_a, [\phi_b, \phi_c]] = \widehat{D}_a \widehat{F}_{bc}$. Using this result, one can derive the equations of motion and the Bianchi identity of NC $U(1)$ gauge fields from Eqs. (1) and (2), respectively. For these field variables, the Lie algebra homomorphism reads as $\mathcal{A}_\theta \to \mathfrak{D} : [\phi_a, [\phi_b, \phi_c]] \mapsto [\widehat{V}_a, [\widehat{V}_b, \widehat{V}_c]]$. In the commutative limit, we thus get the following correspondence [1, 2]:

$$\widehat{D}_b \widehat{F}_{ab} = 0 \quad \overset{|\theta| \to 0}{\Longrightarrow} \quad [V_b, [V_a, V_b]] = 0, \qquad \widehat{D}_{[a} \widehat{F}_{bc]} = 0 \quad \overset{|\theta| \to 0}{\Longrightarrow} \quad [V_{[a}, [V_b, V_{c]}]] = 0. \tag{6}$$

Note that the torsion and curvature are multi-linear differential operators defined by $T(X, Y) = \nabla_X Y - \nabla_Y X - [X, Y]$ and $R(X, Y)Z = [\nabla_X, \nabla_Y]Z - \nabla_{[X,Y]}Z$, where X, Y and Z are vector fields on M. Therefore they satisfy the relations $T(V_a, V_b) = \lambda^2 T(E_a, E_b)$ and $R(V_a, V_b)V_c = \lambda^3 R(E_a, E_b)E_c$. After imposing the torsion free condition $T(E_a, E_b) = 0$, it is easy to derive the identity $R(E_{[a}, E_b)E_{c]} = \lambda^{-3} R(V_{[a}, V_b)V_{c]} = \lambda^{-3}[V_{[a}, [V_b, V_{c]}]]$. Consequently we see that the Bianchi identity for NC $U(1)$ gauge fields in the commutative limit is equivalent to the first Bianchi identity for the Riemann curvature tensors, i.e.,

$$\widehat{D}_{[a} \widehat{F}_{bc]} = 0 \quad \overset{|\theta| \to 0}{\Longrightarrow} \quad R(E_{[a}, E_b)E_{c]} = 0. \tag{7}$$

The mission for the equations of motion is more involved. But, from the experience on the Bianchi identity (7), we basically expect that it will be reduced to the Einstein equations

$$\widehat{D}_b \widehat{F}_{ab} = 0 \quad \stackrel{|\theta|\to 0}{\Longrightarrow} \quad R_{ab} = 8\pi G\Big(T_{ab} - \frac{1}{2}\delta_{ab}T\Big). \tag{8}$$

After a straightforward but tedious calculation, we get a remarkably simple but cryptic result [2]

$$R_{ab} = -\frac{1}{\lambda^2}\Big(g_d^{(+)i} g_d^{(-)j}(\eta^i_{ac}\bar{\eta}^j_{bc} + \eta^i_{bc}\bar{\eta}^j_{ac}) - g_c^{(+)i} g_d^{(-)j}(\eta^i_{ac}\bar{\eta}^j_{bd} + \eta^i_{ac}\bar{\eta}^j_{bd})\Big). \tag{9}$$

To get the above result, we have taken the canonical decomposition of the structure equation as

$$g_{abc} = g_c^{(+)i}\eta^i_{ab} + g_c^{(-)i}\bar{\eta}^i_{ab}. \tag{10}$$

First it is convenient to decompose the energy-momentum tensor into two parts

$$8\pi G T^{(M)}_{ab} = -\frac{1}{\lambda^2}\Big(g_{acd}g_{bcd} - \frac{1}{4}\delta_{ab}g_{cde}g_{cde}\Big), \tag{11}$$

$$8\pi G T^{(L)}_{ab} = \frac{1}{2\lambda^2}\Big(\rho_a\rho_b - \Psi_a\Psi_b - \frac{1}{2}\delta_{ab}(\rho_c\rho_c - \Psi_c\Psi_c)\Big), \tag{12}$$

where $\rho_a \equiv g_{bab}$ and $\Psi_a = -\frac{1}{2}\varepsilon^{abcd}g_{bcd}$. A close inspection reveals that the first one is the Maxwell energy-momentum tensor given by

$$T^{(M)}_{ab} = \frac{\hbar^2 c^2}{g^2_{YM}}\Big(F_{ac}F_{bc} - \frac{1}{4}\delta_{ab}F_{cd}F_{cd}\Big) \tag{13}$$

but the second one seems to be mystic.

The reason is the following. The curvature tensor R_{abcd} in general relativity can be decomposed according to the canonical split of Lie algebra $so(4) = su(2)_L \oplus su(2)_R$ and the canonical decomposition of two-forms $\Omega^2(M) = \Omega^2_+ \oplus \Omega^2_-$ as

$$R = \begin{pmatrix} W^+ + \frac{1}{12}s & B \\ B^T & W^- + \frac{1}{12}s \end{pmatrix}. \tag{14}$$

A notable point is that the Ricci scalar s appears in the blocks (++) and (−−) while the traceless Ricci tensors B and B^T show up in the blocks (+−) and (−+). Since the Ricci tensors (9) emergent from NC $U(1)$ gauge fields belong to the mixed sector (+−) and (−+), they should be traceless according the decomposition (14). However the Ricci tensor (9) has a nontrivial Ricci scalar given by $s = \frac{1}{2\lambda^2}(\rho_a\rho_a - \Psi_a\Psi_a)$. Hence the Liouville energy-momentum tensor (12) cannot be realized in the context of general relativity. In order to descry closer aspects of the second energy-momentum tensor, let us separate the scalar and tensor perturbations as $\rho_a\rho_b = \frac{1}{4}\delta_{ab}\rho_c\rho_c + \big(\rho_a\rho_b - \frac{1}{4}\delta_{ab}\rho_c\rho_c\big)$ and $\Psi_a\Psi_b = \frac{1}{4}\delta_{ab}\Psi_c\Psi_c + \big(\Psi_a\Psi_b - \frac{1}{4}\delta_{ab}\Psi_c\Psi_c\big)$. In a long wavelength limit where the quadruple modes can be ignored, the Liouville energy-momentum tensor (12) behaves like a cosmological constant [3]

$$T^{(L)}_{\mu\nu} \approx -\frac{s}{32\pi G}g_{\mu\nu}. \tag{15}$$

In order to get a corresponding result in (3+1)-dimensional Lorentzian spacetime, let us take the analytic continuation defined by $x^4 = ix^0$. Under this Wick rotation, $g_{\mu\nu} \to g_{\mu\nu}$, $\rho_\mu \to \rho_\mu$ and $\Psi_\mu \to i\Psi_\mu$, so the Liouville energy-momentum tensor in the Lorentzian signature is given by

$$T^{(L)}_{\mu\nu} = \frac{1}{16\pi G\lambda^2}\Big(\rho_\mu\rho_\nu + \Psi_\mu\Psi_\nu - \frac{1}{2}g_{\mu\nu}(\rho^2_\lambda + \Psi^2_\lambda)\Big). \tag{16}$$

Note that ρ_μ and Ψ_μ are four-vectors and random fluctuations in nature and the Lorentzian four-vectors have their own causal structure. They are classified into three classes:

(A) (ρ_μ, Ψ_μ) are space-like vectors, i.e. $g^{\mu\nu}\rho_\mu\rho_\nu > 0$, etc.
(B) (ρ_μ, Ψ_μ) are time-like vectors, i.e. $g^{\mu\nu}\rho_\mu\rho_\nu < 0$, etc.
(C) (ρ_μ, Ψ_μ) are null vectors, i.e. $g^{\mu\nu}\rho_\mu\rho_\nu = 0$, etc.

which is in sharp contrast to the Riemannian space where every vectors are positive-definite. We will see that the nature of gravitational interactions depends on their causal structure.

In macroscopic scales where the scalar fluctuations are dominant compared to tensor perturbations, the Raychaudhuri equation can be approximated as

$$\dot{\Theta} \approx \frac{s}{4} \tag{17}$$

where $s = \frac{1}{2\lambda^2} g^{\mu\nu}(\rho_\mu\rho_\nu + \Psi_\mu\Psi_\nu)$ is the Ricci scalar. Thus the behavior of spacetime expansion/contraction crucially depends on the causal structure of random fluctuations in the classes (A), (B) and (C):

(A) $\dot{\Theta} \approx \frac{s}{4} > 0$ when (ρ_μ, Ψ_μ) are space-like vectors,
(B) $\dot{\Theta} \approx \frac{s}{4} < 0$ when (ρ_μ, Ψ_μ) are time-like vectors,
(C) $\dot{\Theta} \approx \frac{s}{4} = 0$ when (ρ_μ, Ψ_μ) are null vectors.

As a result, space-like fluctuations give rise to the repulsive gravitational force while time-like fluctuations generate the attractive gravitational force. Let us forget about the null case (C) since it does not contribute to the spacetime expansion/contraction.

It should be important to notice that the NC Coulomb branch is generated by the Planck energy condensate in vacuum, i.e., $\rho_{vac} = \frac{1}{4g_{YM}^2}|B_{\mu\nu}|^2 \sim 10^{-2}M_P^4$ [2, 3]. An important point is that the fluctuations over the coherent vacuum (3) should be subject to the spacetime uncertainty relation since the NC Coulomb branch satisfies the Heisenberg algebra. The spacetime uncertainty relation is realized as the UV/IR mixing of vacuum fluctuations. This means that UV fluctuations in the NC Coulomb branch are always paired with IR fluctuations and these UV/IR fluctuations can be extended to macroscopic scales [3]. Hence let us consider the vacuum fluctuations whose IR fluctuations have a macroscopic wavelength $L_H = M_H^{-1}$. Having this in mind, one can estimate the energy scale of the Liouville energy-momentum tensor (16) by observing that the vacuum fluctuations represented by $\widehat{F}_{\mu\nu}(y)$ are not localized but extended to a macroscopic scale L_H [2, 3]. Actually the vacuum fluctuation energy

$$\delta\rho = -\frac{1}{2g_{YM}^2} B_{\mu\nu}\widehat{F}^{\mu\nu} \sim \frac{1}{L_P^2 L_H^2} \tag{18}$$

is the leading contribution to the curvature by the vacuum fluctuations and coming from the boundary term on a hypersurface of radius L_H. This implies that the Ricci scalar in Eq. (15) has the curvature scale given by $|s| \sim \frac{1}{L_H^2}$, i.e.,

$$T_{\mu\nu}^{(L)} \sim \pm \frac{1}{L_P^2 L_H^2} g_{\mu\nu}. \tag{19}$$

A few remarks are in order. If the macroscopic scale L_H is identified with the size of cosmic horizon of our observable universe, $L_H = 1.3 \times 10^{26} m$, the extended (nonlocal) energy in Eq. (18) or (19) is in good agreement with the observed value of current dark energy $\rho_{DE} := \delta\rho \approx (10^{-3}eV)^4$. Moreover one can determine the total energy within the hypersurface of radius L_H, which is given by $\delta E = \frac{4\pi L_H}{3L_P^2}$. Thus the corresponding total entropy $\delta S = \frac{\delta E}{T_H}$ is determined as $\delta S = \frac{A_H}{4G}$ since the de Sitter temperature of the cosmological horizon is given by $T_H = \frac{1}{2\pi L_H}$, where $A_H = 4\pi L_H^2$ and $8\pi G = L_P^2$. Of course the numerical factor is a wishful thinking. This argument shows that the dark energy/matter in our Universe would be a holographic manifestation of a microscopic physics, a.k.a. quantum gravity. We showed before that space-like fluctuations give rise to the repulsive gravitational force while time-like fluctuations generate the attractive gravitational force. When considering the fact that the fluctuations are random in nature and we are living in the (3+1)-dimensional spacetime, the ratio of the repulsive and attractive components will end in $\frac{3}{4} : \frac{1}{4} = 75 : 25$ and this ratio curiously coincides with the dark composition of our current Universe [3]. Note that the dark energy in (19) sets the current Hubble parameter $H_0 = \frac{c}{L_H}$ and the Hubble parameter induces a characteristic acceleration scale $a_0 = cH_0 = \frac{c^2}{L_H}$. Since the dark matter is the third of the dark energy, the dark matter will give rise to the attractive acceleration scale $\frac{a_0}{3} = \frac{c^2}{3L_H}$. This attractive force will compete with ordinary matters depending on their characteristic scales. The inclusion of ordinary matters definitely changes the previous ratio as 75↓ : 25↑, that will cause a better match with the current observation. Therefore it is expected that the emergent gravity can explain the dark sector of our Universe more precisely after including ordinary matters in this scheme.

The emergent gravity picture requires to unify geometry and matters on an equal footing. But it is not yet understood what matter is from the emergent gravity picture although a tough idea was suggested in [1–3]. So we simply assume the matter fields that obey the law of Standard Model. Suppose that ordinary matters are added in the background of dark matter and dark energy given by (19). Since the ordinary matter acts as the attractive gravitational force which decays as the $\frac{1}{r^2}$-law, two attractive forces due to the dark matter and ordinary matters will compete each other with their own characteristic scales. The gravitational force due to ordinary matters will dominate at small scales while the gravitational acceleration due to dark matter dominates at large scales. To illuminate this aspect, let us consider a (spiral or disk) galaxy and $M(r)$ be the mass contained inside an orbit of radius r. The mass distribution of the galaxy gives rise to the acceleration $a_M = \frac{GM(r)}{r^2}$ which decreases as $1/r^2$ as r increases if there is no mass outside this radius. Thus there is a crossover where the acceleration a_M becomes equal to the acceleration $\frac{a_0}{3} = \frac{c^2}{3L_H}$ due to the dark matter:

$$a_M = \frac{GM}{r^2} \lesssim \frac{a_0}{3} = \frac{c^2}{3L_H}. \tag{20}$$

One can see from Eq. (20) that this crossover arises at the distance $r_c = \frac{\sqrt{3GML_H}}{c}$ from the center of the galaxy, over which the acceleration $a_0/3$ due to the dark matter dominates. For example, the crossover distances for M33 ($M = 5 \times 10^{10} M_\odot$, $R = 9\,\text{kpc}$) and the Milky Way ($M = 10^{12} M_\odot$, $R = 65\,\text{kpc}$) are $r_c \approx 5.6\,\text{kpc}$ ($1.7 \times 10^{20} m$) and $29.8\,\text{kpc}$ ($9.2 \times 10^{20} m$), respectively, which are roughly half the size of the galaxies. This implies that the flattening of the galaxy rotation curve may be explained by the dark matter given by Eq. (19).

Acknowledgments

This work was supported by the National Research Foundation of Korea (NRF) grant funded by the Korea government (MOE) (No. NRF-2015R1D1A1A01059710).

References

[1] H. S. Yang, Int. J. Mod. Phys. A **30**, 1550016 (2015).
[2] H. S. Yang, J. High Energy Phys. **05**, 012 (2009).
[3] H. S. Yang, J. Phys. Conf. Ser. **343**, 012132 (2012).

Study of GSLT in Curvature-Matter Coupling Gravity

Muhammad Sharif[1,*] and *Ayesha* Ikram[1,**]

[1] *Department of Mathematics, University of the Punjab, Quaid-e-Azam Campus, Lahore-54590, Pakistan.*

Abstract. In this work, we study the first and generalized second laws of thermodynamics at the apparent horizon of homogeneous and isotropic universe model in the context of $f(\mathcal{G}, T)$ gravity ($\mathcal{G}$ and T represent the Gauss-Bonnet invariant and trace of the energy-momentum tensor, respectively). We formulate the corresponding field equations as well as determine the radius, temperature and entropy to analyze these laws. An extra term associated with entropy production is appeared in the first law due to the non-equilibrium treatment of thermodynamics. It is found that the universal condition is obtained to preserve the generalized second law of thermodynamics.

1 Introduction

Thermodynamics has been a subject of great interest to explore the fascinating characteristics of matter variables in general relativity (GR) as well as in modified gravitational theories. For isotropic and homogeneous universe, Einstein field equations can be expressed in terms of first law of thermodynamics (FLT) [1]. Akbar and Cai [2] found that Friedmann equations evaluated at the apparent horizon can be rewritten in the form $d\mathcal{E} = \tau dS + WdV$ ($\mathcal{E}$, τ, S, V and W are the energy, temperature, entropy, volume inside the horizon and work density, respectively) in the background of GR, Gauss-Bonnet (GB) gravity and the general Lovelock theory. It is found that an auxiliary entropy production term corresponding to the non-equilibrium treatment of thermodynamics is appeared in Clausius relation in modified theories of gravity while no such additional term is obtained in braneworld, GB and Lovelock gravitational theories [3]. Wu et al. [4] formulated the universal condition to check the validity of generalized second law of thermodynamics (GSLT) in the context of modified theories of gravity. Sadjadi [5] explored the validity of second law and GSLT for power-law solution as well as de Sitter universe in $f(R, \mathcal{G})$ gravity. Abdolmaleki and Najafi [6] investigated GSLT in $f(\mathcal{G})$ gravity for isotropic and homogeneous universe filled with matter and radiation enclosed by apparent horizon with Hawking temperature.

In this paper, we explore the first and GSLT at the apparent horizon in $f(\mathcal{G}, T)$ gravity. This modified gravitational theory deals with the non-minimal coupling between quadratic curvature invariant (a linear combination of Ricci scalar (R), Ricci ($R_{\alpha\beta}$) and Riemann ($R_{\alpha\beta\xi\eta}$) tensors) and matter. The paper has the following format. In the next section, we construct the corresponding field equations for isotropic and homogeneous universe with any spatial curvature while section **3** investigates the laws of thermodynamics at the apparent horizon of universe model. We summarize the results in the last section.

*e-mail: msharif.math@pu.edu.pk
**e-mail: ayeshamaths91@gmail.com

2 $f(\mathcal{G},T)$ Gravity

The action of $f(\mathcal{G},T)$ gravity is given by [7]

$$\mathcal{S}_{f(\mathcal{G},T)} = \int d^4x\sqrt{-g}\left[\frac{1}{16\pi G}(R+f(\mathcal{G},T))+\mathcal{L}_m\right], \tag{1}$$

where $\mathcal{G} = R_{\alpha\beta\xi\eta}R^{\alpha\beta\xi\eta} - 4R_{\alpha\beta}R^{\alpha\beta} + R^2$, g, $\mathcal{L}_m$ and G denote determinant of the metric tensor $(g_{\alpha\beta})$, Lagrangian density associated with matter configuration and gravitational constant, respectively. The suitable form of generic function $f(\mathcal{G},T)$ describing the non-minimal coupling between curvature and matter is of the form

$$f(\mathcal{G},T) = f_1(\mathcal{G}) + f_2(\mathcal{G})f_3(T). \tag{2}$$

The variation of the action (1) with respect to $g_{\alpha\beta}$ for the above model in the presence of pressureless fluid yields the following field equations

$$G_{\alpha\beta} = 8\pi\mathbb{G}T^{(\text{eff})}_{\alpha\beta} = 8\pi\mathbb{G}T_{\alpha\beta} + T^{(\mathcal{G}T)}_{\alpha\beta}, \tag{3}$$

where

$$\begin{aligned}
T^{(\mathcal{G}T)}_{\alpha\beta} &= \frac{1}{2}g_{\alpha\beta}(f_1(\mathcal{G})+f_2(\mathcal{G})f_3(T)) - (2R_\alpha^{\xi\eta\delta}R_{\beta\xi\eta\delta} - 4R_{\alpha\xi\beta\eta}R^{\xi\eta} - 4R^\xi_\alpha R_{\xi\beta} + 2RR_{\alpha\beta})\\
&\times (f_1'(\mathcal{G}) + f_2'(\mathcal{G})f_3(T)) - (4R_{\alpha\xi\beta\eta}\nabla^\xi\nabla^\eta - 4g_{\alpha\beta}R^{\xi\eta}\nabla_\xi\nabla_\eta + 4R^\xi_\alpha\nabla_\beta\nabla_\xi + 4R^\xi_\beta\nabla_\alpha\nabla_\xi\\
&- 4R_{\alpha\beta}\nabla^2 - 2R\nabla_\alpha\nabla_\beta + 2Rg_{\alpha\beta}\nabla^2)(f_1'(\mathcal{G}) + f_2'(\mathcal{G})f_3(T)),\\
\mathbb{G} &= G\mathfrak{B}, \quad \mathfrak{B} = 1 + \frac{f_2(\mathcal{G})f_3'(T)}{8\pi G},
\end{aligned}$$

$\nabla^2 = \nabla_\alpha\nabla^\alpha$ (∇_α is a covariant derivative) and prime represents derivative with respect to the corresponding variable.

The line element for homogeneous and isotropic universe is

$$ds^2 = dt^2 - \frac{a^2(t)}{1-\mathcal{K}r^2}dr^2 - \chi^2 d\theta^2 - \chi^2\sin^2\theta d\phi^2, \tag{4}$$

where $\chi = a(t)r$, $a(t)$ and $\mathcal{K}$ represent the scale factor and spatial curvature parameter associated with open, flat and closed cosmic geometries for $\mathcal{K} = -1,\ 0$ and 1, respectively. Using Eqs.(3) and (4), we obtain

$$3\left(H^2 + \frac{\mathcal{K}}{a^2}\right) = 8\pi\mathbb{G}\rho^{(\text{tot})} = 8\pi\mathbb{G}(\rho + \rho^{(\text{hc})}), \tag{5}$$

$$-2\left(\dot{H} - \frac{\mathcal{K}}{a^2}\right) = 8\pi\mathbb{G}(\rho^{(\text{tot})} + p^{(\text{tot})}) = 8\pi\mathbb{G}(\rho + \rho^{(\text{hc})} + p^{(\text{hc})}), \tag{6}$$

where $\rho^{(\text{hc})}$ and $p^{(\text{hc})}$ are the higher order curvature terms given by

$$\begin{aligned}
\rho^{(\text{hc})} &= \frac{\rho^{(\mathcal{G}T)}}{8\pi\mathbb{G}} = \frac{1}{8\pi G\mathfrak{B}}\left[\frac{1}{2}(f_1(\mathcal{G}) + f_2(\mathcal{G})f_3(T)) - \frac{1}{2}\mathcal{G}\Delta + 12H\left(H^2 + \frac{\mathcal{K}}{a^2}\right)\left(\Delta_\mathcal{G}\dot{\mathcal{G}} + \Delta_T\dot{T}\right)\right],\\
p^{(\text{hc})} &= \frac{p^{(\mathcal{G}T)}}{8\pi\mathbb{G}} = \frac{1}{8\pi G\mathfrak{B}}\left[-\frac{1}{2}(f_1(\mathcal{G}) + f_2(\mathcal{G})f_3(T)) + \frac{1}{2}\mathcal{G}\Delta - 8H(H^2+\dot{H})\left(\Delta_\mathcal{G}\dot{\mathcal{G}} + \Delta_T\dot{T}\right)\right.\\
&- \left.4\left(H^2 + \frac{\mathcal{K}}{a^2}\right)\left(\Delta_\mathcal{G}\ddot{\mathcal{G}} + \Delta_T\ddot{T} + 2\Delta_{\mathcal{G}T}\dot{\mathcal{G}}\dot{T} + \Delta_{\mathcal{G}\mathcal{G}}\dot{\mathcal{G}}^2 + \Delta_{TT}\dot{T}^2\right)\right],
\end{aligned}$$

where $\Delta = f_1'(\mathcal{G}) + f_2'(\mathcal{G})f_3(T)$, $\mathcal{G} = 24(H^2+\dot{H})\left(H^2 + \frac{\mathcal{K}}{a^2}\right)$ and $H = \frac{\dot{a}}{a}$ is a Hubble parameter. The subscripts $\mathcal{G}$ and T denote derivatives of Δ with respect to $\mathcal{G}$ and T, respectively whereas dot represents time derivative.

3 Laws of Thermodynamics

In this section, we discuss the first as well as GSLT at the apparent horizon of homogeneous and isotropic universe in the background of $f(\mathcal{G}, T)$ gravity.

3.1 First Law

Here, we construct the FLT which is based on the concept of energy conservation for model (2). The relation

$$h^{\alpha\beta}\partial_\alpha \chi \partial_\beta \chi = 0,$$

where $h_{\alpha\beta} = \text{diag}(1, \frac{-a^2(t)}{1-\mathcal{K}r^2})$ is a two-dimensional line element, provides the radius of apparent horizon for the FRW universe model as

$$\chi_{(\text{ah})} = \left(H^2 + \frac{\mathcal{K}}{a^2}\right)^{-\frac{1}{2}}.$$

To measure the infinitesimal change in apparent horizon radius, we take the derivative of the above equation with respect to time and using Eq.(6), it follows that

$$d\chi_{(\text{ah})} = 4\pi G\left(\rho^{(\text{tot})} + p^{(\text{tot})}\right)\chi^3_{(\text{ah})} H\mathfrak{B}dt, \tag{7}$$

where dt represents the corresponding small time interval. The temperature on the apparent horizon is given by [1]

$$\tau = \frac{|K_{sg}|}{2\pi}, \tag{8}$$

where $K_{sg} = \frac{1}{2\sqrt{-h}}\partial_\alpha\left(\sqrt{-h}h^{\alpha\beta}\chi_{,\beta}\right)$ (h is the determinant of $h_{\alpha\beta}$) is the surface gravity. For homogeneous and isotropic universe model, we have

$$K_{sg} = \frac{1}{\chi_{(\text{ah})}}\left(1 - \frac{\dot{\chi}_{(\text{ah})}}{2\chi_{(\text{ah})}H}\right) = \frac{1}{2}\chi_{(\text{ah})}\left(H^2 + \dot{H} + \frac{\mathcal{K}}{a^2}\right). \tag{9}$$

Bekenstein-Hawking entropy is measured in units of Newton's gravitational constant defined as one fourth of area of apparent horizon ($A = 4\pi\chi^2_{(\text{ah})}$) [8]. The entropy of stationary black hole solutions with bifurcate Killing horizons in the context of modified gravitational theories is a Noether charge entropy also dubbed as Wald entropy [9]. Brustein et al. [10] presented that this entropy is equal to quarter of apparent horizon area in units of effective gravitational coupling in these modified theories. Wald entropy in $f(\mathcal{G}, T)$ gravity is given by

$$S = \frac{A}{4G\mathfrak{B}}\left(1 - \frac{4\Delta}{\chi^2_{(\text{ah})}}\right). \tag{10}$$

It is worth mentioning here that the entropy in $f(\mathcal{G})$ gravity is obtained for $f_3(T) = 0$ while this formula corresponds to GR for $\Delta = 0$ with $\mathfrak{B} = 1$ [11]. Using Eqs.(8)-(10), it follows that

$$\begin{aligned} \tau dS &= 4\pi\chi^3_{(\text{ah})}\left(\rho^{(\text{tot})} + p^{(\text{tot})}\right)Hdt - 2\pi\chi^2_{(\text{ah})}\left(\rho^{(\text{tot})} + p^{(\text{tot})}\right)d\chi_{(\text{ah})} - \frac{4\pi\tau}{G\mathfrak{B}}d\Delta \\ &+ \frac{\pi\chi^2_{(\text{ah})}\tau}{G}\left(1 - \frac{4\Delta}{\chi^2_{(\text{ah})}}\right)d\left(\frac{1}{\mathfrak{B}}\right). \end{aligned} \tag{11}$$

The total energy inside $\chi_{(ah)}$ for homogeneous and isotropic universe model is

$$\mathcal{E} = V\rho^{(tot)} = \frac{4}{3}\pi\chi^3_{(ah)}\rho^{(tot)} = \frac{3V}{8\pi\mathbb{G}}\left(H^2 + \frac{\mathcal{K}}{a^2}\right). \tag{12}$$

Using Eqs.(11) and (12), we obtain

$$\tau dS = -d\mathcal{E} + WdV - \frac{4\pi\tau}{\mathfrak{B}G}d\Delta + \frac{\chi_{(ah)}}{2G}\left[1 + 2\pi\chi_{(ah)}\tau\left(1 - \frac{4\Delta}{\chi^2_{(ah)}}\right)\right]d\left(\frac{1}{\mathfrak{B}}\right), \tag{13}$$

where the infinitesimal change $d\mathcal{E}$ is caused by the small displacement in horizon radius $\chi_{(ah)}$ and $W = \frac{\rho^{(tot)} - p^{(tot)}}{2}$ measures the work done by the system. The FLT for model (2) can be expressed as

$$\tau(dS + d_iS) = -d\mathcal{E} + WdV, \tag{14}$$

where

$$d_iS = \frac{4\pi}{\mathfrak{B}G}d\Delta - \frac{\chi_{(ah)}}{2G\tau}\left[1 + 2\pi\chi_{(ah)}\tau\left(1 - \frac{4\Delta}{\chi^2_{(ah)}}\right)\right]d\left(\frac{1}{\mathfrak{B}}\right),$$

is the entropy production term which demonstrates the non-equilibrium behavior of thermodynamics. This shows that the field equations for model (2) do not meet with the universal form of FLT ($d\mathcal{E} = \tau dS + WdV$) due to the presence of this auxiliary term.

3.2 Generalized Second Law

The GSLT is associated with the total entropy of the system and is defined as

$$\dot{S} + \dot{S}_{(tot)} + d_i\dot{S} \geq 0, \tag{15}$$

where $d_i\dot{S} = \partial_t(d_iS)$ and $S_{(tot)}$ is associated with the entropy due to all matter contents present inside the horizon. The relationship between total entropy, energy density and pressure in a form of differential equation (Gibbs equation) is given by [4]

$$\tau_{(tot)}dS_{(tot)} = d(\rho^{(tot)}V) + p^{(tot)}dV, \tag{16}$$

where $\tau_{(tot)}$ measures the total temperature of all contents within the horizon and is related to the temperature at the apparent horizon as

$$\tau_{(tot)} = \lambda\tau, \quad 0 < \lambda < 1.$$

This shows that the apparent horizon temperature is always greater than the total temperature inside the horizon. Susbtituting Eqs.(14) and (16) in (15), we have

$$\dot{S} + \dot{S}_{(tot)} + d_i\dot{S} = \left(\frac{24 + \chi^4_{(ah)}\mathcal{G}}{96\pi\lambda\chi_{(ah)}}\right)\left[\left(\rho^{(tot)} + p^{(tot)}\right)\left(1 - \frac{\lambda}{2}\right)\dot{V} + (1 - \lambda)V\dot{\rho}^{(tot)}\right] \geq 0, \tag{17}$$

Using Eqs.(5) and (6), this inequality becomes

$$\left(\frac{24 + \chi^4_{(ah)}\mathcal{G}}{192\pi\lambda\mathfrak{B}G}\right)\Omega\chi^4_{(ah)} \geq 0, \tag{18}$$

where

$$\Omega = (2-\lambda)H\left(\dot{H}-\frac{\mathcal{K}}{a^2}\right)^2 + \frac{2H(1-\lambda)}{\chi^2_{(\mathrm{ah})}}\left(\dot{H}-\frac{\mathcal{K}}{a^2}\right) + \frac{(1-\lambda)\mathfrak{B}}{\chi^4_{(\mathrm{ah})}}\partial_t\left(\frac{1}{\mathfrak{B}}\right).$$

The GSLT is protected for $\mathcal{G} > 0$, $\mathfrak{B} > 0$ and $\Omega > 0$. For flat FRW universe model, the inequalities $\mathcal{G} > 0$, $H > 0$, $\dot{H} > 0$, $\mathfrak{B} > 0$ and $\partial_t\left(\frac{1}{\mathfrak{B}}\right) > 0$ must be fulfilled for the validity of GSLT. The equilibrium description of thermodynamics is evaluated at $\lambda = 1$ as

$$\chi^4_{(\mathrm{ah})}H\left(\dot{H}-\frac{\mathcal{K}}{a^2}\right)^2\left(\frac{24+\chi^4_{(\mathrm{ah})}\mathcal{G}}{192\pi\lambda\mathfrak{B}G}\right) \geq 0,$$

which holds for positive values of $\mathcal{G}$, H and $\mathfrak{B}$.

4 Concluding Remarks

In this paper, we have explored the first and GSLT in the background of $f(\mathcal{G}, T)$ gravity. These laws are discussed for the non-equilibrium thermodynamical picture at the apparent horizon of isotropic and homogeneous universe model. It is observed that total entropy is a combination of horizon entropy as well as an auxiliary entropy production term in FLT. The non-equilibrium thermodynamical behavior interprets energy exchange between inside and outside the apparent horizon leading to the additional entropy term. To check the validity of GSLT, we have constructed the general expression for model (2) in terms of horizon entropy, total entropy due to all matter contents inside the horizon and entropy production term. It is concluded that viability condition for this law is found to be achieved relative to universal condition in the context of modified gravitational theories [4].

References

[1] Cai, R.G. and Kim, S.P., J. High Energy Phys. **02**, 050 (2005).
[2] Akbar, M. and Cai, R.G., Phys. Rev. D **75**, 084003 (2007).
[3] Eling, C., Guedens, R. and Jacobson, T., Phys. Rev. Lett. **96**, 121301 (2006); Cai, R.G. and Cao, L.M., Phys. Rev. D **75**, 064008 (2007); Akbar, M. and Cai, R.G., Phys. Lett. B **648**, 243 (2007); Sheykhi, A., Wang, B. and Cai, R.G., Nucl. Phys. B **779**, 1 (2007); Cai, R.G. et al., Phys. Rev. D **78**, 124012 (2008).
[4] Wu, S.F. et al., Class. Quantum Grav. **25**, 235018 (2008).
[5] Sadjadi, H.M., Europhys. Lett. **92**, 50014 (2010).
[6] Abdolmaleki, A. and Najafi, T., Int. J. Mod. Phys. D **25**, 1650040 (2016).
[7] Sharif, M. and Ikram, A., Eur. Phys. J. C **76**, 640 (2016).
[8] Bardeen, J.M., Carter, B. and Hawking, S.W., Commun. Math. Phys. **31**, 161 (1973); Bekenstein, J.D., Phys. Rev. D **7**, 2333 (1973); Hawking, S.W., Commun. Math. Phys. **43**, 199 (1975); ibid. **46**, 206 (1976).
[9] Wald, R.M., Phys. Rev. D **48**, R3427 (1993).
[10] Brustein, R., Gorbonos, D. and Hadad, M., Phys. Rev. D **79**, 0444025 (2009).
[11] Sadjadi, H.M., Phys. Scr. **83**, 055006 (2011); Sharif, M. and Fatima, I., Astrophys. Space Sci. **354**, 507 (2014).

Entropy in Spacetime and Topological Hair

Young-Hwan Hyun[1,⋆] and *Yoonbai* Kim[1,⋆⋆]

[1] *Department of Physics and Institute of Basic Science, Sungkyunkwan University, Suwon 16419, Republic of Korea*

Abstract. Global topological soliton of the hedgehog ansatz is added to de Sitter spacetime in arbitrary dimensions larger than three, and then thermodynamic law is checked at the cosmological horizon. All geometric and thermodynamic quantities are varied in the presence of a long-range interacting matter distribution with negative pressure, however the entropy-area relation is satisfied in the exact form. Its geometry involves deficit solid angle but maintains a single horizon which allows unique temperature normalization, different from the case of Schwarzschild-de Sitter spacetime.

1 Introduction

Thermodynamic laws are established as empirical laws through numerous tests. Since the entropy-area relation was proposed as a thermodynamic law in curved spacetime with a black hole horizon [1], it has contributed for establishing the black hole thermodynamics [2, 3], and has been an active area of research towards quantum gravity, particularly for counting the degrees of freedom and information loss. In the middle of the progress of the black hole thermodynamics, the entropy-area relation has been checked from representative idealized systems to realistic dirty situations. The relation was examined in various quantum gravity theories and also in many different systems of mass, spin, charges, or the number of dimensions [4]. Except for some higher-derivative gravity theories, the entropy-area relation remains to be fulfilled. This universality occurs even for black solutions with different horizon topologies and for the cosmological horizon [5]. Although the various methods including string theory repeatedly confirm the entropy-area relation, the universality never have proper explanation in the context of quantum gravity despite many empirical evidences [4].

In this work we add one more intriguing empirical evidence to investigate the problem of universality. As an attractive candidate, we propose non-BPS topological solitons (as additional nontrivial matter distribution) in the de Sitter spacetime (dS) to test the entropy-area relation. Nonlinear interactions are included to form topological solitons in arbitrary spatial dimensions and non-BPS nature implies complicated mutual interaction between multiple solitons. Despite of this complexity, since its matter distribution is controllable and possesses the topological stability, it is tractable to compute the entropy-area relation in the exact form.

This article is organized as follows. In Section 2, we consider a matter distribution of divergent energy $E \sim R^{d-3}$ and test the entropy-area relation on the cosmological horizon of dS with deficit

⋆e-mail: yhhyun@skku.edu
⋆⋆e-mail: yoonbai@skku.edu

solid angle. In Section 3, we consider a scalar field theory of O($d-1$) symmetry and identify the source of this divergent energy as the global topological soliton of a hedgehog ansatz. We conclude in Section 4 with a few discussions.

2 Entropy-area relation with deficit solid angle

Suppose that there is a matter distribution of which energy-momentum tensor is

$$-T^t{}_t = -T^r{}_r = \cdots \sim \frac{d-2}{2}\frac{v^2}{r^2}, \tag{1}$$

where the coefficient is chosen for later convenience. For $d \geq 3$, the energy in flat spacetime is divergent as the spatial volume goes to infinity,

$$E \sim \int^R drr^{d-2}(-T^t{}_t) \sim \int^R drr^{d-2}\frac{1}{r^2} \Rightarrow \left\{ \begin{matrix} \overset{d=3}{\sim} \ln R \\ \overset{d\geq 3}{\sim} R^{d-3} \end{matrix} \right\} \overset{R\to\infty}{\longrightarrow} \infty. \tag{2}$$

This R^{d-3} divergence for $d > 3$ seems not serious in the d-dimensional de Sitter spacetime (dS$_d$) in which the vacuum energy is growing with larger power, $E_\Lambda \sim R^{d-1}$. In fact, the $1/r^2$-term is that of the natural largest power among various terms with different powers lower than the zeroth power (constancy) of the vacuum energy density, and thus seems naturally allowable as a matter distribution in dS$_d$. A question at the moment is to find a possible field theoretic candidate which can be identified as a natural source giving such energy distribution (1) but will be discussed in the next section.

Introduce a metric in dS$_d$ ($d > 3$)

$$ds^2 = -e^{2\Omega(r)}A(r)dt^2 + \frac{dr^2}{A(r)} + r^2 d\Omega^2_{d-2}, \tag{3}$$

and then the solution of Einstein equation coupled to the aforementioned energy-momentum tensor in (1) is

$$\Omega(r) \overset{\text{set}}{=} 0, \quad A(r) = -\left(\frac{r}{\ell}\right)^2 + 1 - \delta - \frac{C}{r^{d-3}}, \tag{4}$$

where $\delta = \dfrac{8\pi G v^2}{d-3}$. Since our main goal is to test the entropy-area relation in the presence of the matter of which distribution is relatively simple, we set $C = 0$ to deal without creating a new black hole horizon. Performing a coordinate transformation, $dt' = \sqrt{1-\delta}\, dt$, $dr' = \frac{dr}{\sqrt{1-\delta}}$, $d\Omega'_{d-2} = (1-\delta)d\Omega_{d-2}$, the metric returns to that of pure dS$_d$ with the different range of solid angle variables, which leads to the deficit solid angle.

To address the question whether or not the entropy-area relation is sustained for non-vanishing δ, we begin by reading the location of the horizon $r_{\rm h} = \sqrt{1-\delta}\,\ell$ and thus the area of the horizon is decreased by the factor $(1-\delta)^{\frac{d-2}{2}}$,

$$\mathcal{A}^{\rm h}_{\delta {\rm dS}} = \ell^{d-2}\Omega_{d-2}(1-\delta)^{\frac{d-2}{2}} = \mathcal{A}^{\rm h}_{\rm dS}(1-\delta)^{\frac{d-2}{2}}, \tag{5}$$

where Ω_{d-2} is the volume of a unit $(d-2)$-sphere. The surface gravity at the horizon $\kappa(r_{\rm h}) = \left|dA/dr\right|_{r_{\rm h}}/2$ fixes the temperature under the standard normalization [6, 7]. The corresponding thermodynamic law with temperature and pressure accompanies the change of volume [6]

$$T_{\delta {\rm dS}} {\rm d}S_{\delta {\rm dS}} = {\rm d}(-E_{\delta {\rm dS}}) + P_{\delta {\rm dS}}{\rm d}(-V_{\delta {\rm dS}}), \tag{6}$$

where $-E_{\delta\mathrm{dS}}$ counts the negative energy and the surplus angle behind the horizon [5]. Note that there is a minus sign in front of the volume as well, since it accounts for the change of the volume of the inaccessible region [6]. According to the thermodynamic law (6), straightforward computation of the energy and the work inside the horizon results in the following entropy difference,

$$\mathrm{d}S_{\delta\mathrm{dS}} = \frac{A_{\mathrm{dS}}}{4G}\left(-\frac{d-2}{2}\right)(1-\delta)^{\frac{d-4}{2}}\mathrm{d}\delta. \tag{7}$$

Though every thermodynamic quantity is changed by the matter distribution and its signal appears through deficit solid angle, the entropy-area relation is exactly satisfied irrespective of the change,

$$S_{\delta\mathrm{dS}} = S_{\mathrm{dS}} + \Delta S_{\delta\mathrm{dS}} = S_{\mathrm{dS}} + \int_{S(\delta=0)}^{S(\delta)} dS_{\delta\mathrm{dS}} = \frac{\mathcal{A}^{\mathrm{h}}_{\mathrm{dS}}}{4G}(1-\delta)^{\frac{d-2}{2}} = \frac{\mathcal{A}^{\mathrm{h}}_{\delta\mathrm{dS}}}{4G}. \tag{8}$$

In this section a matter distribution of long-ranged energy density and pressure is added to dS_d. Then the corresponding geometry includes deficit solid angle but does not create a new black hole horizon in addition to the cosmological horizon. Though all thermodynamic quantities, *e.g.*, area, temperature, energy, pressure, and entropy, are changed in $\delta\mathrm{dS}_d$ from those in pure dS_d, we show that the exact entropy-area relation holds irrespective of the value of the deficit solid angle. This test supports the claimed universality of the entropy-area relation in the presence of interacting non-BPS matter.

3 Global topological soliton of hedgehog ansatz as a source

Once we look at the distribution of energy-momentum tensor in (1), it seems difficult not to ask a question on a possible candidate of such distribution. In this section let us identify a natural source of the matter of which energy-momentum tensor is given by (1). To find the origin in the framework of field theory, the action of the $\mathrm{O}(d-1)$ scalar fields ϕ^i $(i, j, \cdots = 1, 2, \ldots, d-1)$ is considered

$$S_\phi = \int d^d x \sqrt{-g}\left[-\frac{g^{\mu\nu}}{2}\partial_\mu\phi^i\partial_\nu\phi^i - \frac{\lambda}{4}(\phi^2 - v^2)^2\right], \tag{9}$$

where $\phi^i \equiv \hat{\phi}^i\phi$ with $\hat{\phi}^i\hat{\phi}^i = 1$ and a quartic scalar potential $V(\phi)$ with spontaneous symmetry breaking is chosen for simplicity. The configuration under the hedgehog ansatz, $\phi^i = \hat{r}^i\phi(r)$, supports a topologically stable soliton and its topological charge is given by the winding between the S^{d-2} of broken vacua and the spatial S^{d-2} at the horizon r_h.

To solve the scalar equation and the non-vanishing components of the Einstein equation, we apply the following boundary conditions.

$$\phi(0) = 0, \quad \phi(r_\mathrm{h}) = \phi_\mathrm{h}(< v) \overset{\ell \gg 1/\sqrt{\lambda}v}{\longrightarrow} v, \quad A(0) = 1, \quad \Omega(r_\mathrm{h}) = 0. \tag{10}$$

The first condition is for a nonsingular field configuration under the hedgehog ansatz, and the second condition supports the topological soliton. The presumed dS horizon r_h seems subtle because of possible existence of the short scalar hair. When the size of the horizon ($r_\mathrm{h} \sim \ell$) is much larger than that of the core of a topological defect ($r_{\mathrm{core}} \sim 1/\sqrt{\lambda}v$), the scalar amplitude takes the value of the vacuum. Since no additional mass was assumed, that is $C = 0$ in (4), $A(r)$ becomes the metric of flat spacetime at the origin. The last condition is for the correct normalization of temperature at the dS horizon, so the condition is assigned for later convenience without losing generality.

Approximate solution near but inside the dS horizon is attempted under the assumption that r is much larger than the core radius $1/\sqrt{\lambda}v$,

$$\phi(r) \approx v - \frac{d-2}{2\left[1+\dfrac{3-d}{(\sqrt{\lambda}v\ell)^2}\right]}\frac{v}{(\sqrt{\lambda}vr)^2}+\cdots, \tag{11}$$

$$A(r) \approx -\left(\frac{r}{\ell}\right)^2 + 1 - \delta + \cdots, \tag{12}$$

$$\Omega(r) \approx -\frac{(d-2)(d-3)}{4\left[1+\dfrac{3-d}{(\sqrt{\lambda}v\ell)^2}\right]^2}\delta\frac{1}{(\sqrt{\lambda}vr)^4}+\cdots. \tag{13}$$

Since the expansion assumed the scale $\ell \gg 1/\sqrt{\lambda}v$, the formula $1 + (3-d)/\lambda v^2\ell^2$ in (11) and (13) is always positive so that $\phi(r)$ and $\Omega(r)$ become monotonically increasing at interior region of the horizon. Then the non-vanishing component of the Einstein equation leads to the energy-momentum tensor in (1) exactly irrespective of the spacetime dimensions d. Since the deficit solid angle must be smaller than the solid angle itself or equivalently $\delta < 1$, the scale of the vacuum expectation value v is constrained by the d-dimensional Planck scale, $v < \sqrt{\dfrac{d-3}{8\pi G}}$. The radial component of pressure also has a long tail,

$$T^r{}_r = -\frac{d-2}{2}\frac{\phi^2}{r^2} + \frac{A}{2}\left(\frac{d\phi}{dr}\right)^2 - V \underset{r\sim r_h}{\overset{\phi\approx v}{\longrightarrow}} -\frac{d-2}{2}\frac{v^2}{r^2}, \tag{14}$$

which approaches (1) near the cosmological horizon. Negativity of the pressure explains the decrease of the area of the cosmological horizon (5) by adding this global topological soliton.

Intriguingly enough, the origin of divergent energy is identified by the global topological soliton of the hedgehog ansatz. In $d = 4$ dimensions, the obtained topologically stable global soliton is the global monopole with linearly divergent energy [8]. The signature of energy density leads to a repulsive gravitational potential [9], which explains impossibility of formation of a new horizon in addition to the cosmological horizon. Though the energy of a global topological soliton diverges as $E \sim R^{d-3}$, it can easily be produced through a period of cosmological phase transition by consuming a part of vacuum energy [10].

4 Conclusion

Global topological soliton of the hedgehog ansatz is considered in $O(d-1)$ symmetric scalar field theory in the background of dS_d. The entropy-area relation is confirmed in the presence of divergent energy of the global solitonic object through an easier analysis without creating an additional horizon. This result adds a circumstantial evidence on the issue of universality of the entropy-area relation [4]. Since the matter distribution of the global topological solitons is stable and makes the problem tractable, many of them can provide good candidates to test and establish thermodynamic relations in curved spacetime.

We conclude with a few brief discussions:

1. Though our analysis was performed without mass term, $C = 0$ in (4), in order not to create an additional horizon, it is also intriguing to examine the entropy-area relation with both global topological soliton and usual matter of $C \neq 0$, the analysis of the entropy-area relation in δSdS_d including the Nariai limit.

2. Three-dimensional case was excluded because the $1/r^2$ term in dS_3 is nothing but the energy density of the electrostatic field of point charge, $-T^t{}_t \sim \mathbf{E}^2 \sim 1/r^2$, which leads to a logarithmic term $A(r) \propto \ln r$ as in (2). Different from $d > 3$, the deficit angle $2\pi\delta$ is obtained by shifting the constant term in the metric, of which physical origin is a localized point source [11]. The entropy-area relation in δdS_3 by utilizing the deficit angle was checked in Ref. [12].
3. In this work, since the spacetime symmetry is broken in the radial direction and includes non-trivial red-shift factor at the origin, we fixed the temperature with the local Rindler temperature. Since dS_d is not asymptotically flat, the exact entropy-area relation should be reproduced by employing the Bousso-Hawking temperature [13] instead of the standard normalization of the present work. It is also worth trying to reproduce the relation with different thermodynamic definitions such as quasi-local energy, Wald entropy, and so on.
4. The total energy of the non-perfect fluid of our consideration agrees with the horizon energy defined in Ref. [6], so that the thermodynamic law holds in the same form. However, the large deficit solid angle leads to the discrepancy between two energy values and in the large deficit solid angle limit. It is beyond our consideration in this present work.
5. The pressure term in the first law is written only in terms of the vacuum expectation value of the scalar field. Since we did not vary the cosmological constant and fixed it in the theory, its pressure term did not appear as in the first derivation of the thermodynamic law for the cosmological horizon [5]. Even though we include the energy and pressure contributions from the cosmological constant, both effects are exactly canceled.
6. In the Schwarzschild geometry, the first law of black hole thermodynamics with a global monopole can also be checked in the same limit.

Acknowledgments

This research was supported by Basic Science Research Program through the National Research Foundation of Korea (NRF) funded by the Ministry of Science, ICT and Future Planning (Grant No. NRF-2016R1D1A1B03931090).

References

[1] J. D. Bekenstein, Phys. Rev. D **7**, 2333 (1973).
[2] J. M. Bardeen, B. Carter and S. W. Hawking, Commun. Math. Phys. **31**, 161 (1973).
[3] S. W. Hawking, Commun. Math. Phys. **43**, 199 (1975) Erratum: [Commun. Math. Phys. **46**, 206 (1976)].
[4] S. Carlip, Int. J. Mod. Phys. D **23**, 1430023 (2014) [arXiv:1410.1486 [gr-qc]].
[5] G. W. Gibbons and S. W. Hawking, Phys. Rev. D **15**, 2738 (1977).
[6] T. Padmanabhan, Class. Quant. Grav. **19**, 5387 (2002) [gr-qc/0204019].
[7] A. Corichi and A. Gomberoff, Phys. Rev. D **69**, 064016 (2004) [hep-th/0311030].
[8] M. Barriola and A. Vilenkin, Phys. Rev. Lett. **63**, 341 (1989).
[9] D. Harari and C. Lousto, Phys. Rev. D **42**, 2626 (1990).
[10] Y. Kim, K. Maeda and N. Sakai, Nucl. Phys. B **481**, 453 (1996) [gr-qc/9604030].
[11] S. Deser, R. Jackiw and G. 't Hooft, Annals Phys. **152**, 220 (1984).
[12] M. Spradlin, A. Strominger and A. Volovich, hep-th/0110007.
[13] R. Bousso and S. W. Hawking, Phys. Rev. D **54**, 6312 (1996) [gr-qc/9606052].

EPJ Web of Conferences **168**, 03009 (2018) https://doi.org/10.1051/epjconf/201816803009
Joint International Conference of ICGAC-XIII and IK-15 on Gravitation, Astrophysics and Cosmology

Throat quantization of the Schwarzschild-Tangherlini(-AdS) black hole

Hideki Maeda[1,⋆]

[1] *Department of Electronics and Information Engineering, Hokkai-Gakuen University, Sapporo, Japan*

Abstract. By the throat quantization pioneered by Louko and Mäkelä, we derive the mass and area/entropy spectra for the Schwarzschild-Tangherlini-type asymptotically flat or AdS vacuum black hole in arbitrary dimensions. Using the WKB approximation for black holes with large mass, we show that area/entropy is equally spaced for asymptotically flat black holes, while mass is equally spaced for asymptotically AdS black holes. Exact spectra can be obtained for toroidal AdS black holes in arbitrary dimensions including the three-dimensional BTZ black hole.

1 Classical vacuum solution with maximally symmetric base manifold

Let us consider the vacuum Einstein equations with a cosmological constant

$$\mathcal{R}_{\mu\nu} - \frac{1}{2}g_{\mu\nu}\mathcal{R} + \Lambda g_{\mu\nu} = 0 \tag{1}$$

for the $n(\geq 3)$-dimensional spacetime with spherical ($k = 1$), planar ($k = 0$), or hyperbolic ($k = -1$) symmetry $(\mathcal{M}^n, g_{\mu\nu}) \approx (M^2, g_{AB}) \times (K^{n-2}, \gamma_{ab})$ with the general metric

$$\begin{aligned} ds^2 =& g_{\mu\nu}(x)dx^\mu dx^\nu \\ =& g_{AB}(\bar{y})d\bar{y}^A d\bar{y}^B + R(\bar{y})^2\gamma_{ab}(z)dz^a dz^b, \end{aligned} \tag{2}$$

where g_{AB} is an arbitrary two-dimensional Lorentzian metric on M^2 and γ_{ab} is the metric on the $(n-2)$-dimensional maximally symmetric space K^{n-2} with its curvature k. In three dimensions ($n = 3$), K^{n-2} is one-dimensional which corresponds to $k = 0$.

By the Birkhoff's theorem, the most general solution with non-constant areal radius R is the following:

$$ds^2 = -f(\mathbf{m}, r)dT^2 + f(\mathbf{m}, r)^{-1}dr^2 + r^2\gamma_{ab}dz^a dz^b, \tag{3}$$

$$f(\mathbf{m}, r) := k - \frac{2\kappa_n^2 \mathbf{m}}{(n-2)V_{n-2}^{(k)} r^{n-3}} + \frac{r^2}{l^2}, \tag{4}$$

where $\kappa_n^2 := 8\pi G_n$, defined by the n-dimensional Newton constant G_n, and $l := \sqrt{-(n-1)(n-2)/2\Lambda}$ is the AdS radius. This is the topological generalization of the Schwarzschild-Tangherlini solution in the presence of a cosmological constant. $\mathbf{m}$ is the mass parameter and its range that gives a black-hole configuration is summarized in Table 1. The global structures of the resulting black hole with a single horizon for $\Lambda \leq 0$ are drawn in Fig. 1. Hereafter we assume that K^{n-2} is compact and then $V_{n-2}^{(k)}$ represents the surface area of the submanifold K^{n-2}.

⋆e-mail: h-maeda@hgu.jp

Table 1. The range of $\mathbf{m}$ for the black-hole configuration given by the Schwarzshild-Tangherlini-type solution (3), where $\mathbf{m}_{ex}$ is the mass for the extremal horizon.

	$k = 1$	$k = 0$	$k = -1$
$\Lambda = 0$	$\mathbf{m} > 0$	None	None
$\Lambda < 0$	$\mathbf{m} > 0$	$\mathbf{m} > 0$	$\mathbf{m} \geq \mathbf{m}_{ex}(< 0)$

In order to investigate the quantum aspects of this Schwarzshild-Tangherlini-type black hole, one needs to quantize the Einstein equations (1). The ADM formalism of the Einstein equations shows that general relativity is a dynamically constrained system. Reduced phase-space quantization is one possible way to quantize such a system, in which one performs canonical quantization on the constraint surface after solving all the (classical) constraint equations.

Of course, it is hopeless to accomplish this with full generality. However, it is possible in the midisuperspace approach for the symmetric spacetime (2). Throat quantization pioneered by Louko and Mäkelä [1] is the reduced phase-space quantization for the spacetime (2) with a certain set of canonical variables.

2 Reduced phase space

In order to write all the classical constraint equations in solvable forms, one carefully chooses the canonical variables. This is obvious when we choose the following ADM variables for the metric:

$$\begin{aligned} ds^2 =& -N(t,r)^2dt^2 + \bar{\Lambda}(t,r)^2(dr + N_r(t,r)dt)^2 \\ &+ R(t,r)^2\gamma_{ab}dz^a dz^b. \end{aligned} \tag{5}$$

Integrating over K^{n-2}, the Einstein-Hilbert action reduces to the following two-dimensional ADM form:

$$I_{\rm ADM} = \int dt \int dr(\dot{\bar{\Lambda}}P_{\bar{\Lambda}} + \dot{R}P_R - NH - N_r H_r), \tag{6}$$

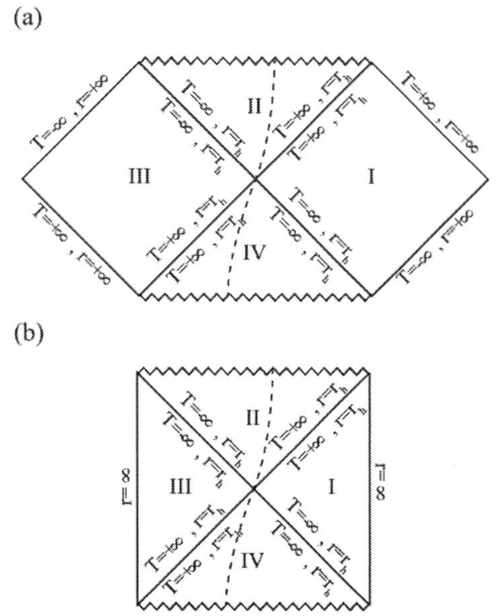

Figure 1. Penrose diagrams of the Schwarzschild-Tangherlini-type black hole (3) with a single Killing horizon for (a) $\Lambda = 0$ and (b) $\Lambda < 0$. A dashed curve represents a constant T timelike hypersurface inside the Killing horizon.

where a dot denotes the derivative with respect to t. While $\bar{\Lambda}$ and R are dynamical variables with their momentum conjugates $P_{\bar{\Lambda}}$ and P_R, N and N_r are Lagrange multipliers providing the constraint equations $H = 0$ ansd $H_r = 0$, respectively. These constraint equations cannot be solved because the super-Hamiltonian H and super-momentum H_r are very complicated functionals of $\bar{\Lambda}$, R, $P_{\bar{\Lambda}}$, and P_R in this ADM formalism.

Thus, in order to solve the constraint equations, one has to find a nice set of the canonical variables which are given by a canonical transformation from the ADM variables $\{\bar{\Lambda}, P_{\bar{\Lambda}}; R, P_R\}$. In the seminal paper [2], Kuchař introduced a set of geometrodynamical variables to solve constraint equations and obtain the reduced phase-space for general spherically symmetric spacetime in four dimensions ($n = 4$) without a cosmological constant ($\Lambda = 0$). In this section, we review his method for $n(\geq 4)$-dimensional spherically symmetric spacetime ($k = 1$).

Kuchař showed that the transformation $\{\bar{\Lambda}, P_{\bar{\Lambda}}; R, P_R\} \to \{M, P_M; S, P_S\}$ defined by

$$S := R, \qquad P_S := P_R - \frac{1}{R'}(\bar{\Lambda} P'_{\bar{\Lambda}} + P_M M'), \tag{7}$$

$$M := \frac{(n-2)V^{(1)}_{n-2}}{2\kappa_n^2} R^{n-3}(1-F), \qquad P_M := -\frac{y\bar{\Lambda}}{F} \tag{8}$$

is canonical, where a prime denotes the derivative with respect to r and y and F are defined by

$$y := N^{-1}(\dot{R} - N_r R'), \qquad F := -y^2 + \bar{\Lambda}^{-2} R'^2. \tag{9}$$

The new canonical variables have clear physical/geometrical meanings; M is the Misner-Sharp quasi-local mass [3] and S is the areal radius. In order to show the canonicality of the transformation, the following boundary conditions are required:

$$N \simeq N_\infty(t) + O(r^{-\epsilon_1}), \qquad N_r \simeq N_r^\infty(t) r^{-(n-3)/2-\epsilon_2}, \tag{10}$$

$$\bar{\Lambda} \simeq 1 + \bar{\Lambda}_1(t) r^{-(n-3)}, \qquad R \simeq r + R_1(t) r^{-(n-4)-\epsilon_4}, \tag{11}$$

where ϵ_1 is a positive number and ϵ_2 and ϵ_4 satisfy $\epsilon_2 > \max[0, -(n-5)/2]$ and $\epsilon_4 > \max[0, -(n-5)]$.

The new action after the canonical transformation with the boundary term is given by

$$I_{\rm geo} = \int dt \int dr (P_M \dot{M} + P_S \dot{S} - N\bar{H} - N_r \bar{H}_r) - \int dt \Big(N_+ M_\infty - N_- M_\infty\Big), \tag{12}$$

where the second term is the boundary term and

$$\bar{H} := \frac{(F + y^2) P_M M'}{yS'} + y P_S, \tag{13}$$

$$\bar{H}_r := P_M M' + P_S S', \tag{14}$$

$$M_\infty(t) := \lim_{r \to \pm\infty} M, \tag{15}$$

$$N_\pm(t) := \pm \lim_{r \to \pm\infty} N. \tag{16}$$

In terms of the new variables, F and y are given by

$$F = 1 - \frac{2\kappa_n^2 M}{(n-2)V^{(1)}_{n-2} S^{n-3}}, \tag{17}$$

$$y^2 = \frac{F^3 P_M^2}{S'^2 - F^2 P_M^2}. \tag{18}$$

The constraint equations $\bar{H} = 0$ and $\bar{H}_r = 0$ in terms of the new variables reduce to $M' = 0$ and $P_S = 0$, of which solution is simply $M = \mathbf{m}(t)$ and $P_S = 0$.

Then, the spatial integration in the new action (12) can be easily performed on the constraint surface to give the following Kuchař's one-dimensional reduced action:

$$I_{\rm K}[\mathbf{m}, \mathbf{p}] = \int dt \Big[\mathbf{p}(t)\dot{\mathbf{m}}(t) - (N_+ - N_-)\mathbf{m}(t)\Big], \tag{19}$$

where $\mathbf{p} := \int_{-\infty}^{\infty} P_M(r,t) dr$. The reduced action (19) depends only $\mathbf{m}(t)$ and its conjugate $\mathbf{p}$. $N_\pm(t)$, the values of the lapse N at either ends of the spatial slice, are prescribed functions and not varied. The Hamilton's equation for $\mathbf{m}$ is simply $\dot{\mathbf{m}} = 0$, whose solution requires $\mathbf{m}$ =constant, which is equal to the ADM mass. This is a different realization of the Birkhoff's theorem, showing that $\mathbf{m}$ carries all the information about the local geometry of the classical solution.

As shown in [4], this Kuchař reduction to the one-dimensional action (19) can be achieved for general k with a negative cosmological constant. In this case, M is replaced by the following generalized Misner-Sharp mass [5]:

$$M = \frac{(n-2)V^{(k)}_{n-2}}{2\kappa_n^2} R^{n-3}\Big\{-\frac{R^2}{l^2} + (1-F)\Big\}. \tag{20}$$

3 Throat quantization

Since the reduced action (19) does not contain any information that the spacetime represents a black hole, canonical quantization of this system does not provide quantum mechanics of the black hole given by Eq. (3). For this purpose, one therefore needs another canonical transformation from the Kuchař action (19). This is what Louko and Mäkelä did for the four-dimensional spherically symmetric spacetime without a cosmological constant [1]. In [6], we generalized their result in the case of general k with $\Lambda(\le 0)$ in arbitrary dimensions. We summarize our result in this section.

3.1 Reduced action with throat variables

Starting from the Kuchař reduce action (19), we first set $N_+ = 1$ and $N_- = 0$. This is a gauge choice at spatial infinities. $N_+ = 1$ means that t coincides with the proper time for an observer at rest at the right-hand spacelike infinity, up to the additive constant. (See Fig. 1.) On the other hand, $N_- = 0$ means that the hypersurface at the left-hand asymptotic infinity is frozen for all t. Thus, the physical meaning of the time t is uniquely deterimined as a proper time of the static observer at one spacelike infinity. Now the action becomes $I_{\rm K} = \int dt(\mathbf{p}\dot{\mathbf{m}} - \mathbf{m})$, for which the Hamiltonian is simply $\mathbf{m}$, the generalized Misner-Sharp mass.

Following Louko and Mäkelä, we take a canonical transformation $(\mathbf{m}, \mathbf{p}) \to (a, p_a)$ from this action, where the relations between old and new variables are given by

$$|\mathbf{p}| = \int_a^{a_{\rm h}(\mathbf{m})} \frac{db}{\sqrt{-f(\mathbf{m}, b)}}, \tag{21}$$

$$p_a = \mathrm{sgn}(\mathbf{p}) \sqrt{-\left(\frac{(n-2)V_{n-2}^{(k)}}{\kappa_n^2}\right)^2 a^{2(n-3)} f(\mathbf{m}, a)}, \tag{22}$$

where the function f is defined by Eq. (4) and $a_{\rm h}$ is defined by $f(\mathbf{m}, a_{\rm h}) = 0$. This is a time-independent canonical transformation and therefore the Hamiltonian $H(a, p_a)$ for the new set of variable (a, p_a) is simply $\mathbf{m}(a, p_a)$. The resulting new action is

$$I_{\rm LM} = \int dt\Big(p_a \dot{a} - H(a, p_a)\Big), \tag{23}$$

where the new Hamiltonian $H(a, p_a)$ $(= \mathbf{m})$ is given by

$$H = \frac{(n-2)V_{n-2}^{(k)} a^{n-3}}{2\kappa_n^2} \times \left[\left(\frac{(n-2)V_{n-2}^{(k)}}{\kappa_n^2}\right)^{-2} p_a^2 a^{-2(n-3)} + k + \frac{a^2}{l^2}\right]. \tag{24}$$

The domains of a and p_a are $0 \le a < \infty$ and $-\infty < p_a < \infty$, respectively.

Hamilton's equations in the system (23) give

$$\frac{d^2 a}{dt^2} = -\frac{(n-3)\kappa_n^2 \mathbf{m}}{(n-2)V_{n-2}^{(k)} a^{n-2}} - \frac{a}{l^2}. \tag{25}$$

Indeed, this is formally equivalent to the dynamics of areal radius r at the wormhole throat in the Schwarzschild-Tangherlini-type black-hole spacetime (3), which is located inside the Killing horizon. This wormhole throat is defined by a radial timelike geodesic with constant T, of which areal radius $r = a(\tau)$ is determined by the following geodesic equation:

$$\frac{d^2 a}{d\tau^2} = -\frac{(n-3)\kappa_n^2 \mathbf{m}}{(n-2)V_{n-2}^{(k)} a^{n-2}} - \frac{a}{l^2}, \tag{26}$$

where τ is the proper time along the geodesic. (A dashed curve in Fig. 1 is an example of such a geodesic.) The (classical) domain of a is $0 \le a \le r_{\rm h}$, where $a = 0$ and $a = r_{\rm h}$ correspond to the curvature singularity and the Killing horizon, respectively.

Equation (25) is identical to Eq. (26) with $t = \tau$. The terminology "throat quantization" comes from this physical interpretation of the dynamics represented by the action (23). However, it is noted that the time t is originally the proper time for an observer at one spatial infinity, not the proper time along the wormhole throat.

3.2 Quantization: Hamiltonian operator and the boundary condition

Finally we canonically quantize the system (23) and perform the quantum mechanics of the Schwarzschild-Tangherlini-type black hole. Replacing p_a in the Hamiltonian (24) by $\hat{p}_a = -i\hbar\partial/\partial a$, we obtain the Schrödinger equation $\hat{H}\Psi = E\Psi$ for the wave function $\Psi = \Psi(a)$ in the stationary state. Under the Laplace-Beltrami operator-ordering, the Schrödinger equation is written as

$$\left(-\frac{\hbar^2}{2m}\frac{\partial^2}{\partial x^2} + V(x)\right)\Psi = E\Psi, \tag{27}$$

where $x := a^{(n-1)/2}$ and the effective mass m and potential $V(x)$ are given by

$$m \equiv \frac{4(n-2)V_{n-2}^{(k)}}{(n-1)^2\kappa_n^2}, \tag{28}$$

$$V(x) \equiv \frac{(n-2)V_{n-2}^{(k)}}{2\kappa_n^2}\left(\frac{x^2}{l^2} + kx^{2(n-3)/(n-1)}\right). \tag{29}$$

The domain of x is $0 \le x < \infty$ and hence this is the quantum mechanics on the half-line. The natural inner product consistent with the Laplace-Beltrami operator-ordering is

$$\langle\Psi|\Phi\rangle := \int_0^\infty \Psi^*\Phi dx. \tag{30}$$

Since the classical Hamiltonian H is the generalized Misner-Sharp mass, we identify E as the mass eigenvalue of the black hole.

Actually, the Hamiltonian operator $\hat{H}$ is not self-adjoint in the square-integrable Hilbert space on the half-line $\mathcal{L}^2([0,\infty))$ but it admits self-adjoint extensions. This self-adjoint extension requires the following boundary condition at the location of the classical singularity $x = 0$:

$$\Psi(0) + L\frac{\partial\Psi}{\partial x}\bigg|_{x=0} = 0, \tag{31}$$

where L is a real constant [7]. $L = 0$ corresponds to the Dirichlet boundary conditions, while $L = \infty$ corresponds to the Neumann boundary condition. Other values of L gives the general Robin boundary conditions. With the boundary condition (31), evolution of this quantum system is unitary ($\partial\langle\Psi|\Phi\rangle/\partial t = 0$).

3.3 Mass spectrum of the black hole

Now we show the mass spectrum E of the black hole under the Dirichlet ($L = 0$) boundary condition at the origin $x = 0$. While an exact spectrum can be obtained for the toroidal AdS black hole ($k = 0$) in arbitrary dimensions, one obtains the mass spectra for black holes with large mass using the WKB approximation in other cases.

The mass spectrum sharply depends on the presence of the cosmological constant. For asymptotically flat black hole ($k = 1$), the spectrum for large E is

$$E^{(n-2)/(n-3)} \simeq (n-2)\hbar\sqrt{\pi}N \times \frac{\Gamma(\frac{n-1}{2(n-3)} + \frac{1}{2})}{\Gamma(\frac{n-1}{2(n-3)})}\left(\frac{(n-2)V^{(1)}_{n-2}}{2\kappa_n^2}\right)^{1/(n-3)}, \tag{32}$$

where N is a large integer. By the following classical relation between the mass E and the Wald entropy S;

$$S = \frac{2\pi}{\kappa_n^2}V^{(1)}_{n-2}\left(\frac{2\kappa_n^2 E}{(n-2)V^{(1)}_{n-2}}\right)^{(n-2)/(n-3)}, \tag{33}$$

Eq. (32) implies that the entropy of the black hole is equally spaced for large E as

$$S \simeq 4\pi^{3/2}\hbar\frac{\Gamma(\frac{n-1}{2(n-3)} + \frac{1}{2})}{\Gamma(\frac{n-1}{2(n-3)})}N =: S_N. \tag{34}$$

Some particular cases are given explicitly as

$$S_N = \begin{cases} 4\pi^{3/2}\hbar N\frac{\Gamma(2)}{\Gamma(3/2)} = 8\pi\hbar N \ (n=4) \\ 4\pi^{3/2}\hbar N\frac{\Gamma(3/2)}{\Gamma(1)} = 2\pi^2\hbar N \ (n=5) \\ 4\pi^{3/2}\hbar\frac{\Gamma(1)}{\Gamma(1/2)}N = 4\pi\hbar N \ (n\to\infty). \end{cases} \tag{35}$$

In contrast, for the asymptotically AdS black hole with general k, the spectrum for large E is

$$E \simeq \frac{(n-1)\hbar}{l}N, \tag{36}$$

where N is a positive integer. The above expression becomes exact in the toroidal case ($k = 0$), which includes the BTZ black hole in three dimensions.

At a glance, two results (32) and (36) look inconsistent in the limit of vanishing cosmological constant $\Lambda \to 0$ ($l \to \infty$) for spherical AdS black holes ($k = 1$). This is understood by noting two different length scales, the Planck length $\ell_p := (\hbar\kappa_n^2/c^3)^{1/(n-2)}$ and the AdS radius l. The metric function (4) shows the relation between the mass of the black hole E and the horizon radius r_h as

$$E = \frac{(n-2)V^{(k)}_{n-2}r_h^{n-3}}{2\kappa_n^2}\left(k + \frac{r_h^2}{l^2}\right). \tag{37}$$

From this relation, we see that (i) a spherical AdS black hole is approximated by an asymptotically flat black hole if $r_h \ll l$ and (ii) it is approximated by the toroidal AdS black hole ($k = 0$) if $r_h \gg l$. Therefore, the spectrum of the spherical AdS black hole is given by Eq. (32) in the region of $\ell_p \ll r_h \ll l$ and by Eq. (36) in the region of $r_h \gg l$, as shown in Fig. 2. In the limit of vanishing cosmological constant ($l \to \infty$), the spacetime becomes asymptotically flat so that the region $r_h \gg l$ disappears in Fig. 2 and the mass spectrum (32) is valid in the whole region satisfying $r_h \gg \ell_p$.

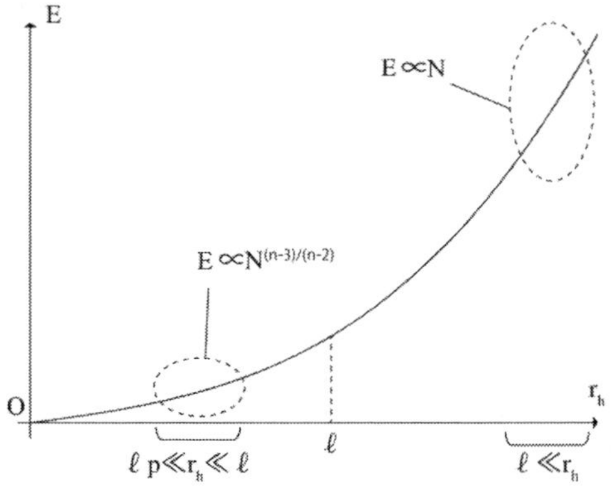

Figure 2. Mass-horizon relation (37) for spherical AdS black holes ($k = 1$). The mass spectrum is given by Eq. (32) in the region of $\ell_p \ll r_h \ll l$ and by Eq. (36) in the region of $r_h \gg l$.

In summary, using the WKB approximation for the Schwarzschild-Tangherlini-type black hole (3) with large mass, we have shown that area/entropy is equally spaced in the asymptotically case, while mass is equally spaced for asymptotically AdS case. This result is unchanged under the Neumann boundary condition ($L = \infty$). In the case of the Robin boundary condition with other values of L, numerical calculations are required to obtain the mass spectrum.

References

[1] J. Louko and J. Mäkelä, Phys. Rev. D **54**, 4982 (1996).
[2] K. Kuchař, Phys. Rev. **D50**, 3961 (1994).
[3] C. W. Misner and D. H. Sharp, Phys. Rev. **136**, B571 (1964).
[4] G.A.S. Dias and J.P.S. Lemos, Phys. Rev. D **79**, 044013 (2009).
[5] H. Maeda and M. Nozawa, Phys. Rev. D **77**, 064031 (2008).
[6] G. Kunstatter and H. Maeda, Class. Quant. Grav. **31**,**31**, 115009 (2014).

[7] M. Reed and B. Simon, *Methods of Modern Mathematical Physics II: Fourier Analysis, Self-adjointness* (Academic Press, New York, 1975).

The orbital stability of a test body motion in the field of two massive bodies

Medeu Abishev[1,2,3,⋆], *Saken* Toktarbay[1,2,⋆⋆] *Aigerim* Abylayeva[1,2,], and *Amanhan* Talkhat[1,2,]

[1] *IETP,*
[2] *Physical and Technical Faculty, Al-Farabi Kazakh National University, al'-Farabi avenue, 71, Almaty 050040, Kazakhstan*
[3] *Institute of Gravitation and Cosmology, RUDN University, Miklukho-Maklaya St.,6, Moscow 117198, Russia*

Abstract. We investigate the orbital stability of a test particle motion in the restricted three-body problem where all bodies have their own rotation. We have shown that it is possible to get some insight into the stability properties of the motion of test particles in restricted three-body problem, without knowing the exact solutions of the relativistic motion equations.

1 Introduction

In the works dedicated to the relativistic three-body problem, some peculiarities of private problems were considered [1–3]. But in some cases, as we believe, cosmogonic general relativistic effects can be substantial, even determining the evolution of a system of bodies, especially when it comes to stable orbital motions of celestial bodies. Therefore, the development of relativistic planetary cosmogony and the study of the evolution of planetary systems in the framework of General relativity (GR) is very important. The orbital stability of two rotating bodies investigated in [4, 5]. As well as, the orbital stability problem of circular motion of a test body in the restricted three-body problem with the relativistic corrections investigated in [6], where the appropriate conditions are

$$U_1 << c^2, \;\; U_2 << U_1, \;\; U_1/U_2 \approx v^2/c^2, \tag{1}$$

where U_1, U_2 are the potentials of central and second bodies in the considered region respectively. In that case, all the bodies do not have their own rotation. The resting position of the central body (the first body) coincides with the reference point of coordinates while the second body moves along the circle around the central body and is not subjected to the disturbance. The test body moves in a perturbed circular orbit. This problem belongs to the class of quasi-keplerian, Therefore, in this case, we can investigate the orbital stability problem by the adiabatic theory of motion in GR mechanics which was developed by academician M.M. Abdildin [4].

⋆e-mail: abishevme@gmail.com
⋆⋆e-mail: saken.yan@yandex.com

2 The equation of motion of a test body

According to the adiabatic theory, the evolutionary motion of the test (third) body describes the average change of its orbital momentum. Therefore, we write down the orbital angular momentum of the test body

$$\vec{M} = [\vec{r}_3, \vec{p}_3], \tag{2}$$

and the time derivative from it

$$\dot{\vec{M}} = \left[\dot{\vec{r}}_3, \vec{p}_3\right] + \left[\vec{r}_3, \dot{\vec{p}}_3\right], \tag{3}$$

where

$$\dot{\vec{r}}_3 = \frac{\partial H}{\partial \vec{p}_3}, \quad \dot{\vec{p}}_3 = -\frac{\partial H}{\partial \vec{r}_3}, \tag{4}$$

are found by means of the equations of Hamilton from the Hamilton function of the system.

In [6], Abishev et al., the relativistic equation of translational motion of a test body in the field of two bodies in the mechanics of general relativity is investigated by the asymptotic methods of the adiabatic theory through the process of averaging the corresponding equations using the vector elements $\vec{M}$ (the orbital moment) and $\vec{A}$ (the Laplace vector). It is shown that in this case the orbit of the test body is stable in the form of circular orbit.

In the present work, we consider the orbital stability problem for the circular motion of a test body in a restricted three-body problem when all bodies have their own rotation. We derived the Lagrange function of a system of bodies up to the second-order terms. It is quite sufficient for the problems of relativistic celestial mechanics that the influence of the internal structure of bodies can generally be neglected and omitted from all terms associated with taking this influence into account. We confine ourselves to the zeroth order terms of the expansion in powers of the ratio of body size to their mutual distances.

This work is organized as follows. In Sec. 2, we present the explicit form of the Lagrangian of the three body, containing an additional rotating term. We also present an equation of motion for the test body. We use the adiabatic theory of motion to analyse the orbital stability of test body. In Sec.3 we will present the averaging equation of motion. Finally, in Sec. 4, we discuss our results.

The Lagrangian for the translational and rotational motion of the three rotating bodies without interior structure can be represented in this form:

$$L = L^{(0)} + L^{(*)}. \tag{5}$$

Consequently, the Hamiltonian function can be represented in this form:

$$H = H^{(0)} + H^{(*)}. \tag{6}$$

According to the equations (3) and (4) the time derivative of orbital angular momentum and equation of motion can be represented in the following form:

$$\dot{\vec{M}} = \dot{\vec{M}}^{(0)} + \dot{\vec{M}}^{(*)}, \tag{7}$$

$$\dot{\vec{r}}_i = \frac{\partial H^0}{\partial \vec{p}_i} + \frac{\partial H^*}{\partial \vec{p}_i} = \dot{\vec{r}}_i^{(0)} + \dot{\vec{r}}_i^{(*)}, \tag{8}$$

and

$$\dot{\vec{p}}_i = \frac{\partial H^{(0)}}{\partial \vec{r}_i} + \frac{\partial H^{(*)}}{\partial \vec{r}_i} = \dot{\vec{p}}_i^{(0)} + \dot{\vec{p}}_i^{(*)}, \tag{9}$$

where (0) term is for the three point like masses and it's explicit form was presented in the work [6] by Abishev at al., and (∗) is the additional term which related to the rotation of the body.

From the general form of the Lagrangian for N bodies [5, 7, 8], we can derive the rotational term of Lagrangian as following:

$$\begin{aligned}
L^{(*)} = & \sum_{i=1}^{3} \omega_i^2 I_i \left(1 + \frac{1}{c^2}\right) - \frac{1}{2c^2}\left[I_2\left(\vec{\omega}_2\dot{\vec{r}}_2\right)^2 + I_3\left(\vec{\omega}_3\dot{\vec{r}}_3\right)^2\right] \\
& - \frac{\gamma}{c^2}\left\{\frac{\vec{r}_2\cdot\dot{\vec{r}}_2}{|\vec{r}_2|^3}\left[3m_1 I_2\omega_2 + 4m_2 I_1\omega_1\right] - \frac{\vec{r}_3\cdot\dot{\vec{r}}_3}{|\vec{r}_3|^3}\left[3m_1 I_3\omega_3 + 4m_3 I_1\omega_1\right]\right. \\
& \left. + \frac{\vec{r}_2-\vec{r}_3}{|\vec{r}_2-\vec{r}_3|^3}\left[\left(3\dot{\vec{r}}_3 - 4\dot{\vec{r}}_2\right)m_2 I_3\omega_3 - \left(3\dot{\vec{r}}_2 - 4\dot{\vec{r}}_3\right)m_3 I_2\omega_2\right]\right\} \\
& - \frac{\gamma}{2c^2}\frac{m_2 I_2 + m_3 I_3}{|\vec{r}_2-\vec{r}_3|^3}\left[\dot{\vec{r}}_2\cdot\dot{\vec{r}}_3 - \frac{3\left((\vec{r}_2-\vec{r}_3)\cdot\dot{\vec{r}}_2\right)\left((\vec{r}_2-\vec{r}_3)\cdot\dot{\vec{r}}_3\right)}{\left|\vec{r}_2-\vec{r}_3\right|^2}\right] \\
& - \frac{3\gamma}{c^2}\left[\frac{1}{|\vec{r}_2|}\left(m_2\omega_2^2 I_2 + m_2\omega_1^2 I_1\right) + \frac{1}{|\vec{r}_3|}\left(m_1\omega_3^2 I_3 + m_3\omega_1^2 I_1\right) + \frac{1}{|\vec{r}_2-\vec{r}_3|}\right. \\
& \left. \times\left(m_2\omega_3^2 I_3 + m_3\omega_2^2 I_2\right)\right] + \frac{\gamma^2}{2c^2}\left\{\frac{m_1 I_2 + m_2 I_1}{|\vec{r}_2|^2} + \frac{m_1 I_3 + m_3 I_1}{|\vec{r}_3|^2}\right. \\
& + \frac{m_2 I_3 + m_3 I_2}{|\vec{r}_2-\vec{r}_3|^2} + m_1 m_2 I_3\frac{\vec{r}_3(\vec{r}_3-\vec{r}_2)}{|\vec{r}_3|^3|\vec{r}_2-\vec{r}_3|^3} + m_2 m_2 I_1\frac{\vec{r}_2\cdot\vec{r}_3}{|\vec{r}_3|^3|\vec{r}_2-\vec{r}_3|^3} \\
& \left. + m_1 m_3 I_2\frac{(\vec{r}_2-\vec{r}_3)\cdot\vec{r}_2}{|\vec{r}_2-\vec{r}_3|^3|\vec{r}_2|^3}\right\} - \frac{\gamma}{c^2}\left\{4\left[\frac{I_1 I_2}{|\vec{r}_2|^3} + \frac{I_1 I_3}{|\vec{r}_3|^3} + \frac{I_2 I_3}{|\vec{r}_2-\vec{r}_3|^3}\right]\right. \\
& -3\left[\frac{(\vec{r}_2\vec{\omega}_1)\cdot(\vec{r}_2\vec{\omega}_2)}{|\vec{r}_2|^3} + \frac{(\vec{r}_3\vec{\omega}_1)(\vec{r}_3\vec{\omega}_3)}{\left|\vec{r}_3\right|^3} + \frac{\left((\vec{r}_2-\vec{r}_3)\vec{\omega}_2\right)\left((\vec{r}_2-\vec{r}_3)\vec{\omega}_3\right)}{\left|\vec{r}_2-\vec{r}_3\right|^3}\right].
\end{aligned} \tag{10}$$

We can find the Hamiltonian using the following well known relation:

$$H = \vec{v}_i\frac{\partial L}{\partial \vec{v}_i} - L, \tag{11}$$

and by taking into account the Eqs. (5) and (6) we can write down the additional term of the Hamiltonian and the derivatives of the canonical variables of the test body.

3 Averaging equation of motion

According to the adiabatic theory, in order to obtain the evolutionary equations of motion one needs to integrate the equation (7) for the repetition period of system configurations T (synodic period of the test body):

$$\overline{\dot{\vec{M}}} = \frac{1}{T}\int_0^T \dot{\vec{M}}dt, \tag{12}$$

where

$$T = \frac{2\pi}{\omega_2 - \omega_3}. \tag{13}$$

In fact, the preliminary results show that it is possible to find the time derivative of the angular momentum, and then integrate the equation to find the average change in the orbital angular momentum. In this case, it can be shown that under certain conditions which satisfy the stability conditions for the orbital motion of the test body we can find solutions for the orbit of test body in the restricted three body problem.

4 Conclusions

In this work, we considered the orbital stability problem for the circular motion of a test body in the restricted three-body problem within the framework of GR, when all bodies have their own rotation. We derived equation of motion from the Lagrange function with the rotational terms. The orbital stability of the test body is investigated by the adiabatic theory of motion in the GR. Moreover, we derived the explicit form of the corresponding motion equation and the time derivative of momentum. The resulting equations was integrated (averaged) by the sinodic period of the test body, which eliminates perturbations, leaving only evolutionary effects. The result of the integral is confirmed that the circular motion will be stable if the spins of all bodies are collinear within a plane. The quasicircular orbits of the test body will be analyzed in a forthcoming works.

Acknowledgments

This work has been partially supported by the Ministry of Education and Science (Kazakhstan), in particular by grant to project "Restricted three-body problem in the General Relativity"

References

[1] Maindl T. I.& Dvorak R. Astron.Astrophys. **290**, 335 (1994)
[2] Imai T1, Chiba T, Asada H. Phys. Lett. B **98**, (2007).
[3] Kei Yamada, Hideki Asada . Phys. Rev. D **82**,104019 (2010).
[4] Abdil'din M. M., Abishev M. E., Beissen N. A. Gravitation and Cosmology **V15**, 141 (2009).
[5] Brumberg A.A. *Relativistic celestial mechanics* (Nauka, 1972), 382 p.
[6] Abishev M. E., Toktarbay, S., Zhami B. A. Gravitation and Cosmology. **V20**, 149 (2014).
[7] Landau L.D., Lifshitz A.I.*The Classical Theory of Fields* (Nauka, 1973), Vol. 2. 402 p.
[8] Abdildin M.M. *Mechanins of Einstein theory of gravity*(Alma-Ata: Nauka, 1988), 198 p.

EPJ Web of Conferences **168**, 04002 (2018) https://doi.org/10.1051/epjconf/201816804002
Joint International Conference of ICGAC-XIII and IK-15 on Gravitation, Astrophysics and Cosmology

THE FIRST ICRANET CATALOG OF BINARY-DRIVEN HYPERNOVAE

G. B. Pisani[1,2], R. Ruffini[1,2,3,4], Y. Aimuratov[1,2,3], C. L. Bianco[1,2], M. Karlica[1,2,3], M. Kovacevic[1,2,3], R. Moradi[1,2], M. Muccino[1,2], A. V. Penacchioni[5,6], D. Primorac[1,2], J. A. Rueda[1,2,4], Y. Wang[1,2]

Draft version September 18, 2017

ABSTRACT

In a series of recent publications, scientists from ICRANet, led by professor Remo Ruffini, have reached a novel comprehensive picture of gamma-ray bursts (GRBs) thanks to their development of a series of new theoretical approaches. Among those, the induced gravitational collapse (IGC) paradigm explains a class of energetic, long-duration GRBs associated with Ib/c supernovae (SN), recently named binary-driven hypernovae (BdHNe).

BdHNe have a well defined set of observational features which allow to identify them. Among them, the main two are: 1) long duration of the GRB explosion, namely larger than 2 s in the rest frame; 2) a total energy, released in all directions by the GRB explosion, larger than 10^{52} ergs.

A striking result is the observation, in the BdHNe sources, of a universal late time power-law decay in the X-rays luminosity after 10^4 s, with typical decaying slope of ~ 1.5. This leads to the possible establishment of a new distance indicator having redshift up to $z \sim 8$.

Thanks to this novel theoretical and observational understanding, it was possible for ICRANet scientists to build the firstst BdHNe catalog, composed by the 345 BdHNe identified up to the end of 2016.

Keywords: supernovae: general — binaries: general — gamma-ray burst: general — stars: neutron

1. TOWARDS A FIRST CATALOG OF BINARY-DRIVEN HYPERNOVAE

The first observations by the BATSE instrument on board the Compton γ-ray Observatory satellite have evidenced what has later become known as the prompt radiation of GRBs. On the basis of their hardness as well as their duration, GRBs were initially classified into short and long in an epoch when their cosmological nature was still being debated (Mazets et al. 1981; Klebesadel 1992; Dezalay et al. 1992; Kouveliotou et al. 1993; Tavani 1998).

The advent of the *BeppoSAX* satellite (Boella et al. 1997) introduced a new approach to GRBs by introducing joint observations in the X-rays and γ-rays thanks to its instruments: the Gamma-ray Burst Monitor (40–700 keV), the Wide Field Cameras (2–26 keV), and the Narrow Field Instruments (2-10 keV). The unexpected discovery of a well separate component in the GRB soon appeared: the afterglow, namely a radiation lasting up to 10^5–10^6 s after the emission of the prompt radiation (see Costa et al. 1997a,b; Frontera et al. 1998, 2000; de Pasquale et al. 2006). Beppo-SAX clearly indicated the existence of a power law behavior in the late X-ray emission (LXRE).

The coming of the *Swift* satellite (Gehrels et al. 2004; Evans et al. 2007, 2010), significantly extending the observation energy band to the X-ray band thanks to its X-ray Telescope (XRT band: 0.3–10 keV), has allowed us for the first time to uncover the unexplored region between the end of the prompt radiation and the power-law late X-ray behavior discovered by *BeppoSAX*: in some long GRBs, a steep decay phase was observed leading to a plateau followed then by a typical LXRE power law behavior (Evans et al. 2007, 2010).

Recently, Pisani et al. (2013) noticed the unexpected result that the LXREs of a "golden sample" (GS) of six long, closeby ($z \lesssim 1$), energetic ($E_{iso} > 10^{52}$ erg) GRBs, when moved in the rest-frame of the sources, were showing a common power-law behavior (see Fig. 1), independently from the isotropic energy E_{iso} coming from the GRB prompt radiation (see Fig 2). More unexpected was the fact that the plateau luminosity and duration before merging in the common LXRE power-law behavior were clearly functions of the E_{iso} (see Fig. 2, and Ruffini et al. 2014c), while the late power-law remains independent from the E_{iso} of the prompt emission (see Fig. 1–2, and Pisani et al. 2013; Ruffini et al. 2014c). For this reason, this striking scaling law has been used as a distance indicator to independently estimate the cosmological redshift of some long GRBs by imposing the overlap of their LXRE (see, e.g., Penacchioni et al. 2012, 2013; Ruffini et al. 2013b,c, 2014a), and also to predict, ten days in advance, the observation of the typical optical signature of the supernova SN 2013cq, associated with GRB 130427A (Ruffini et al. 2015, 2013a; de Ugarte Postigo et al. 2013; Levan et al. 2013).

All these analyses are based on the paradigms introduced in Ruffini et al. (2001a) for the space-time parametrization of the GRB phenomena, in Ruffini et al. (2001b) for the interpretation of the structure of the GRB prompt radiation, and in Ruffini et al. (2001c) for the induced gravitational collapse (IGC) mechanism, further

[1] Dipartimento di Fisica, Sapienza Università di Roma and ICRA, Piazzale Aldo Moro 5, I-00185 Roma, Italy
[2] International Center for Relativistic Astrophysics Network (ICRANet), Piazza della Repubblica 10, I-65122 Pescara, Italy
[3] Université de Nice Sophia-Antipolis, Grand Château Parc Valrose, Nice, CEDEX 2, France
[4] ICRANet-Rio, Centro Brasileiro de Pesquisas Fisicas, Rua Dr. Xavier Sigaud 150, Rio de Janeiro, RJ, 22290-180, Brazil
[5] University of Siena, Department of Physical Sciences, Earth and Environment, Via Roma 56, I-53100 Siena, Italy
[6] ASI Science Data Center, via del Politecnico s.n.c., I-00133 Rome Italy

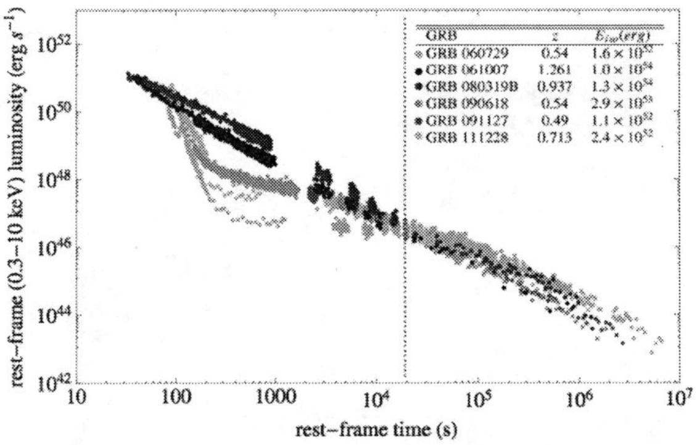

Figure 1. Scaling law found in the isotropic X-ray late times luminosity within the GS by Pisani et al. (2013). Despite the different early behavior, the different light curves join all together the same power law after a rest-frame time of $t_{rf} \sim 2 \times 10^4$ s.

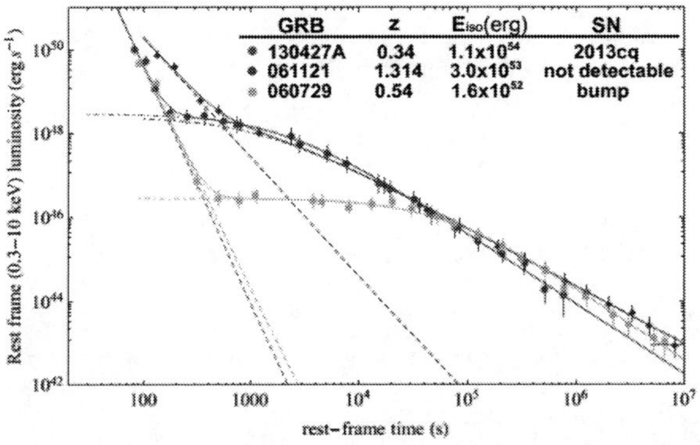

Figure 2. Nested structure of the isotropic X-ray luminosity of the BdHNe. This includes the previously mentioned scaling law of the late power law and leads to an inverse proportionality between the luminosity of the plateau and the rest-frame time delimiting its end and the beginning of the late power law decay Ruffini et al. (2014c).

developed in Ruffini et al. (2007),Rueda & Ruffini (2012), Fryer et al. (2014), and Ruffini et al. (2016). In the present case, the phenomenon points to an IGC occurring when a tight binary system composed of a carbon-oxygen core (CO_{core}) undergoes a supernova (SN) explosion in the presence of a binary neutron star (NS) companion (Ruffini et al. 2001b, 2007; Izzo et al. 2012; Rueda & Ruffini 2012; Fryer et al. 2014; Ruffini et al. 2015). When the IGC leads the NS to accrete enough matter and therefore to collapse to a black hole (BH), the GRB shows a long duration, and its prompt emission overtakes the treshold value of 10^{52} ergs. The overall observed phenomenon is called binary-driven hypernova (BdHN; Fryer et al. 2014; Ruffini et al. 2015, 2016).

A decisive further step has been the identification as a BdHN of GRB 090423 (Ruffini et al. 2014b) at the extremely high redshift of $z = 8.2$ (Salvaterra et al. 2009; Tanvir et al. 2009). On top of that, the LXRE of GRB 090423 overlaps entirely with the ones of the GS (see Fig. 3), extending such a scaling law up to extreme cosmological distances. This result led to the necessity of checking such an common behavior of the LXREs in BdHNe at redshifts larger than $z \sim 1$.

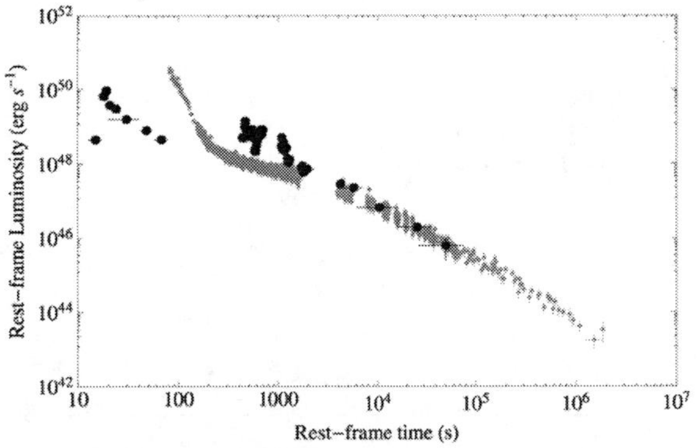

Figure 3. X-ray luminosity of GRB 090423 (black points) compared with the one of GRB 090618 (green points), the prototype BdHN, by Ruffini et al. (2014b).

In Pisani et al. (2016), we present an "enlarged sample" (ES) of 161 BdHNe observed up to the end of 2015. In this work we analysed the signatures contained in the LXREs at $t_{rf} \gtrsim 10^4$ s, where t_{rf} is the rest-frame time after the initial GRB explosion. In particular, we probed a further improvement for the presence of such an LXRE universal behavior of BdHNe by the introduction of a collimation effect within the emission mechanism.

In our recent work (Ruffini et al. 2017), we focused on analyzing the early X-Ray Flares in the GRB flare-plateau-afterglow (FPA) phase observed by Swift/XRT. The FPA occurs only in the BdHNe while is not present in the other subclasses of GRBs, for details see Ruffini et al. (2016). The sample presented in Table 9 of Ruffini et al. (2017), namely an updated version of the ES up to the end of 2016, together with the BdHNe lacking LXRE data and the ones from the pre-Swift-era, counts 345 BdHNe in total. This represents the current BdHNe catalog from ICRANet.

In the following, we present various insightful results which ICRANet scientists gained from this catalog: in Sections 2, 3, and 4, we describe how we built the ES and we study the LXRE features within it; finally, in Section 5 we refer to our up-to-date catalog of BdHNe and we draw our perspectives.

2. THE FIRST ENLARGED SAMPLE OF BDHN

Starting from the GS originally presented in Pisani et al. (2013), in Pisani et al. (2016) we have built a new sample of BdHNe, which we called "enlarged sample" (ES), under the following selection criteria:

- measured redshift z;
- GRB rest-frame duration larger than 2 s;
- isotropic energy E_{iso} larger than 10^{52} erg; and
- presence of associated *Swift*/XRT data lasting at least up to $t_{rf} = 10^4$ s.

We collected 161 sources, satisfying our criteria, which cover 11 years of *Swift*/XRT observations, up to the end of 2015, see Table 2 of Pisani et al. (2016). The E_{iso} of each source has been estimated using the observed redshift z together with the best-fit parameters of the

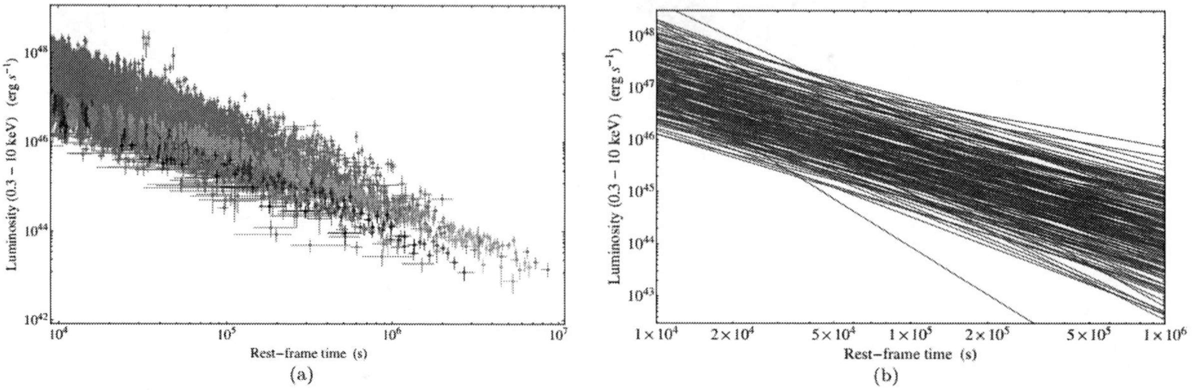

Figure 4. Panel (a): LXRE luminosity light curves of all 161 sources of the ES (gray) compared with the ones of the GS: GRB 060729 (pink), GRB 061007 (black), GRB 080913B (blue), GRB 090618 (green), GRB 091127 (red), and GRB 111228 (cyan), plus GRB 130427A (orange; from Pisani et al. 2016). Panel (b): power laws which best fit the luminosity light curves of the X-ray emissions of all 161 sources of the ES (from Pisani et al. 2016).

γ-ray spectrum published in the GCN circular archive[7]. Most of of the ES sources, 102 out of 161, have γ-ray data observed by *Fermi*/GBM and Konus-WIND, which, with their energy bands being 10–1000 keV and 20–2000 keV, respectively, lead to a solid estimate of the E_{iso}, computed in the "bolometric" 1–10^4 keV band (Bloom et al. 2001). The remaining sources of the ES have had their γ-ray emission provided by *Swift*/BAT only, with the unique exception of one source observed by HETE. The energy bands of these two detectors, being 15–150 keV and 8–400 keV, respectively, lead to an estimate of E_{iso} by extrapolation in the "bolometric" 1–10^4 keV band (Bloom et al. 2001).

3. GOING TO THE REST-FRAME

We compare the *Swift*/XRT isotropic luminosity light curve $L_{iso}(t_{rf})$ for 161 GRBs of the ES in the common rest-frame energy range of 0.3 − 10 keV. We initially adjust the observed *Swift*/XRT flux f_{obs} as if it had been observed in the 0.3 − 10 keV rest-frame energy range. In the detector frame, the 0.3 − 10 keV rest-frame energy band becomes $[0.3/(1+z)] - [10/(1+z)]$ keV, where z is the measured redshift of the GRB. We assume a simple power-law as the best fit for the spectral energy distribution of the *Swift*/XRT data[8]:

$$\frac{dN}{dA\,dt\,dE} \propto E^{-\gamma}\,. \tag{1}$$

Hence, we can calculate the flux light curve in the 0.3−10 keV rest-frame energy band, f_{rf}, multiplying the observed one, f_{obs}, by the k-correctionr:

$$f_{rf} = f_{obs}\frac{\int_{\frac{0.3\,keV}{1+z}}^{\frac{10\,keV}{1+z}} E^{1-\gamma}dE}{\int_{0.3\,keV}^{10\,keV} E^{1-\gamma}dE} = f_{obs}(1+z)^{\gamma-2}\,. \tag{2}$$

Then, to calculate the isotropic X-ray luminosity L_{iso}, we need to multiply f_{rf} by the spherical surface having the luminosity distance as radius

$$L_{iso} = 4\,\pi\,d_l^2(z) f_{rf}\,, \tag{3}$$

[7] http://gcn.gsfc.nasa.gov/gcn3_archive.html
[8] http://www.swift.ac.uk/

where we assume a standard cosmological ΛCDM model, namely $\Omega_m = 0.27$ and $\Omega_\Lambda = 0.73$. In the end, we convert the observed times into rest-frame times t_{rf}:

$$t_{rf} = \frac{t_{obs}}{1+z}\,. \tag{4}$$

After, we fit the isotropic luminosity light-curve late phase with a decaying power-law function defined as:

$$L_{iso}(t_{rf}) = L_0\, t_{rf}^{-\alpha}\,, \tag{5}$$

where α, the power law index, is a positive number, and L_0 is the luminosity at a fixed time $t_{rf} = t_0$ after the GRB initial explosion in the rest-frame of the source. All the power-laws are shown in Fig. 4b. Fig. 5a shows the distribution of the α indexes within the ES. This distribution follows a Gaussian behavior having a mean value of $\mu_\alpha = 1.48$ and a standard deviation of $\sigma_\alpha = 0.32$. The LXRE luminosity light curves of the ES in the 0.3–10 keV rest-frame energy range are plotted in Fig. 4a, together with the curves of the GS. Fig. 4a shows that the power-laws within the ES span around two orders of magnitude in luminosity. The spread of the LXRE light curves in the ES is better shown off by Fig. 5b which display the distribution within the ES of the LXRE integrated energies E_{LT} defined as:

$$E_{LT} \equiv \int_{10^4 s}^{10^6 s} L_{iso}(t_{rf})\,\mathrm{d}t_{rf}\,. \tag{6}$$

The solid red line in Fig. 5b is the Gaussian function that best fits the late integrated energies E_{LT} in logarithmic scale. Its mean value is $\mu_{\mathrm{Log}_{10}(E_{LT})} = 51.40$, and its standard deviation is $\sigma_{\mathrm{Log}_{10}(E_{LT})} = 0.47$.

The LXRE power-law spread, given approximately by $2\sigma_{\mathrm{Log}_{10}(E_{LT})} = 0.94$, is larger than the one of the previous work of Pisani et al. (2013), which results as $2\sigma_{\mathrm{Log}_{10}(E_{LT})} = 0.56$. This is certainly due to the important growth of the number of BdHNe composing the ES (161) in respect to the ones of the GS (6).

Moreover, there is no evidence for a correlation between the LXRE power-law behavior and the isotropic energy radiated by the source during the GRB prompt radiation (for details, see Pisani et al. 2016).

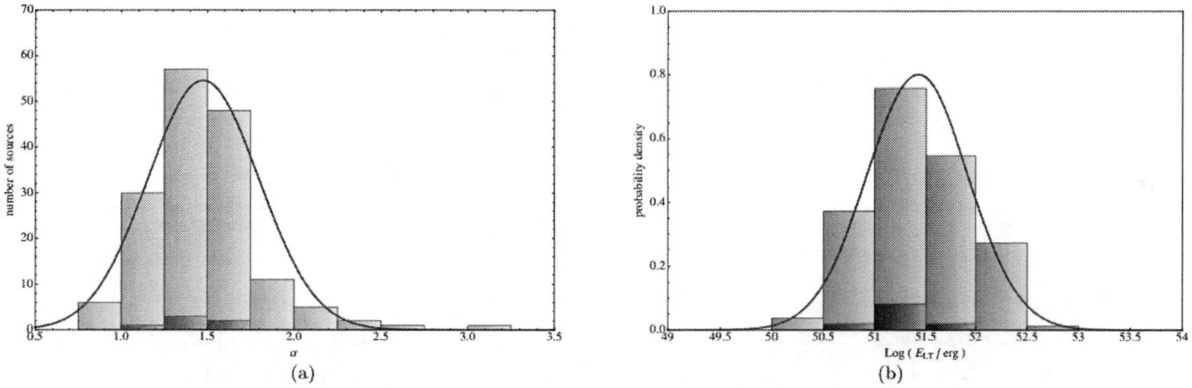

Figure 5. Panel (a): distribution of the LXRE power law indexes α within the ES (cyan) compared to the one of the GS (red). Such a distribution follows a Gaussian behavior (blue line) with a mean value of $\mu_\alpha = 1.48$ and a standard deviation of $\sigma_\alpha = 0.32$ (from Pisani et al. 2016). Panel (b): probability distribution of the LXRE integrated energies within the time interval 10^4–10^6 s in the rest-frame after the initial GRB trigger for all the sources of the ES (in green) compared with the GS (in blue). The solid red line represents the Gaussian function which best fits the ES data in logarithmic scale. Its mean value is $\mu_{\text{Log}_{10}(E_{LT})} = 51.40$, while its standard deviation is $\sigma_{\text{Log}_{10}(E_{LT})} = 0.47$ (from Pisani et al. 2016).

4. COLLIMATION

In Pisani et al. (2016), we also proposed to reduce the spread of the LXRE power laws within the ES by introducing a collimation effect in the emission mechanism. In fact, if such a process is not isotropic, our estimates for the LXRE luminosities are actually overestimations of the intrinsic ones. By introducing a collimation effect, namely assuming that the LXREs are not radiated isotropically but inside a double-cone region having half-opening angle θ, we can compute the intrinsic LXRE luminosity $L_{intr}(t_{rf})$ from the isotropic $L_{iso}(t_{rf})$:

$$L_{intr}(t_{rf}) = L_{iso}(t_{rf})\,(1 - \cos\theta)\,. \qquad (7)$$

From Eq. 7, an angle θ can be computed for each source of the ES if an intrinsec universal LXRE light curve $L_{intr}(t_{rf})$ is given. By assuming GRB 050525A, which has the lowest luminosity within the ES, as our sole "isotropic" LXRE source, we obtain the probability distribution of the half-opening angle θ within the ES showed in Fig. 6a. The blue solid line represents a logarithmic normal distribution, which best fits the data. This distribution has a mode of $Mo_\theta = 17.62°$, a mean of $\mu_\theta = 30.05°$, a median of $Me_\theta = 25.15°$, and a standard deviation of $\sigma_\theta = 19.65°$. In addition, it is possible to verify that, by adjusting the $L_{iso}(t_{rf})$ light curve of each ES source for its corresponding θ, an overlap of the LXRE luminosity light curves as good as the one seen in the GS by Pisani et al. (2013) shown in Fig. 1 is obtained. Since the LXRE follows a power-law behavior, we can evaluate the tightness of the LXREs overlap estimating the correlation coefficient ρ between all the luminosity light-curve data points of the ES sources in log-log scale. Considering the data points of the LXRE power laws within the 10^4–10^6 s time interval (the time interval where we defined E_{LT}), we obtain $\rho = -0.94$ for the GS, $\rho = -0.84$ for the ES before the collimation effect correction, and $\rho = -0.97$ after the collimation correction. Therefore, assuming the collimation not only let the spread of the LXREs within the ES decrease, but makes the LXREs overlap even tighter than the one previously observed in the GS. This leads to the possible establishment of a new distance indicator, eventually useful to test the standard cosmological ΛCDM model.

5. THE CURRENT BDHNE CATALOG

Thanks to the tremendous amount of work from ICRANet scientists in the past years (Ruffini et al. 2001a,b,c, 2007; Rueda & Ruffini 2012; Izzo et al. 2012; Fryer et al. 2014; Ruffini et al. 2015, 2016), today we know that all the observed GRBs having long duration and isotropic energy E_{iso} larger than 10^{52} erg are the observational result of a BdHN phenomenon. Therefore, these two signatures are necessary and sufficient to identify a BdHN source. This holds also in the case it was not possible to observe the other typical features of the BdHNe following the GRB explosion, like: the FPA structure in the X-rays; the LXRE in the X-rays; and the associated Ib/c SN in the optical rays. In our recent work (Ruffini et al. 2017), in order to focus our analysis on the early X-Ray Flares in the FPA phase, we collected all the BdHNe ever observed till the end of 2016, collecting all the GRBs which satisfies the following criteria:

- measured redshift z;
- GRB rest-frame duration larger than 2 s;
- isotropic energy E_{iso} larger than 10^{52} erg.

The updated list presented in Table 9 of (Ruffini et al. 2017) is composed by 345 BdHNe, and represents the current ICRANet catalog of BdHNe. The ES, updated to the end of 2016, counts 182 BdHNe having *Swift*/XRT data up to at least 10^4 s in the rest-frame after the initial GRB explosion. It composes $\sim 53\%$ of the total BdHNe catalog. Since the *Swift* satellite is operating since 2005, we have an average of ~ 15 BdHNe per year having good LXRE observations. Consequently, this representes the expected rate of BdHNe which, in the near future, will be useful to test the standard cosmological ΛCDM model at redshifts up to $z \sim 8$.

This work made use of data supplied by the UK Swift Science Data Center at the University of Leices-

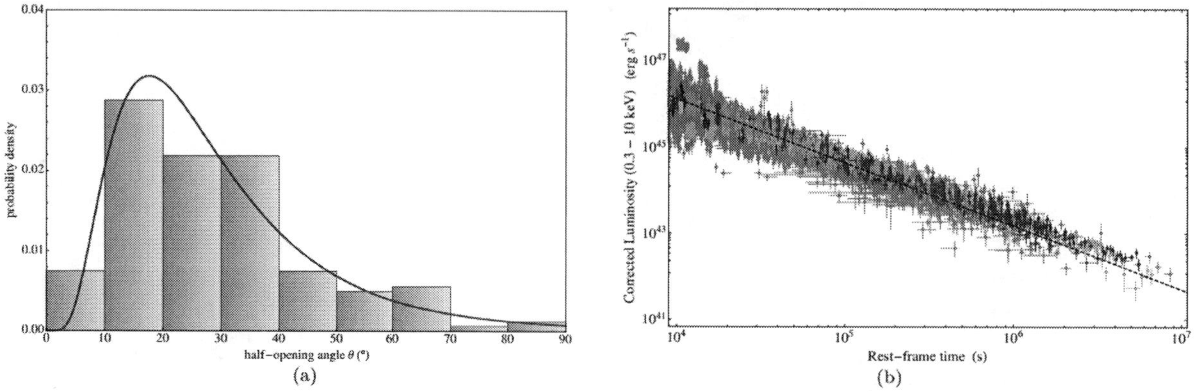

Figure 6. Left panel (a): probability distribution of the half-opening angle θ within the ES. The blue solid line represents a logarithmic normal distribution, which best fits the data. This distribution has a mode of $Mo_\theta = 17.62°$, a mean of $\mu_\theta = 30.05°$, a median of $Me_\theta = 25.15°$, and a standard deviation of $\sigma_\theta = 19.65°$ (from Pisani et al. 2016). Right panel (b): corrected LXRE luminosity light curves of all 161 sources of the ES (gray) compared to the ones of the GS: GRB 060729 (pink), GRB 061007 (black), GRB 080913B (blue), GRB 090618 (green), GRB 091127 (red), and GRB 111228 (cyan), plus GRB 130427A (purple; Pisani et al. 2013; Ruffini et al. 2015). The black dotted line represents the universal LXRE power law, namely the linear fit of the late emission of GRB 050525A (from Pisani et al. 2016).

ter. J. A. R. acknowledges the support by the International Cooperation Program CAPES-ICRANet financed by CAPES-Brazilian Federal Agency for Support and Evaluation of Graduate Education within the Ministry of Education of Brazil. M. M. acknowledges the partial support of the project No. 3101/GF4 IPC-11/2015, and the target program of the Ministry of Education and Science of the Republic of Kazakhstan.

REFERENCES

Bloom, J. S., Frail, D. A., & Sari, R. 2001, AJ, 121, 2879
Boella, G., Butler, R. C., Perola, G. C., et al. 1997, A&AS, 122, 299
Costa, E., Frontera, F., Heise, J., et al. 1997a, Nature, 387, 783
Costa, E., Feroci, M., Piro, L., et al. 1997b, IAU Circ., 6576
de Pasquale, M., Piro, L., Gendre, B., et al. 2006, A&A, 455, 813
de Ugarte Postigo, A., Xu, D., Leloudas, G., et al. 2013, GCN Circ., 14646, 1
Dezalay, J.-P., Barat, C., Talon, R., et al. 1992, in American Institute of Physics Conference Series, Vol. 265, American Institute of Physics Conference Series, ed. W. S. Paciesas & G. J. Fishman, 304–309
Evans, P. A., Beardmore, A. P., Page, K. L., et al. 2007, A&A, 469, 379
Evans, P. A., Willingale, R., Osborne, J. P., et al. 2010, A&A, 519, A102
Frontera, F., Costa, E., Piro, L., et al. 1998, ApJ, 493, L67
Frontera, F., Amati, L., Costa, E., et al. 2000, ApJSS, 127, 59
Fryer, C. L., Rueda, J. A., & Ruffini, R. 2014, ApJ, 793, L36
Gehrels, N., Chincarini, G., Giommi, P., et al. 2004, ApJ, 611, 1005
Izzo, L., Rueda, J. A., & Ruffini, R. 2012, A&A, 548, L5
Klebesadel, R. W. 1992, in Gamma-Ray Bursts - Observations, Analyses and Theories, ed. C. Ho, R. I. Epstein, & E. E. Fenimore (Cambridge University Press), 161–168
Kouveliotou, C., Meegan, C. A., Fishman, G. J., et al. 1993, ApJ, 413, L101
Levan, A. J., Fruchter, A. S., Graham, J., et al. 2013, GCN Circ., 14686, 1
Mazets, E. P., Golenetskii, S. V., Ilinskii, V. N., et al. 1981, Ap&SS, 80, 3
Penacchioni, A. V., Ruffini, R., Bianco, C. L., et al. 2013, A&A, 551, A133
Penacchioni, A. V., Ruffini, R., Izzo, L., et al. 2012, A&A, 538, A58
Pisani, G. B., Izzo, L., Ruffini, R., et al. 2013, A&A, 552, L5
Pisani, G. B., Ruffini, R., Aimuratov, Y., et al. 2016, ApJ, 833, 159
Rueda, J. A., & Ruffini, R. 2012, ApJ, 758, L7
Ruffini, R., Bianco, C. L., Chardonnet, P., Fraschetti, F., & Xue, S.-S. 2001a, ApJ, 555, L107
—. 2001b, ApJ, 555, L113
—. 2001c, ApJ, 555, L117
Ruffini, R., Bernardini, M. G., Bianco, C. L., et al. 2007, ESA SP, 622, 561
Ruffini, R., Bianco, C. L., Enderli, M., et al. 2013a, GCN Circ., 14526, 1
—. 2013b, GRB Coordinates Network, 14888
—. 2013c, GRB Coordinates Network, 15576
—. 2014a, GRB Coordinates Network, 15707
Ruffini, R., Izzo, L., Muccino, M., et al. 2014b, A&A, 569, A39
Ruffini, R., Muccino, M., Bianco, C. L., et al. 2014c, A&A, 565, L10
Ruffini, R., Wang, Y., Enderli, M., et al. 2015, ApJ, 798, 10
Ruffini, R., Rueda, J. A., Muccino, M., et al. 2016, ApJ, 832, 136
Ruffini, R., Wang, Y., Aimuratov, Y., et al. 2017, ArXiv e-prints
Salvaterra, R., Della Valle, M., Campana, S., et al. 2009, Nature, 461, 1258
Tanvir, N. R., Fox, D. B., Levan, A. J., et al. 2009, Nature, 461, 1254
Tavani, M. 1998, ApJ, 497, L21

EPJ Web of Conferences **168**, 04003 (2018) https://doi.org/10.1051/epjconf/201816804003
Joint International Conference of ICGAC-XIII and IK-15 on Gravitation, Astrophysics and Cosmology

Multiphoton amplitude in a constant background field

Aftab Ahmad[1,2,⋆], *Naser* Ahmadiniaz[3,⋆⋆], *Olindo* Corradini[4,5,⋆⋆⋆], *Sang Pyo* Kim[3,6,⋆⋆⋆⋆], and *Christian* Schubert[2,†]

[1] *Department of Physics, Gomal University, 29220 D.I. Khan, K.P.K., Pakistan.*
[2] *Instituto de Física y Matemáticas, Universidad Michoacana de San Nicolás de Hidalgo. Edificio C-3, Ciudad Universitaria, Morelia 58040, Michoacán, México.*
[3] *Center for Relativistic Laser Science, Institute for Basic Science, Gwangju 61005, Korea.*
[4] *Dipartimento di Scienze Fisiche, Informatiche e Matematiche, Universitá di Modena e Reggio Emilia, Via Campi 213/A, I-41125 Modena, Italy.*
[5] *INFN, Sezione di Bologna, Via Irnerio 46, I-40126 Bologna, Italy.*
[6] *Department of Physics, Kunsan National University, Kunsan 54150, Korea.*

Abstract. In this contribution, we present our recent compact master formulas for the multiphoton amplitudes of a scalar propagator in a constant background field using the worldline fomulation of quantum field theory. The constant field has been included non-perturbatively, which is crucial for strong external fields. A possible application is the scattering of photons by electrons in a strong magnetic field, a process that has been a subject of great interest since the discovery of astrophysical objects like radio pulsars, which provide evidence that magnetic fields of the order of 10^{12}G are present in nature. The presence of a strong external field leads to a strong deviation from the classical scattering amplitudes. We explicitly work out the Compton scattering amplitude in a magnetic field, which is a process of potential relevance for astrophysics. Our final result is compact and suitable for numerical integration.

1 Introduction

The "worldline" or "Feynman-Schwinger" [1] representation of the one-loop effective action in scalar QED is

$$\Gamma[A] \;=\; -\int_0^\infty \frac{dT}{T}\,\mathrm{e}^{-m^2T}\int_P Dx(\tau)\,\mathrm{e}^{-\int_0^T d\tau[\frac{1}{4}\dot{x}^2+ie\dot{x}^\mu A_\mu(x(\tau))]}\,. \qquad (1)$$

Here m and T denote the mass and proper-time of the loop scalar, and $\int_P Dx(\tau)$ the path integral over closed loops in (Euclidean) spacetime with periodicity T in the proper-time. The above path integral

⋆e-mail: aftab.gu@gmail.com
⋆⋆e-mail: ahmadiniaz@ibs.re.kr (speaker)
⋆⋆⋆e-mail: olindo.corradini@unimore.it
⋆⋆⋆⋆e-mail: sangkim@kunsan.ac.kr
†e-mail: schubert@ifm.umich.mx

can be converted into the following master formula for the N-photon (off-shell) amplitudes [2, 3]

$$\Gamma(k_1,\varepsilon_1;\ldots;k_N,\varepsilon_N) = -(-ie)^N(2\pi)^D\delta^D\Big(\sum_{i=1}^N k_i\Big)\int_0^\infty \frac{dT}{T}(4\pi T)^{-\frac{D}{2}}\mathrm{e}^{-m^2T}\prod_{i=1}^N\int_0^T d\tau_i$$
$$\times\exp\Big\{\Big[\sum_{i,j=1}^N\Big[\frac{1}{2}G_{Bij}k_i\cdot k_j - i\dot G_{Bij}\varepsilon_i\cdot k_j + \frac{1}{2}\ddot G_{Bij}\varepsilon_i\cdot\varepsilon_j\Big]\Big\}\Big|_{\varepsilon_1\varepsilon_2\cdots\varepsilon_N}. \tag{2}$$

Here G_{Bij}, $\dot G_{Bij}$ and $\ddot G_{Bij}$ are the "bosonic" worldline Green's function and its first and second derivatives with respect to the first variable ($G_{Bij}\equiv G_B(\tau_i-\tau_j)$ etc):

$$G_{Bij} = |\tau_i-\tau_j| - \frac{(\tau_i-\tau_j)^2}{T}\ ,\quad \dot G_{Bij} = \mathrm{sign}(\tau_i-\tau_j) - 2\frac{\tau_i-\tau_j}{T}\ ,\quad \ddot G_{Bij} = 2\delta(\tau_i-\tau_j) - \frac{2}{T}. \tag{3}$$

G_{Bij} is the Green's function for the operator $\frac{d^2}{d\tau^2}$ with the periodic boundary condition, also obeying the "string-inspired" (SI) boundary conditions: $\int_0^T d\tau_i G_{Bij} = \int_0^T d\tau_j G_{Bij} = 0$. The notation $\big|_{\varepsilon_1\varepsilon_2\cdots\varepsilon_N}$ means taking linear terms in each polarization vector after expanding the exponential. This master formula is extremely compact and well-organized with respect to gauge invariance [2–4]. See [3] for its generalization to the spinor QED case, [6, 7] for multi-loop generalizations, and [5] for a review on the worldline formalism. For scalar QED in the presence of a constant background field $F_{\mu\nu}$, a similar "Bern-Kosower type" master formula was obtained in [8, 9]:

$$\Gamma(k_1,\varepsilon_1;\ldots;k_N,\varepsilon_N) = -(-ie)^N(2\pi)^D\delta^D\Big(\sum_{i=1}^N k_i\Big)\int_0^\infty\frac{dT}{T}(4\pi T)^{-\frac{D}{2}}\mathrm{e}^{-m^2T}\det{}^{-\frac{1}{2}}\Big[\frac{\sin(\mathcal{Z})}{\mathcal{Z}}\Big]$$
$$\times\prod_{i=1}^N\int_0^T d\tau_i\exp\Big\{\sum_{i,j=1}^N\Big[\frac{1}{2}k_i\cdot\mathcal{G}_{Bij}\cdot k_j - i\varepsilon_i\cdot\dot{\mathcal{G}}_{Bij}\cdot k_j + \frac{1}{2}\varepsilon_i\cdot\ddot{\mathcal{G}}_{Bij}\cdot\varepsilon_j\Big]\Big\}\Big|_{\varepsilon_1\varepsilon_2\cdots\varepsilon_N}, \tag{4}$$

where $\mathcal{Z} = eFT$. The main differences between this master formula and the vacuum one in Eq. (2) is the additional factor $\det^{-1}\big[\frac{\sin\mathcal{Z}}{\mathcal{Z}}\big]$, and the replacement of the G_{Bij} Green's function with a new one $\mathcal{G}_{Bij}$ that is a function of the external field:

$$\mathcal{G}_{Bij} = \frac{T}{2}\mathcal{Z}^2\Big(\frac{\mathcal{Z}}{\sin(\mathcal{Z})}\mathrm{e}^{-i\mathcal{Z}\dot G_{Bij}} + i\mathcal{Z}\dot G_{Bij} - 1\Big). \tag{5}$$

This Green's function also follows the SI boundary conditions. This master formula, and particularly its extension to spinor QED [5, 10], are generally more efficient for constant field calculations in QED than the standard method based on Feynman diagrams. They have already been applied to photonic processes in a constant field like vacuum polarization [5, 10, 11], photon splitting (in a magnetic field) [5, 12] and the two loop Euler-Heisenberg Lagrangian [5, 9]. Much less has been done for the corresponding amplitudes with an open line (propagator) instead of the closed loop, either in vacuum or in an external field. In 1996 Daikouji et al. [13] obtained the following "Bern-Kosower type" master formula for the N-photon dressed scalar propagator in vacuum (see also [14] for a recent rederivation):

$$D^{pp'}(k_1,\varepsilon_1;\cdots;k_N,\varepsilon_N) = (-ie)^N(2\pi)^D\delta^D\Big(p+p'+\sum_{i=1}^N k_i\Big)\int_0^\infty dT\,\mathrm{e}^{-m^2T}$$
$$\times\prod_{i=1}^N\int_0^T d\tau_i\,\mathrm{e}^{-T[p+\frac{1}{T}\sum_{i=1}^N(k_i\tau_i-i\varepsilon_i)]^2+\sum_{i,j=1}^N[\Delta_{ij}k_i\cdot k_j-2i{}^\bullet\!\Delta_{ij}\varepsilon_i\cdot k_j-{}^\bullet\!\Delta^\bullet_{ij}\varepsilon_i\cdot\varepsilon_j]}\Big|_{\varepsilon_1\varepsilon_2\cdots\varepsilon_N}, \tag{6}$$

where Δ_{ij} is the open line Green's function adopted to Dirichlet boundary conditions (DBC)

$$\Delta(\tau_i,\tau_j) = \frac{\tau_i\tau_j}{T} + \frac{|\tau_i-\tau_j|}{2} - \frac{\tau_i+\tau_j}{2} \quad , \quad \Delta(0,\tau_j) = \Delta(T,\tau_j) = \Delta(\tau_i,0) = \Delta(\tau_i,T) = 0 \,. \tag{7}$$

For this Green's function, we need to distinguish between derivatives with respect to first and second

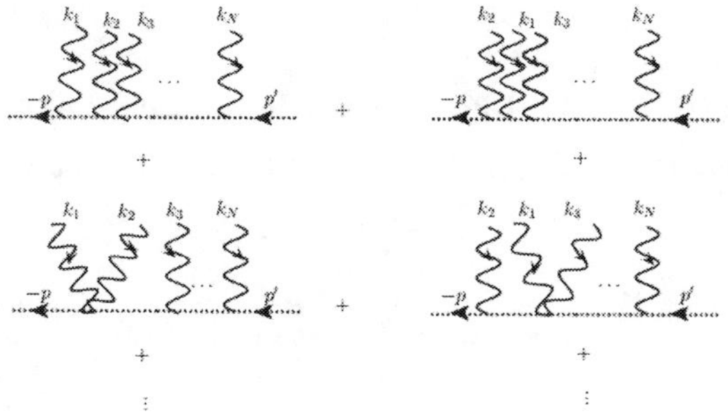

Figure 1. Multi-photon Compton-scattering diagram.

argument, since the DBC break the translation invariance in proper-time. A convenient notation is [15] to use left and right dots to indicate derivatives with respect to the first and the second argument, respectively:

$$\begin{aligned} {}^\bullet\Delta_{ij} &= \frac{\tau_j}{T} + \frac{1}{2}\mathrm{sign}(\tau_i-\tau_j) - \frac{1}{2} \quad , \quad {}^\bullet\Delta^\bullet_{ij} = \frac{1}{T} - \delta(\tau_i-\tau_j)\,, \\ \Delta^\bullet_{ij} &= \frac{\tau_i}{T} - \frac{1}{2}\mathrm{sign}(\tau_i-\tau_j) - \frac{1}{2} \quad , \quad 2\Delta_{ij} = G_{Bij} - G_{Bi0} - G_{B0j}\,, \end{aligned} \tag{8}$$

where the last formula gives a relation between Δ_{ij} and G_{Bij} [16]. The organization of the paper is as follows. In section 2 we include the constant field into the Green's function of a scalar propagator, then dress it with N photons to obtain master formulas in configuration and momentum space. In section 3 we work the momentum space formula out for $N = 2$ to find a compact representation of the Compton scattering cross section in a general constant field. In section 4 we summarize our discussion.

2 Scalar propagator in a constant field

In the previous section we have presented a master formula for an open line in vacuum, in the following we consider a constant external field to be included in this master formula. We first start with a free propagator in a constant field.

2.1 Free propagator

A constant field in the Fock-Schwinger gauge is written as

$$A_\mu(y) = -\frac{1}{2}F_{\mu\nu}(y-x')^\nu \quad , \quad A(y=x') = 0 \;\; (x' \text{ is the reference point of the potential})\,. \tag{9}$$

After decomposing the arbitrary trajectory $x(\tau)$ into a straight-line part and a fluctuation part

$$x(\tau) = x' + \frac{\tau}{T}(x-x') + q(\tau) \quad , \quad q(T) = q(0) = 0 \quad \text{(DBC)}\,, \tag{10}$$

and using (9) and $Q^\mu = \int_0^T d\tau q^\mu(\tau)$, with some manipulations the free scalar propagator can be written as

$$\begin{aligned} D^{xx'}[A] &= \int_0^\infty dT\, \mathrm{e}^{-m^2T} \int_{x(0)=x'}^{x(T)=x} Dx \mathrm{e}^{-\int_0^T d\tau\left[\frac{1}{4}\dot{x}^2+ie\dot{x}\cdot A(x)\right]} \\ &= \int_0^\infty dT\, \mathrm{e}^{-m^2T-\frac{(x-x')^2}{4T}} \int Dq(\tau)\, \mathrm{e}^{-\int_0^T d\tau \frac{1}{4}q\left(-\frac{d^2}{d\tau^2}+2ieF\frac{d}{d\tau}\right)q+\frac{ie}{T}(x-x')FQ}, \end{aligned} \quad (11)$$

which is gaussian. Now besides the free path integral normalization $\int Dq(\tau)\mathrm{e}^{-\int_0^T d\tau q\left(-\frac{1}{4}\frac{d^2}{d\tau^2}\right)q} = (4\pi T)^{-\frac{D}{2}}$ [5] we have to take into account the ratio of the field-dependent and free path integral normalization:

$$\frac{\mathrm{Det}'^{-\frac{1}{2}}_P\left(-\frac{1}{4}\frac{d^2}{d\tau^2}+\frac{1}{2}ieF\frac{d}{d\tau}\right)}{\mathrm{Det}'^{-\frac{1}{2}}_P\left(-\frac{1}{4}\frac{d^2}{d\tau^2}\right)} = \mathrm{Det}'^{-\frac{1}{2}}_P\left(\mathbb{1}-2ieF\left(\frac{d}{d\tau}\right)^{-1}\right) = \det{}^{-\frac{1}{2}}\left[\frac{\sin(\mathcal{Z})}{\mathcal{Z}}\right], \quad (12)$$

where we have eliminated the zero mode which exists in the path integral by putting a 'prime'. Now we have to introduce a new Green's function which is again related to the one with SI boundary conditions in (5) for a constant field:

$$\underset{\smile}{\Delta}_{ij} \equiv \langle\tau_i \mid \left(\frac{d^2}{d\tau^2}-2ieF\frac{d}{d\tau}\right)^{-1} \mid \tau_j\rangle_{\mathrm{DBC}} = \frac{1}{2}\left(\mathcal{G}_{Bij}-\mathcal{G}_{Bi0}-\mathcal{G}_{B0j}+\mathcal{G}_{B00}\right). \quad (13)$$

After using this Green's function in the usual completing-the-square procedure, we get the following well-known representation of the scalar propagator in a constant external field in x-space:

$$\begin{aligned} D^{xx'}(F) &= \int_0^\infty dT \mathrm{e}^{-m^2T}\mathrm{e}^{-\frac{(x-x')^2}{4T}}(4\pi T)^{-\frac{D}{2}}\det{}^{-\frac{1}{2}}\left[\frac{\sin(eFT)}{eFT}\right]\mathrm{e}^{\int_0^T d\tau\int_0^T d\tau' \frac{ie}{T}(x-x')F\underset{\smile}{\Delta}(\tau,\tau')\frac{ie}{T}F(x-x')} \\ &= \int_0^\infty dT\mathrm{e}^{-m^2T}(4\pi T)^{-\frac{D}{2}}\det{}^{\frac{1}{2}}\left[\frac{\mathcal{Z}}{\sin\mathcal{Z}}\right]\mathrm{e}^{-\frac{1}{4T}(x-x')\mathcal{Z}\cot\mathcal{Z}(x-x')}. \end{aligned} \quad (14)$$

After Fourier transforming its representation in momentum space gives

$$D^{pp'}(F) = (2\pi)^D\delta(p+p')D(p,F) \quad , \quad D(p,F) = \int_0^\infty dT\,\mathrm{e}^{-m^2T}\frac{\mathrm{e}^{-Tp\left(\frac{\tan\mathcal{Z}}{\mathcal{Z}}\right)p}}{\det^{\frac{1}{2}}[\cos\mathcal{Z}]}. \quad (15)$$

2.2 Dressed propagator

Now, if we dress the scalar propagator with N photons in addition to the constant field, we write the potential as $A = A_{\mathrm{ext}} + A_{phot}$ which the former is the same as in (9) and the later is written as a sum of plane waves, each photon is represented as a vertex operator:

$$A^\mu_{\mathrm{phot}} = \sum_{i=1}^N \varepsilon_i^\mu \mathrm{e}^{ik_i\cdot x} \quad , \quad V^A[k,\varepsilon] = \int_0^T d\tau\varepsilon\cdot\dot{x}(\tau)\,\mathrm{e}^{ik\cdot x(\tau)} = \int_0^T d\tau\mathrm{e}^{ik\cdot x(\tau)+\varepsilon\cdot\dot{x}(\tau)}\Big|_{\mathrm{lin}(\varepsilon)}. \quad (16)$$

The path integral representation of the scalar propagator in a constant field dressed with N photons is written as

$$D^{xx'}(F|k_1,\varepsilon_1;\cdots;k_N,\varepsilon_N) = (-ie)^N\int_0^\infty dT\mathrm{e}^{-m^2T}\int_P Dx\mathrm{e}^{-\int_0^T d\tau\left[\frac{1}{4}\dot{x}^2+ie\dot{x}\cdot A_{\mathrm{ext}}(x)\right]}V^A[k_1,\varepsilon_1]\cdots V^A[k_N,\varepsilon_N]. \quad (17)$$

After applying the path decomposition (10) and completing the square we get ($x_- = x - x'$)

$$
\begin{aligned}
D^{xx'}(F|k_1,\varepsilon_1;\cdots;k_N,\varepsilon_N) &= (-ie)^N \int_0^\infty dT \mathrm{e}^{-m^2T}(4\pi T)^{-\frac{D}{2}} \det{}^{\frac{1}{2}}\Big[\frac{\mathcal{Z}}{\sin\mathcal{Z}}\Big]\mathrm{e}^{-\frac{1}{4T}x_-\mathcal{Z}\cot\mathcal{Z}x_-} \\
&\quad \times \int_0^T d\tau_1 \cdots \int_0^T d\tau_N \, \mathrm{e}^{\sum_{i=1}^N\left(\varepsilon_i\cdot\frac{x_-}{T}+ik_i\cdot\frac{x_-\tau_i}{T}+ik_i\cdot x'\right)}
\end{aligned}
$$

$$
\times\exp\Big[\sum_{i,j=1}^N\Big(k_i \underset{\smile}{\Delta}_{ij}\, k_j - 2i\varepsilon_i\, {}^{\bullet}\underset{\smile}{\Delta}_{ij}\, k_j - \varepsilon_i\, {}^{\bullet}\underset{\smile}{\Delta}{}^{\bullet}{}_{ij}\, \varepsilon_j\Big) + \frac{2e}{T}x_- \sum_{i=1}^N\Big(F\, {}^{\circ}\underset{\smile}{\Delta}_i\, k_i - iF\, {}^{\circ}\underset{\smile}{\Delta}{}^{\bullet}{}_i\, \varepsilon_i\Big)\Big]\Big|_{\varepsilon_1\varepsilon_2\cdots\varepsilon_N}, \tag{18}
$$

where a left (right) 'open circle' on $\underset{\smile}{\Delta}(\tau,\tau')$ denotes an integral $\int_0^T d\tau$ ($\int_0^T d\tau'$). This x-space master formula was obtained by McKeon and Sherry for a purely magnetic field [17]. Now we can Fourier transform to momentum space to get

$$
\begin{aligned}
D^{pp'}(F\mid k_1,\varepsilon_1;\cdots;k_N,\varepsilon_N) &= (-ie)^N(2\pi)^D\delta\Big(p+p'+\sum_{i=1}^N k_i\Big)\int_0^\infty dT\mathrm{e}^{-m^2T}\frac{1}{\det^{\frac{1}{2}}[\cos\mathcal{Z}]} \\
&\quad\times\int_0^T d\tau_1\cdots\int_0^T d\tau_N\,\mathrm{e}^{\sum_{i,j=1}^N\left(k_i\underset{\smile}{\Delta}_{ij}k_j-2i\varepsilon_i{}^{\bullet}\underset{\smile}{\Delta}_{ij}k_j-\varepsilon_i{}^{\bullet}\underset{\smile}{\Delta}{}^{\bullet}{}_{ij}\varepsilon_j\right)}\mathrm{e}^{-Tb(\frac{\tan\mathcal{Z}}{\mathcal{Z}})b}\Big|_{\varepsilon_1\varepsilon_2\cdots\varepsilon_N}, \\
b &\equiv p+\frac{1}{T}\sum_{i=1}^N\Big[\Big(\tau_i-2ieF\,{}^{\circ}\underset{\smile}{\Delta}_i\Big)k_i-i\Big(1-2ieF\,{}^{\circ}\underset{\smile}{\Delta}{}^{\bullet}{}_i\Big)\varepsilon_i\Big]\,.
\end{aligned} \tag{19}
$$

Diagrammatically, these master formula hold the full information on the set of Feynman diagrams appearing in Fig. 1, with the full scalar propagator in a constant field (usually indicated by a double line).

3 Compton scattering in a constant field

Now, let us apply our momentum space master formula in (19) to work out the Compton scattering cross section, corresponding to $N = 2$. Notice that our master formula describes the *untruncated* dressed propagator so in applying it to physical processes one has to truncate the external scalars, we define the matrix element $\mathcal{T}$ as

$$
\mathcal{T}(F|k_1,\varepsilon_1;\cdots;k_N,\varepsilon_N) = \frac{D^{pp'}(F|k_1,\varepsilon_1;\cdots;k_N,\varepsilon_N)}{D(p,F)D(p',F)}\,. \tag{20}
$$

Moreover, it will be convenient to Wick rotate from Euclidean to Minkowski space (see [18] for our conventions). Expanding out the exponentials in (19) up to terms linear in both polarizations we arrive at

$$
D^{pp'}(F\mid k_1,\varepsilon_1;k_2,\varepsilon_2) = e^2\int_0^\infty dT\frac{\mathrm{e}^{-m^2T}}{\det^{\frac{1}{2}}[\cos\mathcal{Z}]}\int_0^T d\tau_1 d\tau_2\mathrm{e}^{-Tb_0(\frac{\tan\mathcal{Z}}{\mathcal{Z}})b_0+\sum_{i,j=1}^2 k_i\underset{\smile}{\Delta}_{ij}k_j}\varepsilon_1 M_{12}\varepsilon_2\,, \tag{21}
$$

with

$$
\begin{aligned}
b_0 &\equiv p+\frac{1}{T}\sum_{i=1}^2\Big(\tau_i-2ieF\,{}^{\circ}\underset{\smile}{\Delta}_i\Big)k_i\,, \\
M_{12} &\equiv 2\,{}^{\bullet}\underset{\smile}{\Delta}{}^{\bullet}{}_{12}-\frac{2}{T}\Big(1+2ie\,{}^{\circ}\underset{\smile}{\Delta}{}^{\bullet T}_{1}F\Big)\frac{\tan\mathcal{Z}}{\mathcal{Z}}\Big(1-2ieF\,{}^{\circ}\underset{\smile}{\Delta}{}^{\bullet}{}_2\Big) \\
&\quad+4\Big[(1+2ie\,{}^{\circ}\underset{\smile}{\Delta}{}^{\bullet T}_{1}F)\frac{\tan\mathcal{Z}}{\mathcal{Z}}b_0-\sum_{i=1}^2{}^{\bullet}\underset{\smile}{\Delta}_{1i}\,k_i\Big]\Big[b_0\frac{\tan\mathcal{Z}}{\mathcal{Z}}\Big(1-2ieF\,{}^{\circ}\underset{\smile}{\Delta}{}^{\bullet}{}_2\Big)-\sum_{i=1}^2 k_i\underset{\smile}{\Delta}{}^{\bullet}{}_{i2}\Big]\,.
\end{aligned} \tag{22}
$$

To obtain the cross section we square and sum over the photon polarizations via $\sum_{\rm pol} \varepsilon_i^{*\mu} \varepsilon_i^{\nu} \to g^{\mu\nu}$ which finally leads to the following expression for the Compton cross section:

$$\sum_{\rm pol} \mathcal{T}^* \mathcal{T} = \frac{e^4}{|D(p,F)^2 | D(p',F)^2} \int_0^\infty dT' \frac{{\rm e}^{-m^2 T'}}{{\rm det}^{\frac{1}{2}} [\cos \mathcal{Z}']} \int_0^{T'} d\tau_1' \int_0^{T'} d\tau_2' \, {\rm e}^{-T' b_0^* (\frac{\tan \mathcal{Z}'}{\mathcal{Z}'}) b_0^* + \sum_{i,j=1}^2 k_i \underset{\smile}{\Delta}'_{ij} k_j}$$
$$\times \int_0^\infty dT \frac{{\rm e}^{-m^2 T}}{{\rm det}^{\frac{1}{2}} [\cos \mathcal{Z}]} \int_0^{T} d\tau_1 \int_0^{T} d\tau_2 \, {\rm e}^{-T b_0 (\frac{\tan \mathcal{Z}}{\mathcal{Z}}) b_0 + \sum_{i,j=1}^2 k_i \underset{\smile}{\Delta}_{ij} k_j} \, {\rm tr}(M_{12}'^{\dagger} M_{12}) \,. \tag{23}$$

4 Summary

In this short report we have extended our previous master formulas for multiphoton amplitudes in vacuum in scalar QED [14] to include a constant external field using the worldline formalism, see [18] for more details. Our master formula is valid off-shell, and combines the various orderings of the N photons along the scalar line. We have worked out the cross section for the case of $N = 2$, that is linear Compton scattering in the field, reaching a compact integral representation suitable for numerical evaluation. This cross section is, for the magnetic field case, of importance for astrophysics, but appears to have been studied so far only in the strong-field limit [19].

References

[1] R.P. Feynman, Phys. Rev. **80**, 440 (1950).
[2] Z. Bern and D. A. Kosower, Phys. Rev. Lett. **66**, 1669 (1991); Nucl. Phys. **B 379**, 451 (1992).
[3] M. J. Strassler, Nucl. Phys. **B 385**, 145 (1992).
[4] N. Ahmadiniaz, C. Schubert and V.M. Villanueva, JHEP **1301** (2013) 312.
[5] C. Schubert, Phys. Rept. **355**, 73 (2001).
[6] M. G. Schmidt and C. Schubert, Phys. Lett. B **331** 69 (1994), hep-th/9403158.
[7] M.G. Schmidt and C. Schubert, Phys. Rev. D 53, 2150 (1996), hep-th/9410100.
[8] R. Zh. Shaisultanov, Phys. Lett. **B 378**, 354 (1996).
[9] M. Reuter, M. G. Schmidt and C. Schubert, Ann. Phys. (N.Y.) **259**, 313 (1997).
[10] C. Schubert, Nucl. Phys. **B 585**, 407 (2000).
[11] W. Dittrich and R. Shaisultanov, Phys. Rev. **D 62**, 045024 (2000).
[12] S.L. Adler and C. Schubert, Phys. Rev. Lett. **77**.
[13] K. Daikouji, M. Shino, Y. Sumino, Phys. Rev. **D 53**, 4598 (1996).
[14] N. Ahmadiniaz, A. Bashir and C. Schubert, Phys. Rev. **D 93**, 045023 (2016).
[15] F. Bastianelli and P. van Nieuwenhuizen, Cambridge University press, 2006.
[16] D. Fliegner, M.G. Schmidt and C. Schubert, Z. Phys. **C 64**, 111 (1994).
[17] D.G.C. McKeon and T.N. Sherry, Mod. Phys. Lett. **A9**, 2167 (1994).
[18] A. Ahmad, N. Ahmadiniaz, O. Corradini, S. P. Kim and C. Schubert, Nucl. Phys. **B 919**,9 (2017).
[19] J. K. Daugherty and A. K. Harding, Astrophys. J. **309**, 362 (1986).

EPJ Web of Conferences **168**, 04004 (2018) https://doi.org/10.1051/epjconf/201816804004
Joint International Conference of ICGAC-XIII and IK-15 on Gravitation, Astrophysics and Cosmology

Baryonic Force for Accelerated Cosmic Expansion and Generalized U_{1b} Gauge Symmetry in Particle-Cosmology

Mehbub Khan[1,⋆], *Yun* Hao[1,⋆⋆], and *Jong-Ping* Hsu[1,⋆⋆⋆]

[1] *Department of Physics, University of Massachusetts Dartmouth, North Dartmouth, MA 02747, USA*

Abstract. Based on baryon charge conservation and a generalized Yang-Mills symmetry for Abelian (and non-Abelian) groups, we discuss a new baryonic gauge field and its linear potential for two point-like baryon charges. The force between two point-like baryons is repulsive, extremely weak and independent of distance. However, for two extended baryonic systems, we have a dominant linear force $\propto r$. Thus, only in the later stage of the cosmic evolution, when two baryonic galaxies are separated by an extremely large distance, the new repulsive baryonic force can overcome the gravitational attractive force. Such a model provides a gauge-field-theoretic understanding of the late-time accelerated cosmic expansion. The baryonic force can be tested by measuring the accelerated Wu-Doppler frequency shifts of supernovae at different distances.

1 Introduction

In particle-cosmology, apart from the well-known forces in quantum chromodynamics, the electroweak theory, and gravity, there are other possible forces related to quark confinement and the accelerated cosmic expansion. The quark confinement force could be related to a linear potential. The force responsible for the accelerated cosmic expansion should be extremely weak and repulsive between baryon matter. Collecting these together, a total-unified group G_{tu} in particle cosmology [1], consistent with established conservation laws, might be $G_{tu} = T_4 \times [SU_3]_{color} \times \{SU_2 \times U_1\} \times U_{1b} \times U_{1\ell}$, where U_{1b} and $U_{1\ell}$ are associated to baryon and lepton charges (or numbers) respectively. For simplicity, we ignore $U_{1\ell}$ here, for it can be included without difficulty.

Let us concentrate on the baryon sector U_{1b} of the particle-cosmology model based on the following two postulates:

(i) a 'generalized' gauge symmetry [2, 3] for conserved baryon charges with the U_{1b} group,

(ii) the invariant Lagrangian $\propto \partial^\mu B_{\mu\lambda} \partial_\nu B^{\nu\lambda}$ for the baryon dynamics with the gauge fields B_μ.

This particle-cosmology model could provide an understanding of the late-time accelerated cosmic expansion based on the U_{1b} gauge fields in flat space-time rather than the unknown dark energy. Such a particle-cosmology model does not require 68% of dark energy to explain the accelerated expansion. Its prediction of the specific acceleration could be tested in the future.

The generalization of gauge symmetry with exact conservation law has been discussed by introducing vector gauge functions [2, 3] and Hamilton's characteristic (phase) functions [4, 5] into the new

⋆e-mail: mkhan4@umassd.edu
⋆⋆e-mail: yhao@umassd.edu
⋆⋆⋆e-mail: jhsu@umassd.edu

gauge transformations. It differs from the usual gauge transformation, which involves scalar gauge function and ordinary phase function. They lead to new massless gauge fields which satisfy 4th-order filed equations. These 4th-order field equations in the static limit give us a linear static potential and a constant repulsive force between two point-like baryon charges.

In the special case, when the vector gauge functions $\omega_\mu(x)$ can be expressed in terms of the space-time derivative of scalar functions, $\omega_\mu(x) = \partial_\mu\omega(x)$, the generalized gauge symmetry reduces to the usual gauge symmetry [2]. As a result, one has the usual second-order differential equation for massless fields, which has been discussed by Lee and Yang in 1955 [6]. This usual Lee-Yang U_1 gauge field for conserved baryon charge is identical to that of the electromagnetic theory with the conserved electric charge. Eötvös experiments gave a stringent limit of the strength of such a 'Lee-Yang force.' The inverse-square force between baryons in nucleus would violate the equivalence principle and, hence, the force should be roughly a million times weaker than the gravitational force. Thus, such an extremely weak Lee-Yang force does not have any observable results in the laboratory or in the cosmos. However, the new 'generalized' gauge symmetry for baryon charges with the U_{1b} group will have important implications in cosmos and may be called 'general Lee-Yang gauge symmetry.'

2 Generalized U_{1b} gauge symmetry

Let us consider a physical system with a gauge field $B_\mu(x)$ and a quark field $q(x)$ with spin 1/2 . We define the generalized U_{1b} gauge transformations for these two fields [2],

$$B'_\mu(x) = B_\mu(x) + \omega_\mu(x), \qquad \mu = 0,1,2,3, \qquad c = \hbar = 1, \tag{1}$$

$$q'(x) = e^{-iP_\omega(x)}q(x), \qquad \overline{q}'(x) = \overline{q}(x)e^{iP_\omega(x)}, \qquad P_\omega(x) = g_b\left(\int_{x'_o}^{x'_e=x}\omega_\lambda(x')dx'^\lambda\right)_{Le}, \tag{2}$$

where $P_\omega(x)$ is a 'characteristic function'[1] and the subscript Le in (2) implies that the vector gauge function $\omega(x)_\mu$ satisfies the Lagrange equation,

$$d\omega_\lambda(x') - \frac{\partial\omega_\mu(x')}{\partial x'^\lambda}dx'^\mu = 0. \tag{3}$$

We can show that the variation of the action functional P_ω is actually a function of the coordinates. The reasons are that the initial point x_o is fixed so that $\delta x'(x'_o) = 0$, the end point is variable, $x'_e = x$, and that the gauge function $\omega_\lambda(x')$ satisfies (3). Furthermore, in order to have the general gauge invariant Lagrangian for the baryon dynamics, the vector gauge functions $\omega_\mu(x)$ are required to satisfy the constraint,

$$\partial^2\omega_\mu(x) - \partial_\mu\partial^\lambda\omega_\lambda(x) = 0, \qquad \partial^2 = \partial_\mu\partial^\nu\eta^{\mu\nu}, \qquad \eta^{\mu\nu} = (1,-1,-1,-1). \tag{4}$$

The requirement (3) is necessary for the integral in the variation of $P_\omega(x)$ to vanish, i.e., $+g_b\int_{x'_o}^{x}\left[(\partial\omega_\mu(x')/\partial x'^\lambda)dx'^\mu - d\omega_\lambda(x')\right]\delta x'^\lambda = 0$, so that we have [4, 5]

$$\delta P_\omega(x) = g_b\omega_\mu(x)\delta x^\mu, \qquad or \qquad \partial_\mu P_\omega = g_b\omega_\mu(x). \tag{5}$$

This equation (5) for the U_{1b} phase 'characteristic function $P_\omega(x)$ is crucial for a Lagrangian to have the general Lee-Yang symmetry. The U_{b1} covariant derivative $\Delta_{b\mu}$ is defined as usual

$$\Delta_{b\mu} = \partial_\mu + ig_bB_\mu. \tag{6}$$

[1] In the literature, it is called Hamilton's characteristic function, which is well-defined, independent of path, and a local function of x, as discussed in refs. 4 and 5.

In the generalized gauge symmetry, we have the vector gauge functions $\omega_\mu(x)$, which have four functions instead of a mere scalar function in the usual gauge symmetry. These four gauge functions are required to satisfy four partial differential equations in (4), therefore the choice of these four vector gauge functions is not completely arbitrary. However, the gauge functions $\omega_\lambda(x)$ in the generalized gauge transformations still involve infinitely many functions.

One can verify that $\overline{\psi}(\partial_\mu + ig_b B_\mu)\psi$ is invariant under the general U_{1b} transformations (1)-(3),

$$\overline{\psi}'(\partial_\mu + ig_b B'_\mu)\psi' = \overline{\psi}(\partial_\mu + ig_b B_\mu)\psi, \tag{7}$$

where we have used equations (1), (2), (3) and (6).

The general U_{1b} gauge covariant derivative, $\partial_\mu + ig_b B_\mu$, in (6) has the same form as that in the usual U_1 gauge theory QED. The U_{1b} gauge curvature is, as usual, determined by the commutator of the covariant derivative,

$$[\partial_\mu + ig_b B_\mu, \partial_\nu + ig_b B_\nu] = ig_b B_{\mu\nu}, \qquad B_{\mu\nu} = \partial_\mu B_\nu - \partial_\nu B_\mu. \tag{8}$$

Under the general U_{1b} transformation (3), the U_{1b} gauge curvature $B_{\mu\nu}$ turns out not to be invariant, $B'_{\mu\nu} = B_{\mu\nu} + \partial_\mu \Lambda_\nu - \partial_\nu \Lambda_\mu \neq B_{\mu\nu}$. However, the divergence of the gauge curvature, $\partial^\mu B_{\mu\nu}$, is invariant,

$$\partial^\mu B'_{\mu\nu} = \partial^\mu B_{\mu\nu}, \tag{9}$$

because the vector functions $\omega_\mu(x)$ satisfy the constraint equation (4). The result (9) enables us to write down the general U_{1b} gauge invariant Lagrangian,

$$L = \frac{L_s^2}{2}\partial^\mu B_{\mu\lambda}\partial_\nu B^{\nu\lambda} + i\overline{q}[\gamma^\mu(\partial_\mu - ig_b B_\mu) - M]q, \tag{10}$$

where L_s denotes a universal scale with the dimension of length (for a physical system with a general Lee-Yang symmetry). The fourth order gauge field equation can be derived from (10),

$$\partial^2\partial^2 B_\mu = \frac{g_b}{L_s^2}\overline{q}\gamma_\mu q, \tag{11}$$

where we have chosen the gauge condition $\partial_\mu B^\mu = 0$. One can replace the source term in (11) for $\mu = 0$ by a point baryon charge $g_b\delta^3(\mathbf{r})/L_s^2$, and find the static potential $B_0(\mathbf{r}) = -g_b r/(8\pi L_s^2)$. This linear potential B_0 leads to a constant force, i.e., independent of distance between two point baryon charges. The four-dimensional symmetry framework for the baryon dynamics can be formulated in a broader view with the coordinate $x^\mu = (w, x, y, z)$, so that it can be generalized to a general frame (inertial and non-inertial) on the basis of the limiting Lorentz-Poincaré invariance [7, 8].

3 Calculations of cosmic force between baryonic systems

Now let us consider a simple model of particle-cosmology to investigate cosmic baryon force in detail. Although the force between two point baryonic charges is distance-independent, one cannot simply assume that the total baryonic charges in a uniform sphere are concentrated at the center of the sphere and that the resultant force is distance-independent. The situation for baryon force differs from that of the well-known inverse-square force.

The cosmic baryon force can be calculated in two steps: First, we approximate the supernova as a point in a huge cluster of galaxies, which is treated as a big sphere of uniform baryon charges with a very large radius R_o.[2] The resultant force is no longer distance-independent, but the sphere of baryon charges can be considered as concentrated at the center of the sphere. Second, we replace the point-like supernova by a sphere of baryon charge with a radius $R_s \ll R_o$. The result could provide a way to test the cosmic baryon force by measuring the distance-dependence of accelerations of supernovae.

We may picture a supernova having a mass m_s and a radius R_s inhabiting in a sphere of, say, approximately 100 billions of galaxies. These baryonic galaxies can be idealized as points uniformly spread out in the entire sphere with a baryon mass density ρ. The mass of this sphere is $M = \rho 4\pi R_o^3/3$. Each nucleon has three quark charges, i.e., $3g_b$. Suppose the point-like supernova has a mass m and

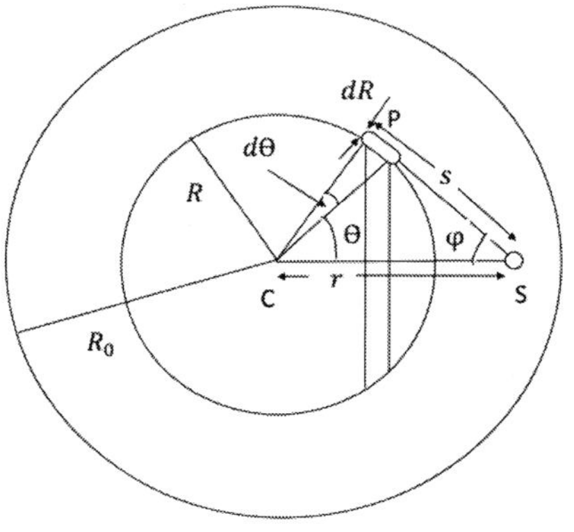

Fig. 1. A schematic diagram for calculations of the cosmic baryon force F_{CS1} between a gigantic uniform sphere of baryon galaxies and a point-like baryon supernova.

baryon charges $3g_b m/m_p$, where m_p stands for the mass of a baryon (proton or neutron). In Fig. 1, we consider a thin uniform shell of radius R , where R is less than R_o . The thickness on this shell is dR . Here the supernova is outside the shell at a distance $r > R$. A point P is on the ring with radius R. The angle SCP in the diagram is θ . The mass dM_r of the ring is

$$dM_r = \rho 2\pi R^2 dR \ \sin\theta d\theta,$$

where $2\pi R \sin\theta$ is the circumference of the ring. At point P , let us consider an infinitesimal element. The effective repulsive force in between this infinitesimal element and the point-like supernova is dF_p. The horizontal component of the force dF_p is given by $dF_p \cos\phi$. The vertical components of the force dF_p are cancelled completely due to the symmetry involved in the geometry we are considering here. We calculate the total force dF_{CS} in the horizontal direction CS from the entire ring on the point-like supernova at S. Using the equation for the constant force between two point-like baryonic charges, we find

$$dF_{CS} = A\int dR \int_0^{\pi} \left[R^2 \cos\phi \right] \sin\theta \, d\theta = A \int \int_{R-r}^{R+r} [dF] \frac{s}{Rr} ds, \qquad A = \frac{(3g_b)^2 m\rho}{4L_s^2 m_p^2}, \qquad (12)$$

$$dF \equiv dR \, R^2 \left[\frac{s^2 + r^2 - R^2}{2sr} \right], \quad s^2 = R^2 + r^2 - 2Rr \cos\theta, \quad R^2 = s^2 + r^2 - 2rs \cos\phi.$$

[2]The observable portion of the universe is approximate by a finite and gigantic sphere for our calculations.

The magnitude of the total repulsive force can be obtained [2] from the integrations of ds and dR in (12) for two cases:

$$(i) \qquad F_{CS1} = A \int_0^r \frac{RdR}{2r^2} \int_{r-R}^{r+R} ds(s^2 + r^2 - R^2), \quad r > R, \tag{13}$$

$$(ii) \qquad F_{CS2} = A \int_r^{R_o} \frac{R'dR'}{2r^2} \int_{R'-r}^{R'+r} ds(s^2 + r^2 - R'^2), \quad R_o > R' > r. \tag{14}$$

In the calculation of F_{CS2}, we modify Fig.1 such that the distance CS is smaller than $R'(= R)$. We obtain the distance-dependence of the cosmic baryon force $F_{CS1} + F_{CS2}$ to be $\propto [r/R_o - r^3/(5R_o^3)]$.

Next, we consider an extended supernova with a radius R_s and baryon mass density ρ_s. We may use Fig. 1 with the modifications: (a) C is the center of the supernova and replace R_o by R_s, and (b) move the point S to outside so that $r > R_s$, and consider all baryon charges of the gigantic sphere, $3g_bM/m_p$, as concentrated at S. After some tedious but straightforward calculations, the total effective cosmic baryon force (cbf) of the gigantic sphere with baryon galaxies exerted on a supernova with radius R_s is

$$F_{cbf} = \left(\frac{(3g_b)^2 mM}{8\pi L_s^2 m_p^2}\right)\left[\frac{r}{R_o} - \frac{r^3}{5R_o^3} - \frac{rR_s^3}{5R_o^3}\right], \qquad m = \frac{4\pi R_s^3 \rho_s}{3}, \tag{15}$$

which turns out to be dominated by a linear force because $R_s < r << R_o$.[3] In the limit $R_s \to 0$, the force in (15) reduces to that for a point-like supernova, i.e., $F_{CS1} + F_{CS2}$ in (13)-(14), as it should be. We may remark that (a) we use a simple model corresponds to a finite R_o for simplicity and (b) if one wishes, one may take the limit $R_o \to \infty$, provided that the density ρ also approaches zero such that $R_o^2\rho$ remains finite. We may remark that the trajectories of two macroscopic baryon objects have been numerically calculated [9]. For the case corresponding to the elliptical orbit in the Kepler problem, one has a 'distorted elliptical orbit' for two baryonic systems.

The dominant linear force in (15) predicts that the measured acceleration of the cosmic expansion should be approximately linear in r because R_o is very much larger than r and R_s. This prediction could be tested by measuring the accelerated frequency shifts[4] of supernovae at different distances.

References

[1] J. P. Hsu and L. Hsu, *Space-Time Symmetry and Quantum Yang-Mills Gravity* (Singapore, World Scientific, 2013) pp. 220-223.

[2] J. P. Hsu, Chinese Phys. C. **41**: 015101, (2017). arXiv:1609.00227 [physics.gen-ph].

[3] J. P. Hsu, Eur. Phys. J. Plus, **129**: 108, (2014). Note that Ω functions in (5)-(6) in this ref. [3] should be replaced by Hamilton's characteristic functions, $P_\omega(x)$, in eq. (2) of the present paper.

[4] W. Yourgrau and S. Mendelstam, *Variational Principles in Dynamics and Quantum Theory*, (Dover, 3rd ed. 1979) p. 50.

[5] L. Landau and E. Lifshitz, *The Classical Theory of Fields* (Addision-Wesley, 1951) p. 29.

[6] T. D. Lee and C. N.Yang, Phys. Rev. **98**:1501 (1955). J. Hsu, Mod. Phys. Lett. A **20**: 2855 (2005).

[7] J. P. Hsu and L. Hsu, Phys. Letters, A **196**: 1 (1994). Such a formulation of space-time transformations based solely on the first postulate is essential for generalization to non-inertial frames.

[8] See ref. 1, ch.2, pp. 23-24, pp. 38-42.

[9] M. Khan, Thesis (Physics Department, Univ. of Massachusetts Dartmouth, 2017).

[3]This new r-dependent force between baryonic systems with general U_{1b} symmetry was called the 'Okubo force' [2].
[4]The accelerated Wu-Doppler effect can be derived from the limiting Lorentz-Poincaré invariance [8].

EPJ Web of Conferences **168**, 04005 (2018) https://doi.org/10.1051/epjconf/201816804005
Joint International Conference of ICGAC-XIII and IK-15 on Gravitation, Astrophysics and Cosmology

Hot Accretion onto Black Holes with Outflow

Myeong-Gu Park[1,⋆] and *Du-Hwan* Han[1]

[1] *Kyungpook National University, Daegu 41566, KOREA*

Abstract. Classic Bondi accretion flow can be generalized to rotating viscous accretion flow. Study of hot accretion flow onto black holes show that its physical charateristics change from Bondi-like for small gas angular momentum to disk-like for Keperian gas angular momentum. Especially, the mass accretion rate divided by the Bondi accretion rate is proportional to the viscosity parameter alpha and inversely proportional to the gas angular momentum divided by the Keplerian angular momentum at the Bondi radius for gas angular momentum comparable to the Keplerian value. The possible presence of outflow will increase the mass inflow rate at the Bondi radius but decrease the mass accretion rate across the black hole horizon by many orders of magnitude. This implies that the growth history of supermassive black holes and their coevolution with host galaxies will be dramatically changed when the accreted gas has angular momentum or develops an outflow.

1 Introduction

The amount of energy produced by the accretion is determined by the mass accretion rate $\dot{M}$ times the energy production efficiency ε. Black holes, however, do not have a hard surface, and the energy production efficiency is not given a priori unlike in accretion onto neutron stars or whited dwarfs. The efficiency is determined only by the detailed physical state and processes during the accretion, which depend critically on the mass accretion rate. The important question, yet often ignored, is what is the mass accretion rate onto black holes for general accretion flow such as rotating one and how does it depend on the physical parameters?

In this talk, the luminosity and the mass accretion rate are, respectively, normalized by the Eddington luminosity (for pure hydrogen accretion) $L_E \equiv 4\pi G M m_p c/\sigma_{Th} = 1.23 \times 10^{38}(M/M_\odot)$ ergs/s and the Eddington mass accretion rate $\dot{M}_E \equiv L_E/c^2 = 2.2 \times 10^{-9}(M/M_\odot)\ M_\odot$/yr where G is the Gravitational constant, c speed of light, M mass of the accreting black hole, m_p proton mass, and σ_{Th} Thomson cross section. The dimensionless luminosity $l \equiv L/L_E$ and the mass accretion rate $\dot{m} \equiv \dot{M}/\dot{M}_E$ along with the radius in units of the Schwarzschild radius $r_{Sch} \equiv 2GM/c^2$ make the problem scale-free with respect to the black hole mass, in most cases.

2 Bondi Flow

Bondi in his seminal work [1] studied spherical accretion of polytropic gas onto a point mass. He showed that four classes of solutions exist: supersonic, subsonic, transonic, and non-physical. Subsonic and transonic ones are the correct physical solutions for the real accretion: accretion flow may

⋆e-mail: mgp@knu.ac.kr

stay subsonic all the way from outer-boundary to the stellar surface or starts subsonic, passes a sonic (critical) point, and becomes supersonic. Since black holes have no hard surface to build up enough pressure to slow down the infalling gas, transonic solutions are the relevant ones for accretion onto black holes. Bondi also realized that for given gas density and temperature at infinity, transonic accretion is possible only for a specific value of the mass accretion rate which is the maximum accretion rate. This mass accretion rate is called 'Bondi accretion rate',

$$\dot{M}_B = \Lambda\gamma^{-3/2} 4\pi r_B^2 \rho_\infty c_{s,\infty} = 4\pi\Lambda \frac{(GM)^2 \rho_\infty}{\gamma^{3/2} c_{s,\infty}^3}, \tag{1}$$

where ρ_∞ is the mass density at infinity, $c_{s,\infty}$ the sound speed of gas at infinity, $r_B \equiv GM/c_{s,\infty}^2$ the Bondi radius, and Λ is an order-of-unity function of the adiabatic index γ [1]. This accretion rate has been widely used in almost any kind of accretion problems, perhaps too widely used.

3 Rotating accretion flow

Bondi's work dealt with accretion of gas with no angular momentum at all. But real cosmic accretion inevitably involves rotation. Park [2] used slim disk approximation (vertical structure of the disk is averaged) to study the behavior of the hot rotating accretion flow. Rotating accretion flow should have some kind of viscosity which transports the angular momentum outward, thereby enabling the inward accretion of gas; Park [2] adopted widely used alpha-viscosity prescription but in a simplified form. The equations contain a critical point (sonic point) as shown in Bondi flow [1] and slim disk [3], and the regularity condition at the critical point made the mass accretion rate an eigenvalue for given gas density, temperature, and angular momentum at the outer boundary. He found that the flow behavior changes from that of Bondi flow to that of disk flow as the gas angular momentum increases. Hot accretion flow constitutes a continuous family of solutions from non-rotating Bondi flow to almost Keplerian advection-dominated accretion flow (ADAF; [5–7]). The mass accretion rate also changed continuously from the Bondi accretion rate for small enough specific gas angular momentum at the outer boundary $J_{out} \lesssim 0.1 J_B$ (J_B is the Keplerian angular momentum at the Bondi radius), to an order-of-magnitude smaller rate for near Keplerian gas angular momentum (Fig. 1) with functional dependence

$$\frac{\dot{M}}{\dot{M}_B} = \frac{\dot{m}}{\dot{m}_B} = 0.09 \left(\frac{\alpha}{0.01}\right) \left(\frac{J_{out}}{J_B}\right)^{-1}. \tag{2}$$

Any gas that is fed to the central supermassive black hole in the real universe is likely to have non-negligible angular momentum, and the mass accretion rate can be much smaller than the classic Bondi accretion rate. This result has very important implications for the growth of black hole and black hole-galaxy coevolution: black holes accreting gas with near Keplerian angular momentum will grow with a factor of 10 smaller accretion rate than the Bondi accretion rate and, accordingly, the feedback from black hole to the host galaxy will be weaker.

Narayan & Fabian [4] subsequently studied similar slowly rotating accretion flow onto a supermassive black hole to describe the cooling flow in a massive elliptical galaxy. They used alpha-prescription viscosity proportional to the angular shear and slim disk approximation. They confirmed the result of [2] that the flow changes from Bondi-like to disk-like and the mass accretion rate decreases as J_{out}/J_B increases. However, their $\dot{M}/\dot{M}_B$ versus J_{out}/J_B has a much shallower slope than that of [2] and, therefore, the mass accretion rate decreases only to a factor of a few less than $\dot{M}_B$ even for a near-Keplerian flow. They suspected that the difference with [2] may be due to the different value of r_B and suggested that a more thorough exploration of solutions in $\alpha - (J_{out}/J_B) - (r_B/r_{Sch})$ is needed [4].

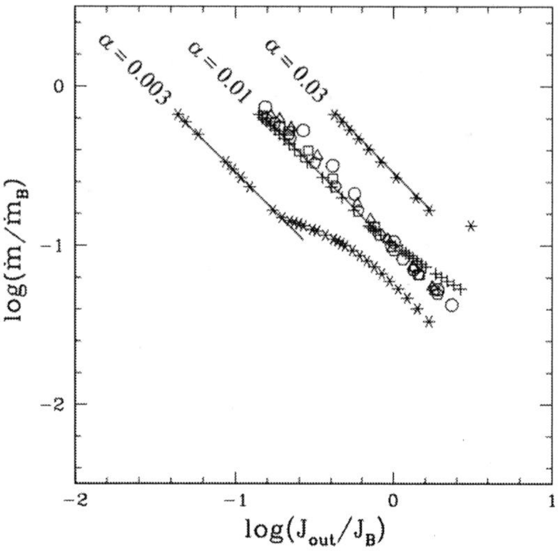

Figure 1. Mass accretion rate (in units of Bondi accretion rate) as a function of gas angular momentum (in units of Keplerian angular momentum) at the Bondi radius [2].

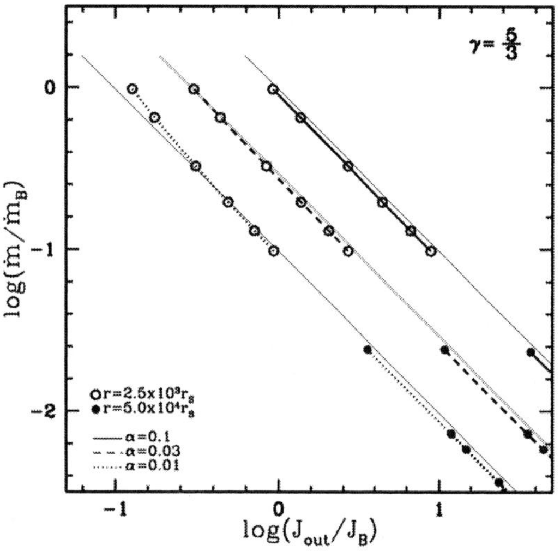

Figure 2. Mass accretion rate versus gas angular momentum for different values of the Bondi radius r_B and the viscosity parameter α.

4 Generalized Bondi flow

Following [2] and [4], we used slim disk approximation with alpha-description viscosity on the generalized Bondi flow, i.e., rotating polytropic gas, accreting onto a black hole. The adoption of polytropic gas equation of state will facilitate comparison with the original Bondi flow. We specifically focused on the dependence of flow characteristics on r_B, α, and J_{out}/J_B as suggested by [4].

Figure 2 shows the dependence of the mass accretion rate (in units of M_B) on the gas angular momentum at the Bondi radius (in units of J_B) for two different Bondi radii ($r_B = 2.5 \times 10^3 r_{Sch}$, and $r_B = 5.0 \times 10^3 r_{Sch}$ and for three different values of viscosity parameter ($\alpha = 0.1, 0.03$, and 0.01). All solutions show $\dot{M}/\dot{M}_B \propto (J_{out}/J_B)^{-1}$ (grey lines in Fig. 2 represent Eq. 2) seen in [2]. Different values of the Bondi radius, i.e., gas temperature, or viscosity parameter do not change the $\dot{M}/\dot{M}_B$-$(J_{out}/J_B)^{-1}$ relation.

We also checked the dependence on the adiabatic index γ of accreting gas. Figure 3 show the $\dot{M}/\dot{M}_B$-$(J_{out}/J_B)^{-1}$ relation for different γ. The mass accretion rate ratio $\dot{M}/\dot{M}_B$ hardly changes for

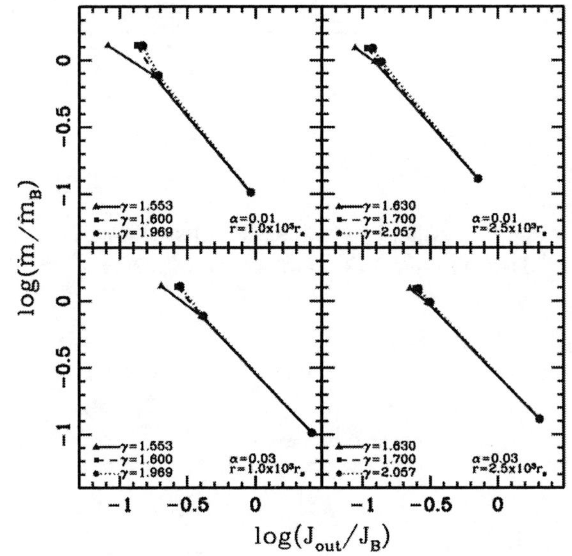

Figure 3. Mass accretion rate versus gas angular momentum for different values of the adiabatic index γ.

different values of γ, showing that $\dot{M}_B$ carries all the dependence on the adiabatic index even for generalized rotating Bondi flow.

So we can conclude that $\dot{M}/\dot{M}_B$-$(J_{out}/J_B)^{-1}$ relation discovered in [2] is valid in an extended parameter space of r_B, α, and γ.

5 ADAF with Outflow

As soon as the ADAF solutions appear, it was immediately realized that the ADAF has a positive Bernoulli constant [5]. This means that hot ADAF flows are very likely to develop an outflow or a wind. Blanford & Begelman [8] proposed simple self-similar advection-dominated inflow-outflow solutions (ADIOS). In this models, hydrodynamic or magnetic wind, or equally outflow, carries off enough of the mass, angular momentum and energy to bind the gas to the black hole and to allow accretion to proceed [8]. That is, outflow helps maintain the accretion. Numerical simulations also show that outflow is common rather than exception, and the mass accretion rate through the inner boundary can be almost 1/100 of the mass inflow rate from large radius [9].

We extended our calculation by implementing self-similar ADIOS of [8] to generalized Bondi flow to answer the question what is the mass accretion rate of the accreting flow in the presence of outflow. The equations contain similar critical structure seen in [2] and solved similarly. Figure 4 shows the resulting mass accretion rate $\dot{M}/\dot{M}_B$ of ADIOS versus ADAF as a function of gas angular momentum $(J_{out}/J_B)^{-1}$. The mass accretion rate of ADIOS are filled circles while those of ADAF open circles. ADIOS has a higher mass accretion rate than ADAF for the same gas conditions, including the gas angular momentum. This is because outflow removes angular momentum and the existence of outflow is equivalent to decreased gas angular momentum, and therefore, yields a higher mass accretion rate.

So we can conclude that outflows increase the mass accretion rate. However, the mass accretion rate here means the mass inflow rate at the outer boundary. In the presence of outflow, the mass accretion rate is not conserved and becomes a function of radius. In self-similar solutions of ADIOS, it is a power-law of radius [8], and the mass accretion rate at the inner boundary can be many factors of magnitude smaller than the mass inflow rate the outer boundary. Although the mass inflow rate increases by a factor of a few in the presence of outflow for given physical conditions of gas that

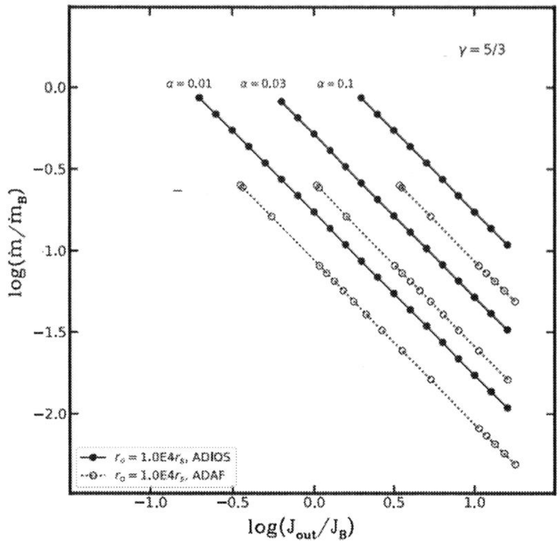

Figure 4. Mass accretion rate of ADIOS versus ADAF as a function of gas angular momentum.

surrounds the black hole, the final mass accretion rate that crosses the black hole horizon is much smaller by many orders of magnitude.

6 Summary

We have studied the structure and physical characteristics of polytropic rotating accretion flow onto black holes. We have confirmed that the mass accretion rate divided by the Bondi accretion rate is proportional to the viscosity parameter alpha and inversely proportional to the gas angular momentum divided by the Keplerian angular momentum at the Bondi radius, for a large parameter space of viscosity parameter alpha, gas temperature, and the adiabatic index. If a self-similar outflow is developed, the mass inflow rate at the outer boundary increases because the outflow carries away angular momentum but the mass accretion rate across the black hole horizon decreases by many orders of magnitude. The decrease in the mass accretion rate due to the gas angular momentum or outflow will have profound implications on the growth of supermassive black holes and their coevolution with host galaxies.

We thank the organizing committee of ICGAC-XIII/IK15 for hospitality. This research was supported by Kyungpook National University Research Fund, 2016.

References

[1] H. Bondi, MNRAS **112**, 195 (1952)
[2] M.-G. Park, ApJ **706**, 637 (2009)
[3] M. A. Abramowicz, B. Czerny, J. P. Lasota, and E. Szuszkiewicz, ApJ, **332** 646 (1988)
[4] R. Narayan and A. C. Fabian, MNRAS, **415** 372 (2011)
[5] R. Narayan and I. Yi, ApJL, **428** L13 (1994)
[6] R. Narayan and I. Yi, ApJ, **444** 231 (1995)
[7] R. Narayan and I. Yi, ApJ, **452** 710 (1995)
[8] R. D. Blandford and M. C. Begelman, MNRAS, **303** L1 (1999)
[9] J. Li, J. Ostriker, and R. Sunyaev, ApJ, **767** 105 (2013)

Particle Dynamics Around a Charged Black Hole

Sehrish Iftikhar[1,⋆]

[1] *Department of Mathematics,*
Lahore College for Women University, Lahore-54000, Pakistan.

Abstract. This paper investigates particle dynamics around the noncommutative Reissner Nordström black hole. We study escape velocity of the particle at innermost stable circular orbit. In order to discuss the stability of orbits we analyze effective potential.We compare our results with the dynamics of particles in Reissner Nordström as well as non-commutative Schwarzschild black hole. We observe that the noncommutative parameter affects the motion of particles.

1 Introduction

It is believed that the concept of continuous spacetime breaks down at short distances (of the order of Planck lenght). The noncommutative (NC) theory proved to be very useful to discuss the dynamics of spacetime at Planck scale. The NC geometry is described by universal minimal length scale, $\sqrt{\Theta}$ where, Θ is a deformation parameter that is equivalent to Planck's length. The noncommutativity of spacetime is based on quantum mechanics. In this regard, the NC spacetime is described by Hermitian operators that are not commutative, i.e., [1]

$$[x^{\alpha}, x^{\beta}] = \iota\Theta^{\alpha\beta},$$

where $\Theta^{\alpha\beta}$ is antisymmetric matrix that has dimension $(\text{lenght})^2$, also it is real valued. It describes a small region in $\alpha\beta$-plane (of the observable domain) as the Planck's constant represents very small fundamental cell of the observable phase space in quantum mechanics. The ordinary spacetime can be obtained in the limit $\Theta \rightarrow 0$.

The existence of black holes (BHs) is one of the most prominent predictions of general relativity (GR). Black holes are the regions in space in which singularities are hidden inside the event horizons. Generally, in the framework of quantum mechanics, the singular sources are smeared out and the spacetime is described by NC geometry. Such modifications to GR have a compelling effect on the BH structure. Many other interesting features of NC spacetime have been studied in literature [2].

The study of geodesics (null or timelike) explains the properties related to the geometry of curved spacetime. Among the different types of geodesics, circular geodesics are of great importance for astrophysical as well as theoretical research. Motion of particles in weakly magnetized Schwarzschild BH was considered by Frolov and shoom [3]. They mainly focused on the bounded trajectories in equatorial plane. Frolov [4] explored the collision of particles for the Schwarzschild BH and found

⋆e-mail: sehrish3iftikhar@gmail.com

that in the strong magnetic field, maximal collision energy can be high. Zahrani et al. [5] studied the escape velocity for particles at the innermost stable circular orbit (ISCO) and found that such motion is chaotic, generally. Some interesting features of neutral as well as charged particle dynamics are discussed in [6]

In this paper, we explore the dynamics of particles around Noncommutative Reissner Nordström (NCRN) BH. The paper is organized as follows. In section **2**, we present the NCRN BH and study the equations of motion. In section **3** we numerically analyze the behavior of effective potential as well as escape velocity of the particle. The results are summarized in the last section.

2 Equations of Motion

In order to discuss the BHs in the context of NC geometry, the corresponding field equations have to be solved. In this regard, the Einstein tensor remains unchanged and the effect of noncommutativity is taken into account only on the matter part (energy momentum tensor) [7–9]. This approach is analogues to the NC modification of quantum field theory [10]. The effect of smearing is carried out by considering energy density of a static, spherically symmetric, particle-like, smeared gravitational source [7–9]

$$\rho_\Theta = \frac{M}{(4\pi\Theta)^{\frac{3}{2}}} e^{-\frac{r^2}{4\Theta}}, \tag{1}$$

where M denotes the mass of the particle diffused throughout the region of linear size $\sqrt{\Theta}$ and Θ represents the NC parameter. The spherically symmetric line element by solving the Einstein field equations with ρ_Θ, is given by [11]

$$ds^2 = -h(r)dt^2 + h(r)^{-1}dr^2 + r^2(d\theta^2 + \sin^2\theta d\phi^2), \tag{2}$$

where

$$h(r) = 1 - \frac{2M_\Theta}{r} + \frac{Q_\Theta^2}{r^2}, \tag{3}$$

here

$$M_\Theta = \frac{2M}{\sqrt{\pi}}\gamma(\frac{3}{2}, \frac{r^2}{4\Theta}), \tag{4}$$

$$Q_\Theta = \frac{Q}{\sqrt{\pi}}\sqrt{\gamma^2(\frac{1}{2}, \frac{r^2}{4\Theta}) - \frac{r}{\sqrt{2\Theta}}\gamma(\frac{1}{2}, \frac{r^2}{2\Theta})}, \tag{5}$$

and

$$\gamma(\frac{a}{b}, x) = \int_o^x \frac{d\mu}{\mu}\mu^{\frac{a}{b}}e^{-\mu}, \tag{6}$$

is the incomplete gamma function and Q is the charge. In the limit $\frac{r}{\sqrt{\Theta}} \to \infty$, the classical RN BH is recovered.

Using $f(r) = 0$, we find the horizons

$$r_h = \frac{M_\Theta \pm \sqrt{M_\Theta^2 - 4Q_\Theta^2}}{2}, \tag{7}$$

This leads to a complicated equation, therefore it is difficult to find the horizons in closed form. A detailed discussion about horizons and other important properties for spacetime (2) is given in [11]. For a spherically symmetric spacetime, there are three integrals of motion. Two of them are related to the symmetries of the BH metric with respect to translations of time and rotations about the symmetry axis and are obtained by the Killing vectors, given by

$$\xi^{\nu}_{(t)}\partial_{\nu} = \partial_t, \quad \xi^{\nu}_{(\phi)}\partial_{\nu} = \partial_{\phi}. \tag{8}$$

here $\xi^{\nu}_{(t)} = (1,0,0,0)$ and $\xi^{\nu}_{(\phi)} = (0,0,0,1)$.

The conserved quantities (the specific energy E and azimuthal angular momentum L_z (per unit mass)) corresponding to Eq.(8) are given by,

$$E = \dot{t}h(r), \quad L_z = \dot{\phi}r^2\sin^2\theta,$$

where the dot represents the differentiation over the affine parameter τ. By using normalization condition $u^{\nu}u_{\nu} = -1$, where u is the four velocity, we obtain

$$\dot{r}^2 = E^2 - (1 - \frac{2M_{\Theta}}{r} + \frac{Q^2_{\Theta}}{r^2})(1 + \frac{L^2}{r^2\sin^2\theta}).$$

At tuning points of particles moving from trajectories $\dot{r} = 0$

$$E^2 = (1 - \frac{2M_{\Theta}}{r} + \frac{Q^2_{\Theta}}{r^2})(1 + \frac{L^2}{r^2\sin^2\theta}) \equiv V_{eff}(r), \tag{9}$$

where V_{eff} represents the effective potential. The ISCO for the NCRN BH in the equatorial plane ($\theta = \frac{\pi}{2}$, $\dot{\theta} = 0$) can be obtained solving $V_{eff} = 0$

$$r_0^4 M'_{\Theta} + r_0^3(-M_{\Theta} - Q_{\Theta}Q'_{\Theta}) + r_0^2(L^2 + Q^2_{\Theta} + L^2 M'_{\Theta}) - r_0(3L^2 M_{\Theta} + L^2 Q_{\Theta}Q'_{\Theta}) = 0,$$

where

$$M'_{\Theta} = -\frac{e^{-\frac{r_0^2}{4\Theta}} M r_0 \sqrt{\frac{r_0^2}{\Theta}}}{2\Theta\sqrt{\pi}},$$

$$Q'_{\Theta} = \frac{Q^2}{\pi}\left[\frac{e^{\frac{r_0^2}{2\Theta}}}{\Theta^{3/2}\sqrt{\frac{r_0^2}{\Theta}}} - \frac{\sqrt{2}r_0^2 e^{\frac{r_0^2}{2\Theta}}}{\Theta^{3/2}\sqrt{\frac{r_0^2}{\Theta}}} + \sqrt{2}\sqrt{\frac{1}{\Theta}}\gamma(\frac{1}{2}, \frac{r_0^2}{4\Theta}) - \frac{2r_0 e^{\frac{r_0^2}{2\Theta}}\gamma(\frac{1}{2}, \frac{r_0^2}{4\Theta})}{\Theta\sqrt{\frac{r_0^2}{\Theta}}} - \frac{\gamma(\frac{1}{2}, \frac{r_0^2}{2\Theta})}{\sqrt{2\Theta}}\right]. \tag{10}$$

The corresponding energy and azimuthal angular momentum are

$$L_{z0} = \sqrt{\frac{M_{\Theta}r_0^3 - Q^2_{\Theta}r^2 - r_0^4 M'_{\Theta} + r_0^3 Q_{\Theta}Q'_{\Theta}}{r^2 - 3M_{\Theta}r_0 + 2Q^2_{\Theta} + r_0^2 M'_{\Theta} - r_0 Q_{\Theta}Q'_{\Theta}}}, E_0^2 = \frac{(r_0^2 - 2M_{\Theta}r_0 + Q^2_{\Theta})^2}{r_0^2(r^2 - 3M_{\Theta}r_0 + 2Q^2_{\Theta} + r_0^2 M'_{\Theta} - r_0 Q_{\Theta}Q'_{\Theta})}.$$

Now we consider that initially the particle is orbiting in ISCO, collides with another particle coming from infinity (initially at rest). There are three possible outcomes for the particle motion as a result of this collision, i.e, either have bounded motion around BH, captured by BH, or escape to infinity. If there is a small change in energy as well as the angular momentum, then the path of the particle is slightly changed. On the other hand, the major change in the energy and angular momentum can cause the particle to deviate from its original path, as a result, it can escape to infinity or attracted towards the BH. After colliding with another particle, there will be have new integrals of motion. We consider a simple case when the initial radial velocity as well as azimuthal angular momentum do not change after the collision. Now angular momentum and energy at the equatorial plane takes the form [5]

$$L^2 = L^2_{z0} + r_0^2 v_{\perp}^2, \quad E = \sqrt{E_0^2 + v_{\perp}^2 h(r_0)}. \tag{11}$$

If $E \geq 1$, the particle will escape to infinity this yields

$$|v_{\perp}| \geq \sqrt{(1 - E_0)h(r_0)^{-1}}. \tag{12}$$

3 Analysis of Effective Potential

1 2

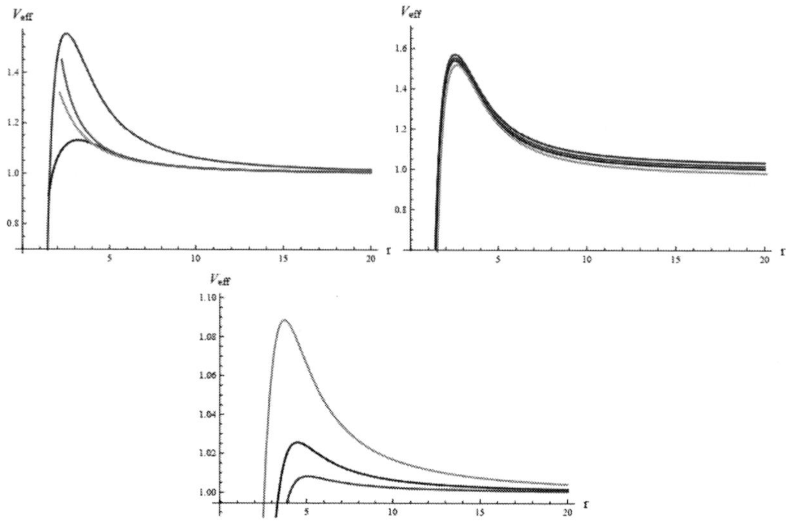

Figure 1. Effective potential with respect to r for $M = 1$ and $L = 1.5$ (upper panel). In the left graph, $Q = 0.5$ and red, blue, green, and pink curves corresponds to $\Theta = 0.7, 0.5, 0.3, 0$. In the right graph, $\Theta = 0.5$ and red, blue, green, and purple curves corresponds to $Q = 0.5, 0.3, 0.2, 0$. In lower panel, $M = 1$, $\Theta = 0.5$ and $Q = 0.5$. Here, red, blue, and green curves corresponds to $L = 2.5, 2, 1.5$.

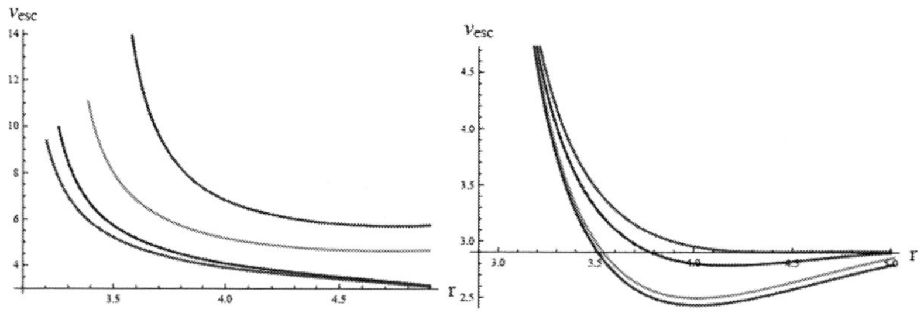

Figure 2. Escape velocity versus r for $M = 1$, and $L = 1.5$. Here, red, blue, green, and purple curves corresponds to $\Theta = 0.7, 0.5, 0.3, 0$ (Left) and $Q = 0.5, 0.3, 0.2, 0$ (Right).

In this section, we examine the behavior of the effective potential as well as the escape velocity of the particle graphically using Eqs.(9) and (12). In Figure **1**, the effective potential is plotted with respect to r. The stable and unstable orbits corresponds to the maximum and minimum values of the effective potential. In the left plot, we study the behavior of V_{eff} by varying the values of Θ. We observe that for the large values of Θ, the effective potential attains higher values. Initially, the orbits are unstable but gradually V_{eff} decreases as r increases which corresponds to stable orbits. The behavior of V_{eff} for NCRN BH is similar to the case of RN BH. The right plot indicates V_{eff} for the different values of Q. These plots indicate that the V_{eff} has smaller values as Q increases, also for the neutral BH ($Q = 0$) the orbits are highly unstable. We also analyze the stability for the different values of the angular momentum. The graph in the lower panel describes that for the small values of angular momentum, the orbits are unstable while the large values of V_{eff} correspond to stable paths.

In Figure **2**, the escape velocity is plotted for the different values of Θ and Q. The left plot shows that the escape velocity is higher for $\Theta = 0$, also the increasing values NC parameter correspond to less chances for the particle to escape from the vicinity of the BH. The right plot indicates that the escape velocity of the particle becomes higher for the large values of charge while for $Q = 0$, it is the smallest.

4 Final Remarks

In this paper we have investigated the particle dynamics orbiting around the NCRN BH. For this purpose, we first formulate the equations of motion. We have examine the behavior of effective potential as well as escape velocity of the particles moving around a charged NC BH. By observing the graphical properties of the effective potential, we found that the path of the particle is effected by the NC parameter. The largest value of NC parameter corresponds to highly unstable orbit, also the change in charge parameter (in the presence of Θ) leads to both unstable and stable orbits. However, the orbits are more stable when the angular momentum is high. The cases of neutral and RN BHs have shown the similar behavior as compared to NCRN BH. We have also studied the effect of noncommutativity as well as charge of the BH on the particle's escape velocity. We observe that the the particle have less chances to escape to infinity in the presence of NC parameter while the increasing values of the BH charge leads to high values of the escape velocity. We conclude that noncommutativity causes the deviation in the original path of the particle.

References

[1] P. Nicolini, Int. J. Mod. Phys. A **24**, 1229 (2009).
[2] H. Garcia-Compean and C. Soto-Campos, Phys. Rev. D **74**, 104028 (2006); K. Nozari and B. Fazlpour, Mod. Phys. Lett. A **22**, 2917 (2007); R. Banerjee, B.R. Majhi and S. Samnta, Phys. Rev. D **77**, 124035 (2008); Y.G. Miao, Z. Xue and S.J. Zhang, Int. J. Mod. Phys. D **21**, 1250018 (2012); M. Sharif and S. Iftikhar, Eur. Phys. J. C **76**, 630 (2016).
[3] V.P. Frolov and A.A. Shoom, Phys. Rev. D **82**, 084034 (2010).
[4] V.P. Frolov, Phys. Rev. D **85**, 024020 (2012).
[5] A.M. Al Zahrani, V.P. Frolov and A.A. Shoom, Phys. Rev. D **87**, 084043 (2013).
[6] B. Majeed, S. Hussain and M. Jamil, Adv. High Energy Phys. **2015**, 671259 (2015); M. Sharif, and S. Iftikhar, Int. J. Mod. Phys. D **26**, 1750091 (2017).
[7] P. Nicolini, A. Smailagic and E. Spallucci, Phys. Lett. B **632**, 547 (2006).
[8] S. Ansoldi, P. Nicolini, A. Smailagic and E. Spallucci, Phys. Lett. B **645**, 261 (2007).
[9] P. Nicolini, J. Phys. A **38**, L631 (2005).
[10] A. Smailagic and E. Spalluci, J. Phys. A **36**, L517 (2003); J. Phys. A **37**, 1 (2004).
[11] S.A., Alavi, Acta Phys. Polon. B **40**, 2679 (2009).

Quintessential inflation: A unified scenario of inflation and dark energy

Md. Wali Hossain[1,⋆]

[1]*Asia Pacific Center for Theoretical Physics, Pohang 37673, Korea*

Abstract. Quintessential inflation unifies inflation and late time acceleration by a single scalar field. Such a scenario, with canonical and non-canonical scalar fields, has been discussed . The scalar field behaves as an inflaton field during inflation and as a quintessence field during late time. Also the predictions of the models has been compared with the recent Planck data.

1 Introduction

Current observations suggest that there are two accelerated phases of our universe, one during very early time known as inflation [1] and the other one is the late time acceleration [2]. These regimes of two accelerated phases are generally treated independently. However, it is interesting to think that there is a unique reason responsible for both phases of acceleration. Such a scenario is known as quintessential inflation [3–9].

The scalar field should not interfere with the thermal history of the universe. It should be invisible during radiation and matter dominated era and reappear only at the recent past giving rise to late-time cosmic acceleration. This requirement demands that the scalar field potential should have a steep region in the potential just after the end of inflation. This steep region results a scalar field kinetic energy dominated regime which has two major effects, one is the fast decay in scalar field energy density which eventually results radiation dominated universe and the other one is the blue spectrum in relic gravitational wave [4, 10, 11]. Apart from these effects steep region of the potential can cause tracker behavior [12] of the scalar field depending upon the fact that the scalar field potential has a shallow region during late time dynamics or there is a special mechanism, like non-minimal coupling, which helps the scalar field to dominate over matter. Since the scalar field must survive till late times the conventional reheating mechanism is not applicable and one needs alternative reheating mechanism. Instant preheating [13] mechanism is one of such efficient mechanisms.

In this article, I shall discuss quintessential inflation models with canonical and non-canonical scalar fields [5, 6, 8, 9]. A non-minimal coupling between scalar field and massive neutrinos has been considered. This non-minimal coupling becomes effective only when the massive neutrinos become non-relativistic *i.e.*, only during the late times. in these scenarios the non-minimal coupling plays an important role in late time acceleration and the dark energy scale is related to the massive neutrino mass scale.

⋆e-mail: wali.hossain@apctp.org

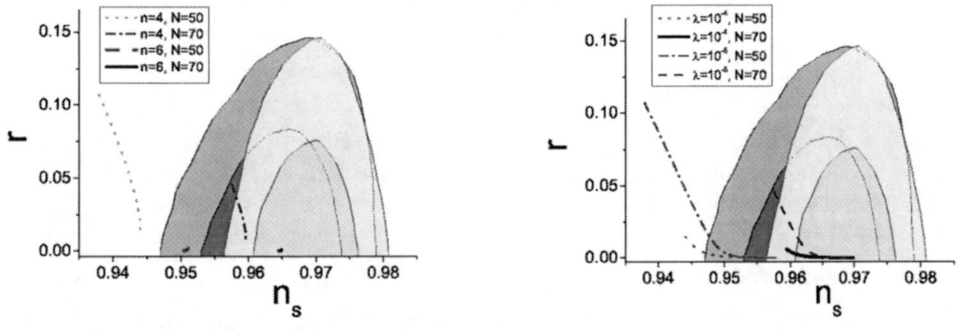

Figure 1. 1σ (yellow) and 2σ (light yellow) contours are for Planck 2015 results ($TT + lowP + lensing + BAO + JLA + H_0$) [14] and gray contours are for Planck 2013 results ($Planck + WP + BAO$) [15]. Predictions of our scenario is shown by varying λ for different n (left figure) and varying n for different λ (right figure) and for 50 and 70 e-foldings. Figures are taken from Ref. [9].

2 Inflation

We shall first analyze the inflationary phase for both canonical and non-canonical scalar fields. We shall also compare the scenario with Planck data.

2.1 Canonical scalar field

Let us consider the following action [9]

$$\mathcal{S} = \int d^4x \sqrt{-g}\Big[\frac{M_{\rm Pl}^2}{2}R - \frac{1}{2}\partial^\mu\phi\partial_\mu\phi - V(\phi)\Big] \tag{1}$$

where $M_{\rm Pl}$ is the Planck mass, ϕ is the scalar field and $V(\phi)$ is the scalar field potential.

We consider the potential

$$V = V_0 e^{-\lambda\phi^n/M_{\rm Pl}^n}, \tag{2}$$

where V_0, λ and n are the constant parameters. $n > 1$ makes the potential steeper than exponential which can affect the scalar field dynamics [9]. For $n > 1$ the potential has a flat region around $\phi = 0$ which can be responsible for the slow-roll of the scalar field. Increasing the values of n increases the flat region in the potential which can give sufficiently low tensor to scalar ratio.

Slow-roll parameters, for $n \neq 2$, are given below [9]

$$\epsilon = \frac{M_{\rm Pl}^2}{2}\left(\frac{1}{V}\frac{{\rm d}V}{{\rm d}\phi}\right)^2 = \frac{1}{2}n^2\lambda^2\, Q(n,\lambda,\mathcal{N})^{2n-2}, \tag{3}$$

$$\eta = \frac{M_{\rm Pl}^2}{V}\frac{{\rm d}^2V}{{\rm d}\phi^2} = n\lambda\, Q(n,\lambda,\mathcal{N})^{n-2}\,\{1 - n + n\lambda\, Q(n,\lambda,\mathcal{N})^n\}, \tag{4}$$

$$Q(n,\lambda,\mathcal{N}) = \left\{n\lambda\left((n-2)\mathcal{N} + n\lambda 2^{\frac{2-n}{2(n-1)}}\left(\frac{1}{n^2\lambda^2}\right)^{\frac{n}{2(n-1)}}\right)\right\}^{1/(2-n)}. \tag{5}$$

Tensor to scalar ratio (r) and spectral index (n_s) are given by

$$r(N) = 16\epsilon(N), \tag{6}$$

$$n_s(N) = 1 - 6\epsilon(N) + 2\eta(N), \tag{7}$$

In figure 1 the predictions of the model is being compared with the Planck 2015 [14] and 2013 [15] results. It can be seen that we can find some parameter space where the predictions of the model agree with the Planck results.

2.2 Non-canonical scalar field

We consider the following non-canonical action for the scalar field along with the standard Einstein-Hilbert term [5, 16],

$$S = \int d^4x\sqrt{-g}\Big[-\frac{M_{\rm Pl}^2}{2}R + \frac{k^2(\phi)}{2}\partial^\mu\phi\partial_\mu\phi + V(\phi)\Big], \tag{8}$$

$$k^2(\phi) = \left(\frac{\alpha^2 - \tilde{\alpha}^2}{\tilde{\alpha}^2}\right)\frac{1}{1+\beta^2 e^{\alpha\phi/M_{\rm Pl}}} + 1, \tag{9}$$

$$V(\phi) = M_{\rm Pl}^4 e^{-\alpha\phi/M_{\rm Pl}}, \tag{10}$$

where α, $\tilde{\alpha}$ and β are the model parameters. The kinetic function $k^2(\phi)$ has been chosen such a way so that we have tracker behavior in scalar field dynamics in post-inflationary periods. Non-canonical scalar field action can be transformed into a canonical form by doing a simple transformation $k^2(\phi) = (\partial\sigma/\partial\phi)^2$ where σ is the canonical scalar field [5]. In terms of the canonical field the potential will behave like exponential potential with slope $\tilde{\alpha}$ during inflation and α during post-inflationary dynamics. So by fixing $\tilde{\alpha}$ very small we can have a sufficiently flat region in the potential which can give rise to inflation with small tensor to scalar ratio. On the other hand α can be fixed from the constraint on the early dark energy. The parameter β is related to the energy scale of inflation.

Tensor to scalar ratio (r) and spectral index (n_s), for this scenario, are given by [5, 6, 9]

$$r(N,\tilde{\alpha}) = 16\epsilon(N) \approx \frac{8\tilde{\alpha}^2}{1-e^{-\tilde{\alpha}^2 N}}, \tag{11}$$

$$n_s(N,\tilde{\alpha}) = 1 - 6\epsilon + 2\eta \approx 1 - \tilde{\alpha}^2\coth\left(\frac{\tilde{\alpha}^2 N}{2}\right), \tag{12}$$

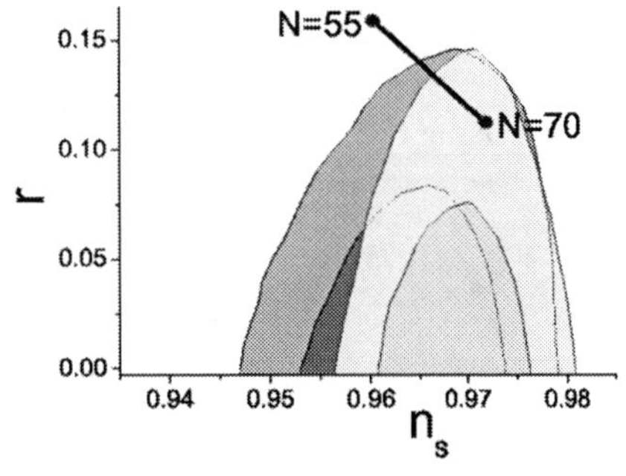

Figure 2. 1σ (yellow) and 2σ (light yellow) contours are for Planck 2015 results ($TT + lowP + lensing + BAO + JLA + H_0$) [14] and gray contours are for Planck 2013 results ($Planck + WP + BAO$) [15]. Predictions of the scenario are shown by varying e-foldings from 55 to 70 for $\tilde{\alpha} \to 0$. The figure is taken from Ref. [9].

In figure 2 we have compared the model with recent observational results of Planck [14, 15] and found that the scenario under consideration can be ruled out by the future experiments as the model already lies in the 2σ region of Planck 2015 results.

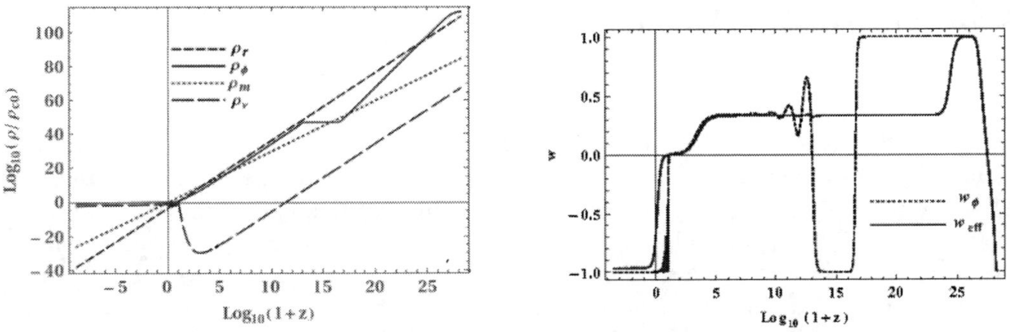

Figure 3. Left: Evolution of matter ($\rho_{\rm m}$, dotted green), radiation ($\rho_{\rm r}$, blue short dashed), scalar field (ρ_ϕ, red solid), and massive neutrinos (ρ_ν, purple long dashed) energy densities. Right: Evolution of the effective ($w_{\rm eff}$, solid red) and scalar field (w_ϕ, dotted blue) equation of states. Figures are taken from Ref. [9]

3 Late time dynamics

For both the cases, canonical and non-canonical, during the late time the potential is very steep. For canonical case the potential is steeper than exponential for $n > 1$. For non-canonical case the potential is steep exponential as the function $k^2(\phi) \rightarrow 1$ for large values of ϕ *i.e.*, during the late time and the slope of the potential is steep as from the early dark energy constraints we get $\alpha \geq 20$. Because of this steep nature of the potential the scalar field energy density follows the background during late time (scaling behavior) but can not take over matter as this behavior is an attractor solution for the potential with exponential nature. Here we should mention that the large field behavior of the potential (2) is similar as exponential potential [9] which gives same late time behavior of the scalar field as exponential potential. So for these scenarios under consideration we need some special mechanism so that the scalar field can take over matter at the recent past. In this regard we consider non-minimal coupling between scalar field and massive neutrinos. The motivation behind this is that the mass scale of massive neutrinos is of similar order of dark energy and in these scenarios we can relate the dark energy scale with the massive neutrinos energy density and the conformal coupling constant[5, 8, 9]. Another reason is that the massive neutrinos are relativistic during radiation and early matter dominated regimes and become non-relativistic only during late times. So the conformal coupling becomes effective only during the recent past as the scalar field is coupled with the trace of the energy momentum tensor of the massive neutrinos. We consider the conformal coupling of the form $e^{2\gamma\phi/M_{\rm Pl}}$ where γ is the coupling constant.

Non-minimal coupling modifies the scalar field equation of motion. This modification along with the potential term forms an effective potential. This effective potential has a minimum for $\gamma > 0$. This minimum depends on the massive neutrinos energy density and the conformal coupling constant. The scalar field, during late times, starts oscillating around the minimum. Now for large values of γ (*e.g.*, $\gamma = 800$) the minimum becomes very narrow and the scalar field eventually settles down at the minimum giving rise to late time acceleration. When the scalar field settles down at the minimum we can relate the dark energy density with the minimum of the effective potential and thereby with the massive neutrinos energy density and the conformal coupling constant. figure 3 shows the evolution of the energy densities of different components of the universe during post-inflationary eras along with the evolution of the effective and scalar field equation of states. These figures are plotted by considering the canonical case. But it is needless to say that for the non-canonical case also we have similar dynamics of the scalar field because of the steep exponential nature of the potential.

4 Summary and Discussions

Quintessential inflation models unify inflation and late time acceleration by a single scalar field. Models based on both canonical and non-canonical scalar fields are being discussed here. For canonical scalar field with potential steeper than exponential we can have low tensor to scalar ratio within the 1σ bound given by Planck 2015 results. But for non-canonical scalar field the tensor to scalar ratio > 0.1 which lies in the 2σ contours of Planck 2015 results. In future, if we have more precise data, the non-canonical scalar field model of quintessential inflation, considered here, will be ruled out.

Non-minimal coupling between the scalar field and massive neutrinos has been considered in both the scenarios. This non-minimal coupling plays an important role in late time acceleration. Since massive neutrinos are relativistic during early times and become non-relativistic only during the recent past the non-minimal coupling becomes effective only from the recent past. So the non-minimal coupling does not affect the radiation era. In both the scenarios, the late time behavior of the scalar field is a tracker behavior. Because of the non-minimal coupling there forms an effective potential with minimum and the scalar field oscillates around the minimum and eventually settles down at the minimum during late times giving rise to late time acceleration.

Here we have presented models which can successfully unify inflation and late time acceleration by a single scalar field. While both of the scenario fit with the Planck 2015 results, at least up to 2σ level, we expect that some of the models will be ruled out by future experiments.

Acknowledgment

I thank C. Q. Geng, R. Myrzakulov, M. Sami and E. N. Saridakis for a fruitful collaboration on this topic. I also thank the organizers of the International Conference on Gravitation : Joint Conference of ICGAC-XIII and IK15, held at Ewha Womans University, Seoul, Korea, where the work was presented.

References

[1] A. D. Linde, Phys. Lett. **129B**, 177 (1983).
[2] E. J. Copeland, M. Sami and S. Tsujikawa, Int. J. Mod. Phys. D **15**, 1753 (2006)
[3] P. J. E. Peebles and A. Vilenkin, Phys. Rev. D **60**,103506 (1999)
[4] V. Sahni, M. Sami and T. Souradeep, Phys. Rev. D **65**, 023518 (2001)
[5] M. W. Hossain, R. Myrzakulov, M. Sami and E. N. Saridakis, Phys. Rev. D **90**, 023512 (2014)
[6] M. W. Hossain, R. Myrzakulov, M. Sami and E. N. Saridakis, Phys. Rev. D **89**, 123513 (2014)
[7] M. W. Hossain, R. Myrzakulov, M. Sami and E. N. Saridakis, Phys. Lett. B **737**, 191 (2014)
[8] M. Wali Hossain, R. Myrzakulov, M. Sami and E. N. Saridakis, Int. J. Mod. Phys. D **24**, no. 05, 1530014 (2015)
[9] C. Q. Geng, M. W. Hossain, R. Myrzakulov, M. Sami and E. N. Saridakis, Phys. Rev. D **92**, no. 2, 023522 (2015)
[10] V. Sahni, Phys. Rev. D **42**, 453 (1990).
[11] M. Giovannini, Phys. Rev. D **58**, 083504 (1998).
[12] P. J. Steinhardt, L. M. Wang and I. Zlatev, Phys. Rev. D **59**, 123504 (1999)
[13] G. N. Felder, L. Kofman and A. D. Linde, Phys. Rev. D **59**, 123523 (1999).
[14] P. A. R. Ade *et al.* [Planck Collaboration], Astron. Astrophys. **594**, A20 (2016)
[15] P. A. R. Ade *et al.* [Planck Collaboration], Astron. Astrophys. **571**, A22 (2014)
[16] C. Wetterich, Phys. Rev. D **89**, no. 2, 024005 (2014)

GRB 110731A within the IGC paradigm

Daria Primorac[1,2,⋆], *Remo* Ruffini[1,2,3,4,⋆⋆], *Giovanni Battista* Pisani[1,2,⋆⋆⋆], *Yerlan* Aimuratov[1,2,3], *Carlo Luciano* Bianco[1,2], *Mile* Karlica[1,2,3], *Julio David* Melon Fuksman[1,2], *Rahim* Moradi[1,2], *Marco* Muccino[1,2], *Ana Virginia* Penacchioni[5,6], *Jorge Armando* Rueda[1,2,4], and *Yu* Wang[1,2]

[1] *Dipartimento di Fisica and ICRA, Sapienza Università di Roma, Piazzale Aldo Moro 5, I–00185 Rome, Italy*
[2] *ICRANet, Piazza della Repubblica 10, I–65122 Pescara, Italy*
[3] *Université de Nice Sophia Antipolis, CEDEX 2, Grand Château Parc Valrose, Nice, France*
[4] *ICRANet-Rio, Centro Brasileiro de Pesquisas Físicas, Rua Dr. Xavier Sigaud 150, 22290–180 Rio de Janeiro, Brazil*
[5] *ASI Science Data Center, Via del Politecnico s.n.c., I-00133 Rome, Italy*
[6] *Dept. of Physical Sciences, Earth and Environment, University of Siena, Via Roma 56, I-53100 Siena, Italy*

Abstract. Bright gamma-ray burst (GRB) 110731A was simultaneously observed by *Fermi* and *Swift* observatories, with a follow up optical observation which inferred the redshift of $z = 2.83$. Thus, available data are spanning from optical to high energy (GeV) emission. We analyze these data within the induced gravitational collapse (IGC) paradigm, recently introduced to explain temporal coincidence of some long GRBs with type Ic supernovae. The case of binary-driven hypernova (BdHN) assumes a close system, which starts as an evolved core - neutron star binary. After the core-collapse event, the new NS - black hole system is formed, emitting the GRB in the process. We performed the time-resolved and time-integrated analysis of the *Fermi* data. Preliminary results gave isotropic energy $E_{\rm iso} = 6.05 \times 10^{53}$ erg and the total P-GRB energy of $E_{\rm P-GRB} = 3.7 \times 10^{52}$ erg. At transparency point we found a Lorentz factor $\Gamma \sim 2.17 \times 10^3$, laboratory radius of 8.33×10^{13} cm, P-GRB observed temperature of 168 keV and a baryon load $B = 4.35 \times 10^{-4}$. Simulated light-curve and prompt emission spectra showed the average circum burst medium density to be $n \sim 0.03$ particles per cm^3. We reproduced the X-ray light-curve within the rest-frame of the source, finding the common late power-law behavior, with $\alpha = -1.22$. Considering these results, we interpret GRB 110731A as a member of a BdHNe group.

1 Introduction

Gamma-ray bursts (GRBs) are known to be the most luminous events in the universe. The data provided by CGRO/BATSE observations showed that the T_{90} distribution of the observed bursts is bimodal in structure [1]. Since then, GRBs are traditionally divided into long/soft and short/hard bursts with the separation of about 2 seconds in the observed frame [2]. This division came prior to the redshift measurements. Once X-ray observations provided more accurate positions, optical telescopes could determine the distance at which the GRBs occurred, confirming their cosmological origin. Thanks to their extreme brightness, GRBs are detectable up to redshift $z \approx 9$ (see, e.g., [3]). With redshifts available, correlations between burst observables could facilitate the use of GRBs as standard candles, constraining the expansion history of the universe. The separation in long and short GRBs was still present when one inquired into their rest-frame properties. For example, the $E_{iso} - E_p$ dependence for long bursts follows Amati relation [4, 5], where E_{iso} is the burst isotropic energy and E_p is the rest-frame peak energy in the $\nu F(\nu)$ spectrum. The short GRBs on the other hand follow a different relation (see, e.g., [6–8]). Therefore, it is generally accepted that long and short GRBs have different progenitors, where the first are the result of a gravitational collapse event (see [9, 10]) while the latter originate from compact mergers [11, 12]. However, there are still doubts regarding the dominant mechanism of long GRBs.

Connection between some long GRBs and type Ib/c supernovae (SNe) has been known since GRB 980425/SN 1998bw observations [13]. The induced gravitational collapse (IGC) paradigm has been introduced in the past years to explain this temporal coincidence in the case of binary-driven hypernovae (BdHNe): long, extremely energetic bursts ($E_{iso} > 10^{52}$ erg) with high rest-frame peak energies ($E_p > 200$ keV) up to some MeV. The model invokes a close binary system consisting of a neutron star (NS) and a carbon-oxygen (CO) core close to a core-collapse event (see, e.g., [14–16]). A well-determined time sequence is implied, where each stage is expected to have distinctive observational properties:

Hypercritical accretion. After the CO core SN explosion, the companion NS gains mass via hypercritical accretion of the ejected material. Spectrum is described with an expanding thermal component ($\sim 10^{9-10}$ cm) plus an ex-

⋆e-mail: Daria.Primorac@icranet.org
⋆⋆e-mail: ruffini@icra.it
⋆⋆⋆e-mail: giovanni.b.pisani@gmail.com

tra power-law (PL). The black body (BB) temperature decreases with time following a broken PL [17].

BH formation and the P-GRB. The companion NS reaches critical mass and collapses into a BH. This leads to a formation of an optically thick e^+e^- plasma – the *fireshell* of total energy E^{tot} [18, 19]. The fireshell expands and self-accelerates due to the e^+e^- annihilation up to ultra-relativistic velocities, until the transparency condition has been reached. The dynamics of the fireshell up to this point are fully described with E^{tot} and the baryon load $B = M_B c^2/E^{tot}$, where M_B is the mass of the baryons engulfed by the plasma. In the case of long GRBs, it is characterized by $10^{-4} \lesssim B < 10^{-2}$. The proper-GRB (P-GRB) is then emitted, with a spectrum defined by the geometry of the pair-creation region. PL with an exponential cutoff is expected in the case of axial symmetry. After the transparency has been achieved, evidence of a BH formation is seen in form of high energy photons (up to few GeV). The BdHN scenario invokes the BH formation. X-ray flashes (XRFs) are less energetic sub-class of long GRBs, where the NS only gains mass, and there is no BH formation.

Canonical GRB emission. After the transparency emission, the optically thin shell of baryons collides with a circum burst medium (CBM) of density n_{CBM}, giving rise to the prompt emission. The CBM is modeled by the filling factor, which takes into account filamentary structures of the medium, $R = A_{eff}/A_{vis}$. Here, A_{eff} is the area covered by the CBM clouds while A_{vis} is the total observable area [8, 15]. The prompt emission occurs at $\sim 10^{15-17}$ cm [17].

X-ray afterglow. Rest-frame X-ray light-curves show precise behavior consisting of three PL segments: starting at the endpoint of the prompt emission, followed by a plateau phase, and ending by a late constant PL decay [20, 21]. While plateau shrinks as a function of the increasing E_{iso}, the late PL decay is common to all sources after $\approx 10^4$ s in the rest-frame, with typical slopes of $-1.7 \lesssim a_x \lesssim -1.3$ [22, 23]. In the case of XRFs, light-curves are less energetic and more chaotic, not exhibiting these properties. The emitting region inferred for some BdHNe sources suggested that the origin of the X-ray emission has the initial size of $\sim 10^{12}$, meaning that the place of origin is SN ejecta, not the GRB.

SN remnant detection in optical. Optical SN emission due to the ^{56}Ni decay is expected after $\sim 10 - 15$ days in the source rest-frame, and it is currently detectable only up to $z \lesssim 1$. In the end, system that began as a CO core - NS binary now consists of a newly born NS (ν-NS) and a BH.

In this work we shortly summarize relevant observations of the energetic GRB 110731A. We use these data to search for the BdHN scenario signatures described above. We perform time-integrated and time-resolved spectral analysis and study the results within the IGC model. From this we derive the baryon load, Lorentz factor at transparency point and P-GRB energy. We proceed to simulate the light-curve and spectra to obtain the average densities of CBM.

2 Analysis of *Fermi/GBM* data

Gamma-Ray Burst Monitor (GBM, [24]) onboard *Fermi* observatory was specially designed for GRB studies. Having 12 sodium iodide (NaI) detectors (8 keV–1 MeV) and 2 bismuth germanate detectors (250 keV–40 MeV), it covers the energy range of 8 keV – 40MeV. Large Area Telescope (LAT, [25]) is the second instrument onboard *Fermi*, able to detect photons with energies up to more than 300 GeV. GRB 110731A triggered the GBM [26] at 11:09:29.94 (UT) on 2011 July 31 (T_0). The GRB was already in the LAT field of view ($\approx 3°$), so both instruments observed the burst from early on. Using LAT data burst was preliminarily localized at (R.A., J2000)=−280°.39, (decl., J2000)=−28°.53. We downloaded the time-tagged event (TTE) data of the triggering detectors NaI-n0 & NaI-n3, and of the corresponding BGO-b0 detector, from the NASA Fermi Web site[1]. Software package RMFIT[2] was used to perform the time-integrated and time-resolved spectral analysis of the TTE data. Data preparation included subtracting the background by fitting it to a low order polynomial and excluding the overflow channels. We reproduced the LAT and GBM (background subtracted) light-curves in various energy bands (see figure 1) using the TTE data (GBM) and 'P8R2_TRANSIENT020'-class events (LAT). The light-curve, starting around $T_0 - 0.4$s can be described as a complex single-peak structure lasting up to $\approx$10 s, with a sudden decrease after $\approx$7.3 s.

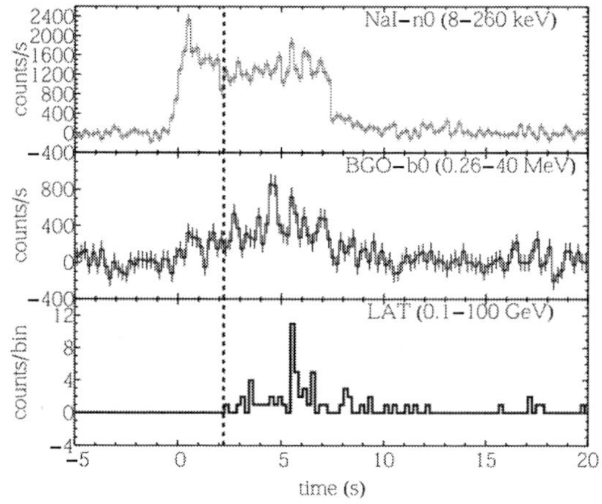

Figure 1. LAT and GBM (NaI & BGO, background subtracted) light-curves of GRB 110731A. Top two panels show NaI-n0 (8–260 keV, top panel) and BGO-b0 (0.26–40 MeV, middle panel) data with 0.2 s binning. The vertical dashed line at $T_0 \sim 2.4$ s indicates the onset of high energy emission observed by LAT (bottom panel, 0.5 s binning).

We performed a time-integrated analysis using the GBM energy range (8 keV – 40 MeV) and approximate GBM-T_{90} interval [0.19 s, 7.9 s]. Three different empirical models were tested: a simple power-law (PL), a

[1] ftp://legacy.gsfc.nasa.gov/fermi/data/gbm/bursts/
[2] fermi.gsfc.nasa.gov/ssc/data/analysis

Table 1. Best-fit parameters of models used for the time-integrated ([0.19 s, 7.9 s]) spectral analysis. The reference energy is fixed to 100 keV. The energy flux is evaluated for the 10–1000 keV range.

Model	E_p (keV)	α	β	Γ_{PL}	F (erg cm^{-2}s^{-1})	C-STAT/DOF
PL	-	-	-	-1.51 ± 0.01	$(2.600 \pm 0.026) \times 10^{-6}$	1921.9/362
Compt	344.5 ± 12.8	-0.86 ± 0.03	-	-	$(2.815 \pm 0.045) \times 10^{-6}$	618.4/361
BAND	324.1 ± 17.3	-0.85 ± 0.03	-2.50 ± 0.15	-	$(2.803 \pm 0.040) \times 10^{-6}$	612.9/360

power-law with the exponential cutoff (Compton) and the Band function [27]. The goodness of the fit was measured using C-statistic [28], requiring an improvement by ΔC-stat=9 for every degree of freedom (DOF). Compton-fit showed a significant improvement over the power-law fit. Band function, having one parameter more, improved the fit by ΔC-stat≈6. The Compton was therefore considered the best-fit model with parameter values of E_p = 344.5(±12.8) keV and $\alpha = -0.86(\pm 0.03)$ (see table 1). In order to calculate the isotropic energy E_{iso}, we fitted the Band function using the entire observed emission interval [-0.38 s, 10.56 s]. This avoids flux underestimation due to the Compton model and detector sensitivity (see, e.g., [29]). The best-fit parameters were: $\alpha = -0.91^{+0.03}_{-0.03}$, $\beta = 2.19^{+0.07}_{-0.09}$, with $E_p = 319.4^{+18.9}_{-16.8}$ keV. E_{iso} was obtained using

$$E_{iso} = \frac{4\pi d_l^2}{(1+z)} S_{bol}, \qquad (1)$$

where d_l^2 is the luminosity distance, z is the redshift and S_{bol} is the bolometric fluence in the $1/(1+z)$ keV to $10^4/(1+z)$ keV frame:

$$S_{bol} = S_{obs} \frac{\int_{1/(1+z)}^{10^4/(1+z)} E\phi(E)dE}{\int_{E_{min}}^{E_{max}} E\phi(E)dE}. \qquad (2)$$

Here, ϕ is the Band model obtained from the spectral fit and $S_{obs} = \Delta t F_{obs}$, where F_{obs} is the mean energy flux in the (E_{min}, E_{max}) detection band and Δt is the time interval of the spectral fit. We find the value of $E_{iso} = 6.05(\pm 0.09) \times 10^{53}$ erg.

We proceed to the time-resolved spectral analysis and P-GRB identification. P-GRB is expected in the form of the first flash of thermal radiation , once the e^+e^- plasma reaches transparency point. This takes place before the evidence of the BH formation, and thus, before the high energy emission (~2.4 s). Expected spectra is thermal (spherically symmetric case) or Compton, produced by a convolution of many instantaneous co-moving spectra with different temperatures (axially symmetric case). Therefore, in order to identify the P-GRB, we selected the times before the high energy emission using a time binning criteria of $S/N = 15$, and fitted six different models to each bin: power-law, Compton, Band and the same three empirical models with added black body (BB) component. We adopted the same criteria of ΔC-stat=9 for every degree of freedom. For every time bin Compton was showed to be the best-fit model (similar as [30], see table 2).

The effective temperature was calculated using $kT = E_p/3.92$, or $kT^{RF} = E_p(1+z)/3.92$ for the rest-frame value. We identified the P-GRB as the emission up to T_0 + 0.68 s, at which point the temperature starts to drop (see figure 2). This coincides with the light-curve maximum. The Compton-fit parameters of this interval are: E_p = 171.9(±17.1) keV, $\alpha = -1.08(\pm 0.08)$ and A = 0.0631(±0.0077) p/s cm^2 keV, where A is the amplitude. They were used to obtain the P-GRB energy $E_{P-GRB} = 3.7 \times 10^{52}$ erg and the rest-frame temperature kT^{RF} = 167.9 keV.

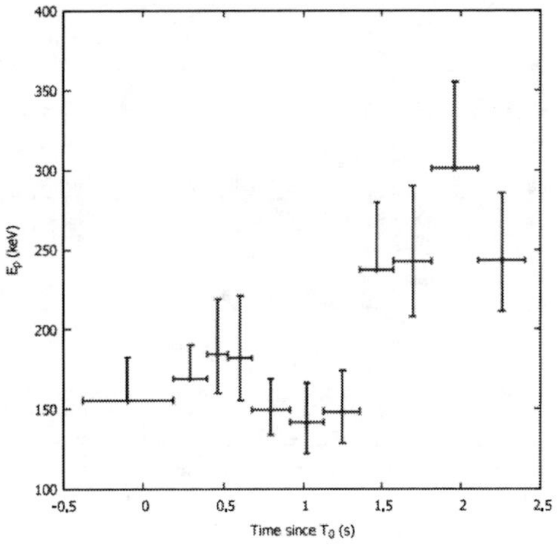

Figure 2. Temporal evolution of the E_p spectral parameter within the Compton model. Bars represent $\pm 1\sigma$ error intervals. Time is given in the observer frame with the binning criteria of S/N=15.

2.1 Light-curve and spectra simulation

The obtained energy values for E_{iso} and E_{P-GRB} served as an input in our numerical code . Calculated value of the baryon load was $B = 4.35 \times 10^{-4}$. At the transparency point, the laboratory radius was 8.33×10^{13} cm and the P-GRB observed temperature 168 keV, with Lorentz factor $\Gamma \sim 2.17 \times 10^3$. We simulated the light-curve and the prompt emission spectrum (see figure 3 and figure 4). The resulting average CBM value was $n \sim 0.03$ particles cm^3.

3 *Swift/XRT* observations

In addition to *Fermi*, GRB 110731A triggered many other observatories, including the burst alert telescope (BAT) onboard *Swift* [31]. Besides BAT, *Swift* spacecraft [32]

Table 2. C-STAT values of 6 models used in the time-resolved spectral analysis, with bin duration given in the first column. E_p value of the best fitting Compton model is given in the second column.

Time bins (s)	Best-fit Compt		C-STAT/DOF other models				
	E_p (keV)	C-STAT/DOF	PL	PL+BB	Compt+BB	Band	Band+BB
$[-0.38, 0.19]$	$155.7^{+26.9}_{-0.0}$	413.2/361	429.7/362	413.8/360	412.5/359	413.2/360	413.1/358
$[0.19, 0.40]$	$168.9^{+21.6}_{-0.0}$	368.3/361	403.0/362	366.7/360	361.9/359	368.3/360	-NaN/358
$[0.40, 0.53]$	$184.7^{+34.5}_{-24.6}$	319.1/361	361.8/362	332.7/360	314.9/359	319.0/360	315.1/358
$[0.53, 0.67]$	$181.9^{+38.9}_{-26.9}$	331.6/361	359.8/362	332.4/360	331.1/359	330.0/360	329.8/358
$[0.67, 0.92]$	$149.3^{+19.8}_{-15.3}$	362.3/361	421.3/362	364.5/360	362.2/359	361.7/360	360.1/358
$[0.92, 1.13]$	$141.7^{+24.8}_{-19.7}$	373.7/361	425.8/362	369.1/360	361.4/359	362.0/360	361.2/358
$[1.13, 1.36]$	$148.5^{+25.8}_{-19.5}$	357.7/361	393.2/362	359.3/360	357.2/359	358.1/360	358.7/358
$[1.36, 1.58]$	$237.3^{+42.3}_{-0.0}$	365.1/361	391.8/362	368.9/360	363.4/359	364.4/360	363.9/358
$[1.58, 1.81]$	$242.8^{+47.3}_{-34.5}$	366.9/361	429.5/362	380.3/360	361.6/359	365.0/360	361.5/358
$[1.81, 2.11]$	$301.0^{+54.2}_{-0.0}$	431.8/361	455.8/362	433.8/360	425.9/359	430.2/360	424.1/358
$[2.11, 2.40]$	$243.2^{+42.4}_{-31.6}$	421.6/361	487.5/362	419.6/360	418.7/359	421.8/360	418.6/358

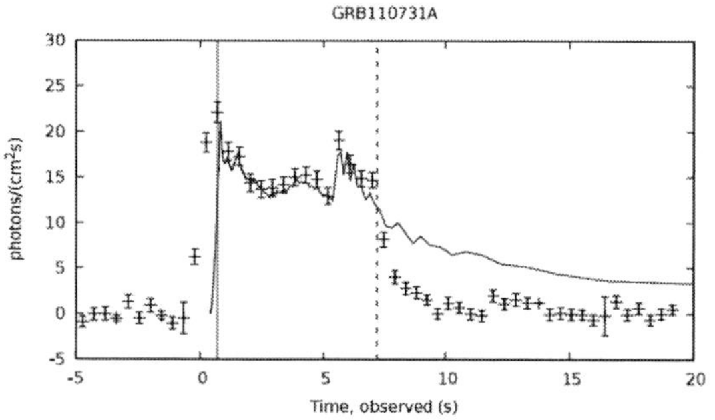

Figure 3. Simulated and NaI-n3 detector (0.2 s binning) light-curve of GRB 110731A. Red vertical line at 0.69 s marks the end of the P-GRB. The dashed line marks the end of the simulation for the spectra.

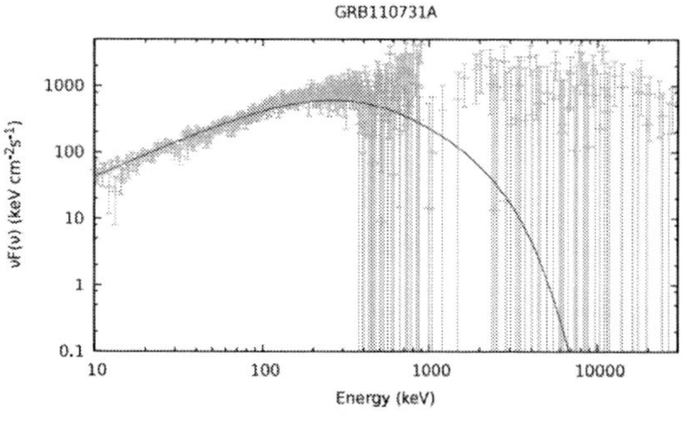

Figure 4. Simulated $\nu F(\nu)$ spectra of the prompt emission, superimposed on the joined NaI-n0, n3 & BGO-b0 observed spectra.

harbors X-ray (XRT) and UV/optical (UVOT) telescopes, intended for the afterglow observations. Available multi-wavelength data make this burst a good candidate to investigate the IGC paradigm. The XRT began observations 56 s after the BAT trigger, where first 8 s were taken while *Swift* was still slewing. The GBM and LAT localizations are consistent with the position provided by *Swift*. Redshift z=2.83 was provided by spectroscopic observations done by GMOS-N [33]. This enabled the XRT light-curve to be transformed to the 0.3–10 keV rest-frame energy range, plotted as a function of the rest-frame time (see figure 5). The GRB 110731A light-curve is superimposed on the X-ray luminosity light-curves of other BdHNe for comparison of the common late power-law behavior, typical for BdHNe. Non-existence of the plateau phase is in agreement with the high E_{iso} of this burst. Furthermore, the PL index $\alpha = -1.22$ is within the limits presented in [22, 23].

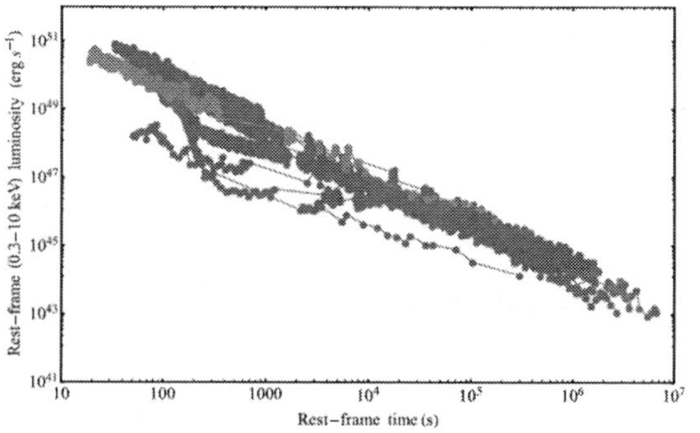

Figure 5. Rest-frame luminosity light-curve of GRB 110731A in the 0.3–10 keV range (orange), superimposed on other BdHNe light-curves (grey): 050525, 060729, 061007, 080319B, 090618, 091127, 100816A, 111228A, 130427A [34]. All GRBs have known redshift.

4 Conclusion and future perspectives

The mechanism responsible for the prompt emission of GRBs is still a debated issue. The origin of the GRB prompt emission in a BdHNe scenario is a consequence of a well-determined sequence of a binary system interaction. Being observed by multiple observatories, GRB

110731A makes a good candidate for investigation of the expected BdHNe signatures. The light-curve showed no evidence of the hypercritical accretion phase, which possibly remains under the detection level, considering the large distance of this GRB ($z = 2.83$). Calculated isotropic energy $E_{\rm iso} = 6.05 \times 10^{53}$ erg is well above the BdHNe lower limit of $\sim 10^{52}$ erg. Using the calculated $E_{\rm iso}$ and P-GRB identified via time-resolved spectral analysis, we inferred the baryon load of $B = 4.35 \times 10^{-4}$, which is within the expected limits for long GRBs. LAT detected a high energy emission $\sim$ 2.4 s after the the GBM trigger, expected in the BdHNe case as a signature of a BH formation. X-ray rest-frame luminosity light-curve showed typical late PL behavior, shared between the other members within the BdHNe group.

This preliminary analysis showed that GRB 110731A is a strong BdHN candidate. Still, in our ongoing analysis we will check the influence of the background selection, bad channels and different binning on the obtained parameters. We will investigate the use of different models and compare these results with the rest of the BdHNe sample.

References

[1] Meegan, C. A., Fishman, G. J., Wilson, R. B., Horack, J. M., Brock, M. N., Paciesas, W. S., Pendleton, G. N., & Kouveliotou, C., Nature **355**, 143 (1992)

[2] Kouveliotou, C., Meegan, C. A., Fishman, G. J., et al., ApJL (413), L101 (1993)

[3] Cucchiara, A., Levan, A. J., Fox, D. B., et al., ApJ (736), 7 (2011)

[4] Amati, L., et al., A&A **390**, 81 (2002)

[5] Amati, L., Guidorzi, C., Frontera, F., Della Valle, M., Finelli, F., Landi, R., & Montanari, E., MNRAS **391**, 577 (2008)

[6] Zhang, F. W., Shao, L., Yan, J. Z., & Wei, D. M., ApJ **750**, 88 (2012)

[7] Calderone, G., Ghirlanda, G., Ghisellini, G., et al. 2015, MNRAS, 448, 403.

[8] Ruffini, R., Rueda, J. A., Muccino, M., et al., ApJ **832**, 136 (2016)

[9] Damour, T., & Ruffini, R., Physical Review Letters **35**, 463 (1975)

[10] Piran, T., Reviews of Modern Physics **76**, 1143 (2004)

[11] Narayan, R., Piran, T., & Shemi, A., ApJL **379**, L17 (1991)

[12] Lee, W. H., Ramirez-Ruiz, E., & Page, D., ApJL **608**, L5 (2004)

[13] Galama, T. J., Vreeswijk, P. M., van Paradijs, J., et al., Nature **395**, 670 (1998)

[14] Fryer, C. L., Rueda, J. A., & Ruffini, R., ApJL **793**, L36 (2014)

[15] Ruffini, R., Bianco, C. L., Fraschetti, F., Xue, S.-S., & Chardonnet, P., ApJ **555**, L117 (2001)

[16] Rueda, J. A., & Ruffini, R., ApJL **758**, L7 (2012)

[17] Izzo, L., Ruffini, R., Penacchioni, A. V., et al., A&A **543**, A10 (2012)

[18] Ruffini, R., Salmonson, J. D., Wilson, J. R., & Xue, S.-S., A&A **350**, 334 (1999)

[19] Ruffini, R., Salmonson, J. D., Wilson, J. R., & Xue, S.-S., A&A **359**, 855 (2000)

[20] Nousek, J. A., Kouveliotou, C., Grupe, D., et al., ApJ **642**, 389 (2006)

[21] Zhang, B., Fan, Y. Z., Dyks, J., et al., ApJ **642**, 354 (2006)

[22] Pisani, G. B., Izzo, L., Ruffini, R., et al., A&A **552**, L5 (2013)

[23] Ruffini, R., Muccino, M., Bianco, C. L., et al., 2014, A&A, 565, L10 (2014)

[24] Meegan, C., Lichti, G., Bhat, P. N., et al., ApJ **702**, 791 (2009)

[25] Atwood, W. B., Abdo, A. A., Ackermann, M., et al., ApJ **679**, 1071 (2009)

[26] Gruber, D., GRB Coordinates Network, 12221 (2011)

[27] Band, D., Matteson, J., Ford, L., et al., ApJ **413**, 281 (1993)

[28] Cash, W., ApJ **228**, 939 (1979)

[29] Virgili, F. J., Qin, Y., Zhang, B., & Liang, E., MNRAS **424**, 2821 (2012)

[30] Ackermann, M., et al., ApJ **763**, 71 (2013)

[31] Oates, S. R., Beardmore, A. P., et al., GRB Coordinates Network, 12215 (2011)

[32] Gehrels, N., et al., ApJ **611**, 1005 (2004)

[33] Tanvir, N. R., Wiersema, K., et al., GRB Coordinates Network, 12225 (2011)

[34] Ruffini, R., Wang, Y., Aimuratov, Y., et al., arXiv:170403821 (2017)

Evolution of an electron-positron plasma produced by induced gravitational collapse in binary-driven hypernovae

J. D. Melon Fuksman[1,2,⋆], *L.* Becerra[1,2], *C. L.* Bianco[1,2], *M.* Karlica[1,2,3], *M.* Kovacevic[1,2,3], *R.* Moradi[1,2], *M.* Muccino[1,2], *G. B.* Pisani[1,2], *D.* Primorac[1,2], *J. A.* Rueda[1,2,4], *R.* Ruffini[1,2,3,4], *G. V.* Vereshchagin[1,2], and *Y.* Wang[1,2]

[1] *Dipartimento di Fisica, Sapienza Università di Roma, Piazzale Aldo Moro 5, 00185 Rome, Italy*
[2] *ICRANet, Piazza della Repubblica 10, 65122 Pescara, Italy*
[3] *Université de Nice Sophia Antipolis, CEDEX 2, Grand Château Parc Valrose, Nice, France*
[4] *ICRANet-Rio, Centro Brasileiro de Pesquisas Físicas, Rua Dr. Xavier Sigaud 150, 22290–180 Rio de Janeiro, Brazil*

Abstract. The binary-driven hypernova (BdHN) model has been introduced in the past years, to explain a subfamily of gamma-ray bursts (GRBs) with energies $E_{iso} \geq 10^{52}$ erg associated with type Ic supernovae. Such BdHNe have as progenitor a tight binary system composed of a carbon-oxigen (CO) core and a neutron star undergoing an induced gravitational collapse to a black hole, triggered by the CO core explosion as a supernova (SN). This collapse produces an optically-thick e^+e^- plasma, which expands and impacts onto the SN ejecta. This process is here considered as a candidate for the production of X-ray flares, which are frequently observed following the prompt emission of GRBs. In this work we follow the evolution of the e^+e^- plasma as it interacts with the SN ejecta, by solving the equations of relativistic hydrodynamics numerically. Our results are compatible with the Lorentz factors estimated for the sources that produce the flares, of typically $\Gamma \lesssim 4$.

1 Introduction

The *induced gravitational collapse* (IGC) (see, e.g., [1], [2], [3]) has been proposed in the past years, as a way to explain a sub-class of long gamma-ray bursts (GRBs) called binary-driven hypernovae (BdHNe), characterised by an isotropic energy $E_{iso} \geq 10^{52}$ erg and a rest-frame spectral peak energy between 0.2 and 2 MeV. The model considers a binary system formed by a carbon-oxygen (CO) core and a neutron star (NS), which undergo a tight orbit. If the core-collapse of the CO star produces a supernova explosion, the ejected material may trigger an hypercritical accretion process onto the NS, due to a copious neutrino emission and the trapping of photons within the accretion flow. This process can cause the NS to collapse as well, thus forming a black hole (BH). It has been proposed in [4] and, e.g., [5], that such collapse can lead to the formation of an e^+e^- plasma, that later expands and interacts with the SN ejecta, finally producing a GRB.

In this scenario, the major portion of the optically thick e^+e^--baryon plasma originating from the collapse expands away from the supernova (SN), giving rise to the canonical GRB *prompt emission*. This emission occurs at approximately 10^{15}-10^{17} cm from the BH, and is measured to come from material that expands at Lorentz factors $\Gamma \sim 10^2$-10^3 (see e.g. [5]). Right after this first stage, that can last up to $\sim$ 100 s, *X-ray flares* are frequently observed, followed by the so-called "plateau" and finally by the late decay of the X-ray afterglow [6]. By studying the time evolution of the thermal component of the X-ray flares, it can be inferred that they originate from regions which move at roughly $\Gamma \lesssim 4$, as pointed out in [6].

These differences in the features of the prompt emission and the flares can be explained in terms of the IGC model. In it, the prompt emission is produced after the interaction of the e^+e^- with the SN ejecta, in a direction that corresponds to lower overall densities along the line of sight of an external observer (see [7] and Fig. 1). On the other hand, as the binary system keeps spinning, the mass density profile along that line changes. If a bigger amount of mass gets between the BH and the observer (see [8] and Fig. 1), an X-ray flare is emitted at the moment the plasma escapes the outermost regions of the SN ejecta, namely, at the shock *breakout*. Due to the deceleration of the shock by its interaction with the surrounding material, the Lorentz factors measured for the flares will be smaller than for the prompt emission.

In this work we describe numerically the evolution of the plasma along different directions, and study the compatibility of the IGC model with some of the above-mentioned observational features.

2 Equations and numerical scheme

We have modeled the evolution of the e^+e^- plasma and the SN ejecta following a single-fluid approach, where all the involved particle species, in this case baryons, photons, electrons and positrons, coming either from the plasma or the SN, are in *local thermodynamic equilibrium* (LTE).

⋆e-mail: david.melon@icranet.org

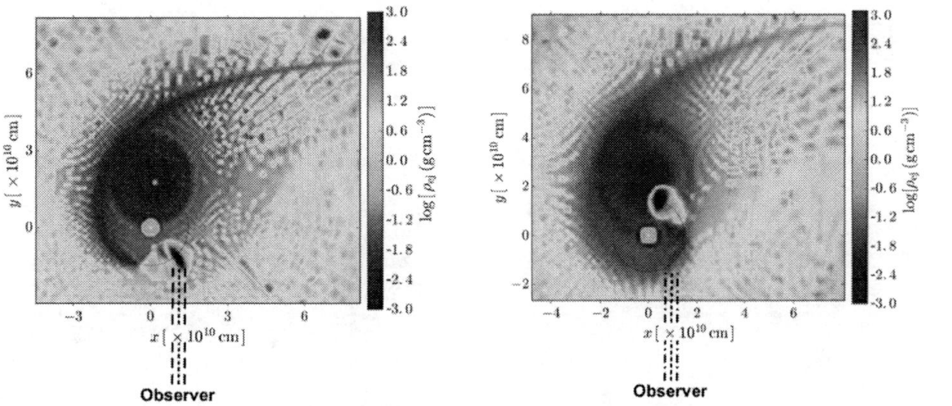

Figure 1. Density profiles calculated in [8] corresponding to the SN ejecta at (left) the moment of the collapse and (right) 100 s later. The line of sight of an observer that sees the initial prompt emission is indicated, to point out the change of the mass profile along it.

Under this assumption, the dynamics of the resulting fluid is governed by the equations of relativistic hydrodynamics (RHD). Throughout this work, we have numerically solved the one-dimensional RHD equations with the additional assumption of spherical symmetry, considering only a dependence of the intensive variables on the radial spherical coordinate. This allows us to study the evolution of the plasma along one selected radial direction at a time, thus to consider the different density distributions as seen through each direction. In the absence of gravity, the resulting equations of motion can be written as follows:

$$\frac{\partial(\rho\Gamma)}{\partial t} + \nabla.(\rho\Gamma\mathbf{v}) = 0, \quad (1)$$

$$\frac{\partial m_r}{\partial t} + \nabla.(m_r\mathbf{v}) + \frac{\partial p}{\partial r} = 0, \quad (2)$$

$$\frac{\partial \mathcal{E}}{\partial t} + \nabla.\,(\mathbf{m} - \rho\Gamma\mathbf{v}) = 0, \quad (3)$$

where ρ is the comoving mass density, $\mathbf{v}$ is the fluid velocity (natural units where $c = 1$ are used), $\Gamma = (1 - \mathbf{v}^2)^{-\frac{1}{2}}$ is the Lorentz factor, $\mathbf{m}$ the total momentum density, and the subscript r indicates the radial component of a vector. The momentum density $\mathbf{m}$ is defined as $\mathbf{m} = h\Gamma^2\mathbf{v}$, where h is the comoving enthalpy density, given by $h = \rho + \epsilon + p$, where p is the gas pressure and ϵ its internal energy density, both measured in the comoving frame. Finally, we denote by $\mathcal{E}$ the value of ϵ in the laboratory (or coordinate) frame. In the above formulation of these equations, we compute this energy density by substracting the coordinate mass density $\rho\Gamma$ to the $(0, 0)$ component of the fluid's energy-momentum tensor $T^{\mu\nu}$, as

$$\begin{aligned}\mathcal{E} &= T^{00} - \rho\Gamma \\ &= h\Gamma^2 - p - \rho\Gamma\,. \end{aligned} \quad (4)$$

Whenever the LTE condition holds, an equation of state relating ϵ, p and ρ can be obtained, thus closing the system defined by equations (1) to (4). In this work, we have used the equation of state of an ideal relativistic gas, which can be expressed in terms of its enthalpy as:

$$h = \rho + \frac{\gamma p}{\gamma - 1}, \quad (5)$$

with $\gamma = 4/3$. Considering the contributions of all the involved particles to the total density, mass and energy, we have verified that equation (5) holds in the full range of parameters used in our simulations, with a value of γ that deviates from $4/3$ with a maximum error of 0.2%. The details of this calculation can be found in [6].

To integrate the above-defined system of equations, we have used the one-dimensional RHD module included in the PLUTO code [9]. The code works by making use of Godunov-type Riemann solvers, of which we have chosen an extension of the HLLC scheme to the equations of RHD (see [10] for the complete details). Among the possible configurations included in PLUTO, we have used a second-order total variation diminishing scheme for spatial reconstruction, and second-order Runge Kutta integration for time evolution. On each time step, the grid was updated in order to better follow total energy density gradients, by means of an adaptive mesh refinement algorithm provided by the CHOMBO library [11].

3 Simulations

3.1 Low density directions

As a verification of this scheme's applicability to our system, we have performed similar simulations to those in [7], which account for the evolution of the plasma in the lower density regions (see Fig. 1). To be more precise about this, we define the *baryon load* as $B = M_B c^2/E_{e^+e^-}$, where M_B is the total integrated mass considering the assumed spherically symmetric distribution, and $E_{e^+e^-}$ is the initial internal energy of the plasma. Namely, B works as an indicator of the mass-to-energy ratio along each direction.

In this part, we will only consider density distributions such that $B < 10^{-2}$. As it is explained in detail in [7], in that case the plasma forms a slab that expands while

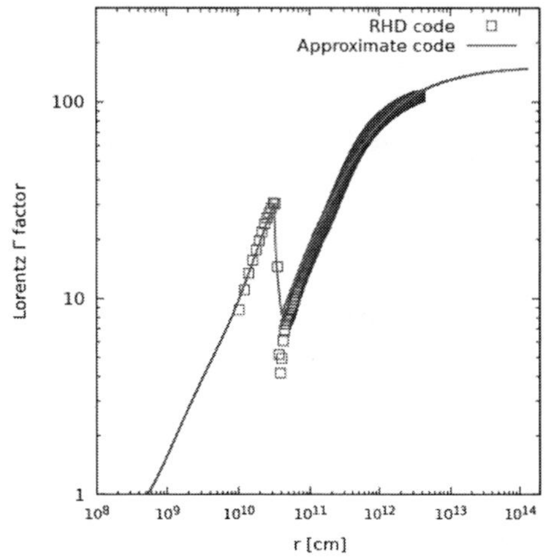

Figure 2. Lorentz factor Γ computed in with the PLUTO code, compared with the one computed with the semi-analytic approximate code. For this plot, the values $E_{e^+e^-} = 1.0 \times 10^{53}$ erg and $B = 6.61 \times 10^{-3}$ have been chosen. Similar good agreement is found for other values of $E_{e^+e^-}$ and B as long as $B < 10^{-2}$. Reproduced from [6].

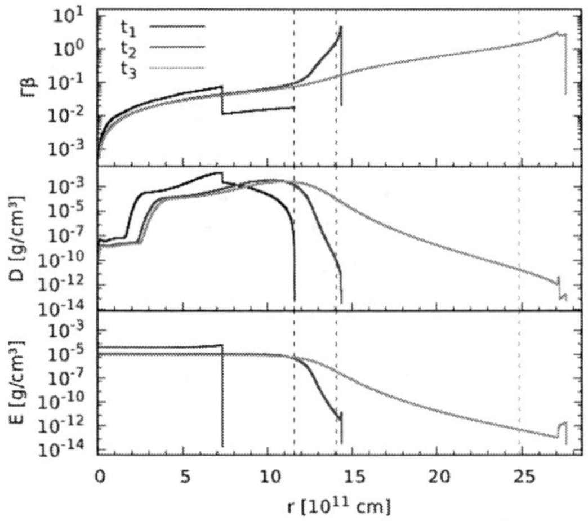

Figure 3. Profiles of $\Gamma v/c$ (up), $D = \rho\Gamma$ (center) and plasma internal energy $\mathcal{E}$ (down) for $B = 200$ at three different times, labeled as t_1 (before the breakout), t_2 (at the breakout) and t_3 (after the breakout). The factor $\Gamma v/c$ is approximately equal to v/c when $v \ll c$, and to Γ when $v \sim c$. Contributions to $\mathcal{E}$ due to the density have been neglected, in order to substract the baryon's kinetic energy, and show the position of the shock. The vertical dashed lines indicate the position of R_{ph} at each time. See also [6].

accelerating, until it reaches a constant Lorentz factor of roughly $\Gamma \sim 1/B$. During its evolution, and still under the condition $B < 10^{-2}$, the slab's width remains constant when measured in the laboratory frame. This process is studied in [7] using an approximate semi-analytic code that assumes this feature, and allows to predict average intensive quantities by means of conservation equations.

In Fig. 2 we show the comparison between the Lorentz factors computed with both the semi-analytic code and PLUTO, for one particular value of $E_{e^+e^-}$ and B. The simulations run with PLUTO show the formation of a slab of constant width, that accelerates accordingly to what is expected from the semi-analytic code. This is in turn consistent with the treatment done in [7], and therefore, the analysis done so far for the prompt emission remains unchanged.

3.2 High density directions

Having applied the current scheme to the already-known region of the parameter space, we proceed to study the case $B > 10^{-2}$, which corresponds to the interaction of the e^+e^- plasma with the SN remnant along the higher density directions, and the subsequent emission of the flares. To this end, we have chosen for each simulation an initial density profile that matches the ones obtained in [8] (see Fig. 1). Hence, all the considered profiles were set in the following way:

$$\rho \propto (R_0 - r)^\alpha , \qquad (6)$$

where R_0 and α, with $2 < \alpha < 3$, are fitting parameters. Each profile of this kind has a single baryon load, that corresponds to the evolution of the fluid along one particular direction. Similarly, we have taken the velocity to depend on the radius as $v_r \propto r$, in order to set an initial profile that corresponds to the homologous expansion of the SN ejecta. Lastly, the plasma is initially contained within a radius of order $10^8 - 10^9$ cm, and has an uniform energy density.

Once the system is let to evolve from these initial conditions, the plasma expands and forms a shock that reaches the outermost part of the SN ejecta. Instead of forming a thin shell with an almost uniform Lorentz factor, the plasma evolves in such a way that the shock is followed by smooth energy and velocity distributions, where the module of the last may differ in several orders of magnitude from one point to another, as it is shown in Fig. 3.

After an initial expansion where Γ may reach values of several tenths, the shock rapidly elgulfs enough mass to decelerate and reach a non relativistic velocity distribution, typically in $t < 1$ s. This is maintained through the whole time the plasma is contained within the SN ejecta, until the breakout, in which the sudden decrease of the density causes the closest areas to the shock to reach relativistic velocities, as it can be seen in Fig. 3. From then on, the matter pushed by the shock -and no longer in homologous expansion- keeps expanding while cooling down.

As the shock propagates inside of the SN ejecta, photons are trapped inside the region occupied by the plasma, since their diffusion timescale is much longer than the dynamical times. Since characteristic equilibrium times are much shorter than both scales (see e.g. [12]), LTE is maintained during the whole evolution of the plasma within the SN material. However, at the breakout, the plasma reaches

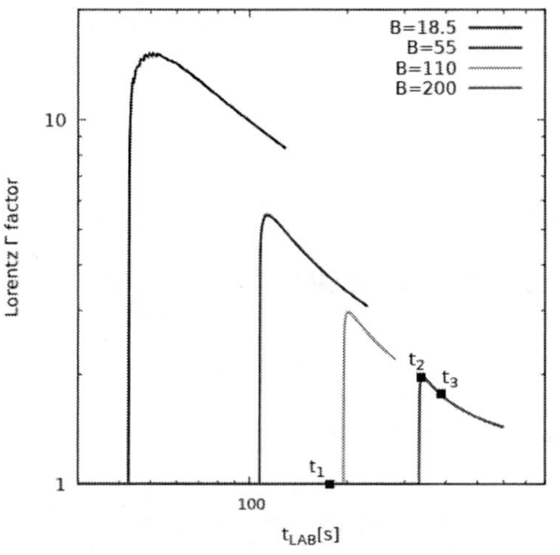

Figure 4. Lorentz gamma at the photospheric radius, calculated for different values of the baryon load, corresponding to different considered directions along the SN ejecta. The three times considered for $B = 200$ in Fig. 3 are indicated. Reproduced from [6].

the *photosphere*, which is an optically thin area located in the outermost regions of the SN ejecta. When this happens, photons begin to escape, and produce what is later observed as a flare.

To perform a comparison between these results and the observed Γ factors, we firstly recall from [13] that, before reaching an external observer, most of the photons are lastly scattered from a region peaked at $\tau \sim 1$, where τ is the optical depth calculated from the observer's line of sight. We hence define the transparency -or photospheric- radius R_{ph} as the value of the r coordinate that satisfies $\tau(R_{ph}) = 1$. Finally, we can give an estimation of the observed Γ by computing its value at $r = R_{ph}$.

Since τ is a Lorentz invariant when the total photon cross section is constant, which we assume as a first approximation, it can be calculated in laboratory coordinates as

$$\tau = \int_{R_{ph}}^{\infty} \mathrm{d}r\, \sigma_T\, n_{e^-}(r), \tag{7}$$

where we set the cross section to $\sigma_T = 6.65 \times 10^{-25}$ cm^2, i.e., the one corresponding to Thomson scattering by electrons. Moreover, we compute the electron density as $n_{e^-} = \rho\Gamma/m_P$, where m_P is the proton mass. In doing so, we neglect the mass of the electrons, and we assume to have an average of one electron per nucleon.

Fig. 4, already shown in [6], shows the time evolution of $\Gamma(R_{ph})$, calculated for four different baryon loads that correspond to four different high-density directions along the SN ejecta. For a high enough B, it can be seen that, indeed, $\Gamma(R_{ph}) \lesssim 4$.

4 Final remarks and future work

The performed simulations of the evolution of an e^+e^- plasma inside of a SN ejecta show as a result the formation, expansion and breakout of a shock. Within the IGC model's parameters, the Lorentz factor at the photospheric radius verifies $\Gamma \lesssim 4$ for a high enough integrated mass across the observer's line of sight. This is consistent with the existence of a thermal emitter expanding at such a Lorentz factor, as it is inferred from the thermal component observed in the X-ray flares.

We have said that RHD holds as long as LTE is granted. However, this is not the case for the thin region close to the shock, from where photons can escape. If this is taken into account, the pressure radiation, which is the dominant one, should be actually smaller, and consequently we should expect the actual Lorentz factors to be even smaller. Therefore, the results of this work must be taken as a superior limit for Γ. To take this effect into account, we are currently working on a scheme that evolves radiation and massive particles separately, which would allow to compute the emitted luminosity as well.

References

[1] Rueda, J. A., & Ruffini, R., ApJL, **758**, L7 (2012)
[2] Izzo, L., Rueda, J. A., & Ruffini, R., A&A, **548**, L5 (2012)
[3] Fryer, C. L., Rueda, J. A., & Ruffini, R., ApJL, **793**, L36 (2014)
[4] Damour, T., Ruffini, R., APS, **35**, **7**, 463 (1975)
[5] Ruffini, R., Rueda, J. A., Muccino, M., et al., **832**, 136 (2016)
[6] Ruffini, R., Wang Y. et al, https://arxiv.org/abs/1704.03821, submitted to ApJ (2017)
[7] Ruffini, R., Salmonson, J. D., Wilson, J. R., Xue, S., A&A, **359**, 855 (2000)
[8] Becerra L., Bianco C. L., Fryer C. L., Rueda J. A., Ruffini R., ApJ **833**, 107 (2016)
[9] Mignone, A., Zanni, C., Tzeferacos, P., et al., The Astrophysical Journal Supplement Series, **198**, 7 (2011)
[10] Mignone, A., Bodo, G., Mon. Not. R. Astron. Soc., **364**, 126 (2005)
[11] Colella, P., Graves, D. T., Johnson, J. N., et al., *Chombo software package for AMR applications design document*, Tech. Rep. LBNL-6615E (2003)
[12] Aksenov A. G. , Ruffini R., Vereshchagin G. V., Phys. Rev. D **79**, 043008 (2009)
[13] Pe'er A., ApJ, **682**, 463 (2008)

Searches for magnetic monopoles with IceCube

Anna Pollmann[1,*] for the IceCube Collaboration[**]

[1] *Dept. of Physics, University of Wuppertal, 42119 Wuppertal, Germany*

Abstract. Particles that carry a magnetic monopole charge are proposed by various theories which go beyond the Standard Model of particle physics. The expected mass of magnetic monopoles varies depending on the theory describing its origin, generally the monopole mass far exceeds those which can be created at accelerators. Magnetic monopoles gain kinetic energy in large scale galactic magnetic fields and, depending on their mass, can obtain relativistic velocities.
IceCube is a high energy neutrino detector using the clear ice at the South Pole as a detection medium. As monopoles pass through this ice they produce optical light by a variety of mechanisms. With increasing velocity, they produce light by catalysis of baryon decay, luminescence in the ice associated with electronic excitations, indirect and direct Cherenkov light from the monopole track, and Cherenkov light from cascades induced by pair creation and photonuclear reactions. By searching for this light, current best limits for the monopole flux over a broad range of velocities was achieved using the IceCube detector. A review of these magnetic monopole searches is presented.

1 Introduction

The existence of magnetic monopoles, particles carrying a single magnetic charge, is motivated by various theories which extend the Standard Model of particles, such as Grand Unified Theories (GUTs), String theory, Kaluza-Klein, and M-Theory [1]. The elementary magnetic charge g_D is derived from basic principles with a value of $g_D = e/2\alpha$ [2] where α is the fine structure constant and e is the elementary electric charge. Other parameters of magnetic monopoles, such as their mass and predicted flux, depend on the details of the models regarding particle creation.

Relic monopoles are predicted to have been created during symmetry breaking shortly after the Big Bang with a mass of the order of 10^{13} GeV to 10^{19} GeV [3]. The proposed inflationary phase of the early universe leads to a dilution of the monopole density. Intermediate (or smaller) mass monopoles are predicted within intermediate steps of symmetry breaking during or after inflation with masses in the order of 10^7 GeV to 10^{13} GeV [4, 5]. Monopoles at relativistic speeds are expected to have intermediate masses since only in this mass range they can be sufficiently accelerated in cosmic magnetic fields [4]. Magnetic monopoles can also be introduced by modest extensions of the electroweak theory of the standard model with masses ranging down to a few tera electronvolts [6].

Although the existence of light monopoles may be tested at accelerator experiments (e.g. MoEDAL) [7, 8], for intermediate and heavier masses the cosmic abundance of monopoles is

* e-mail: anna.pollmann@uni-wuppertal.de
** http://icecube.wisc.edu

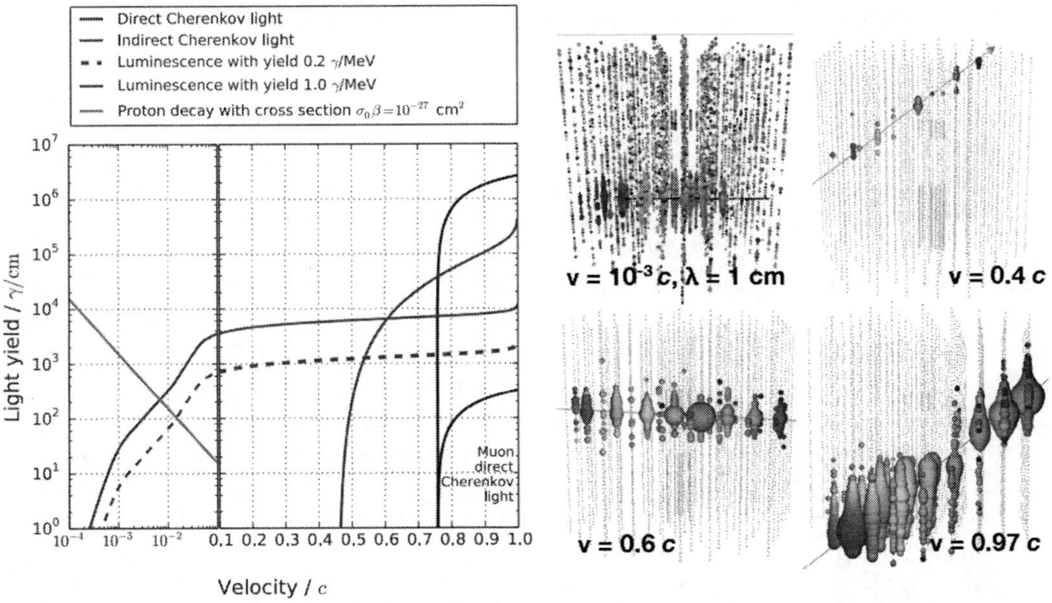

Figure 1. *Left:* Light yield of monopole signatures at different speeds compared to the light yield of a bare muon (black solid). The luminescence yield is varied between $0.2\,\gamma/\mathrm{MeV}$ [11] (green dashed) and $1.0\,\gamma/\mathrm{MeV}$ [12] (green solid). *Right:* Event views of simulated monopole signatures showing light emission due to proton decay (top left) [13], luminescence (top right), indirect Cherenkov (bottom left), and direct Cherenkov radiation (bottom right). The position of the IceCube DOMs are shown with gray spheres. Hit DOMs are visualized with colored spheres, scaling with the recorded charge. The color denotes the time from early (red) to late (blue). The simulated track of the monopole is shown with a solid line. The event view at the top left is simulated using a mean free path of 1 cm for proton decay catalysis and includes hits produced by photomultiplier noise and passing muons during the long event duration over long millisecond event duration.

constrained by astrophysical arguments [9] or by direct searches. Special purpose detectors for monopoles, such as MACRO or MoEDAL [7, 10], use induction, visible damage of plastic targets, and time of flight measurement. Generically, light yield from a monopole is much greater than for particles with electronic charges as shown in Fig. 1 (left). As a result, large aperture general purpose detectors of high energy particles (e.g. IceCube, ANTARES, Baikal) provide the best limits on the flux of monopoles.

2 Monopole signatures

IceCube is a cubic-kilometer neutrino detector installed in the ice at the geographic South Pole between depths of 1450 m and 2450 m [14]. It consists of 86 strings, each with 60 digital optical modules (DOMs). Neutrino reconstruction relies on the optical detection of Cherenkov radiation emitted by secondary particles produced in neutrino interactions in the ice or the nearby bedrock. The DeepCore sub-array, as used in this work, includes the 8 more densely instrumented strings, optimized for low energies, plus 7 adjacent standard strings [15]. The large volume of IceCube and its generic detection method relying on light production is beneficial for rare event searches, such as magnetic monopole detection. Depending on the monopole mass and arrival direction at Earth, magnetic monopoles can have a range of velocities when penetrating IceCube.

Cherenkov light is produced by charged particles exceeding the speed of $0.76\,c$ in ice. The Cherenkov light yield for magnetic monopoles is $\approx 2 \cdot (g_D/e)^2$ times higher than that of electric charges [16]. The event signature of *highly relativistic* monopoles is therefore expected to be a bright and homogeneous track passing through the entire detector. At *ultra-relativistic* speeds ($\gamma \gtrsim 10^3$), photo-nuclear interactions and pair production take over as the dominant light production processes [17].

Below the Cherenkov threshold magnetic monopoles are energetic enough to ionize the surrounding matter and accelerate the ejected δ-electrons above the Cherenkov threshold. This so-called indirect Cherenkov light can be used for monopole detection down to $\approx 0.5\,c$ [18]. At these *mildly relativistic* velocities monopoles have a comparable light yield as a bare muon.

In some GUT theories protons are unstable particles with a long lifetime [19, 20]. A proton coming close to a monopole's core can decay under baryon number violation. The decay products of a 1 GeV proton induce a small particle cascade in which the secondaries emit Cherenkov light. The catalysis cross section is inversely dependent on the monopole speed. IceCube is sensitive to mean free paths from 1 mm to 10 m [13]. Since IceCube's strings are 125 m apart from each other, the resulting signature of a *non-relativistic* monopole is a slow track. The brightness of the signature depends on the catalysis cross section.

The described detection methods do not cover low relativistic velocities, nor non-relativistic velocities if baryon decay does not exist. Therefore a new approach is followed in order to fill this gap. At *low relativistic* velocities magnetic monopoles produce luminescence light by electronic excitation of the surrounding medium. The luminescence yield dN_γ/dx per path length scales linearly with the monopole energy loss dE/dx. The energy loss can be estimated from the heavy ion energy loss by using an effective charge of $\beta \cdot g_D$. The light yield of pure ice luminescence irradiated with α-particles was recently measured to be on the order of $\approx (1.0 \pm 0.3)\,\gamma/\text{MeV}$ at IceCube temperatures, integrated over the wavelengths to wish IceCube is sensitive [12]. Simulations of monopoles with speed above $0.1\,c$ show, that their signatures can be detected by IceCube [21]. However, the effect of impurities and different charges of the projectile are yet to be determined. The light yield of the described light emissions is shown in Fig. 1 (left) and compared with the light yield of a bare muon. Simulated event views of the expected monopole signatures are shown in Fig. 1 (right).

3 Recent IceCube searches

Searches for magnetic monopoles with IceCube have been executed targeting three of the described signatures. All IceCube analyses are based on simulated signal and background, along with 10% of the recorded data, to avoid experimental bias. After an internal review the analysis scheme is fixed and applied on the remaining 90% of data. Since 2011 the IceCube data acquisition at Pole is running a specialized trigger for the collection of long lasting events produced by non-relativistic (i.e. heavy) monopoles catalyzing proton decay. During the required time, up to several milliseconds, tens of muons pass through the detector and the photomultipliers' dark rates cause almost every DOM to detect a hit. The challenge is to reconstruct the monopole track out of the dominant background. The trigger searches for triplets (three pairs of hit DOMs) which fulfill conditions related to timing and opening angle. The further analysis requests monopole candidates to have at least 26 triplets above a reconstructed speed of $10^{-3}\,c$ or 60 triplets below that speed. With the application of this analysis on one year of DeepCore data no monopole candidate was found. Upper flux bounds were derived for catalysis cross sections $\sigma_0\beta$ from $5 \cdot 10^{-26}\,\text{cm}^2$ up to $2 \cdot 10^{-22}\,\text{cm}^2$ on the order of $2 \cdot 10^{-15}\text{cm}^{-2}\text{s}^{-1}\text{sr}^{-1}$

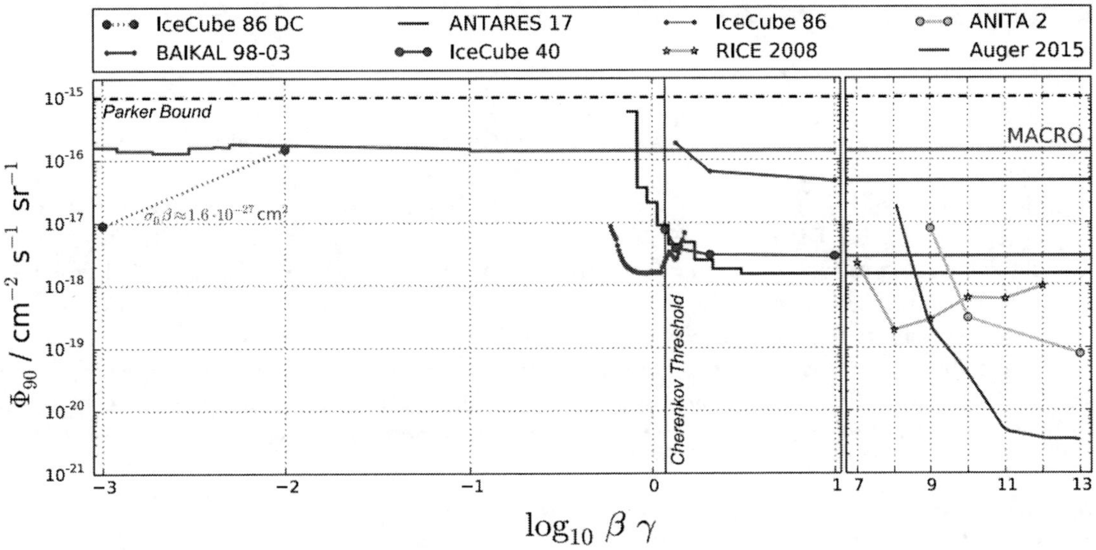

Figure 2. Current upper limits on the flux of magnetic monopoles in dependence on the speed of monopoles at the detector [11, 13, 18, 22–26]. The best limits for cosmic monopoles are achieved by general purpose detectors. However, low relativistic speeds, from $0.1\,c$ to $0.5\,c$, are not covered by them yet so that the latest limit in this region is almost two orders of magnitude higher than for all other speed ranges.

to $2 \cdot 10^{-18}\text{cm}^{-2}\text{s}^{-1}\text{sr}^{-1}$ [13]. Assuming dark matter is dominated by magnetic monopoles, these limits require a monopole mass above $1.22 \cdot 10^{19}$ GeV.

The search for highly relativistic magnetic monopoles of intermediate mass producing direct Cherenkov light [18] used data from 2008 when IceCube was half constructed. Since the monopole signature is as bright as a few hundred tera electronvolt muon, the analysis was based on the extremely high energy filter which is used to select candidate events for astrophysical neutrino searches. Before the detection of astrophysical neutrinos the signature of monopoles would have been distinctive because of its brightness and its upwards going direction. Using this as a main criterion the analysis found three events of which one was an expected muon neutrino and two were of unexpected origin. The latter were identified as partially contained, bright cascades which were possibly early hints of the later detected astrophysical neutrino flux.

IceCube's most recent search for magnetic monopoles targeted mildly relativistic monopoles of intermediate mass emitting indirect Cherenkov light. Data from 2011, when the detector was completed, was used in this analysis. After selecting tracks, reconstructed within the anticipated velocity range and in upwards direction, the main backgrounds were coincident air showers and neutrinos which pass through the 125 m gaps in the geometry of IceCube. Further event selection was done using machine learning algorithms (boosted decision trees). All used variables rely on simple geometrical and timing patterns of the hit DOMs. After the last selection up to 3.61 background events were expected. Three vertical track signatures did pass this criterion. They were identified as muons resulting from neutrino interactions below the IceCube volume. The derived flux bound improves on previous bounds by almost two orders of magnitude [18].

4 Summary and outlook

Due to its unprecedented size IceCube holds the strongest constraints on the flux of non-relativistic and mildly relativistic, i.e. intermediate mass and heavy, magnetic monopoles. Even better sensitivity is expected from further years of data taking with IceCube, or from proposed volume extensions of the detector [27].

Ongoing analyses using IceCube data are expected to improve the current limits and extend over larger velocity ranges. The new analysis searching for slow monopoles introduces a new filter extending data taking to long duration events from DeepCore to whole IceCube. In addition there is an ongoing analysis searching for highly relativistic monopoles discriminating tracks caused by high energetic neutrinos by the homogeneity of the tracks.

Finally there are measurements of light yield from luminescence in ice ongoing, enabling new searches for low and non-relativistic magnetic monopoles. These analyses open a new detection channel enhancing the chance for observation of magnetic monopoles in IceCube.

References

[1] Y.M. Shnir, *Magnetic Monopoles* (Springer, 2005)
[2] P. Dirac, Proc. Roy. Soc. **A 133**, 60 (1931)
[3] A. Rajantie, Contemp. Phys. **53**, 195 (2012)
[4] C. Patrignani et al., Chin. Phys. C **40** (2016)
[5] T.W. Kephart, G.K. Leontaris, Q. Shafi (2017), arXiv:1707.08067
[6] Y.M. Cho, K. Kimm, S. Oh, J.H. Yoon (2017), arXiv:1703.02188
[7] B. Acharya et al., Phys. Rev. Lett. **118**, 061801 (2017)
[8] G. Aad et al., Phys. Rev. D **93**, 052009 (2016)
[9] E.N. Parker, Astrophys. J. **160**, 383 (1970)
[10] M. Ambrosio et al. (MACRO Collaboration), Eur. Phys. J. C **25**, 511 (2002)
[11] V. Aynutdinov et al. (BAIKAL Collaboration), Astrophys. J. **29**, 366 (2008)
[12] A. Pollmann, S. Pieper, *Measurement of water luminescence*, in *ICRC* (2017), POS1060
[13] M.G. Aartsen et al. (IceCube Collaboration), Eur. Phys. J. C **74**, 2938 (2014)
[14] M.G. Aartsen et al. (IceCube Collaboration), JINST **12**, P03012 (2017), arXiv:1612.05093
[15] R. Abbasi et al. (2011), arXiv:1109.6096
[16] D.R. Tompkins, Phys. Rev. **138** (1964)
[17] S.D. Wick, T.W. Kephart, T.J. Weiler, P.L. Biermann, Astropart. Phys. **18**, 663 (2003)
[18] M.G. Aartsen et al. (IceCube Collaboration), Eur. Phys. J. C **76**, 1 (2016)
[19] V.A. Rubakov, Nucl. Phys. B **203**, 311 (1982)
[20] C.G. Callan, Nucl. Phys. B **212**, 391 (1983)
[21] A. Pollmann, *Luminescence of water or ice ...*, in *ICNFP 16* (2016), arXiv:1610.06397
[22] A. Albert et al., J. High Energ. Phys. **2017**, 54 (2017)
[23] R. Abbasi et al. (IceCube Collaboration), Phys. Rev. D **87**, 022001 1 (2013)
[24] M. Detrixhe et al. (ANITA Collaboration), Phys. Rev. D **83**, 023513 (2011)
[25] D.P. Hogan et al., Phys. Rev. D **78**, 075031 (2008)
[26] A. Aab et al. (Pierre Auger Collaboration), Phys. Rev. D **94**, 082002 (2016), arXiv:1609.04451
[27] M.G. Aartsen et al. (IceCube Collaboration), *IceCube-Gen2: A Vision for the Future of Neutrino Astronomy in Antarctica* (2014), arXiv:1412.5106

EPJ Web of Conferences **168**, 04011 (2018) https://doi.org/10.1051/epjconf/201816804011
Joint International Conference of ICGAC-XIII and IK-15 on Gravitation, Astrophysics and Cosmology

Role of strangeness to the neutron star mass and cooling

Chang-Hwan Lee[1,⋆], *Yeunhwan* Lim[2], *Chang Ho* Hyun[3], and *Kyujin* Kwak[4]

[1]*Department of Physics, Pusan National University, Busan 46241, Korea*
[2]*Cyclotron Institute, Texas A&M University, College Station, TX 77843, USA*
[3]*Department of Physics Education, Daegu University, Gyeongsan 38453, Korea*
[4]*Department of Physics, Ulsan National Institue of Science and Technology (UNIST), Ulsan 44919, Korea*

Abstract. Neutron star provides unique environments for the investigation of the physics of extreme dense matter beyond normal nuclear saturation density. In such high density environments, hadrons with strange quarks are expected to play very important role in stabilizing the system. Kaons and hyperons are the lowest mass states with strangeness among meson and bayron families, respectively. In this work, we investigate the role of kaons and hyperons to the neutron star mass, and discuss their role in the neutron star cooling.

1 Introduction

Observations of two massive neutron stars ($\sim 2M_{\odot}$) in neutron star-white dwarf binaries raised questions on the role of strangeness [1, 2]. In general, for a given neutron star equation of state (EOS), strangeness softens the EOS and reduces the maximum mass compared to that without strangeness [3]. Before the observation of $2M_{\odot}$ neutron stars, strangeness is often introduced to make soft EOS and reduce the maximum mass of neutron stars down to $1.5M_{\odot}$ to be consistent with the observed neutron star masses in neutron star-neutron star binaries [3]. Due to the observation of $2M_{\odot}$ neutron stars, the maximum mass of neutron star has to be bigger than $2M_{\odot}$, and those soft EOSs with strangeness inconsistent with $2M_{\odot}$ neutron stars are ruled out. However, the existence of strangeness in the central part of neutron stars still remains as an open problem.

In dense matter, as baryon number density increases, the energy densities of protons and neutrons increase because fermions cannot exist at the same states and higher energy states have to be included. Hence, it is quite natural for the new states with lower energy density to exist at high densities. Possible candidates for the new states are condensed pions, condensed kaons, hyperons, and deconfined quark states. In hadronic phase before the phase transition to the quark states, strangeness might appear as hadrons such as hyperons and kaons which are lowest energy states with strange quark (or strange anti-quark) among baryons and mesons, respectively.

In this work, based on the recent works by Lim *et al.* [4–7], we investigate the neutron star EOS with kaons and hyperons at the core, and check the effect of new EOS on the neutron star cooling. In Sec. 2, neutron star mass with strangeness is discussed. In Sec. 3, neutron star cooling with strangeness is discussed. Final conclusion is summarized in Sec. 4.

⋆e-mail: clee@pusan.ac.kr

2 Neutron star mass with strangeness

In our works [4–6], in order to investigate the role of strangeness to neutron star EOS, we use Skyrme force models for NN, $N\Lambda$ and $\Lambda\Lambda$ interactions and SU(3) chiral effective models for the kaon-nucleon interactions. Even though strangeness reduces the maximum mass of neutron stars, we found that $2M_\odot$ neutron star with strangeness is still possible if the original EOS without strangeness is stiff enough. In Table 1, we list the maximum mass of neutron stars with given EOS which are consistent with $2M_\odot$ observations.

Table 1. Maximum mass of neutron stars with strangeness [4–6].

with Λ hyperons

NN model	$N\Lambda$ model	$\Lambda\Lambda$ model	Maximum neutron star mass
SKI4	YBZ6	S$\Lambda\Lambda$3	2.03 $M_\odot$
SGI	YBZ6	S$\Lambda\Lambda$3	2.04 $M_\odot$

with kaon condensation

NN model	a_3m_s	Σ_{KN}	Maximum neutron star mass
SKI4	−134 MeV	167 MeV	2.06 $M_\odot$
SGI	−134 MeV	167 MeV	2.19 $M_\odot$
SGI	−178 MeV	256 MeV	2.03 $M_\odot$

For the contribution of Λ hyperons, $\Lambda\Lambda$ interaction plays very important role. Without $\Lambda\Lambda$ interaction, EOS with $N\Lambda$ interaction becomes too soft to explain $2M_\odot$ neutron stars. Hence, those models with repulsive $\Lambda\Lambda$ interaction are consistent with current observations. For the contribution of kaons, we found that the smaller values of Σ_{KN} are consistent with observations. Σ_{KN} is responsible for the attraction in the KN scalar interaction and the reduction of the kaon effective mass in medium.

In summary, even though strangeness reduces the maximum mass of neutron stars, original stiffer nuclear EOS (without strangeness) combined with less attractive interactions for the strangeness can be consistent with $2M_\odot$ neutron star observations.

3 Neutron star cooling with strangeness

In Lim *et al.* [7], we investigate the cooling behavior of neutron stars with strangeness for the selected models summarized in Sec. 2. In Figure 1, we show the cooling curves of neutron stars for SGI model without strangeness and SGI+YBZ6+S$\Lambda\Lambda$3 model with hyperons [7]. In both cases, the sudden drop in temperature occurs in a very narrow mass range. These sudden drops are caused by the nucleon direct URCA. For SGI model, the direct URCA starts to occur when the neutron star mass reaches around $1.7 - 1.8M_\odot$, and for SGI+YBZ6+S$\Lambda\Lambda$3 model, the direct URCA starts to occur around $1.5 - 1.6M_\odot$. So, the hyperons can have a significant effect on the neutron stars with masses around $1.5 - 1.8M_\odot$ as far as cooling is concerned. Other models show similar behavior except for the shift in the mass ranges for sudden temperature drop.

For kaons, the cooling curves with and without kaon contribution are almost the same because the critical densities for kaon condensation are larger than the critical densities for direct URCA. This indicates that the kaon condensation occurs only for the neutron stars with masses higher than $1.8M_\odot$ for SGI model. For SKI4 model, the critical mass for kaon condensation is higher than $1.7M_\odot$. Note that the high critical mass for the kaon condensation results in the high maximum mass of neutron star as in Table 1.

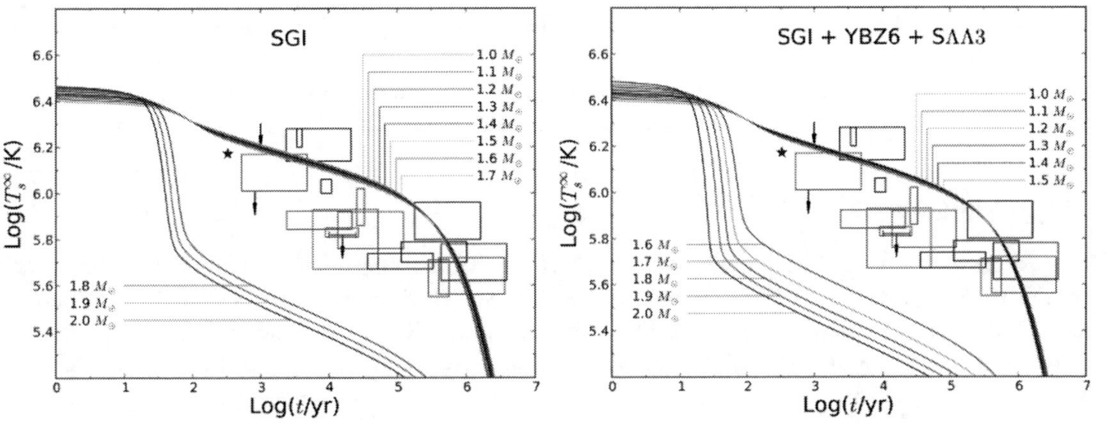

Figure 1. Cooling curves of neutron stars (taken from Lim *et al.* [7]). Left figure corresponds to SGI model (without strangeness) and right figure to SGI+YBZ6+SΛΛ3 model (with hyperons). Cooling curves with SGI model combined with kaon interaction are almost the same as in the left figure.

The sudden drop of the cooling curves in a very narrow mass range in Figure 1 raises a question on the mass distribution function of neutron stars because masses of observed neutron stars are distributed in a wider mass range [3].

4 Discussion

In this article, we discuss the role of strangeness to the maximum mass and cooling behavior of neutron stars. Even though particles with strangeness such as hyperons and kaons soften the EOS and consequently reduce the maximum mass of neutron stars, the specific value of maximum mass strongly depends on the original (without strangeness) nuclear EOS. If the original EOS is stiff enough, even with strangeness, the maximum mass can still be consistent with the $2M_\odot$ neutron star observations [1, 2]. Among Skyrme force models, SKI4 and SGI models satisfy these constraints [4–6]. We found that the repulsive ΛΛ interaction (SΛΛ3) and less attractive KN interaction (corresponding to the smaller $\Sigma_K N$) can be consistent with $2M_\odot$ neutron star observations.

Using these models, we investigate the cooling behavior of neutron stars. With original (without strangeness) SKI4 and SGI nuclear models, the temperature drops very rapidly as soon as the direct URCA occurs. The sudden drop of the cooling curves in the left panel of Figure 1 occurs for a very narrow mass range around $1.7 - 1.8M_\odot$. This behavior remains the same in the right panel except for the shift in mass range around $1.5 - 1.6M_\odot$. The sudden drop of the cooling curve in a very narrow mass range raises a question on the mass distribution, because it may not be consistent with the mass distribution of well-measured neutron star masses [3]. Since our work is focused on the Skyrme force models, we cannot make a firm conclusion. However, we confirm that the strangeness in neutron star is still an open possibility.

Many new developments in the neutron star research have been in progress. First of all, analyses on the low-mass X-ray binaries (LMXBs) have provided constraints on both masses and radii of neutron stars even though the uncertainty is large [8, 9]. New X-ray mission NICER (The Neutron star Interior Composition ExploreR) was launched in June 2017. In near future, NICER is expected to provide new observational data on the neutron stars in LMXBs [10]. Gravitational wave detection from neutron star binaries is another possibility. After the first success of the detection of gravitational waves from black hole binaries [11–13], gravitational waves generated by neutron star-neutron star or neutron star-black hole binaries are expected to be detected in near future. In the merger process, the inner structure of neutron star can be revealed via gravitational waves [14].

Acknowledgments

This article is based on the talk given at the International Conference on Gravitation: Joint Conference of ICGAC-XIII and IK15, July 3-7, 2017, Ewha Womans University, Seoul, Korea. CHL is partially supported by the National Research Foundation of Korea (NRF) grant funded by the Korea government (MSIP) No. 2015R1A2A2A01004238. CHL and KK are partially supported by the NRF grant No. 2016R1A5A1013277.

References

[1] P. Demorest, T. Pennucci, S. Ransom, M. Roberts, and J. Hessels, Nature **467**, 1081 (2010).
[2] J. Antoniadis *et al.*, Science **340**, 448 (2013).
[3] M. Prakash, Proceedings of 8th International Workshop on Critical Point and Onset of Deconfinement, arXiv:1307.0397 (2013).
[4] Y. Lim, K. Kwak, C. H. Hyun, C.-H. Lee, Phys. Rev. C **89**, 055804 (2014).
[5] Y. Lim, C. H. Hyun, K. Kwak, C.-H. Lee, Int. J. of Mod. Phys. E **24**, 1550100 (2015).
[6] Y. Lim, C. H. Hyun, C.-H. Lee, Int. J. of Mod. Phys. E **26**, 1750015 (2017).
[7] Y. Lim, C. H. Hyun, C.-H. Lee, *Strangeness in neutron star cooling*, arXiv:1608.02078.
[8] A. W. Steiner, J. M. Lattimer, and E. F. Brown, Astrophys. J. **722**, 33 (2010).
[9] A. W. Steiner, J. M. Lattimer, and E. F. Brown, Eur. Phys. J. A **52**, 18 (2016).
[10] K. C. Gendreau, *et al.*, *"The Neutron star Interior Composition ExploreR (NICER): Design and Development"*, Proc. SPIE **9905**, Space Telescopes and Instrumentation 2016; Ultraviolet to Gamma Ray, 99051H (22 July 2016).
[11] B. P. Abbott *et al.* (LIGO Scientific and Virgo Collaboration), Phys. Rev. Lett.**116**, 061102 (2016).
[12] B. P. Abbott *et al.* (LIGO Scientific and Virgo Collaboration), Phys. Rev. Lett. **116**, 241103 (2016).
[13] B. P. Abbott *et al.* (LIGO Scientific and Virgo Collaboration), Phys. Rev. Lett. **118**, 221101 (2017).
[14] K. Kiuchi, Y. Sekiguchi, K. Kyutoku, and M. Shibata, Classical and Quantum Gravity **29**, 124003 (2012).

EPJ Web of Conferences **168**, 04012 (2018) https://doi.org/10.1051/epjconf/201816804012
Joint International Conference of ICGAC-XIII and IK-15 on Gravitation, Astrophysics and Cosmology

Equation of state with scale-invariant hidden local symmetry and gravitational waves

Hyun Kyu Lee[1,⋆]

[1] *Department of Physics, Hanyang University, Seoul 04763, Korea*

Abstract. The equation of state (EoS) for the effective theory proposed recently in the frame work of the scale-invariant hidden local symmetry is discussed briefly. The EoS is found to be relatively stiffer at lower density and but relatively softer at higher density. The particular features of EoS on the gravitational waves are discussed. A relatively stiffer EoS for the neutron stars with the lower density induces a larger deviation of the gravitational wave form from the point-particle-approximation. On the other hand, a relatively softer EoS for the merger remnant of the higher density inside might invoke a possibility of the immediate formation of a black hole for short gamma ray bursts or the appearance of the higher peak frequency for gravitational waves from remnant oscillations. It is anticipated that this particular features could be probed in detail by the detections of gravitational waves from the binary neutron star mergers.

1 Introduction

The study of QCD under extreme conditions at high temperature and/or at high baryon number density has been one of the leading research subjects in understanding the theory of strong interaction. The object of this study includes the early Universe at high temperature as well as the core of the compact object with high density. While there has been lots of interesting progress in understanding QCD at high temperature theoretically and experimentally, at RHIC and LHC for example, there has been little development in QCD at high density and it remains as an uncharted area.

There are several experimental projects proposed or being constructed in near future like FAIR (Germany), FRIB (USA), HIAF (China), NICA (Russia), RAON (Korea), RIBF (Japan) as facilities for exploring the dense hadronic matter. The target density depending on the facilities is expected to be more than 2 times (but not much higher) normal nuclear density, $2n_0$. Neutron stars are believed to be compact objects with high density cores inside. The core densities of typical neutron stars of mass up to 1.5 $M_\odot$ is estimated to be roughly 2-3 n_0. On the other hand the core densities of the recently observed two-solar-mass neutron stars [1] are believed to be most likely greater than 5 n_0.

Recent detections of gravitational waves form black hole binaries [2] indicate strongly the possibilities of detecting gravitational waves from coalescing binary-neutron stars (BNS). According to the population analysis based on the designed detection range of BNS gravitational waves of aLIGO $\sim$ 100 Mpc, a number of detections of gravitational waves from BNSs are expected in O2 and O3 runs at aLIGO and aVirgo[3].

⋆e-mail: hyunkyu@hanyang.ac.kr

For a static and spherically symmetric astrophysical compact star, the relativistic hydrodynamic equilibrium is governed by the Tolmann-Oppenheimer-Volkoff(TOV) equation

$$\frac{dm}{dr} = 4\pi r^2\epsilon, \quad \frac{dp}{dr} = -(\epsilon+p)\frac{m+4\pi r^3 p}{r(r-2m)}, \quad \frac{d\Phi}{dr} = -\frac{1}{\epsilon+p}\frac{dp}{dr}. \tag{1}$$

where ϵ and p are energy density and pressure at r respectively. $m(r)$ is the mass enclosed inside the radius r. We can calculate the mass of compact star, M, and its radius, R, by integrating the TOV equation up to $p(R) = 0$ and we get the profile of $\Phi(r), m(r)$ and $p(r)$.

When a compact star is placed in a static external field, a star supposed to be in a spherically symmetric configuration is then deformed by the external field. The asymptotic expansion of the metric at large distances r from the star defines the quadrupole moment, Q_{ij}, and the external tidal field, $\mathcal{E}_{ij}$, as expansion coefficients [4] given by $-\frac{1+g_{00}}{2} = -\left[\frac{m}{r} + \frac{3}{2}\frac{Q_{ij}}{r^3}n^i n^j + \cdots\right] + \frac{1}{2}\mathcal{E}_{ij}r^2 n^i n^j + \cdots$ where $n^i = x^i/r$ and Q_{ij} and $\mathcal{E}_{ij}$ are both symmetric and traceless .The deformability parameter λ is defined by

$$Q_{ij} = -\lambda\mathcal{E}_{ij} \text{ or } \Lambda = \lambda G\left(\frac{c^2}{GM}\right)^5 = 9.51\times 10^2\lambda_{36}\left(\frac{M_\odot}{M}\right)^5 \tag{2}$$

where G is the gravitational constant and $\lambda = \lambda_{36}\times 10^{36}\text{g cm}^2\text{ s}^2$. Deformation parameters[1] obtained from the known EoSs consistent with certain criteria, for example on velocity of sound $v_s < 1$ and on maximum mass $M_{max} \geq 2M_\odot$, are calculated to be in a rather broad range of $10^2 - 10^3$ [5, 7].

The effect of the tidal deformation on the gravitational waves in binary coalescence process is found to be appearing at the order of post newtonian approximation in the analytical formulation. The numerical simulation matched to the analytic expression shows a clear deviation from point particle approximation at high frequency(up to near 1 kHz) component of the gravitational wave: the larger tidal deformability Λ, the more deviation. It also shows the feasibility of distinguishability between EoSs if $\delta\Lambda > 100$ [8]. At the final stage of the binary coalescing process, there are two possibilities depending on the EoS, a prompt emergence of a black hole for soft EoSs or a merger remnant for sufficiently stiffer EoSs. For a sufficiently stiffer EoS, for which prompt collapse into a black hole is unlikely, the results of numerical simulations demonstrate interesting results of the appearance of peaks in their higher frequency components, $f > 1$ kHz [9]. The peaks in the frequency components are considered to be the evidence of the gravitational emission during the remnant oscillations. The location of peaks, f_2, due to the quadrupolar oscillation are found to be sensitive on the EoS such that a larger deformation Λ leads to a smaller peak frequency f_2.

Recently a new EoS has been proposed in the quite different frame work [10, 11] when compared with the presently available EoSs. The EoS is obtained based on the model constructed in responding to the question whether certain fundamental symmetries of QCD, invisible or hidden in matter-free space, can emerge at high baryonic density such as in massive compact stars. Therefore it is very interesting to see whether its particular characteristics can be probed by gravitational waves, which will be discussed briefly in the final section.

2 EoS with scale-invariant hidden local symmetry

The constituents of a neutron stars are mostly neutrons with a finite fraction of protons and electrons, for low mass neutron stars with core density not much higher than $2n_0$. In recent papers [11], it was proposed that both hidden local symmetry and hidden scale symmetry could emerge at densities

[1]For details how to obtain the deformation parameters in combination with TOV, see for example [5, 6].

relevant to $\sim$ 2 solar-mass neutron stars. As density increases it is natural to expect new degrees of freedom to be exited with the explicit emergence of those symmetries. The particle contents entangled with hidden local symmetry are rho meson and omega meson and a scalar particle, dilaton χ as a conformal compensator with scale symmetry. The corresponding lagrangian, $\mathcal{L}_{\rm inv}$, dictated by the symmetries with particle contents can be written down in a straightforward way [11]. The only term which does not obey the scale symmetry, $\mathcal{L}_{\rm SB}$, is the dilaton potential term, $V(\chi)$, which is put such that it produce the QCD scale anomaly properly.

$$\theta^{\mu}_{\mu} = 4V(\chi) - \chi\frac{\partial V(\chi)}{\partial \chi}(= \frac{\beta(\alpha_s)}{4\alpha_s}G^a_{\mu\nu}G^{a\mu\nu}) \tag{3}$$

where $G^{\mu\nu}$ is the gluon field in QCD.

In the medium characterized by density n, the basic premise is that the effect of changes in the vacuum structure caused by density is entirely encoded in the density dependent parameters of the Lagrangian up to the density domain relevant in compact stars. How these density dependencies, intrinsic density dependencies (IDDs), are defined is described in detail in [10]. For the refinement of the IDDs behavior, the sophisticated high order correlation beyond the mean field treatment, the V_{lowk}-RG treatment is adopted.

Up to nuclear matter density $n_0 \simeq 0.16$ fm^{-3}, IDDs can be more or less determined either from experiments or theoretically. It will however be extremely difficult to track them at higher densities going beyond n_0. Here we rely on certain topological structure present in the skyrmion approach. An extremely important observation in the skyrmion approach, when skyrmions are put on crystal lattice, is the changeover from skyrmions to half-skyrmions at a density $n > n_0$ [12]. When the topology change is translated into the bare parameters, it makes a drastic – and novel – change in the IDDs across the transition density denoted $n_{1/2}$. Specifically it gives the prediction that going into the half-skyrmion phase, the in-medium pion (dilaton) decay constant f^*_π (f^*_σ) and the in-medium nucleon mass m^*_N go over to a constant $f^*_\pi/f_\pi \approx f^*_\sigma/f_\sigma \approx m^*_N/m_N \longrightarrow \kappa \sim (0.6-0.9)$.

At the lower density $n \leq n_{1/2}$, the results of the V_{lowk} calculation for $n \lesssim n_{1/2}$ can be well fit for the neutron matter by the fitting functions for energy density and pressure given by

$$\epsilon = m_N n + n_0 A_I\left(\frac{n}{n_0}\right)^2 + n_0 B_I\left(\frac{n}{n_0}\right)^{1+D_I}, \quad p = n_0 A_I(\frac{n}{n_0})^2 + n_0 BD_I(\frac{n}{n_0})^{D_I+1}. \tag{4}$$

where $(A_I, B_I, D_I) = (9.11\ \text{MeV}, 2.14\ \text{MeV}, 4.08)$.

In the density regime $n > n_{1/2}$, there is a topology change that drastically affects the IDDs at $n = n_{1/2}$, parameters changed from those in $n \leq n_{1/2}$ as explained in detail in [11]. It turns out that the results of the V_{lowk} calculation for $n > n_{1/2}$ can be captured very well by the simple formula

$$\epsilon = n_0 B\left(\frac{n}{n_0}\right)^{4/3} + n_0 D, \quad p = \frac{1}{3}n_0 B\left(\frac{n}{n_0}\right)^{4/3} - n_0 D, \tag{5}$$

where $B = 686$ MeV, $D = 253$ MeV for the pure neutron matter.

It is interesting to note that the equation of state looks very similar to the relativistic free gas with a 'bag-constant-like'term $n_0 D$($\sim$ 41MeV/fm^3 for the pure neutron matter). The former is consistent with the conformal symmetric nature of the effective theory considered here and the latter the scale-symmetry-breaking dilaton potential. The observation that the 'bag-constant-like' is found to be a constant independent on the density leads to the conclusion that the velocity of sound is given by the conformal value, $\sqrt{1/3}c$.

Given the equation of state described above, it is a standard procedure to use the TOV equation to calculate the properties of compact stars. In doing this we will take into account of the β stability, thereby including the leptons. The maximum mass of compact star is estimated to be $\simeq 2.05M_\odot$ and the radius $R \simeq 12.19$ km with the maximum central density $n_{max} \simeq 5.1n_0$. The result is consistent with the well-measured values $M \sim 2M_\odot$ [1]. It is fair to conclude that the effective theory provides an evidently respectable EoS for massive compact stars. Up to the predicted central density $\sim 5.1n_0$, there are no other degrees of freedom than the pseudo-NG bosons (π and σ), vector mesons and nucleons.

3 Discussion

A simple comparison with other EoSs in the literature can be given using polytropic fits. The adiabatic indices, Γs, for polytrope fits of EoSs [13] are compared with the highest indices in Eqs.4 and 5. At the lower density $n \leq n_{1/2}$, the highest index, ~ 5, in Eq.4 is found to be larger than adiabatic indices Γs. At the higher density $n > n_{1/2}$ the highest index in Eq. 5 is fund to be 4/3, which is less than those in [13].

During the early phase of coalescing process neutron stars in binary can be treated as point particles as far as gravitational wave is concerned and the gravitational wave depends essentially on their masses of neutron stars such that there is no more information on equation of states than the masses. Since the most abundant NSs are believed to have masses around $1.4M_\odot$, the density to be explored is not expected to be much higher than $\sim 2n_0$ [14]. When they get closer the gravity between neutron stars becomes strong enough to induce tidal deformations on each orbiting neutron star. The point particle approximation is no more valid and the effect of tidal deformation on the gravitational wave should be considered. The tidal deformation depends strongly on the equation of states and therefore the deviation of gravitational waves form the point particle approximation will certainly provide an useful information in differentiating various EoSs, which otherwise cannot be distinguished by the masses alone. At the final stage of merging process, there is a possibility of appearing a black hole or a merger remnant if not prompt collapse into a black hole , which depends on the EoS of merging neutron stars. Numerical simulation demonstrates that the density of the remnants can be as large as more than $\sim 4n_0$ even for the merging neutrons stars of masses, $1.4M_\odot$.

For $1.4M_\odot$ neutron star with core density $2n_0$ with new EoS, the tidal deformation parameter is calculated to be $\lambda_{36} = 4.44$ or $\Lambda = 815$, which is larger than the most of the other known EoSs for $1.4M_\odot$ neutron star. It is because the EoS is stiffer with larger deformation parameter for the density of the $1.4M_\odot$ star. And consequently the larger deviation from point-like-approximation in the gravitational wave form from the coalescing binary is expected. On the other hand, the density of the merger remnants are expected to be higher than $\sim 4n_0$, at which the new EoS is found to become softer. Hence the prompt formation of a black hole after merger might be one of the interesting expectations. The possibility could be explored in numerical simulation when the change over of the EoS from lower density relevant to binary neutron stars into higher density relevant for a remnant is seriously considered. The viable scenario of soft gamma ray bursts is the presence of a black hole at the central region. When the black hole for short GRBs is formed from the binary coalescence, then softer EoS at higher density might be preferred for those GRBs followed by gravitational waves. If it is not soft enough to form a prompt black hole, an oscillating merger remnant might be expected with the peak frequency feature. But the peak frequency for new EoS is likely to be higher than those of stiffer EoSs [9].

In summary, it is observed that the EoS with scale-invariant hidden local symmetry behaves as a stiffer EoS in a sense that it gives a larger radius and larger deformation parameter, which in turn results in the larger deviation of gravitational wave from point-like-approximation. In the higher density regime for the merger remnant it behaves as relatively softer EoS with the possibility of a prompt formation of a black hole for short gamma ray bursts or higher peak frequency for gravitational waves from remnant oscillations. These particular features could be probed in detail by the detections of gravitational waves from the binary neutron star mergers.

The author would like to thank Kyungmin Kim, Tom Kuo, Yong-Liang Ma, Won-Gi Paeng and Mannque Rho for the useful discussions and long-time collaborations. This work was supported in part by a Collaborative Research Grant from KISTI GSDC in 2017.

References

[1] P.B. Demorest *et al.*, Nature **467**, 1081 (2010); J. Antoniadis *et al.*, Science **340**, 6131 (2013).
[2] B.P. Abbot *et al.* , Phys. Rev. Lett. 116, 061102 (2016).
[3] J. Aasi, et al., Class. Quantum Grav. 32, 074001 (2015);F. Acernese, et al., Class. Quantum Grav. 32(2), 024001 (2015).
[4] K. S.Thorne, Phys. Rev. D **58**, 124031 (1998)
[5] T. Hinderer, B.D. Lackey, R.N. Lang and J.S. Read, Phys. Rev. D **81**, 123016 (2010).
[6] K. Kim, H.K. Lee and J. Lee, Int. J. Mod. Phys. E **26**, 1740011 (2017)
[7] J.S. Read *et al.*, Phys. Rev. D **88**, 044042 (2013)
[8] K. Hotokezaka, K. Kyutoku, Y. Sekiguchi and M. Shibata, Phys. Rev. D **93**, 064082 (2016).
[9] A. Bauswein, N.Stergioulas, H-T. Janka, EPJA 52, 56B(2016)
[10] W.G. Paeng, T.T.S. Kuo, H.K. Lee and M. Rho, Phys. Rev. C **93**, 055203 (2016).
[11] W.G. Paeng, T.T.S. Kuo, H. K. Lee, Yong-Liang Ma and M. Rho, Phys. Rev. D 96, 014031 (2017).
[12] H.K. Lee, B.Y. Park and M. Rho, Phys. Rev. C **83**, 025206 (2011) Erratum: [Phys. Rev. C **84**, 059902 (2011)].
[13] J.S. Read, B.D. Lackey, B.J. Owen, J.L. Friedman, Phys. Rev. D79, 124032(2009).
[14] J. Lattimer, Ann. Rev. Nucl. Part. Sci. 62, 485(2012).

EPJ Web of Conferences **168**, 04013 (2018) https://doi.org/10.1051/epjconf/201816804013
Joint International Conference of ICGAC-XIII and IK-15 on Gravitation, Astrophysics and Cosmology

A model for high-energy emission of the Intrabinary shock in pulsar binaries

Hongjun An[1,*] on behalf of the Fermi-LAT collaboration

[1] *Department of Astronomy and Space Science, Chungbuk National University, Cheongju, 28644, Republic of Korea*

Abstract. We present our studies of intrabinary shock emission for astrophysical binary systems with a neutron star. We construct a model for the shock emission and compare the model calculation with the light curve and the spectral energy distribution of the gamma-ray binary 1FGL J1018.6−5856. The model assumes a slow and a fast population of particles accelerated in the shock, and computes the high-energy emission spectra and orbital light curves produced by synchrotron, self-Compton and external Compton processes of the high-energy particles in the shock. The model allows one to study plasma properties and to constrain the binary geometry, most importantly the inclination angle (i). We discuss potential use of this model for other pulsar binaries to determine the inclination angle of the binary hence the mass of the neutron star.

1 Introduction

In some astrophysical binary objects, it is believed that winds of the two objects interact strongly and form shocks, so called intrabinary shocks (IBSs). These shocks can accelerate particles to very high energies, and the particles in the shock radiate high-frequency electromagnetic waves [1]. If one of the stars is a neutron star, the wind of the neutron star is likely composed of cold energetic electrons and positrons. When they are accelerated in the shock, they radiate their energy very efficiently. This radiation can be observed and help to study IBS.

IBS is important for many reasons. Particle acceleration mechanisms in astrophysical shocks are not yet well understood; diffusive shock acceleration (DSA) [2] is a commonly used one but there are others as well. These mechanisms have been studied with particle-in-cell (PIC) simulations [3]; the resulting energy distribution of particles is relatively sensitive to the shock environments which are not well known. Because particles are believed to be accelerated stochastically (e.g., via DSA) in IBS, observations of IBS emission can be used for inferring properties of the shock. IBS emission can also provide us with an independent way of estimating the binary orbital parameters which are important to infer the mass of the neutron star in the binary. In particular, this can supplement the current method of inferring the neutron star mass with optical spectroscopic and photometric observations and modeling.

Formation of shocks in pulsar binaries is similar to that in pulsar wind nebulae [4, 5]: by the interaction of pulsar wind with the ambient medium (stellar wind in pulsar binaries and interstellar gas or supernova remnants in pulsar wind nebulae). A contact discontinuity (CD) is formed when the wind of the companion and the relativistic pulsar wind interact in the orbit. Around the CD, shocks are formed of which the relativistic pulsar wind shock is of interest in regard to the high-energy emission in these systems [6]. In the pulsar wind shock, particles are stochastically accelerated and are believed to be bulk accelerated as they flow [7, 8]; Lorentz beaming is expected which may be observed as spikes in the orbital light curve [9, 10].

Using the IBS scenario, we develop a phenomenological model for the high-energy emission [10] in pulsar binaries. Below we show details of the model components and present the results of application to the gamma-ray binary 1FGL J1018.6−5856. We then discuss plasma properties in the IBS of the source and potential use of the model for other pulsar binaries. Note that similar models have been used for other sources such as the gamma-ray binary LS 5039 [8] and the pulsar binary PSR J2215+5135 [9].

2 The IBS emission Model

Emission in an IBS is very complex. There are relativistic electrons and positrons stochastically accelerated in the shock, hence synchrotron radiation is expected. Then the particles will inverse-Compton upscatter the synchrotron photons (synchrotron-self Compton, SSC) and the stellar optical/UV photons external to the shock (external Compton, EC); these emissions will be in the ~TeV band [6, 10]. Due to these emissions, the energy distribution of the particles evolves in time and space, so does the emitted photon spectrum. Furthermore, the IBS emission will be Lorentz boosted because some of the particles are bulk accelerated in the flow as shown in magnetohydrodynamic simulations [7, 8], and so the Lorentz beaming also needs to be considered. Of course, the stellar and the pulsar emission can contaminate the IBS emission. Therefore, studying IBS requires detailed numerical simulations and comparison of simulations and observations. However, these simulations are yet to be improved for the relativistic pulsar

*e-mail: hjan@chungbuk.ac.kr

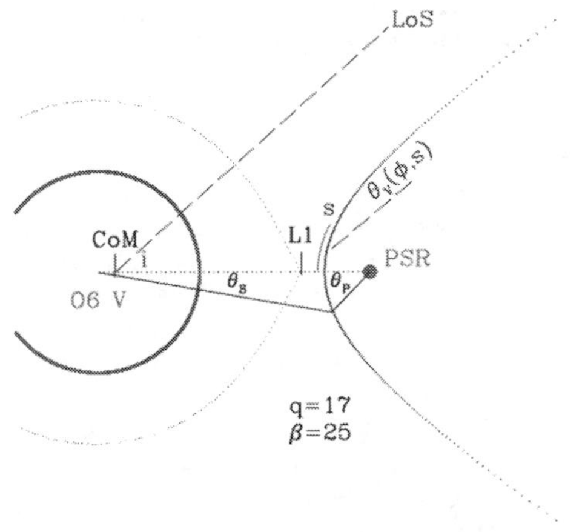

Figure 1. Vertical cross section of the binary system and the IBS. Two stars (pink) and the CD (blue) are shown. The CD should look like a paraboloid in 3D. This plot is generated for the mass ratio $q = 17$ and the wind momentum flux ratio $\beta = 25$. Various orbital parameters are denoted. The figure is taken from [10].

wind, so instead, we use a relatively simple phenomenological model for IBS emission in order to infer the flow properties and the orbital geometry using high-energy observations [10].

2.1 The Shock Geometry

As noted above, IBS emission is highly anisotropic, so properly estimating the shape of an IBS is very important. The geometry of an IBS is determined by the shape of the CD. The CD produced due to isotropic wind interaction is in principle paraboloid like, and the shape can be calculated [11]: $r(\theta_p) = D\sin\theta_s/\sin(\theta_p + \theta_s)$ and $\theta_s\cot\theta_s = 1 + \beta(\theta_p\cot\theta_p - 1)$, where $r(\theta_p)$ is the distance from the compact object to a point in the CD, D is the orbital separation, θ_s and θ_p are angles with respect to the line of centers from the companion star and the compact object, respectively, and β is the wind momentum flux ratio (Figure 1). Once β and D are given, the full locus of the CD can be calculated.

We note that the above calculation is for isotropic winds. Realistically, the wind of a neutron star (and the stellar companion) may not be isotropic, then the CD shape may change. However, isotropic winds may be a good approximation, and the model with the approximation can capture most of the important features of the emission as we show below.

Of course, the shock flow can extend to a large distance and may make other structures [8]. We note that the IBS emission becomes rapidly weaker with distance s due to gradually reducing B and seed photon density (for SSC and EC). So we calculate the IBS emission out to 3× orbital separation ($s \lesssim 3D$). Emission at larger s may explain the low-energy radio emission of pulsar binaries.

2.2 High-energy emission in IBS

In an IBS, particles are stochastically accelerated, perhaps via DSA. Of course, magnetic reconnection may play a role. The energy distribution of the particles can be diverse depending on the shock environment, but generally a power law with an index of $p = 2$ is expected (i.e., $dN/d\gamma_e \propto \gamma_e^{-p}$). Some of the particles are further bulk accelerated in the IBS flow to have a Lorentz factor $\Gamma > 1$, hence there can be multiple populations with different Γ [7, 8]. In hydrodynamic simulations, it is shown that Γ can be modest, but the exact value and the distribution of Γ for relativistic pulsar winds are not known.

As we mentioned earlier, in the IBS scenario e^-/e^+ plasma in the shock makes synchrotron radiation under the influence of B field supplied by the pulsar. Although it is impossible to measure B in the shock directly, we can scale the neutron star's field: the $\sim 10^{12}$ G field of the neutron star will decrease following the dipole formula ($\propto r^{-3}$) out to the light cylinder radius ($R_{LC} = cP/2\pi$, where c is the speed of light and P is the rotation period of the neutron star), and then the stripped pulsar wind will carry the field ($\propto r^{-1}$) to the shock at $r(\theta_p)$ [6]. So once the timing solution (spin period P and the derivative $\dot{P}$) of the pulsar is obtained, dipolar B at the neutron star's pole can be inferred with $B = 3.2 \times 10^{19}\sqrt{P\dot{P}}$ G, and hence B in the shock can be estimated. Note that this estimation is rough and serves only as a guide. The synchrotron loss is $dE/dt \propto -\gamma_e^2$, and the emitted photon energy distribution follows a power law with an index $(p-1)/2$ (i.e., $F_\nu \propto \nu^{-(p-1)/2}$) and lies in the X-ray to the low-energy gamma-ray bands for typical values of shock parameters.

The dominant photon populations in the shock are synchrotron photons emitted by the particles themselves; there are also stellar blackbody photons supplied by the stellar companion. All these photons interact with the particles in the shock via inverse-Compton scattering (i.e., SSC and EC). In these processes, the synchrotron photons gain energy by a factor of $\sim \gamma_e^2$ by SSC; when observed, these are further Lorentz-boosted by $\delta = 1/[\Gamma(1 - \sqrt{1 - 1/\Gamma^2}\cos\theta_V)]$, where θ_V is the viewing angle of the observer (Figure 1). The stellar photons, however, are external to the shock and so there is another $\sim \Gamma^2$ energy gain (with the scattering angle dependence), so the gain factor is $\sim \Gamma^2\gamma_e^2$ for EC. These SSC and EC photons become high-energy gamma-ray photons (>10 GeV) to be observed.

The observed spectral energy distribution (SED) of the IBS-emitted (+reprocessed) photons is characterized with a "double-hump" continuum structure, the low-energy hump being produced by the synchrotron process and the high-energy hump by SSC and EC. Of course, there are contaminations from the stellar blackbody photons at ~ eV and pulsar magnetospheric emission at ~ GeV.

2.3 The IBS model

The model we develop [10] is basically a combination of the above. Using the masses of the two stars and the orbital period measurement, we can determine the orbital separation (D) using the Kepler's law. Of course, orbital eccentricity (e) needs to be assumed, and later can be verified because change in the separation due to e will cause the magnetic field strength to vary over the orbit. This change may be visible in the synchrotron light curve (orbital variation).

At each orbital phase, distance to a point on the CD ($r(\theta_p)$) is calculated using the formula in Section 2.1 for given β. We then assume the shock-accelerated electron distribution (power law), magnetic field strength B, and the bulk Lorentz factor ($\Gamma = 1$) at the apex of the shock. As the particles flow along the shock to a distance $s(\theta_p)$ from the apex (see Figure 1), these quantities vary: $N_e[s(\theta_p)]$, $B[s(\theta_p)]$, and $\Gamma[s(\theta_p)]$. Although the exact functional forms need to be found using magnetohydrodynamic simulations, simple analytic forms can be used for our phenomenological model. For example, $N_e \propto 1-\cos\theta_p$ can be used because pulsar wind particles are injected over the whole shock and at θ_p particles injected from $\theta_p = 0$ to θ_p are accumulated, and $B \propto r(\theta_p)^{-1}$ since $r(\theta_p) >> R_{LC}$ for typical millisecond pulsars. The functional form for Γ may be more complex [7, 8], but a simple linear trend ($\Gamma(s) \propto s$) may be a good starting point.

At a given point s_0 on the shock, there are multiple particle populations injected: a fresh injected one from the pulsar and another that flows from the apex ($s < s_0$). The slope (power-law index) of the distribution is the same between the populations, but those from the apex lose more energy as they already emitted synchrotron/SSC/EC photons (cooling) while they flow to s_0 from the apex at $s = 0$. Because the energy loss is proportional to $\sim \gamma_e^2$ for the synchrotron and the Compton processes, the highest-energy particles lose energy quickly and become low-energy particles. Hence, the flow from the apex lacks high-energy particles. This cooling results in a break in the particle distribution (cooling break at $\gamma_{e,b}$), so the electron energy distribution will be a broken power law: $dN/d\gamma_e \propto \gamma_e^{-p_1}$ for $\gamma_e \leq \gamma_{e,b}$ and $dN/d\gamma_e \propto \gamma_e^{-p_2}$ for $\gamma_e > \gamma_{e,b}$. This will be visible as a break in the observed photon spectrum as well. In addition, the minimum and the maximum electron energies need to be specified. Because we are concerned with high-energy emission, we use a reasonably small γ_e for the minimum such that the emitted photon energy is well below the X-ray band. The maximum γ_e depends on the acceleration process, but in DSA the maximum can be as high as $\sim$ 100 MeV, the radiation reaction limit. This results in a cut off in the photon spectrum at the highest frequency.

With all these prescriptions, we compute the synchrotron and the inverse-Compton radiation at each point of the shock and Lorentz-boost them by δ (for synchrotron and SSC) to get the observed spectrum (see [12] for example). We repeat this for each orbital phase of interest, calculate fluxes, and construct the model light curves. Also, we construct the phase-summed model SED by summing up the spectra over the orbital phases.

3 Application to the gamma-ray binary 1FGL J1018.6−5856

1FGL J1018.6−5856 is identified as a gamma-ray binary based on the gamma-ray modulation [13]. The orbital period is measured to be 16.544 days [14]. Importantly, phase 0 in Figure 2 is identified as the inferior conjunction of the compact object [15]. Because pulsations are not detected from this source yet, it is not clear whether the compact object is a neutron star or a black hole, but previous studies favor a neutron star over a black hole [15–18], so we assume a neutron star in this work.

Figure 2 shows the broadband SED (left) and the high-energy orbital light curves (middle and right) of the source [10]. In the SED, the lowest-frequency radio photons ($\nu \sim 10^{10}$ Hz) are assumed to be emitted in a different region from that for the high-energy photons, so we do not attempt to explain these points. In the $\sim 10^{15}$ Hz band, stellar blackbody emission dominates, hence these data allow an accurate determination of the seeds for EC. The X-ray to low-energy gamma-ray ($\nu \sim 10^{17}-10^{22}$ Hz) photons are produced by the synchrotron process, hence provide us with information on the particle distribution. The slope of the observed SED in this band in Figure 2 left is ~ -0.5, which implies $p \approx 2$ for the electron distribution, and the maximum energy of the synchrotron emitted photons is $\sim$1 GeV which implies that $\delta \approx 10$ ($h\nu_{max} = \delta \times 100$ MeV). Although the spectral break point is not detected by any observatory (no observatory operating in that band), the model expects it to be in the MeV range to match both the X-ray and the $\sim$100 MeV spectra; then B is inferred to be $\sim$a few Gauss. For this magnetic field, the maximum energy of electrons required to emit synchrotron photons at the radiation reaction limit is estimated to be $\gamma_e \approx 10^8$ (see Table 1).

At $\sim$ 1 GeV, the emission is mainly from the pulsar magnetosphere and contaminates the IBS emission (Figure 2 right). So we model the pulsar emission with a power-law with an exponential cut-off spectrum to separate out the IBS emission (see [10] for the pulsar spectral parameters). SSC and EC photons lie in the $\nu \geq 10^{26}$ Hz band; these are also sensitive to the properties of the flow, the electron distribution and Γ. The flow properties we inferred above explain the SSC+EC SED well (see Figure 2 left).

Moving on to the light curve modeling, the X-ray and low-energy gamma-ray light curves are explained by the beamed emission aligned with the observer's line of sight (LoS) for the spike at phase 0 and isotropic emission (broad hump peaking at phase $\sim$0.4) modulating orbitally due to the change in separation between the stars (and so B) with an addition of the orbitally constant pulsar magnetosphere emission in the gamma-ray light curve (Figure 2 right). The models are shown in lines and the parameters are presented in Table 1 (see [10] for more details). Because the model is very complicated and does not allow us to determine a unique set of parameters, we do not formally fit the data and so do not report the uncertainties for the parameter values.

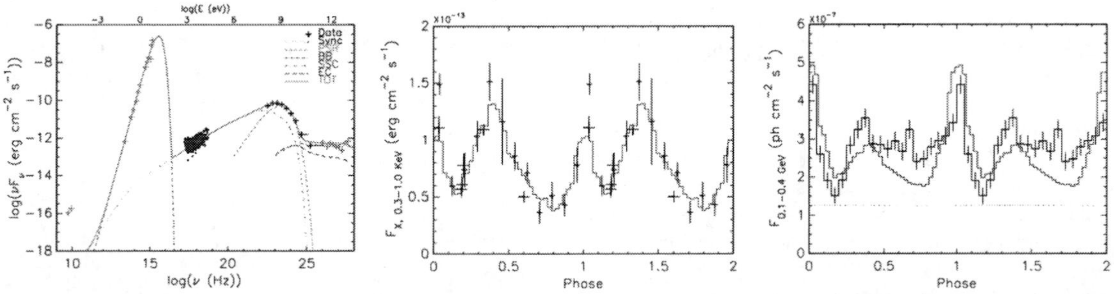

Figure 2. Broadband SED (left), the X-ray (middle) and the gamma-ray (right) light curves of 1FGL J1018.6−5856. The X-ray data are taken with *Swift*, *XMM-Newton* and *NuSTAR* observatories, and the gamma-ray data are taken with the *Fermi* gamma-ray space telescope. The IBS models for this source described in Section 3 are shown in lines. Various emission components are denoted in different colors in the left figure. In the low-energy gamma-ray light curve (right), we also plotted the pulsar emission (green dotted) to isolate the IBS emission (red). The figures are taken from [10].

Table 1. The orbital parameters and the shock properties obtained with the modeling

Parameter	Symbol	Value
Eccentricity	e	0.35
Assumed momentum flux ratio	β	25
Inclination (deg.)	i	50
Low-energy spectral index	p_1	1.93
Break electron energy[a]	$\gamma_{e,b}$	5×10^7
High-energy spectral index	p_2	2.15
Maximum electron energy[a]	$\gamma_{e,max}$	10^8
Maximum bulk Lorentz factor	Γ_{max}	7

[a]These vary along the orbit and the shock distance s.

4 Discussion

We developed an IBS emission model for binary systems in which winds of the two stars interact, producing IBS. In particular, we are concerned with high-energy (>X-ray) emission in astrophysical binary systems which have a relativistic wind and so can produce a relativistic wind shock. The characteristics of the model are the double-hump structure in the SED, and spikes (for proper inclinations) and broad hump (for eccentric orbits) in the high-energy light curve.

We applied the model to the gamma-ray binary 1FGL J1018.6−5856 because it has features that can be easily explained by our IBS model [10]. Although the model parameters are not uniquely determined, there are several things we learned about the shock properties and the orbit. (1) the electron distribution in the shock is very similar to that expected by diffusive shock acceleration theories, $p \approx 2$, (2) in the shock, electrons are maximally accelerated to the radiation reaction limit of DSA, (3) the observed data require two populations of electrons, a slow $\Gamma = 1$ population and a fast bulk-accelerated population $\Gamma \leq 7$, the latter being only a small fraction (a few %) of the former, (4) our LoS is aligned well with the shock tail (asymptotic tangent).

(1) and (2) suggest that particles can be accelerated to very high energies (to tens of TeV) in the IBS which may have interesting implications for the cosmic-ray studies. (3) qualitatively agrees with the limited numerical magnetohydrodynamic simulations [7, 8]. Although these simulations were not done using a realistic pulsar wind with bulk Lorentz factor $\geq 10^4$, they seem to predict some features of the flow: multiple populations with a smaller particle fraction for higher Γ. In our model, we used two populations, but the fast population has multiple values of Γ (linearly growing with s in our simple model) so in principle, we have multiple populations. In addition, our modeling requires that the fast population is only a small fraction of the slow population, which is qualitatively similar to the prediction of the numerical studies [8]. We find that the result is not very sensitive to the choice of the function $\Gamma(s)$, hence more realistic Γ distribution in an IBS flow obtained with the simulations may work; this can be done in the future when both simulation and the IBS model are further improved.

Finally, (4) has very interesting implications for studying pulsar binaries because it can give us a new estimation of the inclination of the systems [9, 19]. Even without detailed modeling, (4) can be concluded based on the amplitude and the width of the spike. This can be done in the following way. In the IBS model, the spike in the light curve is produced because of beaming which is maximum when our LoS is aligned with the accelerated flow, i.e., the shock tail. Suppose an observer is looking at the binary near the normal to the orbit. Then the shock tail will not cross the LoS anywhere in the orbit; the observer sees only the side of the shock. In this case, beaming (strongly dependent on the viewing angle θ_V) will be very weak and the observer will not see any spike in the light curve. On the other hand, if an observer is near the orbital plane, the observer will see two edges of the shock as the binary orbits and hence will detect two spikes. In between, if an observer is near the tail of the shock, the observer will see only one spike. So simply the number of spikes in the light curve can already tell us much about the inclination, and the light curve modeling can give us more detailed information. Of course, the situation may be more complex if the orbital speed of the stars is comparable to the wind speed [9].

Determination of the orbital inclination is particularly important for inferring the mass of the neutron star in pulsar binaries. The mass of the neutron star in a pulsar binary can be estimated with the mass function if the mass ratio of the two stars and the orbital inclination are determined. The former can be determined by the radial velocity measurement, and the latter by the optical light curve modeling (see [9, 20] for example). The mass determined in this way is subject to relatively large uncertainties (to the third power in sin$i$) due to the correction for the mismatch between the centers of light and mass, and an independent determination will be very helpful. Of course, for relativistic pulsar binaries measuring the relativistic delay of pulses can be used for accurate determination of the neutron star mass (see [21] for example).

As we show above, the IBS emission can tell us about the orbital inclination of pulsar binaries, especially when there are spikes in the high-energy light curve [9, 19]. In particular, the orbital inclination can be determined with the synchrotron light curve modeling only which can run reasonably quickly [9] compared to the one with the full SSC+EC emission models. Of course, there is a weakness in this method; measuring the crucial parameter β, the wind momentum flux ratio, for determining the inclination is hard. The way the shock bends can tell us about the relative strength of the two winds, but accurate determination of β (and so i) requires measurements of the momentum flux of both winds. Furthermore, the pulsar wind can be anisotropic, so the angular profile of the pulsar wind momentum needs to be considered. Also note that for relativistic shocks [7] the opening angle may differ from the analytic derivation of [11], hence more realistic magnetohydrodynamic simulations need to be performed. These will benefit from the high-energy modeling presented here for inferring the shock properties which can be compared with the simulation results, helping to improve the simulations further.

The *Fermi*-LAT Collaboration acknowledges support for LAT development, operation and data analysis from NASA and DOE (United States), CEA/Irfu and IN2P3/CNRS (France), ASI and INFN (Italy), MEXT, KEK, and JAXA (Japan), and the K.A. Wallenberg Foundation, the Swedish Research Council and the National Space Board (Sweden). Science analysis support in the operations phase from INAF (Italy) and CNES (France) is also gratefully acknowledged. This work performed in part under DOE Contract DE-AC02-76SF00515. This research was supported by Basic Science Research Program through the National Research Foundation of Korea (NRF) funded by the Ministry of Science, ICT & Future Planning (NRF-2017R1C1B2004566).

References

[1] Blandford, R., Yuan, Y., Hoshino, M. & Sironi, L., SSRv, **207**, 291 (2017)
[2] Drury, L. O., Reports on Progress in Physics, **46**, 973 (1983)
[3] Spitkovsky, A., ApJL, **682**, L5 (2008)
[4] Rees, M. J., & Gunn, J. E., MNRAS, **167**, 1 (1974)
[5] Kennel, C. F., & Coroniti, F. V., ApJ, **283**, 710 (1984)
[6] Dubus, G., A&A, **451**, 9 (2013)
[7] Bogovalov, S. V., Khangulyan, D. V., Koldoba, A. V., Ustyugova, G. V., & Aharonian, F. A., MNRAS, **387**, 63 (2008)
[8] Dubus, G., Lamberts, A., & Fromang, S., A&A, **581**, A27 (2015)
[9] Romani, R. W., & Sanchez, N., ApJ, **828**, 7 (2016)
[10] An, H. & Romani, R. W., ApJ, **838**, 145 (2017)
[11] Canto, J., Raga, A. C., & Wilkin, F. P., ApJ, **469**, 729 (1996)
[12] Finke, J. D., Dermer, C. D., & Böttcher, M., ApJ, **686**, 181 (2008)
[13] Ackermann, M., Ajello, M., Ballet, J., et al., Science, **335**, 189 (2012)
[14] An, H., Bellm, E., Bhalerao, V., et al., ApJ, **806**, 166 (2015)
[15] Strader, J., Chomiuk, L., Cheung, C. C., Salinas, R., & Peacock, M., ApJ, **813**, L26 (2015)
[16] An, H., Dufour, F., Kaspi, V. M., & Harrison, F. A., ApJ, **775**, 135 (2013)
[17] Waisberg, I. R., & Romani, R. W., ApJ, **805**, 18 (2015)
[18] Williams, B. J., Rangelov, B., Kargaltsev, O., & Pavlov, G. G., ApJ, **808**, L19 (2015)
[19] Wadiasingh, Z., Harding, A. K., Venter, C., Böttcher, M., & Baring, M. G., ApJ, **839**, 80 (2017)
[20] Sanchez, N., & Romani, R. W., ArXiv e-prints, , 1706.05467 (2017)
[21] Demorest, P. B., Pennucci, T., Ransom, S. M., nature, **467**, 1081 (2010)

EPJ Web of Conferences **168**, 04014 (2018) https://doi.org/10.1051/epjconf/201816804014
Joint International Conference of ICGAC-XIII and IK-15 on Gravitation, Astrophysics and Cosmology

Colliding scalar pulses in the Einstein-Gauss-Bonnet gravity

Hisaaki Shinkai[1,*] and *Takashi* Torii[2,**]

[1] *Department of Information Systems, Osaka Institute of Technology, Hirakata City, Osaka 573-0196, Japan*
[2] *Department of System Design, Osaka Institute of Technology, Osaka City, Osaka 530-8568, Japan*

Abstract. We numerically investigated how the nonlinear dynamics depends on the dimensionality and on the higher-order curvature corrections in the form of Gauss-Bonnet (GB) terms, with a model of colliding scalar pulses in plane-symmetric space-time. We observed that a collision of large scalar pulses will produce a large-curvature region, of which the magnitude depends on $\alpha_{\rm GB}$. The normal corrections ($\alpha_{\rm GB} > 0$) work for avoiding the appearance of singularity, although it is inevitable.

1 Introduction

The Einstein-Gauss-Bonnet (GB) gravity theory is one of the plausible candidates which describes the early Universe, but so far little is known of its nonlinear dynamical behaviors. We numerically investigated the dynamics in higher-dimensional space-time with and without the GB terms. (Please refer the detail report [1]).

The Einstein-GB gravity is derived from string theory, with additional higher-order-curvature correction terms to GR in the form of the Lagrangian,

$$\mathcal{L}_{\rm GB} = \mathcal{R}^2 - 4\mathcal{R}_{\mu\nu}\mathcal{R}^{\mu\nu} + \mathcal{R}_{\mu\nu\rho\sigma}\mathcal{R}^{\mu\nu\rho\sigma}, \tag{1}$$

where $\mathcal{R}$, $\mathcal{R}_{\mu\nu}$, and $\mathcal{R}_{\mu\nu\rho\sigma}$ are the n-dimensional scalar curvature, the Ricci tensor, and the Riemann curvature, respectively. This particular combination gives us several reasonable properties, as such ghost-free combinations, and a set of equations up to the second derivative in spite of the higher-curvature combinations. The theory is expected to have singularity-avoidance features in the context of gravitational collapses and/or cosmology. However, only a few studies so far have reported on the investigation of nonlinear dynamical features in Einstein-GB gravity (e.g., numerical studies on critical phenomena [2, 3], black hole formation in AdS [4]).

Our investigative model deals with colliding wave packets. Due to the nonlinear features of the theory, in GR, gravitational waves interact with themselves when they pass through each other. Considering a collision of plane gravitational waves is the simplest scenario of this nonlinear interaction problem (see Ref. [5] and references therein). In fact, Penrose [6] pointed out that the future light cone of a plane wave is distorted as it passes through another plane wave. Since these solutions assume a plane-symmetric space-time, this singularity does not have a horizon; it is a "naked" one.

Our attention to this problem focuses on the differences in the growth of curvature, especially the dependences on the dimension and the GB terms.

*e-mail: hisaaki.shinkai@oit.ac.jp
**e-mail: takashi.torii@oit.ac.jp

2 Field Equations and numerical technique

2.1 Action

The Einstein-GB action in n-dimensional space-time $(\mathcal{M}, g_{\mu\nu})$ is described as

$$S = \int_{\mathcal{M}} d^n x \sqrt{-g}\Big[\frac{1}{2\kappa^2}(\mathcal{R} + \alpha_{\rm GB}\mathcal{L}_{\rm GB}) + \mathcal{L}_{\rm matter}\Big], \tag{2}$$

where $\mathcal{L}_{\rm GB}$ is the GB term [Eq. (1)], κ^2 is the n-dimensional gravitational constant, and $\mathcal{L}_{\rm matter}$ is the matter Lagrangian. This action reproduces the standard n-dimensional Einstein gravity, if we set $\alpha_{\rm GB} = 0$. In the actual simulations, we set $\kappa^2 = 1$ and change $\alpha_{\rm GB}$ as a parameter.

The action (2) gives the gravitational equation as

$$G_{\mu\nu} + \alpha_{\rm GB} H_{\mu\nu} = \kappa^2 T_{\mu\nu}, \tag{3}$$

where $G_{\mu\nu} = \mathcal{R}_{\mu\nu} - \frac{1}{2} g_{\mu\nu}\mathcal{R}$, and

$$H_{\mu\nu} = 2\big(\mathcal{R}\mathcal{R}_{\mu\nu} - 2\mathcal{R}_{\mu\alpha}\mathcal{R}^{\alpha}{}_{\nu} - 2\mathcal{R}^{\alpha\beta}\mathcal{R}_{\mu\alpha\nu\beta} + \mathcal{R}_{\mu}{}^{\alpha\beta\gamma}\mathcal{R}_{\nu\alpha\beta\gamma}\big) - \frac{1}{2} g_{\mu\nu}\mathcal{L}_{\rm GB}. \tag{4}$$

We assume massless scalar field, ψ, of which energy-momentum tensor is given by

$$T_{\mu\nu} = \partial_\mu\psi\partial_\nu\psi - g_{\mu\nu}\frac{1}{2}(\nabla\psi)^2, \tag{5}$$

which obeys the Klein-Gordon equations, $\Box\psi = 0$.

2.2 Dual-null formulation

We use dual-null formulation for expressing space-time which has planar symmetry. The use of dual-null coordinates simplifies the treatment of horizon dynamics, enables us to approach close to large-curvature regions, and also clarifies radiation propagation in far regions. We implement our dual-null evolution code which was used for four-dimensional GR [7] so as to follow higher-dimensional space-time with the GB terms.

We adopt the line element

$$ds^2 = -2e^{f(x^+,x^-)}dx^+dx^- + r^2(x^+, x^-)\gamma_{ij}dz^i dz^j, \tag{6}$$

where the coordinates (x^+, x^-) are along the null propagation directions, and $\gamma_{ij}dx^i dx^j$ is the metric of the flat $(n-2)$-dimensional space.

For writing down the field equations, we introduce the variables

$$\Omega \equiv \frac{1}{r}, \qquad \vartheta_\pm \equiv (n-2)\partial_\pm r, \qquad \nu_\pm \equiv \partial_\pm f, \qquad \pi_\pm \equiv \partial_\pm\psi \tag{7}$$

where $\partial_\pm \equiv \partial/\partial x^\pm$, and these are the conformal factor, expansions, and inaffinities, respectively. Please refer [1] for a set of equations.

2.3 Numerical integration scheme

The basic idea of numerical integration is as follows. We prepare our numerical integration range as drawn in Fig. 1. We give initial data on a surface Σ_0, where $x^+ = x^- = 0$, and the two null hypersurfaces $\Sigma_\pm$, generated from it, where $x^\mp = 0$ and $x^\pm > 0$. Due to the dual-null decomposition, the causal region of a grid is clear, and there are in-built accuracy checks: the integrability conditions or consistency conditions $\partial_-\partial_+ u = \partial_+\partial_- u$.

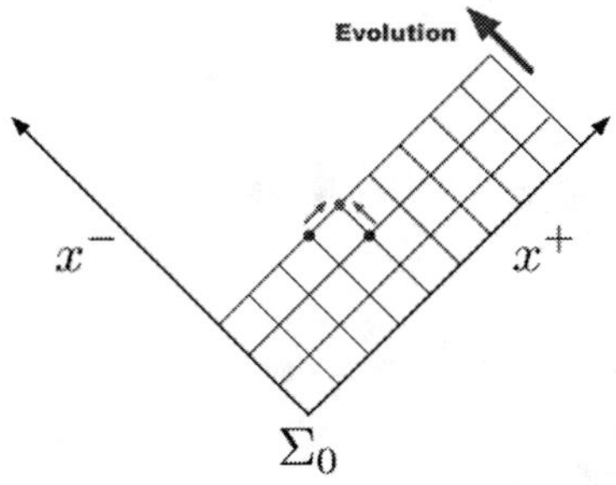

Figure 1. Numerical grid structure. Initial data are given on null hypersurfaces $\Sigma_\pm$ ($x^\mp = 0$, $x^\pm > 0$) and their intersection Σ_0. Generally, the initial data have to be given as $(\Omega, f, \vartheta_\pm, \psi)$ on Σ_0, and $(\nu_\pm, \pi_\pm)$ on $\Sigma_\pm$. We then evolve the data $u = (\Omega, \vartheta_\pm, f, \nu_\pm, \psi, \pi_\pm)$ on a constant-x^- slice to the next.

3 Numerical evolutions of the collision of scalar pulses

There are several exact solutions of the colliding plane waves, which produce curvature singularity after their collisions. We prepare a similar situation in our code and examine such a strong curvature effect in higher-dimensional GR and in Einstein-GB gravity. We first note that in the construction of exact solutions, the wave fronts are assumed to be a step function, while in our simulations the wave fronts are a continuous function. We have found that we can compare the behaviors more easily when we place colliding matter rather than colliding gravitational waves.

We put a perturbed scalar field (ψ) in the flat background on the initial surfaces and evolve it. The initial scalar field is set as $\psi = 0$ and has momentum

$$\begin{cases} \pi_+ = a\exp(-b(x^+/\sqrt{2}-c)^2) \\ \pi_- = 0 \end{cases} \text{on } \Sigma_+, \qquad \begin{cases} \pi_+ = 0 \\ \pi_- = a\exp(-b(x^-/\sqrt{2}-c)^2) \end{cases} \text{on } \Sigma_-. \tag{8}$$

3.1 Evolutions in GR

The two typical evolutions are shown in Fig. 2. We plot the behaviors of the scalar field and the Kretschmann scalar, $\mathcal{I}^{(5)}$, for five-dimensional GR. We set $a = 0.2$ and 0.4, and $b = 10$, $c = 2$ for these plots. For small pulses [Fig. 2(a)], we see that two pulses just pass through each other and the curvature $\mathcal{I}^{(5)}$ turns back to the flat again. On the contrary, for large pulses [Fig. 2(b)], the nonlinear curvature evolution appears after the collision of pulses. The latter behavior is similar to the exact solutions of the plane-wave collision (see, e.g., figures in Ref. [8]). We actually find that in all blow-up regions, both expansions are $\vartheta_\pm < 0$. In four-dimensional plane-symmetric space-time, if the curvature blows up, then it means the appearance of a naked singularity, since there is no chance to form a horizon. However, in higher dimensions, we expect such a blow-up will be hidden in a horizon as the expansions suggest.

3.2 Evolutions in Einstein-GB

We also evolved the same initial data by the set of evolution equations with nonzero α_{GB}.

Figure 3(a) displays the Kretschmann scalar, $\mathcal{I}^{(5)}(= R^{ijkl}R_{ijkl})$, for both $\alpha_{GB} = +1$ and $\alpha_{GB} = -1$ cases for the same initial data with the large-amplitude case ($a = 0.4$) in Fig. 2(b). We see that the local peak of $\mathcal{I}^{(5)}$ at the collision of two pulses (at $x^+ = x^- = 2\sqrt{2}$) is smaller (larger) when $\alpha_{GB} > 0$ ($\alpha_{GB} < 0$) than that in GR.

In Fig. 3(b), we plot the "evolution" behavior of the Kretschmann scalar, $\mathcal{I}^{(5)}$, at the origin ($x^+ = x^-$) where two pulses collide. At later times, we see that the curvature will diverge for all the cases (GR and Einstein-GB) due to the large amplitude of the initial pulses, but these growing behaviors are

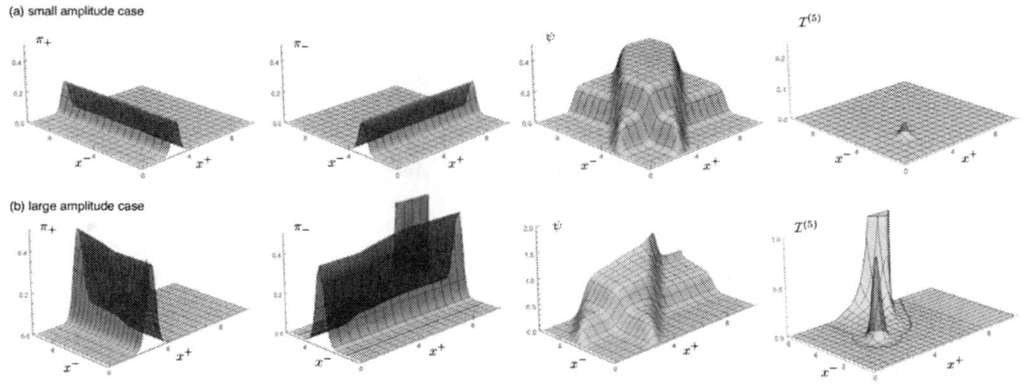

Figure 2. Evolutions of colliding two scalar pulses in five-dimensional GR: (a) the small-amplitude case [$a = 0.2$ in Eqs. (8)], and (b) the large-amplitude case ($a = 0.4$). The scalar momentum $\pi_{\pm}$, scalar field ψ, and the Kretschmann scalar $\mathcal{I}^{(5)}$ are plotted in the (x^+, x^-) coordinates. Initial data were set at both Σ_- ($x^+ = 0$, $x^- > 0$) and Σ_+ ($x^+ > 0$, $x^- = 0$) and evolved. For small pulses, we see that they just cross, and space-time turns back toward flat again, while for large pulses, we see that nonlinear curvature evolution appears after the collision of pulses. The latter behavior is similar to the exact solutions of the plane-wave collision.

again ordered by $\alpha_{\rm GB}$. Supposing that the curvature singularity will be formed at the final phase of this evolution (analogues to the plane-wave collision), then we can say that introducing the GB terms cannot stop the formation of the singularity, but it will shift its appearance later if $\alpha_{\rm GB} > 0$.

Figure 3(right) shows the magnitude of the Kretschmann scalar, $\mathcal{I}^{(n)}$, at the moment of the collision of scalar pulses (at the first peak of $\mathcal{I}^{(n)}$). We plot the cases $\alpha_{\rm GB} = 0, \pm 0.1, \pm 0.5, \pm 1.0$ and the dimensions $n = 4, 5, 6$, and 7. We see that for $n = 4$, all three cases have the same magnitude, which is consistent with the fact that the GB correction does not appear at $n = 4$. For larger dimensions, the magnitude becomes lower. We also find that introducing positive $\alpha_{\rm GB}$ (i.e. the normal higher-curvature correction) reduces its magnitude.

4 Summary and Discussions

We investigated fully nonlinear dynamics in n-dimensional Gauss-Bonnet gravity using a model of colliding scalar pulses. We observed that curvature (Kretschmann scalar) evolves more mildly in the presence of the normal GB terms ($\alpha_{\rm GB} > 0$) and in higher-dimensional space-time. The appearance of the singularity is inevitable in our model, but the basic feature is matched with the expected effect of the cosmologists; i.e., the avoidance (or lower possibility) of the appearance of the singularity.

As is shown in other models (e.g., Refs. [1, 9]), in higher-dimensional GR, the chance of appearances of naked singularities is suppressed compared to the four-dimensional GR cases. This is suggested by the existence of many freedoms in gravity which suppresses the growth of curvature and makes the formation of horizons less eccentric. The introduction of the GB terms seems to work for this direction. We hope that these results will be used as a guiding principle for understanding the fundamental dynamical features of the Einstein-GB gravity.

This work was supported in part by the Grant-in-Aid for the Scientific Research Fund of the JSPS (C) No. 25400277, and also by MEXT KAKENHI Grants No. 17H06357 and No. 17H06358.

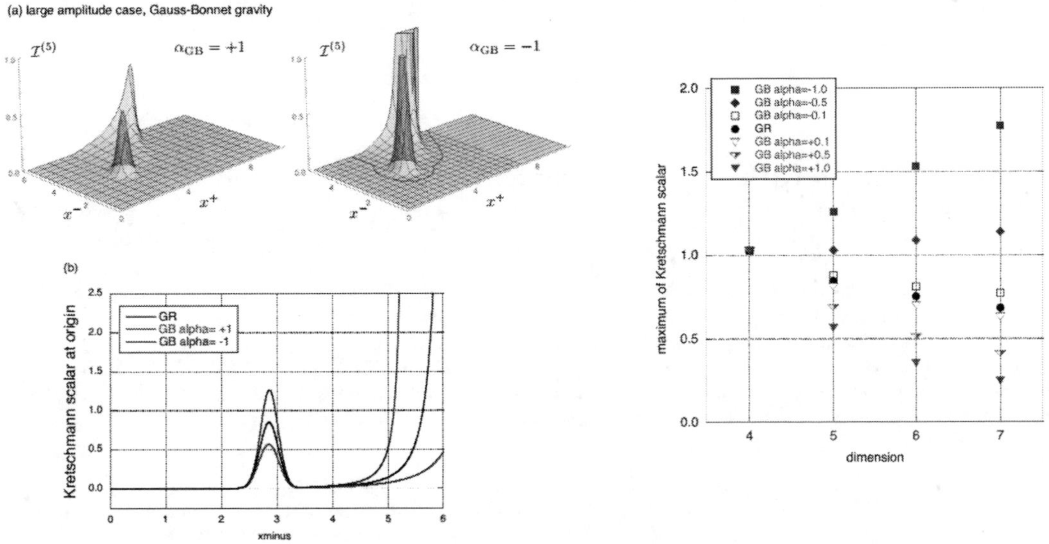

Figure 3. (a) Kretschmann scalar, $\mathcal{I}^{(5)}$, of the evolutions of colliding two scalar pulses in five-dimensional Einstein-GB gravity with $\alpha_{\rm GB} = \pm 1$. The initial data are the same with the large-amplitude case in Fig. 2(b). We see that the local peak of $\mathcal{I}^{(5)}$ at the collision of two pulses (at $x^+ = x^- = 2\sqrt{2}$) is smaller (larger) when $\alpha_{\rm GB} > 1$ ($\alpha_{\rm GB} < 1$). (b) Kretschmann scalar, $\mathcal{I}^{(5)}$, at the origin ($x^+ = x^-$), of these evolutions together with one with $\alpha_{\rm GB} = 0$ (i.e. GR). (Right) The Kretschmann scalar, $\mathcal{I}^{(n)}$, at the moment of the collision of scalar pulses (at $x^+ = x^- = 2\sqrt{2}$). We plot for the models with $\alpha_{\rm GB} = 0, \pm 0.1, \pm 0.5, \pm 1.0$ and for the dimensions $n = 4, 5, 6, 7$. For larger dimensions, the magnitude becomes lower in GR. We also find that introducing positive $\alpha_{\rm GB}$ (i.e. the normal higher-curvature correction) reduces its magnitude.

References

[1] H. Shinkai and T. Torii, Phys. Rev. D **96**, 044009 (2017).
[2] S. Golod and T. Piran, Phys. Rev. D **85**, 104015 (2012).
[3] N. Deppe, *et al.* Phys. Rev. D **86**, 104011 (2012).
[4] N. Deppe, *et al.* Phys. Rev. Lett. **114**, 071102 (2015); JHEP10, 087 (2016).
[5] J. B. Griffiths, *Colliding Plane Waves in General Relativity* (Clarendon Press, Oxford, 1991)
[6] R. Penrose, Phys. Rev. Lett. **14**, 57 (1965); Rev. Mod. Phys. **37**, 215 (1965).
[7] H. Shinkai and S. A. Hayward, Phys. Rev. D **66**, 044005 (2002).
[8] R.A. Matzner and F. J. Tipler, Phys. Rev. D **29**, 1575 (1984).
[9] Y. Yamada and H. Shinkai, Class. Quant. Grav. **27**, 045012 (2010); Phys. Rev. D **83**, 064006 (2011).

EPJ Web of Conferences **168**, 04015 (2018) https://doi.org/10.1051/epjconf/201816804015
Joint International Conference of ICGAC-XIII and IK-15 on Gravitation, Astrophysics and Cosmology

Universal relations with fermionic dark matter

A. Krut[1,2,3,⋆], C. R. Argüelles[3,5], J. A. Rueda[1,3,4], and R. Ruffini[1,3,4]

[1] *Dip. di Fisica, Sapienza Università di Roma, Piazzale Aldo Moro 5, I-00185 Rome, Italy*
[2] *University of Nice-Sophia Antipolis, 28 Av. de Valrose, 06103 Nice Cedex 2, France*
[3] *ICRANet, Piazza della Repubblica 10, I-65122 Pescara, Italy*
[4] *ICRANet-Rio, Centro Brasileiro de Pesquisas Físicas, Rua Dr. Xavier Sigaud 150, 22290–180 Rio de Janeiro, Brazil*
[5] *Grupo de Astrofísica, Relatividad y Cosmología, Facultad de Ciencias Astronómicas y Geofísicas, UNLP and CONICET Paseo del Bosque S/N 1900, La Plata, Buenos Aires, Argentina*

Abstract. We have recently introduced a new model for the distribution of dark matter (DM) in galaxies, the Ruffini-Argüelles-Rueda (RAR) model, based on a self-gravitating system of massive fermions at finite temperatures. The RAR model, for fermion masses above keV, successfully describes the DM halos in galaxies, and predicts the existence of a denser quantum core towards the center of each configuration. We demonstrate here, for the first time, that the introduction of a cutoff in the fermion phase-space distribution, necessary to account for galaxies finite size and mass, defines a new solution with a compact quantum core which represents an alternative to the central black hole (BH) scenario for SgrA*. For a fermion mass in the range $48\text{keV} \leq mc^2 \leq 345\text{keV}$, the DM halo distribution fulfills the most recent data of the Milky Way rotation curves while harbors a dense quantum core of $4 \times 10^6 M_\odot$ within the S2 star pericenter. In particular, for a fermion mass of $mc^2 \sim 50\text{keV}$ the model is able to explain the DM halos from typical dwarf spheroidal to normal elliptical galaxies, while harboring dark and massive compact objects from $\sim 10^3 M_\odot$ tp to $10^8 M_\odot$ at their respective centers. The model is shown to be in good agreement with different observationally inferred universal relations, such as the ones connecting DM halos with supermassive dark central objects. Finally, the model provides a natural mechanism for the formation of supermassive BHs as heavy as few $\sim 10^8 M_\odot$. We argue that larger BH masses (few $\sim 10^{9-10} M_\odot$) may be achieved by assuming subsequent accretion processes onto the above heavy seeds, depending on accretion efficiency and environment.

1 Introduction

The problem of the distribution of stars in globular clusters, and more general in galactic systems, has implied one of the results of most profound interest in classical astronomy. In particular, in the pioneering works of [1] and [2], they considered the effects of collisional relaxation and tidal cutoff by studying solutions of the Fokker-Planck equation. There, it was shown that stationary solutions of this kind can be well described by lowered isothermal sphere models, based on simple Maxwellian energy distributions with a constant subtracting term interpreted as an energy cutoff. An extension of this statistical analysis with thermodynamic considerations, which includes the effects of violent (collisionless) relaxation, has been studied in [3], with important implications to the problem of virialization in galaxies which are still of actual interest. In a series of works, [4–6] changed the emphasis from self-gravitating systems of classic stars (which verify Maxwellian distributions) to systems of fermionic particles with the aim of describing galactic dark matter halos. It was there considered a quantum fermionic distribution taking into account the possible presence of a cutoff in the energy as well as in the angular momentum. A remarkable contribution in the understanding of these issues was given in [7], based on the study of generalized kinetic theories accounting for collisionless relaxation processes, and leading to a class of generalized Fokker-Planck equation for fermions. It was there explicitly shown the possibility to obtain, out of general thermodynamic principles, a generalized Fermi-Dirac distribution function including an energy cutoff, extending the former results by [1] and [2] to quantum particles. More recently, it was shown that quantum particles fulfilling fermionic quantum statistics and gravitational interactions are able to successfully describe the distribution of galactic dark matter halos when contrasted with observations [8–11]. In particular, [11] proposed a new model (hereafter RAR model), based on such fundamental physical pillars to attack this problem, presenting a novel and more general *dense quantum core - classical*

⋆E-mail: e-mail: andreas.krut@icranet.org (ICRANet)

EPJ Web of Conferences **168**, 04015 (2018) https://doi.org/10.1051/epjconf/201816804015
Joint International Conference of ICGAC-XIII and IK-15 on Gravitation, Astrophysics and Cosmology

halo distribution. Thus, given the apparent ubiquity of massive black holes at the center of galaxies, in [11] the RAR model was proposed as a viable possibility to establish a link between the dark central cores to dark matter halos within a unified approach.

In this work we extend the RAR model by introducing a cutoff in the momentum distribution to account for finite galaxy sizes (analogously as previously done in [6]) and to account for more realistic galaxy relaxation mechanisms as indicated above and in [7]. This approach provides a new family of solutions with an overall redistribution of the bounded fermions. Consequently, the more stringent outer halo constraints of these novel configurations allow a higher compactness of the central cores. A detailed description of the presented results, the RAR model, its application to galactic structures and the comparison with universal relations is provided in [12].

2 Novel constraints from Milky Way Observables

We consider the extended high resolution rotation curve data of the Galaxy in [13], ranging from pc scales up to $\sim 10^2$ kpc, together with the orbital data of the eight best resolved S-cluster stars taken from [14]. Our analysis will thus cover in total more than nine orders of magnitude of radial extent.

As we will show below, the key result of our work is that there is a continuous underlying dark matter distribution covering the whole observed Galactic extent, which not only governs the dynamics of the outer halo above $r \gtrsim 10$ kpc, but also the central pc, being the intermediate region dominated by the baryonic components (bulge+disk).

The objective now is to solve the metric potential $\nu(r)$ for given regular initial conditions at the center (β_0, θ_0, W_0) and different dark matter particle mass m, in order to find the mass solution $M(r)$ consistent with the observationally constrained dark matter halo of the Galaxy. In detail, a dark matter halo described by $(r_1, M_1) = (40\ \text{kpc}, 2 \times 10^{11} M_\odot)$ consistent with the dynamics of the outer dark matter halo as shown recently in [15]. This is equivalently consistent with $(r_2, M_2) = (12\ \text{kpc}, 5 \times 10^{10} M_\odot)$ as constrained in [13]. Simultaneously, we require a quantum core of mass $M(r_c) = M_c = 4.2 \times 10^6 M_\odot$ enclosed within a radius $r_c \leq r_{p(S2)} = 6 \times 10^{-4}$ pc, the S2 star pericenter [14]. The results of this full boundary condition problem for the Milky Way observables are summarized in fig. 1.

The Milky Way outermost dark matter halo behavior is subjected to the cutoff conditions $W(r_b) \approx 0$ when $\rho(r_b) = 10^{-5} M_\odot/\text{pc}^3$ at the boundary radius $r_b = 50$ kpc (see fig. 1). Here, $\rho(r_b)$ is the Local Group density as constrained in [17][1]. The limiting behavior of such a dark matter density profile is also consistent with a dark matter halo mass of $M(40\ \text{kpc}) = 2 \times 10^{11} M_\odot$ as required above, further implying a total Galactic mass (dark + baryonic) at r_b of $M_{\text{all}} \approx 3 \times 10^{11} M_\odot$, of which 80% is dark according to our model (i.e. $M(r_b) = 2.4 \times 10^{11} M_\odot$). It is clear that such a dark matter mass distribution must be also in agreement with the dynamical constraints set by the Galactic satellite dwarf observations, e.g. the Sagittarius (Sgr) dwarf satellite. Indeed, such observational constraints have been recently considered in [15], who showed that their fulfillment requires a total mass of the Galaxy (at $\sim 80\%$ confidence level)

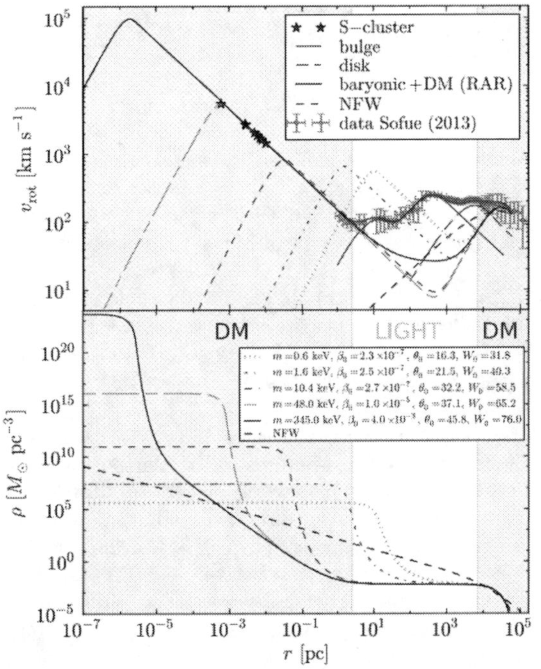

Figure 1. (Colour online, with permission of [12]) Theoretical rotation curves (upper panel) and density profiles (lower panel) for different DM fermion masses in the keV region, in agreement with all the Milky Way observables from $\sim 10^{-3}$ pc to $\sim 10^5$ pc. The continuous thick-red curve that fit the observed data (blue points) is the total rotation curve given by the RAR model plus the baryonic (bulge + disk) component without the need for a central BH, i.e. using any RAR profile for $mc^2 = 48$–345 keV. We also show, for the sake of comparison, the contribution of the NFW DM profile to the total rotation curve (not shown here) which produces an equally good fit to the data in the halo region [see 13, for details]. The stars-symbols represent the eight best resolved S-cluster stars [14], whose positions in the plot indicates the *effective* circular velocity at pericenter (i.e. without considering the ellipticity of the orbits). In the inner bulge region ($3 \lesssim r \lesssim 10^2$) pc, the large velocity error bars of about ±20–30% are mainly due to non-circular motions, while in the halo region there are larger observational errors bars of up to $\sim 50\%$ due to systematics [13]. The DM contribution to the Galactic halo becomes necessary above ~ 7 kpc, in agreement with [16].

[1]Note that the exact $W(r) = 0$ cutoff condition, is fulfilled in the limiting case $\rho(r) = 0$ achieved for $r \gtrsim r_b$.

$M_{\rm all}(r \gtrsim 50$ kpc)$\approx 3 \times 10^{11} M_\odot$, in agreement with our results here.

The different regimes in the $\rho(r)$ profiles are also manifest in the dark matter rotation curve. We define the first maxima as the core radius r_c and the second maxima as the halo (or the one-halo scale length) r_h. Following then the standard assumption in the literature that baryonic and dark matter do not interact each other, we have theoretically calculated the total rotation curve as $v^2_{\rm rot} = v^2_b(r) + v^2_d(r) + v^2_{\rm DM}(r)$. With $v_b(r)$ and $v_d(r)$ the baryonic circular velocities for bulge and disk. We calculated the total (inner + main) bulge circular velocity using the same mass model parameters as in [13]. For the disk, we have performed the calculations with mass models parameters slightly changed (within 12%) with respect to those given in [13], where the NFW dark matter profile was assumed. We do this change to improve the fit of the observational data when adopting our dark matter profile. Finally, $v^2_{\rm DM}(r)$ is the dark matter contribution given by the RAR model.

Varying the particle mass produces configurations which are classified in three groups. The fermion mass range $mc^2 \leq 10.4$ keV is firmly ruled out by the present analysis because the corresponding rotation curve exceeds the total velocity observed in the baryonic (bulge+disk) dominated region, between ~ 1 pc and 10^2 pc (including upper bound in error bars, see fig. 1).

In the intermediate range 10.4 keV$\lesssim mc^2 \lesssim$48 keV, the theoretical rotation curve is not in conflict with any of the observed data and dark matter inferences in [13], but the compactness of the quantum core is not enough to be an alternative to the central BH scenario.

There is a fermion mass range 48 keV$\lesssim mc^2 \lesssim$345 keV with corresponding accompanying parameters (β_0, θ_0, W_0), whose associated solutions explain the Galactic dark matter halo while providing at the same time an alternative for the central BH. The lower bound in m is imposed by the dynamics of the stellar S-cluster, while the upper bound corresponds to the last stable configuration before reaching the critical mass for gravitational collapse ($M_c^{cr} \sim m^3_{\rm Planck}/m^2$, see also [18]). The critical configuration has a core radius $r_c \approx 4.75 r_\bullet$ with $r_\bullet$ the Schwarzschild radius associated to a BH of $4.2 \times 10^6 M_\odot$ (see also [19]).

3 Comparing observed galaxy correlations

We turn now to analyze if the RAR model agrees with the $M_{\rm BH} - M_{\rm tot}$ relation [20–22] where $M_{\rm tot}$ is the total dark matter halo mass and $M_{\rm BH}$ is the mass of the compact dark object at the center of galaxies. Traditionally, that dark object is assumed as SMBHs but here interpreted as dark matter quantum cores in the case of inactive galaxies. In the following we consider $M_{\rm BH} = M_c$, being $M_c = M(r_c)$ the quantum core mass. In order to show this, we use the full family of RAR solutions for typical dSphs ($r_h = 0.26$ kpc, $M_h = 2.7 \times 10^7 M_\odot$), spiral ($r_h = 48$ kpc, $M_h = 1 \times 10^{12} M_\odot$) and elliptical galaxies ($r_h = 90$ kpc, $M_h = 5 \times 10^{12} M_\odot$) calculated for the particle mass $mc^2 = 48$ keV and spanning the maximal free parameter space (β_0,θ_0,W_0) in each galaxy case. See [12] for details. These solutions cover correspondingly a well defined window of predicted masses $(M_c, M_{\rm tot})$ summarized in fig. 2.

For typical spiral and elliptical galaxy types the family of predicted $(M_c, M_{\rm tot})$ covers the (horizontal) spread of the observed correlation for each galaxy type. In addition, it extends out of it indicating a window of predicted masses by the RAR model, not yet observed. The limiting values of $(M_c, M_{\rm tot})$, with observationally given halo radius r_h and halo mass M_h, are automatically established by the gravitational core collapse at the critical point (intrinsic physical condition) and the necessity to have a dark matter density contribution above the galaxy group-scale (astrophysical condition). The Milky Way RAR-solution is also plotted for completeness showing a good agreement as well. The (M_c, M_{tot}) RAR-predictions are displayed together with the observationally inferred best-fit relations [20, 22], only in the region where data supports (i.e. the so-called Ferrarese window in fig. 2). The case of typical dwarf galaxies is located at the lower end of the $(M_c, M_{\rm tot})$ plane. It is worth to stress that no observational data exist yet in that part of the correlation and thus the blue curve at the bottom of fig. 2 is a prediction of the RAR model. Additional verification of the above predictions of the RAR model needs the observational filling of the gaps in the $(M_c, M_{\rm tot})$ plane from dwarfs all the way up to ellipticals. This is a work in progress within our group.

It is appropriate at this point to recall that the majority of the values of $M_{\rm BH}$ in the observed $M_{\rm BH} - M_{\rm tot}$ relation have been obtained through the so-called $M_{\rm BH} - \sigma_*$ relation, with σ_* the bulge dispersion velocity [23–25]. The $M_{\rm tot}$ values were there calculated at the virial radius within a Navarro-Frenk-White dark matter model [20] while in our case they were obtained at the boundary radius r_b of the RAR model equilibrium configurations. In the case of dwarf galaxies the observational inference of the mass of the central dark region via the dispersion velocity is unclear [see, e.g., 21, 26, and references therein]. However, [26] attempted to give an estimate of the dark central object mass in dwarf galaxies obtaining $M_{\rm BH} \sim 10^3$–10^4 $M_\odot$ (often called intermediate massive BHs), in agreement with the values of M_c for dSph analyzed in this work. Special attention has to be given to our dwarf galaxy predictions in view of the recent observational result reported in [27] on the putative massive black holes of few $\sim 10^6 M_\odot$ detected in ultra-compact dwarf galaxies of total mass of few $\sim 10^7 M_\odot$. Interestingly, these (rather evolved) compact dwarf galaxies show no (observational) hints of being dominated by a dark matter halo, remarkably in agreement with

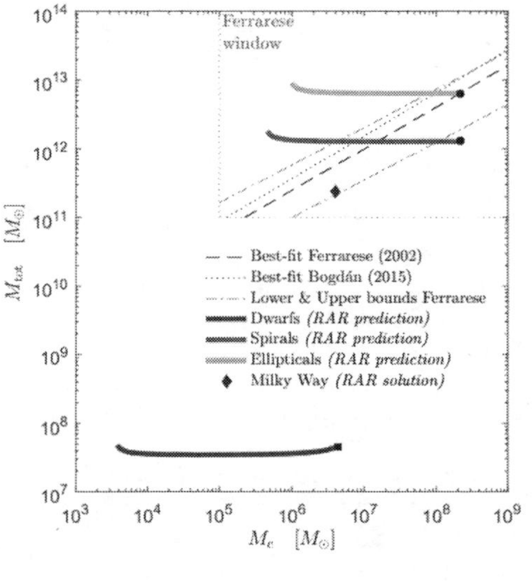

Figure 2. (Colour online, with permission of [12]) The different *predicted* colour lines read for each galaxy type showing the ability of the three parametric (β_0, θ_0, W_0) RAR solutions for a particle masse $mc^2 = 48$ keV to be in agreement with the different $M_{\rm BH} - M_{\rm tot}$ relations considered in the literature (see text) and explicited in the picture box. While the red and yellow RAR prediction lines, together to the Milky Way solution (i.e. spiral and elliptical galaxies) lay within the *observable Ferrarese window*, the blue RAR prediction for dwarfs, is located at the lower end of the M_c-$M_{\rm tot}$ plane, where data do not support. The black dots correspond to the critical core masses M_c^{cr}, the black square indicates the limiting maximum core mass for dwarfs M_c^{max}.

our *halo deficit*[2] solution, see fig. 2. It is interesting that this observational inference was published few weeks after the completion of our predictions. This is a potentially important prediction of the RAR model which deserves further analysis.

Moving to the larger elliptical galaxies, it is interesting to note that the maximum quantum core mass $M_c^{cr} \sim 2\times10^8 M_\odot$ (for $mc^2 = 48$ keV) predicted by our model, is in striking consistency with the uppermost (sample-representative) central mass $M_{\rm BH}$ obtained from an X-ray imaging analysis of more than 3000 isolated and inactive elliptical galaxies [22]. These results, when viewed through our theoretical $M_c - M_{\rm tot}$ relation, give support to our idea that normal elliptical galaxies may harbor dark central objects (not yet BHs) without showing AGN-like activity, while larger SMBHs masses do show AGN properties and thus reach the upper end of the $M_{\rm BH} - M_{\rm tot}$ relation.

[2]This has to be understood in a technical context rather than in an astrophysical.

4 Conclusion

It is now clear from our results that gravitationally bounded systems based on fermionic phase-space distributions including for escape velocity effects and central degeneracy, can explain the DM content in the Galaxy while providing a natural alternative for the central BH scenario in SgrA*. A key point of the present RAR model is the ability to predict entire DM halo configurations which fulfil the observed properties of galaxies, such as the $M_{\rm BH}$-M_{tot}, for a unique DM fermionic mass. At the same time it provides, on astrophysical basis, possible clues on the formation of super massive BHs in galactic nuclei.

Our general results on the DM distribution in galaxies can be considered as complementary to those based on standard cosmological simulations. However, the latter being based on N-body purely Newtonian simulations, contrast with our semi-analytical four-parametric approach which has the chance to include more rich physical ingredients, such as quantum statistics (arising from specific phase-space relaxation mechanisms), thermodynamics and gravity. We have shown that our density profiles successfully agree with a large variety of galactic observables and universal laws.

An important aspect of the particle mass range of few 10–100 keV obtained here from galactic observables is that it produces basically the same behavior in the power spectrum (down to Mpc scales) from that of standard ΛCDM cosmologies, thus providing the expected large-scale structure (see for details [28]). In addition, it is not *too warm* (i.e. our masses are larger than $mc^2 \sim 1$–3 keV) to enter in tension with current Lyα forest constraints [29, 30] and the number of Milky Way satellites [31] as in standard ΛWDM cosmologies.

All the above arguments offer significant support for our keV-scale fermions as dark matter, which may well co-exist harmonically with other DM species in the universe. These aspects will have to interplay with the physics of elementary particles regarding the nature of these fermions (Majorana neutrinos, supersymmetric particles, sterile neutrinos, etc.) as well as with the possible detection through decaying processes involving weak interactions. Indeed, DM fermion masses within the relatively narrow window obtained here, 48 keV$\lesssim mc^2 \lesssim$ 345 keV, have also arisen within different microscopic models based on extensions of the standard model and consistent with all cosmological (i.e. large scale structure) and X-ray constraints as the ones considered in [29].

Acknowledgments

A. K. is supported by the Erasmus Mundus Joint Doctorate Program by Grants Number 2014-0707 from the agency EACEA of the European Commission.

References

[1] R.W. Michie, MNRAS**125**, 127 (1963)
[2] I.R. King, AJ**71**, 64 (1966)
[3] D. Lynden-Bell, MNRAS**136**, 101 (1967)
[4] R. Ruffini, L. Stella, A&A**119**, 35 (1983)
[5] J.G. Gao, M. Merafina, R. Ruffini, A&A**235**, 1 (1990)
[6] G. Ingrosso, M. Merafina, R. Ruffini, F. Strafella, A&A**258**, 223 (1992)
[7] P.H. Chavanis, Physica A Statistical Mechanics and its Applications **332**, 89 (2004), cond-mat/0304073
[8] H.J. de Vega, P. Salucci, N.G. Sanchez, MNRAS**442**, 2717 (2014), 1309.2290
[9] C.R. Argüelles, R. Ruffini, I. Siutsou, B. Fraga, Journal of Korean Physical Society **65**, 801 (2014), 1402.0700
[10] I. Siutsou, C.R. Argüelles, R. Ruffini, Astronomy Reports **59**, 656 (2015), 1402.0695
[11] R. Ruffini, C.R. Argüelles, J.A. Rueda, MNRAS**451**, 622 (2015), 1409.7365
[12] C.R. Argüelles, A. Krut, J.A. Rueda, R. Ruffini, ArXiv e-prints (2017), 1606.07040
[13] Y. Sofue, PASJ**65**, 118 (2013), 1307.8241
[14] S. Gillessen, F. Eisenhauer, T.K. Fritz, H. Bartko, K. Dodds-Eden, O. Pfuhl, T. Ott, R. Genzel, ApJ**707**, L114 (2009), 0910.3069
[15] S.L.J. Gibbons, V. Belokurov, N.W. Evans, MNRAS**445**, 3788 (2014), 1406.2243
[16] F. Iocco, M. Pato, G. Bertone, Nature Physics **11**, 245 (2015), 1502.03821
[17] Y. Sofue, PASJ**64**, 75 (2012), 1110.4431
[18] C.R. Argüelles, R. Ruffini, International Journal of Modern Physics D **23**, 1442020 (2014), 1405.7505
[19] C.R. Argüelles, R. Ruffini, B.M.O. Fraga, Journal of Korean Physical Society **65**, 809 (2014), 1402.1329
[20] L. Ferrarese, ApJ**578**, 90 (2002), astro-ph/0203469
[21] J. Kormendy, R. Bender, Nature**469**, 377 (2011), 1101.4650
[22] Á. Bogdán, A.D. Goulding, ApJ**800**, 124 (2015), 1502.05043
[23] L. Ferrarese, D. Merritt, ApJ**539**, L9 (2000), astro-ph/0006053
[24] K. Gebhardt, R. Bender, G. Bower, A. Dressler, S.M. Faber, A.V. Filippenko, R. Green, C. Grillmair, L.C. Ho, J. Kormendy et al., ApJ**539**, L13 (2000), astro-ph/0006289
[25] K. Gültekin, D.O. Richstone, K. Gebhardt, T.R. Lauer, S. Tremaine, M.C. Aller, R. Bender, A. Dressler, S.M. Faber, A.V. Filippenko et al., ApJ**698**, 198 (2009), 0903.4897
[26] M. Valluri, L. Ferrarese, D. Merritt, C.L. Joseph, ApJ**628**, 137 (2005), astro-ph/0502493
[27] C.P. Ahn, A.C. Seth, M. den Brok, J. Strader, H. Baumgardt, R. van den Bosch, I. Chilingarian, M. Frank, M. Hilker, R. McDermid et al., ApJ**839**, 72 (2017), 1703.09221
[28] A. Boyarsky, O. Ruchayskiy, M. Shaposhnikov, Annual Review of Nuclear and Particle Science **59**, 191 (2009), 0901.0011
[29] A. Boyarsky, J. Lesgourgues, O. Ruchayskiy, M. Viel, Physical Review Letters **102**, 201304 (2009), 0812.3256
[30] M. Viel, G.D. Becker, J.S. Bolton, M.G. Haehnelt, Phys. Rev. D**88**, 043502 (2013), 1306.2314
[31] E.J. Tollerud, J.S. Bullock, L.E. Strigari, B. Willman, ApJ**688**, 277-289 (2008), 0806.4381

EPJ Web of Conferences **168**, 05001 (2018) https://doi.org/10.1051/epjconf/201816805001
Joint International Conference of ICGAC-XIII and IK-15 on Gravitation, Astrophysics and Cosmology

Gravitational waves from the first order electroweak phase transition in the Z_3 symmetric singlet scalar model [*]

Toshinori Matsui[1,**]

[1] *School of Physics, KIAS, Seoul 02455, Korea*

Abstract. Among various scenarios of baryon asymmetry of the Universe, electroweak baryogenesis is directly connected with physics of the Higgs sector. We discuss spectra of gravitational waves which are originated by the strongly first order phase transition at the electroweak symmetry breaking, which is required for a successful scenario of electroweak baryogenesis. In the Z_3 symmetric singlet scalar model, the significant gravitational waves are caused by the multi-step phase transition. We show that the model can be tested by measuring the characteristic spectra of the gravitational waves at future interferometers such as LISA and DECIGO.

1 Introduction

In the scenario of electroweak baryogenesis (EWBG) [2, 3], the strongly first order electroweak phase transition (SFOEWPT) is required to satisfy the condition of the departure from thermal equilibrium

$$\langle h \rangle_* / T_* \gtrsim 1, \tag{1}$$

with T_* being the temperature of EWPT and $\langle h \rangle_*$ the vacuum expected value (VEV) of the SM Higgs field h at T_*. In order to satisfy this condition, the extended Higgs sector from standard model (SM) is required. These extensions could help to build a barrier between the EW vacuum and a metastable vacuum at tree or loop level [3, 4]. The mechanism to generate a thermal cubic term for h by a tree level barrier is most easily implemented in the extended Higgs sectors by a singlet S, containing effective tree-level cubic terms $\sim S^3 + S|H|^2$ with H the SM Higgs doublet [5–11].

If the extended Higgs sector respects some symmetry such as Z_2, under which $S \to -S$ and $H \to H$, an alternative way to the desired tree level barrier is available in the symmetric limit where S does not acquire VEV at the present universe [7, 12–16]. Such a scenario is associated with multi-step PT's. The universe may have been once in the intermediate phase $\Omega_{\rm meta}$ and then tunneled through a tree level barrier to the phase $\Omega_{\rm EW}$, recovering the Z_2 symmetry.

We expect that gravitational wave (GW) is available to explore the nightmare scenario which is a case that the model cannot be tested at colliders. In principle, EWPT of $T_* \simeq 100$ GeV can be detectable at the GW observation experiments [17]. The space-based interferometers: LISA [18], DECIGO [19] and BBO [20], designed to be sensitive to GW density $\Omega_{\rm GW}h^2 \gtrsim 10^{-16} - 10^{-10}$ (depending on frequency $\simeq 10^{-3} - 10^{-1}$ Hz), will be launched in the near future [17].

[*]This proceeding paper is based on Ref. [1] in collaborated with Zhaofeng Kang and Pyungwon Ko.
[**]e-mail: matsui@kias.re.kr

2 Z_3 symmetric singlet scalar model

We introduce an isospin complex singlet scalar S transforming as $S \to e^{i2w}S$ with $w = \pi/3$ under Z_3, while the SM fields including the SM Higgs doublet H are neutral under Z_3. The most general renormalizable and Z_3-symmetric scalar potential $V(H,S)$ is given by

$$V_0(H,S) = -\mu_h^2|H|^2 - \mu_s^2|S|^2 + \lambda_h|H|^4 + \lambda_s|S|^4 + \lambda_{sh}|H|^2|S|^2 + \sqrt{2}\left(\frac{A_s}{3}S^3 + h.c.\right). \quad (2)$$

Compared to the Z_2-symmetric model, there is just one more parameter describing the cubic term A_sS^3. [1] After EWSB, two scalar fields are parametrized as $H = (G^+, (v + h^0 + iG^0)/\sqrt{2})$ and $S = (s^0 + ia_s^0)/\sqrt{2}$. There appear two physical degrees of freedom h and s in addition to Nambu-Goldstone (NG) modes $G^\pm$ and G^0 that are absorbed by the W- and Z-bosons. The vacuum stability condition reads as $\lambda_s > 0$, $\lambda_h > 0$ and $4\lambda_s\lambda_h > \lambda_{sh}^2$. At zero temperature $T = 0$, the model parameters are fixed to be $\lambda_h = m_h^2/(2v^2)$, $\mu_h^2 = m_h^2/2$ and $\mu_s^2 = \lambda_{sh}v^2/2 - m_s^2$ up to radiative corrections with v which is the VEV of h. Here, m_h and m_s are the physical masses of h and s. We use $v = 246$ GeV, $m_h = 125$ GeV, $m_s, \lambda_s, \lambda_{sh}$ and A_s as the input parameters.

Expanding the scalar fields around their classical backgrounds, $\langle H\rangle = (0, \varphi_h/\sqrt{2})$ and $\langle S\rangle = \varphi_s/\sqrt{2}$, the one-loop effective potential at finite temperature is given by

$$V_{\rm eff}(\varphi_h,\varphi_s,T) = V_0(\langle H\rangle,\langle S\rangle) + \sum_i n_i\frac{M_i^4(\varphi_h,\varphi_s,T)}{64\pi^2}\left(\ln\frac{M_i^2(\varphi_h,\varphi_s,T)}{Q^2} - c_i\right) + \sum_i n_i\frac{T^4}{2\pi^2}I_{B,F}\left(\frac{M_i^2(\varphi_h,\varphi_s,T)}{T^2}\right), \quad (3)$$

where Q is the renormalization scale, which is set at v in our analysis. Here, n_i and $M_i(\varphi_\Phi, \varphi_S, T)$ denote the degrees of the freedom and the field-dependent masses for particles i, respectively. We consider loop contributions from the fields $i = h^0, s^0, a_s^0, G^\pm, G^0, W_{T,L}^\pm, Z_{T,L}, \gamma_{T,L}, t$ and b. We take the $\overline{\rm MS}$ scheme, where the numerical constants c_i are set at $3/2$ $(5/6)$ for scalars and fermions (gauge bosons). The contribution of the finite temperature is defined by $I_{B,F}(a^2) = \int_0^\infty dx\, x^2 \ln\left[1 \mp \exp\left(-\sqrt{x^2+a^2}\right)\right]$ for boson and fermions, respectively. The thermally corrected field-dependent masses for the CP-even/odd, Goldstone, the weak gauge bosons and top quarks are given by, for example, Ref. [1, 22].

3 Multi-step phase transitions with first order electroweak phase transition

For a given scalar potential $V_{\rm eff}(\vec{\varphi}, T)$ with $\vec{\varphi}$ denoting a vector of real scalar fields in the multi dimensional fields space, the (critical) bubble can be found by extremizing the Euclidean action $S_E(T) \equiv S_3(T)/T$ where $S_3(T)$ is defined as $S_3(T) \equiv \int d^3x\left[(\partial\vec{\varphi})^2/2 + V_{\rm eff}(\vec{\varphi},T)\right]$. Then, the bubble nucleation rate per unit volume per unit time will be given by $\Gamma(t) = \Gamma_0(t)\exp[-S_E(t)]$ with the pre-factor $\Gamma_0 \sim T^4$. In order for the nucleated vacuum bubbles to percolate through the whole Universe, the nucleation rate per Hubble volume per Hubble time should reach the unity $\Gamma/H^4|_{T=T_*} \simeq 1$, which determines the transition temperature T_*.

The GW spectrum from first order phase transition (FOPT) can be parameterized by several parameters, with the most crucial two, α and β, which capture the main features of FOPT dynamics and largely determine the features of GW spectrum. We will follow the conventions in Ref. [17]. The parameter $\alpha \equiv \epsilon/\rho_{\rm rad}$ is the total energy budget of FOPT normalized by the radiative energy $\rho_{\rm rad} = (\pi^2/30)g_*T_*^4$ with $g_*(= 108.75)$ being the relativistic degrees of freedom in the plasma at the PT temperature T_*. The liberated latent heat $\epsilon = -(\Delta V + T\partial V/\partial T)|_{T_*}$, with ΔV the vacuum energy gap between two vacua. Another parameter β is defined by $\beta \equiv -dS_E/dt|_{t_*}$. We use the dimensionless parameter $\widetilde{\beta} \equiv \beta/H_*$, where $H_* \equiv 1.66\sqrt{g_*}\,T_*^2/m_{\rm pl}$ is the Hubble constant.

[1] In this paper we do not consider the possibility that S makes the dark matter (DM) candidate [21], because we failed in finding viable parameter space with $\lambda_{sh} \sim O(0.01)$ that is necessary to accommodate correct DM phenomenology.

4 Numerical results

4.1 Parameter space with various transition pattern

In order to study the vacuum structure at finite temperature, we use the code `cosmoTransitions` [23] for numerical studies on PT in the Z_3 symmetric scalar Higgs sector. Each path of the transition pattern and the metastable vacua at the intermediate stage of the model are shown in Fig. 1. At $T = 0$, we are interested in the case where the EWSB but Z_3-preserving vacuum $\Omega_h \equiv (\langle h \rangle = v, 0)$ is the ground state, which may be accompanied by a metastable vacuum $\Omega_s \equiv (0, \langle s \rangle \neq 0)$ or $\Omega_{sh} \equiv (\langle h \rangle \neq 0, \langle s \rangle \neq 0)$. The presence of Ω_{sh} is a new aspect in the Z_3-symmetric model compared to the Z_2-symmetric model, and it will make possible three-step PT's in our model.

We summarize the parameter region of multi-step PT in Fig. 1, where two-step PT and three-step PT are plotted [2]. In the $\mu_s^2 > 0$ region, we find that the two-step PT ($\Omega_0 \to \Omega_s \to \Omega_h$) can happen, with the first-step either second or first order, depending on the relevant parameters.

Two step (second order - first oder) PT case is basically corresponding to the Z_2-symmetric model in the $A_s \to 0$ limit. For the $\lambda_s = 1$ example, A_s is restricted to be smaller than tens of GeV and thus the resulting deviations as expected are not significant. But it can still increase or decrease T_h^* with appreciate amount, see the green and blue points in Fig. 2 (left).

Two step (first order - first oder) PT for finite A_s, the first-step PT significantly becomes the FOPT. For a large $\lambda_s = 3$, the metastable Ω_s can be accommodated for much larger $A_s \sim O(100)$ GeV. That large A_s, by contrast, is able to change the nature of transition $\Omega_0 \to \Omega_s$, into the first order type; furthermore, the strength of the second-step can be significantly enhanced and then reopens the smaller λ_{sh} region with $\lambda_{sh} \sim O(0.1)$; see Fig. 2 (middle). We can find that the requirement $T_s^* \gtrsim T_h$ yields an upper bound on $|A_s| \lesssim 300$ GeV in this example. Note that the figures indicate that for a given A_s, the region for λ_{sh} is restricted and within this region increasing λ_{sh} could lead to lower T_h^*.

The three-step (first order - second order- first oder) PT ($\Omega_0 \to \Omega_s \to \Omega_{sh} \to \Omega_h$) cases are shown in Fig. 1 (left) for $\mu_s^2 < 0$ and Fig. 1 (right) for $\mu_s^2 > 0$. In Fig. 2 (right), we display the allowed region for $\lambda_{sh} = 0.24$ by taking the feasible values of (λ_s, m_s, A_s) in which we can find a point of Fig. 1 (left). Increasing λ_s lowers T_s^* and it will eventually go below T_h, thus shutting down the three-step PT. On the other hand, when λ_s becomes fairly small (thus for a much larger v_s), then T_s^* (T_h^*) is getting higher (smaller), FOPT is enhanced in this limit.

4.2 Detectability of gravitational waves in the Z_3-symmetric model

We display the results on the $(\alpha, \widetilde{\beta})$ plane in the Fig. 3, with the experimental sensitivities of eLISA [17, 24] and DECIGO [19] labelled by the shaded regions. The sensitivity regions of four eLISA detector configurations described in Table I in Ref. [17] are denoted by "C1", "C2", "C3" and "C4". The expected sensitivities for the future DECIGO stages are labeled by "Correlation", "1 cluster" and "Pre" following Ref. [19]. The transition temperature T_* depends on the model parameters (see, Fig. 2) and the velocity of the bubble wall v_b is uncertain. Although the experimental sensitivities on the $(\alpha, \widetilde{\beta})$ depend on T_* and v_b, we take $T_* = 50$ GeV and $v_b = 0.95$ as a reference for the purpose of illustration. It is seen that typically one needs $\alpha \gtrsim O(0.01)$ for the near future detection.

However, the first source from $\Omega_0 \to \Omega_s$ with FOPT turns out to be undetectable since it always gives $\alpha \lesssim 0.01$. On the other hand, in particular in the three-step PT case, most of the parameter space can be covered for the other source of EWPT. One of the main reasons causing this difference is that the first-step happened at a relatively high temperature $T_s^* \gtrsim 160$ GeV, which typically is rather higher than the EWPT temperature $T_h^* \lesssim 100$ GeV; recalling that $\alpha \propto 1/T^4$, thus the first source is suppressed. A lower T_h^* also leads to smaller $\widetilde{\beta}$, which is determined by the PT temperature.

[2] The one-step EWPT ($\Omega_0 \to \Omega_h$) is the second order for the range in Fig. 1. The one-step FOEWPT is realized for $m_s \gtrsim$ 400 GeV with large λ_{sh} by the non-decoupling thermal loop effects even for $A_s = 0$ as discussed in Refs. [13, 15, 16, 22, 25].

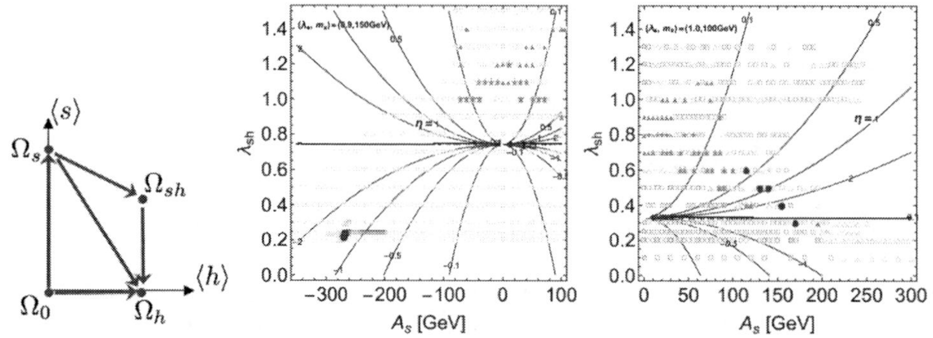

Figure 1. Each path of the transition pattern and the metastable vacua at the intermediate stage in the Z_3 model (left). Global picture of multi-step PT in the (A_s, λ_{sh}) plane for $(\lambda_s, m_s[\text{GeV}]) = (0.9, 150)$ (left) and $(1.0, 100)$ (middle). PT of three-step (red, circle), two-step (green, triangle for the second-first order PT or star for the first-first order PT) and one-step (blue, square) are plotted. Filled plots satisfy the condition of SFOEWPT in Eq. (1). In $\mu_s^2 < 0$ region, the three-step PT can happen only in a very narrow space, consistent with Fig. 2 (right).

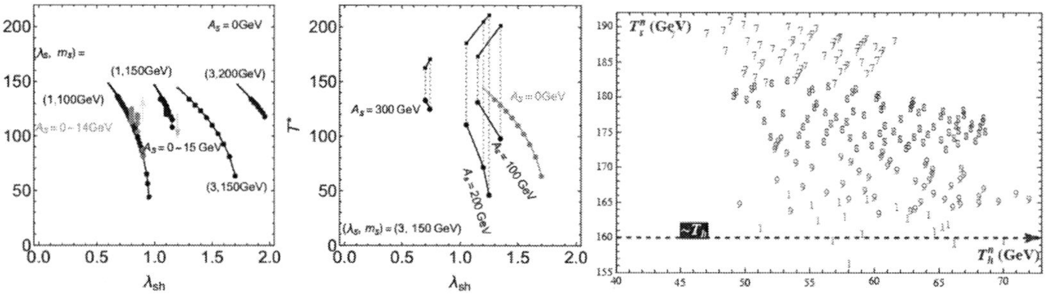

Figure 2. (Left/middle) The two-step PT in the $\mu_s^2 > 0$ region as the function of λ_{sh}. For the second-first order PT (left), we show $A_s = 0$ (Z_2-like) case (black lines) and $A_s \neq 0$ cases by fixing λ_{sh} (green and blue lines), for four cases $(\lambda_s, m_s[\text{GeV}]) = (1, 100), (1, 150), (3, 100), (3, 150)$. The first-first order PT (middle) arises are shon for $(\lambda_s, m_s[\text{GeV}]) = (3, 150)$ by taking $A_s[\text{GeV}] = 100, 200, 300$ (blue lines). For each dashed line, the upper and the lower ends denote T_s^* and T_h^*, respectively. In these plots we just keep the points which give FOPT. (Right) The three-step EWPT in the $\mu_s^2 < 0$ region with $\lambda_{sh} = 0.24$, varying $\lambda_s = 0.7, 0.8, 0.9, 1.0$ which is labelled by numbers 7, 8..., respectively. Distributions of two FOPT temperatures, T_s^* and T_h^*; the red dashed line denotes T_h, the typical second order PT temperature for $\Omega_0 \to \Omega_h$.

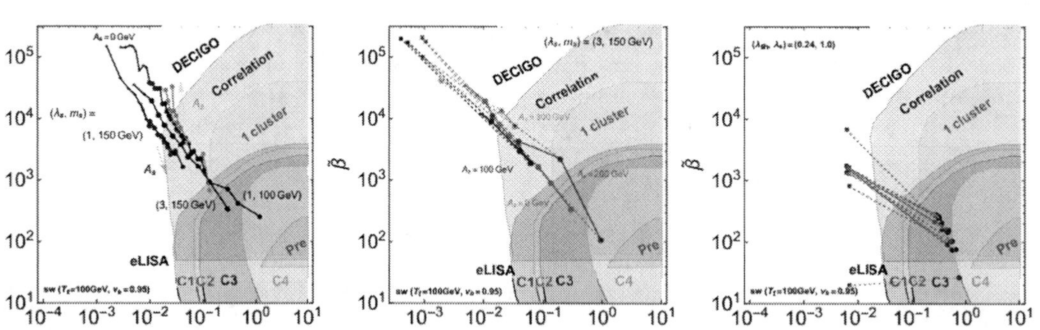

Figure 3. Detectability of GWs in the $(\alpha, \widetilde{\beta})$ plan from the two-step (second-first/first-first order) PT (left/middle) and the three-step (first-second-first order) PT (right) which are corresponding to Fig. 2. In middle/left plane, the two FOPTs are labelled respectively by the square and circle points, connected by a dashed line. The expected sensitivities of eLISA and DECIGO are set by using the sound wave contribution for $T_* = 100$ GeV and $v_b = 0.95$.

5 Conclusion

A potential barrier can be created during EWPT by the tree level effects due to a doublet-singlet mixing [5–11]. As a result, such models can be tested by the synergy between the measurements of various Higgs boson couplings at future collider experiments and the observation of GWs at future space-based interferometers as discussed in Refs. [10, 11]. In another implementation imposing unbroken discrete symmetry like Z_2 [7, 12–16], multi-step PT could utilize a tree level barrier. But generically the absence of mixing renders the tests at colliders difficult without taking enough large λ_{sh} coupling as discussed in Refs. [13, 15, 16, 22, 25]. In this paper, we have focused on such the nightmare scenario in the Z_3 symmetric single scalar model. Especially, the three-step PT produces two sources of GW in the model. Despite of the undetectability from the first-step in the near future, the other source from EWPT basically can be completely covered by LISA and DECIGO.

Acknowledgements

This work is based on the collaboration with Zhaofeng Kang and Pyungwon Ko. I would like to thank them for their support.

References

[1] Z. Kang, P. Ko and T. Matsui, arXiv:1706.09721 [hep-ph].

[2] V. A. Kuzmin, V. A. Rubakov, and M. E. Shaposhnikov, Phys. Lett. B155 (1985) 36; M. E. Shaposhnikov, Nucl. Phys. B287 (1987) 757-775.

[3] D. E. Morrissey and M. J. Ramsey-Musolf, New J. Phys. **14**, 125003 (2012).

[4] D. J. H. Chung, A. J. Long and L. T. Wang, Phys. Rev. D **87**, no. 2, 023509 (2013).

[5] S. Profumo, M. J. Ramsey-Musolf and G. Shaughnessy, JHEP **0708**, 010 (2007).

[6] A. Ashoorioon and T. Konstandin, JHEP **0907**, 086 (2009).

[7] J. R. Espinosa, T. Konstandin and F. Riva, Nucl. Phys. B **854**, 592 (2012).

[8] K. Fuyuto and E. Senaha, Phys. Rev. D **90**, no. 1, 015015 (2014).

[9] S. Profumo, M. J. Ramsey-Musolf, C. L. Wainwright and P. Winslow, Phys. Rev. D **91**, no. 3, 035018 (2015).

[10] P. Huang, A. J. Long and L. T. Wang, Phys. Rev. D **94**, no. 7, 075008 (2016).

[11] K. Hashino, M. Kakizaki, S. Kanemura, P. Ko and T. Matsui, Phys. Lett. B **766**, 49 (2017).

[12] J. M. Cline and K. Kainulainen, JCAP **1301**, 012 (2013).

[13] D. Curtin, P. Meade and C. T. Yu, JHEP **1411**, 127 (2014).

[14] V. Vaskonen, Phys. Rev. D **95**, no. 12, 123515 (2017).

[15] A. Beniwal, M. Lewicki, J. D. Wells, M. White and A. G. Williams, arXiv:1702.06124 [hep-ph].

[16] G. Kurup and M. Perelstein, arXiv:1704.03381 [hep-ph].

[17] C. Caprini *et al.*, JCAP **1604**, no. 04, 001 (2016).

[18] P. A. Seoane *et al.* [eLISA Collaboration], arXiv:1305.5720 [astro-ph.CO].

[19] S. Kawamura *et al.*, Class. Quant. Grav. **28**, 094011 (2011).

[20] V. Corbin and N. J. Cornish, Class. Quant. Grav. **23**, 2435 (2006).

[21] G. Belanger, K. Kannike, A. Pukhov and M. Raidal, JCAP **1301**, 022 (2013).

[22] K. Hashino, M. Kakizaki, S. Kanemura and T. Matsui, Phys. Rev. D **94**, no. 1, 015005 (2016).

[23] C. L. Wainwright, Comput. Phys. Commun. **183**, 2006 (2012).

[24] Data sheet by A. Petiteau, http://www.apc.univ-paris7.fr/Downloads/lisa/eLISA/Sensitivity/Cfgv1/StochBkgd/

[25] M. Kakizaki, S. Kanemura and T. Matsui, Phys. Rev. D **92**, no. 11, 115007 (2015).

Event Rates of Gravitational Waves from merging Intermediate-mass Black Holes: based on a Runaway Path to a SMBH

Hisaaki Shinkai[1,⋆]

[1] *Department of Information Systems, Osaka Institute of Technology, Hirakata City, Osaka 573-0196, Japan*

Abstract. Based on a dynamical formation model of a supermassive black hole (SMBH), we estimate the expected observational profile of gravitational wave at ground-based detectors, such as KAGRA or advanced LIGO/VIRGO. Noting that the second generation of detectors have enough sensitivity from 10 Hz and up, we are able to detect the ring-down gravitational wave of a BH with the mass $M < 2 \times 10^3 M_\odot$. This enables us to check the sequence of BH mergers to SMBHs via intermediate-mass BHs. We estimate the number density of galaxies from the halo formation model and estimate the number of BH mergers from the giant molecular cloud model assuming hierarchical growth of merged cores. At the designed KAGRA (and/or advanced LIGO/VIRGO), we find that the BH merger of its total mass $M \sim 60M_\odot$ is at the peak of the expected mass distribution. With its signal-to-noise ratio $\rho = 10(30)$, we estimate the event rate $R \sim 200(20)$ per year in the most optimistic case, and we also find that BH mergers in the range $M < 150M_\odot$ are $R > 1$ per year for $\rho = 10$. Thus, if we observe a BH with more than $100M_\odot$ in future gravitational-wave observations, our model naturally explains its source.

1 Introduction

The LIGO group reported three detections of gravitational waves (GW150914, GW151226, GW170104) and one transient event (LVT151012) so far (see Table. 1), all three of which are regarded as the events of coalescence of binary black holes (BBHs). These announcements were not only valuable on the point of the direct detections of the gravitational wave, but also the first results of confirming the existence of BHs, the existence of BHs of this mass range, and the existence of BBHs.

In this article, as a possible sources of BHs of $30M_\odot$–$60M_\odot$, we consider simple sequences of mergers of BHs, which start from $10M_\odot$ BHs and end with a supermassive BH (SMBH) via intermediate-mass BHs (IMBHs). After counting the numbers of galaxies in the Universe from a halo formation model, and modeling the numbers of BHs in a galaxy from a giant molecular cloud scenario, we estimate their observational detectability at ground-based gravitational-wave detectors. More details are reported in [1].

The formation process of an SMBH is one of the unsolved problems in galaxy evolution history. Many possible routes were suggested by Rees [5] long ago, but we still debate a plausible route. In this article, we take the scenario that SMBHs are formed by accumulations of BHs. This route was came to be believed when an IMBH (10^2–$10^3 M_\odot$) was first discovered in a starburst galaxy M82. So far, many IMBHs have been found in the center of galaxies.

⋆e-mail: hisaaki.shinkai@oit.ac.jp

Table 1. GW events (as of July 2017)

	Primary & Secondary		Final BH			SNR	Mpc	deg^2
	$m_1/M_\odot$	$m_2/M_\odot$	$M_{\rm final}/M_\odot$	$\Delta M/M_\odot$	$a/M_\odot$	ρ	z	
GW150914	36.2	29.1	62.3	−3	0.68	24	410Mpc	600
ref [2]	$^{+5.2}_{-3.8}$	$^{+3.7}_{-4.4}$	$^{+3.7}_{-3.1}$	(4.59%)			0.09	
LVT151012	23	13	35	−1	0.66			
	$^{+18}_{-6}$	$^{+4}_{-5}$	$^{+14}_{-4}$	(2.78%)				
GW151226	14.2	7.5	20.8	−0.9	0.74	13	440Mpc	
ref [3]	$^{+8.3}_{-3.7}$	$^{+2.3}_{-2.3}$	$^{+6.1}_{-1.7}$	(4.15%)			0.09	
GW170104	31.2	19.4	48.7	−1.9	0.64	13	880Mpc	1300
ref [4]	$^{+8.4}_{-6.0}$	$^{+5.3}_{-5.9}$	$^{+5.7}_{-4.6}$	(3.75%)			0.18	

This runaway path was first proposed by Ebisuzaki *et al.*[6]. The scenario consists of three steps: (1) formation of IMBHs by runaway mergers of massive stars in dense star clusters, (2) accumulations of IMBHs at the center region of a galaxy due to sinkages of clusters by dynamical friction, and (3) mergers of IMBHs by multibody interactions and gravitational radiation. Successive mergers of IMBHs are likely to form an SMBH with a mass of heavier than $> 10^6 M_\odot$. Numerical simulations support the above first step and the second step is also confirmed in a realistic mass-loss model, while the third step is not yet investigated in detail. The discovery of an SMBH binary system, together with a simulation of an eccentric evolution of SMBH binaries, supports this formation scenario through merging of IMBHs. (See references in [1]).

In Matsubayashi *et al.*(2004)[7], we pointed out that gravitational waves from IMBHs can be a trigger to prove this process. Later, Fregeau *et al.*[8] discussed the event rates of IMBH–IMBH binary observations at advanced LIGO and VIRGO and concluded that we can expect ~ 10 mergers per year. This work was followed by Gair *et al.*[9], including the Einstein Telescope project. Amaro-Seoane & Santamaría [10] also discussed the IMBH–IMBH system, including the pre-merger phase.

2 Black Hole Merger Model

Ring-down Frequency from BHs

The gravitational waveform of binary-star mergers which ends up with a single BH, has three typical phases: inspiral phase, merging phase, and ring-down phase. The ring-down frequency is simply estimated by the quasi-normal frequency of BHs, which is determined from the mass and spin of the final BH and is estimated to be higher frequency than in its inspiral phase.

Supposing that advanced GW interferometers can detect $f_{\rm qnm}$ above 10 Hz, then BHs less than 1200 $M_\odot$ are within the target if BHs are nonrotating ($a = 0$), while BHs less than $2500 M_\odot$ are in the detectable range for highly rotating cases ($a = 0.98$).

Number of Galaxies in the Universe

In order to model the typical mass of galaxies and its distribution, we apply the halo mass function given by Vale & Ostriker [11], in which they discuss an empirically based, nonparametric model for galaxy luminosities with halo/subhalo masses. They apply the Sheth-Tormen mass function for halo number density. Integrating it by the volume as a function of redshift z (we use the standard cosmology model), we can derive the number density of halos (Figure 1(left)). Combining the average number of galaxies and the number density of halos, we get the number density of galaxies $n_{\rm galaxy}(M, z)$, which we show in Fig.1(right).

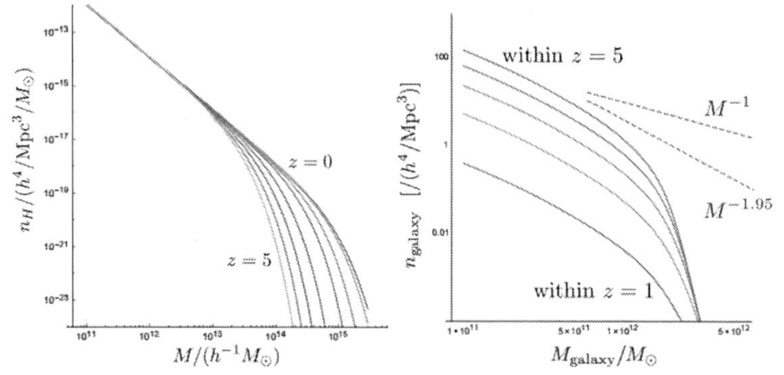

Figure 1. (left) Global mass functions for halos (halo and subhalo), $n_H(M)$, for $z = 0, 0.1, 0.5, 2, 4, 5$. $n_H(M)$ is in units of $h^4/\mathrm{Mpc}^3/M_\odot$. M is in units of $h^{-1}M_\odot$, with $h = 0.7$. (right) Number density of galaxies, $n_{\mathrm{galaxy}}(M)$.

Number of BHs in a Galaxy

We next estimate the number of BH candidates in a galaxy. Recently, Inutsuka *et al.* [12] developed a scenario of galactic-scale star formation from a giant molecular cloud. Their model includes both the growth of molecular clouds and the destruction of magnetized molecular clouds by radiation. Simulations and steady-state analysis show that the mass density function of molecular clouds, $n_{\mathrm{cl}}(M_g)$, converges at the Schechter-like function, $n_{\mathrm{cl}}(M_{\mathrm{cl}}) \sim M_{\mathrm{cl}}^{-1.7}\exp(-M_{\mathrm{cl}}/M_{\mathrm{cut}})$ where the cutoff mass $M_{\mathrm{cut}} = 10^6 M_\odot$. On the other hand, many N-body simulations report that there is a simple relation between the mass of the most massive cluster m_{max} and the total mass of the molecular cloud M_{cl}, $m_{\mathrm{max}} = 0.20 M_{\mathrm{cl}}^{0.76}$.

We therefore combine these results, and we suppose that each molecular cloud forms a single BH in its core if it is more than $10 M_\odot$, and we suppose that these BHs become "building blocks" for forming stellar-sized and intermediate-mass BHs. Many N-body simulations suggest that massive objects will accumulate in the center of a galaxy owing to dynamical friction, so that we modeled that these seed BHs accumulate and merge repeatedly (as we model below), resulting in IMBHs and SMBHs. We do not specify where these mergers occur, but we count our BH mergers after we set up the initial seeds. We show the number density of BHs in a galaxy, $n_{\mathrm{BH}}(M_{\mathrm{BH}})$ in Fig. 2(left).

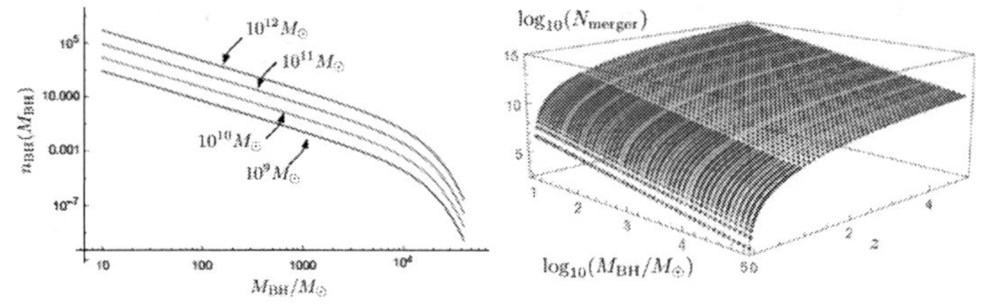

Figure 2. (left) Number density of BHs per galaxy as a function of BH mass for different total mass of galaxies $M_{\mathrm{galaxy}} = 10^9 M_\odot$–$10^{12} M_\odot$. (right) Cumulative distribution function of the number of BH mergers $N_{\mathrm{merger}}(M_{\mathrm{BH}})$ as a function of the redshift z. We express number with binned one, of which we binned 20 for one order in M_{BH}.

Number of BH Mergers in a Galaxy

We simply assume that BHs formed at cores of clouds will accumulate each other hierarchically, i.e. the mass and the number of BHs at steps from k to $k+1$ can be expressed simply by $M_{k+1} = 2M_k$, and $N_{k+1} = N_k/2$. The mass of a BH merger, then, obeys the distribution M^{-1}. On the other hand, we know empirically that the mass of the central BH of the galaxy, M_{SMBH}, and the total mass of the galaxy, M_{galaxy}, has a relation $M_{\mathrm{SMBH}} = 2 \times 10^{-4} M_{\mathrm{galaxy}}$ (or equal to 10^{-3} of the bulge mass).

Combining these facts, for a certain galaxy with M_{galaxy}, we pick up BHs with total mass M_{SMBH} (equation above), obeying the mass distribution of Fig.2 (left). We suppose that picked-up BHs will form an SMBH in its series of mergers in the hierarchical model. Together with galaxy distribution function $n_{\text{galaxy}}(M, z)$, we are able to count the possible events of BH mergers, $N_{\text{merger}}(M_{\text{BH}}, z)$, in the universe, which we show in Fig.2 (right).

3 Signal-to-noise Ratio, Detectable Distance, and Event Rate

The standard procedure for the detection is judged by the optimal S/N ratio, ρ, which is given by

$$\rho = 2\left[\int_0^\infty \frac{\tilde{h}(f)\tilde{h}^*(f)}{S_n(f)} df\right]^{1/2}, \tag{1}$$

where $\tilde{h}(f)$ is the Fourier-transformed quantity of the wave, and $S_n(f)$ is the (one-sided) power spectral density of strain noise of the detector. For KAGRA (bKAGRA), we use a fitted function

$$\sqrt{S_n(f)} = 10^{-26}\left(6.5 \times 10^{10} f^{-8} + 6 \times 10^6 f^{-2.3} + 1.5 f\right), \tag{2}$$

where f is measured in Hz, as was used in [13].

Following Flanagan & Hughes [14], we use the energy spectrum formula for the ring-down wave which results in $E_{\text{ringdown}} \approx \frac{1}{8}A^2 M^2 f_{\text{ringdown}} Q$, where M is the total mass of the binary, $M = m_1 + m_2$.

Let $\epsilon_r(a) \equiv E_{\text{ringdown}}/M$, which expresses the energy fraction of the emitted gravitational wave to the total mass. As we cited in the introduction, three GW events suggest us that the 4.0% of the total mass is emitted before the merger. The associate numerical simulations show that the ring-down part emits the energy around 0.6 % of the total mass. If we use $A \sim 0.4$, then we recover the ratio $\epsilon_r(0.67) = 0.58\%$. The magnitude of this A is also consistent with the quadrupole formula.

The S/N is, then, expressed using the inertial mass $\mu = m_1 m_2/M$ and the redshift of the source z,

$$\rho^2 = \frac{8}{5}\frac{\epsilon_r(a)}{f_R^2}\frac{(1+z)M}{S_h(f_R/(1+z))}\left(\frac{(1+z)M}{d_L(z)}\right)^2\left(\frac{4\mu}{M}\right)^2. \tag{3}$$

By specifying the BH mass and spin, together with ρ, we can then find the distance d_L that satisfies eq. (3). We call this distance the detectable distance, $D(M, a, \rho)$. We show them in Fig.3(left) for KAGRA for S/R=10 and 100. Using $D(M, a, \rho)$, we set the upper limit of z for integrating the number of galaxies, N_{galaxy}, and then obtain the number of BH mergers, N_{merger}. We show N_{merger} in Figures 3 (a1) and (b1) for S/R=10 and 30, respectively.

The event rate R, then, is estimated by $R[/\text{yr}] = N_{\text{merger}}(z)/V(D/2.26)$, where the factor 2.26 is for averaging the distance for all directions. We show them in Figs 3 (a2) and (b2). These figures are for specifying the BH spin parameter a, but if we assume that a is homogeneously distributed, then the averaged R is estimated as its peak $R \sim 7.13[/\text{yr}]$ at $M \sim 59.1 M_\odot$ (200–375 [Hz] for $a = 0$–0.9). The mergers of the range above $R > 1$ [/yr] have mass $40 M_\odot < M < 150 M_\odot$. The total number of events above $R > 1$ [/yr] is ~ 211.

4 Summary

Our event rate sounds similar to that of other groups. For example, the LIGO-Virgo group updated their estimated event rates after the detection of GW150914 as 2–600 Gpc^3 yr^{-1} assuming BH mass distribution models as flat or power law ($\sim M^{-2.35}$; [15]). Kinugawa *et al.* [16] estimate as 70–140

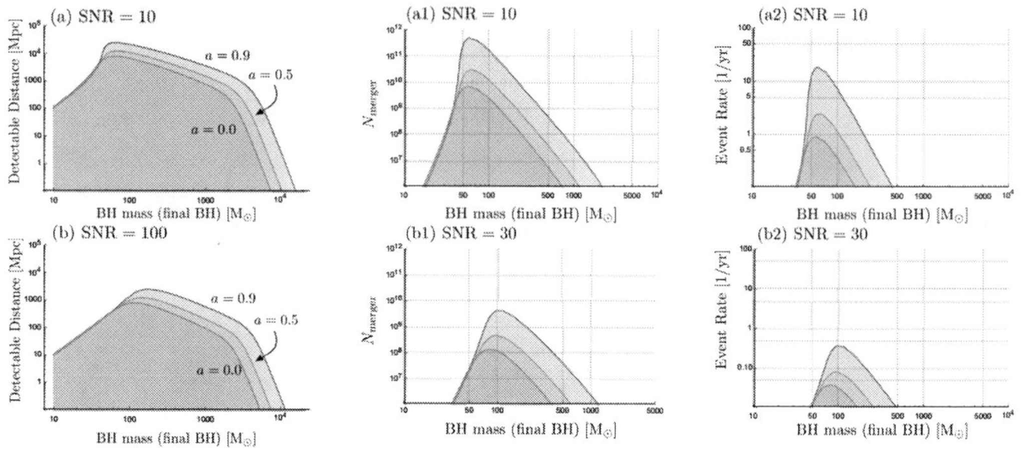

Figure 3. (left) Detectable distance D of the ring-down signal at KAGRA. S/R is set to (a) 10 and (b) 100. (middle and right) Number of BH mergers within the detectable distances (a1, b1) and event rate R (a2, b2) as a function of BH mass M with S/N ratio ρ = 10 and 30 for KAGRA. Three distributions for each figure are of $a = 0.9, 0.5$, and 0.0 (from largest to lowest), respectively.

yr^{-1} from their Population III model. Inoue *et al.*[17] estimate as $< 60\ \text{yr}^{-1}$ from their BH merger model inferred from the luminosity function of ultraluminous X-ray sources. However, our model predicts BH mergers with $M > 100M_{\odot}$, which will be a key to test our model in the future.

This work was supported in part by the Grant-in-Aid for Scientific Research Fund of the JSPS (C) No. 25400277, MEXT KAKENHI Grants No. 17H06357 and No. 17H06358.

References

[1] Shinkai, H., Kanda, N., and Ebisuzaki, T., Astrophys. J. 835 (2017) 276.
[2] Abbott, B. *et al.* Phys. Rev. Lett. 116 (2016) 061102.
[3] Abbott, B. *et al.* Phys. Rev. Lett. 116 (2016) 241103.
[4] Abbott, B. P. *et al.* Phys. Rev. Lett. 118 (2017) 221101.
[5] Rees, M. J. Observatory 98(1978) 210; Annu. Rev. Astron. Astrophys. 22(1984) 471.
[6] Ebisuzaki, T., *et al.*, 2001, Astrophys. J. Lett. 562, L19.
[7] Matsubayashi, T., Shinkai, H., & Ebisuzaki, T., Astrophys. J. 614 (2004) 864.
[8] Fregeau J. M., *et al.* Astrophys. J. 646 (2006) L135.
[9] Gair, J. R., Gen. Rel. Grav. 43 (2011) 485.
[10] Amaro-Seoane, P., & Santamaría, L., Astrophys. J. 722 (2010) 1197.
[11] Vale, A., & Ostriker, J.P., Mon. Not. Roy. Astro. Soc. 371 (2006) 1173.
[12] Inutsuka, S., Inoue, T., Iwasaki, K. & Hosokawa, T., Astro. Astrophys. 580 (2015) A49.
[13] Nakano, H., Tanaka, T., & Nakamura, T., Phys. Rev. D 92 (2015) 064003.
[14] Flanagan, É. É., & Hughes. S. A., Phys. Rev. D 57 (1998) 4535.
[15] Abbott, B. *et al.* Astrophys. J. Lett. 833 (2016) 1.
[16] Kinugawa, T., *et al.* Mon. Not. Roy. Astro. Soc. 442 (2014) 2963.
[17] Inoue, Y., Tanaka, Y.T., & Isobe, N., Mon. Not. Roy. Astro. Soc. 461 (2016) 4329.

EPJ Web of Conferences **168**, 06001 (2018) https://doi.org/10.1051/epjconf/201816806001
Joint International Conference of ICGAC-XIII and IK-15 on Gravitation, Astrophysics and Cosmology

Flattening the inflaton potential beyond minimal gravity

Hyun Min Lee[1,*]

[1] *Department of physics, Chung-Ang University, Seoul 06974, Korea.*

Abstract. We review the status of the Starobinsky-like models for inflation beyond minimal gravity and discuss the unitarity problem due to the presence of a large non-minimal gravity coupling. We show that the induced gravity models allow for a self-consistent description of inflation and discuss the implications of the inflaton couplings to the Higgs field in the Standard Model.

1 Introduction

It was precisely five years ago since the discovery of Higgs boson on July 4, 2012 [1] that I gave this talk in the 2017 International Conference on Gravitation. Together with the detection of gravitational waves coming from the merger of binary black holes on September 14, 2015 [2], we have the fortune to encounter the memorable moments of triumphs of the Standard Model (SM) and General Relativity in our century. Even with such a huge success of our understanding fundamental interactions from subatomic scales to cosmological distances, there remain open questions on the validity of our theories at high energies and in the early Universe.

Cosmological inflation solves problems of horizon, homogeneity, flatness, cosmic relics, and structure formation, requiring the period of an exponential expansion with a scalar field "inflaton" just after Big Bang. From the measurement of Cosmic Microwave Background (CMB) anisotropies by Planck satellite [3], almost scale-invariant and Gaussian scalar perturbations are needed, meaning that canonical single field inflation models are favored. Inflation could also lead to primordial gravity waves in the CMB polarizations at the detectable level in the future CMB experiments.

Almost scale-invariant CMB anisotropies require the inflation potential to be flat for a long period of time, i.e. $N = 60$ e-foldings. In most of large field inflation models, the inflaton field makes a trans-Planckian excursion during inflation. The question is then whether inflation potential remains flat for entire range of inflaton fields such that UV physics can be ignored.

In this article, we review on a class of inflation models beyond minimal gravity and discuss the unitarity problems due to a large non-minimal coupling. We show that in induced gravity models, the inflaton potential is justified against the quantum corrections due to physics below the Planck scale.

*e-mail: hminlee@cau.ac.kr

2 Inflation beyond minimal gravity

In Einstein gravity, the Lagrangian for a real scalar inflaton ϕ is described by

$$\mathcal{L}_{\rm E} = \sqrt{-g}\left(\frac{1}{2}\mathcal{R} - \frac{1}{2}(\partial\phi)^2 - V(\phi)\right). \tag{1}$$

In order to realize a sufficiently long period of inflation, we need to introduce small masses or couplings in the inflaton potential. For instance, we meed $m \sim 10^{13}$ GeV for quadratic potential, $V(\phi) = \frac{1}{2}m^2\phi^2$; $\lambda \sim 10^{-12}$ for quartic potential, $V(\phi) = \frac{1}{4}\lambda\phi^4$.

In the case of scalar-tensor gravity, the general Lagrangian for a real scalar inflaton contains a non-minimal coupling $F(\phi)$ as below,

$$\mathcal{L}_{\rm J} = \sqrt{-g}\left(\frac{1}{2}\mathcal{R} + \frac{1}{2}F(\phi)\mathcal{R} - \frac{1}{2}(\partial\phi)^2 - U(\phi)\right). \tag{2}$$

Then, if the effective Planck mass is field-dependent as $M_P^2(\phi) \sim F(\phi)$ with $F^2(\phi) \sim U(\phi)$ during inflation, gravity becomes weaker and makes the effective inflation potential $V \sim U/F^2$ asympotically constant, without a need of small couplings.

It is known that higher curvature terms can be introduced to drive inflation without a scalar field, because the degree of freedom in gravity becomes then larger than two. The Starobinsky model [4] introduces an $\mathcal{R}^2$ term in the Lagrangian as below,

$$\mathcal{L}_{\rm Star} = \sqrt{-g}\left(\frac{1}{2}\mathcal{R} + \frac{1}{16}\xi^2\mathcal{R}^2\right). \tag{3}$$

Then, it can be shown that the model is dual to a scalar-tensor gravity at the classical level, after the introduction of a Lagrange multiplier ϕ, as

$$\mathcal{L}_{\rm dual} = \sqrt{-g}\left(\frac{1}{2}\mathcal{R} + \frac{1}{2}\xi f(\phi)\mathcal{R} - U(\phi)\right) \tag{4}$$

for $f(\phi), U(\phi)$ satisfying $U(\phi) = f^2(\phi)$, which is called the Starobinsky condition. Thus, there are a class of scalar-tensor theories, that are equivalent to the Starobinsky model. As discussed previously, the dual theory automatically satisfies the condition for the asymptotic flat effective potential of the inflaton, due to the functional relation, $U(\phi) = f^2(\phi)$. For instance, for $U(\phi) = \frac{1}{4}\lambda\phi^4$, we would need $f(\phi) \sim \phi^2$.

As a result, after rescaling the metric and the scalar field for canonical kinetic terms by $g_{\mu\nu} \to g_{\mu\nu}/\Omega$ with $\Omega = 1 + \xi f(\phi)$ and $\frac{d\chi}{d\phi} = \sqrt{\frac{3}{2}}\frac{\Omega'}{\Omega}$, respectively, the inflaton Lagrangian in the scalar dual theory becomes the following form,

$$\mathcal{L}_{\rm dual} = \sqrt{-g_E}\left(\frac{1}{2}\mathcal{R} - \frac{1}{2}(\partial\chi)^2 - \frac{1}{\xi^2}\left(1 - e^{-2\chi/\sqrt{6}}\right)^2\right). \tag{5}$$

The Starobinsky model leads to inflationary predictions for the spectral index and tensor-to-scalar ratio, $n_s = 1 - \frac{2}{N}, r = \frac{12}{N^2}$, with $\xi \sim 10^4$ for the normalization of the CMB anisotropies. Therefore, the results are well consistent with the Planck data [3]. On the other hand, in the SM Higgs inflation with non-minimal coupling [5], the Higgs kinetic term can be ignored during inflation, so it is classically equivalent to the Starobinsky model, leading to the similar predictions for inflation. However, the consistency of the Starobinsky model and the Higgs inflation below the Planck scale might be questioned at the quantum level [6, 7], due to the presence of a large non-minimal coupling ξ.

3 Induced gravity and unitarity problem

We consider a general gravity Lagrangian including the general inflaton kinetic term in the following [8],

$$\mathcal{L}_{\rm gen} = \sqrt{-g}\left(\frac{1}{2}\Omega(\phi)\mathcal{R} - \frac{1}{2}K(\phi)(\partial\phi)^2 - U(\phi)\right). \tag{6}$$

Then, after rescaling the metric by $g_{\mu\nu} \to g_{\mu\nu}/\Omega$, the above Lagrangian becomes in Einstein frame as

$$\mathcal{L}_{\rm gen} = \sqrt{-g_E}\left(\frac{1}{2}\mathcal{R} - \frac{1}{2}\left(\frac{K}{\Omega} + \frac{3\Omega'^2}{\Omega^2}\right)(\partial\phi)^2 - V\right) \tag{7}$$

with $V = U/\Omega^2$. Identifying the canonical inflation field by $\frac{d\chi}{d\phi} = \sqrt{K/\Omega + 3\Omega'^2/2\Omega^2}$, the slow-roll condition is ensured for $\epsilon = \frac{1}{2}(dV/\chi/V) = \frac{1}{3}(\Omega U'/(\Omega' U) - 2)^2 \ll 1$, constraining $U(\phi) \sim \Omega^2[1 + O(\sqrt{\epsilon})]$. Furthermore, for Starobinky-like inflation, we take

$$\Omega'^2/\Omega^2 \gg K \qquad \text{and} \qquad \frac{1}{2}\Omega' R = U'. \tag{8}$$

As a result, we get $\Omega = 1 + \frac{R}{4V_I}$ and $U = \frac{R^2}{16V_I}$, fixing the potential completely [8] to

$$U = V_I(\Omega - 1)^2 \tag{9}$$

with V_I being the vacuum energy during inflation. Then, during inflation, the Lagrangian (7) with eq. (9) becomes the same form as eq. (5) in the Starobinsky model, but it takes a more general form due to the general kinetic terms for inflation, thus leading to Starobinsky-like models.

Choosing $K = 1$ and the frame function Ω and the potential [8] as follows,

$$\Omega = \xi f(\phi), \quad U = \lambda\left(f(\phi) - \xi^{-1}\right)^2, \tag{10}$$

we can generate the Planck mass by the inflaton VEV such that $f(\langle\phi\rangle) = \xi^{-1}$, as in the induced gravity model [9]. Then, for $f(\phi) = \phi^n$, the inflaton kinetic term in Einstein frame becomes in vacuum [8]

$$\frac{\mathcal{L}_{\rm kin}}{\sqrt{-g_E}} = -\frac{1}{2}\left(\frac{K}{\Omega} + \frac{3\Omega'^2}{2\Omega^2}\right)\bigg|_{\phi=\langle\phi\rangle}(\partial\phi)^2 = -\frac{3}{4}n^2\xi^{2/n}(\partial\phi)^2. \tag{11}$$

In this case, defining the canonical inflaton field χ in vacuum by $\phi = \sqrt{\frac{2}{3}\frac{1}{n}}\,\xi^{-1/n}\chi$, we obtain the inflaton potential in Einstein frame, $V = U/\Omega^2$, with $f(\phi) = \phi^n$, as

$$V = \frac{\lambda}{\xi^2}\left[1 - (\xi\phi^n)^{-1}\right]^2 = \frac{\lambda}{\xi^2}\left[1 - \left(\frac{3}{2}\right)^{n/2} n^n\chi^{-n}\right]^2. \tag{12}$$

Therefore, after expanding the inflaton around $\langle\phi\rangle = \xi^{-1/n}$, we find that the interaction terms for the canonical inflaton field χ are suppressed by the Planck scale, so there is no violation of unitarity below the Planck scale [8]. The induced gravity model with $f(\phi) = \phi^2$ and a mixing quartic coupling between ϕ and Higgs field has been proposed as a unitarization of the original Higgs inflation with non-minimal coupling [10]. According to the above discussion, we note that there are a general class of induced gravity models unitarizing the Higgs inflation. The discussion with a large inflaton VEV has been extended to the case with a general form of the kinetic term $K(\phi)$ [11].

On the other hand, in universal attractor models [12] where $K = 1, \Omega = 1 + \xi f(\phi)$ and $U = \lambda f^2(\phi)$ are taken, a small or vanishing inflation VEV is assumed. In this case, although the inflationary predictions are the same as in induced gravity models, the non-minimal coupling does not rescale the inflaton field in vacuum such that the interaction terms in the inflaton potential are suppressed by $\Lambda_{UV} = M_P/\xi^{1/(n-1)}$, which hints at the premature violation of unitarity below the Planck scale [8]. This class of models contains the Higgs inflation with non-minimal coupling for $n = 2$ [6, 7]. We note that there is an exception to the low-scale violation of unitarity for $n = 1$ [13, 14], in which case the inflaton VEV is not physical because it can be always absorbed by the shift of the inflation field.

Finally, we remark the implications of the inflaton coupling in induced gravity model for vacuum stability and reheating. In the presence of both the inflaton ϕ and the SM Higgs scalar h, we can choose the non-minimal couplings and the scalar potential in eq. (10) [10] as follows,

$$\xi f(\phi, h) = \xi_\phi \phi^2 + \xi_h h^2, \tag{13}$$

$$U(\phi, h) = \frac{1}{4}\lambda_\phi(\phi^2 - w^2)^2 + \frac{1}{4}\lambda_h(h^2 - v^2)^2 + \frac{1}{2}\lambda_{H\phi}(h^2 - v^2)(\phi^2 - w^2), \tag{14}$$

with the inflation VEV given by $w \sim 1/\sqrt{\xi_\phi} \gg v$. Then, it has been shown that a nonzero mixing quartic coupling $\lambda_{H\phi}$ improves the stability of electroweak vacuum by inducing a tree-level shift in the effective quartic coupling, $\lambda_{\rm eff} = \lambda_h - \frac{\lambda_{H\phi}^2}{\lambda_\phi}$ [15, 16], which is inferred from the measured Higgs mass, $m_h = \sqrt{2\lambda_{\rm eff}}\, v$. Furthermore, the inflaton decay into a pair of Higgs bosons reheats the Universe after inflation, thanks to the mixing quartic coupling $\lambda_{H\phi}$, leading to the reheating temperature in the range of $10^9\,{\rm GeV} \lesssim T_R \lesssim 10^{13}\,{\rm GeV}$ [17].

4 Conclusions

We have shown that the inflaton potential can be made flat without a small coupling, as gravity becomes weaker due to an inflaton-dependent effective gravity coupling during inflation. We have revisited the Starobinsky model and its scalar-dual theories as well as the Higgs inflation with a large non-minimal coupling, all of which are consistent with Planck data. Generalizing to Starobinsky-like models in scalar-tensor theories, we have identified the induced gravity models to be consistent at the quantum level below the Planck scale. We have also discussed the implications of the coupling between the inflaton and the Higgs field for vacuum stability and reheating.

Acknowledgments

The author would like to thank Cliff Burgess, Jose Espinosa, Gian Giudice and Mike Trott for collaboration and discussion. The work is supported in part by Basic Science Research Program through the National Research Foundation of Korea (NRF) funded by the Ministry of Education, Science and Technology (NRF-2016R1A2B4008759).

References

[1] G. Aad *et al.* [ATLAS Collaboration], Phys. Lett. B **716** (2012) 1 doi:10.1016/j.physletb.2012.08.020 [arXiv:1207.7214 [hep-ex]]; S. Chatrchyan *et al.* [CMS Collaboration], Phys. Lett. B **716** (2012) 30 doi:10.1016/j.physletb.2012.08.021 [arXiv:1207.7235 [hep-ex]].

[2] B. P. Abbott *et al.* [LIGO Scientific and Virgo Collaborations], Phys. Rev. Lett. **116** (2016) no.6, 061102 doi:10.1103/PhysRevLett.116.061102 [arXiv:1602.03837 [gr-qc]].

[3] P. A. R. Ade *et al.* [Planck Collaboration], Astron. Astrophys. **594** (2016) A20 doi:10.1051/0004-6361/201525898 [arXiv:1502.02114 [astro-ph.CO]].

[4] A. A. Starobinsky, Phys. Lett. **91B** (1980) 99. doi:10.1016/0370-2693(80)90670-X

[5] F. L. Bezrukov and M. Shaposhnikov, Phys. Lett. B **659** (2008) 703 doi:10.1016/j.physletb.2007.11.072 [arXiv:0710.3755 [hep-th]].

[6] C. P. Burgess, H. M. Lee and M. Trott, JHEP **0909** (2009) 103 doi:10.1088/1126-6708/2009/09/103 [arXiv:0902.4465 [hep-ph]]; C. P. Burgess, H. M. Lee and M. Trott, JHEP **1007** (2010) 007 doi:10.1007/JHEP07(2010)007 [arXiv:1002.2730 [hep-ph]].

[7] J. L. F. Barbon and J. R. Espinosa, Phys. Rev. D **79** (2009) 081302 doi:10.1103/PhysRevD.79.081302 [arXiv:0903.0355 [hep-ph]]. M. P. Hertzberg, JHEP **1011** (2010) 023 doi:10.1007/JHEP11(2010)023 [arXiv:1002.2995 [hep-ph]].

[8] G. F. Giudice and H. M. Lee, Phys. Lett. B **733** (2014) 58 doi:10.1016/j.physletb.2014.04.020 [arXiv:1402.2129 [hep-ph]].

[9] A. Zee, Phys. Rev. Lett. **42** (1979) 417; doi:10.1103/PhysRevLett.42.417 L. Smolin, Nucl. Phys. B **160** (1979) 253. doi:10.1016/0550-3213(79)90059-2

[10] G. F. Giudice and H. M. Lee, Phys. Lett. B **694** (2011) 294 doi:10.1016/j.physletb.2010.10.035 [arXiv:1010.1417 [hep-ph]].

[11] H. M. Lee, Eur. Phys. J. C **74** (2014) no.8, 3022 doi:10.1140/epjc/s10052-014-3022-0 [arXiv:1403.5602 [hep-ph]].

[12] R. Kallosh, A. Linde and D. Roest, Phys. Rev. Lett. **112** (2014) no.1, 011303 doi:10.1103/PhysRevLett.112.011303 [arXiv:1310.3950 [hep-th]].

[13] A. Kehagias, A. M. Dizgah and A. Riotto, Phys. Rev. D **89** (2014) no.4, 043527 doi:10.1103/PhysRevD.89.043527 [arXiv:1312.1155 [hep-th]].

[14] J. L. F. Barbon, J. A. Casas, J. Elias-Miro and J. R. Espinosa, JHEP **1509** (2015) 027 doi:10.1007/JHEP09(2015)027 [arXiv:1501.02231 [hep-ph]].

[15] J. Elias-Miro, J. R. Espinosa, G. F. Giudice, H. M. Lee and A. Strumia, JHEP **1206** (2012) 031 doi:10.1007/JHEP06(2012)031 [arXiv:1203.0237 [hep-ph]].

[16] O. Lebedev, Eur. Phys. J. C **72** (2012) 2058 doi:10.1140/epjc/s10052-012-2058-2 [arXiv:1203.0156 [hep-ph]].

[17] H. M. Lee, Phys. Lett. B **722** (2013) 198 doi:10.1016/j.physletb.2013.04.024 [arXiv:1301.1787 [hep-ph]].

Higgs Portal Inflation with Fermionic Dark Matter

Aditya Aravind[1,2,⋆], *Minglei* Xiao[1,⋆⋆], and *Jiang-Hao* Yu[1,3,⋆⋆⋆]

[1] *Department of Physics and Texas Cosmology Center, The University of Texas at Austin, TX 78712, USA*
[2] *Asia Pacific Center for Theoretical Physics, Pohang 37673, Korea*
[3] *Amherst Center for Fundamental Interactions, Department of Physics, University of Massachusetts, Amherst, MA 01003, USA*

Abstract. We discuss the inflationary model presented in [1], involving a gauge singlet scalar field and fermionic dark matter added to the standard model. Either the Higgs or the singlet scalar could play the role of the inflaton, and slow roll is realized through its non-minimal coupling to gravity. The effective scalar potential is stabilized by the mixing between the scalars as well as the coupling with the fermionic field. Mixing of the two scalars also provides a portal to dark matter. Constraints on the model come from perturbativity and stability, collider searches and dark matter constraints and impose a constraining relationship on the masses of dark matter and scalar fields. Inflationary predictions are generically consistent with current Planck data.

1 Introduction

The paradigm of cosmic inflation is currently very well established as a possible description of the early history of the observable universe. It is interesting from both the quantum gravity as well as the particle phenomenology viewpoints. While the simplest inflationary scenarios such as single field slow roll (SFSR) inflation, where exponential expansion of the universe is driven by the potential energy of a slowly rolling scalar (inflaton) field, are still consistent with observations, the picture remains unsatisfactory due to the lack of knowledge about the nature of the inflaton field and its connection with the familiar standard model fields.

A few years ago it was shown [2] that the standard model Higgs (with nonminimal coupling to gravity) could play the role of the inflaton field. While the idea is very attractive, it has been observed that the model may be unviable due to the instability of the Higgs potential after accounting for loop corrections [3]. Without accounting for new physics, the only way to avoid this is by assuming a top quark pole mass about 3σ below its central value [4]. Even after doing this, inflationary predictions could still be highly sensitive to the exact values of parameters [5].

In the present work we describe our model introduced in [1] that attempts to address the Higgs instability problem by adding two new fields to the standard model. The model aims to enable inflation to proceed successfully and also provide a plausible dark matter candidate.

⋆e-mail: aditya.aravind@apctp.org
⋆⋆e-mail: jerryxiao@physics.utexas.edu
⋆⋆⋆e-mail: jhyu@physics.umass.edu

2 Model Description

The model consists of two fields in addition to the standard model field content: a gauge singlet scalar S which couples to the Higgs, and a gauge singlet fermion ψ which plays the role of dark matter. We introduce a Z_2 symmetry under which S and one of the Weyl components of ψ (which we call ψ_1) are odd, while the other Weyl component (ψ_2) and all the standard model fields are even. Both scalar fields can be nonminimally coupled to gravity.

The Jordan frame Lagrangian is

$$\mathcal{L} = \sqrt{-g}\left[-\frac{M_{\rm pl}^2 + 2\xi_h H^\dagger H + \xi_s S^2}{2} R + \partial_\mu H^\dagger \partial^\mu H + \frac{1}{2}(\partial_\mu S)^2 - V(H,S) + \mathcal{L}_{\rm DM}\right], \tag{1}$$

where $M_{\rm pl}$ is the reduced Planck mass and $H = \begin{pmatrix} \pi^+ \\ \frac{1}{\sqrt{2}}\left(\phi + i\pi^0\right)\end{pmatrix}$ is the Higgs doublet. The two-field scalar potential and dark matter lagrangian are defined as follows:

$$V(H,S) = -\mu_h^2 H^\dagger H + \lambda_h (H^\dagger H)^2 - \frac{1}{2}\mu_s^2 S^2 + \frac{1}{4}\lambda_s S^4 + \frac{1}{2}\lambda_{sh} H^\dagger H S^2 + \kappa S, \tag{2}$$

$$\mathcal{L}_{\rm DM} = i\bar{\psi}\gamma^\mu \partial_\mu \psi - y_\psi S \bar{\psi}\psi. \tag{3}$$

The soft breaking term κS (ignored from here on) raises the degeneracy of the Z_2 symmetry thus helping to avoid the domain wall problem.

After symmetry breaking, in general, both S and ϕ (the neutral component of the Higgs doublet H) in the tree-level potential develop vacuum expectation values. The values of these fields in the electroweak minimum can be denoted as

$$v \equiv \langle\phi\rangle, \quad u \equiv \langle S\rangle. \tag{4}$$

The mass matrix of the two scalars can be written as

$$\mathcal{M}^2_{\rm scalar}(\phi, S) \equiv \begin{pmatrix} m^2_{\phi\phi} & m^2_{s\phi} \\ m^2_{s\phi} & m^2_{ss} \end{pmatrix}, \tag{5}$$

which on diagonalizing gives us the mixing angle φ between the two scalars whose present day value is given by

$$\varphi = \frac{1}{2}\arctan\frac{\lambda_{sh} v u}{\lambda_s u^2 - \lambda_h v^2} \tag{6}$$

3 Constraints

We consider three types of constraints that the model needs to satisfy in the interest of self-consistency and agreement with observations. These constraints are discussed in the following subsections.

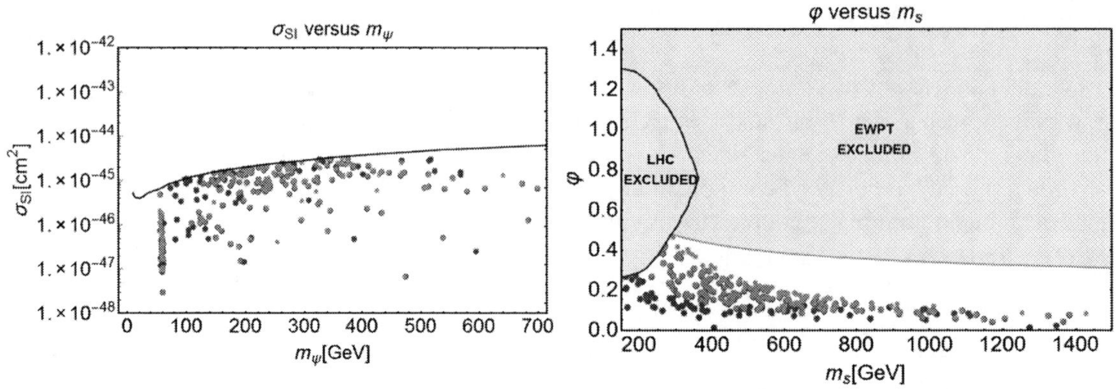

Figure 1. The figure on the left shows the spin independent direct detection cross section σ_{SI} plotted as a function of dark matter mass. The black line here corresponds the the LUX bound. The figure on the right shows mixing angle φ as a function of mass of the scalar field at its low energy vacuum, m_s. The orange line corresponds to the EWPT upper bound on φ and the blue line corresponds to LHC physics lower bound on m_s. The orange and blue shaded regions are excluded by these bounds respectively. In both figures, the green and red points correspond to h-inflation and s-inflation respectively. Both figures are from [1].

3.1 Inflation Constraints

In this model, slow-roll inflation can happen either along the ϕ-axis ($h-$inflation) or the S-axis ($s-$inflation), which means that either ϕ or S would take large field values (typically $O(10^{16}\,\text{GeV})$ or higher). The mechanism of inflation is very similar to the standard Higgs inflation scenario [2], with a large non-minimal coupling of the inflaton helping to flatten the Einstein-frame potential at large field values. For ease of analysis, the nonminimal coupling of the other field is taken to be small ($O(1)$).

The loop corrected inflaton potential can be obtained after accounting for the running of the various couplings, especially the self-coupling of the inflaton field. For the model to be consistent, the couplings should remain perturbative at all scales, starting from their electroweak scale values (relevant for late-time physics) all the way up to inflatonary scale. Moreover, the potential obtained in this manner must be positive all through the relevant scales and and sloping in the correct direction so that the inflaton is likely to eventually settle at the electroweak minimum. Additionally, the inflationary axis must always be a valley and the mass-squared in the transverse direction must be positive and sufficiently large ($m_T^2 \gg H^2$ to avoid isocurvature perturbations).

These conditions impose various constraints on the starting (near electroweak scale) values of the various couplings such as y_ψ, λ_s, λ_{sh}, etc., most notably imposing an upper bound on all of their values and constraining λ_{sh} to be positive. The range of values of couplings that produce satisfactory running behaviour can be further tested against other constraints coming from dark matter physics and collider phenomenology.

3.2 Dark Matter Constraints

At the end of inflation, the potential energy of the inflaton is converted into kinetic energy and eventually transferred into excitations of the various standard model fields. A detailed analysis of this kind was done in the context of standard Higgs inflation in [6]. We expect a similar process (corrected for the existence of S and ψ fields) to occur for our model for typical values of input parameters. As

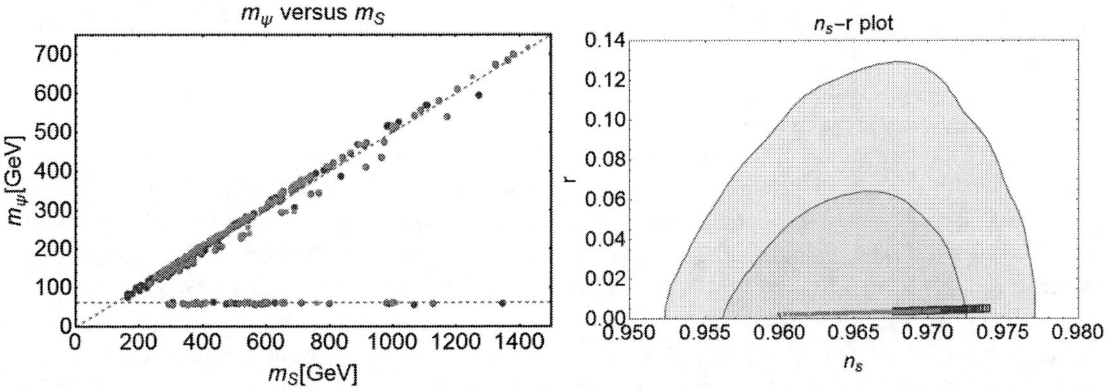

Figure 2. In the figure on the left, dark matter mass m_ψ is plotted against the scalar mass m_s. The dashed lines correspond to $m_\psi = (1/2)m_s$ and $m_\psi = (1/2)m_h$ respectively. In the figure on the right, $n_s - r$ values for $h-$inflation and $s-$inflation are shown with allowed region from Planck 2015 [12] shown in the background. In both figures, the green and red points correspond to $h-$inflation and $s-$inflation respectively. Both figures are from [1].

long as the Yukawa coupling y_ψ and the mixing angle φ are not unnaturally small, we expect dark matter production to follow the usual WIMP scenario. Therefore, the relic abundance of dark matter can be calculated by computing dark matter interaction cross sections from the values of the various couplings. The obtained relic density is then compared to the Planck 2015 TT,TE,EE+lowP data $\Omega_c h^2 = 0.1198 \pm 0.0015$ [7] to obtain an additional constraint on the model parameter space.

Apart from producing the correct relic density of dark matter, the model should also satisfy the constraints on dark matter interaction with the standard model imposed by the direct detection experiments. The spin-independent cross section of dark matter for successful data points is compared to the LUX bounds [8] in the first plot in Figure 1.

3.3 Collider Constraints

Collider constraints for the model primarily arise from the fact that there is a mixing between the Higgs and the S fields. The mixing is parametrized by the mixing angle φ, which is therefore constrained by collider experiments. We consider constraints coming from Electroweak Precision Tests (EWPT) [9] and also constraints from LHC [10] translated into a constraint in the $m_S - \varphi$ plane in [11]. The constraints are shown in the second plot in Figure 1.

4 Results and Conclusions

The first plot in Figure 2 shows dark matter mass m_ψ as a function of scalar mass m_s for data points which successfully pass all the constraints. It is clear that the dark matter mass (m_ψ) tends to take values corresponding to the resonance regions, i.e, nearly half the scalar mass (m_s) or half the Higgs mass (m_h). This is chiefly a consequence of combining the upper bounds on the couplings coming from requiring appropriate running behaviour with the constraints coming from relic density and direct detection.

The second plot in Figure 2 gives the $n_s - r$ predictions from the model. It is clear that the inflationary predictions of $h-$ and $s-$ inflation are not markedly different. This is not surprising since at inflationary scales, both types of inflation involve a scalar field with nearly quartic potential and quadratic nonminimal coupling to gravity, with some modifications coming from the exact running behaviour of the couplings. It is also clear that the model generically predicts a low tensor to scalar ratio, similar to standard (tree level) Higgs inflation and therefore the predictions are generally well within the Planck-selected region.

The original motivation of the paper was to come up with a model that would address the potential instability of the Higgs field and enable to have inflation. An additional objective was to explore the possibility of connecting dark matter physics with inflation. In this model involving adding a gauge singlet scalar and singlet fermionic dark matter to the standard model, both objectives could be achieved, albeit in a limited region of parameter space where dark matter mass is nearly half the mass of one of the scalars (resonance regions). It was discovered that inflation could be achieved along either axis and will generically yield low tensor-to-scalar ratio (r) and a relatively large spread of scalar spectral tilt (n_s) that could still lie within the Planck-allowed region.

In the coming years, we expect to have more restrictive constraints from dark matter direct detection and colliders. Therefore, there is promise that the model parameter space may be further constrained. Moreover, upcoming CMB B-mode searches are expected to detect or further constrain the tensor-to-scalar ratio, which could improve the distinguishing power between various inflationary models and also potentially verify the inflationary predictions of this model.

Acknowledgments

We would like to thank C. Kilic, J. Kim, D. Lorshbough, S. Paban, T. Prokopec, and J. Ren for helpful discussions. This material is based upon work supported by the National Science Foundation under Grant Number PHY-1316033. The work of JHY was supported in part by DOE Grant de-sc0011095.

References

[1] A. Aravind, M. Xiao and J. H. Yu, Phys. Rev. D **93**, no. 12, 123513 (2016) doi:10.1103/PhysRevD.93.123513 [arXiv:1512.09126 [hep-ph]].

[2] F. L. Bezrukov and M. Shaposhnikov, Phys. Lett. B **659**, 703 (2008) doi:10.1016/j.physletb.2007.11.072 [arXiv:0710.3755 [hep-th]].

[3] G. Degrassi, S. Di Vita, J. Elias-Miro, J. R. Espinosa, G. F. Giudice, G. Isidori and A. Strumia, JHEP **1208**, 098 (2012) doi:10.1007/JHEP08(2012)098 [arXiv:1205.6497 [hep-ph]].

[4] A. Salvio, Phys. Lett. B **727**, 234 (2013) doi:10.1016/j.physletb.2013.10.042 [arXiv:1308.2244 [hep-ph]].

[5] K. Allison, JHEP **1402**, 040 (2014) doi:10.1007/JHEP02(2014)040 [arXiv:1306.6931 [hep-ph]].

[6] F. Bezrukov, D. Gorbunov and M. Shaposhnikov, JCAP **0906**, 029 (2009) doi:10.1088/1475-7516/2009/06/029 [arXiv:0812.3622 [hep-ph]].

[7] P. A. R. Ade *et al.* [Planck Collaboration], arXiv:1502.01589 [astro-ph.CO].

[8] D. S. Akerib *et al.* [LUX Collaboration], arXiv:1512.03506 [astro-ph.CO].

[9] S. Baek, P. Ko, W. I. Park and E. Senaha, JHEP **1211**, 116 (2012) doi:10.1007/JHEP11(2012)116 [arXiv:1209.4163 [hep-ph]].

[10] S. Chatrchyan *et al.* [CMS Collaboration], Eur. Phys. J. C **73**, 2469 (2013) doi:10.1140/epjc/s10052-013-2469-8 [arXiv:1304.0213 [hep-ex]].

[11] M. L. Xiao and J. H. Yu, Phys. Rev. D **90**, no. 1, 014007 (2014) Addendum: [Phys. Rev. D **90**, no. 1, 019901 (2014)] doi:10.1103/PhysRevD.90.014007, 10.1103/PhysRevD.90.019901 [arXiv:1404.0681 [hep-ph]].

[12] P. A. R. Ade *et al.* [Planck Collaboration], [arXiv:1502.02114 [astro-ph.CO]].

Inflationary attractors from nonminimal coupling

Zhu Yi[1,⋆] and *Yungui* Gong[1,⋆⋆]

[1] *School of Physics, Huazhong University of Science and Technology, Wuhan, Hubei 430074, China*

Abstract. We show explicitly how the E model attractor is obtained from the general scalar-tensor theory of gravity with arbitrary conformal factors in the strong coupling limit. By using conformal transformations, any attractor with the observables n_s and r can be obtained. The existence of attractors imposes a challenge to distinguish different models.

1 Introduction

The Planck measurements of the cosmic microwave background anisotropies give the scalar spectral tilt $n_s = 0.968 \pm 0.006$ and the tensor to scalar ratio $r_{0.002} < 0.11(95\%\text{C.L.})$[1, 2]. The central value of n_s suggests the relation $n_s = 1 - 2/N$ with $N = 60$, here N is the number of e-folds when the pivotal scale exits the horizon before the end of inflation. To the leading order of $1/N$, many inflation models have the universal result $n_s = 1 - 2/N$, they include the R^2 inflation[3] and the hilltop inflation[4], the Higgs inflation with the nonminimal coupling $\xi\phi^2 R$[5, 6] also gives the same n_s when $\xi \gg 1$, which is a special case of the more general nonminimal coupling $\xi f(\phi)R$ with the potential $\lambda^2 f^2(\phi)$ for an arbitrary function $f(\phi)$ in the strong coupling limit $\xi \gg 1$[7]. The above relation between n_s and N can be generalized to the parametrization of the observables in terms of N and used to reconstruct inflationary potential [8–19].

In this talk, we show that in the strong coupling limit $\xi \gg 1$, any attractor is possible for the nonminimal coupling $\xi f(\phi)R$ with an arbitrary coupling function $f(\phi)$ [20]. The E-model attractor is used as an example to support the claim. The talk is organized as follows. In section 2, we discuss the Higgs inflation. The construction of attractor behaviours is discussed in section 3. The conclusions are drawn in section 4.

2 Higgs inflation

The simplest and easiest way to realize the cosmic inflation is to use scalar fields. Until now, the only scalar field observed is Higgs field. The Higgs potential is

$$V(\phi) = \frac{\lambda}{4}\left(\phi^2 - v^2\right)^2, \tag{1}$$

⋆e-mail: yizhu92@hust.edu.cn, Talk given by this author
⋆⋆e-mail: yggong@mail.hust.edu.cn

where we set $c = \hbar = 1/8\pi G = 1$. Because $v \sim 100\text{GeV}$ and $\phi \sim 10^{19}\text{GeV}$, we can neglect the parameter v and consider the simple power-law potential [21]

$$V_h(\phi) = \frac{\lambda}{4}\phi^4. \tag{2}$$

The action for the Higgs field minimally coupled to gravity is

$$S = \int d^4x\sqrt{-g}\left[\frac{1}{2}R - \frac{1}{2}g^{\mu\nu}\nabla_\mu\phi\nabla_\nu\phi - V_h(\phi)\right], \tag{3}$$

the scalar spectral tilt n_s and the tensor to scalar ratio r can be calculated as

$$n_s = 0.951, \quad r_{0.002} = 0.260, \tag{4}$$

with $N = 60$. Obviously, this inflation model has been ruled out by the observation.

If we insist on using the Higgs field to realize the inflation, then we need to introduce the nonminimal coupling. The action for the Higgs field nonminimally coupled to gravity is [5, 6]

$$S = \int d^4x\sqrt{-\tilde{g}}\left[\frac{1+\xi\phi^2}{2}\tilde{R}(\tilde{g}) - \frac{1}{2}\tilde{g}^{\mu\nu}\nabla_\mu\phi\nabla_\nu\phi - V_h(\phi)\right]. \tag{5}$$

By taking the conformal transformation, in the strong coupling limit $\xi \gg 1$, this action in Einstein frame becomes

$$S = \int d^4x\sqrt{-g}\left[\frac{1}{2}R - \frac{1}{2}g^{\mu\nu}\nabla_\mu\psi\nabla_\nu\psi - U_h(\psi)\right], \tag{6}$$

where

$$\psi = \sqrt{\frac{3}{2}}\ln\left(1+\xi\phi^2\right), \tag{7}$$

$$U_h(\psi) = \frac{\lambda}{4\xi^2}\left[1 - \exp\left(-\sqrt{\frac{2}{3}}\psi\right)\right]^2, \tag{8}$$

the scalar spectral tilt n_s and the tensor to scalar ratio r are

$$n_s \approx 1 - \frac{2}{N}, \quad r \approx \frac{12}{N^2}. \tag{9}$$

If we take $N = 60$, then these results are consistent with the observation.

3 The inflationary attractor

The action for a general scalar-tensor theory of gravity in Jordan frame is

$$S_J = \int d^4x\sqrt{-\tilde{g}}\left[\frac{1}{2}\Omega(\phi)\tilde{R}(\tilde{g}) - \frac{1}{2}\omega(\phi)\tilde{g}^{\mu\nu}\nabla_\mu\phi\nabla_\nu\phi - V_J(\phi)\right], \tag{10}$$

where $\Omega(\phi) = 1 + \xi f(\phi)$, ξ is a dimensionless constant, and $f(\phi)$ is an arbitrary function. Taking the conformal transformations

$$g_{\mu\nu} = \Omega(\phi)\tilde{g}_{\mu\nu}, \tag{11}$$

$$d\psi^2 = \left[\frac{3}{2}\frac{(d\Omega/d\phi)^2}{\Omega^2(\phi)} + \frac{\omega(\phi)}{\Omega(\phi)}\right]d\phi^2, \tag{12}$$

the action (10) in Einstein frame becomes

$$S = \int d^4x \sqrt{-g} \left[\frac{1}{2} R(g) - \frac{1}{2} g^{\mu\nu} \nabla_\mu \psi \nabla_\nu \psi - U(\psi) \right], \tag{13}$$

where $U(\psi) = V_J(\phi)/\Omega^2(\phi)$. From the actions (10) and (13), we see that depending on the choices of the coupling $\Omega(\phi)$, different potentials $V_J(\phi)$ in Jordan frame become the same potential $U(\psi)$ in Einstein frame, so we can obtain any attractor from different $V_J(\phi)$ and $\Omega(\phi)$ [20].

In reference [7], the authors take the potential

$$V_J(\phi) = \lambda^2 f(\phi)^2 = \frac{\lambda^2}{\xi^2} \left[\Omega(\phi) - 1 \right]^2, \tag{14}$$

and $\omega(\phi) = 1$. They point out that under the strong coupling condition

$$\frac{3(d\Omega(\phi)/d\phi)^2}{2\omega(\phi)} \gg \Omega(\phi), \tag{15}$$

the scalar spectral tilt n_s and the tensor to scalar ratio r of this model are independent of $\Omega(\phi)$, and the results are

$$n_s \approx 1 - \frac{2}{N}, \tag{16}$$

$$r \approx \frac{12}{N^2}. \tag{17}$$

So an attractor is reached. If we set $f(\phi) = \phi^2$, this model reduces to the Higgs inflation model (6).

The idea discussed above can be easily extended to more general cases. For any given potential $U(\psi)$ in Einstein frame, we can obtain n_s and r. The same n_s and r can also be obtained from different potentials $V_J(\phi)$ and couplings $\Omega(\phi)$. However, the analytical relation between ψ and ϕ is hard to find, so the explicit form of the potential $V_J(\phi)$ cannot be obtained. To overcome this problem, we take the strong coupling limit for the purpose of giving analytical relation between ψ and ϕ. The limit on ξ from the strong coupling condition (15) was discussed in [18].

Under the strong coupling condition (15), we have

$$\psi = \sqrt{\frac{3}{2}} \ln \Omega(\phi), \tag{18}$$

and the potential in Jordan frame is

$$V_J(\phi) = \Omega^2(\phi) U \left[\sqrt{\frac{3}{2}} \ln \Omega(\phi) \right]. \tag{19}$$

Therefore, different potentials $V_J(\phi)$ in Jordan frame accompanied by $\Omega(\phi)$ result in the same potential $U(\psi)$ in Einstein frame, and the same observables n_s and r, so the attractor is obtained.

As an example, we take the E-model[22, 23]

$$U(\psi) = V_0 \left[1 - \exp\left(-\sqrt{\frac{2}{3\alpha}} \psi \right) \right]^{2n}, \tag{20}$$

to show how to obtain the attractor. Combining Eqs. (19) and (20), we find that

$$V_J(\phi) = V_0\Omega^2(\phi)\left[1 - \Omega^{-1/\sqrt{\alpha}}(\phi)\right]^{2n}. \tag{21}$$

The scalar spectral tilt n_s and the tensor to scalar ratio r for the potential (20) are [20]

$$n_s = 1 + \frac{8n}{3\alpha\left[g(N,n,\alpha)+1\right]} - \frac{8n(n+1)}{3\alpha\left[g(N,n,\alpha)+1\right]^2}, \tag{22}$$

$$r = \frac{64n^2}{3\alpha\left[g(N,n,\alpha)+1\right]^2}, \tag{23}$$

where

$$g(N,n,\alpha) = W_{-1}\left[-\left(\frac{2n}{\sqrt{3\alpha}}+1\right)\exp\left(\frac{-4nN}{3\alpha} - \frac{2n}{\sqrt{3\alpha}} - 1\right)\right], \tag{24}$$

for $n > 1$ and $n/3(2n-1) < \alpha < 4n^2/3(n-1)^2$ or $1/3 < n < 1$ and $\alpha > 4n^2/3(3n-1)^2$;

$$g(N,n,\alpha) = W_{-1}\left[-\left(\frac{2u}{3\alpha} - \frac{2n}{3\alpha} + 1\right)\exp\left(-1 - \frac{2u + 2n(2N-1)}{3\alpha}\right)\right], \tag{25}$$

for $n > 1$ and $\alpha > 4n^2/3(n-1)^2$;

$$g(N,n,\alpha) = W_{-1}\left[-\left(\frac{2n}{3\alpha} + \frac{2v}{3\alpha} + 1\right)\exp\left(-1 - \frac{2v + 2n(2N+1)}{3\alpha}\right)\right], \tag{26}$$

for other case; $u = \sqrt{6\alpha n^2 + n^2 - 3\alpha n}$, $v = \sqrt{n(3\alpha - 6\alpha n + n)}$ and W_{-1} is the lower branch of the Lambert W function.

To show that the attractor (22) and (23) can be reached for arbitrary conformal factor $\Omega(\phi) = 1 + \xi f(\phi)$, we take $\omega(\phi) = 1$, $n = 2$, $\alpha = 1$, and the power-law functions $f(\phi) = \phi^k$ with $k = 1/4$, $1/3$, $1/2$, 1, 2 and 3. From Eqs. (22), (23) and (24), we get the attractor $n_s = 0.9673$ and $r = 0.0031$ for $N = 60$. We vary the coupling constant ξ and choose $N = 60$ to calculate n_s and r for the models with the potential (21), and the results are shown in Fig. 1. From Fig. 1, we see that the attractor is reached in the strong coupling limit $\xi \gtrsim 100$.

4 Conclusion

Under the conformal transformation and the strong coupling condition (15), the potentials in the Jordan and Einstein frames have the relationship $V_J(\phi) = \Omega^2(\phi)U[\sqrt{3/2}\ln\Omega(\phi)]$. For any non-singular function $\Omega(\phi)$, these models will have the same observables n_s and r as those given by the potential $U(\psi)$ in Einstein frame. Therefore, these $\Omega(\phi)$ and $V_J(\phi)$ give the same attractors n_s and r. By this method, we may get any attractor we want, and the existence of these ξ attractors imposes a challenge to distinguish different models [20].

Acknowledgements: This research was supported in part by the National Natural Science Foundation of China under Grant No. 11475065.

References

[1] R. Adam et al. (Planck), Astron. Astrophys. **594**, A1 (2016), 1502.01582
[2] P.A.R. Ade et al. (Planck), Astron. Astrophys. **594**, A20 (2016), 1502.02114

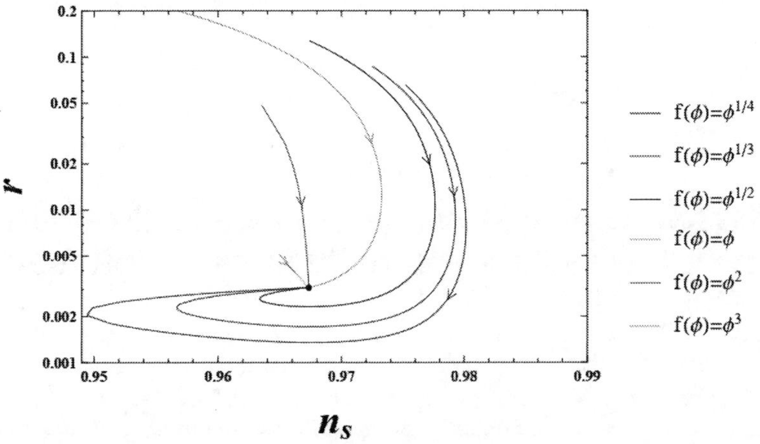

Figure 1. The numerical results of n_s and r for the scalar-tensor theory with the potential (21). We take $\omega(\phi) = 1$, $n = 2$, $\alpha = 1$, the power-law functions $f(\phi) = \phi^k$ with $k = 1/4$, $1/3$, $1/2$, 1, 2 and 3, and $N = 60$. The coupling constant ξ increases along the direction of the arrow in the plot. The E-model attractor (22) and (23) is reached if $\xi \gtrsim 100$.

[3] A.A. Starobinsky, Phys. Lett. B. **91**, 99 (1980)
[4] L. Boubekeur, D. Lyth, JCAP **0507**, 010 (2005), hep-ph/0502047
[5] F.L. Bezrukov, M. Shaposhnikov, Phys. Lett. B **659**, 703 (2008), 0710.3755
[6] D.I. Kaiser, Phys. Rev. D **52**, 4295 (1995), astro-ph/9408044
[7] R. Kallosh, A. Linde, D. Roest, Phys. Rev. Lett. **112**, 011303 (2014), 1310.3950
[8] V. Mukhanov, Eur. Phys. J. C **73**, 2486 (2013), 1303.3925
[9] D. Roest, JCAP **1401**, 007 (2014), 1309.1285
[10] J. Garcia-Bellido, D. Roest, Phys. Rev. D **89**, 103527 (2014), 1402.2059
[11] L. Barranco, L. Boubekeur, O. Mena, Phys. Rev. D **90**, 063007 (2014), 1405.7188
[12] L. Boubekeur, E. Giusarma, O. Mena, H. Ramírez, Phys. Rev. D **91**, 083006 (2015), 1411.7237
[13] T. Chiba, Prog. Theor. Exp. Phys. **2015**, 073E02 (2015), 1504.07692
[14] P. Creminelli, S. Dubovsky, D. López Nacir, M. Simonović, G. Trevisan, G. Villadoro, M. Zaldarriaga, Phys. Rev. D **92**, 123528 (2015), 1412.0678
[15] R. Gobbetti, E. Pajer, D. Roest, JCAP **1509**, 058 (2015), 1505.00968
[16] J. Lin, Q. Gao, Y. Gong, Mon. Not. Roy. Astron. Soc. **459**, 4029 (2016), 1508.07145
[17] Q. Fei, Y. Gong, J. Lin, Z. Yi, JCAP (2017), 1705.02545
[18] Q. Gao, Y. Gong (2017), 1703.02220
[19] Q. Gao, Sci. China Phys. Mech. Astron. **60**, 090401 (2017), 1704.08559
[20] Z. Yi, Y. Gong, Phys. Rev. D **94**, 103527 (2016), 1608.05922
[21] A.D. Linde, Phys. Lett. B **129**, 177 (1983)
[22] R. Kallosh, A. Linde, JCAP **1310**, 033 (2013), 1307.7938
[23] J.J.M. Carrasco, R. Kallosh, A. Linde, Phys. Rev. D **92**, 063519 (2015), 1506.00936

EPJ Web of Conferences **168**, 06004 (2018) https://doi.org/10.1051/epjconf/201816806004
Joint International Conference of ICGAC-XIII and IK-15 on Gravitation, Astrophysics and Cosmology

On the effective Stefan-Boltzmann law and the thermodynamic origin of the initial radiation density in warm inflation

Yongwan Gim[1,2,⋆] and *Wontae* Kim[1,⋆⋆]

[1] *Department of Physics, Sogang University, 35 Baekbeom-ro, Mapo-gu, Seoul 04107, Korea*
[2] *Research Institute for Basic Science, Sogang University, 35 Baekbeom-ro, Mapo-gu, Seoul 04107, Korea*

Abstract. In this presentation, we are going to explain the thermodynamic origin of warm inflation scenarios by using the effective Stefan-Boltzmann law. In the warm inflation scenarios, radiation always exists to avoid the graceful exit problem, for which the radiation energy density should be assumed to be finite at the starting point of the warm inflation. To find out the origin of the non-vanishing initial radiation energy density, we derive an effective Stefan-Boltzmann law by considering the non-vanishing trace of the total energy-momentum tensors. The effective Stefan-Boltzmann law successfully shows where the initial radiation energy density is thermodynamically originated from. And by using the above effective Stefan-Boltzmann law, we also study the cosmological scalar perturbation, and obtain the sufficient radiation energy density in order for GUT baryogenesis at the end of inflation. This proceeding is based on Ref. [1]

1 Introduction

Inflation is an elegant solution to the interesting problems such as the horizon and flatness problems in the big bang cosmology [2–4], during which the rapid expansion lays the universe in a supercooled phase. And thereafter the reheating process should be assumed for the graceful exit problem. In contrast to the assumption of the supercooled universe after inflation, another way to approach this issue without reheating process has been studied in Ref. [5, 6], which is called as a warm inflation scenario. There are two important ingredients to describe the warm inflation. First one is a damping term during warm inflation. Since the interactions of the inflaton and radiation are inevitable during inflation, the damping term describing the decay rate of the inflaton into other fields should be introduced during warm inflation. Second one is the large-scale initial radiation energy density. It can be naturally assumed to be nonzero, $\rho_{\rm r}(t_{\rm i}) \neq 0$ [7], which is compatible with the Stefan-Boltzmann law of $\rho_r = 3\gamma T^4$ in the hot thermal bath at the initial point of inflation, $t = t_{\rm i}$. Now, one might wonder how to exist the large-scale initial radiation energy density and what is the origin of it in the warm inflation scenario.

2 Effective Stefan-Boltzmann law in warm inflation

The radiation energy density is related to the temperature through the Stefan-Boltzmann law in the standard warm inflation models. As compared to this, if one were to treat the inflaton and radiation

⋆e-mail: yongwan89@sogang.ac.kr
⋆⋆e-mail: wtkim@sogang.ac.kr

on an equal footing in equilibrium, then one would encounter generically non-vanishing trace of the total energy-momentum tensor due to the inflaton part while the radiation part is still traceless. Now, it should be emphasized that the usual Stefan-Boltzmann law commonly rests upon the traceless condition of the energy-momentum tensor, and thus we have to modify the Stefan-Boltzmann law in order to incorporate the non-vanishing trace of the total energy-momentum tensor.

Let us start with the first law of thermodynamics $dE = TdS - PdV$, where E, T, S, P, and V are the energy, temperature, entropy, pressure, and volume of a thermal system, respectively. The relevant energy-momentum tensor is assumed to be perfect fluid written as $T_{\mu\nu} = (\rho + p)u_\mu u_\nu + g_{\mu\nu}p$, where u^μ is the four-velocity of radiation flow satisfying $u^\mu u_\mu = -1$. Assuming that the trace of the energy-momentum tensor is non-vanishing generically, the trace relation is obtained as $-\rho + 3p = T^\mu_\mu$ where $\rho = E/V$, the differential equation for the energy density is obtained as $T\partial\rho/\partial T - 4\rho = T^\mu_\mu - T\partial T^\mu_\mu/\partial T$, so that the effective Stefan-Boltzmann law to incorporate the non-vanishing trace of the energy-momentum tensor can be obtained as

$$\rho(T) = 3C_0T^4 - \frac{1}{4}T^\mu_\mu - \frac{3}{4}T^4 \int^T \frac{1}{T^4}\frac{\partial T^\mu_\mu}{\partial T}dT, \tag{1}$$

$$p(T) = C_0T^4 + \frac{1}{4}T^\mu_\mu - \frac{1}{4}T^4 \int^T \frac{1}{T^4}\frac{\partial T^\mu_\mu}{\partial T}dT, \tag{2}$$

where the integration constant C_0 can be fixed from an initial condition. The relations (1) and (2) naturally reduce to the usual Stefan-Boltzmann law for the traceless case, so that $C_0 = \gamma$. However, C_0 will be fixed for the case of the non-vanishing trace for our purpose later by imposing a different boundary condition. In fact, such a modified Stefan-Boltzmann law induced by conformal anomalies had been applied to SU(3) lattice gauge theory in particle physics in the Minkowski spacetime [8] and the recent black hole physics in connection with the information loss problem [9, 10].

From the cosmological point of view, let us assume that the total system of the early universe consists of inflaton and radiation in thermal equilibrium. Then the total energy density $\rho_{\rm tot}$ and pressure $p_{\rm tot}$ are written as [11]

$$\rho_{\rm tot} = \rho_\phi + \rho_{\rm r} = \frac{1}{2}\dot{\phi}^2 + V_{\rm eff}(\phi, T) + \rho_{\rm r}, \tag{3}$$

$$p_{\rm tot} = p_\phi + p_{\rm r} = \frac{1}{2}\dot{\phi}^2 - V_{\rm eff}(\phi, T) + p_{\rm r}, \tag{4}$$

where $\rho_{\rm r}$, $p_{\rm r}$ and ρ_ϕ, p_ϕ denote the energy density and pressure of radiation and inflaton, respectively. Specifically, the temperature dependent effective potential $V_{\rm eff}$ for the inflaton is expressed by [12–14]

$$V_{\rm eff}(\phi, T) = -\gamma T^4 + \frac{1}{2}(\delta m_T)^2\phi^2 + V_0(\phi), \tag{5}$$

where $\gamma = \pi^2 g_*/90$ and g_* is an effective particle number. $V_0(\phi)$ is the zero-temperature potential for the scalar field ϕ, and $\delta m_T(\phi, T)$ denotes a thermal correction which will be neglected for simplicity along the lines of Ref. [15].

The trace for the total energy-momentum tensor appears non-vanishing due to the effective potential for the inflaton as

$$T^\mu_\mu = -\rho_{\rm tot} + 3p_{\rm tot} = -4V_{\rm eff}(\phi, T), \tag{6}$$

where the kinetic energy is assumed to be very small as compared to the potential energy from now on. By plugging Eq. (6) into Eqs. (1) and (2), the explicit forms of the pressure and energy density are obtained as

$$\rho_{\rm tot} = V_{\rm eff} + 3\gamma T^4 \ln\left(\frac{T_{\rm GUT}}{T}\right)^4, \quad p_{\rm tot} = -V_{\rm eff} + \gamma T^4 \ln\left(\frac{T_{\rm GUT}}{T}\right)^4. \tag{7}$$

by using the initial condition of $C_0 = 4\gamma \ln T_{\rm GUT}$ from the assumption that there exists only the inflaton field at the initial temperature of our universe $T_{\rm GUT}$, *i.e.*, $\rho_{\rm tot}(T_{\rm GUT}) = \rho_\phi$ and $p_{\rm tot}(T_{\rm GUT}) = p_\phi$. It is worth mentioning that we take T_0 to be the GUT temperature as the maximum temperature of our universe $T_0 = T_{\rm GUT} = 10^{16}$GeV, since all perturbative interactions can be frozen out and ineffective in maintaining or establishing thermal equilibrium for $T > 10^{16}$GeV, and thus the known interactions are not capable of thermalizing the universe at temperature greater than the GUT scale [11]. Comparing Eq. (7) with Eqs. (3) and (4), we can immediately find the effective Stefan-Boltzmann law for the radiation as

$$\rho_{\rm r} = 3\gamma T^4 \ln\left(\frac{T_{\rm GUT}}{T}\right)^4, \qquad p_{\rm r} = \gamma T^4 \ln\left(\frac{T_{\rm GUT}}{T}\right)^4. \tag{8}$$

The radiation energy density (8) starts from zero with the GUT temperature as an initial condition of our universe, and, subsequently, it increases. And eventually it gives the adequate initial radiation energy density for warm inflation and the sufficient temperature after inflation, which will be calculated in Sec. 3.

3 Temperatures at the horizon crossing and at the end of inflation

One of the most important ingredients in warm inflation is that the decreasing radiation energy density during inflation is replenished in such a way that the energy of the inflaton field is transferred to that of radiation in virtue of dissipation. Now, the energy conservation law, $\dot\rho_{\rm tot} + 3H(\rho_{\rm tot} + p_{\rm tot}) = 0$, can be separated into the inflaton and radiation parts as [15]

$$\dot\rho_\phi + 3H(\rho_\phi + p_\phi) = -\Gamma\dot\phi(t)^2, \qquad \dot\rho_{\rm r} + 3H(\rho_{\rm r} + p_{\rm r}) = \Gamma\dot\phi(t)^2, \tag{9}$$

where $H = \dot a/a$ means the Hubble parameter, and $\Gamma\dot\phi^2$ is the friction term adopted phenomenologically to describe the decay of the inflaton field and its energy transfers into the radiation bath. And, the Friedmann equation is also given as $H^2 - \rho_{\rm tot}/(3m_{\rm p}^2) = 0$. Based on the slow-roll approximations [15], $\dot\phi^2 \ll V_{\rm eff}$, $\ddot\phi \ll \Gamma\dot\phi$, $\dot\rho_{\rm r} \ll 4H\rho_{\rm r}$, $\rho_{\rm r} \ll \rho_\phi$, one can get the following equations,

$$3Hr\dot\phi + \partial_\phi V_{\rm eff} = 0, \quad 3H(\rho_{\rm r} + p_{\rm r}) - \Gamma(\phi)\dot\phi^2 = 0, \quad H^2 - \frac{1}{3m_{\rm p}^2}V_{\rm eff} = 0, \tag{10}$$

where $\Gamma/(3H) \gg 1$ in the warm inflationary regime.

From now on, we adopt the power-law potential V_0 and damping term Γ [15] as

$$V_0(\phi) = \lambda\phi^n, \qquad \Gamma(\phi) = \Gamma_0\left(\frac{\phi}{\phi_0}\right)^m \tag{11}$$

in order to perform the specific calculations, where the coefficients Γ_0, ϕ_0 and λ are constants, and the power n and m are fixed as $n = 2$, $m = 2$ for simplicity. In this specific model, the number of e-folds during warm inflation is obtained as

$$N_{\rm inf} = \int_{t_{\rm HC}}^{t_{\rm end}} H(t)dt = \frac{\Gamma_0(\lambda\phi_{\rm HC}^2 - \gamma_{\rm HC}T_{\rm HC}^4)^{\frac{3}{2}}}{6\sqrt{3}m_{\rm p}\lambda^2\phi_0^2} \tag{12}$$

by assuming that $\phi_{\rm end} \ll \phi_{\rm HC}$, where $\phi_{\rm HC}$ and $\phi_{\rm end}$ are the values of the inflaton field corresponding to the horizon-crossing time $t_{\rm HC}$ and the end time of warm inflation $t_{\rm end}$, respectively.

Next, we are going to determine the temperature bounds at the end of inflation via cosmological perturbation. The thermal fluctuations produce the power spectrum P_ζ for the comoving curvature ζ, $P_\zeta = \pi^{1/2}H^{5/2}\Gamma^{1/2}T/(2\dot{\phi}^2)$, and the power spectral index n_s for the scalar perturbation is defined as $n_s - 1 = d\ln|P_\zeta|/d\ln k$. After some tedious calculations with the slow-roll equations (10) , the spectral index is finally obtained as

$$n_s - 1 = \frac{1}{N_{\rm inf} + N_{\rm inf}^2 \ln\left(\frac{T_{\rm GUT}}{T_{\rm HC}}\right)^4} - \frac{1}{12N_{\rm inf}\left(1 - \ln\left(\frac{T_{\rm GUT}}{T_{\rm HC}}\right)^4\right)} - \frac{7}{4N_{\rm inf}}, \tag{13}$$

where the number of e-folds is assumed to be $N_{\rm inf} = 60$ in order for solving the horizon problem. The spectral index can respect the data of Planck 2015 when the temperature at the horizon crossing $T_{\rm HC}$ lies in the interval of

$$8.026 \times 10^{15}{\rm GeV} \leq T_{\rm HC} \leq 9.985 \times 10^{15}{\rm GeV}. \tag{14}$$

In order to evaluate the temperature at the end of warm inflation, we will use the procedure presented in Ref. [16]. By using Eq. (10), the total number of e-folds $N_{\rm tot}$ from the scale at the horizon crossing $a_{\rm HC}$ to the scale at the present time a_0 is written as $N_{\rm tot} = \ln(a_0/a_{\rm HC}) = \ln\left((\sqrt{3}k_0 m_p)^{-1}\sqrt{\lambda\phi_{\rm HC}^2 - \gamma_{\rm HC}T_{\rm HC}^4}\right)$, where the scale of the present time is fixed as $a_0 = 1$ and the scale at the horizon crossing is given as $a_{\rm HC} = k_0/H(t_{\rm HC})$. Next, the relation of $T_{\rm rec} = (1 + z_{\rm rec})T_{\rm CMB}$, where $z_{\rm rec}$ is the red-shift factor given as $1 + z_{\rm rec} = a_0/a_{\rm rec}$, indicates that the temperature diminishes from the recombination era to present universe due to the expansion of the universe as $N_0 = \ln(a_0/a_{\rm rec}) = \ln(T_{\rm rec}/T_{\rm CMB})$. For the radiation-dominated era, the adiabatic expansion of the universe is assumed as $dS = 0$ [11], then the number of e-folds can be rewritten in the radiation-dominated era $N_{\rm rad}$ as

$$N_{\rm rad} = \ln\left(\frac{a_{\rm rec}}{a_{\rm end}}\right) = \frac{1}{3}\ln\left(\frac{s_{\rm end}}{s_{\rm rec}}\right) = \frac{1}{3}\ln\left(\frac{4\gamma_{\rm end}T_{\rm end}^3 \ln\left(\frac{T_{\rm GUT}}{T_{\rm end}}\right)^4}{4\gamma_{\rm rec}T_{\rm rec}^3}\right), \tag{15}$$

where the entropy at the end of inflation is $S_{\rm end} = 4\gamma_{\rm end}a_{\rm end}^3 T_{\rm end}^3 \ln(T_{\rm GUT}/T_{\rm end})^4$. By the way, $S_{\rm rec} = 4\gamma_{\rm rec}a_{\rm rec}T_{\rm rec}^3$ since the radiation only consists of photons without the inflaton, so that the usual Stefan-Boltzmann law is used.

To perform the specific calculations, we choose the effective particle number at the electroweak energy scale as $g_{\rm HC} = g_{\rm end} = 106.75$ and at the recombination era as $g_{\rm rec} = 2$ [11]. The temperature of CMB is known as $T_{\rm CMB} = 2.725K$, and the spectral index for $k_0 = 0.05{\rm Mpc}^{-1}$ is $n_s = 0.9655 \pm 0.0062$ from Planck 2015 [17, 18]. By solving the equation of $N_{\rm tot} = N_0 + N_{\rm rad} + N_{\rm inf}$, the range of $T_{\rm end}$ is, finally, obtained as

$$2.409 \times 10^{13}\ {\rm GeV} \leq T_{\rm end} \leq 2.216 \times 10^{14}\ {\rm GeV}, \tag{16}$$

where this range lies below the well-known upper bound of the temperature of the universe to avoid monopole proliferation [11] and above the lower bounds in Refs. [19–21]. In addition, the corresponding energy density for radiation is consequently $2.852 \times 10^{56}{\rm GeV}^4 \leq \rho_{\rm end} \leq 1.291 \times 10^{60}{\rm GeV}^4$, which is a sufficient radiation energy density to accommodate the GUT baryogenesis at the end of inflation [22].

4 Conclusion

Motivated by the non-zero initial radiation energy density in warm inflation scenario, we performed thermodynamic analysis for the warm inflation model by using the definitions for the inflaton and radiation energy density presented in Ref. [11]. And then we obtained the effective Stefan-Boltzmann law to show that the zero radiation energy density (8) at the Grand Unification epoch just prior to starting inflation became finite when inflation starts, which gives the adequate radiation energy density for warm inflation. By using the effective Stefan-Boltzmann law for the radiation energy density, we studied the number of e-folds and the spectral index of the scalar perturbation under the slow-roll approximations in the power-law potential and damping terms, so that the temperature (16) at the end of warm inflation was successfully calculated, and it satisfies the upper bound lower than the GUT scale [11], and lower bound of the big bang nucleosynthesis [19, 20] by the CMB data [21]. Additionally, we confirmed that a sufficient radiation energy density could be produced for GUT baryogenesis at the end of inflation [22].

As a matter of fact, we have assumed the simplest setting described by the perfect fluid and the power law potential and the damping term as a toy model. The effective Stefan-Boltzmann law can be applied to other models which have the non-vanishing trace, and it might give interesting results. We hope that this issue will be elaborated in the near future.

References

[1] Y. Gim, W. Kim, JCAP **1611**, 022 (2016), 1608.07466
[2] A.A. Starobinsky, Phys. Lett. **B91**, 99 (1980)
[3] K. Sato, Mon. Not. Roy. Astron. Soc. **195**, 467 (1981)
[4] A.H. Guth, Phys. Rev. **D23**, 347 (1981)
[5] A. Berera, Phys. Rev. Lett. **75**, 3218 (1995), astro-ph/9509049
[6] A. Berera, Phys. Rev. **D54**, 2519 (1996), hep-th/9601134
[7] A. Berera, I.G. Moss, R.O. Ramos, Rept. Prog. Phys. **72**, 026901 (2009), 0808.1855
[8] G. Boyd, J. Engels, F. Karsch, E. Laermann, C. Legeland, M. Lutgemeier, B. Petersson, Nucl. Phys. **B469**, 419 (1996), hep-lat/9602007
[9] Y. Gim, W. Kim, Eur. Phys. J. **C75**, 549 (2015), 1508.00312
[10] M. Eune, Y. Gim, W. Kim, Eur. Phys. J. **C77**, 244 (2017), 1511.09135
[11] E.W. Kolb, M. Turner, Westview press (1990)
[12] L. Dolan, R. Jackiw, Phys. Rev. **D9**, 3320 (1974)
[13] S. Weinberg, Phys. Rev. **D9**, 3357 (1974)
[14] A.D. Linde, Rept. Prog. Phys. **42**, 389 (1979)
[15] L.M.H. Hall, I.G. Moss, A. Berera, Phys. Rev. **D69**, 083525 (2004), astro-ph/0305015
[16] J. Mielczarek, Phys. Rev. **D83**, 023502 (2011), 1009.2359
[17] P.A.R. Ade et al. (Planck) (2015), 1502.02114
[18] P.A.R. Ade et al. (Planck) (2015), 1502.01589
[19] M. Kawasaki, K. Kohri, N. Sugiyama, Phys. Rev. **D62**, 023506 (2000), astro-ph/0002127
[20] S. Hannestad, Phys. Rev. **D70**, 043506 (2004), astro-ph/0403291
[21] J. Martin, C. Ringeval, Phys. Rev. **D82**, 023511 (2010), 1004.5525
[22] M. Bellini, Phys. Rev. **D63**, 123510 (2001), gr-qc/0101062

Brief History of Ultra-light Scalar Dark Matter Models

Jae-Weon Lee[1,★]

[1]*Department of renewable energy, Jungwon university, 85 Munmu-ro, Goesan-eup, Goesan-gun, Chungcheongbuk-do, 367-805, Korea*

Abstract. This is a review on the brief history of the scalar field dark matter model also known as fuzzy dark matter, BEC dark matter, wave dark matter, or ultra-light axion. In this model ultra-light scalar dark matter particles with mass $m = O(10^{-22})eV$ condense in a single Bose-Einstein condensate state and behave collectively like a classical wave. Galactic dark matter halos can be described as a self-gravitating coherent scalar field configuration called boson stars. At the scale larger than galaxies the dark matter acts like cold dark matter, while below the scale quantum pressure from the uncertainty principle suppresses the smaller structure formation so that it can resolve the small scale crisis of the conventional cold dark matter model.

Despite long efforts dark matter (DM) remains a great mystery in physics and astronomy [1]. While numerical results of the cold dark matter (CDM) model are remarkably successful in explaining the large scale structure of the universe, it encounters some problems at the scale of galactic or sub-galactic structures. For example, the numerical simulations usually predict cusped central halo density and too many sub-halos and small galaxies, which seem to be in contradiction with observations [2–5].

Recently, there is a growing interest in the idea that DM particles are ultra-light scalars in Bose-Einstein condensate (BEC). (For a review, see Refs. 6–13) In this model, due to the tiny DM particle mass $m = O(10^{-22})eV$, the DM particle number density is very high and hence the inter-particle distance is much smaller than the de Broglie wave length of the DM particles. Therefore, the particles are in BEC and move collectively as a wave rather than incoherent particles. At the scale larger than galaxies the DM perturbation behaves like that of CDM, while below the scale quantum pressure from the uncertainty principle suppresses the small structure formation, which makes it a viable alternative to CDM. This property helps us to resolve the small scale problems of the CDM model such as the missing satellite problem or the cusp/core problem [14].

Before 2000 there were only a few groups of scientists working on this topic, without much communication even between them. Being unaware of precedent works, many researchers in this field independently proposed similar ideas with various names such as BEC DM, scalar field DM (SFDM), fuzzy DM, ultra-light axion (ULA), ultra-light axion like particle (ALP), wave DM, ψ DM, repulsive DM or (super)fluid DM among many, although the basic physics of these models is quite similar. (Henceforth, I will use the term 'SFDM' for the model.) This unfortunate situation has brought some confusions among researchers in this field. Therefore, at this point, it is desirable to summarize what have already attempted in this exciting field. [1]

The hypothesis that galactic DMs are ultra-light scalar particles in BEC has a long history. In Ref. [15] Baldeschi et al. considered the galactic halo model of self-gravitating bosons with mass $m \simeq 10^{-24}eV$, which was obtained by comparing the de Broglie wave length of DM to the typical galaxy size. In [16] Membrado et al. calculated the rotation curve of self-gravitating boson halos using the ground state of the bosons, which later turned out to be adequate for dwarf galaxies [17]. In [18] Sin suggested that the halos are like giant atoms made of ultra light BEC DM particles such as pseudo Nambu-Goldstone boson (PNGB). He tried to explain the observed flat rotation curves (RCs) using the excited states of the BEC DM of the Non-linear Schrödinger equation and emphasized the ripple structures of the RCs as a smoking gun of the theory. The particle mass $m \simeq 3 \times 10^{-23}eV$ was first obtained by fitting the observed RC of galaxy NGC2998. (See also [19, 20]) Following his work, Lee and Koh [21, 22] suggested that DM halos are giant boson stars (BSs) which can be described by the relativistic scalar field theory with gravity, i.e., Einstein Klein-Gordon equation (EKG). In this model DM particles are represented by scalar field ϕ with a typical action

$$S = \int \sqrt{-g}d^4x[\frac{-R}{16\pi G} - \frac{g^{\mu\nu}}{2}\phi^*_{;\mu}\phi_{;\nu} - U(\phi)], \quad (1)$$

where the potential $U(\phi) = \frac{m^2}{2}|\phi|^2 + \frac{\lambda}{4}|\phi|^4$ was considered as an example. This model can be reduced to the BEC model in the Newtonian limit. The field is often assumed to be complex but can be real and have more general potentials. From the maximum stable central density and

★e-mail: scikid@jwu.ac.kr

[1]The list in the references of this paper is by no means exhaustive.

the maximum halo mass ($M_h \simeq 10^{12} M_\odot$) the mass bound $10^{-28} \leq m \leq 10^{-22} eV$ was obtained for $\lambda = 0$ case (now often called fuzzy DM) from the BS theory. It was emphasized that the repulsive self-interaction among DM particles, if there is, provides an additional pressure against the gravitational collapse and drastically changes the length scale to $O(\sqrt{\lambda} m_P/m^2)$ even for a tiny λ, where m_P is the planck mass. An approximate analytic solution of KGE, $|\phi|^2 \sim sin(kr)/kr$, was obtained for $\Lambda \equiv \lambda m_P^2/m^2 \gg 1$ (Thomas-Fermi limt) [21, 23]. If $\lambda < 0$ (as in the Taylor-expanded potential for ULA) the halo can be unstable [21] as argued in [24].

On the other hand Press et al. [25], consider a 'soft boson' with somewhat longer Compton wavelength $\lambda_c = 30kpc$ than that for the typical SFDM. Widrow and Kaiser [26] proposed a numerical technique for the evolution of collisionless matter using the Schrödinger-Poisson equations (SPE).

Similar ideas were independently investigated by many authors [23, 26–40, 40–45]. For example, a model with $m = 0$ was suggested in [27], which has a stability issue. The SFDM without self-interaction (i.e., Fuzzy DM) with $m \simeq 10^{-22} eV$ was shown to be able to solve the small scale issues [31]. Guzman and Matos et al. [46, 47] extensively studied various astrophysical aspects of the SFDM. Other attempts are based, for example, on a non-minimal coupling [28], quintessence [29, 48], Repulsive DM [30], fluid DM [32], nontopological soliton [33], *cosh* potential [35, 36], and ULA [34, 37, 38] among many [39–44, 49]. The BEC nature of this DM was extensively studied in [9, 23, 50] by Harko, Chavanis and their colleagues. More recently cosmology of the ULA is thoroughly studied in [51, 52] by Marsh et al. The ULA has usually a cosine potential, which is often approximated to be a quadratic potential. In that case the calculation results are almost same to those of the fuzzy DM. Other works are described in the followings.

In the SFDM model the macroscopic wave function of the halo DM is often described by the following SPE,

$$\begin{aligned} i\hbar\partial_t\psi &= -\frac{\hbar^2}{2m}\nabla^2\psi + mV\psi, \\ \nabla^2 V &= 4\pi G\rho_d, \end{aligned} \tag{2}$$

which can be derived from the EKG for weak gravity [53] using $\psi = \sqrt{m}\phi$ or from the many-body formalism. Here, the DM density is $\rho_d = m|\psi|^2 = m^2|\phi|^2$ and V is the self-gravitation potential. Once we get ψ from the SPE we can predict the astrophysical observables. For example, for a spherical halo the rotation velocity at radius r is $v_{rot}(r) = \sqrt{\frac{GM(r)}{r}}$, where $M(r) = 4\pi\int_0^r r'^2\rho_d(r')dr'$ is the mass within r.

To study the cosmological structure formation in the fluid approach, it is useful to reduce the Schrödinger equation to the Madelung euations [7, 10] using

$$\psi(r,t) = \sqrt{\rho(r,t)}e^{iS(r,t)}. \tag{3}$$

Substituting Eq. (3) into the Schrödinger equation gives a continuity equation

$$\frac{\partial\rho}{\partial t} + \nabla\cdot(\rho\mathbf{v}) = 0 \tag{4}$$

and an Euler-like equation

$$\frac{\partial\mathbf{v}}{\partial t} + (\mathbf{v}\cdot\nabla)\mathbf{v} + \nabla V + \frac{\nabla p}{\rho} - \frac{\nabla Q}{m} = 0 \tag{5}$$

with a quantum potential $Q \equiv \frac{\hbar^2}{2m}\frac{\Delta\sqrt{\rho}}{\sqrt{\rho}}$, a fluid velocity $\mathbf{v} \equiv \nabla S/2m$, and the pressure p from a self-interaction (if any). The quantum pressure term $\nabla Q/m$ is absent in the CDM models. We have ignored the cosmic expansion for simplicity. Perturbing the above equations around $\rho = \bar\rho$, $\mathbf{v} = 0$, and $V = 0$ gives an equation for density perturbation $\delta\rho \equiv \rho - \bar\rho$,

$$\frac{\partial^2\delta\rho}{\partial t^2} + \frac{\hbar^2}{4m^2}\nabla^2(\nabla^2\delta\rho) - c_s^2\nabla^2\delta\rho - 4\pi G\bar\rho\delta\rho = 0, \tag{6}$$

where c_s is the classical sound velocity from p, and $\bar\rho$ is the average background matter density (See, for example, Refs. 50, 54 for details.).

The Fourier transformed equation of the density contrast $\delta \equiv \delta\rho/\bar\rho = \sum_k \delta_k e^{ik\cdot r}$ with a wave vector k is

$$\frac{d^2\delta_k}{dt^2} + \left[(c_q^2 + c_s^2)k^2 - 4\pi G\bar\rho\right]\delta_k = 0, \tag{7}$$

where $c_q = \hbar k/2m$ is a quantum velocity. For a small k (at a large scale) the c_s dependent term dominates and SFDM behaves like the CDM (with pressure) while for a large k (at a small scale) the c_q dependent term dominates and quantum pressure prevents the structure formation [31, 55–57]. This interesting behavior of the SFDM is confirmed by the precise numerical studies [58] and makes the SFDM an ideal alternative for the CDM. This property alleviates the small scale problems of the CDM model [32, 35, 56, 59]. Equating $c_q^2k^2$ with $4\pi G\bar\rho$ gives the quantum Jeans length scale [31, 60, 61] at the redshift z,

$$\lambda_Q(z) = \frac{2\pi}{k} = \left(\frac{\pi^3\hbar^2}{m^2 G\bar\rho(z)}\right)^{1/4} \tag{8}$$

which leads to the minimum length scale and the mass scale of galactic halos formed at z [62–64].

Dwarf galaxies are the smallest dark-matter-dominated objects and, hence, are ideal objects for understanding the nature of dark matter. The Jeans length above provides a natural cut-off in the mass power spectrum preventing the overproduction of subhalos and the dwarf satellite galaxies [31, 32, 35, 56, 59, 65]. The SFDM could solve the cusp problem [17, 31, 32, 32, 66, 66–68] because of the boundary condition for the SPE at the halo center. There are many works explaining the rotation curves of dwarf [17, 23, 69], and large galaxies [29, 43, 70–78] in this model. By fitting RCs one can obtain the scalar field mass $m \sim 10^{-22} eV$, and interestingly the SFDM with this mass range can resolve many of the small scale problems of CDM models. One exception is the Lyman alpha tension [79, 80] which is under debate [81]. This model can be used to explain the puzzling minimum mass scale [63, 64] and the minimum length scale of galaxies [82, 83] regardless of their luminosity [84] and the observed size evolution [62] of the most massive galaxies [85–87].

The rotating ellipsoidal gravitational potentials of the SFDM halos was shown to induce spiral arms and shell structures in the visible matter of galaxies [44]. The effect of embedding a supermassive black hole in a SFDM halo was considered in [88]. It was also argued that this model can explain the M-sigma relation of supermassive black holes [89].

It is suggested that the SFDM can explain the contradictory behaviors of DM in collisions of galaxy clusters [90–94]. For a fast collision as observed in the Bullet cluster two dark matter halos pass each other in a soliton-like way, while for a slow collision as observed in the Abell 520 the halos merge due to the wave nature of the SFDM. This idea might explain the origin of the dark galaxy and the galaxy without dark matter [90] .

The SFDM model reproduces the evolution of the cosmological densities [95] and the spectrum of the cosmic microwave background [37, 96] if $m > 10^{-24}eV$.

The typical QCD axion has some difficulties to be a SFDM. It has $cos(\phi)$ potential which is effectively attractive to $O(\phi^4)$, and its mass $m_a \simeq 10^{-5}eV$ is too heavy, and it is a real field giving oscillatons instead of boson stars [21]. On the other hand ultra-light axion (ULA) with mass $m_a \simeq 10^{-22}eV$ recently attracts much interest, but if we consider only the quadratic potential for ULA, ULA is almost identical to the typical SFDM or Fuzzy DM. Therefore, we can use the cosmological constraints of ULA for SFDM and vice versa in many situations. There is an idea that a vector field in a modified gravity action could be identified with a BEC [97]. SFDM can be dark energy [98–101] and have quantum entanglement [102].

It is also suggested that the time dependent potential induced by oscillating SFDM influences the pulse arrival residual which can be observed by the Square Kilometre Array (SKA) experiment [103, 104] or by laser interferometer gravitational wave detectors [105]. The oscillation of SFDM might resonate with binary pulsars [106].

In summary, SFDM with mass about $10^{-22}eV$ satisfies many cosmological constraints [107] and seems to be a viable alternative to CDM. Therefore, this model deserves careful considerations. Especially, it is desirable to find a good particle physics model for SFDM [13, 108].

acknowledgments

This work was supported by the Jungwon University Research Grant (2016-040).

References

[1] J. Silk, *Challenges in Cosmology from the Big Bang to Dark Energy, Dark Matter and Galaxy Formation*, in *14th International Symposium on Nuclei in the Cosmos (NIC-XIV) Niigata, Japan, June 19-24, 2016* (2016), 1611.09846

[2] P. Salucci, F. Walter, A. Borriello, Astron. Astrophys. **409**, 53 (2003), astro-ph/0206304

[3] J.F. Navarro, C.S. Frenk, S.D.M. White, *Astrophys. J.* **462**, 563 (1996)

[4] W.J.G. de Blok, A. Bosma, S.S. McGaugh, astro-ph/0212102 (2002)

[5] A. Tasitsiomi, International Journal of Modern Physics D **12**, 1157 (2003)

[6] J.W. Lee, Journal of Korean Physical Society **54**, 2622 (2009), 0801.1442

[7] A. Suárez, V.H. Robles, T. Matos, Astrophysics and Space Science Proceedings **38**, 107 (2014), 1302.0903

[8] T. Rindler-Daller, P.R. Shapiro, Modern Physics Letters A **29**, 1430002 (2014), 1312.1734

[9] T. Harko, Phys. Rev. **D 89**, 084040 (2014), 1403.3358

[10] P.H. Chavanis, Phys. Rev. **D 84**, 043531 (2011), 1103.2050

[11] K. Huang, C. Xiong, X. Zhao, International Journal of Modern Physics A **29**, 1450074 (2014), 1304.1595

[12] D.J.E. Marsh, Phys. Rept. **643**, 1 (2016), 1510.07633

[13] L. Hui, J.P. Ostriker, S. Tremaine, E. Witten (2016), 1610.08297

[14] J.R. Primack, New Journal of Physics **11**, 105029 (2009), 0909.2247

[15] M.R. Baldeschi, G.B. Gelmini, R. Ruffini, Physics Letters B **122**, 221 (1983)

[16] M. Membrado, A.F. Pacheco, J. Sañudo, Phys. Rev. **A 39**, 4207 (1989)

[17] T. Matos, D. Nunez, astro-ph/0303455 (2003), astro-ph/0303455

[18] S.J. Sin, Phys. Rev. **D50**, 3650 (1994), hep-ph/9205208

[19] S.U. Ji, S.J. Sin, Phys. Rev. D **50**, 3655 (1994)

[20] C. Lee, S. Sin, J.Korean Phys.Soc. **28**, 16 (1995)

[21] J.W. Lee, I.G. Koh, Phys. Rev. **D53**, 2236 (1996), hep-ph/9507385

[22] J.W. Lee, I.G. Koh, *Galactic halo as a soliton star* , Abstracts, bulletin of the Korean Physical Society, 10 (2) (1992)

[23] C.G. Boehmer, T. Harko, JCAP **0706**, 025 (2007), 0705.4158

[24] A.H. Guth, M.P. Hertzberg, C. Prescod-Weinstein, Phys. Rev. **D92**, 103513 (2015), 1412.5930

[25] W.H. Press, B.S. Ryden, D.N. Spergel, Physical Review Letters **64**, 1084 (1990)

[26] L.M. Widrow, N. Kaiser, *Astrophys. J. Lett.* **416**, L71 (1993)

[27] F.E. Schunck, astro-ph/9802258 (1998)

[28] U. Nucamendi, M. Salgado, D. Sudarsky, Phys. Rev. Lett. **84**, 3037 (2000), gr-qc/0002001

[29] A. Arbey, J. Lesgourgues, P. Salati, Phys. Rev. D **64**, 123528 (2001)

[30] J. Goodman, New Astronomy Reviews **5**, 103 (2000)
[31] W. Hu, R. Barkana, A. Gruzinov, Phys. Rev. Lett. **85**, 1158 (2000)
[32] P. Peebles, *Astrophys. J.* **534**, L127 (2000)
[33] E.W. Mielke, F.E. Schunck, Phys. Rev. D **66**, 023503 (2002)
[34] E.W. Mielke, J.A.V. Perez, Physics Letters B **671**, 174 (2009)
[35] V. Sahni, L. Wang, Phys. Rev. D **62**, 103517 (2000)
[36] M. Alcubierre, F.S. Guzman, T. Matos, D. Nunez, L.A. Urena-Lopez, P. Wiederhold, Class. Quant. Grav. **19**, 5017 (2002), gr-qc/0110102
[37] C.G. Park, J.c. Hwang, H. Noh, Phys. Rev. **D 86**, 083535 (2012), 1207.3124
[38] P. Sikivie, Q. Yang, Physical Review Letters **103**, 111301 (2009), 0901.1106
[39] B. Fuchs, E.W. Mielke, Mon. Not. Roy. Astron. Soc. **350**, 707 (2004), astro-ph/0401575
[40] T. Matos, F.S. Guzman, L.A. Urena-Lopez, D. Nunez, astro-ph/0102419 (2001)
[41] M.P. Silverman, R.L. Mallett, Classical and Quantum Gravity **18**, L103 (2001)
[42] U. Nucamendi, M. Salgado, D. Sudarsky, Phys. Rev. D **63**, 125016 (2001)
[43] A.A. Julien Lesgourgues, P. Salati, New Astronomy Reviews **46**, 791 (2002)
[44] H.L. Bray, ArXiv:1212.5745 (2012), 1212.5745
[45] J. Eby, C. Kouvaris, N.G. Nielsen, L.C.R. Wijewardhana, arXiv:1511.04474 (2015)
[46] T. Matos, F.S. Guzman, Class. Quant. Grav. **17**, L9 (2000), gr-qc/9810028
[47] F.S. Guzman, T. Matos, H. Villegas-Brena, Rev. Mex. Astron. Astrofis. **37**, 63 (2001), astro-ph/9811143
[48] A. Arbey, J. Lesgourgues, P. Salati, Phys. Rev. D **65**, 083514 (2002)
[49] V. Sahni, L.M. Wang, Phys. Rev. **D62**, 103517 (2000), astro-ph/9910097
[50] P.H. Chavanis, Astronomy and Astrophysics **537**, A127 (2012), 1103.2698
[51] D.J.E. Marsh, P.G. Ferreira, Phys. Rev. **D82**, 103528 (2010), 1009.3501
[52] R. Hlozek, D. Grin, D.J.E. Marsh, P.G. Ferreira, Phys. Rev. **D91**, 103512 (2015), 1410.2896
[53] R. Friedberg, T.D. Lee, Y. Pang, Phys. Rev. D **35**, 3640 (1987)
[54] A. Suarez, T. Matos, Mon. Not. Roy. Astron. Soc. **416**, 87 (2011), 1101.4039
[55] M. Alcubierre, F.S. Guzman, T. Matos, D. Nunez, L.A. Urena-Lopez, P. Wiederhold, *Scalar field dark matter and galaxy formation*, in *Dark matter in astro- and particle physics. Proceedings, 4th Heidelberg International Conference, DARK 2002, Cape Town, South Africa, February 4-9, 2002* (2002)
[56] T. Matos, L. Arturo Ureña López, Phys. Rev. D **63**, 063506 (2001)
[57] T. Harko, Mon. Not. Roy. Astron. Soc. **413**, 3095 (2011), 1101.3655
[58] H.Y. Schive, T. Chiueh, T. Broadhurst, Nature Physics **10**, 496 (2014), 1406.6586
[59] T. Matos, L.A. Urena-Lopez, Class. Quant. Grav. **17**, L75 (2000)
[60] M.I. Khlopov, B.A. Malomed, I.B. Zeldovich, Mon. Not. Roy. Astron. Soc. **215**, 575 (1985)
[61] D. Grasso, Phys.Rev. **D41**, 2998 (1990)
[62] J.W. Lee, Phys. Lett. **B681**, 118 (2009), 0805.2877
[63] J.W. Lee, Phys. Lett. **B756**, 166 (2016), 1511.06611
[64] J.W. Lee, S. Lim, JCAP **1001**, 007 (2010), 0812.1342
[65] A. Suárez, T. Matos, Mon. Not. Roy. Astron. Soc. **416**, 87 (2011), 1101.4039
[66] A. Riotto, I. Tkachev, Phys. Lett. **B484**, 177 (2000), astro-ph/0003388
[67] K.Y. Su, P. Chen, JCAP **1108**, 016 (2011), 1008.3717
[68] T. Matos, L.A. Urena-Lopez, Gen. Rel. Grav. **39**, 1279 (2007)
[69] A. Arbey, J. Lesgourgues, P. Salati, Phys. Rev. D **68**, 023511 (2003)
[70] J. Lesgourgues, A. Arbey, P. Salati, New Astronomy Reviews **46**, 791 (2002)
[71] V.H. Robles, T. Matos, Mon. Not. Roy. Astron. Soc. **422**, 282 (2012), 1201.3032
[72] H.Y. Schive, M.H. Liao, T.P. Woo, S.K. Wong, T. Chiueh, T. Broadhurst, W.Y.P. Hwang, Phys. Rev. Lett. **113**, 261302 (2014), 1407.7762
[73] T. Harko, Mon. Not. Roy. Astron. Soc. **413**, 3095 (2011), 1101.3655
[74] V.H. Robles, T. Matos, Mon. Not. Roy. Astron. Soc. **422**, 282 (2012), 1201.3032
[75] T. Bernal, L.M. Fernández-Hernández, T. Matos, M.A. Rodríguez-Meza, ArXiv e-prints (2017), 1701.00912
[76] F.S. Guzman, T. Matos, Class. Quant. Grav. **17**, L9 (2000)
[77] J.P. Mbelek, Astron. Astrophys. **424**, 761 (2004), gr-qc/0411104
[78] T.H. Lee, B.J. Lee, Phys. Rev. D **69**, 127502 (2004)
[79] V. Iršič, M. Viel, M.G. Haehnelt, J.S. Bolton, G.D. Becker, Phys. Rev. Lett. **119**, 031302 (2017), 1703.04683
[80] E. Armengaud, N. Palanque-Delabrouille, D.J.E. Marsh, J. Baur, C. Yeche, Mon. Not. Roy. Astron. Soc. (2017), 1703.09126
[81] J. Zhang, J.L. Kuo, H. Liu, Y.L.S. Tsai, K. Cheung, M.C. Chu (2017), 1708.04389
[82] M. Mateo, Ann. Rev. Astron. Astrophys. **36**, 435 (1998), astro-ph/9810070
[83] G. Gilmore, D. Zucker, M. Wilkinson, R.F.G. Wyse, V. Belokurov, J. Kleyna, A. Koch, N.W. Evans, E.K.

Grebel, eprint:arXiv.org:0804.1919 (2008)

[84] L.E. Strigari, J.S. Bullock, M. Kaplinghat, J.D. Simon, M. Geha, B. Willman, M.G. Walker, Nature **454**, 1096 (2008), 0808.3772

[85] E. Daddi et al., Astrophys. J. **631**, L13 (2005), astro-ph/0507504

[86] P.G. van Dokkum, M. Kriek, M. Franx, Nature **460**, 717 (2009), 0906.2778

[87] I. Trujillo, C.J. Conselice, K. Bundy, M.C. Cooper, P. Eisenhardt, R.S. Ellis, **382**, 109 (2007)

[88] L.A. Urena-Lopez, A.R. Liddle, Phys. Rev. **D66**, 083005 (2002), astro-ph/0207493

[89] J.W. Lee, J. Lee, H.C. Kim, arXiv:1512.02351 (2015)

[90] J.W. Lee, S. Lim, D. Choi, arXiv:0805.3827 (2008)

[91] D. Castaneda Valle, E.W. Mielke, Annals Phys. **336**, 245 (2013)

[92] A. Paredes, H. Michinel, Phys. Dark Univ. **12**, 50 (2016), 1512.05121

[93] B. Schwabe, J.C. Niemeyer, J.F. Engels, Phys. Rev. **D94**, 043513 (2016), 1606.05151

[94] F.S. Guzman, J.A. Gonzalez, J.P. Cruz-Perez, Phys. Rev. **D93**, 103535 (2016), 1605.04856

[95] T. Matos, A. Vázquez-González, J. Magaña, Mon. Not. Roy. Astron. Soc. **393**, 1359 (2009), 0806.0683

[96] I. Rodríguez-Montoya, J. Magaña, T. Matos, A. Pérez-Lorenzana, *Astrophys. J.* **721**, 1509 (2010), 0908.0054

[97] J.W. Moffat, astro-ph/0602607 (2006)

[98] A. Arbey, J. Lesgourgues, P. Salati, Phys. Rev. **D64**, 123528 (2001), astro-ph/0105564

[99] T. Matos, arXiv:0909.3634 (2009), 0909.3634

[100] K. Huang, Int. J. Mod. Phys. **A28**, 1330049 (2013), [,15(2014)], 1309.5707

[101] M. Gogberashvili, A.S. Sakharov, arXiv (2017), 1702.05757

[102] J.W. Lee, arXiv:1510.07968 (2015), 1510.07968

[103] A. Khmelnitsky, V. Rubakov, JCAP **1402**, 019 (2014), 1309.5888

[104] A. Aoki, J. Soda, Phys. Rev. **D93**, 083503 (2016), 1601.03904

[105] A. Aoki, J. Soda, arXiv:1608.05933 (2016)

[106] D. Blas, D.L. Nacir, S. Sibiryakov, Phys. Rev. Lett. **118**, 261102 (2017), 1612.06789

[107] B. Li, T. Rindler-Daller, P.R. Shapiro, Phys. Rev. **D89**, 083536 (2014), 1310.6061

[108] J.E. Kim, D.J.E. Marsh, Phys. Rev. **D93**, 025027 (2016), 1510.01701

New Views on Dark Matter from Emergent Gravity

Sichun Sun[1,*] and *Yun-Long* Zhang[2,]

[1] *Center for Theoretical Sciences, National Taiwan University, Taipei 10617, Taiwan*
[2] *Asia Pacific Center for Theoretical Physics, Pohang 790-784, Korea*

Abstract. We discuss a scenario that apparent dark matter comes from the induced gravity in the (3+1)-dimensional spacetime, which can be embedded into one higher dimensional flat spacetime. The stress tensor of dark energy and dark matter is identified with the Brown-York stress tensor on the hypersurface, and we find an interesting constraint relation between the dark matter and dark energy density parameter and baryonic density parameter. Our approach may show a new understanding for Verlinde's emergent gravity from higher dimensions. We also comment on some phenomenological implications, including gravitational wave solutions and MOND limit.

1 A Toy Model for the late Universe

Inspired by the emergent gravity models proposed by Verlinde in [1, 2], we propose the induced gravity from higher dimensional spacetime which gives rise to the similar mechanism. We will modify the Einstein field equation in 3+1 dimensional spacetime at the cosmological scale,

$$R_{\mu\nu} - \frac{1}{2}Rg_{\mu\nu} + \frac{H_0}{c}\left(\mathcal{K}g_{\mu\nu} - \mathcal{K}_{\mu\nu}\right) = \frac{8\pi G}{c^4}T_{\mu\nu}, \qquad (1)$$

where H_0 is the Hubble constant. G is the newton gravitational constant. c is the speed of light. $T_{\mu\nu}$ is the stress tensor for normal matters. Alternatively, we can also put the extra terms on the right hand side, which adds the extra contribution to the stress tensor

$$\mathcal{T}_{\mu\nu} \equiv -\frac{H_0 c^3}{8\pi G}\left(\mathcal{K}g_{\mu\nu} - \mathcal{K}_{\mu\nu}\right). \qquad (2)$$

It is just the Brown-York stress tensor induced from higher dimensional space time, and we explain it as the apparent dark sectors. We are going to show that stress tensor $\mathcal{T}_{\mu\nu}$, which is holographic in nature [4] can provide dark energy and dark matter effects.

We consider the ΛCDM Universe, in which the universe contains a positive cosmological constant Λ contribution to the dark energy with component Ω_Λ, cold dark matter density parameter Ω_D, and Baryon density parameter Ω_B. They satisfy $\Omega_D + \Omega_B + \Omega_\Lambda \simeq 1$. Based on eq.(1) and constraints from the consistent embedding, we obtain an interesting constraint relation of the uniform dark matter density,

$$\text{CSZ:} \quad \Omega_D^2 = \frac{1}{2}\Omega_\Lambda(\Omega_D - \Omega_B). \qquad (3)$$

We will show exactly how to derive this equation in the next section. Let us compare with the constraint relation in Verlinde's emergent gravity [2],

$$\text{Verlinde:} \quad \Omega_D^2 = \frac{4}{3}\Omega_B. \qquad (4)$$

We take some parameters from the current ΛCDM universe from observation[3], with a bit priori choice as following

$$\Omega_\Lambda = 0.685, \quad \Omega_D = 0.265, \quad \Omega_B = 0.050. \qquad (5)$$

Comparing our formula (3) with Verlinde's (4), we obtain

$$\Delta_V \equiv \Omega_D^2 - \frac{4}{3}\Omega_B \simeq 0.36\%, \qquad (6)$$

$$\Delta_{CSZ} \equiv \Omega_D^2 - \frac{1}{2}\Omega_\Lambda(\Omega_D - \Omega_B) \simeq -0.34\%. \qquad (7)$$

We will show that although our relation hold as well as Verlinde's, there are still some differences in approximation.

Constraints From Consistent Embedding— Similar to the formula in (1), let us write down the Einstein equation in d dimensional spacetime as

$$R_{\mu\nu} - \frac{1}{2}R\, g_{\mu\nu} = \kappa_d(T_{\mu\nu} + \mathcal{T}_{\mu\nu}), \qquad (8)$$

with $\mu, \nu = 0, 1, ..., (d-1)$, and $\kappa_d = 8\pi G_d/c^4$. $T_{\mu\nu}$ is the stress tensor of normal matters, and $\mathcal{T}_{\mu\nu}$ is the effective dark sectors of our universe, which can include both of the dark energy and dark matters. The trace of above equation gives the Ricci scalar

$$R = -\frac{2\kappa_d}{d-2}(T + \mathcal{T}). \qquad (9)$$

Now we assume that the geometry with metric $g_{\mu\nu}$ can be embedded into one higher dimensional spacetime, as a

*Corresponding Author and Speaker: sichunssun@gmail.com. Proceeding for the 13th International Conference on Gravitation, Astrophysics, and Cosmology the 15th Italian-Korean Symposium on Relativistic Astrophysics: A Joint Meeting. Parallel Talk on July 4, 2017, Seoul, Korea.

hypersurface with the normal vector $\mathcal{N}^M$, and the indices $M, N = 0, 1, ..., d$. We can define the induced metric on the hypersurface $g^M{}_N = \eta^M{}_N - \mathcal{N}^M \mathcal{N}_N$ as well as the extrinsic curvature

$$\mathcal{K}_{\mu\nu} \equiv g_\mu{}^M g_\nu{}^N \nabla_{(M}\mathcal{N}_{N)}. \tag{10}$$

μ, ν are the indices on the hypersurface, which depend on the coordinate choices.

On the other hand, the Hamiltonian constraint equations of the hypersurface lead to

$$R = \mathcal{K}^2 - \mathcal{K}_{\mu\nu}\mathcal{K}^{\mu\nu} - 2\,G^{(d+1)}_{MN}\mathcal{N}^M\mathcal{N}^N. \tag{11}$$

Now we introduce the following Brown-York stress tensor,

$$\mathcal{T}_{\mu\nu} = -\frac{1}{\kappa_{d+1}}\left(\mathcal{K}g_{\mu\nu} - \mathcal{K}_{\mu\nu}\right), \tag{12}$$

and κ_{d+1} is the Einstein's constant in $d+1$ dimension. Notice that in the above definition, there is a minus sign comparing to the usual Brown-York formula [4]. Replacing the external curvature by the Brown-York stress tensor, the Hamiltonian constraint relation (11) becomes

$$\frac{\mathcal{T}^2}{d-1} - \mathcal{T}_{\mu\nu}\mathcal{T}^{\mu\nu} = \frac{R + 2\,G^{(d+1)}_{MN}\mathcal{N}^M\mathcal{N}^N}{(\kappa_{d+1})^2}. \tag{13}$$

Even though there are matters in the late universe, we require them to be localized on the hypersurface, such that we still have $G^{(d+1)}_{MN}\mathcal{N}^M\mathcal{N}^N = T^{(d+1)}_{MN}\mathcal{N}^M\mathcal{N}^N = 0$ in (11). Then plug (9) into (13), we have

$$\frac{\mathcal{T}^2}{d-1} - \mathcal{T}_{\mu\nu}\mathcal{T}^{\mu\nu} = -\frac{2\kappa_d}{(\kappa_{d+1})^2}\frac{1}{d-2}(T + \mathcal{T}). \tag{14}$$

Pure De Sitter Spacetime. — Firstly we set that the normal matters in the Einstein equation (8) vanish, $T_{\mu\nu}$=0. As a warm up, let us consider the d dimensional de-Sitter spacetime,

$$\mathrm{d}s^2 = -\,c^2\mathrm{d}t^2 + e^{2(ct/L)}\left[\mathrm{d}r^2 + r^2\mathrm{d}\,\Omega_{d-2}\right]. \tag{15}$$

which can be be embedded into the $d+1$ dimensional flat spacetime

$$\mathrm{d}s^2_{d+1} = \eta_{MN}\mathrm{d}X^M\mathrm{d}X^N = -\mathrm{d}X_0^2 + \mathrm{d}X_i^2, \tag{16}$$

with $i = 1, 2, ..., d$. The vacuum Einstein equation associated with the (d+1)-dimensional flat metric (16) turns out to be $G^{(d+1)}_{MN} = 0$.

Let us study the embedding of de-Sitter hypersurface in more details. It is a hyperbolid spacetime with radius L,

$$L^2 = -T^2 + X_i^2, \qquad \mathcal{N}^M = \frac{1}{L}(X_0, X_i). \tag{17}$$

where $\mathcal{N}^M$ is the normal vector of the hypersurface pointing outwards. The cosmological constant $\Lambda = \frac{(d-1)(d-2)}{2L^2}$ will play the role of the dark energy. Notice that to balance the Einstein equations with the induced de-Sitter metric $g_{\mu\nu}$ in (15), one requires either the cosmological constant or the apparent dark energy term.

Interestingly, for the pure de-Sitter spacetime (15), after considering (10) with extrinsic curvature $\mathcal{K}_{\mu\nu} = \frac{1}{L}g_{\mu\nu}$, the Brown-York stress tensor (12) turns out to be $\mathcal{T}_{\mu\nu} = \bar{\mathcal{T}}_{\mu\nu} = -\frac{1}{\kappa_{d+1}}\frac{d-1}{L}g_{\mu\nu}$. Then we arrive at the stress tensor of apparent dark energy,

$$\bar{\mathcal{T}}_{\mu\nu} = -\frac{\Lambda_d}{\kappa_d}g_{\mu\nu}, \quad \text{when } \frac{\kappa_{d+1}}{\kappa_d} = \frac{2L}{d-2}. \tag{18}$$

From (18) we read out the dark energy density formula

$$\rho_\Lambda = \bar{\mathcal{T}}_{\mu\nu}\frac{u^\mu u^\nu}{c^4} = \frac{\Lambda_d}{\kappa_d c^2}. \tag{19}$$

After considering (14) with $T = 0$, we have the identity

$$\frac{\bar{\mathcal{T}}^2}{d-1} - \bar{\mathcal{T}}^\mu{}_\nu\bar{\mathcal{T}}^\nu{}_\mu = -\frac{\rho_\Lambda c^2}{d-1}\bar{\mathcal{T}}. \tag{20}$$

Thus, assuming $\mathcal{T}_{\mu\nu} = \bar{\mathcal{T}}_{\mu\nu} \equiv -\frac{\Lambda}{\kappa_d}g_{\mu\nu}$ in the constraint equation (13), the pure de-Sitter spacetime satisfies the above identity automatically. Notice here the Brown-York stress tensor plays the role of dark energy and there is no matter or dark matter yet in the set-up.

Uniform Matter Perturbations. — Next we consider to add small amount of normal matters in with uniform distribution, such that we treat the de-Sitter metric as background. It describes dark energy dominated universe like today.

We consider that our university is uniform at large scale, and take the FLRM metric in d dimension,

$$\mathrm{d}s^2 = -\,c^2\mathrm{d}t^2 + a(t)^2\left[\frac{\mathrm{d}r^2}{1-kr^2} + r^2\mathrm{d}\,\Omega_{d-2}\right]. \tag{21}$$

In the ΛCDM model, $k = 0$, and

$$H(a)^2 = H_0^2\left[\Omega_\Lambda + (\Omega_D + \Omega_B)a^{-3} + \Omega_R a^{-4}\right] \tag{22}$$

with $H(a) \equiv \frac{\dot{a}}{a}$, and H_0 is the Hubble constant today. Considering (18)(19), our assumption for the constraint relation (14) becomes

$$\frac{\mathcal{T}^2}{d-1} - \mathcal{T}_{\mu\nu}\mathcal{T}^{\mu\nu} = -\frac{\rho_\Lambda c^2}{d-1}(T + \mathcal{T}). \tag{23}$$

This is the main constraint relation in this paper. Since in Einstein equation (8), $\mathcal{T}_{\mu\nu}$ is the Brown-York stress tensor playing the role of dark matter and dark energy, and $T_{\mu\nu}$ is the baryonic visible matter with mass density $\rho_B \ll \rho_\Lambda$. The baryonic matter and radiation are with energy density,

$$T^B_{\mu\nu} = \rho_B u_\mu u_\nu, \quad T^R_{\mu\nu} = \rho_R u_\mu u_\nu + p_R h_{\mu\nu}. \tag{24}$$

Now assume that the dark matter is induced with mass density $\rho_D \ll \rho_\Lambda$. The dark energy and cold dark matters are all assumed to be related to the extrinsic curvature.

$$\mathcal{T}_{\mu\nu} = \mathcal{T}^\Lambda_{\mu\nu} + \mathcal{T}^D_{\mu\nu}, \quad T_{\mu\nu} = T^B_{\mu\nu} + T^R_{\mu\nu}, \tag{25}$$

where $\mathcal{T}^\Lambda_{\mu\nu} = -(\rho_\Lambda c^2)g_{\mu\nu}$ and $\mathcal{T}^D_{\mu\nu} = \rho_D u_\mu u_\nu$. Putting them back into the constraint equation (23), and subtracting equation (20), we arrive at,

$$\rho_D^2 = \frac{\rho_\Lambda}{d-2}\left[\rho_D - \rho_B - \rho_R + (d-1)\frac{p_R}{c^2}\right]. \tag{26}$$

When $d = 4$, the stress tensor of radiation is traceless $-\rho_R c^2 + 3p_R = 0$. If setting $\rho_D \simeq 5\rho_B$, $\rho_\Lambda \simeq \rho_c - \rho_D - \rho_B$, we can recover Verlinde's constraint relation (4) approximately. Considering that the critical mass density of the university as $\rho_c = \frac{3H_0^2}{8\pi G}$, for which the spatial geometry is kept flat from the observation, then dividing both side of equation (26) by by ρ_c^2, we have

$$\Omega_D^2 = \frac{1}{2}\Omega_\Lambda(\Omega_D - \Omega_B). \tag{27}$$

If we further consider $\Omega_\Lambda + \Omega_D + \Omega_B \simeq 1$, then we derive our main result in (3),

$$\Omega_D^2 = \frac{1}{3}(\Omega_D - \Omega_B + \Omega_B^2). \tag{28}$$

Considering $d = 4$, $\Omega_D \simeq 5\Omega_B$, as well as $\Omega_B = 0.05 \ll 1$, we can also arrive at Verlinde's $\Omega_D^2 \simeq \frac{4}{3}\Omega_B$ in (4).

On the other hand, since $\Omega_B + \Omega_D \lesssim \Omega_\Lambda$, although not so precise, our de sitter background is still a good approximation. However, if we consider the dark matter in smaller scales around the galaxies and compare with rotational curves, we need to consider the effects of back-reaction of the normal matters, with new derivations.

Constraints from Gravitational Waves. — Recently it is argued that two relativistic models of modified Newtonian dynamics seem inconsistent with observations [5]. Modified gravity models face constraints from two aspects: one is the constraint of the energy loss rate from ultra high energy cosmic rays, which indicates that gravitational waves should propagate at the speed of light; the other is the observed gravitational waveforms from LIGO, which are consistent with Einstein's gravity and suggest that the gravitational wave should satisfy linear equations of motion in the weak-field limit.

Although E. Verlinde suggested similar modifications of Newtonian dynamics as in MOND theories [2], which emerges with different underlining physical origin, there is no covariant equations of motion for the gravitational waves. While for our toy model in the previous sections, we have shown that the Einstein equations are not necessarily modified in our scenarios. The induced dark sectors play the same role as that in the ΛCDM models, so it can pass the above mentioned constraints [5]. To be more specific, the extra apparent dark matter as extra terms in Einstein equation fills the space as dark medium and modifies the the speed of the light. In our induced gravity, gravitational field equation is

$$\mathcal{R}_{\mu\nu} - \frac{1}{2}g_{\mu\nu}\mathcal{R} = \frac{8\pi G}{c^4}\left(T_{\mu\nu} + \mathcal{T}_{\mu\nu}\right), \tag{29}$$

which leads to $0 = \nabla^\mu G_{\mu\nu} = \nabla^\mu T_{\mu\nu} + \nabla^\mu \mathcal{T}_{\mu\nu}$. Thus it is similar to the effects of cold dark matter, and it does not conflict with the observations from LIGO so far.

The dark matter scenarios provide one extra effect on top of the extra dimensional modification in our setups. However, if we indeed take a higher dimensional point of views [6, 7], we expect one extra breezing mode on top of two polarized propagating modes. The extra breezing mode is constraint by the current experiments. There are also multiple massive Kaluza-Klein gravitational modes associated with the extra dimensions. Although those massive modes decay fast and may not reach the gravitational waves detector, they are constrained as well by the gravitational waves signal templates from the binary black hole signals.

2 Connection with other Scenarios

We compare our approach with two well studied scenarios here. One is the holographic scenario of our universe, and the other is the brane world models. Naively they look similar, in a way that they all embed our universe into higher dimensional spacetime. However, the holographic models is a dual equivalent description, while in the brane world models, there exist real extra dimensions.

Holographic Universe. — Let us first compare with the holographic universe scenario. There is an effective contribution from the holographic stress tensor, which can be identified as the stress tensor of dark energy and dark matter. Let us start with the Einstein-Hilbert action in $(d+1)$-dimension, with the Gibbons-Hawking-York boundary term

$$S_{d+1} = \frac{1}{2\kappa_{d+1}}\int_M d^{d+1}x\sqrt{-\tilde{g}}\,(R_{d+1} - 2\Lambda_{d+1}) + \frac{1}{\kappa_{d+1}}\int_{\partial M} d^d x\sqrt{-g}\mathcal{K}_d, \tag{30}$$

with $\tilde{g}_{MN}$ being the metric in $d+1$ dimension. After the variation, we have

$$(2\kappa_{d+1})\delta S_{d+1} = \left[R_{d+1}^{MN} - \frac{1}{2}R_{d+1}\tilde{g}^{MN} + \Lambda_{d+1}\tilde{g}^{MN}\right]\delta\tilde{g}_{MN} + 2\left(\mathcal{K}^{\mu\nu} - \mathcal{K}g^{\mu\nu}\right)\delta g_{\mu\nu}. \tag{31}$$

$\mathcal{K}_{\mu\nu}$ is the external curvature of the time-like brane, which motivates our Brown-York stress tensor in (12). In the usual paradigm of holography, we do not consider the back reaction of the holographic stress tensor on the hypersurface. In our toy model which modifies the Einstein equations into (1), we do consider it.

In our toy model, the holographic screen is embedded in one higher dimensional flat spacetime, which is quite similar to the holography in Rindler spacetime. It is one successful realization of holography in flat spacetime, where the bulk background is flat spacetime, and the induced Brown-York stress tensor gives the dual Rindler fluid [8]. Rindler fluid has the energy density $\rho_U = 0$, and pressure $p_U = a$, where a is the acceleration of the Rindler observer. The vacuum Einstein equation also yields the equation of states $p_U = T_U s$, with T_U being the Unruh temperature, and s being the Bekenstein entropy density of the horizon. Especially, it is found that the Petrov Type I conditions give rise to additional constraint relation of the holographic stress tensor. [9]. It is expected to be generalized to our case as another candidate of constraint equations that is similar to (13).

Mathematically, the de-Sitter metric (15) or the more general FRW metric (21) can also be embedded into the one higher dimensional AdS spacetime, which leads to the holographic models of the universe. Although the holographic correspondence is more clear there, one needs to deal with another problem on the origin of the cosmological constant. That is one of the reasons that we choose the flat embedding in this paper instead of the AdS. The latter one is interesting for further study.

What is more, it is natural to consider that the entanglement between two cosmological horizons may have an impact on the gravity as suggested by Verlinde [2]. In the 4+1 dimensional space time, it is more clear to see this relation through the embeddings of wormholes. It is show in [10], that the EPR pair in 3+1 can be described by the worm hole in 4+1 dimensional AdS spacetime.

Brane World Scenario. — Now let us compare our model to the well studied brane world models. Except for the constraint equations, we also have the dynamical Einstein equation in higher dimensions. We analyze the equations induced on the hypersurface

$$R^{(d)}_{\mu\nu} = (\mathcal{K}g_{\mu\sigma} - \mathcal{K}_{\mu\sigma})\mathcal{K}^{\sigma}_{\ \nu} + \mathcal{M}_{\mu\nu}, \quad (32)$$
$$\mathcal{M}_{\mu\nu} \equiv g_{\mu}^{\ M} g_{\nu}^{\ N} R^{(d+1)}_{MN} + g_{\mu}^{\ M} N^{P} g_{\nu}^{\ N} N^{Q} R^{(d+1)}_{MPNQ}.$$

Although $R^{(d+1)}_{MN} = 0$ in flat spacetime, it is not necessary true for $R^{(d+1)}_{MPNQ}$, which depends on the coordinate choices. After including the normal matters $T_{\mu\nu}$, in principle we can also define the effective stress tensor from brane world models

$$R_{\mu\nu} - \frac{1}{2}Rg_{\mu\nu} = T_{\mu\nu} + \mathcal{T}^{\mathcal{M}}_{\mu\nu}, \quad (33)$$
$$\mathcal{T}^{\mathcal{M}}_{\mu\nu} \equiv (\mathcal{K}g_{\mu\sigma} - \mathcal{K}_{\mu\sigma})\mathcal{K}^{\sigma}_{\ \nu} + \mathcal{W}_{\mu\nu} - \frac{1}{2}\left(\mathcal{K}^2 - \mathcal{K}_{\rho\sigma}\mathcal{K}^{\rho\sigma}\right) g_{\mu\nu},$$

which includes the description of the the hyper surface evolution. In the de Sitter spacetime, $\mathcal{K}_{\mu\sigma} = \frac{1}{L}g_{\mu\nu}$ leads to $\mathcal{T}^{\mathcal{M}}_{\mu\nu} = -\Lambda g_{\mu\nu}$, and $\mathcal{W}_{\mu\nu}$ is associated with the bulk Weyl tensor. However, we have tried the perturbation based on this formula, but the constraint relations is not as good as our equation (3).

Verlinde's Emergent Gravity. — Following Verlinde's model for apparent dark matter, there seem to be three crucial conditions for his construction so far. First, there is the background entropy, that is what he calls the elastic medium occupying the (3+1)-dimensional spacetime. Second, the positive cosmological constant, provides the thermal bath with the Gibbons-Hawking temperature $H/2\pi$. The last, the apparent dark matter is only the response to the presence of the normal matters.

The braneworld scenarios can offer similar above mentioned conditions and become a natural playground for Verlinde's emergent gravity. The elastic medium full of entropy can be explained from higher dimensions, by treating our (3+1) dimensional spacetime as the boundary of the bulk theory. Cosmological constant and the standard model can be easily implemented with braneworld in many literatures [11].Actually that is one of main motivations that braneworld was widely discussed before. Most interestingly, the branes with tensions and dynamics, can react to the matter fields we put in, with extra terms we introduced in the last section. Especially the exterior curvature, a concept valid only from higher dimensional spacetime, describes the "elastic response" nature of the apparent dark matter from Verlinde's theory.

Summary. — In summary, We give a new viewpoint on the dark components of our universe, which originates from the induced stress tensor of higher dimensional flat spacetime. There are only normal matters on the brane, and the dark energy and dark matter are considered as the response of normal matters due to the geometric effects.

In our approaches, the toy and the more developed ones, partly borrowed from the braneworld scenario and combined with holographic universe models, which resembles Verlinde's emergent gravity in a subtle way. We do have a covariant relativistic form of Lagrangian. Our ansatz produces the late universe component data from ΛCDM model, and is also expected to be consistent with early universe evolution from its braneworld like features, although we have not produced the CMB prediction of our scenarios in this note. We are going to study this CMB prediction in the near future.

From our observation, it is still quite interesting to ask whether there are real extra dimensions, although there is no evidence from experiments so far. One recent study may come from gravitational waves physics in [7], in addition to the long searching constraints from colliders and precision measurements of gravity.

Acknowledgements — S. Sun thanks the supported from NCTS for the conference. We thank R. G. Cai, H. Tye, S. J. Wang for helpful conversions.

References

[1] E. P. Verlinde, JHEP **1104**, 029 (2011) [arXiv:1001.0785].

[2] E. P. Verlinde, SciPost Phys. **2**, 016 (2017) [arXiv:1611.02269 [hep-th]].

[3] P. A. R. Ade *et al.* [Planck Collaboration], Astron. Astrophys. **594**, A13 (2016) [arXiv:1502.01589 [astro-ph.CO]].

[4] J. D. Brown and J. W. York, Jr., Phys. Rev. D **47**, 1407 (1993) [gr-qc/9209012].

[5] P. M. Chesler and A. Loeb, Phys. Rev. Lett. **119**, 031102 (2017) [arXiv:1704.05116 [astro-ph.HE]].

[6] T. Kobayashi and T. Tanaka, Phys. Rev. D **73**, 044005 (2006) [hep-th/0511186].

[7] D. Andriot and G. Lucena Gómez, JCAP **1706**, 048 [arXiv:1704.07392 [hep-th]].

[8] I. Bredberg, C. Keeler, V. Lysov and A. Strominger, JHEP **1207**, 146 (2012) [arXiv:1101.2451 [hep-th]].

[9] V. Lysov and A. Strominger, arXiv:1104.5502 [hep-th]. R. G. Cai, L. Li, Q. Yang and Y. L. Zhang, JHEP **1304**, 118 (2013) [arXiv:1302.2016 [hep-th]]. R. G. Cai, Q. Yang and Y. L. Zhang, Phys. Rev. D **90**, 041901 (2014) [arXiv:1401.7792 [hep-th]].

[10] K. Jensen and A. Karch, Phys. Rev. Lett. **111**, 211602 (2013) [arXiv:1307.1132]. J. Sonner, Phys. Rev. Lett. **111**, 211603 (2013) [arXiv:1307.6850]. J. W. Chen, S. Sun and Y. L. Zhang, arXiv:1612.09513 [hep-th].

[11] J. Frieman, M. Turner and D. Huterer, Ann. Rev. Astron. Astrophys. **46**, 385 (2008) [arXiv:0803.0982 [astro-ph]]. G. R. Dvali, G. Gabadadze and M. Porrati, Phys. Lett. B **485**, 208 (2000) [hep-th/0005016]. D. B. Kaplan and S. Sun, Phys. Rev. Lett. **108**, 181807 (2012) [arXiv:1112.0302 [hep-ph]].

Atomki anomaly and the Secluded Dark Sector

Yasuhiro Yamamoto[1,*]

[1] *Department of Physics and IPAP, Yonsei University, Seoul 03722, Republic of Korea*

Abstract. The Atomiki anomaly can be interpreted as a new light vector boson. If such a new particle exists, it could be a mediator between the Standard Model sector and the dark sector including the dark matter. We discussed some simple effective models with these particles. In the models, the secluded dark matter models are good candidates to satisfy the thermal relic abundance. In particular, we found that the dark matter self-interaction can be large enough to solve the small scale structure puzzles if the dark matter is a fermion.

1 Introduction

A nuclear experiment in Hungary has reported the observation of a new resonance in the decay of excited ^{8}Be nuclei [1]. This is called the Atomki anomaly named after the institute where it is observed. The resonance can be simply interpreted as a new weakly interacting light particle of 16.7 MeV.

It is elucidated by Ref. [2] that the new particle should be a vector boson. They have also pointed out the coupling of the vector with proton is much smaller than that of neutron because of the strong constraint by the π^0 decay experiment [3]. Hence, the vector boson is also referred as the protophobic light vector boson. The vector boson is not only the protophobic but also the leptophobic. In particular, the coupling with the neutrinos should be smaller than about 10^{-5}, instead, the coupling with the electron is $O(10^{-4})$. Because of this property, the model building is difficult; for the details, see also Refs. [4].

If we suppose that the vector is the gauge boson of a broken U(1) symmetry, the Lagrangian can be written as

$$\begin{aligned}\mathcal{L}_X &= -\frac{1}{4}X^{\mu\nu}X_{\mu\nu} + (D_\mu S)^\dagger(D^\mu S) + \mu_S^2|S|^2 - \frac{\lambda_S}{2}|S|^4 \\ &= -\frac{1}{4}X^{\mu\nu}X_{\mu\nu} + \frac{m_X^2}{2}X^\mu X_\mu + \frac{1}{2}(\partial_\mu s)^2 - \frac{m_s^2}{2}s^2 + g_X m_X s X^\mu X_\mu - \frac{g_X m_s^2}{2m_X}s^3 + \cdots \end{aligned} \tag{1}$$

where $X^{\mu\nu}$ is the field strength tensor of the new gauge boson X. The scalar field S is expanded as $S = (v_s + s)/\sqrt{2}$ in the unitarity gauge, and $D_\mu S = (\partial_\mu + ig_X X_\mu)S$. Some terms irrelevant in our computation are suppressed here. The other parameters are defined as $v_s^2 = 2\mu_S^2/\lambda_S$, $m_s^2 = \lambda_S v_s^2$, and $m_X = g_X v_s$. We studied simple dark matter models including the above Lagrangian [5].

If the dark matter directly interacts with X, it is difficult to satisfy the thermal relic scenario because the interaction strength of the new vector is comparable to the weak interaction, namely,

[*] e-mail: yamayasu@yonsei.ac.kr

$g_X/m_X = 1/O(100)$ GeV; see Refs. [6] for exceptions. In the rest of this article, we discuss the models of the real scalar and the Majorana fermion dark matter. Since they are neutral under the new gauge symmetry $U(1)_X$, the models also require other particles which mediate the interaction between the dark matter and X. They can be easily compatible with the thermal relic scenario. These models are called the secluded dark matter [7]. The given models are easily implemented in not only the explicit models of the protophobic light vector but also other models since the dark matter properties are mostly determined in the dark sector.

2 Real scalar dark matter

The Lagrangian of the real scalar dark matter model is

$$\mathcal{L}_\phi = \frac{1}{2}(\partial^\mu \phi)^2 - \frac{m_\phi^2}{2}\phi^2 - \frac{\lambda_{\phi S} m_X}{2g_X} s\phi^2 - \frac{\lambda_{\phi S}}{4} s^2\phi^2 - \frac{\lambda_\phi}{4!}\phi^4 + \mathcal{L}_X, \tag{2}$$

where we impose a Z_2 symmetry ($\phi \leftrightarrow -\phi$) to stabilize the dark matter. The coupling $\lambda_{\phi S}$ is given by the scalar four point interaction. The dark Higgs boson s play a role of the mediator.

The annihilation of the dark matter is dominated by the s-wave processes: $\phi\phi \to XX$ and $\phi\phi \to ss$. Their final states are two and four e^+e^- pairs, respectively. The direct detection is induced by the 1-loop diagram mediated by the sXX vertex. The experimental bounds on the models are shown in Fig. 1.

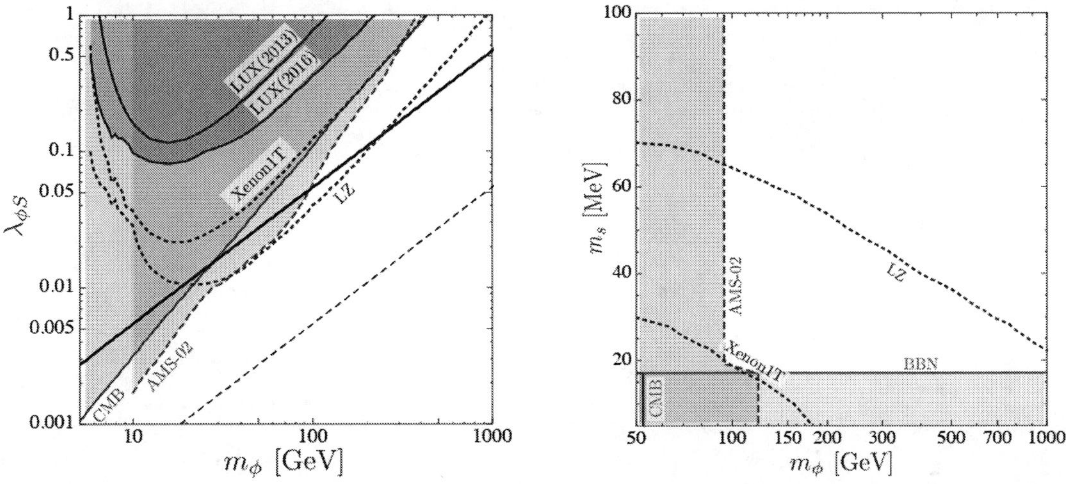

Figure 1. Experimental bounds on the real scalar dark matter. The left figure is the bounds on the typical dark matter coupling $\lambda_{\phi S}$ as a function of the dark matter mass with m_s = 50 MeV. The blue/red region is bounded by the current direct/indirect dark matter searches. The projected direct detection bounds by the XENON1T and the LZ experiments are shown with the blue dotted lines. On the solid black line, the dark matter satisfies the observed thermal relic abundance. On the dotted black line, the relic abundance is 2 orders of magnitude larger than the observed one. The dark Higgs mass dependence of the expected direct detection bounds is shown in the right figure. The coupling $\lambda_{\phi S}$ is chosen to satisfy the thermal relic abundance, i.e. on the black solid line of the left one. The blue dotted lines are the projected direct detection bounds. The red region is excluded by the indirect detections. In the green region, the lifetime of the dark Higgs is larger than 1 sec.

The left figure is the bounds for $m_s = 50$ MeV. The thermal relic abundance can be satisfied if the dark matter is heavier than about 100 GeV. The indirect detection bounds with the e^+e^- spectrum by AMS-02 and Planck exclude the scenario up to there [8]. The current LUX and the future XENON1T bounds are much weaker than the indirect bounds [9]. The current strongest bound by PandaX-II is the middle of them [10]. However, the prospected LZ bound is stronger than them [11].

The right figure shows us the m_s dependence. The coupling λ_{phS} is fixed to obtain the thermal relic abundance. We find that the experimental reach of the dark matter mass strongly depends on the mass of the dark Higgs.

3 Majorana fermion dark matter

The Lagrangian of the Majorana fermion dark matter model is

$$\mathcal{L}_\chi = \bar{\psi}(i\partial\!\!\!/ - g_X X\!\!\!/ - m_\psi)\psi + \frac{1}{2}\bar{\chi}(i\partial\!\!\!/ - m_\chi)\chi - y(S\bar{\psi}\chi + S^\dagger\bar{\chi}\psi) + \mathcal{L}_X, \tag{3}$$

where the Yukawa coupling y can be chosen as real and positive without loss of generality. The Dirac fermion ψ is the additional mediator. After the spontaneous symmetry breaking of $\mathrm{U}(1)_X$, the model includes three Majorana fermions; the lightest one is the dark matter.

The leading contribution of the annihilation is given by the p-wave, so that, the indirect bounds are negligibly weak. The direct detection is given by the same diagram with the real scalar case. However, as shown in Fig. 2, the experimental bounds are significantly stronger than the previous one.

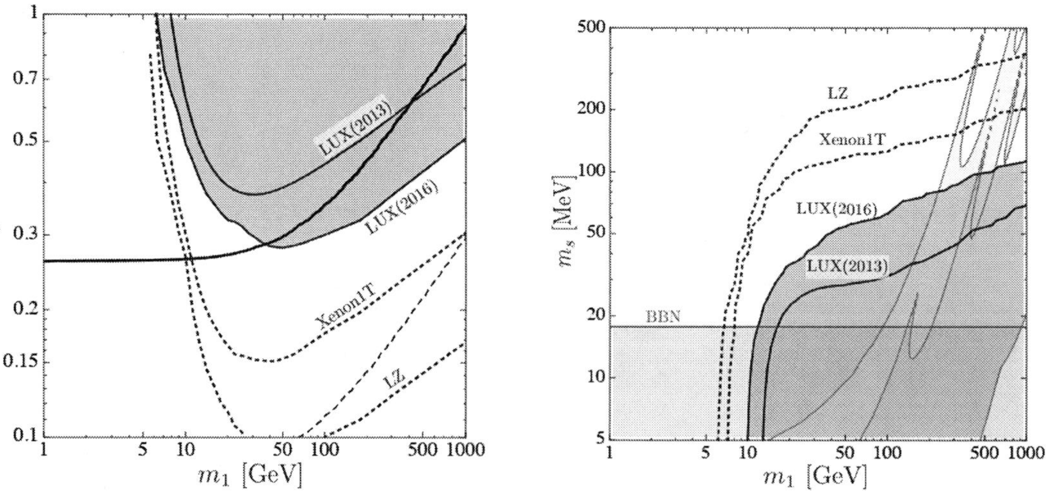

Figure 2. Experimental bounds on Majorana fermion dark matter. The lines and regions are similar to Fig. 1. In the right figure, the small scale structure puzzles can be solved in two yellow regions at the middle and the corner, where $g_X = 10^{-3}$ and 10^{-2} respectively.

In these figures, the additional Majorana fermions are 100 GeV heavier than the dark matter. If $m_s = 50$ MeV, the left figure tell us that the dark matter should be lighter than a few tens GeV. The upper bound becomes about 10 GeV in the near future.

The m_s dependence is shown in the right figure. Since the indirect detection bounds can be neglected, it is difficult to exclude the dark matter lighter than about 5 GeV. If the dark matter is O(100)

GeV or heavier, the dark Higgs should be heavier than about 100 MeV. The bound reaches up to 200–400 MeV by the prospected LZ bound.

Instead, the dark matter annihilates through the p-wave, the dark matter self- scattering is the s-wave. Besides, the cross section is largely enhanced by the Sommerfeld enhancement. The large self-interaction of the dark matter is required to solve the small scale structure puzzles [12]. If $g_X = 10^{-3}$, the self- interaction cross section becomes large enough to explain the structure of the dark matter of O(100) GeV.

4 Summary

A clear peak in the e^+e^- spectrum of excited ^{8}Be has been observed by a Hungarian nuclear experimental group. It looks not a statistical fluctuation like many tail events but a physical event. The peak can originate from the weakly interacting light new vector boson. Such a new particle is an attractive candidate of new physics. As a fascinating scenario, we have studied the thermal relic dark matter models in which the vector boson is a mediator between the dark sector and the Standard Model, see Fig. 3. We have discussed a real scalar and a Majorana fermion as the dark matter. These models are similar each other. However, they have clearly different behavior.

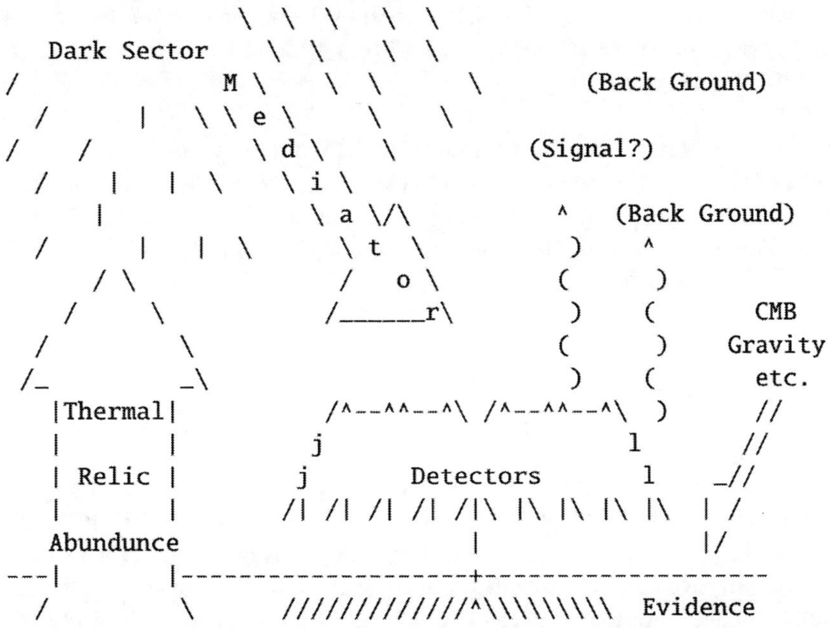

Figure 3. The schematic figure of the models discussed in this paper. The astrophysical observations clarify the dark components in our universe. The dark sector including the dark matter could be thermally produced. We suppose this sector interacts with the Standard Model via the weakly interacting light mediator, namely, the protophobic vector boson of 16.7 MeV. This figure is inspired by the sketch of photosynthesis in Ref. [13]

For the real scalar model, the thermal relic scenario is excluded by the indirect detections if the dark matter is lighter than about 100 GeV. The direct detection bound can be stronger than the indirect bounds for the future LZ result. The reach of the LZ bound is sensitive to the dark Higgs mass.

For the Majorana fermion model, the indirect detection bounds can be negligible because the annihilation is the p-wave. The direct detection bounds exclude the dark matter of heavier than several tens GeV, while it is impossible to constrain the region where the dark matter is lighter than about 5 GeV. In order to evade these experimental bounds for the heavier dark matter, we need to consider the dark Higgs of a few hundred MeV. With the self-interaction, the thermal relic dark matter solves the small scale structure problem if its mass is several hundred GeV.

References

[1] A. J. Krasznahorkay *et al.*, Phys. Rev. Lett. **116** (2016) no.4, 042501 doi:10.1103/PhysRevLett.116.042501 [arXiv:1504.01527 [nucl-ex]]; A. J. Krasznahorkay *et al.*, EPJ Web Conf. **142** (2017) 01019. doi:10.1051/epjconf/201714201019 .

[2] J. L. Feng, B. Fornal, I. Galon, S. Gardner, J. Smolinsky, T. M. P. Tait and P. Tanedo, Phys. Rev. Lett. **117** (2016) no.7, 071803 doi:10.1103/PhysRevLett.117.071803 [arXiv:1604.07411 [hep-ph]].

[3] J. R. Batley *et al.* [NA48/2 Collaboration], Phys. Lett. B **746** (2015) 178 doi:10.1016/j.physletb.2015.04.068 [arXiv:1504.00607 [hep-ex]].

[4] J. L. Feng, B. Fornal, I. Galon, S. Gardner, J. Smolinsky, T. M. P. Tait and P. Tanedo, Phys. Rev. D **95** (2017) no.3, 035017 doi:10.1103/PhysRevD.95.035017 [arXiv:1608.03591 [hep-ph]]; B. Fornal, Int. J. Mod. Phys. A **32** (2017) 1730020 doi:10.1142/S0217751X17300204 [arXiv:1707.09749 [hep-ph]].

[5] T. Kitahara and Y. Yamamoto, Phys. Rev. D **95** (2017) no.1, 015008 doi:10.1103/PhysRevD.95.015008 [arXiv:1609.01605 [hep-ph]].

[6] L. B. Jia and X. Q. Li, Eur. Phys. J. C **76** (2016) no.12, 706 doi:10.1140/epjc/s10052-016-4561-3 [arXiv:1608.05443 [hep-ph]]; O. Seto and T. Shimomura, Phys. Rev. D **95** (2017) no.9, 095032 doi:10.1103/PhysRevD.95.095032 [arXiv:1610.08112 [hep-ph]].

[7] M. Pospelov, A. Ritz and M. B. Voloshin, Phys. Lett. B **662** (2008) 53 doi:10.1016/j.physletb.2008.02.052 [arXiv:0711.4866 [hep-ph]].

[8] L. Accardo *et al.* [AMS Collaboration], Phys. Rev. Lett. **113** (2014) 121101. doi:10.1103/PhysRevLett.113.121101; M. Aguilar *et al.* [AMS Collaboration], Phys. Rev. Lett. **113** (2014) 121102. doi:10.1103/PhysRevLett.113.121102; P. A. R. Ade *et al.* [Planck Collaboration], Astron. Astrophys. **594** (2016) A13 doi:10.1051/0004-6361/201525830 [arXiv:1502.01589 [astro-ph.CO]]; M. Kawasaki, K. Nakayama and T. Sekiguchi, Phys. Lett. B **756** (2016) 212 doi:10.1016/j.physletb.2016.03.005 [arXiv:1512.08015 [astro-ph.CO]].

[9] D. S. Akerib *et al.* [LUX Collaboration], Phys. Rev. Lett. **118** (2017) no.2, 021303 doi:10.1103/PhysRevLett.118.021303 [arXiv:1608.07648 [astro-ph.CO]]; E. Aprile *et al.* [XENON Collaboration], JCAP **1604** (2016) no.04, 027 doi:10.1088/1475-7516/2016/04/027 [arXiv:1512.07501 [physics.ins-det]].

[10] X. Cui *et al.* [PandaX-II Collaboration], arXiv:1708.06917 [astro-ph.CO].

[11] D. S. Akerib *et al.* [LZ Collaboration], arXiv:1509.02910 [physics.ins-det].

[12] S. Tulin, H. B. Yu and K. M. Zurek, Phys. Rev. D **87** (2013) no.11, 115007 doi:10.1103/PhysRevD.87.115007 [arXiv:1302.3898 [hep-ph]]; M. Kaplinghat, S. Tulin and H. B. Yu, Phys. Rev. Lett. **116** (2016) no.4, 041302 doi:10.1103/PhysRevLett.116.041302 [arXiv:1508.03339 [astro-ph.CO]].

[13] https://en.wikipedia.org/wiki/Sexy_Commando_Gaiden:_Sugoi_yo!!_Masaru-san

EPJ Web of Conferences **168**, 06008 (2018) https://doi.org/10.1051/epjconf/201816806008
Joint International Conference of ICGAC-XIII and IK-15 on Gravitation, Astrophysics and Cosmology

Non-minimal dark matter search in dark matter colliders

Seodong Shin[1,2,3,⋆]

[1] *Department of Physics & IPAP, Yonsei University, Seoul 03722, Republic of Korea*
[2] *Enrico Fermi Institute and Department of Physics, University of Chicago, Chicago, IL 60637, USA*
[3] *Physics Department, Indiana University, Bloomington, IN 47405, USA*

Abstract. In the scenarios of dark matter (DM) with a non-minimal dark sector, we revisit a new detection strategy of observing two or three simultaneous signals from inelastic scattering of a boosted DM [1]. The relativistically incoming DM can scatter off inelastically to a heavier unstable dark sector particle which decays back in to the DM associated with visible Standard Model particles inside large volume neutrino detectors. The existence of the secondary procedure renders us to separate it from conventional neutrino scattering background. The relativistically incoming DM can come from the universe by the annihilation of heavy DM component in an inelastic boosted DM scenario or produced by the beam bombardments in fixed target experiments.

1 Introduction

Dark matter (DM) is a key element constituting our universe with energy density about five times larger than that of the Standard Model (SM) particles. Reminding that the SM sector is composed of seventeen elementary particles with various mass spectrum, it is natural to imagine that the dark sector may be also flavorful containing many particles with a variety of mass spectrum. Furthermore, one may consider the scenarios with multiple DM components (with multiple symmetries for the stability of each component) where the right thermal relic can be expected from mechanisms different from that in conventional scenario of weakly interacting massive particle (WIMP).[1]

As a promising strategy to examine non-minimal dark sector directly, we focus on the possibility of observing *relativistic* DM scattering signal with targets, inspired from ordinary collider experiments [1]. Such relativistic scattering may arise either from a DM component which is produced with a relativistic velocity in the universe in a boosted DM (BDM) scenario or from a direct production of light DM inside the laboratory in fixed target experiments. In the framework of BDM, there are at least two DM components (heavy and light) which are stable each other [8–11] and the right thermal relic is obtained via assisted freeze-out mechanism [7]. The mass difference of the two DM particles is huge so that the light DM pair produced from the annihilation of the heavy DM pair in the *present* universe have relativistic velocities (the ratio of the heavy DM mass to the light one is the boost factor). The heavy DM particles do not interact with the SM particles directly but only through the light

⋆e-mail: shinseod@indiana.edu

[1]Null signal observation in DM detection experiments has motivated various non-conventional DM scenarios such as secluded DM [2–6] and non-minimal dark sector models including assisted freeze-out [7], boosted DM (BDM) [8–11], and DM transporting mechanism for cosmic-ray excesses [12].

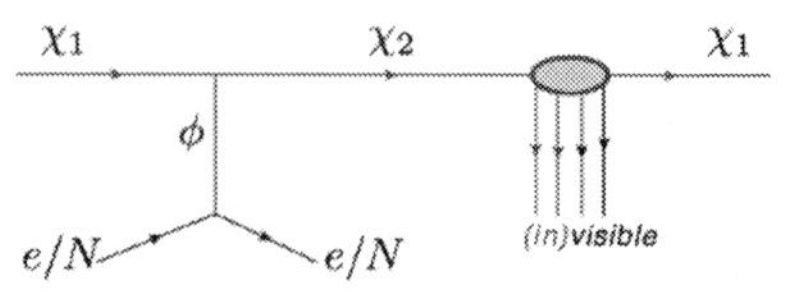

Figure 1. Inelastic boosted DM direct detection scenarios considered in Ref. [1].

DM component so what we can observe in the terrestrial experiments are (in)direct signals of (heavy) light DM. The flux of the relativistic light DM produced from the annihilation of the heavy DM is as small as $O(10^{-7}\,\text{cm}^{-2}\,s^{-1})$ for the heavy DM mass of $O(10\,\text{GeV})$ so large volume neutrino detectors are usually considered to observe the signals [8]. On the other hand, a direct production of light DM particles inside the laboratory is possible for any kind of DM scenarios as long as they include a light DM below the center of mass energy of the fixed target experiments.

In this proceeding based on Ref. [1], we discuss the channel that the incoming boosted DM (χ_1) scatters off inelastically to a heavier (unstable) dark sector particle (χ_2) which decays back into the DM χ_1 associated with visible SM particles by the decay of the mediator ϕ inside the detectors of our interest, as depicted in Fig. 1. The decay of the excited state χ_2 into χ_1 and SM particles, resembling typical cascade decay signatures in collider experiments, provides a unique secondary signature separated from conventional neutrino scattering signals.

2 Model framework and kinematic features

A variety of DM model frameworks can be considered to realize the process in Fig. 1. As a toy model for a concrete analysis, we consider a model of Dirac fermionic DM χ_1 which interacts with target SM electron or nucleus via a vector mediator ϕ. The vector mediator is assumed to be a dark gauge boson X_μ from the toy model Lagrangian satisfying a dark $\text{U}(1)_\text{X}$ gauge symmetry.

$$\mathcal{L}_X \supset -\frac{\sin\epsilon}{2}F_{\mu\nu}X^{\mu\nu} + g_{12}\bar{\chi}_2\gamma^\mu\chi_1 X_\mu + h.c.\,, \tag{1}$$

where the kinetic mixing between $\text{U}(1)_\text{X}$ and $\text{U}(1)_\text{EM}$ [2, 13–15] is parameterized by ϵ. The off-diagonal gauge interaction $\chi_1\chi_2X_\mu$ with coupling g_{12} can be obtained, e.g., from the mixing in the dark sector with different $\text{U}(1)_\text{X}$ charge assignments to χ_1 and χ_2 (see Ref. [16] as an example for the SM quark sector). The red-circled blob in Fig. 1 shows a sequential cascade process of χ_2. For simplicity, we consider a single-step cascade of $\chi_2 \to \chi_1\phi$.

The maximum mass of χ_2 is $\sqrt{s} - m_T$ where m_T is the target mass and $\sqrt{s}$ is the overall center-of-mass energy ($s = m_T^2 + 2E_{\chi_1}m_T + m_{\chi_1}^2$). Then, we get

$$m_{\chi_2} \le \sqrt{m_T^2 + 2E_{\chi_1}m_T + m_{\chi_1}^2} - m_T\,. \tag{2}$$

For χ_1 much heavier than the target ($m_{\chi_1} \gg m_T$) along with a decent boost factor γ_{χ_1}, the above is approximately rewritten as

$$m_{\chi_2} \lesssim m_{\chi_1} + (\gamma_{\chi_1} - 1)m_T\,, \tag{3}$$

which is the case of our e-scattering. On the other hand, if $m_{\chi_1} \ll m_T$, we obtain

$$m_{\chi_2} \lesssim \gamma_{\chi_1}m_{\chi_1}\,, \tag{4}$$

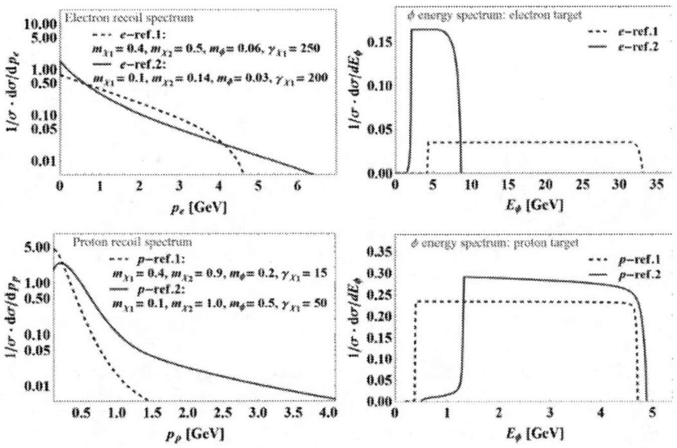

Figure 2. Expected unit-normalized energy spectra of the recoiling target particles from the primary vertex (left panels) and outgoing mediators from the secondary vertex (right panels) for *e*-scattering (top panels) and *p*-scattering (bottom panels). The reference masses are in unit of GeV. See Ref. [1] for the details.

Exp.	Volume [Mt]	p_e^{th} [GeV]	p_p^{th} [GeV]	θ_e^{res} [°]	θ_p^{res} [°]
SK	0.0224	0.1	1.07	3	3
HK	0.56	0.1	1.07	3	3
DUNE	0.04	0.03	0.05	1	5

Table 1. The volume, threshold (kinetic) energy and angular resolution of SK [19], HK [20] and DUNE [21].

which makes it possible that χ_2 is much heavier than χ_1, corresponding to our *p*-scattering.

The expected unit-normalized recoil energy spectra for our four reference points (*e*-ref.1, *e*-ref.2, *p*-ref.1, and *p*-ref.2), written in the plots, are shown in the left panels of Fig. 2. Note that the recoiling energy in large volume neutrino experiments corresponds to the magnitude of the spatial momentum (equivalently, the kinetic energy) of the recoiling target, i.e., $p_T = \sqrt{E_T^2 - m_T^2}$. The reference parameters are chosen to satisfy various experimental bounds [17]. The right panels show the expected (unit-normalized) energy spectra of ϕ produced in the cascade process for our reference points.

Throughout this proceeding and in Ref. [1], we assume that ϕ predominantly decays into e^+e^-. Then the decay length of $\chi_2 \to \chi_1\phi \to \chi_1 e^+e^-$ is at most $\lesssim 1$ cm for our reference points. This is below the detector resolutions (e.g., a few cm for Deep Underground Neutrino Experiment (DUNE) [18]) so the secondary process will be observed as a prompt process in those detectors for the reference points. Note that an appreciable displaced vertex can e also expected for a different choice of parameters of m_ϕ and ϵ.

3 Detection prospects

In order for our signal to be sensitive even with small flux, we choose large volume neutrino detectors such as Super-Kamiokande (SK), Hyper-Kamiokande (HK) and DUNE. The latter two experiments are future proposals which have much attention. Table 1 summarizes the volume, threshold (kinetic) energy and angular resolution of each experiment. Note that our process depicted in Fig. 1 can also

Exp.	Run time	e-ref.1	e-ref.2	p-ref.1	p-ref.2
SK	13.6 yr	170	7.1	3500	5200
HK	1 yr	88	3.7	1900	2800
HK	13.6 yr	6.7	0.28	140	210
DUNE	1 yr	190	9.0	150	1600
DUNE	13.6 yr	14	0.69	11	120

Table 2. Required fluxes in 10^{-7} cm^{-2} s^{-1} in order to make our reference points sensitive in various experiments.

exist in any kind of model with a light DM χ_1 and additional heavier unstable dark sector state χ_2 in fixed target experiments such as LBNF/DUNE [18], SHiP [22] and T2HKK [23]). Note that the threshold energy for p-scattering is much higher that that for e-scattering in Cherenkov light detectors, i.e., SK and HK. This is because a large amount of momentum transfer is needed from χ_1 to boost proton with mass $m_p = 0.938$ GeV. However, this condition is enormously relaxed in liquid Ar TPC detectors like DUNE. From Fig. 2, we see that the differential cross section gets larger for smaller momentum transfer so that the e-scattering is preferred over p-scattering in SK and HK. We also impose an upper limit $p_p \lesssim 1.8$ GeV in order to avoid more concrete analysis on deep inelastic scattering, for simplicity.

The angular resolution separating the primary and secondary signal is also an important feature to observe our signal. For the e-scattering, a huge energy is needed to produce an exited state χ_2 from the initial scattering. Thus, the primary (recoiled e target) and secondary signals can be highly collimated. The angular separation levels between these are shown in Fig. 3 of Ref. [1] for our reference parameters. The results imply that a detector with the angular resolution about 3°, such as SK/HK, can separate the primary and secondary signals for our reference point e-ref.2 with momentum values of the recoiling electron $p_e \in [0.1, 0.3]$ GeV.[2]

On the other hand, much clearer angular separation is possible for p-scattering because the incoming energy of χ_1 only about m_p is enough to produce χ_2 so that the produced particles are not so much boosted. Also, the opening angle of ϕ decay products is large enough as estimated in Ref. [1].

Because the observation of two or three separate (but correlated) signals, expected from our e-scattering and p-scattering processes, is usually not expected in neutrino scattering, it is fair to assume there is no background events. So, in Table 2, we show the experimental sensitivities by requiring three signal events which correspond to the 95% C.L. upper limit under the assumption of a null observation over a null background with Poisson statistics (see also Refs. [24, 25] for useful information). Considering the fact that the typical flux of χ_1 in the minimal BDM scenario is $O(10^{-7})$ cm^{-2} s^{-1} for $E_{\chi_1} = O(20\,\text{GeV})$ [8], we see that e-ref. 2 is very promising. Note that the p-scattering processes are much more sensitive in DUNE than SK/HK because of low p_p^{th} of DUNE. The flux can be increased in a modified BDM scenario (e.g., up to $O(10^{-4})$ cm^{-2} s^{-1} [10, 11]) or probed in fixed target experiments with much higher intensity.

Conclusions

In this proceeding, we revisited the novel proposal of DM detection strategy for the scenarios with non-minimal dark sector: relativistic and inelastic scattering of DM with the targets in terrestrial experiments [1]. The existence of primary and secondary signals provides a unique feature of this process separating the DM signal events from large number of background events (e.g., conventional

[2] For e-ref.1, detecting events with separable signatures is very challenging, which requires future detailed analysis.

neutrino scattering in large volume neutrino experiments). The source of relativistically incoming DM can be the light DM component produced from the annihilation of heavy DM components at the current universe in BDM scenarios or a light DM (in any kind of DM scenarios) produced inside the laboratory in fixed target experiments. A phenomenologically promising feature comes from the secondary cascade process which makes us utilizing the knowledge in collider phenomenology.

Acknowledgments

This work is supported in part by DOE Grant DE-SC0010120 and the National Research Foundation of Korea (NRF-2017R1D1A1B03032076).

References

[1] D. Kim, J. C. Park and S. Shin, arXiv:1612.06867 [hep-ph].
[2] J. H. Huh, J. E. Kim, J. C. Park and S. C. Park, Phys. Rev. D **77**, 123503 (2008) [arXiv:0711.3528 [astro-ph]].
[3] M. Pospelov, A. Ritz and M. B. Voloshin, Phys. Lett. B **662**, 53 (2008) [arXiv:0711.4866 [hep-ph]].
[4] Y. G. Kim, K. Y. Lee and S. Shin, JHEP **0805**, 100 (2008) [arXiv:0803.2932 [hep-ph]].
[5] Y. G. Kim and S. Shin, JHEP **0905**, 036 (2009) [arXiv:0901.2609 [hep-ph]].
[6] Y. G. Kim, K. Y. Lee, C. B. Park and S. Shin, Phys. Rev. D **93**, no. 7, 075023 (2016) [arXiv:1601.05089 [hep-ph]].
[7] G. Belanger and J. C. Park, JCAP **1203**, 038 (2012) [arXiv:1112.4491 [hep-ph]].
[8] K. Agashe, Y. Cui, L. Necib and J. Thaler, JCAP **1410**, no. 10, 062 (2014) [arXiv:1405.7370 [hep-ph]].
[9] J. Berger, Y. Cui and Y. Zhao, JCAP **1502**, no. 02, 005 (2015) [arXiv:1410.2246 [hep-ph]].
[10] K. Kong, G. Mohlabeng and J. C. Park, Phys. Lett. B **743**, 256 (2015) [arXiv:1411.6632 [hep-ph]].
[11] H. Alhazmi, K. Kong, G. Mohlabeng and J. C. Park, arXiv:1611.09866 [hep-ph].
[12] D. Kim, J. C. Park and S. Shin, arXiv:1702.02944 [hep-ph].
[13] L. B. Okun, Sov. Phys. JETP **56**, 502 (1982) [Zh. Eksp. Teor. Fiz. **83**, 892 (1982)].
[14] P. Galison and A. Manohar, Phys. Lett. B **136**, 279 (1984).
[15] B. Holdom, Phys. Lett. B **166**, 196 (1986).
[16] J. E. Kim, M. S. Seo and S. Shin, Phys. Rev. D **83**, 036003 (2011) [arXiv:1010.5123 [hep-ph]].
[17] R. Essig *et al.*, arXiv:1311.0029 [hep-ph].
[18] R. Acciarri *et al.* [DUNE Collaboration], arXiv:1601.02984 [physics.ins-det].
[19] M. Fechner *et al.* [Super-Kamiokande Collaboration], Phys. Rev. D **79**, 112010 (2009) [arXiv:0901.1645 [hep-ex]].
[20] E. Kearns *et al.* [Hyper-Kamiokande Working Group Collaboration], arXiv:1309.0184 [hep-ex].
[21] R. Acciarri *et al.* [DUNE Collaboration], arXiv:1512.06148 [physics.ins-det].
[22] M. Anelli *et al.* [SHiP Collaboration], arXiv:1504.04956 [physics.ins-det].
[23] K. Abe *et al.* [Hyper-Kamiokande proto- Collaboration], arXiv:1611.06118 [hep-ex].
[24] R. Dermisek, E. Lunghi and S. Shin, JHEP **1610**, 081 (2016) [arXiv:1608.00662 [hep-ph]].
[25] R. Dermisek, J. P. Hall, E. Lunghi and S. Shin, JHEP **1412**, 013 (2014) [arXiv:1408.3123 [hep-ph]].

SIMP dark matter and its cosmic abundances

Soo-Min Choi[1,*], *Hyun Min* Lee[1,**], and *Min-Seok* Seo[2,***]

[1] *Department of Physics, Chung-Ang University, Seoul 06974, Korea.*
[2] *Center for Theoretical Physics of the Universe, Institute for Basic Science, 34051 Daejeon, Korea.*

Abstract. We give a review on the thermal average of the annihilation cross-sections for $3 \to 2$ and general higher-order processes. Thermal average of higher order annihilations highly depend on the velocity of dark matter, especially, for the case with resonance poles. We show such examples for scalar dark matter in gauged Z_3 models.

Prepared for the proceedings of the 13th International Conference on Gravitation, Ewha Womans University, Korea, 3-7 July 2017.

1 Introduction

Dark matter has been hinted by missing masses in galaxies and galaxy clusters and supported by indirect evidences such as Cosmic Microwave Background(CMB) or gravitational lensing effect. So many experiments have been carried out to search the dark matter, through direct detection, indirect detection and collider search. Although we do not know characteristics of dark matter yet, we know the non-gravitational interaction of dark matter is getting more constrained.

The most popular candidate for dark matter is Weakly Interacting Massive Particle(WIMP), and WIMP has $2 \to 2$ annihilation channels into the Standard Model particles. WIMP relic density is determined when the $2 \to 2$ annihilation channels are frozen out. There have been a lot of experiments undertaken to find the WIMP, but we have not found any experimental evidence to date.

However, to solve the small scale problems of collisionless cold dark matter, people consider the self-interaction between dark matters as one of the solutions. Small scale problems stem from the differences between observation and simulation of the dark matter profiles, the density of sub-halos, and the number of sub-halos. The popular candidate for self-interacting dark matter is Strongly Interacting Massive Particle(SIMP) [1]. SIMP has a $3 \to 2$ self-annihilation process between dark matter particles determining the current relic density when it is frozen-out. Unlike the WIMP, SIMP doesn't need to have a large coupling with the Standard Model particles. The mass of SIMP is order of 100 MeV scale so it is much lighter than WIMP.

In order to accurately determine the relic density of dark matter, a thermal averaged cross-section of the annihilation channel is required. For the non-relativistic dark matter, the cross-section can be

*e-mail: soominchoi90@gmail.com
**e-mail: hminlee@cau.ac.kr
***e-mail: minseokseo57@gmail.com

expanded as powers of dark matter relative velocity and the lowest order terms of relative velocity are the most important. This applies equally to high order annihilations such as SIMP, especially when there is a higher partial wave or resonance pole.

In this article, we discuss the thermal averaged cross-section of high order annihilations for SIMP and perform a valid thermal average of the velocity dependent cross-section such as in the case with resonance pole.

2 Resonance and thermal average

In order to obtain the relic density of dark matter, we need to calculate a thermal average of cross-section with the integration of a momenta for initial particles. For the non-relativistic dark matter, if the cross-section has a resonance of Breit-Wigner form and the sum of masses for initial particles are almost the same as the resonance mass, the thermal averaged cross-section is very sensitive to the velocities of initial dark matter particles. So we need to consider exactly the momentum dependence of the cross-section. In the paper of Gondolo and Gelmini [2], they showed the exact results of thermal average only for $2 \to 2$ cross-section.

Recently, SIMP and higher order annihilating models have shed light on solving the small scale problems [4–10]. To get the right relic density, the exact calculation of thermal average for $n \to 2$ ($n \geq 3$) cross-section is also very important. So in this article, we show the results of more general thermal average of both non-resonant and resonant cross-sections.

3 Thermal average of SIMP

An interesting way for producing the self-interacting dark matter is through higher order annihilations, namely, the initial particles are three or more in annihilation channels. A good example of self-interacting dark matter with higher order annihilation is SIMP. The relic density of SIMP is governed by the Boltzmann equation,

$$\dot{n}_\chi + 3Hn_\chi = -\langle\sigma v^2\rangle(n_\chi^3 - n_\chi n_{eq}^2) \tag{1}$$

where

$$\sigma v^2 = \frac{1}{2E_1 2E_2 2E_3}\int \frac{d^4p_4}{(2\pi)^3 2E_4}\frac{d^4p_5}{(2\pi)^3 2E_5}|\mathcal{M}|^2(2\pi)^4\delta^4(p_1+p_2+p_3-p_4-p_5). \tag{2}$$

The cross-section with 5-point interactions should be thermal averaged to get the exact relic density. Thermal averaged $3 \to 2$ cross-section is very similar with $2 \to 2$ cross-section. We just need to consider one more initial dark matter. For the initial particles with the same masses, the thermal averaged $3 \to 2$ cross-section is given by [9],

$$\langle\sigma v^2\rangle = \frac{\int d^3v_1 d^3v_2 d^3v_3\,\delta^3(\vec{v}_1+\vec{v}_2+\vec{v}_3)(\sigma v^2)\,e^{-\frac{1}{2}x(v_1^2+v_2^2+v_3^2)}}{\int d^3v_1 d^3v_2 d^3v_3\,\delta^3(\vec{v}_1+\vec{v}_2+\vec{v}_3)\,e^{-\frac{1}{2}x(v_1^2+v_2^2+v_3^2)}}. \tag{3}$$

where delta function comes from center of mass frame of initial particles. If the 5-point interaction is non-resonant, then the cross-section can be expanded by the velocities of initial particles like,

$$(\sigma v^2) = a_0 + a_1(v_1^2+v_2^2+v_3^2) + a_2^{(1)}(v_1^2+v_2^2+v_3^2)^2 + a_2^{(2)}(v_1^4+v_2^4+v_3^4) + \cdots. \tag{4}$$

So there are SO(9) invariant terms and general terms of the form, $(v_1^2)^n(v_2^2)^m(v_3^2)^l$. If the $3 \to 2$ cross-section has only SO(9) invariant terms, the corresponding thermal average is

$$\langle \sigma v^2 \rangle = \frac{1}{2}x^3 \sum_{l=0}^{\infty} \frac{a_l}{l!} \int_0^{\infty} d\eta\, \eta^{l+2} e^{-x\eta} \\ = a_0 + 3a_1 x^{-1} + 6a_2 x^{-2} + \cdots . \tag{5}$$

with $\eta \equiv \frac{1}{2}(v_1^2 + v_2^2 + v_3^2)$. Otherwise, if the cross-section depends on the general terms, then,

$$\langle (v_1^2)^n (v_2^2)^m (v_3^2)^l \rangle = \frac{3\sqrt{3}x^3}{\pi} \int_0^{\infty} dv_1 v_1^2 \int_0^{\infty} dv_2 v_2^2 (v_1^2)^n (v_2^2)^m \times \\ \times \int_{-1}^{+1} d\cos\theta_{12} (v_1^2 + v_2^2 + 2v_1 v_2 \cos\theta_{12})^l \, e^{-x(v_1^2+v_2^2+v_1 v_2 \cos\theta_{12})},$$

especially,

$$\langle (v_1^2)^n \rangle = \left(\frac{4}{3}\right)^n \frac{\Gamma(n+\frac{3}{2})}{\Gamma(\frac{3}{2})} x^{-n} = \langle (v_2^2)^n \rangle = \langle (v_3^2)^n \rangle.$$

When the $3 \to 2$ process is due to resonance, then the resulting cross-section has also a Breit-Wigner form like,

$$\langle \sigma v^2 \rangle = \left\langle \sum_{l=0}^{\infty} \frac{b_R^{(l)}}{l!} \eta^l \frac{\gamma_R}{(\epsilon_R - \frac{2}{3}\eta)^2 + \gamma_R^2} \right\rangle \\ = \frac{3}{4}\pi x^3 \sum_{l=0}^{\infty} \frac{b_R^{(l)}}{l!} G_l(z_R; x), \tag{6}$$

where

$$G_l(z_R; x) = \mathrm{Re}\left[\frac{i}{\pi} \int_0^{\infty} d\eta \frac{\eta^{l+2} e^{-x\eta}}{\frac{3}{2}z_R - \eta}\right] \\ = (-1)^l \frac{\partial^l}{\partial x^l} G_0(z_R; x), \tag{7}$$

$\epsilon_R \equiv \frac{m_R^2 - 9m_{\rm DM}^2}{9m_{\rm DM}^2}$, $\gamma_R \equiv \frac{m_R \Gamma_R}{9m_{\rm DM}^2}$ and $z_R \equiv \epsilon_R + i\gamma_R$. When $\gamma_R \ll 1$, $G_0(z_R; x) \approx \frac{9}{4}\epsilon_R^2 e^{-\frac{3}{2}x\epsilon_R}\theta(\epsilon_R)$. So the thermal averaged cross-section is given by

$$\langle \sigma v^2 \rangle_R \approx \frac{27}{16}\pi \epsilon_R^2 x^3 e^{-\frac{3}{2}x\epsilon_R}\theta(\epsilon_R) \sum_{l=0}^{\infty} \frac{b_R^{(l)}}{l!} \left(\frac{3}{2}\right)^l \epsilon_R^l. \tag{8}$$

Because the phase space integral in the thermal average for SIMP depends on the higher order of dark matter velocity than in the case of WIMP, SIMP case highly depends on the resonance mass (ϵ_R^2) than WIMP case ($\epsilon_R^{1/2}$).

We can generalize the above results of the thermal average of SIMP processes for non-degenerate initial particle masses and even higher order annihilation like $n \to 2$, where $n > 3$. Even if the masses of the initial particles are different, we can calculate the thermal averaged cross-section. Especially, when the cross-section is s-wave, the result can be obtained by $m \to (m_1 + m_2 + m_3)/3$ in the definition of x in eq.(12), (13) and (14). For even higher order annihilations, we can get the general p-wave results by using SO(3n) symmetry.

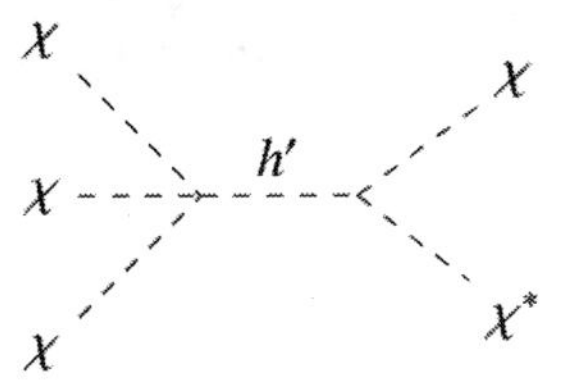

Figure 1. One example of the dark Higgs resonance channel of the discrete Z_3 gauged symmetry dark matter model.

4 Model for resonant SIMP

As a concrete model, we show the discrete Z_3 gauged model for scalar dark matter. In this model, there are dark Higgs and dark matter which are charged under the dark U(1) gauge symmetry. A dark matter triple coupling κ is generated after the dark U(1) is broken down to Z_3. The details of the model are in the references [7, 10]. Because there are dark Higgs resonance channels as shown in figure 1, we should use the results near resonance. In the non-relativistic limit, the effective $3 \to 2$ annihilation cross-section with a dark Higgs resonance has the form,

$$(\sigma v^2)_{Z_3} = C_3 \frac{\gamma_R}{(\epsilon_R - \frac{2}{3}\eta)^2 + \gamma_R^2}, \tag{9}$$

and C_3 is given by

$$C_3 = \frac{\sqrt{5}\kappa^2}{12\beta_\chi m_\chi^5}\left(1 + \frac{\lambda_{\phi\chi} v'^2}{m_\chi^2}\right)^2. \tag{10}$$

Definition of β_χ is $\sqrt{1 - 4m_\chi^2/m_{h'}^2}$. Because the three-body decay ($h' \to \chi\chi\chi$) is suppressed by extra phase space with $\epsilon_R^2/(4\pi^2)$, the decay width of dark Higgs is determined mainly by two-body decay ($h' \to \chi\chi^*$). From eq.(6), the thermal averaged cross-section of eq.(9) is given by

$$\langle \sigma v^2 \rangle_{Z_3} = \frac{3}{4} C_3 \pi x^3 G_0(z_R; x). \tag{11}$$

By using the narrow width approximation,

$$\langle \sigma v^2 \rangle_{Z_3} \approx \frac{27}{16} C_3 \pi \epsilon_R^2 x^3 e^{-\frac{3}{2} x \epsilon_R} \theta(\epsilon_R). \tag{12}$$

In figure 2, we show the parameter space for the relic density of dark matter as a function of C_3 and ϵ_R in the left panel. We choose dark matter mass as 100 MeV. Right panel is the parameter space for dark matter cubic coupling κ and dark matter mass satisfying the relic density of dark matter. In this plot, we show the value of $\epsilon_R = 0.01, 0.02, 0.06$. For both plots in figure 2, we assume narrow width approximation (NWA).

5 Conclusions

We have shown the general results of the thermal averaged cross-sections for SIMP and higher order annihilations. When the cross-section has high dependence on velocity and resonance pole, the exact

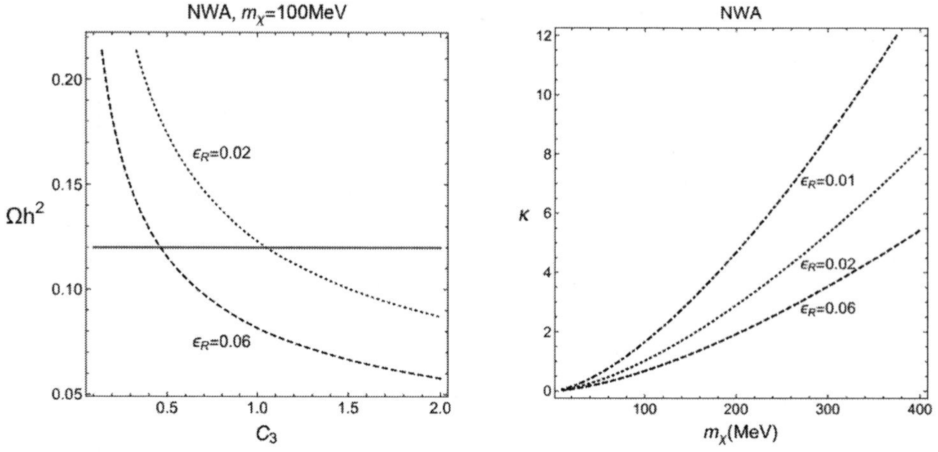

Figure 2. Relic density of Z_3 dark matter as a function of C_3 and ϵ_R in the left. Dark matter cubic coupling satisfying a correct relic density as a function of dark matter mass and ϵ_R in the right. The blue line is the relic density of dark matter obtained by Planck.

calculation of thermal averaged cross-section is very important for the relic density of dark matter. Moreover, the results can be generalized for the non-degenerate initial particle masses and even higher annihilation channels.

6 Acknowledgments

The work is supported in part by Basic Science Research Program through the National Research Foundation of Korea (NRF) funded by the Ministry of Education, Science and Technology (NRF-2016R1A2B4008759). The work of SMC is supported in part by TJ Park Science Fellowship of POSCO TJ Park Foundation. MS is supported by IBS under the project code, IBS-R018-D1.

References

[1] Y. Hochberg, E. Kuflik, T. Volansky and J. G. Wacker, Phys. Rev. Lett. **113** 171301 (2014)
[2] P. Gondolo and G. Gelmini, Nucl. Phys. B **360** 145. (1991)
[3] K. Griest and D. Seckel, Phys. Rev. D **43** 3191. (1991)
[4] Y. Hochberg, E. Kuflik, H. Murayama, T. Volansky and J. G. Wacker, Phys. Rev. Lett. **115** 2, 021301 (2015)
[5] N. Bernal, C. Garcia-Cely and R. Rosenfeld, JCAP **1504** 04, 012 (2015)
[6] H. M. Lee and M. S. Seo, Phys. Lett. B **748** 316 (2015)
[7] S. M. Choi and H. M. Lee, JHEP **1509** 063 (2015)
[8] S. M. Choi and H. M. Lee, Phys. Lett. B **758** 47 (2016)
[9] S. M. Choi, H. M. Lee and M. S. Seo, JHEP **1704** 154 (2017)
[10] S. M. Choi, Y. J. Kang and H. M. Lee, JHEP **1612** 099 (2016)

EPJ Web of Conferences **168**, 06010 (2018) https://doi.org/10.1051/epjconf/201816806010
Joint International Conference of ICGAC-XIII and IK-15 on Gravitation, Astrophysics and Cosmology

Forbidden Channels and SIMP Dark Matter

Soo-Min Choi[1,⋆], *Yoo-Jin* Kang[1,2,⋆⋆], and *Hyun Min* Lee[1,⋆⋆⋆]

[1] *Department of Physics, Chung-Ang University, Seoul 06974, Korea*
[2] *Center for Theoretical Physics of the Universe, Institute for Basic Science, Daejeon 34051, Korea*

Abstract. In this review, we focus on dark matter production from thermal freeze-out with forbidden channels and SIMP processes. We show that forbidden channels can be dominant to produce dark matter depending on the dark photon and / or dark Higgs mass compared to SIMP.

1 Introduction

In the Universe, there are diverse evidences for dark matter such as galaxy rotation curves, gravitational lensing and Cosmic Microwave Background (CMB). Especially, CMB from Planck satellite data infer that the averaged relic density of dark matter is $\Omega_{\rm DM}h^2 = 0.1198 \pm 0.0015$ [1].

Various dark matter candidates have been introduced including Weakly Interacting Massive Particles (WIMPs). WIMPs are well-motivated but carry too small self-interaction to explain small-scale problems. Small-scale problems come from difference between numerical simulations with collisionless dark matter and observations from galaxies. They include core-cusp problem, missing satellite problem and too-big-to-fail problem. Solutions to small-scale problems require dark matter as self-interaction strong as $\sigma_{\rm self}/m_{\rm DM} = 0.1 - 10{\rm cm}^2/{\rm g}$ [2, 3]. In this sense, dark matter annihilating with forbidden channels can be important. In the forbidden channels, masses of final states are larger than dark matter masses. These channels are forbidden at zero temperature but open at high temperature in the early Universe [4]. The forbidden channel needs dark matter with sizable self-interaction, since they have Boltzmann suppression factor. It is discussed in section 3.1 in this review.

Strongly Interacting Massive Particles (SIMP) [5] have started drawing some attention recently because they can allow large self-interacting of light dark matter. The relic density of SIMP is determined by $3 \to 2$ self annihilations in the thermal freeze-out [6]. However, the Boltzmann suppression factor makes SIMP be on the verge of violating unitarity or perturbativity. Also it makes SIMP be in a tension with the bounds from Bullet cluster and halo shapes in the most of the parameter space [7].

In this review, we consider a complex scalar dark matter, dark photon and dark Higgs. We discuss dark matter can be produced from thermal freeze-out with forbidden channels and/or SIMP processes. In forbidden channels, annihilation cross section can be suppressed by a Boltzmann factor which depends on mass difference, $\Delta_i = (m_i - m_{\rm DM})/m_{\rm DM}$, between dark matter and dark photon or dark Higgs. We suggest a model with gauged Z_3 symmetry for forbidden and SIMP dark matter. Specific

⋆e-mail: soominchoi90@gmail.com
⋆⋆e-mail: yoojinkang91@gmail.com
⋆⋆⋆e-mail: hmlee71@gmail.com

range of mass differences allow dark matter to have a smaller self-interaction being compatible with the relic density, as compared to the SIMP case. In our model, the self-coupling contributes to both SIMP and forbidden channels.

2 Model for self-interacting dark matter

We consider a complex scalar dark matter χ and dark Higgs ϕ in dark local $U(1)_d$ model with charge $q_\chi = +1$ and $q_\phi = +3$ each. Local $U(1)_d$ symmetry is spontaneously broken to Z_3 which makes dark matter stable. Then, the dark Higgs ϕ has non-zero vacuum expectation value $\langle\phi\rangle = v_d/\sqrt{2}$ [8, 9].

The Lagrangian for SM singlet scalars, χ and ϕ, and the SM Higgs doublet H, is given by

$$\mathcal{L} = -\frac{1}{4}V_{\mu\nu}V^{\mu\nu} - \frac{1}{2}\sin\xi V_{\mu\nu}B^{\mu\nu} + |D_\mu\phi|^2 + |D_\mu\chi|^2 + |D_\mu H|^2 - V(\phi,\chi,H) \tag{1}$$

where $V_{\mu\nu} = \partial_\mu V_\nu - \partial_\nu V_\mu$, $D_\mu\phi = (\partial_\mu - iq_\phi g_d V_\mu)\phi$, $D_\mu\chi = (\partial_\mu - iq_\chi g_d V_\mu)\chi$, with dark gauge coupling g_d, and $D_\mu H = (\partial_\mu - ig'Y_H B_\mu - \frac{1}{2}igT^a W^a_\mu)H$. The second term means mixing between dark photon Z' and SM gauge boson Z. This $V(\phi,\chi,H) = V_{\rm DM} + V_{\rm SM}$ is the scalar potential and $V_{\rm SM}$ is the Standard Model Higgs potential. $V_{\rm DM}$ is

$$\begin{aligned} V_{\rm DM} &= -m_\phi^2|\phi|^2 + m_\chi^2|\chi|^2 + \lambda_\phi|\phi|^4 + \lambda_\chi|\chi|^4 + \lambda_{\phi\chi}|\phi|^2|\chi|^2 + \left(\frac{\sqrt{2}}{3!}\kappa\phi^\dagger\chi^3 + \text{h.c.}\right) \\ &\quad + \lambda_{\phi H}|\phi|^2|H|^2 + \lambda_{\chi H}|\chi|^2|H|^2. \end{aligned} \tag{2}$$

After the dark Higgs is expanded as $\phi = (v_d + h_d)/\sqrt{2}$, a triple coupling for χ is obtained. This coupling makes dark matter annihilate by $3 \to 2$ process.

3 Thermal freeze-out

We discuss the thermal production of light dark matter from process in combination with dark Higgs quartic coupling and Z' gauge coupling with the 2→2 forbidden channels and the 3→2 annihilation SIMP both by dark matter self-interactions. From that, we can find which process would be dominant process to determine the relic density depending on the mass difference between dark matter and dark photon and/or dark Higgs. Moreover, depending on dark photon and dark Higgs masses, annihilation into a pair of the SM particles also can be dominant. The SM-annihilation is discussed in the original paper [10]. In this review, we focus on comparison between forbidden channels and SIMP channels.

3.1 Forbidden channels

Although the dark photon and/or dark Higgs are heavier than dark matter, they can still contribute to the relic density for dark matter through the forbidden channels. We consider the case where dark photon and / or dark Higgs are light. We assume that dark photon and/or dark Higgs are in kinetic equilibrium during freeze-out process.

If we consider the annihilation cross section for forbidden channels for example $\chi\chi^* \to Z'Z'$ and $\chi\chi^* \to Z'\chi^*$, the detailed balance condition at high temperature are

$$\begin{aligned} \langle\sigma v\rangle_{\chi\chi^*\to Z'Z'} &= \frac{4(n_{Z'}^{\rm eq})^2}{(n_{\rm DM}^{\rm eq})^2}\langle\sigma v\rangle_{Z'Z'\to\chi\chi^*} \\ &= 9(1+\Delta_{Z'})^3 e^{-2\Delta_{Z'}x}\langle\sigma v\rangle_{Z'Z'\to\chi\chi^*} \end{aligned} \tag{3}$$

and similarly

$$\langle\sigma v\rangle_{\chi\chi\to Z'\chi^*} \quad = \quad 3(1+\Delta_{Z'})^3 e^{-2\Delta_{Z'}x}\langle\sigma v\rangle_{Z'\chi^*\to\chi\chi} \tag{4}$$

with $\Delta_{Z'} \equiv (m_{Z'} - m_\chi)/m_\chi$ and $x = m_\chi/T$.

From (4) and (5), we can see that the annihilation cross sections for forbidden channels depend on $e^{-\Delta_{Z'}x}$. Therefore, in the case with $m_\chi < m_{Z'} \ll m_{h_1}$, dark Higgs can't contribute to the relic density for dark matter. Similarly, in the case with $m_\chi < m_{h_1} \ll m_{Z'}$, dark photon can't contribute to the relic density for dark matter. However, the both dark photon and dark Higgs can contribute to the forbidden channels at the same time in the case with $m_\chi < m_{Z'} \sim m_{h_1}$. The first case are discussed below. The other casse are discussed with the first case in the original paper [10].

In the case with $m_\chi < m_{Z'} \ll m_{h_1}$, the Boltzmann equation is approximated to

$$\begin{aligned}\frac{dn_{\rm DM}}{dt} + 3Hn_{\rm DM} \quad &\approx \quad = -\frac{1}{2}\langle\sigma v\rangle_{\chi\chi^*\to Z'Z'} n_{\rm DM}^2 + 2\langle\sigma v\rangle_{Z'Z'\to\chi\chi^*}(n_{Z'}^{\rm eq})^2 \\ &\quad -\frac{1}{2}\langle\sigma v\rangle_{\chi\chi\to Z'\chi^*} n_{\rm DM}^2 + \langle\sigma v\rangle_{Z'\chi^*\to\chi\chi} n_{Z'}^{\rm eq} n_{\rm DM}\end{aligned} \tag{5}$$

It can be rewritten with (4) and (5). Then, the relic density is determined to be

$$\Omega_{\rm DM}h^2 = 5.20\times 10^{-10}{\rm GeV}^{-2}\left(\frac{g_*}{10.75}\right)^{-1/2}\left(\frac{x_f}{20}\right)e^{\Delta_{Z'}x_f}g(\Delta_{Z'},x_f) \tag{6}$$

and the $g(\Delta_{Z'}, x_f)$ is written in the original paper [10].

We put $\langle\sigma v\rangle_{Z'Z'\to\chi\chi^*} = a$ and $\langle\sigma v\rangle_{Z'\chi^*\to\chi\chi} = bv^2 = 6b/x$. Due to $e^{\Delta_{Z'}x_f}$, the $2\to 2$ annihilation cross section can be large and the self-scattering cross section of dark matter also can be large.

3.2 SIMP processes

Z' and dark Higgs are heavier than dark matter and they have small couplings to dark matter and SM particles, the $3\to 2$ annihilation process for dark matter becomes dominant.

Then, the Boltzmann equation is approximated to

$$\frac{dn_{\rm DM}}{dt} + 3Hn_{\rm DM} \approx -\langle\sigma v^2\rangle_{3\to 2}(n_{\rm DM}^3 - n_{\rm DM}^2 n_{\rm DM}^{\rm eq}). \tag{7}$$

Here, the effective $3\to 2$ annihilation cross section is obtained as

$$\langle\sigma v^2\rangle_{3\to 2} = \frac{1}{4}(\langle\sigma v^2\rangle_{\chi\chi\chi^*\to\chi^*\chi^*} + \langle\sigma v^2\rangle_{\chi\chi\chi\to\chi\chi^*}) \equiv \frac{\alpha_{\rm eff}^3}{m_\chi^5}. \tag{8}$$

As a result, the dark matter relic density is given by

$$\Omega_{\rm DM}h^2 = 1.41\times 10^{-8}{\rm GeV}^{-2}\left(\frac{g_*}{10.75}\right)^{-3/4}\left(\frac{x_f}{20}\right)^2\left(\frac{\alpha_{\rm eff}}{M_P^{1/3}m_\chi}\right)^{-3/2}. \tag{9}$$

From the result, the relic density depends on the ratio, $m_\chi/\alpha_{\rm eff}$. The $3\to 2$ self-annihilation cross sections are directly related with the self-scattering cross section, $\sigma_{\rm self} \sim \alpha_{\rm eff}^2/m_\chi^2$. Therefore, the self-scattering cross section can be large enough.

Especially, in SIMP process, we define the effective $2\to 2$ annihilation cross section by $\langle\sigma v\rangle_{{\rm eff},2\to 2} \equiv n_{\rm DM}\langle\sigma v^2\rangle_{3\to 2}$. Then, we can rewrite as below

$$\langle\sigma v\rangle_{{\rm eff},2\to 2} \approx g(x/2\pi)^3 T^3 e^{-x}\langle\sigma v^2\rangle_{3\to 2}. \tag{10}$$

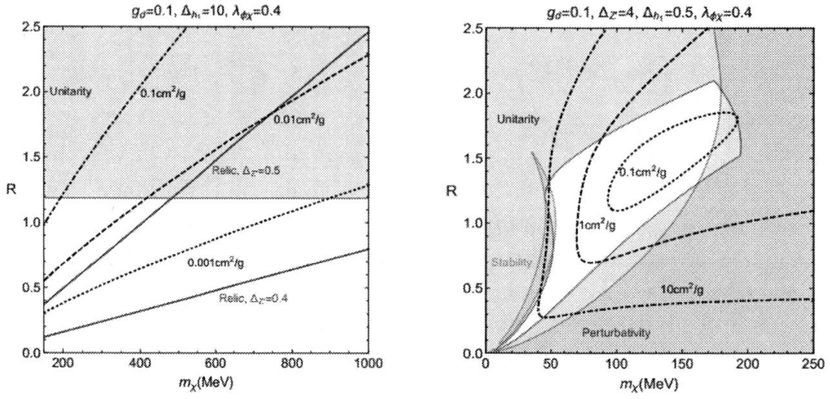

Figure 1. Parameter space of R vs m_χ, satisfying the relic density for forbidden channel (on left) and for SIMP process (on right) with $R = \sqrt{2}\kappa v_d/6m_\chi$. The black lines correspond to self-scattering cross sections, $\sigma_{\rm self}/m_\chi = 0.001, 0.01$ and 0.1 cm^2/g for left, 0.1, 1 and 10 cm^2/g for right.

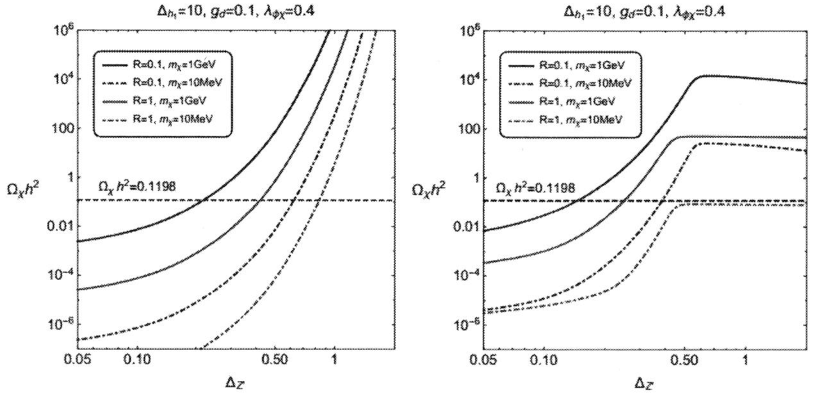

Figure 2. Dark matter relic density as a function of $\Delta_{Z'}$ for only forbidden channels (on left) for both SIMP and forbidden channels (on right). The central value of relic density, $\Omega_\chi h^2 = 0.1198$, is given by black dashed lines. We have taken $R = 0.1$ and $R = 1.0$ with $R = \sqrt{2}\kappa v_d/6m_\chi$.

In Fig.1, we note that SIMP process (on right) requires larger dark matter self-scattering cross section than the case with forbidden channels (on left). Besides, the Boltzmann supression factor in (10) is comparable with those in (4) and (5) for the forbidden channels. From this, we can infer that the forbidden channels can be dominant until $\Delta_{Z'} \sim 0.5$ and we check it exactly in the right side of Fig. 2 which is in the case with $m_\chi < m_{Z'} \ll m_{h_1}$.

4 Conclusion

We have considered the thermal production from freeze-out for self-interacting dark matter in the gauged Z_3 model with dark photon and dark Higgs. We discussed that the forbidden channels ($\chi\chi^* \rightarrow$

$Z'Z'$ and/or h_1h_1, $\chi\chi \to Z'\chi^*$ and/or $h_1\chi^*$), $3 \to 2$ self-annihilation and the Standard Model $2 \to 2$ annihilation can contribute to determining the relic density. We found that forbidden channels can produce light dark matter with sizable self-interation of dark matter to explain the small-scale problem. But the SIMP process favors lighter dark matter and larger self-interaction than forbidden channels. We identified the relevant thermal production mechanisms for self-interacting dark matter.

Acknowledgments

The work is supported in part by Basic Science Research Program through the National Research Foundation of Korea (NRF) funded by the Ministry of Education, Science and Technology (NRF-2016R1A2B4008759). The work of SMC is supported in part by TJ Park Science Fellowship of POSCO TJ Park Foundation. The work of YJK was supported by IBS under the project code, IBS-R018-D1.

References

[1] P. A. R. Ade *et al.* [Planck Collaboration], Astron.Astrophys. **594** A13 (2016)
[2] D. N. Spergel and P. J. Steinhardt, Phys. Rev. Lett. **84** 3760 (2000); W. J. G. de Blok, Adv. Astron **2010** 789293 (2010); M. Boylan-Kolchin, J. S. Bullock and M. Kaplinghat, Mon. Not. Roy. Astron. Soc. **415** L40 (2011)
[3] M Rocha, A. H. G. Peter, J. S. Bullock, M. Kaplinghat, S. Garriwon-Kimmel, J. Onorbe and L. A. Moustakas, Mon. Not. Roy. Astron. Soc. **430** 81 (2013)
[4] R. D'Agnolo and J. T. Ruderman, Phys.Rev.Lett. **115** 061301 (2015)
[5] Y. Hochberg, E. Kuflik, T. Volansky and J. G. Wacker, Phys. Rev. Lett. **113** 171301 (2014)
[6] E. D. Carlson, M. E. Machacek and L. J. Hall, Astrophys. J. **398** 43 (1992)
[7] H. M. Lee, M. S. Seo, ALP Conf. Proc. **1743** 060003 (2016)
[8] S. M. Choi and H. M. Lee, JHEP **1509** 063 (2015)
[9] P. Ko and Y. Tang, JCAP **1405** 04 (2014)
[10] S. M. Choi, Y. J. Kang and H. M. Lee, JHEP **1612** 099 (2016)
[11] L. Bouchet E. Jourdain, J. P. Roques, A. W. Strong, R. Diehl, F. Lebrun and R. Terrier, Astrophys. J. **670** 1315 (2008); A. W. Strong, I. V. Moskalenko and O. Reimer, Astrophys. J. **613** 962 (2004); M. Ackermann *et al.* [Fermi-LAT Collaboration], Astrophys. J. **750** 3 (2012); R. Essig E. Kufilik, S. D. McDermott, T. Volansky and K. M Zurek, JHEP **1311** 193 (2013)

Bell Inequality from Holographic Gravity

Sichun Sun[1] and *Yun-Long* Zhang[2,⋆]

[1] *Center for Theoretical Sciences, National Taiwan University, Taipei 10617, Taiwan*
[2] *Asia Pacific Center for Theoretical Physics, Pohang 790-784, Korea*

Abstract. We study a holographic model of an EPR pair at the boundary of bulk gravity, and use Bell inequality as a sharp test of entanglement. By revealing how Bell inequality is violated by gravity in the bulk, our study sheds light on the entanglement of the original ER=EPR conjecture.

1 Introduction

This proceeding is mainly base on [1], where the Bell inequality from holographic gravity is constructed. Correlations in local classical theories are bounded by the Bell inequality, however, it can be violated by the presence of the non-local entanglement in quantum mechanics [2]. The violation of Bell inequality in the entangled Einstein-Podolsky-Rosen (EPR) pair [3] indicates that two particles have an "instant interaction", in contrast to the hidden variable theories that preserve strict locality [2–6]. In addition to the Bell's tests in laboratories, there are also discussions of detections in cosmological scales, to confirm that whether the inflated primordial fluctuations are quantum mechanical [7–9]. Recently, the ER=EPR conjecture is proposed by Maldacena and Susskind [10, 11]. It stated that the quantum entanglement of the EPR pair can be attributed to the non-traversable Einstein-Rosen (ER) bridge that connects the pair [12]. This conjecture was proposed to resolve the black hole information paradox without resorting to a firewall [13] surround the black hole horizon.

In this work, instead of working with the original ER=EPR setup, we employ a concrete holographic model of the EPR pair proposed by Jensen and Karch in Ref [14] based on the Anti-de Sitter/Conformal Field Theory(AdS/CFT) correspondence [15] and the analytical solution in [16]. The two particles of the boundary EPR pair are connected by a string in the AdS background of the bulk, with an effective ER bridge on the string worldsheet. Therefore, this is a holographic realization of ER=EPR conjecture.

We demonstrate that the Bell inequality violated by the EPR pair living at the boundary can also be violated by the gravitational theory with an ER bridge living in the bulk. Although the holographic setting is different from the original ER=EPR conjecture in which both ER and EPR live in the same spacetime dimensions, our study does shed light on how entanglement can be captured by a classical theory.

2 Holographic EPR Pair

In this section, we review the holographic model of the EPR pair in Ref. [14]. The model proposed that an entangled color singlet quark anti-quark (q-$\bar{q}$) pair in $\mathcal{N}$=4 supersymmetric Young-Mills theory (SYM) can be described by an open string with both of its endpoints attached to the boundary of AdS_5. There is no confinement in this theory and the pair can separate arbitrarily far away from each other. Relevant studies of this model can also be found in [17–23].

There are numerical solutions of the string shapes with different boundary behaviors [24–26]. We will work with the analytic solution for an accelerating string treated in the probe limit at the AdS background [16]. In the analytic solution the open string is accelerated on the Poincáre patch of the AdS_5 with radius L,

$$ds^2 = \frac{L^2}{w^2}[-dt^2 + dw^2 + (dx^2 + dy^2 + dz^2)]. \qquad (1)$$

The string solution in the bulk gravity is given by

$$z^2 = t^2 + b^2 - w^2. \qquad (2)$$

The quark and anti-quark live on the AdS boundary $w = 0$. They are accelerating along the $\pm z$ direction, respectively, with the solution $z = \pm\sqrt{t^2 + b^2}$. Therefore, the two entangled particles are out of causal contact with each other the whole time.

String fluctuations.— To consider the string fluctuations, we transform the solution to the co-moving spacetime $(\tilde{\tau}, \tilde{r}, x, y, \tilde{z})$ of the accelerating quarks via

$$|z| = b\sqrt{1-\tilde{r}e^{\tilde{z}}}\cosh\tilde{\tau},$$
$$t = b\sqrt{1-\tilde{r}e^{\tilde{z}}}\sinh\tilde{\tau},$$

⋆Corresponding Author and Speaker: yunlong.zhang@apctp.org.
Proceeding for the 13th International Conference on Gravitation, Astrophysics, and Cosmology the 15th Italian-Korean Symposium on Relativistic Astrophysics: A Joint Meeting. Parallel Talk on July 6, 2017, Seoul, Korea.

$$w = b\sqrt{\tilde r}e^{\tilde z}, \quad x = b\,\tilde x, \quad y = b\,\tilde y. \tag{3}$$

These two frames, which cover the regions $z \geq 0$ and $z \leq 0$ separately, are accelerating frames with a constant acceleration $a = 1/b$ along opposite directions of z. And (3) only maps the upper part of the string ($0 < w < b$) into the proper frames of the accelerating quarks with $0 < \tilde r < 1$. Plug this transformation (3) in the string solution (2), the string configuration becomes $\tilde z = 0$ for both frames of the quark and anti-quark. Under this transformation (3), the metric (1) becomes

$$\mathrm{d}s^2 = \frac{L^2}{\tilde r}\Big[-f(\tilde r)\mathrm{d}\tilde\tau^2 + \frac{1}{4\tilde r}\frac{\mathrm{d}\tilde r^2}{f(\tilde r)} + e^{-2\tilde z}\left(\mathrm{d}\tilde x^2 + \mathrm{d}\tilde y^2\right) + \mathrm{d}\tilde z^2\Big], \tag{4}$$

where $f(\tilde r) = 1 - \tilde r$. Furthermore, with respect to the time $\tau = b\tilde\tau$, the Hawking temperature $T_H = \frac{1}{2\pi b}$ matches with the Unruh temperature $T_a = \frac{a}{2\pi}$ [16, 23] and we have set the reduced Planck constant $\hbar$ and Boltzmann constant k_B to be unit. Therefore, there is an event horizon at $\tilde r = 1$ associated with the quark and another event horizon also at $\tilde r = 1$ associated with the anti-quark. As shown in Fig. 1, the two horizons are connected by part of the string which can be seen as ER bridge. Hence it is suggested to be a holographic realization of the ER=EPR conjecture [10, 11].

The measurement of the quarks' states will introduce fluctuations to the world lines of quarks which set the boundary conditions for the worldsheet of the string fluctuations. Let $(\tilde\tau, \tilde r)$ be the new worldsheet coordinates in the current frame, then the string fluctuation is $X^\mu(\tilde\tau, \tilde r) = (\tilde\tau, \tilde r, \tilde q_i(\tilde\tau, \tilde r))$, with $i = (\tilde x, \tilde y, \tilde z)$. When $\tilde q_i \ll 1$, the Nambu-Goto action of string with tension T_s becomes

$$S \simeq -T_s L^2 \int \frac{d\tilde\tau d\tilde r}{2\tilde r^{3/2}}\left\{1 + \Big[2\tilde r f(\tilde r)\tilde q_i'\tilde q_j' - \frac{1}{2f(\tilde r)}\dot{\tilde q}_i\dot{\tilde q}_j\Big]h^{ij}\right\}, \tag{5}$$

where $\tilde q_i' \equiv \frac{\partial \tilde q_i}{\partial \tilde r}$, $\dot{\tilde q}_i \equiv \frac{\partial \tilde q_i}{\partial \tilde\tau}$ and $h^{ij} = \mathrm{diag}[1, e^{2\tilde z}, e^{2\tilde z}]$. The equations of motion for the fluctuations on the string are

$$\partial_{\tilde r}\Big(\frac{2f\tilde q_i'}{\tilde r^{1/2}}\Big) - \partial_{\tilde\tau}\Big(\frac{\dot{\tilde q}_i}{2f\tilde r^{3/2}}\Big) = 0. \tag{6}$$

Focusing on the transverse fluctuations $i = \tilde x, \tilde y$,

$$\tilde q_i(\tilde\tau, \tilde r) = \int \frac{d\omega}{2\pi}e^{-i\omega\tilde\tau}\tilde q_i(\omega)Y_\omega(\tilde r), \tag{7}$$

where $\tilde q_i(\omega)$ is the Fourier transform of fluctuation on the boundary. With these we can study the Green's function holographically.

3 Constructing Bell's inequality

The retarded Green's function of the quark under effective random force $\mathcal{F}^i(\tau)$ can be defined as $\mathbf{i}G^{ij}_{\mathcal{R}}(\tau) = \theta(\tau)\langle[\mathcal{F}^i(\tau), \mathcal{F}^j(0)]\rangle$ [16]. In the AdS/CFT correspondence, $\mathcal{F}^i(\tau)$ is the operator conjugate to the fluctuations $q_i(\tau)$, where the dimensionful quantities are $\tau = b\tilde\tau, q_i = b\tilde q_i$. In the low frequency limit $\omega \to 0$, it can be obtained analytically as

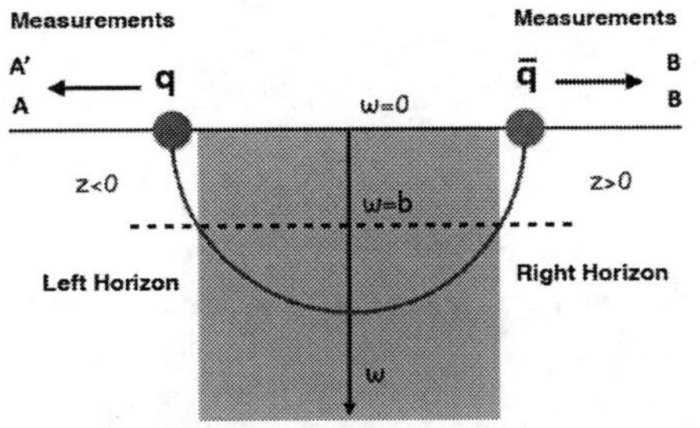

Figure 1. Schematic diagram for the quark(q)-antiquark($\bar q$) EPR pair in holography at a fixed time $t > 0$. The trajectory of left(right) worldsheet horizon is on $w = b$, which depicts the intersection of the string with the world volume horizons seen by $q(\bar q)$ in its co-moving frame. The world volume horizons are the two vertical boundary of the gray region. The Bell inequality test is performed by the measurements at Alice and Bob's locations.

$$\begin{aligned} G^{ij}_{\mathcal{R}}(\omega) &= -\frac{2T_s L^2}{b^2\tilde r^{1/2}}f(\tilde r)Y_{-\omega}(\tilde r)\partial_{\tilde r}Y_\omega(\tilde r)\delta^{ij}\Big|_{\tilde r\to 0} \\ &= -\frac{a^2\sqrt{\lambda}}{2\pi}\mathbf{i}\omega\delta^{ij} + O(\omega^2), \end{aligned} \tag{8}$$

and we have used the fact that $T_s L^2 = \frac{\sqrt{\lambda}}{2\pi}$.

What we need for the Bell's test is the contour time ordered Schwinger-Keldysh (SK) Green's function,

$$\mathbf{i}G^{ij}_{AB}(\tau, x) = \langle\mathcal{F}^i_A(\tau, x)\mathcal{F}^j_B(0)\rangle, \tag{9}$$

where $\mathcal{F}^i_A$ and $\mathcal{F}^j_B$ are separately defined on the causally disconnected left and right wedges of the Penrose diagram, corresponding to the boundaries of different patches of the AdS space. This off-diagonal SK propagator is examined in the Supplemental Material and found to be related to the holographic retarded Green's function

$$G^{ij}_{AB}(\omega) = \frac{2\mathbf{i}e^{-\omega/(2T_a)}}{1 - e^{-\omega/T_a}}\mathrm{Im}G^{ij}_{\mathcal{R}}(\omega), \tag{10}$$

similar to what was found in Refs. [27–29] but with different settings.

For fluctuations coming from two causally separated quarks of an EPR pair along x, y directions, and in the low frequency limit $\omega \to 0$,

$$\mathbf{i}G^{xx}_{AB} = \mathbf{i}G^{yy}_{AB} = \frac{\sqrt{\lambda}a^3}{2\pi^2}, \quad \mathbf{i}G^{xy}_{AB} = \mathbf{i}G^{yx}_{AB} = 0, \tag{11}$$

which indicates that the spatial correlator $G^{ij}_{AB} \propto \delta^{ij}$. The $\sqrt{\lambda}$ factor is consistent with the observation that the entanglement entropy of the entangled pair is of order $\sqrt{\lambda}$ [14]. It is also interesting that this SK correlator does not vanish when the quarks are separated at long distance. This is consistent with the non-local nature of entanglement.

To study the correlators, we normalize the operators such that only the dependence on the spin wave function remains:

$$A_{\mathcal{F}} = (\cos\theta_A\mathcal{F}^x_A + \sin\theta_A\mathcal{F}^y_A)/\langle\mathcal{F}^x_A\mathcal{F}^x_B\rangle^{1/2},$$

$$B_{\mathcal{F}} = (\cos\theta_B \mathcal{F}_B^x + \sin\theta_B \mathcal{F}_B^y)/\langle \mathcal{F}_A^x \mathcal{F}_B^x \rangle^{1/2}, \qquad (12)$$

The mixed measurements for correlators in CHSH correlation formulation become

$$\langle A_{\mathcal{F}} B_{\mathcal{F}} \rangle = \cos(\theta_A - \theta_B) \equiv \cos\theta_{AB}. \qquad (13)$$

Together with the similar normalization of the operators $A'_{\mathcal{F}}$ and $B'_{\mathcal{F}}$, the CHSH correlation formulations becomes

$$\begin{aligned}\langle C_{\mathcal{F}} \rangle &= \langle A_{\mathcal{F}} B_{\mathcal{F}} \rangle + \langle A_{\mathcal{F}} B'_{\mathcal{F}} \rangle + \langle A'_{\mathcal{F}} B_{\mathcal{F}} \rangle - \langle A'_{\mathcal{F}} B'_{\mathcal{F}} \rangle \\ &= \cos\theta_{AB} + \cos\theta_{AB'} + \cos\theta_{A'B} - \cos\theta_{A'B'}. \qquad (14)\end{aligned}$$

For example, when $\theta_{AB} = \theta_{AB'} = \theta_{A'B} = \pi/4$, and $\theta_{A'B'} = 3\pi/4$, we can reach the maximum value $2\sqrt{2}$.

In this derivation, we see the bulk string fluctuations, which come from classical gravity, reproduce the quantum entanglement of an EPR pair on the boundary. The equation of motion of the classical string, Eq.(6), has no coupling between $\tilde{q}_x$ and $\tilde{q}_y$ such that Eq.(11) follows. This can be obtained as long as the string does not experience a force to propagate the fluctuation in the x-direction to the y-direction which breaks parity in general. It seems once these two conditions are satisfied, it does not matter whether there is an ER bridge in the bulk. Hence it is conceivable Bell inequality can still be violated in a holographic model where the EPR pair does not accelerate, similar to how holographic entanglement entropy is computed in a static system [30, 31].

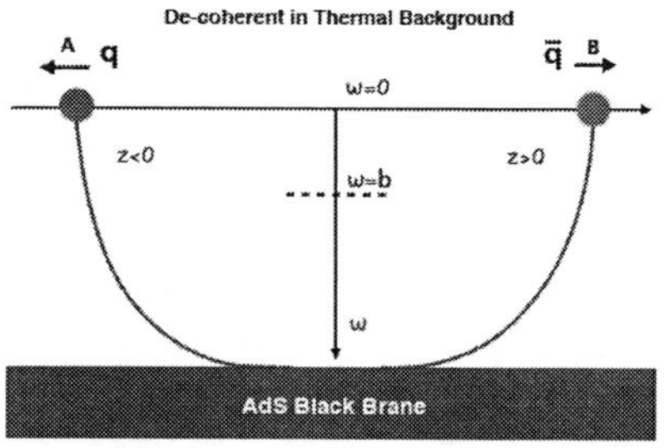

Figure 2. Schematic diagram for the decoherence of quark(q)-antiquark($\bar{q}$) EPR pair in holography at late time. The string breaks after it fell into the black brane. This process correspondends to the decoherence of the quark and anti quak pairs induced by thermal background.

4 Time Evolution and Discussion

It is interesting to consider the decoherence of the EPR pair in the environment. If the environmental effect can be described by thermal fluctuations, then we can add a black brane to the bulk of our model. When the distance of the EPR pair increases with time, the ER bridge also approaches the black brane horizon and then enters the horizon [32]. We expect the ER bridge breaks after it enters the horizon which might shed light on the decoherence process in the boundary field theory. As shown in Fig. 2,

We would also like to comment on the finite temperature effects on the accelerating string solution in AdS_5 black brane spacetime. Although there is no analytical solution of the string expansion, qualitatively the string will touch the horizon of the black hole at late time. There is a semi-quantitative discussion in the small temperature limit [32], that the Hawking temperature of the black hole is much smaller that the Unruh temperature of the accelerating particle $T_H \ll T_a$. An external electric field is imposed to accelerate the quark-antiquark pair from $t = 0$ to $t = \infty$. It is expected at late time energy is flowing into the system from the top of the string to the bottom part, and then being absorbed by the black brane. As time increases, the acceleration slows down and the string trajectory starts to evolve towards the trailing string solution. This corresponds to that the Einstein-Rose bridge is broken up by the horizon of AdS black brane. And for the static string solution in the AdS black brane where two straight strings is hanging down to the horizon, there is no entanglement in the dual state [24]. Other approach to describe quantum operations in holography can also be found in [31].

For future work, it is interesting to consider the back reaction by the measurements and see whether the ER bridge is broken due to the energy injected by measurements. This might provide an opportunity to study the "wave function collapse" typically used to describe how measurements change the states. For our case in pure AdS, if we consider the back reaction after the measurements, the ER bridge might be broken due to the energy injected by measurements, similar to the case with black brane. This is aligned with the fact that the wave function of the EPR pair will collapse after the measurements. Since the string is in probe limit, it is not explicit that the bridge will break due to the string perturbations.

The ER=EPR conjecture is proposed to resolve the black hole information paradox without introducing a firewall in the black hole. The conjecture implies that entanglement of the EPR pair, which is thought to be a quantum mechanical effect, can be captured by classical gravity through the ER bridge. Using Bell inequality as a sharp test of entanglement, we study a holographic model with an EPR pair at the boundary and an ER bridge in the bulk. By revealing how Bell inequality is violated by classical gravity in the bulk, our study sheds light on the possible conditions needed for the entanglement of the original ER=EPR.

Acknowledgements — Y. L. Zhang thanks the supported by APCTP for the conference. We thank D. Berenstein, J. W. Chen, A. Karch, Y. Neiman for helpful communication.

References

[1] J. W. Chen, S. Sun and Y. L. Zhang, "Holographic Bell Inequality," arXiv:1612.09513 [hep-th].

[2] J. S. Bell, "On the Einstein-Podolsky-Rosen paradox," Physics **1**, 195 (1964).

[3] A. Einstein, B. Podolsky and N. Rosen, "Can quantum mechanical description of physical reality be considered complete?," Phys. Rev. **47**, 777 (1935).

[4] J. F. Clauser, M.A. Horne, A. Shimony, R.A. Holt, "Proposed experiment to test local hidden-variable theories," Phys. Rev. Lett. **23**, 880 (1969).

[5] B. S. Cirelson, "Quantum generalizations of Bell's inequality," Lett. Math. Phys. **4**, 93 (1980).

[6] J. B. Hartle, "Space-time quantum mechanics and the quantum mechanics of space-time," gr-qc/9304006.

[7] J. Maldacena, "A model with cosmological Bell inequalities," Fortsch. Phys. **64**, 10 (2016) [arXiv:1508.01082 [hep-th]].

[8] S. Choudhury, S. Panda and R. Singh, "Bell violation in the Sky," Eur. Phys. J. C **77**, no. 2, 60 (2017) [arXiv:1607.00237 [hep-th]].

[9] J. W. Chen, S. H. Dai, D. Maity, S. Sun and Y. L. Zhang, "Towards Searching for Entangled Photons in the CMB Sky," arXiv:1701.03437 [quant-ph].

[10] J. Maldacena and L. Susskind, "Cool horizons for entangled black holes," Fortsch. Phys. **61**, 781 (2013) [arXiv:1306.0533 [hep-th]].

[11] L. Susskind, "Copenhagen vs Everett, Teleportation, and ER=EPR," Fortsch. Phys. **64**, no. 6-7, 551 (2016) [arXiv:1604.02589 [hep-th]].

[12] A. Einstein and N. Rosen, "The Particle Problem in the General Theory of Relativity," Phys. Rev. **48**, 73 (1935).

[13] A. Almheiri, D. Marolf, J. Polchinski and J. Sully, "Black Holes: Complementarity or Firewalls?," JHEP **1302**, 062 (2013) [arXiv:1207.3123 [hep-th]].

[14] K. Jensen and A. Karch, "Holographic Dual of an Einstein-Podolsky-Rosen Pair has a Wormhole," Phys. Rev. Lett. **111**, no. 21, 211602 (2013) [arXiv:1307.1132 [hep-th]].

[15] J. M. Maldacena, "The Large N limit of superconformal field theories and supergravity," [Adv. Theor. Math. Phys. **2**, 231 (1998)] [hep-th/9711200].

[16] B. W. Xiao, "On the exact solution of the accelerating string in AdS(5) space," Phys. Lett. B **665**, 173 (2008) [arXiv:0804.1343 [hep-th]].

[17] J. Sonner, "Holographic Schwinger Effect and the Geometry of Entanglement," Phys. Rev. Lett. **111**, no. 21, 211603 (2013) [[arXiv:1307.6850 [hep-th]].

[18] K. Jensen, A. Karch and B. Robinson, "Holographic dual of a Hawking pair has a wormhole," Phys. Rev. D **90**, no. 6, 064019 (2014) [arXiv:1405.2065 [hep-th]].

[19] K. Jensen and J. Sonner, "Wormholes and entanglement in holography," Int. J. Mod. Phys. D **23**, no. 12, 1442003 (2014) [arXiv:1405.4817 [hep-th]].

[20] M. Chernicoff, A. Güijosa and J. F. Pedraza, "Holographic EPR Pairs, Wormholes and Radiation," JHEP **1310**, 211 (2013) [arXiv:1308.3695 [hep-th]].

[21] A. Karch and S. Sun, "Matrix Flavor Brane and Dual Wilson Line," Phys. Rev. D **89**, no. 6, 066008 (2014) [arXiv:1312.2694 [hep-th]].

[22] T. Hirayama, P. W. Kao, S. Kawamoto and F. L. Lin, "Unruh effect and Holography," Nucl. Phys. B **844**, 1 (2011) [arXiv:1001.1289 [hep-th]].

[23] E. Caceres, M. Chernicoff, A. Guijosa and J. F. Pedraza, "Quantum Fluctuations and the Unruh Effect in Strongly-Coupled Conformal Field Theories," JHEP **1006**, 078 (2010) [arXiv:1003.5332 [hep-th]].

[24] C. P. Herzog, A. Karch, P. Kovtun, C. Kozcaz and L. G. Yaffe, "Energy loss of a heavy quark moving through N=4 supersymmetric Yang-Mills plasma," JHEP **0607**, 013 (2006) [hep-th/0605158].

[25] M. Chernicoff and A. Guijosa, "Acceleration, Energy Loss and Screening in Strongly-Coupled Gauge Theories," JHEP **0806**, 005 (2008) [arXiv:0803.3070 [hep-th]].

[26] P. M. Chesler, K. Jensen and A. Karch, "Jets in strongly-coupled N = 4 super Yang-Mills theory," Phys. Rev. D **79**, 025021 (2009) [arXiv:0804.3110 [hep-th]].

[27] D. T. Son and A. O. Starinets, "Minkowski-space correlators in AdS/CFT correspondence: Recipe and applications," JHEP **0209**, 042 (2002). [arXiv:hep-th/0205051].

[28] C. P. Herzog and D. T. Son, "Schwinger-Keldysh propagators from AdS/CFT correspondence," JHEP **0303**, 046 (2003). [arXiv:hep-th/0212072].

[29] D. T. Son and A. O. Starinets, "Viscosity, Black Holes, and Quantum Field Theory," Ann. Rev. Nucl. Part. Sci. **57**, 95 (2007) [arXiv:0704.0240 [hep-th]].

[30] S. Ryu and T. Takayanagi, "Holographic derivation of entanglement entropy from AdS/CFT," Phys. Rev. Lett. **96**, 181602 (2006) [hep-th/0603001].

[31] T. Numasawa, N. Shiba, T. Takayanagi and K. Watanabe, "EPR Pairs, Local Projections and Quantum Teleportation in Holography," JHEP **1608**, 077 (2016) [arXiv:1604.01772 [hep-th]].

[32] F. Dominguez, C. Marquet, A. H. Mueller, B. Wu and B. W. Xiao, Nucl. Phys. A **811**, 197 (2008) [arXiv:0803.3234 [nucl-th]].

EPJ Web of Conferences **168**, 07002 (2018) https://doi.org/10.1051/epjconf/201816807002
Joint International Conference of ICGAC-XIII and IK-15 on Gravitation, Astrophysics and Cosmology

Exact Holography of Massive M2-brane Theories and Entanglement Entropy

Dongmin Jang[1,*], *Yoonbai* Kim[1,**], *O-Kab* Kwon[1,***], and *D. D.* Tolla[1,2,****]

[1] *Department of Physics, BK21 Physics Research Division, Institute of Basic Science, Sungkyunkwan University, Suwon 440-746, South Korea*
[2] *University College, Sungkyunkwan University, Suwon 440-746, South Korea*

Abstract. We test the gauge/gravity duality between the $\mathcal{N} = 6$ mass-deformed ABJM theory with $U_k(N) \times U_{-k}(N)$ gauge symmetry and the 11-dimensional supergravity on LLM geometries with $SO(4)/\mathbb{Z}_k \times SO(4)/\mathbb{Z}_k$ isometry. Our analysis is based on the evaluation of vacuum expectation values of chiral primary operators from the supersymmetric vacua of mass-deformed ABJM theory and from the implementation of Kaluza-Klein (KK) holography to the LLM geometries. We focus on the chiral primary operator (CPO) with conformal dimension $\Delta = 1$. The non-vanishing vacuum expectation value (vev) implies the breaking of conformal symmetry. In that case, we show that the variation of the holographic entanglement entropy (HEE) from it's value in the CFT, is related to the non-vanishing one-point function due to the relevant deformation as well as the source field. Applying Ryu Takayanagi's HEE conjecture to the 4-dimensional gravity solutions, which are obtained from the KK reduction of the 11-dimensional LLM solutions, we calculate the variation of the HEE. We show how the vev and the value of the source field determine the HEE.

*e-mail: dongmin@skku.edu
**e-mail: yoonbai@skku.edu
***e-mail: okab@skku.edu
****e-mail: ddtolla@skku.edu

1 Introduction

We consider a non conformal quantum field theory and test the gauge/gravity duality away from the conformal fixed point in the large N limit. We start with the $\mathcal{N} = 6$ $U_k(N)\times U_{-k}(N)$ Aharony-Bergman-Jafferis-Maldacena (ABJM) theory with Chern-Simons level k [1], as the CFT at the UV fixed point. The ABJM theory allows the supersymmetry preserving mass deformation and the deformed theory (mABJM) [2, 3] has discrete Higgs vacua presented by the Gomis, Rodriguez-Gomez, Van Raamsdonk, Verlinde (GRVV) matrices [3]. It was known that the vacua of the mABJM theory have one-to-one correspondence [4, 5] with the half BPS Lin-Lunin-Maldacena (LLM) geometries [6, 7] having $SO(2,1)\times SO(4)/\mathbb{Z}_k\times SO(4)/\mathbb{Z}_k$ isometry in 11-dimensions [5, 8]. Since the mABJM theory is obtained by a relevant deformation from the ABJM theory at the UV fixed point, the dual geometry should be asymptotically $AdS_4 \times S^7/\mathbb{Z}_k$. We test this gauge/gravity duality, by calculating the the vacuum expectation values (vevs) of the chiral primary operator with conformal dimension $\Delta = 1$ using both sides of the duality.

The novelty of gauge/gravity duality is that, quantities which are difficult to calculate on one side of the duality can easily be obtained on the other side. One such case is the calculation of entanglement entropy in dimension higher than two, which is very difficult to obtain from the quantum field theory. However, Ryu and Takanayagi (RT) suggested a holographic method of entanglement entropy calculation, for d-dimensional conformal field theory from the AdS_{d+1} solution of the dual gravity [9]. In this letter, we extend RT conjecture to non-conformal case and calculate the entanglement entropy for mABJM from an asymptotically AdS_4 solution of 4-dimensional supergravity, which is obtained by the KK reduction of the LLM solutions.

2 Construction of 4-dimensional Gravity

A 4-dimensional gravity theory on an asymptotically AdS_4 background can be constructed from the KK reduction of 11-dimensional gravity on the LLM geometries. The KK reduction procedure involves, expansion of the 11-dimensional fields in terms of the spherical harmonics on S^7 and then projecting the 11-dimensional equations of motion on the appropriate spherical harmonic elements. The resulting equations are diagonalized to obtain the equations for infinite towers of gauge invariant KK modes. In general, this procedure results in the field equations which contain higher derivative terms. In order to absorb the higher derivative terms and write the canonical 4-dimensional equations of motion for those KK modes, we need to introduce some field redefinitions. Such field redefinitions are some times called the KK maps. See [11, 12] for details.

Following the above procedure and considering the small mass parameter expansion, only the equations of motion of few KK modes, which includes the 4-dimensional graviton mode $H_{\mu\nu}$, one scalar mode Ψ, and one pseudo scalar mode T, are non trivial at quadratic order in the mass parameter. These and all the other 4-dimensional fields are related to the 11-dimensional fields by the KK maps. The equations of motion of those three modes are

$$\begin{aligned}
&\left(\Box - M_T^2\right)T = 0, \qquad \left(\Box - M_\Psi^2\right)\Psi = 0, \quad \text{with } M_T^2 = M_\Psi^2 = -\frac{8}{L^2}\\
&\frac{1}{2}\left(-\Box H_{\mu\nu} + \nabla^\rho\nabla_\mu H_{\nu\rho} + \nabla^\rho\nabla_\nu H_{\mu\rho} - \nabla_\mu\nabla_\nu H\right) + \frac{12}{L^2}H_{\mu\nu}\\
&+ 8\pi G_N A_T\left(\nabla_\mu T\nabla_\nu T + \frac{M_T^2}{2}g^{(0)}_{\mu\nu}T^2\right) + 8\pi G_N A_\Psi\left(\nabla_\mu\Psi\nabla_\nu\Psi + \frac{M_\Psi^2}{2}g^{(0)}_{\mu\nu}\Psi^2\right) = 0, \qquad (1)
\end{aligned}$$

where $g^{(0)}_{\mu\nu}$ is metric of the pure AdS_4, L is the radius of S^7 in 11-dimensions, and G_N is the Newtion's constant in 4-dimensions. For the detailed definitions of parameters, see [11, 12].

Asymptotically AdS_4 solutions of the 4-dimensional equations of motion are obtained from the asymptotic expansion of the LLM solutions in 11-dimensions. Performing the expansion we obtain the following results

$$H_{ij} = \left[-\frac{(L\mu_0)^2}{180}\left(30+\beta_3^2\right)+O\left(\mu_0^4\right)\right]\eta_{ij} \quad H_{zz} = -\frac{(L\mu_0)^2}{1440}\left(960+29\beta_3^2\right)+O\left(\mu_0^4\right),$$
$$\Psi = -24\beta_3\mu_0 z + O(\mu_0^3), \quad T = 16\sqrt{3}\,\mu_0 z + O(\mu_0^3), \tag{2}$$

where μ_0 is the mass parameter of the LLM solutions and β_3 parametrizes all LLM solutions. We have verified that Ψ is dual to CPO of conformal dimension one, while T is dual to a gauge invariant operator of conformal dimension two. Using these 4-dimensional gravity results, we examine the vev of chiral primary operators and holographic entanglement entropy in mABJM theory.

3 Exact Holography

In mABJM theory, there is a unique CPO with conformal dimension one ($O^{(1)}$) that have a non-vanishing vev [11, 12]. Following the standard perturbative expansion in mABJM theory, we can calculate the one point function for such CPO and express the result as an expansion in $1/N$. In the large N limit, only the leading terms which is completely determined by the discrete Higgs vacua of the mABJM theory is relevant. See [11, 12] for details. The result is

$$\langle O^{(1)}\rangle_m = \frac{k\mu}{4\sqrt{2}\pi}\sum_{n=0}^{2N_B+1} n(n+1)(N_n - N_n'), \tag{3}$$

where $\{N_n, N_n'\}$ are the occupation numbers of the discrete Higgs vacua, and $\mu = 4\mu_0$ is the mass parameter in the mABJM theory. This result is identified with the coefficient of the asymptotic expansion of the dual scalar field Ψ in (2), by using the one-to-one correspondence between the discrete Higgs vacua and the LLM solutions. Then, the vev is given by

$$\langle O^{(1)}\rangle = \frac{N^{\frac{3}{2}}\mu_0}{3\sqrt{2}\pi}\beta_3. \tag{4}$$

4 Holographic Entanglement Entropy

According to the RT conjecture, the HEE with a subsystem A on the boundary of $(d+1)$-dimensional AdS geometry is given by $S_A = \frac{\mathrm{Min}(\gamma_A)}{4G_N}$, where G_N is the Newton constant in the $(d+1)$-dimensional gravity theory and γ_A is an area of the surface stretched to the bulk direction, which has the same boundary with the subsystem A [9]. The 4-dimensional metric obtained in section 2 is a perturbation of pure AdS_4, $g_{\mu\nu} = g^{(0)}_{\mu\nu} + H_{\mu\nu}$. Then the induced metric is given by

$$\tilde{g}_{ij} = \frac{\partial x^\mu}{\partial\sigma^i}\frac{\partial x^\nu}{\partial\sigma^j}g_{\mu\nu} = \frac{\partial x^\mu}{\partial\sigma^i}\frac{\partial x^\nu}{\partial\sigma^j}\left(g_{\mu\nu}+H_{\mu\nu}\right) = \bar{g}_{ij} + \tilde{H}_{ij}. \tag{5}$$

The deviation $\tilde{H}_{ij}$ of the induced metric from its AdS_4 value results in the variation of the area for the surface whose boundary is a disk of radius l,

$$\delta\gamma_A = \frac{1}{2}\int d^2\sigma\sqrt{\det\bar{g}_{ij}}\,\bar{g}^{ij}\tilde{H}_{ij} = \pi\int_0^l d\rho\sqrt{\det\bar{g}_{ij}}\,\bar{g}^{ij}\tilde{H}_{ij}. \tag{6}$$

Inserting the results in (2) into (5), we obtain the expansions of $\tilde{H}_{ij}$ up to μ_0^2-order. Using those expansions in (6), we obtain the variation of the minimum area,

$$\delta\gamma_A = -\frac{\pi L^2\mu_0^2}{1440}\int_0^l d\rho \frac{\rho}{\sqrt{1+z'^2}}\left[\left(1200+37\beta_3^2\right)z'^2+16\left(30+\beta_3^2\right)\right] = -\frac{\pi L^2(\mu_0 l)^2}{48}\left(32+\beta_3^2\right).$$

Therefore, the HEE up to μ_0^2-order is given by

$$S_A = S_A^{(0)} + \delta S_A = \frac{\pi L^2}{8G_N}\left[\frac{l}{\epsilon} - 1 - \frac{4}{3}\left(1+\frac{\beta_3^2}{32}\right)(\mu_0 l)^2\right]. \tag{7}$$

We see that the solutions of the 4-dimensional metric (2) for all possible droplets exactly reproduce the HEE which was calculated from the 11-dimensional LLM solutions [10].

Since the metric deformation in (2) is due to the presence of the matter fields Ψ and T, we conclude that the β_3-term in δS_A in (7) is originated from the *vevs* of the CPO $\mathcal{O}^{(1)}$, which is dual to the scalar field Ψ, while the constant term in δS_A is originated from the *source* of a gauge invariant operator $\tilde{\mathcal{O}}^{(2)}$, which is dual to the pseudo scalar field T. Therefore, we see the HEE up to μ_0^2-order in the large N limit is the function of the source and the vevs, $\delta S_A = \delta S_A(J_{\tilde{\mathcal{O}}^{(2)}}, \langle \mathcal{O}^{(1)}\rangle)$, where $J_{\tilde{\mathcal{O}}^{(2)}}$ denotes the source of $\tilde{\mathcal{O}}^{(2)}$.

5 Conclusion

In this letter, we carried out the KK reduction and the holographic renormalization procedure for the mABJM theory and the LLM geometry in 11-dimensional supergravity. By calculating the vevs of CPO with $\Delta = 1$ in field theory side and the corresponding asymptotic coefficients in gravity side, we found a supporting evidence for an exact gauge/gravity duality with $k = 1$ in the large N limit. We could test the duality since discrete Higgs vacua exist in the mABJM theory and they correspond one-to-one with the LLM geometries. We also extended the exact holographic relation to the case of any k for LLM geometries represented by rectangular-shaped Young-diagrams.

It seems that the Higgs vacua of the mABJM theory are parametrized by the vevs of CPOs and those are nonrenormalizable due to the high supersymmetry. This is similar to the case of the Coulomb branch in large N limit in $\mathcal{N} = 4$ super Yang-Mills theory [13, 14]. Though our quantitative results for the gauge/gravity correspondence involve infinite examples, we need to accumulate more analytic evidences for CPOs with Δ (≥ 2) and k (≥ 1) to define supersymmetric vacua. One should also test the dictionary of the gauge/gravity duality for one point functions of vector and tensor fields. For instance, it is important to verify that one point functions of the energy-momentum tensor vanish for all possible supersymmetric vacua, since the mABJM theory is a supersymmetric theory. We leave these issues for future study.

Acknowledgements

This work was supported by the National Research Foundation of Korea(NRF) grant with the grant number NRF-2016R1D1A1B03931090 (Y.K.), NRF-2017R1D1A1A09000951 (O.K.), NRF-2017R1D1A1B03032523 (D.T.).

References

[1] O. Aharony, O. Bergman, D. L. Jafferis and J. Maldacena, JHEP **0810**, 091 (2008).

[2] K. Hosomichi, K. M. Lee, S. Lee, S. Lee and J. Park, JHEP **0809**, 002 (2008).
[3] J. Gomis, D. Rodriguez-Gomez, M. Van Raamsdonk and H. Verlinde, JHEP **0809**, 113 (2008).
[4] H. C. Kim and S. Kim, Nucl. Phys. B **839**, 96 (2010).
[5] S. Cheon, H. C. Kim and S. Kim, arXiv:1101.1101 [hep-th].
[6] I. Bena and N. P. Warner, JHEP **0412**, 021 (2004).
[7] H. Lin, O. Lunin and J. M. Maldacena, JHEP **0410**, 025 (2004).
[8] R. Auzzi and S. P. Kumar, JHEP **0910**, 071 (2009).
[9] S. Ryu and T. Takayanagi, Phys. Rev. Lett. **96**, 181602 (2006); JHEP **0608**, 045 (2006).
[10] K. K. Kim, O. K. Kwon, C. Park and H. Shin, Phys. Rev. D **90**, no. 4, 046006 (2014); Phys. Rev. D **90**, no. 12, 126003 (2014); C. Kim, K. K. Kim and O. K. Kwon, Phys. Lett. B **759**, 395 (2016).
[11] D. Jang, Y. Kim, O. K. Kwon and D. D. Tolla, Eur. Phys. J. C **77**, no. 5, 342 (2017); JHEP **1704**, 104 (2017).
[12] D. Jang, Y. Kim, O. K. Kwon and D. D. Tolla, in preparation.
[13] K. Skenderis and M. Taylor, JHEP **0605**, 057 (2006).
[14] K. Skenderis and M. Taylor, JHEP **0608**, 001 (2006).

Thermodynamic Volume in AdS/CFT

Kyung Kiu Kim[1,⋆] and *Byoungjoon* Ahn[2,⋆⋆]

[1] *Department of Physics and Astronomy, Sejong University, Seoul 05006, Korea*
[2] *Department of Physics, College of Science, Yonsei University, Seoul 03722, Korea*

Abstract. In this note, we study on extended thermodynamics of AdS black holes by varying cosmological constant. We found and discussed pressure and volume of both bulk and boundary physics through AdS/CFT correspondence. In particular, we derive the relation between thermodynamic volume and a chemical potential for M2 brane dual to four dimensional AdS space. In addition, we show that thermodynamic volume of hyperbolic black hole is related to 'entanglement pressure' coming from a generalized first law of entanglement entropy.

1 Introduction and Summary

Thermodynamics of black hole is the starting point of all controversies in quantum gravity. It has given rise to many puzzles, new approaches and fusion researches among various areas in physics. One of the approaches supported by most people is the holographic principle [1, 2]. Maldacena found that this principle can be realized exactly with various backgrounds in string theory[3]. This correspondence was extended to the black holes immediately[4]. Thus the thermodynamics of black holes is identified with thermodynamics of conformal field theories through AdS/CFT correspondence.

Recently there have been meaningful attempts to extend thermodynamics of black holes. It is simply to consider a pressure of the bulk thermodynamics [5]. To this end, the cosmological constant can be regarded as a thermodynamic variable, pressure. It is quite natural, since the cosmological constant plays a role of energy-momentum in the bulk. Under this generalization, the mass of black hole is interpreted as enthalpy rather than internal energy. Moreover, a van der Waals like critical point appears and provides a richer phase structure. [6–8]

Smarr relation shows the effect of the pressure directly even for the fixed cosmological constant, while The first law of black hole thermodynamics doesn't give us any sign of the pressure without varying cosmological constant. Furthermore, it reveals the conjugate (thermodynamic) volume to the pressure and gives new interpretation of the black hole mass as an enthalpy. In AdS space, there are three kinds of non-rotating black holes, such as black brane, black hole and black hole with hyperbolic spatial section. They share a universal form of Smarr relation [9].

On the other hand, these three sorts of black holes are dual to finite temperature conformal field theories in flat space, spherical space and hyperbolic space, respectively, through AdS/CFT correspondence. Nonvanishing curvature of space corrects the energy-momentum tensor of the dual field theory

⋆e-mail: kimkyungkiu@sejong.ac.kr
⋆⋆e-mail: bjahn@yonsei.ac.kr

with Casimir contribution. Smarr relation in the bulk can be written down in terms of on-shell action and boundary quantities shch as energy density, entropy, temperature, charge density and chemical potential. From this relation one can notice that the gravity on-shell action is nothing but the grand potential of the dual boundary system. By varying this relation and adopting on-shell condition, the first law of the dual theory can be derived. Also, the pressure of the dual system can be read off from this. we will provide summary of the results in the next section.

Physical interpretation of the thermodynamic volume depends on how the cosmological constant changes physically. For the black holes with spherical and flat spatial sections, varying cosmological constant corresponds to taking different large N limits because the cosmological constant depends on N. This is a very nontrivial task which contains $1/N$ effect and extended AdS/CFT correspondence. Usually AdS/CFT correspondence is defined for a fixed large N limit as a correspondence between low energy effective theories of coincident N D branes. Thus varying N means adding D branes to the original N D brane system. Also thermodynamic volume is a response from changing number of D branes and it can be interpreted as a chemical potential of branes in various situations. We present an explicit expression for the ABJM theory in section 3.

In addition there is another interesting interpretation of thermodynamics volume for black holes with hyperbolic spatial section, so called, hyperbolic black holes. In this case, the entropy of the black holes can be mapped to an entanglement entropy with ball shape entangling regions whose radius is given by the cosmological constant. So physical meaning of varying N is changing the size of entangling region. This generalizes the thermodynamics of entanglement entropy and we introduce new quantity, 'entanglement pressure' to explain an additional term in the generalized thermodynamics. We will show details in section 3.

2 Thermodynamics of Bulk and Boundary

In this section we summarize and compare thermodynamics of a gravitational system and corresponding boundary theory. Let us start with a following gravity system.

$$S = \frac{1}{16\pi G}\int d^Dx\sqrt{-g}\left(R + \frac{(D-1)(D-2)}{L^2} - \frac{1}{4}F^{MN}F_{MN}\right) , \tag{1}$$

where F is a Maxwell field strength given by $F = dA$. This action admits an ansatz as follows:

$$ds^2 = -f(r)dt^2 + r^2 d\Omega_\kappa^2 + \frac{dr^2}{f(r)} , \tag{2}$$

where the subscript κ of the spatial section Ω_κ can be $1, 0$ and -1 for spherical, flat and hyperbolic spatial section, respectively. Then one can obtain the following charged black hole solution,

$$f(r) = \frac{r^2}{L^2} + \kappa - \frac{16\pi\tilde{m}}{(D-2)r^{D-3}} + \frac{G^{\frac{2(D-3)}{D-2}}\tilde{q}}{2(D-2)(D-3)r^{2(D-3)}} \tag{3}$$

$$A = \left(16G^{\frac{1}{D-2}}\mu - \frac{G^{\frac{D-3}{D-2}}}{(D-3)r^{D-3}}\tilde{q}\right)dt , \tag{4}$$

where $\tilde{m}$ and $\tilde{q}$ are mass and charge density over Ω_k.

Now, one can define thermodynamic variables in the bulk and construct the thermodynamics of the gravitational system. First the Hawking temperature and the entropy are defined as follows:

$$T_H = \frac{1}{4\pi}e^{A(r_h)}f'(r_h) , \quad S = \frac{1}{4G}r_h^{D-2}Vol_{\Omega_\kappa} , \tag{5}$$

where r_h denotes the horizon of the black holes. The bulk pressure is obtained by contribution of the cosmological constant to the energy-momentum tensor.

$$P_B = \frac{(D-1)(D-2)}{16\pi G} \ . \tag{6}$$

Then the Smarr relation for this system can be derived by considering $f(r_h) = 0$ as follows:

$$(D-3)M = (D-2)ST_H - 2P_B V_{th} + (D-3)\mu Q \ , \tag{7}$$

where M and Q are the total mass and charge for Ω_k. Here we defined a quantity V_{th} as

$$V_{th} = \frac{r_h^{D-1}}{D-1} Vol_{\Omega_\kappa} \ . \tag{8}$$

Also, one can obtain the first law of thermodynamics from variation of the Smarr relation (7). It can be written in the following form:

$$dM = T_H dS + V_{th} dP_B + \mu dQ \ . \tag{9}$$

From this, one can see that V_{th} is indeed playing a role of the thermodynamic volume and the mass M can be identified with an enthalpy of the system.

Now, one can consider corresponding dual thermodynamics of the boundary system. Thermodynamic variables of the dual conformal field theory can be obtained by holographic renormalization [10–13]. The metric of the boundary system is easily read off from the asymptotic behavior of the geometries. It is given by

$$ds_b^2 = g_{\mu\nu}^b dx^\mu dx^\nu = -dt^2 + L^2 d\Omega_\kappa^2 \ . \tag{10}$$

Then the volume of the space is

$$\mathcal{V}_b = L^{D-2} Vol_{\Omega_\kappa} \ , \tag{11}$$

thus varying L correponds to varying the size of space in the dual field theory. Derivative of the on-shell gravity action with respect to this boundary metric gives energy momentum tensor as follows:

$$\langle T^{\mu\nu} \rangle = \epsilon u^\mu u^\nu + \frac{\epsilon}{D-2}\left(g_b^{\mu\nu} + u^\mu u^\nu\right) \ , \tag{12}$$

where ϵ and u^μ are energy density and a timelike normal vector of the boundary system. The energy density is $\tilde{m}$ for even D. Otherwise, for odd D, it contains Casimir contributions as follows:

$$\epsilon = \tilde{m} + \frac{D-2}{32} L^{D-3} \kappa^{\frac{D-1}{2}} \begin{pmatrix} \frac{1}{2} \\ \frac{D-1}{2} \end{pmatrix} \ . \tag{13}$$

The Smarr relation (7) gives an expression of the euclidean on-sell action in terms of other thermodynamic variables,

$$S^{on-shell} \frac{T_H}{\mathcal{V}_b} = \epsilon - sT_H - \mu\rho \ , \tag{14}$$

where ρ is the energy density of the boundary system and it can be obtained by $\langle \rho \rangle = \frac{\delta S^{on-shell}}{\delta \mu}$ and s is the entropy density of the boundary system.

Also, the first law (9) can be written in terms of boundary quantities as follows:

$$d\epsilon = T_H ds + \mu d\rho - \frac{1}{D-2}\frac{(\epsilon - \mu\rho)}{\mathcal{V}_b} d\mathcal{V}_b \tag{15}$$

With extensive quantities, this becomes

$$dE_b = T_H dS_b + \mu dQ_b - \left(\frac{D-3}{D-2}(\epsilon - \mu\rho) - sT_H\right) d\mathcal{V}_b \tag{16}$$

where E_b adnd Q_b are total energy and total charge of the boundary system. Finally, we may conclude that the pressure is

$$\mathcal{P}_b = \left(\frac{D-3}{D-2}(\epsilon - \mu\rho) - sT_H\right) \tag{17}$$

and the grand potential is given by the euclidean on-shell action,

$$\mathcal{W} = E_b - ST_H - \mu Q_b \ . \tag{18}$$

In addition, there is a difference from the bulk thermodynamics. The energy of the boundary system still plays a role of internal energy rather than enthalpy. This is confirmed by the first law. Lastly we would like to point out there is an invariant quantity between bulk and boundary theories, which is pressure times volume:

$$P_B V_{th} = \mathcal{P}_b \mathcal{V}_b \ . \tag{19}$$

Therefore, the thermodynamic volum in the bulk is given by ratio of the pressures times the boundary volume. The meaning of this relation is not clear so far. However, it must be an interesting relation for the bulk-boundary duality.

3 Physical Interpretation and Discussion

In this final section we move on to physical interpretation of the extended thermodynamics. As we explained in section 1. The thermodynamic volume is related to a kind of chemical potential to the number of branes for black branes and black holes. In particular, we know dual field theories for some cases exactly. $D = 4$ case is for $\mathcal{N}$=4 super Yang-Mills theory and $D = 3$ case can describe the ABJM theory. We consider the 4 dimensional case.

The radius of 4 dimensional AdS spacetime can be written in terms of parameters of the ABJM theory as follows:

$$L = \frac{l_p}{2}(2^5\pi^2 Nk)^{1/6} \ , \tag{20}$$

where l_p, N and k are the Planck length, the rank of gauge group and the Chern-Simons level of the ABJM theory. Then the first law of the dual theory is given by

$$dE_b = T_H dS + \mu dQ + \mu_N dN \ , \tag{21}$$

where μ_N is the chemical potential of M2 brane and it is related to thermodynamic volume. The expression is

$$\mu_N = -\mathcal{P}_b\left(\frac{\partial \mathcal{V}_b}{\partial N}\right) = -P_B V_{th}\left(\frac{\partial \log \mathcal{V}_b}{\partial N}\right) \ . \tag{22}$$

Now we discuss on the hyperbolic black hole case, i.e, black hole with hyperbolic spatial section. The relevant background is

$$f(r) = \frac{r^2}{L^2} - 1 \ , \tag{23}$$

which is given by setting $\tilde{m} = \tilde{q} = 0$. Then the black hole entropy corresponds to the entanglement entropy with a ball shape entangling region [14]. The radius of the entangling region is given by L. Thus varying L is nothing but changing the size of the region for trace in entanglement entropy calculation. Thus the thermodynamic volume is related to a field theory quantity which is defined in the following generalized first law of entanglement entropy S_{EE} over a region $\mathcal{A}$.

$$\delta S_{EE} = \delta \langle H_{\mathcal{A}} \rangle + \mathcal{P}_{EE} \delta \mathcal{A} \ , \tag{24}$$

Where $H_{\mathcal{A}}$ is modular hamiltonian over $\mathcal{A}$ and $\mathcal{P}_{EE}$ is a response of entanglement entropy to varying $\mathcal{A}$.

The extended black hole thermodynamics by varying cosmological constant has many applications in the dual field theory. In this paper, we introduce several examples. More details and study would be presented in our next paper [15]. As a future direction, we may consider the thermodynamic volume in an inflation model or de Sitter space. In these cases, the thermodynamic volume is related to the slow roll parameter of inflation models and it is interesting to see if the thermodynamic volume plays a role of an effective volume describing dark matter [16] .

Acknowledgement

K.K. was supported by NRF Grant 2015R1D1A1A01058220.

References

[1] G. 't Hooft, "Dimensional reduction in quantum gravity," Salamfest 1993:0284-296 [gr-qc/9310026].

[2] L. Susskind, "The World as a hologram," J. Math. Phys. **36**, 6377 (1995) [hep-th/9409089].

[3] J. M. Maldacena, "The Large N limit of superconformal field theories and supergravity," Int. J. Theor. Phys. **38**, 1113 (1999) [Adv. Theor. Math. Phys. **2**, 231 (1998)] [hep-th/9711200].

[4] E. Witten, "Anti-de Sitter space, thermal phase transition, and confinement in gauge theories," Adv. Theor. Math. Phys. **2**, 505 (1998) [hep-th/9803131].

[5] D. Kastor, S. Ray and J. Traschen, "Enthalpy and the Mechanics of AdS Black Holes," Class. Quant. Grav. **26**, 195011 (2009) [arXiv:0904.2765 [hep-th]].

[6] B. P. Dolan, "The cosmological constant and the black hole equation of state," Class. Quant. Grav. **28**, 125020 (2011) [arXiv:1008.5023 [gr-qc]].

[7] B. P. Dolan, "Pressure and volume in the first law of black hole thermodynamics," Class. Quant. Grav. **28**, 235017 (2011) [arXiv:1106.6260 [gr-qc]].

[8] B. P. Dolan, "Black holes and Boyle's law — The thermodynamics of the cosmological constant," Mod. Phys. Lett. A **30**, no. 03n04, 1540002 (2015) [arXiv:1408.4023 [gr-qc]].

[9] D. Kubiznak, R. B. Mann and M. Teo, "Black hole chemistry: thermodynamics with Lambda," Class. Quant. Grav. **34**, no. 6, 063001 (2017) [arXiv:1608.06147 [hep-th]].

[10] V. Balasubramanian and P. Kraus, "A Stress tensor for Anti-de Sitter gravity," Commun. Math. Phys. **208**, 413 (1999) [hep-th/9902121].

[11] M. Henningson and K. Skenderis, "The Holographic Weyl anomaly," JHEP **9807**, 023 (1998) [hep-th/9806087].

[12] S. de Haro, S. N. Solodukhin and K. Skenderis, "Holographic reconstruction of space-time and renormalization in the AdS / CFT correspondence," Commun. Math. Phys. **217**, 595 (2001) [hep-th/0002230].

[13] K. Skenderis, "Lecture notes on holographic renormalization," Class. Quant. Grav. **19**, 5849 (2002) [hep-th/0209067].

[14] H. Casini, M. Huerta and R. C. Myers, "Towards a derivation of holographic entanglement entropy," JHEP **1105**, 036 (2011) [arXiv:1102.0440 [hep-th]].

[15] B. Ahn, K. Kim and S. Hyun, in progress.

[16] E. P. Verlinde, "Emergent Gravity and the Dark Universe," SciPost Phys. **2**, no. 3, 016 (2017) [arXiv:1611.02269 [hep-th]].

Gravitational Scattering Amplitudes and Closed String Field Theory in the Proper-Time Gauge

Taejin Lee[1]*

[1] *Department of Physics, Kangwon National University, Chuncheon 24341 Korea*

Abstract. We construct a covariant closed string field theory by extending recent works on the covariant open string field theory in the proper-time gauge. Rewriting the string scattering amplitudes generated by the closed string field theory in terms of the Polyakov string path integrals, we identify the Fock space representations of the closed string vertices. We show that the Fock space representations of the closed string field theory may be completely factorized into those of the open string field theory. It implies that the well known Kawai-Lewellen-Tye (KLT) relations of the first quantized string theory may be promoted to the second quantized closed string theory. We explicitly calculate the scattering amplitudes of three gravitons by using the closed string field theory in the proper-time gauge.

1 Introduction

Construction of a finite quantum theory of gravity was the main motivation for theorists to turn towards the string theory [1] as it has become clear that the perturbative theory of gravity based on the quantum field theory [2–4] is not renormalizable [5, 6]. Given that the spectrum of a free closed string contains the spin two massless particle state, which corresponds to the graviton and the perturbative string theory is expected to be free from ultraviolet divergences, it stands to reason that the string theory should receive attention from theorists as a consistent framework for unifying all quantum theories of fundamental forces, including gravity. In view of this, the closed string field theory is anticipated to be a passage to the finite quantum theory of gravity. However, despite years of effort, construction of a consistent perturbative closed string field theory has not been yet completed.

In this work, we construct the covariant closed string field theory by extending recent works on open string field theory in the proper-time gauge [7–11]. Following the general study on the classical limit of quantum gravity [12], the closed string field theory as a consistent quantum theory of gravity should reduce to the classical Einstein gravity in the low energy domain. For a comparison of the proposed closed string field theory with the Einstein gravity, we will calculate the gravitational scattering amplitudes at tree level and show that the closed string field theory correctly reproduces that of the conventional Einstein gravity. In the process of evaluating the string scattering, we also find that the Fock space representations of the closed string field theory are completely factorized into those of the open string field theory in general: This factorization indicates that the KLT [13] relations of the first quantized string theory may be fully extended to the second quantized string theory.

*e-mail: taejin@kangwon.ac.kr

2 Closed string field theory in the proper-time gauge

The closed string field theory in the proper-time may be formally described by the action

$$S = \langle\Phi|\mathcal{K}\Phi\rangle + \frac{g}{3}\Big(\langle\Phi|\Phi\circ\Phi\rangle + \langle\Phi\circ\Phi|\Phi\rangle\Big)$$

where $\mathcal{K} = L_0 - i\epsilon$ and the second term denotes the closed string interaction as depicted in Fig. 1 In the proper-time gauge we fix the length parameters of three strings as $\alpha_1 = \alpha_2 = 1$ and $\alpha_3 = -2$. Because the length parameters can be written in terms of the two dimensional metric on the string world-sheet, fixing the length parameters is equivalent to fixing the metric on the world-sheet by using the reparametrization invariance. The kinetic term is indpendent of the length parameters if we adopt the Fock space representation for the the string field [9]. Although the string field action, describing the off-shell dynamics, may depend on the length parameters, the physical scattering amplitudes, which are generated by the string field action, may be independent of the length parameters.

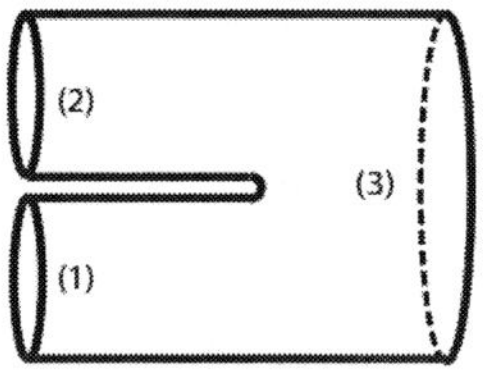

Figure 1. Closed string interaction in the proper-time gauge.

3 Fock space representation of the closed string field theory

Construction of the Fock space representation of the closed string field theory begins with the Schwarz-Christoffel (SC) mapping from the string world-sheet to the complex plane [9]:

$$\rho = \ln(z-1) + \ln z. \tag{1}$$

The local coordinates $\zeta_r = \xi_r + i\eta_r$, $r = 1,2,3$ defined on invidual string world sheet patches are related to z as follows,

$$e^{-\zeta_1} = e^{\tau_0}\frac{1}{z(z-1)}, \tag{2a}$$

$$e^{-\zeta_2} = -e^{\tau_0}\frac{1}{z(z-1)}, \tag{2b}$$

$$e^{-\zeta_3} = -e^{-\frac{\tau_0}{2}}\sqrt{z(z-1)}. \tag{2c}$$

Then, by using the SC mapping, we may write the the Green's function on the world sheet as follows:

$$\begin{aligned} G_C(\rho_r,\rho'_s) &= \ln|z_r - z'_s| \\ &= -\delta_{rs}\left\{\sum_{n=1}\frac{e^{-n\Delta}}{2n}\left(e^{in(\eta'_s-\eta_r)} + e^{-in(\eta'_s-\eta_r)}\right) - \max(\xi,\xi')\right\} \\ &\quad + \sum_{n,m}\bar{C}^{rs}_{nm} e^{|n|\xi_r + |m|\xi'_s} e^{in\eta_r} e^{im\eta'_s}, \end{aligned} \tag{3}$$

where $\bar{C}^{rs}_{nm}$ is the Fourier components of the Green's function on the string world sheet.

The closed string field theory action generates perturbative string Feynmann diagrams corresponding to the multi-string scattering amplitudes

$$\mathcal{W}_N = \langle \mathbf{P}| \exp\left(\sum_r \xi_r L_0^{(r)}\right)|V[N]\rangle. \tag{4}$$

For three-string scattering $\mathcal{W}_3$, by employing the oscillatory basis, we obtain the Fock space representation of the three-string-vertex $|V[3]$:

$$\begin{aligned} |V[3]\rangle = & \exp\left\{\sum_{r,s}\Big(\sum_{n,m\geq 1}\frac{1}{2}\bar{N}^{rs}_{nm}\frac{\alpha_n^{(r)\dagger}}{2}\cdot\frac{\alpha_m^{(r)\dagger}}{2} + \sum_{n\geq 1}\bar{N}^{rs}_{n0}\frac{\alpha_n^{(r)\dagger}}{2}\cdot\frac{p^{(s)}}{2}\Big)\right\} \\ & \exp\left\{\tau_0\sum_r\frac{1}{\alpha_r}\left(\frac{1}{2}\left(\frac{p^{(r)}}{2}\right)^2-1\right)\right\} \\ & \exp\left\{\sum_{r,s}\Big(\sum_{n,m\geq 1}\frac{1}{2}\bar{N}^{rs}_{nm}\frac{\tilde{\alpha}_n^{(r)\dagger}}{2}\cdot\frac{\tilde{\alpha}_m^{(r)\dagger}}{2} + \sum_{n\geq 1}\bar{N}^{rs}_{n0}\frac{\tilde{\alpha}_n^{(r)\dagger}}{2}\cdot\frac{p^{(s)}}{2}\Big)\right\} \\ & \exp\left\{\tau_0\sum_r\frac{1}{\alpha_r}\left(\frac{1}{2}\left(\frac{p^{(r)}}{2}\right)^2-1\right)\right\}|0\rangle, \end{aligned} \tag{5}$$

where $\bar{N}^{rs}_{nm}$, $r, s = 1, 2, 3$ are the Neumann functions of the open string theory. From the explicit expression of the Fock space representation of the three-closed-string vertex, it is evident that the closed string vertex may be completely factorized into those of open string vertices. This observation suggests that the KLT relations may be fully extended to the second quantized string theory.

4 Three-graviton scattering amplitude

Having obtained the Fock space representation of the three-closed-string vertex, we shall calculate the three-graviton-scattering amplitude to confirm that the closed string field theory, as a consistent quantum theory of gravity reduces to the Einstein gravity in the low energy domain. We may decompose the masless closed string states $h_{\mu\nu}\alpha^{\mu}_{-1}\tilde{\alpha}^{\nu}_{-1}|0\rangle$ into the graviton, the anti-symmetric ternsor, and the scalar field states:

$$h_{\mu\nu} = \left\{\frac{1}{2}\left(h_{\mu\nu}+h_{\nu\mu}\right) - \eta_{\mu\nu}\frac{d}{d-2}h^{\sigma}{}_{\sigma}\right\} + \left\{\frac{1}{2}\left(h_{\mu\nu}-h_{\nu\mu}\right)\right\} + \eta_{\mu\nu}\left\{\frac{1}{d-2}h^{\sigma}{}_{\sigma}\right\}. \tag{6}$$

For the three-graviton scattering, we choose the symmetric traceless part and write the external string state as

$$|\Psi_{3G}\rangle = \prod_{r=1}^{3}\left\{h_{\mu\nu}(p^r)\alpha^{(r)\mu}_{-1}\tilde{\alpha}^{(r)\nu}_{-1}\right\}|0\rangle. \tag{7}$$

The three-graviton scattering amplitude follows from the explicit expression of the three-string vertex Eq.(5) with the three-graviton state Eq. (7):

$$\begin{aligned}
\mathcal{A}_{3G} &= \int \prod_{r=1}^{3} dp^{(r)} \delta\left(\sum_{r=1}^{3} p^{(r)}\right) \frac{2g}{3} \langle \Psi_{3G} | V[3] \rangle \\
&= \left(\frac{2g}{3}\right) e^{-2\tau_0 \sum_{r=1}^{3} \frac{1}{\alpha_r}} \int \prod_{i=1}^{3} dp^{(i)} \delta\left(\sum_{i=1}^{3} p^{(i)}\right) \\
& \langle 0 | \left\{ \prod_{i=1}^{3} h_{\mu\nu}(p^{(i)}) a_1^{(i)\mu} \cdot \tilde{a}_1^{(i)\nu} \right\} \frac{1}{2^5} \left(\sum_{r,s=1}^{3} \bar{N}_{11}^{rs} a_1^{(r)\dagger} \cdot a_1^{(s)\dagger} \right) \\
& \left(\sum_{t=1}^{3} \bar{N}_1^t a_1^{(t)\dagger} \cdot \boldsymbol{p} \right) \frac{1}{2^5} \left(\sum_{l,m=1}^{3} \bar{N}_{11}^{lm} \tilde{a}_1^{(l)\dagger} \cdot \tilde{a}_1^{(m)\dagger} \right) \left(\sum_{n=1}^{3} \bar{N}_1^n \tilde{a}_1^{(n)\dagger} \cdot \boldsymbol{p} \right) |0\rangle,
\end{aligned} \tag{8}$$

where $\boldsymbol{p} = p^{(1)} - p^{(2)}$. Making use of the Neumann functions of the open string given in Ref. [9],

$$\bar{N}_{11}^{11} = \frac{1}{2^4}, \quad \bar{N}_{11}^{22} = \frac{1}{2^4}, \quad \bar{N}_{11}^{33} = 2^2, \tag{9a}$$

$$\bar{N}_{11}^{12} = \bar{N}_{11}^{21} = \frac{1}{2^4}, \quad \bar{N}_{11}^{23} = \bar{N}_{11}^{32} = \frac{1}{2}, \quad \bar{N}_{11}^{31} = \bar{N}_{11}^{13} = \frac{1}{2}, \tag{9b}$$

$$\bar{N}_1^1 = \bar{N}_1^2 = \frac{1}{4}, \quad \bar{N}_1^3 = -1, \tag{9c}$$

we are able to evaluate the three-graviton amplitude

$$\begin{aligned}
\mathcal{A}_{3G} &= \left(\frac{2g}{3}\right) 2^6 \left(\frac{1}{2^5}\right)^2 \int \prod_{i=1}^{3} dp^{(i)} \delta\left(\sum_{i=1}^{3} p^{(i)}\right) h_{\mu_1\nu_1}(p^{(1)}) h_{\mu_2\nu_2}(p^{(2)}) h_{\mu_3\nu_3}(p^{(3)}) \\
& \left\{ -\frac{1}{2^4} \eta^{\mu_1\mu_2} \boldsymbol{p}^{\mu_3} + \frac{1}{2^3} \eta^{\mu_1\mu_3} \boldsymbol{p}^{\mu_2} + \frac{1}{2^3} \eta^{\mu_2\mu_3} \boldsymbol{p}^{\mu_1} - \frac{1}{2^4} \eta^{\mu_2\mu_1} \boldsymbol{p}^{\mu_3} + \frac{1}{2^3} \eta^{\mu_3\mu_1} \boldsymbol{p}^{\mu_2} \right. \\
& \left. + \frac{1}{2^3} \eta^{\mu_3\mu_2} \boldsymbol{p}^{\mu_1} \right\} \left\{ -\frac{1}{2^4} \eta^{\nu_1\nu_2} \boldsymbol{p}^{\nu_3} + \frac{1}{2^3} \eta^{\nu_1\nu_3} \boldsymbol{p}^{\nu_2} + \frac{1}{2^3} \eta^{\nu_2\nu_3} \boldsymbol{p}^{\nu_1} - \frac{1}{2^4} \eta^{\nu_2\nu_1} \boldsymbol{p}^{\nu_3} \right. \\
& \left. + \frac{1}{2^3} \eta^{\nu_3\nu_1} \boldsymbol{p}^{\nu_2} + \frac{1}{2^3} \eta^{\nu_3\nu_2} \boldsymbol{p}^{\nu_1} \right\}.
\end{aligned} \tag{10}$$

Repeatedly making use of the covariant gauge condition $\partial^\mu h_{\mu\nu} = 0$, which is equivalent to the de Donder gauge condition for the graviton sector, $\partial^\mu h_{\mu\nu} - \frac{1}{d-2} \partial_\nu h^\sigma{}_\sigma = 0$, we find

$$\begin{aligned}
\mathcal{A}_{3G} &= \kappa \int \prod_{i=1}^{3} dp^{(i)} \delta\left(\sum_{i=1}^{3} p^{(i)}\right) h_{\mu_1\nu_1}(p^{(1)}) h_{\mu_2\nu_2}(p^{(2)}) h_{\mu_3\nu_3}(p^{(3)}) \\
& \left\{ \eta^{\mu_1\mu_2} p^{(1)\mu_3} + \eta^{\mu_2\mu_3} p^{(2)\mu_1} + \eta^{\mu_3\mu_1} p^{(3)\mu_2} \right\} \left\{ \eta^{\nu_1\nu_2} p^{(1)\nu_3} + \eta^{\nu_2\nu_3} p^{(2)\nu_1} + \eta^{\nu_3\nu_1} p^{(3)\nu_2} \right\},
\end{aligned} \tag{11}$$

where $\kappa = \frac{g}{2^7 \cdot 3} = \sqrt{32\pi G_{10}}$. $\mathcal{A}_{3G}$ is precisely the three-graviton interaction term of the conventional perturbative Einstein gravity [14].

5 Conclusions

By extending recent works on the open string field theory in the proper-time gauge [8–11], we constructed a covariant closed string field theory. The closed string field theory generates perturbative three string diagrams, which may be re-expressed in terms of the Polyakov string path-integral. By mapping the three-string world sheet diagram onto the complex plane, we evaluate explicitly the three-string-scattering amplitude and obtain the Fock space representation of the three-string-vertex. For the three-string diagram, we may map the positions of the external strings onto the real line by using the $SL(2, C)$ invariance. Then, it follows that the Fourier components of the closed string Green's function on the string world sheet may be written in terms of those of the open string, *i.e.*, the Neumann functions of the open string. We also find that the Fock-space representation of the closed three-string-vertex may be completely factorized into those of the open string in the proper-time gauge. This factorization of the closed string vertex may be a string field theoretical extension of the KLT relations, previously studied in the first quantized string theory. To confirm that the closed string field theory in the proper-time gauge correctly reduces to the perturbative Einstein gravity, we calculate the three-graviton-scattering amplitude.

More work remains to be done to complete the construction of the covariant closed string field theory in the proper-time gauge: 1) Fock space representations of the general closed string amplitudes should be constructed. 2) We may explicitly evaluate the four-graviton scattering amplitudes. 3) Full extensions of the KLT relations at the level of the second quantized string theory would be certainly among the remaining tasks to be done. We will discuss these subjects in papers sequel to this work.

Acknowledgments This work was supported by Basic Science Research Program through the National Research Foundation of Korea (NRF) funded by the Ministry of Education (2017R1D1A1A02017805).

References

[1] J. Scherk and J. H. Schwarz, Nucl. Phys. B **81**, 118 (1974).
[2] B. S. DeWitt, Phys. Rev. **160**, 1113 (1967).
[3] B. S. DeWitt, Phys. Rev. **162**, 1195 (1967).
[4] B. S. DeWitt, Phys. Rev. **162**, 1239 (1967).
[5] G. 't Hooft and M. J. Veltman, Ann. Inst. Henri Poincare, A **20**, 69 (1974).
[6] M. J. Veltman, in R. Balian and J. Zinn-Justin, eds., *Methods in Field Theory, Proceedings of the Les Houches Summer School 1975*, (North-Holland, Amsterdam, 1976) 265–327.
[7] T. Lee, Ann. Phys. **183**, 191 (1988).
[8] T. Lee, Phys. Lett. B **768**, 248 (2017).
[9] T. Lee, *Covariant open bosonic string field theory on multiple D-branes in the proper-time gauge*, to be published in Jour. Kor. Soc. (2017), arXiv:1609.01473 [hep-th].
[10] T. Lee, arXiv:1703.06402 [hep-th].
[11] S. H. Lai, J. C. Lee, Yi Yang, and T. Lee, arXiv:1706.08025 [hep-th].
[12] D. G. Boulware and S. Deser, Ann. Phys. **89**, 193 (1975).
[13] H. Kawai, D. C. Lewellen, and S. H. Tye, Nucl. Phys. B **269**, 1 (1986).
[14] S. Sannan, Phys. Rev. D **34**, 1749 (1986).

EPJ Web of Conferences **168**, 08001 (2018) https://doi.org/10.1051/epjconf/201816808001
Joint International Conference of ICGAC-XIII and IK-15 on Gravitation, Astrophysics and Cosmology

Conformal (In)Equality

Young-Hwan Hyun[1,⋆], *Yoonbai* Kim[1,⋆⋆], and *Seokcheon* Lee[2,⋆⋆⋆]

[1] *Department of Physics and Institute of Basic Science, Sungkyunkwan University, Suwon 16419, Republic of Korea*
[2] *The Research Institute of Natural Science, Gyeongsang National University, Jinju 52828, Republic of Korea*

Abstract. The current accelerating expansion of the Universe is explained either by dark energy or by modified gravity theories. Both of them can explain exactly the same background evolution of the Universe, however this degeneracy may be broken when the observation of large scale structure formation is taken into account. Two observables are parameterized by the so-called dark energy equation of state, ω and the growth index parameter, γ. From these observed parameters, one may reconstruct the model parameters of the so-called scalar-tensor gravity theory, one of the modified gravity theories. Especially, the scalar-tensor gravity theory is described both in Jordan frame and in Einstein frame. If cosmological observations are interpreted in one frame, then all of the observables should also be interpreted in that frame. This explicitly shows conformal inequality of cosmological observables.

1 Introduction

Cosmological observations support the current Universe in the phase of accelerating expansion. In the theory side, two descriptions compete. One is to adopt general relativity as the correct gravity theory and introduce the dark energy (DE) in addition to the known matter components. The other is to invoke modified gravity theories (MG) and tries to replace the DE with the modifications of the Friedmann equations as an ejective dark energy.

In this manuscript, we shall deal with one of the simplest but long-standing MG, the so-called scalar-tensor gravity (STG) which is natural in the viewpoint of string theory. When the evolution of the Universe is depicted in the context of STG, two model functions are introduced. Comparison to the observable quantities determines these two theoretically introduced functions which specifically are reconstructed in terms of the DE equation of state (EOS), ω [1, 2] and the growth index parameter (GIP), γ [3].

Once STG is considered, the question on the frame-dependence should be addressed. Since no conformal symmetry is possessed in STG coupled to matter fields even at the classical level, every conformal transformation from a chosen frame defines a new frame, which results in the set of infinitely-many physically-distinctive frames. Theoretically favored frames are two extremes, i.e., the Einstein frame (EF) and the Jordan frame (JF). The question on the frame that "where we live (now)"

⋆e-mail: younghwan.hyun@gmail.com
⋆⋆e-mail: yoonbai@skku.edu
⋆⋆⋆e-mail: skylee2@gmail.com

has been a long-standing debate, it seems beyond our scope at the present stage. Instead, we explicitly show that some cosmological observables are frame-dependent and if one obtains model parameters in one frame, then one should interpret all the other observables in that frame.

For this purpose, we express two functions of STG in both the JF and the EF in terms of frame-independent observable parameters, the redshift z in the next section. We show the physical (in)equalities of cosmological observables in two frames such as the Hubble parameter, H and the effective DE EOS, ω. Thus, we show explicitly that one should keep using the only one frame for probing cosmological observables.

2 Reconstruction of Model Functions

We begin this section with the action of the STG in the JF [4]

$$S = \frac{1}{16\pi G_*} \int d^4x \sqrt{-\overline{g}} \left[F(\phi)\overline{R} - \nabla^\mu\phi\nabla_\mu\phi - 2U(\phi) \right] + S_{\rm m}(\overline{g}_{\mu\nu}, \psi_{\rm m}), \tag{1}$$

where G_* is the bare gravitational coupling constant different from the Newton's constant, $\overline{g}_{\mu\nu}$ is the JF metric, $F(\phi)$ is dimensionless function of scalar field ϕ, $U(\phi)$ is its potential, and $S_{\rm m}(\overline{g}_{\mu\nu})$ is the matter action for $\psi_{\rm m}$ representing any matter field. $F(\phi)$ is required to be positive definite to ensure attractive nature of gravity. However its specific functional form and that of $U(\phi)$ are not determined yet but should be restricted by observations. In this manuscript, all the physical variables in the JF are denoted by bars on top of them. We summarize the Friedmann equations in both frames in table 1. In this table $\overline{H} \equiv d\ln\overline{a}/d\overline{t}$ means the Hubble parameter in the JF, $\overline{\rho}_{\rm m}$ and $\overline{P}_{\rm m}$ define the matter energy density and the pressure, respectively. They are conserved quantities by the Bianchi identity. We also introduce the so-called mis-interpreted frame where the Hubble parameter is defined by $H = d\ln a/dt$ which is not the observable quantity. Some of literatures use this quantity to interpret the Hubble parameter in the EF and this is mis-interpreted [5].

	the frist Friedmann Eq	the second Friedmann Eq
JF	$3F\overline{H}^2 = 8\pi G_*\overline{\rho}_{\rm m} + \frac{1}{2}\overline{H}^2\phi'^2 - 3\overline{H}^2F' + U$	$-2F\overline{H}\,\overline{H}' = 8\pi G_*(\overline{\rho}_{\rm m} + \overline{P}_{\rm m}) + \overline{H}^2\phi'^2 + \overline{H}\overline{H}'F' + \overline{H}^2F'' - \overline{H}^2F'$
EF	$3H_{\rm E}^2 = 8\pi\frac{G_*}{F^2}\overline{\rho}_{\rm m} + \left[\frac{1}{2}\left(\frac{\phi'}{\sqrt{F}}\right)^2 - 3\frac{F'}{F} + \frac{U}{F\overline{H}^2}\right]\frac{\overline{H}^2}{F}$	$-2H_{\rm E}\frac{dHE}{d\ln a} = 8\pi\frac{G_*}{F^2}(\overline{\rho}_{\rm m} + \overline{P}_{\rm m}) + \left[\frac{1}{2}\left(\frac{\phi'}{\sqrt{F}}\right)^2 + \frac{1}{2}(\frac{F'}{F})^2 + \frac{F''}{F}\right]\frac{\overline{H}^2}{F}$
MF	$3H^2 = 8\pi\frac{G_*}{F^2}\overline{\rho}_m + \frac{1}{F}\left[\frac{1}{2}\left(\frac{\phi'}{\sqrt{F}}\right)^2 + \frac{3}{4}(\frac{F'}{F})^2\right]\overline{H}^2 + \frac{U}{F^2}$	$-2\frac{dH}{dt} = 8\pi\frac{G_*}{F^2}(\overline{\rho}_{\rm m} + \overline{P}_{\rm m}) + \frac{2}{F}\left[\frac{1}{2}\left(\frac{\phi'}{\sqrt{F}}\right)^2 + \frac{3}{4}(\frac{F'}{F})^2\right]\overline{H}^2$

Table 1. Friedmann equations in JF, EF, and mis-interpreted frame (MF). Primes mean the differentiation with respect to $\ln\overline{a}$.

The evolution equations for the background and the matter perturbation in the flat Friedmann-Lemaitre-Robertson-Walker (FLRW) metric are given by [6]

$$\Omega_{\rm m}(\bar{z}) = \frac{\Omega_{\rm m0}}{\Omega_{\rm m0} + (1-\Omega_{\rm m0})(1+\bar{z})^{3(\omega_0+\omega_a)}e^{-3\omega_a\bar{z}/(1+\bar{z})}} \quad (2)$$

$$\frac{\overline{H}^2(\bar{z})}{\overline{H}_0^2} = \Omega_{\rm m0}(1+\bar{z})^3 + (1-\Omega_{\rm m0})(1+\bar{z})^{3(1+\omega_0+\omega_a)}e^{-3\omega_a\bar{z}/(1+\bar{z})}\,, \quad (3)$$

$$\frac{\overline{H}'(\bar{z})}{\overline{H}(\bar{z})} = -(1+\bar{z})\frac{\overline{H}_{,\bar{z}}}{\overline{H}} = -\frac{3}{2}\left[1 + \omega(\bar{z})\frac{(1-\Omega_{\rm m0})(1+\bar{z})^{3(\omega_0+\omega_a)}e^{-3\omega_a\bar{z}/(1+\bar{z})}}{\Omega_{\rm m0} + (1-\Omega_{\rm m0})(1+\bar{z})^{3(\omega_0+\omega_a)}e^{-3\omega_a\bar{z}/(1+\bar{z})}}\right], \quad (4)$$

$$\overline{\omega}(\bar{z}) = -1 + 2(1+\bar{z})\frac{(1+\bar{z})\frac{\phi_{,\bar{z}}^2}{F_0} + (1+\bar{z})\frac{F_{,\bar{z}\bar{z}}}{F_0} + \left(2 + (1+\bar{z})\frac{\overline{H}_{,\bar{z}}}{\overline{H}}\right)\frac{F_{,\bar{z}}}{F_0} - 2(\frac{F}{F_0}-1)\frac{\overline{H}_{,\bar{z}}}{\overline{H}}}{(1+\bar{z})^2\frac{\phi_{,\bar{z}}^2}{F_0} + 6(1+\bar{z})\frac{F_{,\bar{z}}}{F_0} - 6(\frac{F}{F_0}-1) + \frac{2U}{F_0\overline{H}_0^2}\frac{\overline{H}_0^2}{\overline{H}^2}} \quad (5)$$

$$\frac{\phi_{,\bar{z}}}{\sqrt{F_0}} = -\sqrt{-\frac{F_{,\bar{z}\bar{z}}}{F_0} - \frac{1}{(1+\bar{z})}\left(2 + (1+\bar{z})\frac{\overline{H}_{,\bar{z}}}{\overline{H}}\right)\frac{F_{,\bar{z}}}{F_0} + \frac{2}{(1+\bar{z})}\frac{\overline{H}_{,\bar{z}}}{\overline{H}}\frac{F}{F_0} - 3\frac{\Omega_{\rm m}(\bar{z})}{(1+\bar{z})^2}} \quad (6)$$

$$\frac{F_{,\bar{z}}}{F_0} = \left(\frac{3}{(1+\bar{z})} - 2\frac{\overline{H}_{,\bar{z}}}{\overline{H}} - \frac{P_{,\bar{z}}}{P}\right)\frac{F}{F_0}\,, \quad (7)$$

$$\frac{F_{,\bar{z}\bar{z}}}{F_0} = \left(\left[\frac{3}{(1+\bar{z})} - 2\frac{\overline{H}_{,\bar{z}}}{\overline{H}} - \frac{P_{,\bar{z}}}{P}\right]^2 - \frac{3}{(1+\bar{z})^2} - 2\left[\frac{\overline{H}_{,\bar{z}}}{\overline{H}}\right]_{,\bar{z}} - \left[\frac{P_{,\bar{z}}}{P}\right]_{,\bar{z}}\right)\frac{F}{F_0}\,, \quad (8)$$

where $\bar{H}_0 \equiv \bar{H}(\bar{z}=0)$, subscript z means the differentiation with respect to redshift which related to $F_{,\bar{z}} = -F'/(1+\bar{z})$ and DE EOS is parametrized as the so-called CPL parametrization

$$\omega(\bar{z}) = \omega_0 + \omega_a\frac{\bar{z}}{1+\bar{z}}\,. \quad (9)$$

F/F_0 is given by

$$\frac{F(\bar{z})}{F_0} = \frac{3}{2}\frac{\Omega_{\rm m}(\bar{z})}{P(\bar{z})}\,, \quad (10)$$

$$P(\bar{z}) = \Omega_{\rm m}(\bar{z})^{\gamma}\left(\Omega_{\rm m}(\bar{z})^{\gamma} - (1+\bar{z})\gamma_{,\bar{z}}\ln\Omega_{\rm m}(\bar{z}) - \gamma\left[3 - 2(1+\bar{z})\frac{\overline{H}_{,\bar{z}}}{\overline{H}}\right] + 2 - (1+\bar{z})\frac{\overline{H}_{,\bar{z}}}{\overline{H}}\right), \quad (11)$$

$$\frac{U(\bar{z})}{F_0\overline{H}_0^2} = \frac{1}{2}\frac{\overline{H}^2}{\overline{H}_0^2}\left((1+\bar{z})^2\frac{F_{,\bar{z}\bar{z}}}{F_0} + (1+\bar{z})\frac{F_{,\bar{z}}}{F_0} - (1+\bar{z})\left[5 - (1+\bar{z})\frac{\overline{H}_{,\bar{z}}}{\overline{H}}\right]\frac{F_{,\bar{z}}}{F_0}\right.$$
$$\left. + 2\left[3 - (1+\bar{z})\frac{\overline{H}_{,\bar{z}}}{\overline{H}}\right]\frac{F}{F_0} - 3\Omega_{\rm m}(\bar{z})\right), \quad (12)$$

where GIF is given by $\gamma = \gamma_0 + \gamma_a\frac{\bar{z}}{(1+\bar{z})}$ and we express all observables in terms of frame-independent variable, redshift $\bar{z} = z$.

We summarize the frame-(in)dependent quantities in table 2. As shown, the redshift, z is a frame-independent measurable quantity. Thus, if one express frame-dependent observables, the Hubble parameter, (luminosity distance, angular diameter distance) in terms of z, then one can compare measured quantities between two frames.

Now, we compare the observed quantities both in the JF and in the EF. We show the evolutions of Hubble parameters, $\overline{H}$ and $H_{\rm E}$ in both frames in Fig. 1. We assume that both the background evolution

Table 2. Frame-(in)dependence of (im)measurable quantities. In measurable column, ✗-marks mean unmeasurable and ✓-marks denote measurable. In equality columns, ✗-mark imply inequality and ✓-marks mean equality.

Quantity	Jordan Frame	Einstein Frame	Measurable	Equality
Metric	$\bar{g}_{\mu\nu} = F(\phi) g_{\mu\nu}$	$g_{\mu\nu}$	✗	✗
Time Interval	$d\bar{t} = \frac{dt}{\sqrt{F(\phi)}}$	dt	✗	✗
Scale Factor	$\bar{a}(\bar{t}) = \frac{a(t)}{\sqrt{F(\phi)}}$	$a(t)$	✗	✗
Density (Pressure)	$\bar{\rho}\,(\bar{P}) = F^2 \rho(P)$	$\rho\,(P)$	✗	✗
Hubble Parameter	$\bar{H}(\bar{t}) = \frac{d \ln \bar{a}(\bar{t})}{d\bar{t}}$ $= \sqrt{F} H_E(t)$	$H_E(t) = \frac{d(-\ln \sqrt{F} + \ln a)}{dt}$ $= -\frac{1}{2F}\frac{dF}{dt} + H(t)$	✓	✗
Redshift	$1 + \bar{z} = \frac{\bar{\nu}_s(\bar{t}_s)}{\bar{\nu}_{obs}(\bar{t}_{obs})}$ $= \frac{\bar{a}_{obs}}{\bar{a}_s} = \frac{\sqrt{F(t_s)} a(t_{obs})}{\sqrt{F(t_{obs})} a(t_s)}$	$1 + z = \frac{\nu_s(t_{obs})}{\nu_{obs}(t_{obs})}$ $= 1 + \bar{z}$	✓	✓
Luminosity Distance	$\bar{d}_L = \sqrt{\bar{L}_s / 4\pi \bar{f}_{obs}}$ $= \bar{a}_{obs}(1 + \bar{z}) \sin_K \chi$ $\chi = \int_{\bar{t}_s}^{\bar{t}_{obs}} \frac{d\bar{t}}{\bar{a}}$	$d_L = \sqrt{\frac{L_s}{4\pi f_{obs}} (\frac{e^{\alpha_{obs}}}{e^{\alpha_s}})^2}$ $= a_{obs}(1 + z) \sin_K(\chi)$ χ	✓	✗
Angular Diameter Distance	$\bar{d}_A = \frac{\bar{D}}{\theta}$ $= \bar{a}_{obs}(1 + \bar{z})^{-1} \sin_K \chi$	$d_A = \frac{e^{a_s}}{e^{a_{obs}}} a_s \sin_K \chi$ $= e^{-a_{obs}} \bar{d}_A$	✓	✗

and the large scale structure formation provide $\omega = -1.1 + 0.2\frac{z}{1+z}$ and $\gamma = 0.6$ [7]. The solid line describes the $\overline{H}/\overline{H}_0$ in the JF and the dashed line depicts the $H_{\rm E}/H_{\rm E0}$ in the EF. The difference of Hubble parameters between two frames is as large as 9 % at present. We also show the evolutions of DE EOSs in both frames at Fig. 2. The solid line and the dashed line describe that of the JF and of the EF, respectively. The JF EOS is exactly same as the measured one, $\overline{\omega} = \omega = -1.1 + 0.2\frac{z}{1+z}$ as expected. However, the EF EOS differs from that of the JF $0\%(z = 1.4) \leq \delta\omega \leq 30\%(z = 0)$.

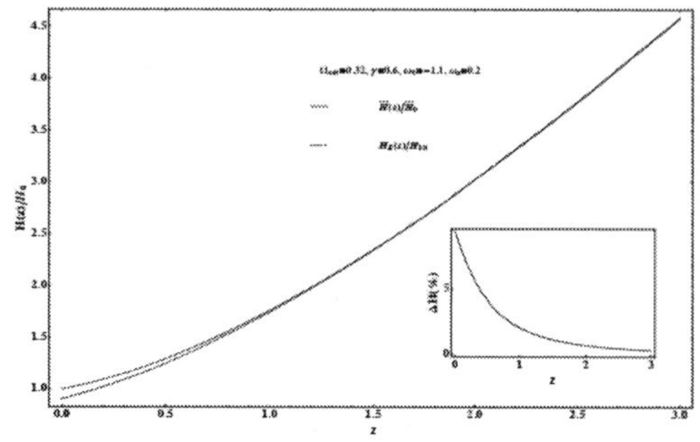

Figure 1. Evolutions of Hubble parameter both in the JF and in the EF. The solid line and the dashed line depict $\overline{H}^2/\overline{H}_0^2$ and $H_{\rm E}^2/H_{\rm E0}^2$, respectively.

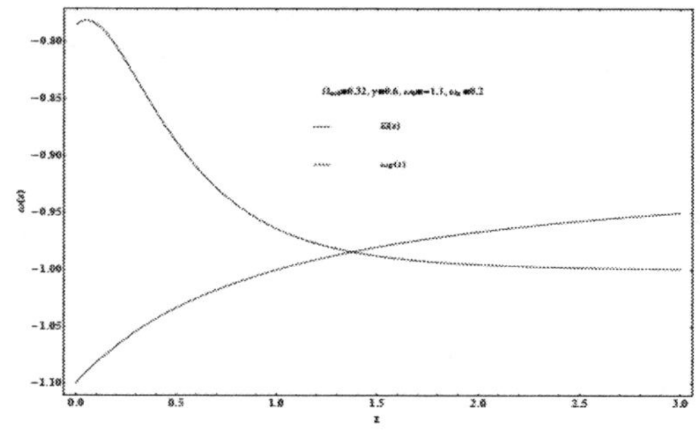

Figure 2. Evolutions of DE EOS both in the JF and in the EF. The solid line and the dashed line depict $\overline{\omega}$ and $\omega_{\rm E}$, respectively.

3 Conclusions

In this manuscript, we explicitly show how to reconstruct the model functions in scalar-tensor gravity theory from the observed parameters, ω and γ. Thus, if observational measurements on both parameters are very accurate, then the dynamics of model functions is determined. From this reconstruction, one can compare the Jordan frame cosmological observables like Hubble parameter and luminosity function with those in the Einstein frame. We explicitly show that the cosmological observables are frame-dependent under conformal transformation. Thus, one should choose a frame first before analysis of any observable and should keep using the same frame in the interpretation of other observable quantities. This demonstrates the explicit conformal inequality for cosmological observables.

Acknowledgments

SL and YK are supported by Basic Science Research Program through the National Research Foundation of Korea (NRF) funded by the Ministry of Science, ICT and Future Planning (Grant No. NRF-2017R1A2B4011168 and NRF-2016R1D1A1B03931090, respectively).

References

[1] M. Chevallier and D. Polarski, Int. J. Mod. Phys. D **10**, 213 (2001) [arXiv:gr-qc/0009008].

[2] E. V. Linder, Phys. Rev. Lett. **90**, 091301 (2003) [arXiv:astro-ph/0208512].

[3] S. Lee and K.-W. Ng, Phys. Rev. D **82**, 043004 (2010) [arXiv:0907.2108].

[4] G. Esposito-Farèse and D. Polarski, Phys. Rev. D **63**, 063504 (2001) [arXiv:gr-qc/0009034].

[5] T. Chiba and M. Yamaguchi, J. Cosmol. Astropart. Phys. **1310**, 040 (2013) [arXiv:1308.1142].

[6] S. Lee, J. Cosmol. Astropart. Phys. **1103**, 021 (2011) [arXiv:1012.2646].

[7] S. Lee, [arXiv:1205.6304].

EPJ Web of Conferences **168**, 08002 (2018) https://doi.org/10.1051/epjconf/201816808002
Joint International Conference of ICGAC-XIII and IK-15 on Gravitation, Astrophysics and Cosmology

Hartle-Hawking wave function and large-scale power suppression of CMB*

Dong-han Yeom**

[1] *Leung Center for Cosmology and Particle Astrophysics, National Taiwan University, Taipei 10617, Taiwan*

Abstract. In this presentation, we first describe the Hartle-Hawking wave function in the Euclidean path integral approach. After we introduce perturbations to the background instanton solution, following the formalism developed by Halliwell-Hawking and Laflamme, one can obtain the scale-invariant power spectrum for small-scales. We further emphasize that the Hartle-Hawking wave function can explain the large-scale power suppression by choosing suitable potential parameters, where this will be a possible window to confirm or falsify models of quantum cosmology. Finally, we further comment on possible future applications, e.g., Euclidean wormholes, which can result in distinct signatures to the power spectrum.

1 Introduction

In order to resolve and understand the initial singularity of our universe, we need to rely on quantum gravity. There are various approaches, but the most conservative approach is to quantize space and time by following the canonical approach. The corresponding master equation for the wave function of the universe is called by the Wheeler-DeWitt equation [1].

The Hartle-Hawking wave function [2] is a proposal of the boundary condition of the Wheeler-DeWitt equation, where this can be presented by the Euclidean path integral

$$\Psi\left[\partial g_{\mu\nu}, \partial\Phi\right] = \int \mathcal{D}g_{\mu\nu}\mathcal{D}\Phi\; e^{-S_{\mathrm{E}}[g_{\mu\nu},\Phi]}, \tag{1}$$

where S_{E} is the Euclidean action as a functional of the metric $g_{\mu\nu}$ and the inflaton field Φ. This path integral is over all geometries and fields that have the boundary $\partial g_{\mu\nu}$ and $\partial\Phi$. This path integral can be further approximated by using the steepest-descent approximation, i.e., this integral is presented by the sum over on-shell solutions:

$$\Psi\left[\partial g_{\mu\nu}, \partial\Phi\right] \simeq \sum e^{-S_{\mathrm{E}}^{\text{on-shell}}}, \tag{2}$$

where these on-shell solutions are called by instantons. Since several years, dynamical and complexified instantons have been extensively investigated [3].

*Proceedings of the 13th International Conference on Gravitation, Astrophysics, and Cosmology & the 15th Italian-Korean Symposium on Relativistic Astrophysics: A Joint Meeting. Talk on July 7, 2017, Seoul, Republic of Korea

**e-mail: innocent.yeom@gmail.com

The Hartle-Hawking wave function not only resolves the initial singularity problem but also gives implications for inflationary cosmology [4]. In this presentation, we first revisit the investigations of Halliwell and Hawking [5]. Then by using this, we can investigate physical implications of the Hartle-Hawking wave function.

2 Halliwell-Hawking revisited

This section is a brief review of the historical work of Halliwell and Hawking [5].

We consider the Einstein gravity with a scalar field:

$$S_{\mathrm{E}} = \int \sqrt{g} d^4x \left[\frac{R}{16\pi} - \frac{1}{2}\nabla_\mu \Phi \nabla^\mu \Phi - V(\Phi) \right], \tag{3}$$

where R is the Ricci scalar and $V(\Phi)$ is the potential term of the inflaton field Φ. For simplicity, we consider the massive scalar field (with the rescaled scalar field $\phi = \Phi\sqrt{4\pi/3}$)

$$V(\Phi) = \frac{8\pi\sigma^2}{3}\left(1 + m^2\phi^2\right), \tag{4}$$

where σ is a constant and m is the mass parameter of the scalar field. It is useful to define the effective Hubble parameter $H_0 = \sqrt{8\pi/3}$.

In order to consider perturbations to the homogeneous solution, we introduce the following metric and field ansatz ($i, j = 1, 2, 3$ and we assume the summation convention for the spherical coordinate indices n, l, m):

$$\begin{aligned}
ds^2 &= \sigma^2\left[\left(N^2 - N_i N^i\right)d\tau^2 + 2N_i dx^i d\tau + a^2\gamma_{ij}dx^i dx^j\right], && (5)\\
N &= N_0\left[1 + \frac{1}{\sqrt{6}} g_{nlm} Q_{nlm}\right], && (6)\\
N_i &= a\left[\frac{1}{\sqrt{6}} k_{nlm}\left(P_i\right)_{nlm} + \sqrt{2} j_{nlm}\left(S_i\right)_{nlm}\right], && (7)\\
\gamma_{ij} &= \left[1 + \frac{\sqrt{6}}{3} q_{nlm} Q_{nlm}\right] d\Omega_3^2 + \Big[\sqrt{6} b_{nlm}\left(P_{ij}\right)_{nlm} \\
&\quad + \sqrt{2} c^o_{nlm}\left(S^o_{ij}\right)_{nlm} + \sqrt{2} c^e_{nlm}\left(S^e_{ij}\right)_{nlm} + 2d^o_{nlm}\left(G^o_{ij}\right)_{nlm} + 2d^e_{nlm}\left(G^e_{ij}\right)_{nlm}\Big], && (8)\\
\Phi &= \sqrt{\frac{3}{4\pi}}\phi + \frac{3\pi}{2} f_{nlm} Q_{nlm}, && (9)
\end{aligned}$$

where N_0 is an arbitrary constant, $a = H_0^{-1}\sin H_0\tau$ and $\phi = $ const. are the background solutions of the de Sitter space (or an approximate solution of the slowly rolling scalar field), g, k, j, q, b, $c^{o,e}$, $d^{o,e}$, f are time dependent mode functions and Q, P_i, S_i, P_{ij}, $S^{o,e}_{ij}$, $G^{o,e}_{ij}$ are space dependent expansion basis. The detailed definitions for the expansion basis are in [5].

In this work, we are focusing on the scalar field perturbations. In the slow-roll limit, the scalar field perturbation f_{nlm} is determined by the following equation [5, 6]:

$$\ddot{f}_{nlm} + 3H_0\dot{f}_{nlm} + \left(m^2 + \frac{n^2 - 1}{a^2}\right) f_{nlm} = 0. \tag{10}$$

From this, we can calculate the power spectrum $P(n)$ as a function of n (after summing over l and m)

$$P(n) = \frac{3n^2}{4\pi}\langle f_n^2\rangle, \tag{11}$$

where we simply denote f_n since it only depends on n.

In order to calculate the expectation value of f_n^2, we need to present the wave function for each perturbation f_n. By plugging the on-shell condition of f_n, the wave function will look like

$$\Psi[f_n] \propto \exp\left[-\frac{1}{2}a^3 f_n \dot{f}_n\right]. \tag{12}$$

There would be many ways to write $\dot{f}_n$ as a function of f_n. However, if we give the Euclidean vacuum condition following Laflamme [7], one can obtain the proper form:

$$\Psi[f_n] \propto \exp\left[-\frac{\dot{\tilde{f}}_n}{2\tilde{f}_n}a^3 f_n^2\right], \tag{13}$$

where $\tilde{f}_n$ and $\dot{\tilde{f}}_n$ are f_n and $\dot{f}_n$ which are evaluated at the horizon crossing time, respectively. By using this wave form, one can finally obtain a closed form of the power spectrum $P(n)$:

$$P(n) = \frac{3n^2}{4\pi}\frac{\tilde{f}_n}{2a^3\dot{\tilde{f}}_n}. \tag{14}$$

3 Suppression of CMB power spectrum

By solving equation (10), we can obtain $\tilde{f}_{nlm}$. There are two linearly independent solutions, but we need to impose the correct boundary condition at $\tau = 0$. The physically viable solution can be presented by the following analytic form [7]:

$$\tilde{f}_n = D\left(z - z^2\right)^{(n-1)/2} {}_2F_1\left(n - \nu, n + \nu + 1; n + 1; z\right), \tag{15}$$

where

$$z = \frac{1 - \cos H_0\tau}{2}, \tag{16}$$

$$\nu = -\frac{1}{2} \pm \sqrt{\frac{9}{4} - \frac{m^2}{H_0^2}}, \tag{17}$$

$$D = \frac{2^{n+1}\Gamma(1 + n/2 + \nu/2)}{n! H_0 \Gamma(1 - n/2 + \nu/2)}. \tag{18}$$

The time parameter τ will be Wick-rotated to $\tau = \pi/2H_0 + it$. The function $\tilde{f}_n$ should be evaluated at the horizon crossing time, i.e., $n = aH = \dot{a} = \sinh H_0 t$ which is the analog relation of $k/aH \simeq 1$ for the flat topology case.

The final result of the power spectrum $P(n)$ is plotted in figure 1 (more detailed discussions are in [6]). If the mass term is zero, then the large-scale power spectrum is enhanced (upper), which is consistent with quantum field theoretical expectations [8]. However, as one turns on the mass term, the large-scale power spectrum can be partly suppressed (lower). Of course, the detailed choices of parameters depend on the detailed inflation model, but it is enough to show that quantum cosmological calculations based on the Hartle-Hawking wave function can explain the large-scale power suppression. Note that for both cases, the power spectrum is scale-invariant for small-scales as we expected.

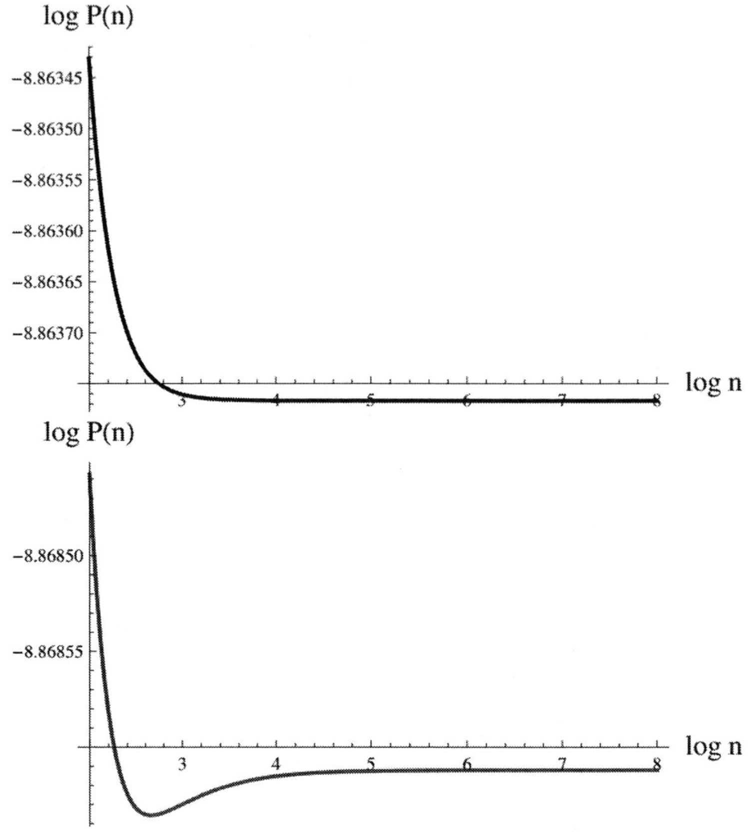

Figure 1. The power spectrum $P(n)$ for $H_0 = 0.01$ and $m = 0$ (upper) and $m = 0.001$ (lower). If the mass term is zero, then the power spectrum is enhanced for small n. However, as m increases, there is a possibility that the power spectrum is suppressed for small n.

4 Future perspectives

Up to now, the observed CMB data has a tension with the theoretical expectations (based single field inflation and the ΛCDM model), especially for large length scales [9]. It is not clear whether this tension will be confirmed to be real or not for future experiments. However, if it is the case, then this tension with theoretical expectations is a good window to test quantum gravity.

In this paper, we show that the Hartle-Hawking wave function can easily explain this power suppression behavior. The Hartle-Hawking wave function is not the only way to explain the large-scale power suppression (e.g., [10]), but we can say that this approach is very well-established and conservative compared to the others.

We can extend this investigation to the other inflationary models [11] and also improve approximations for dynamics of the inflaton field [12]. Perhaps, we may study not only compact instantons but also non-compact instantons, e.g., Euclidean wormholes [13]. In the end, we hope that future experiments can confirm or falsify several models of quantum cosmology. Then, quantum gravity will be established on the experimental ground.

Acknowledgment

The author would like to thank Pisin Chen and Yu-Hsiang Lin for the stimulated collaborations of this project. The work is supported by Leung Center for Cosmology and Particle Astrophysics (LeCosPA) of National Taiwan University (103R4000).

References

[1] B. S. DeWitt, Phys. Rev. **160**, 1113 (1967).

[2] J. B. Hartle and S. W. Hawking, Phys. Rev. D **28**, 2960 (1983).

[3] D. Hwang, H. Sahlmann and D. Yeom, Class. Quant. Grav. **29**, 095005 (2012) [arXiv:1107.4653 [gr-qc]];
D. Hwang, B. H. Lee, H. Sahlmann and D. Yeom, Class. Quant. Grav. **29**, 175001 (2012) [arXiv:1203.0112 [gr-qc]];
P. Chen, T. Qiu and D. Yeom, Eur. Phys. J. C **76**, no. 2, 91 (2016) [arXiv:1503.08709 [gr-qc]].

[4] D. Hwang, S. A. Kim, B. H. Lee, H. Sahlmann and D. Yeom, Class. Quant. Grav. **30**, 165016 (2013) [arXiv:1207.0359 [gr-qc]];
D. Hwang, S. A. Kim and D. Yeom, Class. Quant. Grav. **32**, no. 11, 115006 (2015) [arXiv:1404.2800 [gr-qc]].

[5] J. J. Halliwell and S. W. Hawking, Phys. Rev. D **31**, 1777 (1985).

[6] P. Chen, Y. H. Lin and D. Yeom, arXiv:1707.01471 [gr-qc].

[7] R. Laflamme, Phys. Lett. B **198**, 156 (1987).

[8] A. A. Starobinsky, astro-ph/9603075.

[9] N. Aghanim *et al.* [Planck Collaboration], Astron. Astrophys. **594**, A11 (2016) [arXiv:1507.02704 [astro-ph.CO]].

[10] A. Ashtekar and B. Gupt, Class. Quant. Grav. **34**, no. 1, 014002 (2017) [arXiv:1608.04228 [gr-qc]].

[11] D. Hwang and D. Yeom, JCAP **1406**, 007 (2014) [arXiv:1311.6872 [gr-qc]].

[12] D. Hwang, B. H. Lee, E. D. Stewart, D. Yeom and H. Zoe, Phys. Rev. D **87**, no. 6, 063502 (2013) [arXiv:1208.6563 [gr-qc]].

[13] P. Chen, Y. C. Hu and D. Yeom, JCAP **1707**, no. 07, 001 (2017) [arXiv:1611.08468 [gr-qc]];
S. Kang and D. Yeom, arXiv:1703.07746 [gr-qc];
P. Chen and D. Yeom, arXiv:1706.07784 [gr-qc].

EPJ Web of Conferences **168**, 08003 (2018) https://doi.org/10.1051/epjconf/201816808003
Joint International Conference of ICGAC-XIII and IK-15 on Gravitation, Astrophysics and Cosmology

On Gauge Invariant Cosmological Perturbations in UV-modified Hořava Gravity: A Brief Introduction*

Mu-In Park[1]**

[1] *Research Institute for Basic Science, Sogang University, Seoul, 121-742, Korea*

Abstract. We revisit gauge invariant cosmological perturbations in UV-modified, $z = 3$ Hořava gravity with one scalar matter field, which has been proposed as a renormalizable gravity theory without the ghost problem in four dimensions. We confirm that there is no extra graviton modes and general relativity is recovered in IR, which achieves the consistency of the model. From the UV-modification terms which break the detailed balance condition in UV, we obtain scale-invariant power spectrums for *non*-inflationary backgrounds, like the power-law expansions, without knowing the details of early expansion history of Universe. This could provide a new framework for the Big Bang cosmology.

"All truth passes trough three stages. First, it is ridiculed. Second, it is violently opposed. Third, it is accepted as being self-evident." (A. Schopenhauer)

1 Introduction

(1). As a particle physicist's point of view, a surprise of cosmology is as follows. In particle physics, we need (fundamental) scalar fields (called Higgs fields) for "theoretical" reasons (*i.e.*, to give masses to fundamental particles with renormalizability), but we have waited for a long time (about 40 years, after 1964 papers) for an experimental confirmation at LHC (4 July, 2012). This is (thought to be) the first elementary scalar particle discovered in Nature.

However, in cosmology, it has been known for a long time that there *is* scalar (cosmological fluctuation mode) in the sky after the discovery of CMB (1964, the same year that Higgs field has been proposed !). But, the existence of a scalar mode in the sky is a big mystery from the following reasons. The small fluctuations in CMB are thought to be due to (space-time) fluctuations in the early epoch of our (Big Bang) Universe, which should be described by General Relativity (GR). However, GR alone can *not* have scalar (fluctuation) mode but only the (usual) tensor mode (2 polarizations, called gravitational waves) ! The cosmological/primordial tensor modes have not been discovered yet, though astrophysical tensor modes (gravitational waves) from black hole mergers have been detected in LIGO. So, it would not be quite strange (at least to me) even though we discover the cosmological/primordial tensor modes (called B-mode) in the near future. Rather, the existence of the scalar mode is much more surprising !

*Proceeding of International Joint Conference of ICGAC-XIII and IK15 (July 3-7, 2017, Seoul, Republic of Korea)

**e-mail: muinpark@gmail.com

The simplest way to explain the data is introduce a "cosmic scalar" field and "assume" some peculiar behaviors in the early epoch (known as the inflationary epoch). But, there are huge numbers of explicit models and we still do not know what is the right one and its origin–Is Higgs the remnant of the primordial scalar ? This situation is opposite to that of particle physics: In cosmology, we have discovered scalar mode (CMB) long ago but we do not have its theory yet !

Another way to explain the scalar mode is to modify GR, like $f(R)$, massive gravity, etc., so that the gravity itself has additional scalar modes. But, in that case the scalar (gravitation) could be disaster since it could affect the known (or, well-established) GR test in solar system (,*i.e.,* low energy (IR) test of GR), unless we decouple it from low energy GR sector. And also, this could affect the dark matter and dark energy problems as well. Sometimes, the scalar could be ghost as well, which should be avoided. Today, I will consider another modified gravity theory but without the scalar gravitation mode so that the usual scalar matters are needed in explaining the cosmology data.

(2). One the other hand, our universe is considered to be created from "quantum (vacuum) fluctuations". Actually, since the scalar power spectrum in inflationary cosmology contains Planck constant $\hbar$ as well as Newton's constant G (H is the Hubble parameter), $\Delta^2(k) = 8\pi G\hbar(2H^2/\pi^2)$, one can consider the power spectrum as a "quantum gravity effect" and so does the inflationary cosmology as a "quantum cosmology" ! Moreover, recent LIGO's detections of gravitational waves from merging black holes seems to imply that we need to consider quantum gravity more seriously. Actually, we open the strong gravity test era of GR, beyond the weak gravity test in solar system.

(3). Do we have quantum gravity then ? For quantum theory of particle's interactions, "renormalization" has been a powerful constraint and Higgs particle is its natural consequence. Now, what if we require renormalizability in quantum gravity ? But, it is well-known that the renormalizable quantum gravity can not be realized in Einstein's gravity or its (relativistic) higher-derivative generalizations: There are "ghosts" which have kinetic terms with a wrong sign, in addition to massless gravitons.

Recently, Hořava (or Hořava-Lifshitz (HL)) gravity has been proposed as a renormalzable gravity [1]. There is no ghosts (in the tensor modes) by abandoning the equal-footing treatment of space and time (*i.e.,* Lorentz symmetry) in the higher energy regime (UV). It is power-counting renormalizable but no proof of renormalizability yet. This resembles the situations of Yang-Mills theory when it first appeared.

In cosmology, this theory *may* provide inflationary effect without "inflationary phase (*i.e.*, early de Sitter or accelerating phase)" so that scale-invariant (scalar/tensor) spectrum can be also produced. This has been first argued by Mukohyama [2] but there is no rigorous analysis about this yet. On the contrary, in the works of Gao *et. al.* [3] and Gong *et. al.* [4], scalar spectrum are not scale-invariant though tensor spectrum does. This is today's topic and it provides a natural formulation of cosmology, known as "effective field theory (EFT)", by construction.

2 Hořava (-Lifshitz) Gravity: Basic Idea

In 2009, Hořava proposed a renormalizable, higher-derivative gravity theory, without ghost problems, by abandoning Lorentz symmetry in UV but keeping, so called, "foliation preserving" diffemorphism ($FPDiff$) [1].

The basic ingredients of the Hořava are (1) Einstein gravity (with a Lorentz deformation parameter λ), (2) non-covariant deformations with higher-spatial derivatives (up to six orders), and (3) "detailed balance" in the coefficients (with five constant parameters, $\kappa, \lambda, \nu, \mu, \Lambda_W$), in contrast to the Einstein gravity case with the three physical parameters c, G, Λ and $\lambda = 1$. Here, six orders of spatial derivatives ($z = 3$) came from the power counting renormalizability in $3 + 1$ dimensions. (For the limitation of space and time, I can not explain all the details of the construction. So, please read the original work of Hořava [1] about this.)

From the Hořava's construction, some improved UV behaviors, without ghosts, are expected so that the gravity theory "could" be renormalizable and one might have predictable quantum gravity theory. But, in cosmology applications, it seems that the detailed-balance condition is too strong to get the required, scale invariant cosmological perturbations. For example, tensor spectrum is scale invariant but scalar spectrum is not ! [3, 4]. So, we need to break the detailed balancing in some way but without altering the improved UV behaviors of scale invariant tensor modes in the previous works [3, 4] and we called it as "UV modifications" [5].

3 Cosmological Perturbations in Hořava gravity

We start by considering the four-dimensional, UV-modified Hořava gravity action with $z = 3$, which is power-counting renormalizable [1],

$$S_g = \int d\eta d^3x\sqrt{g}N\left[\frac{2}{\kappa^2}\left(K_{ij}K^{ij}-\lambda K^2\right)-\mathcal{V}\right], \tag{1}$$
$$-\mathcal{V} = \sigma+\xi R+\alpha_1 R^2+\alpha_2 R_{ij}R^{ij}+\alpha_3\frac{\epsilon^{ijk}}{\sqrt{g}}R_{il}\nabla_j R^l{}_k+\alpha_4\nabla_i R_{jk}\nabla^i R^{jk}+\alpha_5\nabla_i R_{jk}\nabla^j R^{ik}+\alpha_6\nabla_i R\nabla^i R,$$

where (the prime ($'$) denotes the derivative with respect to η) $K_{ij} = \frac{1}{2N}\left(g_{ij}{}'-\nabla_i N_j-\nabla_j N_i\right)$, $ds^2 = -N^2c^2d\eta^2+g_{ij}\left(dx^i+N^id\eta\right)\left(dx^j+N^jd\eta\right)$. With the detailed balance condition, the number of independent coupling constants can be reduced to six, *i.e.*, $\kappa, \lambda, \mu, \nu, \Lambda_W, \omega$ for a viable model in IR [8, 9], $\sigma = \frac{3\kappa^2\mu^2\Lambda_W^2}{8(3\lambda-1)}$, $\xi = \frac{\kappa^2\mu^2(\omega-\Lambda_W)}{8(3\lambda-1)}$, $\alpha_1 = \frac{\kappa^2\mu^2(4\lambda-1)}{32(3\lambda-1)}$, $\alpha_2 = -\frac{\kappa^2\mu^2}{8}$, $\alpha_3 = \frac{\kappa^2\mu}{2\nu^2}$, $\alpha_4 = -\frac{\kappa^2}{82\nu^4} = -\alpha_5 = -8\alpha_6$, But, we do not restrict to this case only, at least for the UV couplings $\alpha_4, \alpha_5, \alpha_6$ so that the power-counting renormalizable and scale-invariant cosmological scalar fluctuations can be obtained.

For the power-counting renormalizable matter action, we consider $z = 3$ scalar field action [6, 7],

$$S_m = \int d\eta d^3x\sqrt{g}N\left[\frac{1}{2N^2}\left(\phi'-N^i\partial_i\phi\right)^2-V(\phi)-Z(\partial_i\phi)\right], \tag{2}$$

where $Z(\partial_i\phi) = \sum_{n=1}^{3}\xi_n\partial_i^{(n)}\phi\partial^{i(n)}\phi$ with the superscript (n) denoting n-th spatial derivatives, and $V(\phi)$ is the matter's potential without derivatives.

The actions, S_g and S_m are invariant under the foliation preserving *FPDiff* [1],

$$\delta x^i = -\zeta^i(\eta,\mathbf{x}),\ \delta\eta = -f(\eta),\ \delta g_{ij} = \partial_i\zeta^k g_{jk}+\partial_j\zeta^k g_{ik}+\zeta^k\partial_k g_{ij}+fg'_{ij}, \tag{3}$$
$$\delta N_i = \partial_i\zeta^j N_j+\zeta^j\partial_j N_i+\zeta'^j g_{ij}+fN'_i+f'N_i,\ \delta N = \zeta^j\partial_j N+fN'+f'N,\ \delta\phi = \zeta^j\partial_j\phi+f\phi'.$$

In order to study the cosmological perturbations around the homogeneous and isotropic backgrounds (as seen in CMB), we expand the metric and the scalar field as,

$$N = a(\eta)[1+\mathcal{A}(\eta,\mathbf{x})],\ N_i = a^2(\eta)\mathcal{B}(\eta,\mathbf{x})_{,i},\ g_{ij} = a^2(\eta)[\delta_{ij}+h_{ij}(\eta,\mathbf{x})],\ \phi = \phi_0(\eta)+\delta\phi(\eta,\mathbf{x}), \tag{4}$$

by considering spatially flat ($k = 0$) backgrounds and the conformal (or *comoving*) time η, for simplicity. By substituting the metric and scalar field of (4) into the actions one can obtain the linear-order perturbation part of the total action $S = S_g + S_m$ which gives the Friedman's equations for the background,

$$\mathcal{H}^2 = -\frac{\kappa^2}{6(1-3\lambda)}\left(\frac{1}{2}\phi_0'^2+a^2(V_0-\sigma)\right),\ \mathcal{H}^2+2\mathcal{H}' = \frac{\kappa^2}{2(1-3\lambda)}\left(\frac{1}{2}\phi_0'^2-a^2(V_0-\sigma)\right),$$
$$\phi_0''+2\mathcal{H}\phi_0'+a^2V_{\phi_0} = 0, \tag{5}$$

with the comoving Hubble parameter $\mathcal{H} \equiv a'/a$, $V_0 \equiv V(\phi_0)$, $V_{\phi_0} \equiv (\partial V/\partial\phi)_{\phi_0}$, and $h \equiv h^i_i$. Here, it important to note that there is *no* higher-derivative corrections to the Friedman's equations for spatially flat case and so the background equations are the same as those of GR [8, 9]. However, even in this case, the higher-derivative effects can *re*appear in the perturbed parts.

The quadratic part of the total perturbed action is given by

$$\begin{aligned}\delta_2 S &= \int d\eta d^3x \left\{ \frac{2a^2}{\kappa^2}\left[(1-3\lambda)\mathcal{H}\left(3\mathcal{H}\mathcal{A}^2 + \mathcal{A}(2\partial\mathcal{B}^i - h')\right) + (1-\lambda)(\partial_i\mathcal{B}^i)^2 + \frac{1}{2}\partial_i\mathcal{B}_j\partial^i\mathcal{B}^j \right.\right.\\ &\left. -\partial_i\mathcal{B}_j h^{ij\prime} + \frac{1}{4}h_{ij}{}' h^{ij\prime} + \lambda\left(\partial_i\mathcal{B}^i h' - \frac{1}{4}h'^2\right)\right] + a^2\xi\left(\mathcal{A} + \frac{1}{2}h\right)\left(\partial_i\partial_j h^{ij} - \Delta h\right)\\ &+a^2\left[\frac{1}{2}\delta\phi'^2 - \mathcal{A}\phi_0'\delta\phi' + \frac{1}{2}\mathcal{A}^2\phi_0'^2 + \partial_i\mathcal{B}^i\phi_0'\delta\phi - \frac{a^2}{2}V_{\phi_0\phi_0}\delta\phi^2 - a^2 V_{\phi_0}\delta\phi\mathcal{A} - \frac{1}{2}\phi_0'\delta\phi h'\right]\\ &\left. -a^4\left(\mathcal{V}^{(2)} + \delta Z\right)\right\}, \qquad (6)\end{aligned}$$

where $\mathcal{V}^{(2)}$ is the quadratic part of the potential $\mathcal{V}$ and $\Delta \equiv \delta^{ij}\partial_i\partial_j$, $\delta Z = \sum_{n=1}^{3}\xi_n\,\partial_i^{(n)}\delta\phi\partial^{i(n)}\delta\phi$.

Now, in order to separate the scalar, vector, and tensor contributions, we consider the most general (SVT) decompositions ($\partial_i S^i = \partial_i F^i = \tilde{H} = \partial_i\tilde{H}^i_j = 0$), $\mathcal{B}_i = \partial_i\mathcal{B} + S_i$, $h_{ij} = 2\mathcal{R}\delta_{ij} + \partial_i\partial_j\mathcal{E} + \partial_{(i}F_{j)} + \tilde{H}_{ij}$ .Then, the pure tensor, vector, and scalar parts of the total action are given by, respectively,

$$\begin{aligned}\delta_2 S^{(t)} &= \int d\eta d^3x\, a^2\left[\frac{2}{\kappa^2}\tilde{H}'_{ij}\tilde{H}^{ij\prime} + \xi\tilde{H}_{ij}\Delta\tilde{H}^{ij} + \frac{\alpha_2}{a^2}\Delta\tilde{H}_{ij}\Delta\tilde{H}^{ij} + \frac{\alpha_3}{a^3}\epsilon^{ijk}\Delta\tilde{H}_{il}\Delta\partial_j\tilde{H}^l_k\right.\\ &\left.\quad -\frac{\alpha_4}{a^4}\Delta\tilde{H}_{ij}\Delta^2\tilde{H}^{ij}\right], \qquad (7)\\ \delta_2 S^{(v)} &= \frac{1}{\kappa^2}\int d\eta d^3x\, a^2\partial_i\left(S^j - F^{j\prime}\right)\partial_i\left(S_j - F'_j\right), \qquad (8)\\ \delta_2 S^{(s)} &= \int d\eta d^3x\, a^2\left\{\frac{2(1-3\lambda)}{\kappa^2}\left[3\mathcal{R}'^2 - 6\mathcal{H}\mathcal{A}\mathcal{R}' + 3\mathcal{H}^2\mathcal{A}^2 - 2(\mathcal{R}' - \mathcal{H}\mathcal{A})\Delta(\mathcal{B} - \mathcal{E}')\right]\right.\\ &\quad +\frac{2(1-\lambda)}{\kappa^2}\left[\Delta(\mathcal{B} - \mathcal{E}')\right]^2 - 2\xi(\mathcal{R} + 2\mathcal{A})\Delta\mathcal{R} + \frac{2}{a^2}(8\alpha_1 + 3\alpha_2)(\Delta\mathcal{R})^2\\ &\quad -\frac{2}{a^4}(3\alpha_4 + 2\alpha_5 + 8\alpha_6)\Delta\mathcal{R}\Delta^2\mathcal{R} - a^2 V_{\phi_0}\mathcal{A}\delta\phi - \frac{1}{2a^2}V_{\phi_0\phi_0}\delta\phi^2 - \delta Z\\ &\quad \left. +\frac{1}{2}\delta\phi'^2 - \phi_0'\delta\phi'\mathcal{A} + \frac{1}{2}\phi_0'^2\mathcal{A}^2 + [\Delta(\mathcal{B} - \mathcal{E}') - 3\mathcal{R}']\phi_0'\delta\phi\right\}. \qquad (9)\end{aligned}$$

Here, it is important to note that sixth-order-derivative terms in the gravity action contribute to scalar as well as tensor perturbations, through the specific combination of '$3\alpha_4 + 2\alpha_5 + 8\alpha_6$' for the former but through only 'α_4' for the latter. This is what we need for renormalizability and scale-invariant spectrums (as we can see shortly), for both scalar and tensor. In the detailed-balanced case, the combination '$3\alpha_4 + 2\alpha_5 + 8\alpha_6$' vanishes though α_4 does not. So in that case, the theory would not be renormalizable and nor scale invariant for scalar part, which is in contradict to observational data !

Now, in order to exhibit the true dynamical degrees of freedom we consider the Hamiltonian reduction method [10], for the cosmologically perturbed actions [4, 11] and, after some computation, the action reduces to

$$\delta_2 S^{(s)}_\star = \int d\eta d^3x \frac{1}{2}\left\{u'^2 - u\left[\frac{1}{2}\left(\mathcal{G}_1'\mathcal{G}_1^{-1}\right)' - \frac{1}{4}\left(\mathcal{G}_1'\mathcal{G}_1^{-1}\right)^2 - \mathcal{G}_1\left(\mathcal{G}_2\mathcal{G}_1^{-1}\right)' - \mathcal{G}_2^2 + 4\mathcal{G}_1\mathcal{G}_3\right]u\right\} \qquad (10)$$

for a true scalar degree of freedom u. In UV limit ($3\tilde{\alpha}_4 \equiv 3\alpha_4 + 2\alpha_5 + 8\alpha_6$, $z \equiv a\phi_0'/\mathcal{H}$), its equations of motion reduce to

$$u'' = -\omega_{u(UV)}^2 u, \quad \omega_{u(UV)}^2 = \frac{-6\xi_3\tilde{\alpha}_4}{a^2 z^2}\left[2 + \frac{\kappa^2(1-\lambda)}{4(1-3\lambda)}\frac{z^2}{a^2}\right]\Delta^3. \tag{11}$$

Here, it is important to note that there are sixth-spatial derivatives, as required by the scale invariance of the observed power spectrum [2] as well as the (power-counting) renormalizability [1]. This occurs only when there are sixth-derivative terms in the starting scalar action as well as some breaking of the detailed balance condition in sixth-derivative terms for the gravity action (*i.e.*, $\tilde{\alpha}_4 \neq 0$)

Regarding the scale invariance of the power spectrums, it has been noted that Hořava gravity could provide an alternative mechanism for the early Universe without introducing the hypothetical inflationary epoch [2]: The basic reason of the alternative mechanism comes from the momentum-dependent speeds of gravitational perturbations which could be much larger than the current, low energy (*i.e.*, IR) speed c so that the exponentially expanding early space-time could be mimicked. In order to see this explicitly in our case, we consider the power-law expansions [12] (t is the physical time, defined by $dt = ad\eta$), $a = a_0 t^p$, $(1/3 < p < 1)$.Then we can produce the scale-invariant power spectrums for the quantum field $\hat{\zeta}$ of the ζ (scalar) perturbation, as follows,

$$\left\langle 0|\hat{\zeta}_{\mathbf{k}}(\eta)\hat{\zeta}_{\mathbf{k}'}(\eta)|0\right\rangle = (2\pi)^3\delta(\mathbf{k}+\mathbf{k}')\frac{2\pi^2}{k^3}\Delta_\zeta^2(k), \quad \Delta_\zeta^2 = \frac{k^3}{2\pi^2}|\zeta_{\mathbf{k}}|^2 = \frac{\hbar}{8\pi^2}\sqrt{\frac{(3\lambda-1)[1-(1-\lambda)/4]}{6\kappa^2\xi_3\tilde{\alpha}_4}}, \tag{12}$$

without knowing the details of the history of the early Universe and the form of the (non-derivative) potential $V(\phi)$.

This result shows that one can achieve the "inflation without inflation" picture, as argued by Mukohyama [2]. The basic reason of this is that, in Hořava gravity, due to the momentum-dependent, superluminal speeds of fluctuations is possible in the early Universe, which is assumed to be UV region, so that one can mimic "the inflationary scenario without inflationary epoch" ! (See Fig. 1 in [5].)

Acknowledgments

This was supported by Basic Science Research Program through the National Research Foundation of Korea (NRF) funded by the Ministry of Education, Science and Technology (2016R1A2B401304).

References

[1] P. Hořava, Phys. Rev. D **79**, 084008 (2009) [arXiv:0901.3775 [hep-th]].
[2] S. Mukohyama, Class. Quant. Grav. **27**, 223101 (2010) [arXiv:1007.5199 [hep-th]].
[3] X. Gao, Y. Wang, R. Brandenberger and A. Riotto, Phys. Rev. D **81**, 083508 (2010) [arXiv:0905.3821 [hep-th]].
[4] J. O. Gong, S. Koh and M. Sasaki, Phys. Rev. D **81**, 084053 (2010) [arXiv:1002.1429 [hep-th]].
[5] S. Shin and M. I. Park, arXiv:1701.03844 [hep-th].
[6] G. Calcagni, JHEP **0909**, 112 (2009) [arXiv:0904.0829 [hep-th]].
[7] E. Kiritsis and G. Kofinas, Nucl. Phys. B **821**, 467 (2009) [arXiv:0904.1334 [hep-th]].
[8] A. Kehagias and K. Sfetsos, Phys. Lett. B **678**, 123 (2009) [arXiv:0905.0477 [hep-th]].
[9] M. -I. Park, Class. Quant. Grav. **28**, 015004 (2011) [arXiv:0910.1917 [hep-th]].
[10] L. D. Faddeev and R. Jackiw, Phys. Rev. Lett. **60**, 1692 (1988).
[11] J. Garriga, X. Montes, M. Sasaki and T. Tanaka, Nucl. Phys. B **513**, 343 (1998) [arXiv:astro-ph/9706229].
[12] F. Lucchin and S. Matarrese, Phys. Rev. D **32**, 1316 (1985).

Quintessence in low-energy effective theory

Tae Hoon Lee[1,2,⋆]

[1] *Department of Physics, Soongsil University, Seoul 06978, Korea*
[2] *Research Institute for Origin of Matter and Evolution of Galaxies, Soongsil University, Seoul 06978, Korea*

Abstract. Considering a theory of Brans-Dicke gravity with general couplings of a heavy field, we derive the low-energy effective theory action in the universe of temperature much lower than the heavy field mass. Gravitational equations and the Brans-Dicke scalar field equation including an effective potential of the scalar field are obtained, which is induced through virtual interactions of the heavy field in the late-time universe. We find a deSitter cosmological solution stemming from the inverse power law effective potential of the scalar field and discuss the possibility that the late time acceleration of our universe can be described by means of the solution.

Recent cosmological observations including supernova data [1] suggest that our universe is made up of about 68 % dark energy, about 27 % dark matter, and about 5 % ordinary matter, and they cause an explosion of interest in the origin of dark energy [1] and dark matter [2]. The current cosmic acceleration due to the dark energy and the primordial inflation can be described by using the de Sitter solutions to the gravitational field equations. Various approaches have been used to understand the dark energy, such as the cosmological constant, quintessence [3], and the phantom [4] in general relativity. Other approaches are based on assorted types of modifications to general relativity, such as $f(R)$ gravity and scalar-tensor theories of gravity. We study the late-time acceleration of the Universe, in the Brans-Dicke (BD) theory of gravity coupled to a heavy field.

From the BD gravity theory, we derive the low-energy effective theory [5] action in the universe of temperature much lower than the heavy field mass and obtain an inverse power law effective potential of the BD scalar field, which is stimulated through virtual interactions of the heavy field. With the effective potential we discuss the possibility that the cosmological constant problem can be addressed. Before doing these things, we present the method by which the low-energy effective theory can be obtained from a high energy theory, as follows:

The (effective action or) generating functional (GF) for connected, one-particle-irreducible (1PI) with respect to ϕ- and h-lines Green functions is given by the Legendre transform of a GF for connected Green functions, $W(j, J) = -i \ln Z(j, J)$ with GF for (general) Green functions $Z(j, J)$,

$$\Gamma(\phi, h) = W(j, J) - \int d^4x j\phi - \int d^4x Jh \tag{1}$$

with $\frac{\delta W(j,J)}{\delta j} = \phi, \quad \frac{\delta W(j,J)}{\delta J} = h$. Eq. (1) can be rewritten as

$$\Gamma(\phi, h) = \Gamma(\phi, J) - \int d^4x Jh, \tag{2}$$

⋆e-mail: thlee@ssu.ac.kr

where the GF for connected, 1PI with respect to ϕ-lines Green functions is given by the *partial* Legendre transform

$$\Gamma(\phi, J) = W(j, J) - \int d^4x j\phi. \tag{3}$$

with $\frac{\delta W(j,J))}{\delta j} = \phi$. From (2) we also have the relation

$$\Gamma(\phi, J) = \Gamma(\phi, h) + \int d^4x Jh \tag{4}$$

with

$$\frac{\delta\Gamma(\phi, h)}{\delta h} = -J, \tag{5}$$

which gives us $h = h(\phi, J)$.

When the freedoms associated with matter fields, $h(x^\mu)$, heavier ($m_h > T$) than the temperature of the Universe are hidden from direct observation (*i.e.*, $J = 0$), the effective theory is described by the low-energy effective action $S_{eff}(\phi)$ of light fields, $\phi(x^\mu)$, only. In the tree-level approximation, its action is given by

$$S_{eff}(\phi) \equiv \Gamma^{(0)}(\phi, h(\phi)), \tag{6}$$

where $h(\phi) \equiv h(\phi, J = 0)$ is derived from the equation

$$\frac{\delta\Gamma^{(0)}(\phi, h)}{\delta h} = 0 \tag{7}$$

with the effective action $\Gamma(\phi, h)$ defined in terms of both ϕ and h fields in a high-energy theory, as in Eq. (1).

The BD theory of gravity coupled to a heavy field $h(x^\mu)$ can be described by the action, $S = \int d^4x \sqrt{-g}[\varphi R - \frac{\omega_{bd}}{\varphi} g^{\alpha\beta}\partial_\alpha\varphi\partial_\beta\varphi - V(\varphi)] + S_{(hm)}$, which is rewritten as

$$S = \int d^4x \sqrt{-g}[\phi^2 R - \omega g^{\alpha\beta}\partial_\alpha\phi\partial_\beta\phi - V(\phi)] + S_{(hm)}, \tag{8}$$

with substitution $\varphi = \phi^2$ and $\omega_{bd} = \frac{1}{4}\omega$. The action for the heavy field is assumed by

$$S_{(hm)} = \int d^4x \sqrt{-g}[-\frac{1}{2} g^{\alpha\beta}\partial_\alpha h\partial_\beta h - \bar{V}(h) - V_{int}(\phi, h)], \tag{9}$$

where $\bar{V}$ is the potential of the heavy field and the non-renormalizable interaction potential

$$V_{int}(\phi, h) = \frac{u}{M_d^{k+l-4}}\phi^k h^l \tag{10}$$

with $k + l > 4$.

By applying the formula (7) to Eqs. (9) and (10) and by choosing $\bar{V}(h) = \frac{1}{2}m_h^2 h^2$, we get

$$h(\phi) \simeq (\frac{-ul\phi^k}{m_h^2 M_d^{(k+l-4)}})^{\frac{1}{2-l}}(1 + \frac{1}{2-l}\frac{g^{\alpha\beta}\nabla_\alpha\partial_\beta}{m_h^2}) \simeq (\frac{-ul\phi^k}{m_h^2 M_d^{(k+l-4)}})^{\frac{1}{2-l}}, \tag{11}$$

in the low-energy limit, $|\partial_\beta h| << m_h h$. Substituting Eq. (11) into Eqs. (6) and (8), we have

$$S_{eff}(\phi) \simeq \int d^4x \sqrt{-g}[\phi^2 R - \omega(\phi) g^{\alpha\beta}\partial_\alpha\phi\partial_\beta\phi - V(\phi) - \frac{1}{2} g^{\alpha\beta}\partial_\alpha h(\phi)\partial_\beta h(\phi) - \mathcal{V}^{eff}(\phi)], \tag{12}$$

where the effective potential

$$\mathcal{V}^{eff}(\phi) = d\,\phi^{\frac{-2k}{l-2}} \tag{13}$$

with $d = (1/2l^{\frac{2}{2-l}} - l^{\frac{l}{2-l}})(-u)^{\frac{2}{2-l}} m_h^{\frac{2l}{l-2}} (M_d^{(k+l-4)})^{\frac{2}{l-2}}$ ($u < 0$, $k+l > 4$, and $l > 2$), which gives us an inverse power law potential for a quintessence model. The equation of state is given by

$$\omega_{tot} \equiv \frac{p}{\rho} = \frac{\omega\dot{\phi}^2 + 4(\dot{\phi}^2 + \phi\ddot{\phi} + 2H\phi\dot{\phi}) - V + c\phi^{\frac{2k+2l-4}{2-l}}\dot{\phi}^2 - d\phi^{\frac{-2k}{l-2}}}{\omega\dot{\phi}^2 - 12H\phi\dot{\phi} + V + c\phi^{\frac{2k+2l-4}{2-l}}\dot{\phi}^2 + d\phi^{\frac{-2k}{l-2}}}$$

with $c = \frac{1}{2}\frac{k^2}{(2-l)^2}(\frac{-ul}{m_h^2 M_{dm}^{(k+l-4)}})^{\frac{2}{2-l}}$. Note that $\omega_{tot} > -1$.

In the flat Friedamnn-Robertson-Waker spacetime, we can derive the scalar field equation and the Friedmann equation from Eq. (12).

$$\begin{aligned}&(2\omega + h'^2)\ddot{\phi}\\ &= -(6\omega H + 3Hh'^2 + h'h''\dot{\phi})\dot{\phi} - 2\{V_{,\phi^2} - 6(\dot{H} + 2H^2) + \mathcal{V}^{eff}{}_{,\phi^2}\}\phi\end{aligned}$$

with $h' \equiv \frac{\partial h}{\partial \phi}$, $V_{,\phi^2} \equiv \frac{\partial V}{\partial \phi^2}$, ... and

$$\begin{aligned}&-4\phi^2\dot{H}(= \rho + p)\\ &= 2(\omega + 2)\dot{\phi}^2 + 4\phi\ddot{\phi} + \dot{h}^2(\phi) - 4H\phi\dot{\phi},\end{aligned} \tag{14}$$

subject to the constraint

$$6\phi^2 H^2 = \omega\dot{\phi}^2 - 12H\phi\dot{\phi} + V + \frac{1}{2}\dot{h}^2(\phi) + \mathcal{V}^{eff}(\phi), \tag{15}$$

from which we get a solution for de Sitter universe with $\dot{H} = 0$ and $\ddot{\phi} = 0$, which has following requirements (with $H = const.$ and $\dot{\phi} = const.$):

$$6\phi^2 H^2 = \omega_e\dot{\phi}^2 - 12\phi\dot{\phi}H + V(\phi) + \mathcal{V}^{eff}(\phi), \tag{16}$$

$$(\omega_e + 2)\dot{\phi}^2 - 2H\phi\dot{\phi} = 0,$$

and

$$(6\omega_e H + h'h''\dot{\phi})\dot{\phi} = \{2V_{,\phi^2} - 24H^2 + 2\mathcal{V}^{eff}{}_{,\phi^2}\}\phi$$

with $\omega_e \equiv \omega(1 + \frac{h'^2}{2\omega})$. Linear stability analyses near the above solution are given in [6].

When $\dot{\phi}^2 << 1$ and $V(\phi) << 1$ are assumed, Eq. (16) yields $\Lambda\phi^2 \sim \mathcal{V}^{eff}(\phi)$ with an effective cosmological constant $\Lambda \simeq 10^{-122}M_p^2$ and the Planck mass $M_p \simeq 10^{19}GeV$. We suppose that the BD field ϕ would be an attractor in the asymptotic region with $\phi \sim M_p$. Note that $< \phi^2 > \sim 1/G$ in the BD theory [7] with its Lagrangian density, $\phi^2 R - \omega g^{\alpha\beta}\partial_\alpha\phi\partial_\beta\phi - V(\phi)$. Taking the heavy field mass $m_h \simeq 10^{-17}M_p$ and $M_d \simeq 10^{-x}M_p$ in Eqs. (13) and (16), we have the useful solutions in (possibly smallest) integers l and k:

Interesting cases, [$l = 3, k = 3, x = 5$] and [$l = 3, k = 6, x = 2$], of natural numbers k, l, and x can be related with effective theories, which are derived from a grand unified theory defined at energy higher than $M_d \simeq 10^{14}GeV$ or $M_d \simeq 10^{17}GeV$, respectively. In these cases,

$$\mathcal{V}^{eff}(\phi) \propto m_h^6/M_d^2 \cdot (M_d/\phi)^{2k}. \tag{17}$$

Other cases of [$l = 3$] and [$k =$ any integer, $x = 0$] are obtained with a little different values of $m_h \sim 10^{-20}M_p$ and $\Lambda \sim 10^{-120}M_p^2$, compared to the former cases. In the latter cases, the non-renormalizable interaction term of the BD and the heavy field in Eq. (10) is induced from a high-energy theory which resides at energy near M_p. The effective potential is $\mathcal{V}^{eff}(\phi) \propto m_h^6/M_p^2 \cdot (M_p/\phi)^{2k}$.

We can restate that, in the low-energy effective theory of BD gravity coupled with a heavy field whose non-renormalizable interaction terms are derived from a grand unified theory, we get a petite cosmological constant Λ from the effective potential $\mathcal{V}^{eff}$ in Eq. (16) (with suitable m_h and M_d). The cosmological constant problem thus can be alleviated in this kind of quintessence model with the inverse power law, effective potential of the BD field in the late-time universe, as given in Eqs. (13) and (17).

Acknowledgements

We would like to thank Profs. Y. M. Cho, R. Ruffini, S. W. Kim, S. P. Kim, H. Y. Lee, and G. Kang, who invite us. We thank Profs. C. Lee, H. Min, and P. Y. Pac who helped us learn field theories. We also thank Profs. M-S Kim and H-Y Park who were our colleagues in Soongsil University.

It was supported by the Basic Science Research Program through the National Research Foundation of Korea (NRF) funded by Ministry of Education, Science and Technology (NRF-2017R1D1A1B06032249).

References

[1] Riess A. G. *et al.*, Astron. J. **116**, 1009 (1998); Perlmutter S. J. *et al*, Astrophys. J. **517**, 565 (1999); Perlmutter S. J., Turner M. S., and White M. J., Phys. Rev. Lett. **83**, 670 (1999); Riess A. G. *et al.*, Astrophys. J. **659**, 98 (2007); Spergel D. N. *et al.* [WMAP Collaboration], Astrophys. J. Suppl. **170**, 377 (2007).

[2] Mannheim P. D., Astrophys. J. **479**, 659 (1997); Lee T. H. and Lee B. J., Phys. Rev. D **69**, 127502 (2004).

[3] Ratra B. and Peebles P. J. E., Phys. Rev. D **37**, 3406 (1988); Frieman J. and Waga I., Phys. Rev. D **57**, 4642 (1998); Steinhardt P. J., Wang L., and Zlatev I., Phys. Rev. D **59**, 123504 (1999).

[4] Caldwell R. R., Phys. Lett. B **545**, 23 (2002).

[5] Lee C., Lee T. H., and Min H., Phys. Rev. D **39**, 1681 (1989); 1701 (1989).

[6] Damdinsuren B., Sim J., and Lee T. H., Class. Quant. Grav. **34**, 175012 (2017).

[7] Brans C. and Dicke R., Phys. Rev. **124** , 925 (1961).

EPJ Web of Conferences **168**, 08005 (2018) https://doi.org/10.1051/epjconf/201816808005
Joint International Conference of ICGAC-XIII and IK-15 on Gravitation, Astrophysics and Cosmology

Accelerated cosmological expansion without tension in the Hubble parameter*

Fast evolution of the Hubble parameter $H(z)$

Maurice H.P.M. van Putten[1,**]

[1] *614 Yeongsil-Gwan, Department of Physics and Astronomy, Sejong University, 143-747 Seoul, South Korea*

Abstract. The H_0-tension problem poses a confrontation of dark energy driving late-time cosmological expansion measured by the Hubble parameter $H(z)$ over an extended range of redshifts z. Distinct values $H_0 \simeq 73$ km s^{-1}Mpc^{-1} and $H_0 \simeq 68$ km s^{-1}Mpc^{-1} obtain from surveys of the Local Universe and, respectively, ΛCBM analysis of the CMB. These are representative of accelerated expansion with $H'(0) \simeq 0$ by $\Lambda = \omega_0^2$ and, respectively, $H'(0) > 0$ in ΛCDM, where $\omega_0 = \sqrt{1-q}H$ is a fundamental frequency of the cosmological horizon in a Friedmann-Robertson-Walker universe with deceleration parameter $q(z) = -1 + (1+z)H^{-1}H'(z)$. Explicit solutions $H(z) = H_0\sqrt{1+\omega_m(6z+12z^2+12z^3+6z^4+(6/5)z^5)}$ and, respectively, $H(z) = H_0\sqrt{1-\omega_m+\omega_m(1+z)^3}$ are here compared with recent data on $H(z)$ over $0 \lesssim z \lesssim 2$. The first is found to be free of tension with H_0 from local surveys, while the latter is disfavored at 2.7σ. A further confrontation obtains in galaxy dynamics by a finite sensitivity of inertia to background cosmology in weak gravity, putting an upper bound of $m \lesssim 10^{-30}$eV on the mass of dark matter. A C^0 onset to weak gravity at the de Sitter scale of acceleration $a_{dS} = cH(z)$, where c denotes the velocity of light, can be seen in galaxy rotation curves covering $0 \lesssim z \lesssim 2$. Weak gravity in galaxy dynamics hereby provides a proxy for cosmological evolution.

1 Introduction

Estimates of the Hubble constant $H_0 = H(0)$, where $H(z)$ denotes the Hubble parameter as a function of redshift z, primarily derive from surveys of the Local Universe and fits of power spectra of the Cosmic Microwave Background (CMB) in the framework of ΛCDM. The results $H_0 \simeq 73$ km s^{-1}Mpc^{-1} and, respectively, $H_0 \simeq 68$ km s^{-1}Mpc^{-1} are distinct at a level of confidence better than 3σ [3]. This H_0-tension problem is interesting for its potential implications for dark energy density ρ_Λ, beyond merely $\Lambda = 8\pi\rho_\Lambda > 0$ inferred from a deceleration parameter $q = -\ddot{a}a/\dot{a}^2 = -1 + (1+z)H^{-1}H'$ satisfying

$$q = \frac{1}{2}\Omega_M - \Omega_\Lambda < 0 \tag{1}$$

*Invited talk ICGAC-XIII and IK15, July 3-7, Seoul

**e-mail: mvp@sejong.ac.kr

in ΛCDM based on a classical vacuum in three-flat Friedmann-Robsertson-Walker (FRW) universe with line-element

$$ds^2 = -dt^2 + a(t)^2(dx^2 + dy^2 + dz^2) \tag{2}$$

described by a scale factor $a(t)$, $H = \dot{a}/a$, evolved by general relativity with Λ constant. Here, Ω_M and Ω_Λ refer to baryonic and dark matter density Ω_M and, respectively dark energy density normalization to closure density $\rho = 3H^2/(8\pi G)$, where G denotes Newton's constant.

While $q(0) < 0$ appears relatively secure from surveys of the Local Universe, the relationship (1) derives from classical general relativity, i.e., a covariant embedding of Newton's gravitational potential energy U_N in geodesic motion in a metric of four-dimensional spacetime based on Einstein's principle of equivalence.[1] Applied to galaxy dynamics, we commonly preserve equivalence of geodesic motion to Newton's picture of force balance between gravitational and inertial forces with inertial mass m equal to gravitating mass m_0, given by rest-mass energy m_0c^2, where c denotes the velocity of light. In particular, the latter is assumed to be scale-free, i.e., $m = m_0$ is assumed to hold true at arbitrarily small accelerations α conform Newton's second law (a proportional relation between force and acceleration). It has been suggested that perhaps the latter should be relaxed to account for anomalous galactic dynamics [8, 9].

Our background cosmology introduces a de Sitter scale of acceleration $a_{dS} = cH$, whose present value on the order of $1\text{Å}\,\text{s}^{-2}$ is small but non-zero. If a_{dS} breaks equivalence between m and m_0, galaxy dynamics is expected to be anomalous at distances in weak gravity, where

$$\alpha < a_{dS}. \tag{3}$$

Astronomical evidence for general relativity at low accelerations is limited to verification of gravitational accelerations $\alpha \gtrsim 10^{-6}\text{m}\,\text{s}^{-2}$ (with $m = m_0$), which leaves a window for anomalies in galactic dynamics in weak gravity (3). Recently, we derived inertia in unitary holography [13–15] with the property that $m < m_0$ at accelerations $\alpha < a_{dS}$ supported by high resolution data on galaxy rotation curves (Fig. 1). With invariant kinetic energy E_k and U_N in orbital motion [14], this theory leaves the total energy $H = E_k + U_N$ and in particular the classical Lagrangian

$$L = E_k - U_N \tag{4}$$

unchanged. By volume, weak gravity makes up most of the Universe. *If inertia falls below its Newtonian value in weak gravity, then possibly the Hubble expansion is faster than what is expected in ΛCDM.* For this reason, anomalous galactic dynamics is a potential proxy of novel cosmological evolution.

In (2), we have a cosmological horizon at the Hubble radius

$$R_H = \frac{c}{H}. \tag{5}$$

Defined as an apparent horizon in Cauchy surfaces of constant time t, these horizons are spheres with area $A_H = 4\pi R_H^2$. As a compact surface, these horizons carry a finite fundamental frequency ω_0 of an ordinary differential equation describing geodesic separation of associated null-geodesics,

$$\omega_0 = \sqrt{1-q}H. \tag{6}$$

In cosmological holography, ω_0 is picked up by the induced wave equation of massless fields, notably electromagnetic and gravitational fields, with dispersion relation

$$\omega = \sqrt{k^2 + \omega_0^2} \tag{7}$$

[1] Equivalence of gravitational fields locally around non-inertial observers, whether arising from a massive object or arising from acceleration as seen by Rindler observers [e.g 2].

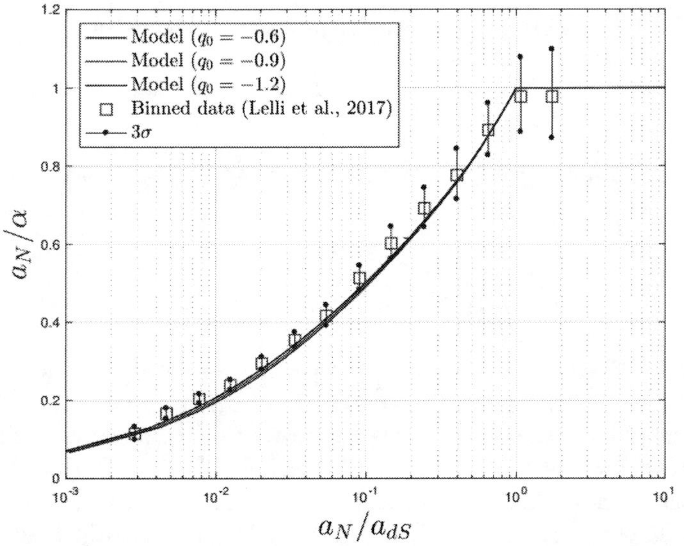

Figure 1. High resolution data of [5] on centripital accelerations α in galaxy rotation curves reveal an onset to weak gravity at $(a_N/a_{dS}, \alpha/a_N) = (1,1)$ in transition to $m/m_0 = a_N/\alpha < 1$. This onset appears to be C^0, identified with a collusion of apparent Rindler and the cosmological horizon at Hubble radius R_H. Binned data shown are accompanied by 3σ uncertainties. Model curves (continuous lines) are included for various values of the deceleration parameter q_0, assuming a Hubble parameter $H_0 = 73$ km s^{-1} Mpc^{-1}. (Reprinted from [15], data from [5]).

for a frequency $\omega(k)$ with associated wave number k. Thus, the cosmological horizon induces a dynamical dark energy

$$\Lambda = \omega_0^2 \tag{8}$$

which, in late time cosmology, is *inherently positive and small.* By (6), Λ is dynamical and includes second time derivatives of $a(t)$. As such, including (8) in the FRW equation describing the Hamiltonian energy constraint,

$$\Omega_M + \Omega_\Lambda = 1, \tag{9}$$

obtains an ordinary differential equations which is second order in time. It defines a *singular perturbation* of (9) which, after all, is first order in time in ΛCMD.

Here, we elaborate on accelerated cosmological expansion by (8) and in ΛCDM, confronted with recent Hubble data $H(z)$ [1, 10] over an extended range of redshifts. This development is facilitated by analytic solutions for both in late time cosmology, parameterized by $H_0 = H(0)$ and $\omega_m = \Omega_M(0)$ of the Hubble parameter and density of (baryonic and dark) matter at the present redshift $z = 0$ (§2). Our model for cosmological evolution has a Hubble parameter H_0 free of tension with estimates from surveys of the Local Universe A further confrontation with galaxy rotation curve data obtains in weak gravity at accelerations $\alpha < a_{dS}$ modeled by inertia of holographic origin (§3). Our model identifies a holographic origin of dark energy and inertia, bringing together theory and data on cosmological evolution and anomalous galaxy dynamics (§4).

2 Accelerated expansion in cosmological holography

Evolution of the FRW scale factor $a(t)$ derives from (9) with either (8) or Λ constant in ΛCDM. Parameterized by H_0 and ω_m, the resulting Hubble parameter satisfies [15]

$$H(z) = H_0\sqrt{1+\omega_m(6z+12z^2+12z^3+6z^4+(6/5)z^5)}/(1+z),\quad H'(0) = H_0(3\omega_m - 1) \simeq 0 \quad (10)$$

and, respectively,

$$H(z) = H_0\sqrt{1-\omega_m+\omega_m(1+z)^3},\quad H'(0) = \frac{3}{2}\omega_m H_0 \simeq 0.5H_0, \quad (11)$$

where we used $\omega_m \simeq 0.3$. According to (8), the Universe is presently at close to a minimum value of $H(z)$, whereas $H(z)$ is decreasing to $H_0\sqrt{1-\omega_m} \simeq 0.83H_0$ of a de Sitter Universe in the distant future. This distinct behavior shows that, in late time cosmology, $H(z)$ will be larger for (8) than in ΛCDM, the latter with a relatively stiff evolution by maintaining $H'(z) > 0$ well into the future.

Table 1 lists estimates of (H_0, ω_m) obtained by nonlinear model regression of (10-11) (Fig. 1) applied to recent data compilations of $(z_k, H(z_k))$. Fig. 1 includes distinct behavior (10-11)) in the qQ-diagram, where $Q(z) = dq(z)/dz$. Table 1 includes estimates of $q_0 = q(0)$ and $Q_0 = Q(0)$ with $1\,\sigma$ uncertainties and fits to a cubic and quartic Taylor series expansion (with no priors on q_0 and Q_0) of $H(z)$.

Table 1. Estimates of $(H_0, q_0, Q_0, \omega_m)$ with $1\,\sigma$ uncertainties by nonlinear model regression applied to the coefficients of the truncated Taylor series of cubic and quartic order, and to (H_0, ω_m) in (10) with $\Lambda = \omega_0^2$ and (11) for ΛCDM. H_0 is expressed in units of km s^{-1}Mpc^{-1}. (Reprinted from [15].)

model	H_0	q_0	Q_0	ω_m	$h'(0)$
Cubic	74.4 ± 4.9	-1.17 ± 0.34	2.49 ± 0.55	-	-0.17
Quartic	74.5 ± 7.3	-1.18 ± 0.67	2.54 ± 1.99		-0.18
$\Lambda = \omega_0^2$	74.9 ± 2.6	-1.18 ± 0.084	2.37 ± 0.073	0.2719 ± 0.028	-0.18
ΛCDM	66.8 ± 1.9	-0.50 ± 0.060	1.00 ± 0.030	0.3330 ± 0.040	0.5

Fig. 1 and Table 1 show a three-fold consistency among model-independent cubic and quartic fits and the model fit to (10). By cubic fit, ΛCDM is inconsistent with data at $2.7\,\sigma$.

Here, $Q_0 \simeq 2.5$ [12] is representative a near-extremal value of $H(z)$ today. The associated relatively high estimate of H_0 from the cosmological data $\{z_k, H(z_k)\}$ is free of tension with $H_0 = 73.24 \pm 1.74$km s^{-1}Mpc^{-1} obtained from surveys of the Local Universe, providing quantitative support for a dynamic dark energy (8). Combining results on H_0, we estimate [15]

$$H_0 \simeq 73.75 \pm 1.44\,\text{km}\,\text{s}^{-1}\text{Mpc}^{-1}. \quad (12)$$

3 Asymptotic behavior in weak gravity

Applying (8) to (9) yields a second order differential equation, we have

$$q = \Omega_M - 2\Omega_\Lambda \quad (13)$$

associated with $\Omega_\Lambda = (1/3)(1-q)$ and $\Omega_M = (1/3)(2+q)$ and $w = (2q-1)/(1-q)$, defined by $p_\Lambda = w\rho_\Lambda$ between dark pressure and dark energy. In the matter dominated era $q = 1/2$ holds true in

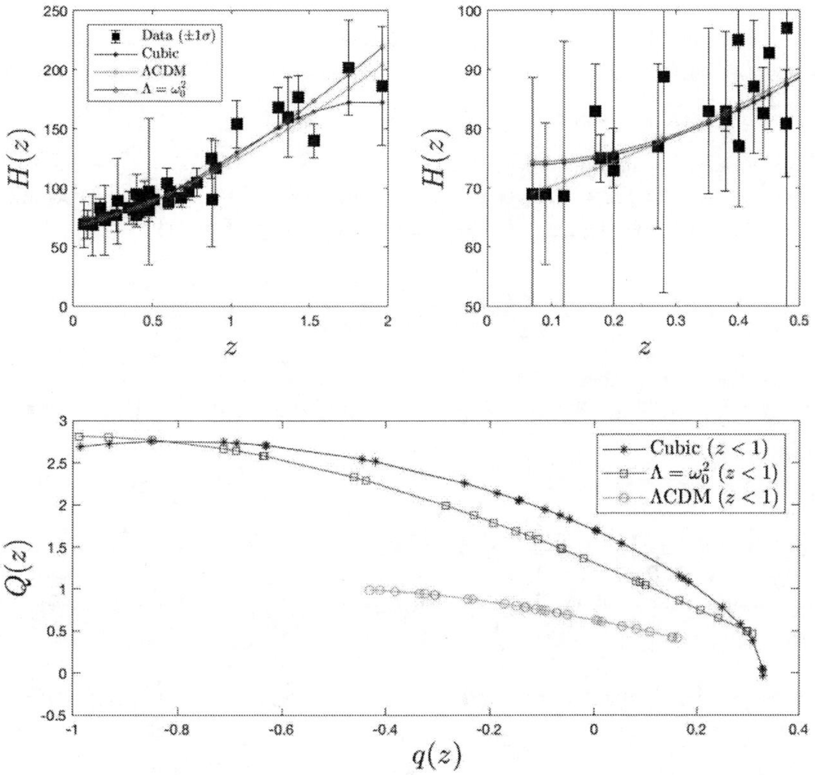

Figure 2. (Top panels.) Fits of (10-11) and a Taylor series expansion to third order by nonlinear model regression to data $(z_k, H(z_k))$ over $0 < z < 2$. Note distinct behavior at $z \sim 0$ with $H'(0) \simeq 0$ for (10) and $H'(0) > 0$ for (11). (Bottom panel.) In the (q, Q)-plane, results for (10) and consistent with the Taylor series expansion. Results for ΛCDM are inconsistent with the latter at $2.7\,\sigma$. (Reprinted from [15], data from [5]).

both (1) and (13), (8). In late time cosmology ($q \lesssim -0.5$), however, q is *twice* the value (1) of ΛCDM, signifying *fast* evolution of the Hubble parameter $H(z)$.

The asymptotic regime of weak gravity $\alpha << a_{dS}$ satisfies the baryonic Tully-Fisher relation [11] or, equivalently, Milgrom's law [7],

$$\alpha = \sqrt{a_0 a_N}. \tag{14}$$

In a background cosmology with dark energy (8), we have [15]

$$a_0 = \frac{\omega_0}{2\pi}. \tag{15}$$

By (6), (15) introduces a sensitivity of galaxy dynamics in the regime (14) to the cosmological parameters (H, q), in addition to sensitivity to a_{dS} at the onset to weak gravity ($a_N = a_{dS}$). A recent sample of galaxy rotation curves at intermediate redshifts $z \sim 2$ clusters close to the onset $a_N = a_{dS}$ but is in the weak gravity regime (3). The transition to (14) is described by holographic inertia, sensitive to background parameters (H, q). By (10), this theory accounts for rotation curves from $z \sim 0$ (Fig. 1) up to $z \sim 2$ [15].

In this process, we encounter a deceleration parameter $q_0 \simeq -1.18 \simeq 1.6 \times 10^{-10}\mathrm{m\,s^{-2}}$ (Table 1) with implied value $a_0 \simeq 1.6 \times 10^{-10}\mathrm{m\,s^{-2}}$. In what follows, we consider the rate of decay to (14) in

$$\alpha = \sqrt{a_0 a_N}\left(1 + O\left(x^k\right)\right) \quad (x << 1) \tag{16}$$

by specific values of $k > 0$, where $x = a_N/a_0$.

Canonical estimates $a_0' \simeq 1.2 \simeq 1.6 \times 10^{-10}\mathrm{m\,s^{-2}}$ derive from fitting an interpolating function to rotation curve data such as Fig. 1 [e.g. 6]. Commonly used is $f(x) = x/(1+x)$, $x = a/a_0'$ satisfying $F_N = mf(x)\alpha$, gives

$$\alpha = \frac{1}{2}a_N\left(1 + \sqrt{1+4x^{-1}}\right) = a_0'\sqrt{x}\left(1 + \frac{1}{2}\sqrt{x} + \frac{1}{8}x + O\left(x^{\frac{3}{2}}\right)\right), \tag{17}$$

showing $k = 1/2$ in (16).

Our value $a_0 \simeq 1.6 \times 10^{-10}\mathrm{m\,s^{-2}}$ is defined by the asymptotic value of $\mu(y) = 2\left\langle B(p)\right\rangle_y$ in (18) in the notation of [15],

$$\alpha = \sqrt{\mu a_{dS} a_N}, \tag{18}$$

defined as a thermal average of the ratio $B(p)$ of dispersion relations on the cosmological horizon and 3+1 spacetime within over momentum space as a function of $y = (a_N/a_{dS})\left((1-q)/2\right)^{-\gamma}$, $\gamma \simeq 0.5$, whereby $y \simeq x/(\sqrt{2}\pi)$ in terms of x above. With $u = p/T_H$, $T_H = (1-q)a_{dS}/(4\pi)$, $A = \Lambda/T_H = 16\pi^2/(1-q)$, we have

$$\left\langle B(p)\right\rangle_y = \frac{1}{W(y)}\int_0^\infty B(p)e^{-s^2}u^2du, \quad W(y) = \int_0^\infty e^{-s^2}u^2du, \quad s^2 = \frac{u^2}{2\sigma^2}, \tag{19}$$

where

$$\sigma^2 \simeq \sqrt{A}\,y \quad (u << 1). \tag{20}$$

With $B(p) = 1 + \frac{1}{2}u^2(1 - 1/A) + O(u^4)$, (14) is reached with $\mu(y) = 1 + O(y)$, i.e., $k = 1$ in (16) of (18).

In a confrontation with data as shown in Fig. 1, we face the fact that x remains finite. In comparing a_0' from a fit using an interpolation function with a_0 associated with (15), discrepancies are expected by distinct decays $k = 1/2$ and $k = 1$ in (16). A fit to a tail of data $y_0 = a_N/a_{dS} = O\left(10^{-2}\right)$ of Fig. 1 involves the first two terms of (17), whereas (14) is attained to order $O(y) = O(x)$ in the latter. Writing $a_0' = (1-\epsilon)a_0$, a_0' and a_0 satisfy

$$\sqrt{a_0 a_N}\left(1 - \frac{1}{2}\epsilon\right)\left(1 + \frac{1}{2}\sqrt{x} + O(x)\right) = \sqrt{a_0 a_N}\left(1 + O(x)\right). \tag{21}$$

With $a_0 = \omega_0/(2\pi) = \sqrt{1-q}H/(2\pi) = a_{dS}\sqrt{1-q}/(2\pi)$, we conclude that a_0' *under*-estimates a_0 by

$$\epsilon \simeq \sqrt{x} \simeq \sqrt{\frac{2\pi}{\sqrt{1-q}}}\sqrt{y_0} \simeq 21\%\sqrt{\frac{y_0}{10^{-2}}}. \tag{22}$$

Indeed, a canonical value $a_0' \simeq 1.18 \times 10^{-10}\mathrm{m\,s^{-2}}$ obtains by fitting the full expression (17) to the data of Fig. 1, whereas a fit of the asymptotic relation (14) restricted to the tail $a_N/a_{dS} \lesssim 10^{-2}$ gives

$$a_0 \simeq 1.41 \times 10^{-10}\mathrm{m,s^{-2}} \tag{23}$$

larger than a_0' by about 19%, as expected based on (22).

Our value $a_0 \simeq 1.6 \times 10^{-10}\mathrm{m\,s^{-2}}$, obtained from (15) based on estimating the cosmological parameters (H, q) in Table 1, agrees with (23) to within about 17%, which is within the $1\,\sigma$ (statistical and systematic) uncertainties of about 20% in rotation curve data [e.g. 4].

4 Conclusions and outlook

The H_0 tension problem points to a discrepancy between accelerated expansion and relatively stiff evolution in ΛCDM. We here present a dynamical dark energy based on a fundamental frequency of the cosmological horizon, that is inherently positive and small. It introduces relatively fast evolution in the Hubble parameter today, satisfying $H'(0) \simeq 0$ with H_0 larger than that expected in ΛCDM. A detailed confrontation with Hubble data covering an extended range in redshifts obtains an estimate of H_0 (Table 1) in full agreement with H_0 obtained from surveys of the Local Universe. Since (10-11) share the same parameters (H_0, ω_m) characterizing late time cosmology, dynamical dark energy and static dark energy can, for the first time, be simultaneously compared with data. The results of Table 1 favor the first and disfavor the second by $2.7\,\sigma$.

Fast evolution (10) arises from novel behavior in the deceleration parameter $q(z)$, that changes the Hamiltonian energy constraint (9) to an ordinary differential equation which is second order in time, rather than first order in time in ΛCDM. As such, (10) is a singular perturbation, disconnected from ΛCDM. On this background, inertia of holographic origin is coevolving in the regime of weak gravity (3) with a specific predictions for anomalous behavior in galaxy dynamics, whose asymptotic behavior parameterized by (15) explicitly expresses sensitivity to background cosmology.

Results on (15) derived from fitting (10) to cosmological data on the Hubble parameter and derived from a direct fit to rotation curve data (Fig. 1) are consistent with rotation curve data within $1\,\sigma$ uncertainties. Our estimates of a_0 are slightly higher than canonical estimates, that we identify with a relatively fast decay of (18) in our theory of weak gravity to the asymptotic behavior (14).

Conceivably, conditions of weak gravity might be reproduced in laboratory (or satellite) experiments. While we cannot escape the presence of the gravitational field of the Earth (or the Sun), perhaps measurements on acceleration along equipotential surfaces in the gravitational field of the Earth (or the Sun) can be realized to test for anomalies $m < m_0$, by observing geodesic separation between particles in free fall, as an extension of Galileo's experiment. Suitable accelerations below a_{dS} may be imparted by gravitational or electrostatic forces.

Acknowledgements. This research is supported in part by the National Research Foundation of Korea (No. 2015R1D1A1A01059793 and 2016R1A5A1013277).

References

[1] Farooq, O., Madiyar, F.R., Crandall, S., & Ratra, B., *ApJ*, **835**, 26 (2017)

[2] Feynman, R.P., Morimigo, F.B., & Wagner, W.G., *Feynman Lectures on Gravitation* (Westview Press, Colorado, 2003), §7.1-2

[3] Freedman, W.L., Nature Astronomy, **1**, 0121 (2017)

[4] Lelli, F., McGaugh, S.S., & Schombert, J.M., 2016, Astron. J., 152, 157

[5] Lelli, F., McGaugh, S.S., Schombert, J.M., & Pawlowski, M.S., 2017, ApJ, 836, 152

[6] McGaugh, S.S., 2012, ApJ, 143, 40

[7] Milgrom, M., 1983, ApJ 270, 365

[8] Milgrom, 1999, M., Phys. Lett. A 253, 273

[9] Smolin, L., 2017, arXiv:1704.00780

[10] Solà, J., Gómez-Valent, A., & de Cruz Pérez, J., 2017, *ApJ*, **836**, 43

[11] Tully, R.B., & Fisher, J.R., 1977, Astron. Astrophys. 54, 661

[12] van Putten, M.H.P.M., MNRAS **405**, L48 (2015)

[13] van Putten, M.H.P.M., ApJ **837**, 22 (2017a)

[14] van Putten, M.H.P.M., MPLA **32**, 1730019 (2017b)

[15] van Putten, M.H.P.M., ApJ, to appear (2017c)

EPJ Web of Conferences **168**, 08006 (2018) https://doi.org/10.1051/epjconf/201816808006
Joint International Conference of ICGAC-XIII and IK-15 on Gravitation, Astrophysics and Cosmology

Stability Analysis of Bulk Viscous Cosmology

M. Sharif[1]* and *Saadia* Mumtaz[1]**

[1] *Department of Mathematics, University of the Punjab, Quaid-e-Azam Campus, Lahore-54590, Pakistan.*

Abstract. In this paper, we study phase space analysis of FRW universe model by taking a power-law model for bulk viscosity coefficient. An autonomous system of equations is developed by defining normalized dimensionless variables. We find corresponding critical points for different values of the parameters to investigate stability of the system. It is found that the presence of power-law model of bulk viscosity appears as an effective ingredient to enhance the stability of the respective universe model.

1 Introduction

Many astronomical observations (type Ia supernova and cosmic microwave background radiation) have suggested that our universe is undergoing an accelerated expansion [1]. An unusual anti-gravitational force, known as dark energy (DE), is considered to be responsible for the current cosmic expansion. There are various alternative candidates for DE to study its mysterious characteristics among which the cosmological constant (Λ) is the simplest one. It has also been suggested that bulk viscous fluids play an important role for accelerated cosmic expansion [2]. Bulk viscosity is one of the favorable dissipative contributions in homogeneous universe scenarios.

A phase space analysis is useful to study different patterns of evolution by transforming the system of equations to an autonomous one. Guo et al. [3] studied phase space analysis for FRW universe model filled with barotropic fluid as well as phantom scalar field and found that phantom dominated solution is a stable late-time attractor. Yang and Gao [4] explored this analysis for k-essence cosmology and obtained that stability of the critical points play a substantial role for the cosmic evolution. Recently, we have discussed the stability of accelerated expansion via phase space analysis for both isotropic [5] as well as anisotropic [6] universe models.

This work is devoted to study stability of FRW universe model by taking bulk viscous fluid through phase space analysis. The plan of the paper is as follows. In Sect. 2, we provide some basic equations and a power-law of bulk viscosity parameter. We study phase space analysis for both matter as well as radiation dominated universe model. We conclude our results in the last section.

2 General Equations and Phase Space Analysis

We consider homogeneous and isotropic universe model defined by the line element

$$ds^2 = -dt^2 + a^2(t)(dr^2 + r^2 d\theta^2 + r^2 \sin^2\theta d\phi^2), \tag{1}$$

*e-mail: msharif.math@pu.edu.pk
**e-mail: sadiamumtaz17@gmail.com

where $a(t)$ is the scale factor. The matter distribution for the cosmic fluid is given by

$$T_{\mu\nu} = (\rho + \tilde{p})u_\mu u_\nu + \tilde{p}g_{\mu\nu}, \tag{2}$$

where ρ and $\tilde{p}$ correspond to the energy density and total pressure, respectively. The effective bulk viscous pressure is defined by the Eckart formalism as [7]

$$\tilde{p} = p - 3\eta H, \tag{3}$$

where η represents bulk viscosity coefficient. The bulk viscosity gives dissipative effects to the cosmic fluid. The Friedmann equations describing the evolution of the respective universe model are

$$3H^2 = \rho, \tag{4}$$

$$2\dot{H} + 3H^2 = -\tilde{p}, \tag{5}$$

where $H = \frac{\dot{a}}{a}$ is the Hubble parameter and dot means derivative with respect to time. The conservation of energy-momentum tensor yields

$$\dot{\rho} + 3(\rho + \tilde{p})H = 0. \tag{6}$$

We consider a power-law model for bulk viscosity [8]

$$\eta = \xi H^\delta, \tag{7}$$

where ξ and δ are free parameters.

2.1 Bulk Viscous Matter Dominated Universe

Her we study stability of FRW model through phase space analysis dominated by matter (neglecting radiation). We define the dimensionless bulk viscous parameters as

$$\tilde{\eta} = \frac{3\eta}{H_0^\delta}, \tag{8}$$

which takes the form

$$\tilde{\eta} = 3\xi\left(\frac{H}{H_0}\right)^\delta, \tag{9}$$

where H_0 defines the current value of Hubble parameter. The constraint and Raychaudhuri equations turn out to be

$$3H^2 = \rho_m, \tag{10}$$

$$HH_0\tilde{\eta} = 2\dot{H} + 3H^2, \tag{11}$$

where ρ_m corresponds to the matter density. The conservation equation for bulk viscous dust becomes

$$\dot{\rho}_m + 3(\rho_m - HH_0\tilde{\eta})H = 0. \tag{12}$$

The deceleration and EoS parameter are defined by

$$q = -1 - \frac{\dot{H}}{H^2}, \quad \omega = -1 - \frac{2\dot{H}}{3H^2}, \tag{13}$$

which, through Eqs.(9) and (11), yield

$$q = \frac{1}{2}\left[1 - 3\xi\left(\frac{H}{H_0}\right)^{\delta-1}\right], \tag{14}$$

$$\omega = -\xi\left(\frac{H}{H_0}\right)^{\delta-1}. \tag{15}$$

To find analytical solution of the evolution equations, we define normalized dimensionless variables as

$$x = \Psi_m = \frac{\rho_m}{3H^2}, \quad z = \frac{1}{1+\frac{H_0}{H}}, \tag{16}$$

which can reduce this system to an autonomous one. The system of Eqs.(11) and (12) in terms of normalized variables become

$$x' = 3\xi(1-x)\left(\frac{1-z}{z}\right)^{1-\delta}, \tag{17}$$

$$z' = \frac{3(1-z)}{2}\left[\xi z^{\delta}(1-z)^{1-\delta} - z\right]. \tag{18}$$

We define a new variable τ for time such that $\frac{dt}{d\tau} = \frac{1}{H^2}$. The corresponding derivative will be represented by prime. In order to determine the critical points (x_c, z_c), we need to solve the respective dynamical system by imposing the condition $x' = z' = 0$. We evaluate two critical points for the autonomous Eqs.(17) and (18).

For $P_1 = (1, 1)$, both the eigenvalues are positive showing unstable past attractor. Here $z = 1$ implies either $H_0 = 0$ or $H \to \infty$. Since H_0 cannot be zero therefore it shows the initial singular cosmic epoch characterized by H. We plot the dynamical behavior of critical points corresponding to different values of ξ and δ as shown in figure 1. In these numerical plots, the white region shows decelerated expansion with $q > 0$ while the green region corresponds to $q < 0$ showing accelerated expansion of the universe. We observe that both eigenvalues are positive indicating the point P_1 as an unstable past attractor. This point lies in the physical phase space outside the green region showing decelerated cosmic expansion for all choices of parameters. The point $P_2 = (1, 0)$ corresponds to a saddle point since the eigenvalues have opposite sign. For $P_3 = \left(1, \left[1+\frac{1}{\xi}\right]^{\frac{1}{\delta-1}}\right)$, the corresponding eigenvalues are negative showing a stable future attractor which lies in green region showing an expanding universe model dominated by viscous matter.

2.2 Bulk Viscous Matter and Radiation

We discuss the phase space structure of the model by adding radiation as a new cosmic constituent to study whether the bulk viscous model indicates a radiation dominated epoch. The constraint and Raychaudhuri equations take the form

$$\rho_m + \rho_r = 3H^2, \tag{19}$$

$$2\dot{H} + 3H^2 = \bar{\eta}HH_0 - \frac{1}{3}\rho_r. \tag{20}$$

The conservation equation for radiation case can be written as

$$\dot{\rho}_r + 4\rho_r H = 0. \tag{21}$$

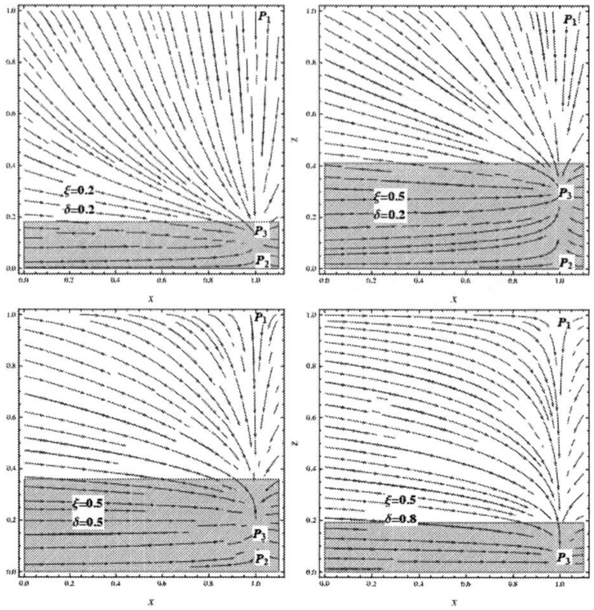

Figure 1. Plots for the phase plane evolution of FRW universe model with different values of ξ and δ.

We generalize the density parameter for radiation by $\Psi_r = \frac{\rho_r}{3H^2}$ such that Eq.(20) becomes

$$\dot{H} = \frac{1}{2}\left[\bar{\eta}HH_0 - H^2\Psi_r - 3H^2\right]. \tag{22}$$

The deceleration and EoS parameters are computed as

$$q = \frac{1}{2}\left[1 + \Psi_r - 3\xi\left(\frac{H}{H_0}\right)^{\delta-1}\right], \tag{23}$$

$$\omega = \frac{\Psi_r}{3} - \xi\left(\frac{H}{H_0}\right)^{\delta-1}. \tag{24}$$

We define dimensionless variables as

$$x = \Psi_m = \frac{\rho_m}{3H^2}, \quad y = \Psi_r = \frac{\rho_r}{3H^2}, \quad z = \frac{1}{1 + \frac{H_0}{H}},$$

through which, the dynamical system of equations for phase space coordinates become

$$x' = 3\xi(1 - x)\left(\frac{1-z}{z}\right)^{1-\delta} + xy, \tag{25}$$

$$y' = y\left[y - 1 - 3\xi\left(\frac{1-z}{z}\right)^{1-\delta}\right], \tag{26}$$

Table 1. Stability Analysis for Matter and Radiation Case

Critical Point	Behavior	Stability	Behavior of q
P_1	Source	Unstable	Deceleration
P_2	Saddle	Unstable	Deceleration
P_3	Sink	Stable	Deceleration

$$z' = \frac{(1-z)}{2}\left[3\xi(1-z)\left(\frac{1-z}{z}\right)^{-\delta} - z(y+3)\right]. \tag{27}$$

We evaluate three critical points by setting $x' = y' = z' = 0$. For $P_1 = (x_c, y_c, z_c) = (0, 1, 1)$, the corresponding eigenvalues show an unstable past attractor. This point lies in decelerated expanding phase. The point $P_2 = (u, 0, 1)$ corresponds to a saddle node showing decelerated cosmic expansion. For $P_3 = \left(1, 0, \frac{1}{1+\xi^{\frac{1}{\delta-1}}}\right)$, the corresponding eigenvalues are negative indicating stable late attractor in deceleration phase for different choices of parameters. The summary of the corresponding results is given in table 1.

3 Summary

We have studied phase space analysis for homogeneous and isotropic universe model by taking a mixture of viscous dust and radiating fluid. We have considered a power-law model for the bulk viscous coefficient. Firstly, we have discussed stability of the universe dominated by bulk viscous matter through their eigenvalues corresponding to different values of ξ and δ (figure 1). We have found an unstable initial matter dominated state undergoing decelerated expansion for P_1 with all choices of parameters. The point P_2 is a saddle point in matter dominated universe. The point P_3 corresponds to a stable matter dominated universe undergoing accelerated expansion. It is observed that the green region (accelerated expansion) tends to increase by increasing ξ while this region gets smaller by increasing δ. It is worth mentioning here that stable solutions (undergoing accelerated expansion) exist in the presence of power-law model of bulk viscosity.

Secondly, we have studied stability of the universe model with bulk viscous radiation and matter to check whether it also indicates a prior radiation dominance. It is found that the critical points P_1 and P_2 are saddle/unstable showing initially the matter dominated phase (without acceleration). The point P_3 is a stable attractor showing decelerated cosmic expansion. We conclude that no stable point (undergoing accelerated expansion) exists with bulk viscous radiation and matter.

References

[1] Riess, A.G. et al., Astron. J. **116**, 1009(1998); Perlmutter, S.J. et al., Astrophys. J. **517**, 565(1999); Bennett, C.L. et al., Astrophys. J. Suppl. **148**, 1(2003).
[2] Zimdahl, W., Phys. Rev. D **53**, 5483(1996).
[3] Guo, Z.K. et al., Phys. Lett. B **608**,177(2005).
[4] Yang, R.J. and Gao, X.T., Class. Quantum Grav. **28**, 065012(2011).
[5] Sharif, M. and Mumtaz, S., Eur. Phys. J. C **77**, 136(2017).
[6] Sharif, M. and Mumtaz, S., arXiv:1706.07659.
[7] Eckart, C., Phys. Rev. **58**, 919(1940).
[8] Wang, D. and Meng, X.H., arXiv:1611.01614.

EPJ Web of Conferences **168**, 09001 (2018) https://doi.org/10.1051/epjconf/201816809001
Joint International Conference of ICGAC-XIII and IK-15 on Gravitation, Astrophysics and Cosmology

Superradiant Phenomena for Spinor Fields in Rotating Black Hole Geometry

Masakatsu Kenmoku[1,⋆], *Yongmin* Cho[2,3,⋆⋆], *Kazuyasu* Shigemoto[4,⋆⋆⋆], and *Jong-Hyuk* Yoon[2,⋆⋆⋆⋆]

[1] *Nara Women's University and Nara Science Academy*
[2] *Konkuk University*
[3] *Seoul National University*
[4] *Tezukayama University*

Abstract. We derive the results (i) the ortho-normal and completeness relations for normal modes and (ii) non-existence of zero mode for spinor fields in rotating black hole geometry. From these results, we show that superradiant phenomena for spinor fields should be type 2: positive momentum on the horizon ($p_H > 0$) and negative frequency at infinity ($\omega < 0$).

1 Introduction

Matter interactions with black holes (BHs) are essential for their existence and observation. As the black hole geometry, rotating BH geometry is important for superradiant phenomena (incident intensity < scattered intensity). Especially Anti-de Sitter (AdS) space-time plays the role of reflection mirror and successive superradiant phenomena lead to instability of BHs. Spinors are fundamental as matter fields because multiple spinors can represent Bosons and Fermions, which can be realized by the Bargmann-Wigner formulation.

The organization of this short report is as follows. In sect. 2, spinor fields in Kerr-AdS space-time is studied. In sect. 3, normal modes are studied to derive the ortho-normal and completeness relations. In sect. 4, non-existence of zero mode ($p_{\mathbf{H}} = \omega - \Omega_{\mathbf{H}} m = 0$) is shown with ω, m as frequency and azimuthal quantum number of spinor fiedls, and Ω_{H} as angular velocity of BH. In sect. 5, we show that the type 2 superradiant modes ($\omega < 0, p_{\mathbf{H}} > 0$) are consistent. Summary of the obtained new results is given in the final section.

This report is based on our paper [1] including the detailed logic and calculation.

⋆e-mail: m.kenmoku@cc.nara-wu.ac.jp
⋆⋆e-mail: ymcho7@konkuk.ac.kr
⋆⋆⋆e-mail: shigemot@tezukayama-u.ac.jp
⋆⋆⋆⋆e-mail: yoonjh@konkuk.ac.jp

2 Spinor fields in Kerr-AdS space-time

The line element of Kerr-AdS space-time with Boyer-Lindquist coordinate is

$$ds^2 = \frac{\Delta_r}{\rho^2}\left(dt - \frac{a\sin^2\theta}{\Xi}d\phi\right)^2 + \frac{\Delta_\theta \sin^2\theta}{\rho^2}\left(adt - \frac{r^2+a^2}{\Xi}d\phi\right)^2 \frac{\rho^2}{\Delta_r}dr^2 + \frac{\rho^2}{\Delta_\theta}d\theta^2 ,$$

with

$$\begin{aligned}&\Delta_r = (r^2+a^2)(1+r^2\ell^{-2}) - 2Mr\,,\ \Delta_\theta = 1 - a^2\ell^{-2}\cos^2\theta\,,\\ &\rho^2 = r^2 + a^2\cos^2\theta \quad , \quad \Xi = 1 - a^2\ell^{-2}\,,\end{aligned} \tag{1}$$

where $\ell = \sqrt{-3/\Lambda}$ is the cosmological parameter and $a = J/M$ is the rotation parameters of BH. The event horizon is defined as the outer zero of Δ_r as $0 = \Delta_r \;\Rightarrow\; r = r_+$.

The Dirac equation for spinor fields in curved spacetime is defined through local Minkowski spacetime, where the Dirac matrices are treated as spacetime independent (Greek suffix $(\mu, \nu ...)$ as curved spacetime and Latin suffix (i,j , ...) as local Cartesian spacetime):

$$\begin{aligned}0 &= (\gamma^\mu(\partial_\mu + \frac{1}{4}\omega_\mu^{ij}\Gamma_{ij}) + \bar{\mu})\Psi(x)\\ &= (\gamma^i b_i^\mu \partial_\mu + \frac{1}{2\sqrt{-g}}\gamma^i\partial_\mu(\sqrt{-g}b_i^\mu) + \frac{1}{4}b_i^\mu b_j^\nu \partial_\mu b_{\nu k}\Gamma^{ijk} + \bar{\mu})\Psi(x),\end{aligned} \tag{2}$$

where ω_μ^{ij}, b_i^μ are denoted as spin connection, vierbein, Γ_{ij}, Γ_{ijk} as anti-symmetric product of Dirac gamma matrix, and $\bar{\mu}$ as mass of spinor field.

We put the spinor fields as separation of variables in the polar coordinate:

$$\Psi(x) = \frac{\exp(-i\omega t + im\phi)}{(\Delta_r\Delta_\theta\rho^2\sin^2\theta)^{1/4}}F(r,\theta). \tag{3}$$

Next we define the chirality eigen-states as

$$F(r,\theta) = (\frac{\chi^*}{\chi})^{1/4}f_+ + (\frac{\chi}{\chi^*})^{1/4}f_-,\ \chi = r + ia\cos\theta,\ \gamma^5 f_\pm = \pm f_\pm, \tag{4}$$

where four component spinors are expressed in two component ones:

$$f_+(r,\theta) = \frac{1}{\sqrt{2}}\begin{pmatrix}\zeta\\ -\zeta\end{pmatrix},\ f_-(r,\theta) = \frac{1}{\sqrt{2}}\begin{pmatrix}\eta\\ \eta\end{pmatrix} \tag{5}$$

with

$$\eta = \begin{pmatrix}R_1(r)S_1(\theta)\\ R_2(r)S_2(\theta)\end{pmatrix},\ \zeta = \begin{pmatrix}R_2(r)S_1(\theta)\\ R_1(r)S_2(\theta)\end{pmatrix}. \tag{6}$$

Then the Dirac equation reduces to the four ordinary differential equations:

$$\begin{aligned}&(\sqrt{\Delta_\theta}\partial_\theta + \frac{a\omega\sin\theta^2 - \Xi m}{\sqrt{\Delta_\theta}\sin\theta})S_1(\theta) = (-a\bar{\mu}\cos\theta + \kappa)S_2(\theta),\\ &(\sqrt{\Delta_\theta}\partial_\theta - \frac{a\omega\sin\theta^2 - \Xi m}{\sqrt{\Delta_\theta}\sin\theta})S_2(\theta) = (-a\bar{\mu}\cos\theta - \kappa)S_1(\theta),\\ &(\sqrt{\Delta_r}\partial_r - i\frac{(r^2+a^2)\omega - a\Xi m}{\sqrt{\Delta_r}})R_1(r) = (\kappa - i\bar{\mu}r)R_2(r),\\ &(\sqrt{\Delta_r}\partial_r + i\frac{(r^2+a^2)\omega - a\Xi m}{\sqrt{\Delta_r}})R_2(r) = (\kappa + i\bar{\mu}r)R_1(r),\end{aligned} \tag{7}$$

where κ is the separation parameter.

3 Normal modes: orthonormal and completeness relations

The local conservation law of the bi-field current $k^\mu_{\rm spinor}$ for Ψ and Ψ' holds using the Dirac equations:

$$\partial_\mu k^\mu_{\rm spinor} = 0, \quad k^\mu_{\rm spinor} = i\sqrt{-g}\,\bar{\Psi}(t,x)\gamma^\mu\Psi'(t,x). \tag{8}$$

The inner product defining the integral of $k^\mu_{\rm spinor}$ is shown to be constant in time:

$$(\Psi \mid \Psi') = \int d^3x k^t_{\rm spinor}\,, \quad \frac{d}{dt}(\Psi \mid \Psi') = 0\,, \tag{9}$$

where the surface term should vanish:

$$\begin{aligned} & k^r(x)\,|^{r=\infty}_{r=r_{\rm H}} \\ = & \frac{1}{\sqrt{\Delta_\theta}} {\rm e}^{-i(\omega-\omega')t}{\rm e}^{i(m-m')\phi}(R^*_{1,\lambda}R_{1,\lambda'} - R^*_{2,\lambda}R_{2,\lambda'})(S^*_{1,\lambda}S_{1,\lambda'} + S^*_{1,\lambda}S_{1,\lambda'})\,|^{r=\infty}_{r=r_{\rm H}} \\ = & \ 0. \end{aligned} \tag{10}$$

The boundary condition on R_1, R_2 are required for vanishing the surface term:

$$R_1 = \pm\gamma R_2 \quad (r = r_{\rm H}), \quad \text{and} \quad R_1, R_2 \to 0 \quad (r \to \infty), \tag{11}$$

where γ is a phase factor.

The orthogonal relation for the separation parameter κ is obtained by using the spinor equations and the boundary condition:

$$\begin{aligned} & (\Psi_{\omega,m,\kappa} \mid \Psi_{\omega,m,\kappa'}) \\ = & \ i\int d^3x \frac{1}{\Xi}\left(\frac{r^2+a^2}{\Delta_r\sqrt{\Delta_\theta}}(R^*_{1,\lambda}R_{1,\lambda'} + R^*_{2,\lambda}R_{2,\lambda'}) + i\frac{a\sin\theta}{\sqrt{\Delta_r\Delta_\theta}}(R^*_{2,\lambda}R_{1,\lambda'} - R^*_{1,\lambda}R_{2,\lambda'})\right)(S^*_{1,\lambda}S_{1,\lambda'} + S^*_{2,\lambda}S_{2,\lambda'}) \\ = & \ 0. \end{aligned} \tag{12}$$

The orthonormal relations for the frequency ω and the azimuthal quantum number m are obtained straightforward. Summarizing there results, orthonormal relations with respect to ω, m and κ are obtained:

$$(u_\alpha \mid u_{\alpha'}) = -(v_\alpha \mid v_{\alpha'}) = \delta_{\alpha,\alpha'}, \quad (u_\alpha \mid v_{\alpha'}) = (v_\alpha \mid u_{\alpha'}) = 0, \tag{13}$$

with the eigenfunctions

$$u_\alpha = \frac{\exp(-i\omega t + im\phi)}{(\Delta_r\Delta_\theta\rho^2\sin^2\theta)^{1/4}}F(r,\theta)_\alpha, \quad v_\alpha = u^c_\alpha = i\gamma^2\gamma^0\bar{u}^T_\alpha \tag{14}$$

where $\alpha = (\omega, m, \kappa)$ and $\alpha' = (\omega', m', \kappa')$ denote sets of quantum numbers.

Any spinor fields can be expanded by eigenfunctions u_α and v_α:

$$\Psi(t,x) = \Sigma_\alpha(c_\alpha u_\alpha + d^\dagger_\alpha v_\alpha), \tag{15}$$

with the coefficient functions $c_\alpha = (u_\alpha \mid \Psi)$, $d^\dagger_\alpha = -(v_\alpha \mid \Psi)$. Inserting the coefficient functions back into the spinor field expression, the completeness relations are obtained:

$$i\sqrt{-g}\,\Sigma_\alpha\left(u_\alpha(t,x)\bar{u}_\alpha(t,x') - v_\alpha(t,x)\bar{v}_\alpha(t,x')\right)\gamma^t = \delta^{(3)}(x-x'). \tag{16}$$

4 Non-existence of zero mode and the spectrum condition

The radial equations for zero mode, defining as $p_{\mathbf{H}} = \omega - \Omega_{\mathbf{H}} m = 0$, become

$$\begin{aligned}
(\sqrt{\Delta_r}\partial_r - i\frac{(r^2 - r_{\mathrm{H}}^2)\omega_*}{\sqrt{\Delta_r}})R_{1\mathrm{zero}}(r) &= (\kappa - i\bar{\mu} r)R_{2\mathrm{zero}}(r), \\
(\sqrt{\Delta_r}\partial_r + i\frac{(r^2 - r_{\mathrm{H}}^2)\omega_*}{\sqrt{\Delta_r}})R_{2\mathrm{zero}}(r) &= (\kappa + i\bar{\mu} r)R_{1\mathrm{zero}}(r),
\end{aligned} \tag{17}$$

where $\omega_* = \Omega_{\mathrm{H}} m$ denotes the zero mode frequency. Near the horizon ($r = r_{\mathbf{H}}$), zero mode equations become

$$\sqrt{r - r_{\mathrm{H}}}\partial_r R_{1\mathrm{zero}}(r) \simeq \beta R_{2\mathrm{zero}}(r), \quad \sqrt{r - r_{\mathrm{H}}}\partial_r R_{2\mathrm{zero}}(r) \simeq \beta^* R_{1\mathrm{zero}}(r), \tag{18}$$

where $\beta = (\kappa - i\bar{\mu} r_{\mathrm{H}})/\sqrt{X(r_{\mathrm{H}})}$ with $X(r) = \partial_r \Delta_r$.

General solutions to eq.(18) are given by

$$\begin{aligned}
R_{1\mathrm{zero}}(r) &\simeq d_1 \exp(2|\beta|\sqrt{r - r_{\mathrm{H}}}) + d_2 \exp(-2|\beta|\sqrt{r - r_{\mathrm{H}}}), \\
R_{2\mathrm{zero}}(r) &\simeq (\beta^*/\beta)^{1/2}(d_1 \exp(2|\beta|\sqrt{r - r_{\mathrm{H}}}) - d_2 \exp(-2|\beta|\sqrt{r - r_{\mathrm{H}}})),
\end{aligned} \tag{19}$$

where d_1 and d_2 are integration constants. For the boundary condition

$$R_1 = (\beta/\beta^*)^{1/2} R_2, \quad \text{or} \quad R_1 = -(\beta/\beta^*)^{1/2} R_2, \tag{20}$$

where the phase factor is $\beta = (\kappa - i\bar{\mu} r_{\mathrm{H}})/\sqrt{X(r_{\mathrm{H}})}$ with $X(r) = \partial_r \Delta_r$, zero mode solutions cannot satisfy the boundary condition in eq.(20). This means that the zero mode does not exist as physical states,i.e. "non-existence of zero mode".

Next we give the statement on the spectrum in the following.

(1) In case without rotation ($a = J/M = 0$), the physical modes satisfy the condition $\omega > 0$.

(2) In case with rotation ($a = J/M \neq 0$), any mode cannot across the zero mode line ($\omega - \Omega_{\mathrm{H}} m = 0$).

(3) The analyticity with respect to the rotation parameter a is required under the conditions:
(i) the well-separated zero of Δ_r as outer horizon ($r_{\mathbf{H}}$) and
(ii) the well-defined value of Δ_θ as $a < \ell$ (smaller value of rotation parameter than cosmological parameter).

Then the set of quantum number $\alpha = (\omega, m, \kappa)$ should satisfy the spectrum condition $\omega - \Omega_{\mathrm{H}} m > 0$.

5 Superradiant modes

General superradiant condition is derived under the conditions (a) the area law of BH and (b) the conservation law of energy and angular momentum by Bekenstein [2] and others [3, 4] as

$$1 - \frac{\Omega_{\mathbf{H}} m}{\omega} < 0 \quad or \quad \omega(\omega - \Omega_{\mathbf{H}} m) < 0, \tag{21}$$

from which there are two possible types of superradiance:
type 1 superradiance: $\omega > 0$ and $\omega - \Omega_{\mathbf{H}} m < 0$ or type 2 superradiance: $\omega < 0$ and $\omega - \Omega_{\mathbf{H}} m > 0$.

Our spectrum condition $\omega - \Omega_{\mathbf{H}} m > 0$, derived in sect. 4, , requires that type 1 superradiance is not consistent and type 2 superradiance is consistent to occur as stable superradiant modes.

6 Summary

As the summary of this report, we classify the superradiant modes as follows. ($p_{\mathrm{H}} \equiv \omega - \Omega_{\mathrm{H}} m$ denotes the momentum of matter fields on the horizon):

For spinor fields,

type 1($\omega > 0$ and $p_{\mathrm{H}} < 0$) is not possible but type 2 ($\omega < 0$ and $p_{\mathrm{H}} > 0$) is possible by our analysis. (Note that both types are not possible by other analysis [3, 5, 6].)

For scalar fields, which is described in our paper [1] though not in this report,

type 1 ($\omega > 0$ and $p_{\mathrm{H}} < 0$) is not possible but type 2 ($\omega < 0$ and $p_{\mathrm{H}} > 0$) is possible by our analysis. (Note that the results are opposite by other analysis.)

The result of our analysis is obtained from the completeness relations in sect.3 and the spectrum condition in sect.4. For spinor and scalar fields, one of superradiance should be possible, because of the completeness relation. Among two types, type 2 can occur because of the spectrum condition. We have obtained the same type 2 superradiance both for spinors and scalars, which is supported by Bargmann-Wigner formulation by our previous work [7].

References

[1] M. Kenmoku, Y. M. Cho, K. Shigemoto and J. H. Yoon, arXiv:1609.00798, to appear in Classical and Quantum Gravity.

[2] J. D. Bekenstein, Phys. Rev. **D7**, 949 (1973).

[3] W. G. Unruh, Phys. Rev. Lett. **31**, 1265 (1973), Phys. Rev. **D10**, 3194 (1974).

[4] S. Mano and E. Takasugi, Prog. Theor. Phys. **97**, 213 (1997).

[5] S. Meda, Prog. Theor. Phys. **55**, 1677 (1976).

[6] S. M. Wagh and N. Dadhich, Phys. Rev. D32, 1863 (1985).

[7] M. Kenmoku and Y. M. Cho, Int. J. Mod. Phys. **A30**, 1550052 (2015).

EPJ Web of Conferences **168**, 09002 (2018) https://doi.org/10.1051/epjconf/201816809002
Joint International Conference of ICGAC-XIII and IK-15 on Gravitation, Astrophysics and Cosmology

Upper Limit of Radiation from Coalescence of Rotating Hayward Black Holes

Bogeun Gwak[1,*]

[1] *Department of Physics and Astronomy, Sejong University, Seoul 05006, Republic of Korea*

Abstract. We review the upper limit of the gravitational radiation thermally allowed under the collision of two rotating Hayward black holes. The upper limit is dependent on the gravitational spin interaction between black holes and is useful to determine one of the parameters for consistency with Laser Interferometer Gravitational-Wave Observatory (LIGO) data.

1 Introduction

Colliding black holes have been an expected source of gravitational waves. The first gravitational wave, GW150914, was successfully detected at the Laser Interferometer Gravitational-Wave Observatory (LIGO)[1]. Since then, LIGO announced two more detections of gravitational waves[2, 3]. In this proceeding, we will review the upper limit of the gravitational radiation thermally allowed under the collision of two Hayward black holes[4]. Black hole thermodynamics are studied using the laws of black-hole mechanics, which are analogous to the laws of thermodynamics. They suggest that information about a black hole can be retrieved using Hawking radiation and Bekenstein-Hawking entropy. [5–8]. Particularly, the irreducibility of the entropy is important in the Penrose process[9, 10] as well as the weak cosmic censorship conjecture[11–20]. Since the collision of black holes is a natural process, we can expect an increase in the entropy of the system based on the second law of thermodynamics[21]. Fundamentally, the upper limit of the radiation depends on the interaction between black holes[22–24] and the instability of the resultant black hole formed by the collision[25–28]. The Hayward black hole is a regular black hole[29], given a non-zero energy-momentum tensor in Einstein's equations. Based the weak cosmic censorship conjecture, the singularity should be hidden by the event horizon of the black hole, but the Hayward black hole has no singularity within the horizon. Using the Newman-Janis transformation[30], the rotating Hayward black hole is obtained. The metric of the Hayward black hole is also found in the gravity theory coupled with non-linear electrodynamics[31].

We *review* the upper limit of the gravitational radiation under the collision of two rotating Hayward black holes and the possible parameter range using LIGO data[4].

*e-mail: rasenis@sejong.ac.kr

2 Hayward Black Hole

We consider the coalescence of two rotating Hayward black holes. [29, 30]. Given a Hayward black hole of mass M and angular momentum $J = Ma$, the metric is in Boyer-Lindquist coordinates

$$ds^2 = -\left(1 - \frac{2mr}{\Sigma}\right)dt^2 - \frac{4amr\sin^2\theta}{\Sigma}dtd\phi + \frac{\Sigma}{\Delta}dr^2 + \Sigma d\theta^2 + \left(r^2 + a^2 + \frac{2a^2mr\sin^2\theta}{\Sigma}\right)\sin^2\theta d\phi^2\,, \quad (1)$$

where

$$\Delta = r^2 - 2mr + a^2\,, \quad \Sigma = r^2 + a^2\cos^2\theta\,, \quad m = M\frac{r^{3+\rho}}{r^{3+\rho} + g^3\Sigma^{\rho/2}}\,. \quad (2)$$

The mass function m has two free parameters, ρ and g, and depends on the coordinate θ based on the function Σ. The parameter ρ mainly affects the shape of the event horizon of the black hole[4]. We will use the Bekenstein-Hawking entropy to find the upper limit of the radiation. The entropy is proportional to the area of the horizon A_H, which is obtained using a numerical approach. Thus,

$$S_{\rm BH} = \frac{1}{4}A_H\,, \quad A_{\rm H} = 2\pi\int_0^{\pi}\left(r_{\rm h}^2(\theta) + a^2\right)\sin\theta d\theta\,, \quad (3)$$

which are slightly different values from that of a Kerr black hole due to the effect of additional parameters.

3 Coalescence of Rotating Hayward Black Holes

We will review the general process for the coalescence of black holes to obtain the upper limit of the gravitational radiation[21, 22]. The upper limit will be obtained using the increased entropy of the system. This results in an inequality between the initial mass of the system and the mass of the resultant black hole. Thus, we can derive the upper limit of the radiation from the minimum mass of the resultant black hole.

3.1 Upper Limit of Radiation

We begin by assuming that the two Hayward black holes are far enough away from each other to eliminate the effects of interactions[4, 23–25]. Then, these black holes slowly come together and merge to form a single, resultant black hole. This is a natural process; hence, the entropy of the system will increase. We denote subscripts 1 and 2 for the initial black holes and 3 for the resultant black hole. The increase in the Bekenstein-Hawking entropy is given by the inequality

$$A_{\rm H}(M_1, J_1) + A_{\rm H}(M_2, J_2) \le A_{\rm H}(M_3, J_3)\,. \quad (4)$$

where M_1, M_2 and J_1, J_2 are, respectively, the masses and angular momenta of the initial black holes. M_3 is the mass of the resultant black hole. The final angular momentum J_3 is assumed to be the total angular momentum of the system due to conservation of angular momentum; therefore,

$$J_3 = \sqrt{J_1^2 + J_2^2 + 2J_1J_2\cos\psi}\,, \quad (5)$$

where ψ is the angle between the angular momentum vectors of the initial black holes. In Eq. (4), there is only one variable: M_3; therefore, we can find $M_{\rm min}$, the minimum value of M_3, by solving the inequality. The mass difference between the initial and final black holes is presumably due to gravitational radiation, which has an energy of $M_{\rm rad}$, such that

$$M_{\rm rad} = M_1 + M_2 - M_3 \,. \tag{6}$$

We can now estimate the minimum mass of the resultant black hole and obtain the upper limit of the radiation, $M_{\rm r}$.

$$M_{\rm r} = M_1 + M_2 - M_{\rm min} \,, \tag{7}$$

Therefore, the actual radiation will not exceed this value for identical initial conditions.

3.2 Parameter Range from LIGO Observations

The rotating Hayward black hole includes several parameters. In particular, g has the greatest effect on the black hole solution and upper bound of the radiation. A possible range for g can be estimated using the following LIGO data: GW150914, GW151226, and GW170104[1–3]. To do so, we fix the following parameters: $\rho = 0.9$ and $a = 0$, since they are not as significant as g. The actual observations for a have an error; therefore, setting $a = 0$ will not affect our results for g. In Fig. 1, we estimate the range of g using the extremal spin parameter, minimum mass, and the upper limit of the radiation. The final spin parameters of black holes are estimated; hence, the spin parameters

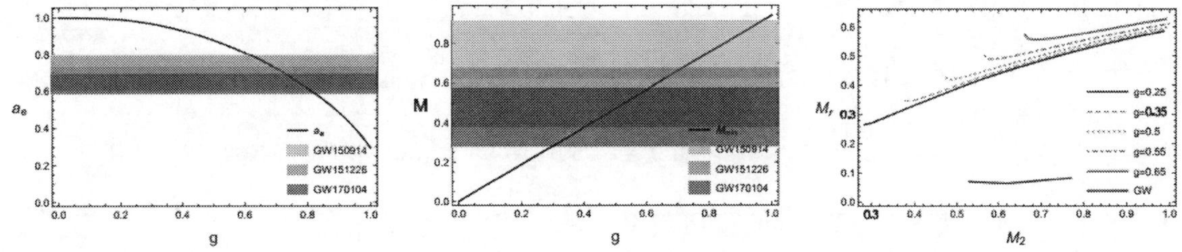

(a) The extremal spin parameter under $\rho = 0.9$ and $M = 1$.

(b) The minimum mass under $\rho = 0.9$, $M = 1$, and $a = 0$.

(c) The upper bounds on the radiation for $\rho = 0.9$, $M_1 = 1$, $a_1 = 0$, and $a_2 = 0$, and the slope between LIGO data.

Figure 1. The upper bounds thermally allowed and detections of the LIGO.

should be within the extremal spin parameter to exist stably. In Fig. 1 (a), we show the dimensionless extremal spin parameter for a given g, and the straight area represents the specific observation for a_3. Accordingly, the straight area should be below the solid line to exist in our Universe. The upper bound of g is approximately 0.75. The mass ratio of the second initial black hole is greater than the extremal mass, if we set the mass of the first initial black hole to unity. In Fig. 1 (b), the minimum mass of the black hole is given in terms of each value of g. The upper bound of g is approximately 0.99. This value is too large to extract any meaning from it. . Using the upper limit of the radiation, as shown in Fig. 1 (c), we can see that the lines for the upper limit should span the entirety of the red line, which represents the LIGO data. Hence, the upper value of g is approximately 0.5 and the possible values for g are within 0 and 0.5.

4 Summary

We have reviewed the upper limit of the gravitational radiation thermally allowed under the collision of two rotating Hayward black holes. We suppose an initial state in which two black holes begin far enough from each other, where the effects of interaction can be ignored. Then, the two black holes slowly approach one another and, finally, merge to form a resultant black hole. The entropy of the system increases, as is required by the second law of thermodynamics. Using the second law, we determine the possible minimum mass of the resultant black hole. Knowing that the total mass of the initial state is greater than that of final black hole and that mass should be conserved in spacetime, the difference in mass is presumably released as a form of gravitational radiation. Using the minimum possible mass, the upper limit of the radiation can be obtained for a given initial state. The most important parameter for the upper limit of the radiation of a rotating Hayward black hole is g. The upper limit can be useful to determine the possible range of g using LIGO data. By using an extremal spin parameter, the minimum mass, and the upper limit of the radiation, the possible values of g are within 0 and 0.5.

Acknowledgments
This work was supported by Basic Science Research Program through the National Research Foundation of Korea (NRF) funded by the Ministry of Science, ICT & Future Planning (NRF-2015R1C1A1A02037523).

References

[1] B. P. Abbott *et al.* [LIGO Scientific and Virgo Collaborations], Phys. Rev. Lett. **116**, no. 6, 061102 (2016).
[2] B. P. Abbott *et al.* [LIGO Scientific and Virgo Collaborations], Phys. Rev. Lett. **116**, no. 24, 241103 (2016).
[3] B. P. Abbott *et al.* [LIGO Scientific and VIRGO Collaborations], Phys. Rev. Lett. **118**, no. 22, 221101 (2017).
[4] B. Gwak, Eur. Phys. J. C **77**, no. 7, 482 (2017).
[5] S. W. Hawking, Commun. Math. Phys. **43**, 199 (1975).
[6] S. W. Hawking, Phys. Rev. D **13**, 191 (1976).
[7] J. D. Bekenstein, Phys. Rev. D **7**, 2333 (1973).
[8] J. D. Bekenstein, Phys. Rev. D **9**, 3292 (1974).
[9] J. M. Bardeen, Nature **226**, 64 (1970).
[10] R. Penrose and R. M. Floyd, Nature **229**, 177 (1971).
[11] D. Christodoulou, Phys. Rev. Lett. **25**, 1596 (1970).
[12] D. Christodoulou and R. Ruffini, Phys. Rev. D **4**, 3552 (1971).
[13] L. Smarr, Phys. Rev. Lett. **30**, 71 (1973).
[14] B. Gwak, Phys. Rev. D **95**, no. 12, 124050 (2017).
[15] B. Gwak and B. H. Lee, Phys. Lett. B **755**, 324 (2016).
[16] B. Gwak and B. H. Lee, JCAP **1602**, 015 (2016).
[17] B. Gwak and B. H. Lee, Class. Quant. Grav. **29**, 175011 (2012).
[18] B. Gwak and B. H. Lee, Phys. Rev. D **84**, 084049 (2011).
[19] B. Gwak, arXiv:1707.09128 [gr-qc].
[20] B. Gwak, arXiv:1709.08665 [gr-qc].
[21] S. W. Hawking, Phys. Rev. Lett. **26**, 1344 (1971).
[22] R. M. Wald, Phys. Rev. D **6**, 406 (1972).
[23] B. Gwak and D. Ro, Eur. Phys. J. C **77**, no. 8, 554 (2017).
[24] B. Gwak and D. Ro, arXiv:1610.04847 [gr-qc].

[25] B. Gwak and B. H. Lee, JHEP **1607**, 079 (2016).
[26] B. Gwak, B. H. Lee and D. Ro, Phys. Lett. B **761**, 437 (2016).
[27] W. K. Ahn, B. Gwak, B. H. Lee and W. Lee, Eur. Phys. J. C **75**, no. 8, 372 (2015).
[28] B. Gwak and B. H. Lee, Phys. Rev. D **91**, no. 6, 064020 (2015).
[29] S. A. Hayward, Phys. Rev. Lett. **96**, 031103 (2006).
[30] C. Bambi and L. Modesto, Phys. Lett. B **721**, 329 (2013).
[31] Z. Y. Fan and X. Wang, Phys. Rev. D **94**, no. 12, 124027 (2016).

EPJ Web of Conferences **168**, 09003 (2018) https://doi.org/10.1051/epjconf/201816809003
Joint International Conference of ICGAC-XIII and IK-15 on Gravitation, Astrophysics and Cosmology

Wormholes in Einstein-Born-Infeld Gravity

Jin Young Kim[1,⋆] and *Mu-In* Park[2,⋆⋆]

[1] *Department of Physics, Kunsan National University, Kunsan 54150, Korea*
[2] *Research Institute for Basic Science, Sogang University, Seoul 04107, Korea*

Abstract. We introduce a new approach to construct wormholes without introducing exotic matters in Einstein-Born-Infeld gravity with a cosmological constant. Contary to the conventional approach, the throat is located at the place where the solutions can be joined smoothly. The metric and its derivatives are continuous so that the exotic matters are not needed there. The exoticity of the energy-momentum tensor is not essential to sustain the wormhole. We also present a method to check the stability of wormholes in the new approach.

1 Introduction

Wormholes are non-singular space-time structures connecting two different universes or parts of the same universe [1–3]. In the conventional approaches to construct traversable wormholes, although wormholes are exact solutions of Einstein equation, the crucial problem is the existence of exotic matter which violates energy conditions. Also the usual cuts and pastes of the throats and subsequent putting of the hypothetical matters to the throats are too artificial to be considered as a natural process.

To circumvent the obstacle by exotic matters, one may consider other extended theories of gravity so that the exotic matters are naturally generated. For example, one may consider non-minimally coupled scalar-tensor gravity theories, higher curvature gravities, higher dimensional gravities, and brane-worlds theories. A new approach for the wormhole construction has been proposed recently [4, 5]. In the new approach, the throat is defined as the place where the solutions are smoothly joined. There, the metric and its derivatives are continuous so that the exotic matters are not introduced. In this sense the new type wormholes were named as natural wormholes to distinguish from the conventional wormholes [5].

The new approach is based on Hořava (or Hořava-Lifshitz) gravity which was proposed as a renormalizable quantum gravity based on different scaling dimensions between the space and time coordinates [6]. This new approach can be applied to other gravity theories as well when the condition for the existence of the natural throat is satisfied. Recently we have shown that this new approach is also possible in Einstein-Born-Infeld gravity with a cosmological constant [7]. In this paper, we review the geometric aspect of this new approach and introduce a method to see the stability of this new wormhole solution by computing the quasinormal modes.

⋆e-mail: jykim@kunsan.ac.kr
⋆⋆e-mail: muinpark@gmail.com

2 Black hole solutions in Einstein-Born-Infeld gravity

We consider the EBI gravity action with a cosmological constant Λ in $D = 3 + 1$ dimensions,

$$S = \int d^4x\sqrt{-g}\left[\frac{(R-2\Lambda)}{16\pi G} + L(F)\right], \quad L(F) = 4\beta^2\left(1 - \sqrt{1 + \frac{F_{\mu\nu}F^{\mu\nu}}{2\beta^2}}\right). \tag{1}$$

Here, the parameter β is a coupling constant with dimensions $[\text{length}]^{-2}$ which needs to flow to infinity to recover the usual Maxwell electrodynamics with $L(F) = -F_{\mu\nu}F^{\mu\nu} \equiv -F^2$ at low energies. The correction terms from a finite β represent the effect of the non-linear BI fields.

Taking $16\pi G = 1$ for simplicity, the equations of motion are

$$\nabla_\mu\left(\frac{F^{\mu\nu}}{\sqrt{1 + \frac{F^2}{2\beta^2}}}\right) = 0, \tag{2}$$

$$R_{\mu\nu} - \frac{1}{2}Rg_{\mu\nu} + \Lambda g_{\mu\nu} = \frac{1}{2}T_{\mu\nu}, \tag{3}$$

where the energy momentum tensor for BI fields is given by

$$T_{\mu\nu} = g_{\mu\nu}L(F) + \frac{4F_{\rho\mu}F^{\rho}_{\nu}}{\sqrt{1 + \frac{F^2}{2\beta^2}}}. \tag{4}$$

Consider a static and spherically symmetric solution with the metric ansatz

$$ds^2 = -N^2(r)dt^2 + \frac{1}{f(r)}dr^2 + r^2(d\theta^2 + \sin^2\theta d\phi^2). \tag{5}$$

For the static electrically charged case where the only non-vanishing component of the field strength tensor is $F_{rt} \equiv E$, one can find $N^2(r) = f(r)$. Then, the general solution for the BI electric field is obtained from the equation (2) as

$$E(r) = \frac{Q}{\sqrt{r^4 + \frac{Q^2}{\beta^2}}}, \tag{6}$$

where Q is the integration constant that represents the electric charge localized at the origin.

With (6), the Einstein equation (3) reduces to

$$\frac{d}{dr}(rf) = 1 - \Lambda r^2 + 2\beta^2\left(r^2 - \sqrt{r^4 + \frac{Q^2}{\beta^2}}\right), \tag{7}$$

and a simple integration over r gives

$$f = 1 - \frac{2C}{r} - \frac{\Lambda}{3}r^2 + \frac{2\beta^2}{r}\int_0^r r^2\left(1 - \sqrt{1 + \frac{Q^2}{\beta^2 r^4}}\right)dr, \tag{8}$$

where C is an integration constant. The solution can be expressed in terms of the hypergeometric function [8] as

$$f(r) = 1 - \frac{2M}{r} - \frac{\Lambda}{3}r^2 + \frac{2}{3}\beta^2 r^2\left(1 - \sqrt{1 + \frac{Q^2}{\beta^2 r^4}}\right) + \frac{4}{3}\frac{Q^2}{r^2}\,{}_2F_1\left(\frac{1}{2}, \frac{1}{4}; \frac{5}{4}; \frac{-Q^2}{\beta^2 r^4}\right). \tag{9}$$

Here, M is another mass parameter, defined by

$$M \equiv C + M_0, \quad M_0 = \frac{2}{3}\sqrt{\frac{\beta Q^3}{\pi}}\Gamma\left(\frac{1}{4}\right)\Gamma\left(\frac{5}{4}\right). \tag{10}$$

The total mass M, which is the sum of the intrinsic mass C and the (finite) self-energy of a point charge M_0, is the usual ADM mass.

The solution (9) has a curvature singularity at the origin, but there is no curvature singularity at the horizon as expected. The solution can have two horizons generally and one can classify the black holes in terms of the values of βQ and the cosmological constant Λ [7].

(i) $\beta Q > 1/2$: In this case, the black hole may have one non-degenerate horizon (type I), or two non-degenerate (non-extremal) horizons or one degenerate (extremal) horizons (type II), depending on the mass for the asymptotically AdS ($\Lambda < 0$) or flat ($\Lambda = 0$) case. There can be a phase transition from Schwarzschild (Sch)-like type I black holes to Reissner-Nodstrom(RN)-like type II black holes when the mass becomes smaller than the marginal mass M_0. When the mass is smaller than the extremal mass M^*, the singularity at $r = 0$ becomes naked. In the limit $\beta \to \infty$, only the RN-like type II black holes are possible. On the other hand, for asymptotically dS ($\Lambda > 0$) case, the situation is more complicated due to the cosmological horizon. In this case, there can exist maximally three horizons (two smaller ones for black holes and the largest one for the cosmological horizon), depending on the mass and the cosmological constant.

(ii) $\beta Q = 1/2$: In this case, only the Sch-like type I black holes are possible. The horizon shrinks to zero size for the marginal case $M = M_0$, even though its mass has still a non-vanishing value. This does not mean that the metric is a flat Minkowskian. This rather means that all the gravitational energy is stored in the BI electric field as a self-energy. When the mass is smaller than the marginal mass M_0, the singularity at $r = 0$ becomes naked.

(iii) $\beta Q < 1/2$: This case is similar to the case (ii), except that the marginal case has no (even a point) horizon so that its singularity is naked always.

3 Construction of natural wormhole solutions

We consider a four dimensional spherically symmetric wormhole connecting two remote parts of the same universe (or mathematically equivalently, the reflection symmetric two universes). The metric is described by

$$ds^2 = -N_\pm(r)^2 c^2 dt^2 + \frac{dr^2}{f_\pm(r)} + r^2\left(d\theta^2 + \sin^2\theta d\phi^2\right) \tag{11}$$

with two coordinate patches, each one covering the range $[r_0, +\infty)$. In the new approach, the radius of throat r_0 is defined as

$$\left.\frac{dN_\pm}{dr}\right|_{r_0} = \left.\frac{df_\pm}{dr}\right|_{r_0} = 0 \tag{12}$$

with the usual matching condition $N_+(r_0) = N_-(r_0)$, $f_+(r_0) = f_-(r_0)$ by demanding that the metric and the Schwarzschild coordinates be continuous across the throat. If there exists a coordinate patch $\mathcal{M}_+$ in which the singularities are absent for all values of $r \geq r_0$, one can construct a smooth regular wormhole-like geometry, which may or may not be traversable depending on the location of r_0, by joining $\mathcal{M}_+$ and its mirror patch $\mathcal{M}_-$ at the throat r_0. Here, it is important to note that, in the new approach, $f_\pm(r_0)$ needs not to be vanished in contrast to the Morris-Thorne approach [2], while the quantities $dN_\pm(r_0)/dr$, $df_\pm(r_0)/dr$ in (12) need not to be vanished in both the Morris-Thorne approach [2] and the cuts and pastes approach of Visser [3].

The first step to construct a natural wormhole is to find the throat radius r_0 defined by (12) with the matching condition [4, 5]. For the case (i) of $\beta Q > 1/2$, there is no throat for the Sch-like type I black holes with $M > M_0$ down to the marginal case of $M = M_0$, where a point-like throat is generated at the origin. The zero-size ($r_0 = 0$) throat can grow as the black hole loses its mass. But, down to the extremal case of $M = M^*$ ($< M_0$), the throat is inside the outer horizon and is non-traversable. As the black hole loses its mass further, there emerges and grows a traversable throat from the extremal black hole horizon. For the AdS case, the throat grows and persists until its mass M vanishes. However, for the cases of flat and dS, the throat does not grow to an indefinitely large size. There is a maximum size at a certain mass $\widetilde{M}$ before M vanishes, *i.e.*, $M^* > \widetilde{M} > 0$. If we further reduce the mass below $\widetilde{M}$, only the space-time with a naked singularity is a possible solution. So, the usual naked singularity solution for $M \leq M^*$ may be replaced by the non-singular wormhole geometry until a certain (non-negative) mass $\widetilde{M}$ is reached in flat and dS cases. This may suggest the generalized cosmic censorship [4, 5] by the existence of a wormhole-like structure around the origin, which has been thought to be a singular point of black hole solutions where all the usual physical laws could break down and might be regarded as an indication of the incompleteness of general relativity.

For the case (ii) of $\beta Q = 1/2$, a point-like throat can be generated at the origin contrary to the generation of a finite-size throat for the case (i). But its growing process is the same as the case (i), showing the existence of a maximum size of the throat with the mass $\widetilde{M} > 0$ for flat and dS cases, and $\widetilde{M} = 0$ for AdS case.

For the case (iii) of $\beta Q < 1/2$, the natural wormhole geometry exists only for AdS case and its growing behavior is similar to the case (ii). On the other hand, for the flat and dS cases, only the Sch-like type I black hole solution for $M > M_0$ or the naked singularity solution for $0 < M < M_0$ can exist.

Here, we have considered only the geometric aspects in constructing the natural wormhole. The analysis on the Born-Infeld field shows that the Born-Infeld eauation is not modified [7].

4 Stability of natural wormhole solutions

One easy way to see the stability of black hole or wormhole is to study the scalar perturbation. We consider the minimally coupled scalar wave equation

$$\nabla^2\Phi = m^2\Phi, \tag{13}$$

where m is mass of the scalar field. The mode solution can be studied with the following form

$$\Phi(t, r, \theta, \phi) = \frac{\psi(r)}{r} Y(\theta, \phi) e^{-i\omega t}, \tag{14}$$

where $Y(\theta, \phi)$ is the spherical harmonics on S^2. In general. it is not easy to solve the equation analytically. The stability of this new type of wormhole can be considered in analogy with the quasinormal modes in black hole physics. There are several ways to compute quasinormal modes in the literature. Here, we follow the technique developed by Horowitz and Hubeny [9].

Since we are interested in the mode solution of the form $e^{-i\omega(t+r_*)}$, with a new radial variable $dr_* = dr/f(r)$, near the throat, it is convenient to consider the ingoing Eddington coordinates, $v = t+r_*$,

$$ds^2 = -f dv^2 + 2dvdr + r^2 d\Omega_2^2. \tag{15}$$

Then the scalar wave equation (13) reduces to the following equation for the radial equation $\psi(r)$

$$f(r)\frac{d^2}{dr^2}\psi(r) + (f' - 2i\omega)\frac{d}{dr}\psi(r) - V(r)\psi(r) = 0, \quad V(r) = \frac{f'}{r} + \frac{\ell(\ell+1)}{r^2} + m^2, \tag{16}$$

where $\ell(\ell+1)$ is the laplacian on S^2 and prime denotes the derivative with respect to r.

The stability can be studied by finding the complex ω such that equation (16) has a solution with only ingoing modes near the throat and vanishing at infinity. If we introduce a new variable $x \equiv 1/r$, equation (16) becomes

$$s(x)\frac{d^2\psi}{dx^2} + \frac{t(x)}{x-x_0}\frac{d\psi}{dx} + \frac{u(x)}{(x-x_0)^2}\psi = 0, \tag{17}$$

with

$$s(x) = -\frac{x^4 f(x)}{x-x_0}, \quad t(x) = -2x^3 f - x^4\dot{f} - 2i\omega x^2, \quad u(x) = (x-x_0)[\ell(\ell+1)x^2 - x^3\dot{f} + m^2,$$

where $\dot{f} = df/dx$. One can expand s, t, and u in powers of $(x-x_0)$ about the throat x_0. For example, $s(x) = \sum_{n=-1}^{\infty} s_n(x-x_0)^n$, where $s_{-1} = -x_0^4 f_0$, $s_0 = -4x_0^3 f_0$, $s_1 = -(1/2)(12x_0^2 f_0 + x_0^4\ddot{f}_0)$, $\cdots$.

To see the leading behavior of the solution near the throat, we set $\psi = (x-x_0)^\alpha$ and put into equation (17). To the leading order, we have $\alpha(\alpha-1) = 0$, which have two solutions $\alpha = 0$ and $\alpha = 1$. Since $\alpha = 0$ corresponds to the ingoing mode, we consider a solution of the form

$$\psi(x) = \sum_{n=0}^{\infty} a_n(x-x_0)^n. \tag{18}$$

Substituting equation (18) into (17) and equating the coefficients of $(x-x_0)$ for each n to zero, we have the recursion relation

$$a_n = -\frac{1}{n(n-1)s_{-1}}\sum_{k=0}^{n-1}[k(k-1)s_{n-k-1} + kt_{n-k-1} + u_{n-k-1}]a_k. \tag{19}$$

One can find the quasinormal modes from the boundary condition $\psi \to 0$ as $x \to 0$ ($r \to \infty$). This means that we solve $\sum_{n=0}^{\infty} a_n(-x_0)^n = 0$ in the complex ω plane. This can be done by truncating the series and computing the partial sum as a function of ω. Then one can discuss the stability of wormhole from the sign of the imaginary part of ω.

Acknowledgments

MIP was supported by Basic Science Research Program through the National Research Foundation of Korea (NRF) funded by the Ministry of Education, Science and Technology (2016R1A2B401304).

References

[1] J. A. Wheeler, Ann. Phys. (N.Y.) **2**, 604 (1957).
[2] M. S. Morris and K. S. Thorne, Am. J. Phys. **56**, 395 (1988).
[3] M. Visser, Nucl. Phys. B **328**, 203 (1989); *Lorenzian wormholes* (AIP Press, 1995).
[4] M. B. Cantcheff, N. E. Grandi, and M. Sturla, Phys. Rev. D **82**, 124034 (2010).
[5] S. W. Kim and M.-I. Park, Phys. Lett. B **751** , 220 (2015).
[6] P. Hořava, Phys. Rev. D **79**, 084008 (2009).
[7] J. Y. Kim and M.-I. Park, Eur. Phys. J. C, **76**, 621 (2016).
[8] T. K. Dey, Phys. Lett. B **595**, 484 (2004).
[9] G. T. Horowitz and V. E. Hubeny, Phys. Rev. D **62**, 024027 (2000).

EPJ Web of Conferences **168**, 09004 (2018) https://doi.org/10.1051/epjconf/201816809004
Joint International Conference of ICGAC-XIII and IK-15 on Gravitation, Astrophysics and Cosmology

Dilatonic dyon black hole solutions in the model with two Abelian gauge fields

Medeu Abishev[1,2,4,⋆], *Dmitry* Ivashchuk[3,4,] *Kuantay* Boshkayev[1,2,], and *Algys* Malybayev[1,2,]

[1] *IETP,*
[2] *Physical and Technical Faculty, Al-Farabi Kazakh National University, al'-Farabi avenue, 71, Almaty 050040, Kazakhstan*
[3] *Center for Gravitation and Fundamental Metrology, VNIIMS, Ozyornaya St., 46, Moscow 119361, Russia*
[4] *Institute of Gravitation and Cosmology, RUDN University, Miklukho-Maklaya St.,6, Moscow 117198, Russia*

Abstract. Dilatonic black hole dyon-like solutions in the gravitational $4d$ model with a scalar field, two 2-forms, two dilatonic coupling constants $\lambda_i \neq 0$, $i = 1, 2$, obeying $\lambda_1 \neq -\lambda_2$ and sign parameter $\varepsilon = \pm 1$ for scalar field kinetic term are considered. Here $\varepsilon = -1$ corresponds to ghost scalar field. These solutions are defined up to solutions of two master equations for two moduli functions, when $\lambda_i^2 \neq 1/2$ for $\varepsilon = -1$. A set of bounds on gravitational mass and scalar charge are presented by using a certain conjecture on parameters of solutions, when $1 + 2\lambda_i^2 \varepsilon > 0$, $i = 1, 2$.

1 Introduction

Here we give a brief extension of our previous work [1] devoted to dilatonic dyon black hole solutions. We consider a subclass of dilatonic black hole solutions with electric and magnetic charges Q_1 and Q_2, respectively in $4d$ model with metric g, scalar field φ, two 2-forms $F^{(1)}$ and $F^{(2)}$, corresponding to two dilatonic coupling constants λ_1 and λ_2, respectively. For coinciding dilatonic couplings $\lambda_1 = \lambda_2 = \lambda$ we get a trivial non-composite generalisation of dilatonic dyon black hole solutions in the model with one 2-form which was considered in ref. [1]. The dilatonic scalar field may be either an ordinary one or a phantom (or ghost) one.

Here we present lower bounds on gravitational mass M and scalar charge Q_φ. As in ref. [1] this problem is solved here up to a conjecture, which states one to one (smooth) correspondence between the pair (Q_1^2, Q_2^2), where Q_1 is electric charge and Q_2 is magnetic charge, and the pair of positive parameters (P_1, P_2), which appear in decomposition of moduli functions at large distances. This conjecture is believed to be valid for all $\lambda_i \neq 0$ in the case of ordinary scalar field and for $0 < \lambda_i^2 < 1/2$ for the case of phantom scalar field (in both cases the inequality $\lambda_1 \neq -\lambda_2$ is assumed).

⋆e-mail: abishevme@gmail.com

2 Black hole dyon solutions

Let us consider a model governed by the action

$$S = \frac{1}{16\pi G} \int d^4x \sqrt{|g|} \Big\{ R[g] - \varepsilon(\partial\varphi)^2 - \sum_{i=1}^{2} \frac{1}{2} e^{2\lambda_i \varphi} (F^{(i)})^2 \Big\}, \tag{2.1}$$

where $g = g_{\mu\nu}(x) dx^\mu \otimes dx^\nu$ is metric, φ is the scalar field, $F^{(i)} = dA^{(i)} = \frac{1}{2} F^{(i)}_{\mu\nu} dx^\mu \wedge dx^\nu$ is the 2-form with $A^{(i)} = A^{(i)}_\mu dx^\mu$, $\varepsilon = \pm 1$, G is the gravitational constant, $(\partial\varphi)^2 = g^{\mu\nu} \partial_\mu \varphi \partial_\nu \varphi$, $(F^{(i)})^2 = F^{(i)}_{\mu\nu} F^{(i)\mu\nu}$, $i = 1, 2$. Here $\lambda_1, \lambda_2 \neq 0$ are coupling constants obeying $\lambda_1 \neq -\lambda_2$ and $|g| = |\det(g_{\mu\nu})|$. We put $\lambda_i^2 \neq 1/2$, $i = 1, 2$, for $\varepsilon = -1$.

We consider a family of dyonic-like black hole solutions to the field equations corresponding to the action (2.1) which are defined on the manifold

$$\mathcal{M} = (2\mu, +\infty) \times S^2 \times \mathbb{R}, \tag{2.2}$$

and have the following form

$$ds^2 = g_{\mu\nu} dx^\mu dx^\nu = H_1^{h_1} H_2^{h_2} \Big\{ -H_1^{-2h_1} H_2^{-2h_2} \left(1 - \frac{2\mu}{R}\right) dt^2 \tag{2.3}$$

$$+ \frac{dR^2}{1 - \frac{2\mu}{R}} + R^2 d\Omega_2^2 \Big\},$$

$$\exp(\varphi) = H_1^{h_1 \lambda_1 \varepsilon} H_2^{-h_2 \lambda_2 \varepsilon}, \tag{2.4}$$

$$F^{(1)} = \frac{Q_1}{R^2} H_1^{-2} H_2^{-A_{12}} dt \wedge dR, \tag{2.5}$$

$$F^{(2)} = Q_2 \tau. \tag{2.6}$$

Here Q_1 and Q_2 are (colored) charges - electric and magnetic, respectively, $\mu > 0$ is the extremality parameter, $d\Omega_2^2 = d\theta^2 + \sin^2\theta d\phi^2$ is the canonical metric on the unit sphere S^2 ($0 < \theta < \pi, 0 < \phi < 2\pi$), $\tau = \sin\theta d\theta \wedge d\phi$ is the standard volume form on S^2,

$$h_i = K_i^{-1}, \qquad K_i = \frac{1}{2} + \varepsilon\lambda_i^2, \tag{2.7}$$

$i = 1, 2$, and

$$A_{12} = (1 - 2\lambda_1\lambda_2\varepsilon) h_2. \tag{2.8}$$

Functions $H_s > 0$ obey the equations

$$R^2 \frac{d}{dR} \left(R^2 \frac{\left(1 - \frac{2\mu}{R}\right)}{H_s} \frac{dH_s}{dR} \right) = -K_s Q_s^2 \prod_{l=1,2} H_l^{-A_{sl}}, \tag{2.9}$$

with the following boundary conditions imposed

$$H_s \to H_{s0} > 0 \tag{2.10}$$

for $R \to 2\mu$, and

$$H_s \to 1 \tag{2.11}$$

for $R \to +\infty$, $s = 1, 2$.

In (2.9) we denote

$$(A_{ss'}) = \begin{pmatrix} 2 & A_{12} \\ A_{21} & 2 \end{pmatrix}, \tag{2.12}$$

where A_{12} is defined in (2.8) and

$$A_{21} = (1 - 2\lambda_1\lambda_2\varepsilon)h_1. \tag{2.13}$$

These solutions may be obtained just by using general formulae for non-extremal (intersecting) black brane solutions from [2]. The composite analogs of the solutions with one 2-form and $\lambda_1 = \lambda_2$ were presented in ref. [1].

The first boundary condition (2.10) guarantees (up to a possible additional demand on analicity of $H_s(R)$ in the vicinity of $R = 2\mu$) the existence of (regular) horizon at $R = 2\mu$ for the metric (2.3). The second condition (2.11) ensures an asymptotical (for $R \to +\infty$) flatness of the metric.

3 Bounds on mass and scalar charge

For ADM gravitational mass we get from (2.3)

$$GM = \mu + \frac{1}{2}(h_1P_1 + h_2P_2), \tag{3.1}$$

where the parameters P_s appear in asymptotical relations $H_s = 1 + P_s/R + o(1/R)$, as $R \to +\infty$. The scalar charge just follows from (2.4)

$$Q_\varphi = \varepsilon(\lambda_1h_1P_1 - \lambda_2h_2P_2). \tag{3.2}$$

Here we outline the following hypothesis, which is supported by certain numerical calculations [1, 3]. For $h_1 = h_2$ this conjecture was proposed in ref. [1].

Conjecture. *For any $h_1 > 0$, $h_2 > 0$, $\varepsilon = \pm 1$, $Q_1 \neq 0$, $Q_2 \neq 0$ and $\mu > 0$: (A) the moduli functions $H_s(R)$, which obey (2.9), (2.10) and (2.11), are uniquely defined and hence the parameters P_1, P_2, the gravitational mass M and the scalar charge Q_φ are uniquely defined too; (B) the parameters P_1, P_2 are positive and the functions $P_1 = P_1(Q_1^2, Q_2^2)$, $P_2 = P_2(Q_1^2, Q_2^2)$ define a diffeomorphism of $\mathbb{R}_+^2$ ($\mathbb{R}_+ = \{x|x > 0\}$); (C) in the limiting case we have: (i) for $Q_2^2 \to +0$: $P_1 \to -\mu + \sqrt{\mu^2 + K_1Q_1^2}$, $P_2 \to +0$ and (ii) for $Q_1^2 \to +0$: $P_1 \to +0$, $P_2 \to -\mu + \sqrt{\mu^2 + K_2Q_2^2}$.*

The conjecture could be readily verified for the case $\varepsilon = 1$, $\lambda_1\lambda_2 = 1/2$. Another integrable case $\varepsilon = 1$, $\lambda_1 = \lambda_2 = \lambda$, $\lambda^2 = 3/2$ is more involved [3].

The conjecture implies the following proposition.

Proposition. *For $h_s > 0$, $Q_s \neq 0$, $\lambda_s \neq 0$ $(s = 1, 2)$ and $\lambda_1 + \lambda_2 \neq 0$ we have the following bounds on the gravitational mass M and the scalar charge Q_φ which are valid for all $\mu > 0$:*

$$\frac{1}{2}\sqrt{h_{min}(Q_1^2 + Q_2^2)} < GM, \tag{3.3}$$

$$|Q_\varphi| < |\lambda|_{max}\sqrt{h_{min}(Q_1^2 + Q_2^2)}, \tag{3.4}$$

for $\varepsilon = +1$ $(0 < h_s < 2)$, and

$$\sqrt{\frac{1}{2}(Q_1^2 + Q_2^2)} < GM, \tag{3.5}$$

$$|Q_\varphi| < |\lambda|_{max}\sqrt{h_{max}(Q_1^2 + Q_2^2)}, \tag{3.6}$$

for $\varepsilon = -1$ $(h_s > 2)$. *Here* $h_{min} = \min(h_1, h_2)$, $h_{max} = \max(h_1, h_2)$, *and* $|\lambda|_{max} = \max(|\lambda|_1, |\lambda|_2)$; $h_{min} = (\frac{1}{2} + |\lambda|^2_{max})^{-1}$ *for* $\varepsilon = +1$ *and* $h_{max} = (\frac{1}{2} - |\lambda|^2_{max})^{-1}$ *for* $\varepsilon = -1$.

In ref. [1] Proposition was proved for the case $\lambda_1 = \lambda_2$. In this case the bound (3.3) is coinciding (up to notations) with that from ref. [4], which was proved there by using certain spinor techniques.

4 Conclusions

Here a family of non-extremal black hole dyon-like solutions in a 4d gravitational model with a scalar field and two Abelian vector fields is considered. The scalar field is either ordinary ($\varepsilon = +1$) or phantom one ($\varepsilon = -1$). The model contain two dilatonic coupling constants $\lambda_s \neq 0$, $s = 1, 2$, obeying $\lambda_1 \neq -\lambda_2$.

The solutions are defined up to two moduli functions $H_1(R)$ and $H_2(R)$, which obey two differential equations of second order with boundary conditions imposed. For $\varepsilon = +1$ these equations are integrable for four cases, corresponding to Lie algebras $A_1 + A_1$, A_2, $B_2 = C_2$ and G_2. In the first case $(A_1 + A_1)$ we have $\lambda_1 \lambda_2 = 1/2$, while in the second one (A_2) we get $\lambda_1 = \lambda_2 = \lambda$ and $\lambda^2 = 3/2$.

Here we have presented lower bounds on the gravitational mass and upper bounds on the scalar charge for $1 + 2\lambda_s^2 \varepsilon > 0$, which are based on the conjectur on the parameters of solutions $P_1 = P_1(Q_1^2, Q_2^2)$, $P_2 = P_2(Q_1^2, Q_2^2)$. For $\varepsilon = +1$ the lower bound on the gravitational mass is in agreement for $\lambda_1 = \lambda_2$ with that obtained earlier by Gibbons et al. by using certain spinor techniques.

Acknowledgment. The authors acknowledge the support from the Program of target financing of the Ministry of Education and Science of the Republic of Kazakhstan Grant No. F.0755. The paper was also funded by by the Ministry of Education of the Russian Federation (Agreement Number 02.a03.21.0008 of 24 June 2016).

References

[1] M.E. Abishev, K.A. Boshkayev, V.D. Dzhunushaliev and V.D. Ivashchuk, Dilatonic dyon black hole solutions, *Class. Quantum Grav.* **32**, No. 16, 165010 (2015).

[2] V.D. Ivashchuk and V.N. Melnikov, P-brane black holes for general intersections, *Grav. Cosmol.* **5**, No 4, 313-318 (1999); gr-qc/0002085.

[3] M.E. Abishev, K.A. Boshkayev, V. D. Ivashchuk, A. Malybaev, in preparation.

[4] G.W. Gibbons, D. Kastor, L.A.J. London, P.K. Townsend and J. Traschen, Supersymmetric Self-Gravitating Solitons, Nucl. Phys. B **416**, 850-880 (1994); arxiv: hep-th/9310118.

Quasilocal angular momentum of gravitational fields in (2+2) formalism

Seung Hun Oh[1,⋆] and *Jong Hyuk* Yoon[1,⋆⋆]

[1] *School of Physics, Konkuk University, 120 Neungdong-ro, Gwangjin-gu, Seoul 05029, Korea*

Abstract. Recently the Poisson algebra of a quasilocal angular momentum of gravitational fields $L(\xi)$ in (2+2) formalism of Einstein's theory was studied in detail [1]. In this paper, we will briefly review the definition of $L(\xi)$ and its remarkable properties. Especially, it will be discussed that $L(\xi)$ satisfies the Poisson algebra $\{L(\xi), L(\eta)\}_{\text{P.B.}} = L([\xi,\eta]_{\text{L}})$, up to a constant normalizing factor, and this algebra reduces to the standard SO(3) algebra at null infinity. It will be also argued that our angular momentum is a quasilocal generalization of A. Rizzi's geometric definition.

1 Formalism

In (2+2) formalism of Einstein's theory [2], the 4-dimensional spacetime E_4 is regarded as a fiber bundle that consists of a 2-dimensional base space M_{1+1} of the Lorentzian signature and a 2-dimensional spacelike fiber N_2 at each point on M_{1+1}. Then, the Einstein's theory can be interpreted as a gauge theory defined on M_{1+1} with *diff*(N_2) as a gauge symmetry, the diffeomorphism group of N_2 [3]. Let us introduce a coordinate system $\{u, v, y^a : a = 2, 3\}$ on E_4, where $\{u, v\}$ and $\{y^a\}$ are coordinates on M_{1+1} and N_2, respectively. The most general line element is given by [4],

$$ds^2 = 0 \cdot dv^2 - 2f du dv - 2h du^2 + \phi_{ab}(dy^a + A_+^a du + A_-^a dv)(dy^b + A_+^b du + A_-^b dv). \tag{1}$$

All the metric components are functions of u, v, and y^a, since we don't assume any spacetime isometry. Throughout this paper, the subscripts + and − denote u and v, respectively, and N_2 is assumed to be compact. The horizontal lifts $\hat{\partial}_\pm$ of the tangent vector fields $\partial_\pm$ are defined by

$$\hat{\partial}_+ = \partial_+ - A_+^a \partial_a, \quad \hat{\partial}_- = \partial_- - A_-^a \partial_a \tag{2}$$

where $A_\pm^a$ are the gauge connections of *diff*(N_2). The *diff*(N_2)-covariant derivative of a *diff*(N_2)-tensor density $T_{ab\cdots}{}^{cd\cdots}$ with weight w is defined by

$$D_\pm T_{ab\cdots}{}^{cd\cdots} := \partial_\pm T_{ab\cdots}{}^{cd\cdots} - [A_\pm, T]_{ab\cdots}{}^{cd\cdots}, \tag{3}$$

where $[A_\pm, T]_{Lab\cdots}{}^{cd\cdots}$ denotes the Lie derivative of $T_{ab\cdots}{}^{cd\cdots}$ with respect to $A_\pm := A_\pm^a \partial_a$. It would be helpful to define the following in- and out-going null vector fields for understanding geometrical meanings of quasilocal angular momentum;

$$n = k\Big(\hat{\partial}_+ - \frac{h}{f}\hat{\partial}_-\Big), \quad l = \frac{k}{f}\hat{\partial}_-, \tag{4}$$

⋆e-mail: shoh.physics@gmail.com
⋆⋆e-mail: yoonjh@konkuk.ac.kr

with a normalization condition

$$< n, l > = -k^2, \quad k = \text{constant} > 0. \tag{5}$$

The positive-definite metric ϕ_{ab} on the two-surface N_2 defined by u, v = constant can be factored into a product of the area element e^σ and the conformal two-metric ρ_{ab},

$$\phi_{ab} = e^\sigma \rho_{ab}, \quad \det \rho_{ab} = 1. \tag{6}$$

When we identify v with a physical time in this formalism, the induced metric on the hypersurface Σ of v = constant is given by

$$ds^2|_\Sigma = -2hdu^2 + e^\sigma \rho_{ab}(dy^a + A_+^a du)(dy^b + A_+^b du). \tag{7}$$

Therefore, the phase variables on Σ consist of $q^I = (h, \sigma, A_+^a, \rho_{ab})$ and their conjugate momenta $\pi_I = (\pi_h, \pi_\sigma, \pi_a, \pi^{ab})$. Here "0", f, and A_-^a in the metric (1) are turned out to be the Lagrange multipliers. After some lengthy calculations, the first-order form of the Einstein-Hilbert action is found to be

$$S = \int dv du d^2y \{\pi_h \partial_- h + \pi_\sigma \partial_- \sigma + \pi_a \partial_- A_+^a + \pi^{ab} \partial_- \rho_{ab} - 0 \cdot C_+ - f\, C_- - A_-^a C_a\}, \tag{8}$$

where the constraints are given by

$$\begin{aligned} C_+ &:= \pi^{ab} D_+ \rho_{ab} + \pi_\sigma D_+ \sigma - h D_+ \pi_h - \partial_+ (h\pi_h + 2e^\sigma D_+ \sigma) \\ &\quad + \partial_a (h\pi_h A_+^a + 2A_+^a e^\sigma D_+ \sigma + 2he^{-\sigma} \rho^{ab} \pi_b + 2\rho^{ab} \partial_b h), \end{aligned} \tag{9}$$

$$C_- := \mathcal{H} - \partial_+ \pi_h + \partial_a (A_+^a \pi_h + e^{-\sigma} \rho^{ab} \pi_b), \tag{10}$$

$$\begin{aligned} C_a &:= -\partial_+ \pi_a + \partial_b (A_+^b \pi_a) + \pi_b \partial_a A_+^b + \pi_\sigma \partial_a \sigma - \partial_a \pi_\sigma \\ &\quad + \pi_h \partial_a h + \pi^{bc} \partial_a \rho_{bc} - \partial_b (\pi^{bc} \rho_{ac}) - \partial_c (\pi^{bc} \rho_{ab}) + \partial_a (\pi^{bc} \rho_{bc}), \end{aligned} \tag{11}$$

and the functional $\mathcal{H}$ in (10) is defined as

$$\begin{aligned} \mathcal{H} &:= -\frac{1}{2} e^{-\sigma} \pi_\sigma \pi_h + \frac{1}{4} h e^{-\sigma} \pi_h^2 - \frac{1}{2} e^{-2\sigma} \rho^{ab} \pi_a \pi_b - e^\sigma R_2 + \frac{1}{2h} e^{-\sigma} \rho_{ac} \rho_{bd} \pi^{ab} \pi^{cd} \\ &\quad + \frac{1}{8h} e^\sigma \rho^{ab} \rho^{cd} (D_+ \rho_{ac})(D_+ \rho_{bd}) + \frac{1}{2h} \pi^{ab} D_+ \rho_{ab} + \frac{1}{2} \pi_h D_+ \sigma. \end{aligned} \tag{12}$$

Notice that the conjugate momentum π^{ab} is traceless, $\rho_{ab}\pi^{ab} = 0$, since the determinant of ρ_{ab} is constant. It was shown that the constraints (9), (10), and (11) are the first-class constraints [5]. The variations over the Lagrange multipliers yield

$$C_+ = C_- = C_a = 0. \tag{13}$$

The 12 Hamilton's equations together with these constraint equations are equivalent to the vacuum Einstein' equations $R_{AB} = 0$ for $A, B = \pm, a$. In the present paper, however, we concern the constraint equation $C_a = 0$ only. Multiplying (11) with an arbitrary tangent vector field $\xi = \xi^a \partial_a$ on N_2, we have

$$\begin{aligned} \xi^a C_a &= \pi^{ab} \mathcal{L}_\xi \rho_{ab} + \pi_\sigma \mathcal{L}_\xi \sigma + \pi_h \mathcal{L}_\xi h + \pi_a \mathcal{L}_\xi A_+^a + \pi_a \partial_+ \xi^a \\ &\quad - \partial_+ (\xi^a \pi_a) + \partial_a (-\xi^a \pi_\sigma - 2\pi^{ab} \xi^c \rho_{bc} + A_+^a \xi^b \pi_b) = 0, \end{aligned} \tag{14}$$

where $\mathcal{L}_\xi$ is the Lie derivative with respect to ξ. Thus, if we integrate the constraint equation (14) over N_2, then we obtain the following integro-differential equation:

$$\frac{\partial}{\partial u} L(u, v; \xi) = \frac{1}{16\pi} \oint d^2y (\pi^{ab} \mathcal{L}_\xi \rho_{ab} + \pi_\sigma \mathcal{L}_\xi \sigma + \pi_h \mathcal{L}_\xi h + \pi_a \mathcal{L}_\xi A_+^a + \pi_a \partial_+ \xi^a), \tag{15}$$

where $L(u, v; \xi)$ is *quasilocal angular momentum of gravitational fields* defined by

$$L(u, v; \xi) := \frac{1}{16\pi} \oint d^2y (\xi^a \pi_a) + L_0. \tag{16}$$

Here, L_0 is an *undetermined integration constant*, and we simply set $L_0 = 0$ throughout this paper. The equation (15) is the balance equation that relates the rate of change in L to the corresponding flux as the parameter u changes [2, 6]. It should be emphasized that the quasilocal angular momentum can be also written in terms of geometric quantities as

$$L(u, v; \xi) = \frac{1}{16\pi} \oint d\mu \frac{< \xi, [n, l]_{\rm L} >}{< n, l >} + L_0, \tag{17}$$

where $d\mu = d^2y\ e^{\sigma}$ is the invariant measure on N_2. It was proved that the sufficient and necessary condition that quasilocal angular momentum $L(\xi)$ of a two-surface along a given vector field ξ^a is insensitive to the ways of labeling the surface is

$$\tilde{\nabla}_a \xi^a = 0, \tag{18}$$

where $\tilde{\nabla}_a$ is the covariant derivative on N_2 [1]. Therefore, the definition of $L(\xi)$ makes sense only if ξ^a is a *divergence-free* vector field on N_2.

2 Poisson algebra of quasilocal angular momentum

By integrating the equation (15), we have another expression of $L(\xi)$ given by

$$L(u, v; \xi) = \frac{1}{16\pi} \int_\Sigma du\, d^2y\, (\pi^{ab} \mathcal{L}_\xi \rho_{ab} + \pi_\sigma \mathcal{L}_\xi \sigma + \pi_h \mathcal{L}_\xi h + \pi_a \mathcal{L}_\xi A^a_+ + \pi_a \partial_+ \xi^a) \quad (\tilde{\nabla}_a \xi^a = 0). \tag{19}$$

Using this form of 3-dimensional volume integral over hypersurface Σ, we can define a Poisson algebra of $L(\xi)$ on Σ by

$$\{L(\xi), L(\eta)\}_{\rm P.B.} = \int du \oint d^2y \left(\frac{\delta L(\xi)}{\delta q^I} \frac{\delta L(\eta)}{\delta p_I} - \frac{\delta L(\eta)}{\delta q^I} \frac{\delta L(\xi)}{\delta p_I} \right). \tag{20}$$

Now we state a significant theorem for the Poisson algebra of $L(\xi)$ as follows:

Theorem. Let ξ and η be two arbitrary vector fields tangent to N_2. The Poisson bracket of the 3-dimensional quasilocal angular momentum flux integrals associated with ξ and η is given by the 3-dimensional integral of the corresponding flux associated with the Lie bracket of ξ and η up to a constant factor. Precisely, it is given by

$$\{L(\xi), L(\eta)\}_{\rm P.B.} = \frac{1}{16\pi} L([\xi, \eta]_{\rm L}). \tag{21}$$

This theorem implies that there exists a homomorphism between the Poisson algebra of quasilocal angular momentum and the Lie algebra of tangent vectors on N_2. We can also define an invariant quasilocal angular momentum L^2 by

$$L^2(u, v) := \frac{1}{96\pi} \left(\frac{\mathcal{A}}{4\pi}\right)^2 \oint d^2y\, e^{-2\sigma} \rho^{ab} \pi_a \pi_b, \tag{22}$$

where $\mathcal{A}$ is the area of N_2, as an analog of the Casimir invariant of SO(3) algebra in the sense that

$$\{L^2, L(\xi)\}_{\rm P.B.} = 0, \tag{23}$$

for any divergence-free vector field ξ on N_2. By using this quasilocal quantity, we can avoid the problems related to the singular nature of the divergence-free tangent vector field ξ^a on a compact N_2.

3 Asymptotic limit at null infinity of asymptotically flat spacetime

At null infinity of asymptotically flat spacetimes, N_2 becomes S_2. Let $y^a = (\theta, \phi)$ be the usual angular coordinates on S_2. The generators $\bar{\xi}_{(\alpha)}$ $(\alpha = 1, 2, 3)$ of the SO(3) isometry of S_2 are given by

$$\bar{\xi}_{(1)} = -\sin\phi\frac{\partial}{\partial\theta} - \cot\theta\cos\phi\frac{\partial}{\partial\phi}, \quad \bar{\xi}_{(2)} = \cos\phi\frac{\partial}{\partial\theta} - \cot\theta\sin\phi\frac{\partial}{\partial\phi}, \quad \bar{\xi}_{(3)} = \frac{\partial}{\partial\phi}, \tag{24}$$

which satisfy the commutation relations $[\bar{\xi}_{(\alpha)}, \bar{\xi}_{(\beta)}]_{\mathrm{L}} = -\epsilon_{\alpha\beta}{}^{\gamma}\bar{\xi}_{(\gamma)}$. By continuity, there exist tangent vector fields $\xi_{(\alpha)}$ $(\alpha = 1, 2, 3)$ on N_2 in the asymptotic zone of asymptotically flat spacetimes, which approach $\bar{\xi}_{(\alpha)}$ as N_2 approaches S_2 in the limit $v \to \infty$. Let $\bar{L}_\alpha$ be the gravitational angular momentum at null infinity defined by the limit

$$\bar{L}_\alpha := \lim_{v\to\infty} L(u, v; \xi_{(\alpha)}), \tag{25}$$

for such tangent vector fields $\xi_{(\alpha)}$. One can easily show that $\bar{\xi}_{(\alpha)}$ is divergence-free, that is, $\bar{\nabla}_a\bar{\xi}^a_{(\alpha)} = 0$ for $\alpha = 1, 2, 3$, and show that the following commutation relation holds

$$\{\bar{L}_\alpha, \bar{L}_\beta\}_{\mathrm{P.B.}} = -\frac{1}{16\pi}\epsilon_{\alpha\beta}{}^{\gamma}\bar{L}_\gamma. \tag{26}$$

as a direct consequence of the theorem proven in Section 2. This means that the algebra of gravitational angular momentum at null infinity realizes the SO(3) Lie algebra under the Poisson bracket.

The angular momentum of the Kerr spacetime can be obtained by taking the limit (25). Let us notice that the Kerr metric in the asymptotic zone can be written as

$$\begin{aligned} ds^2 = &-2dudv - (1 - \frac{2m}{v} + \cdots)du^2 + (\frac{4ma\sin^2\theta}{v} - \frac{4ma^3\sin^2\theta\cos^2\theta}{v^3} + \cdots)dud\phi \\ &+v^2(1 + \frac{a^2\cos^2\theta}{v^2} + \cdots)d\theta^2 + v^2\sin^2\theta(1 + \frac{a^2}{v^2} + \cdots)d\phi^2 \\ &+\sin^2\theta(\frac{4ma^3}{v^3} + \frac{8m^2a^3}{v^4} + \cdots)dvd\phi - (\frac{a^2\sin^2\theta}{v^2} + \cdots)dv^2, \end{aligned} \tag{27}$$

where m and a are the mass and the specific angular momentum of Kerr spacetime, respectively. It can be readily shown that

$$\bar{L}_1 = \bar{L}_2 = 0, \qquad \bar{L}_3 = ma, \qquad \lim_{v\to\infty} L^2 = (ma)^2 \tag{28}$$

which is precisely the angular momentum of the Kerr spacetime at null infinity.

Now let us show that the asymptotic limit of our quasilocal angular momentum coincides with the angular momentum defined by A. Rizzi [7]. His approach is based on an affine foliation of the asymptotic zone of a asymptotically flat spacetime with families of S_2 [8]. He introduced the in- and out-going null vector fields n and l with the normalization $< n, l >= -2$. Together with an orthonormal frame e_a $(a = 1, 2)$ on S_2, n and l form a tetrad system

$$\{\, e_a, e_3, e_4 : a = 1, 2,\ e_3 = n, e_4 = l\,\}. \tag{29}$$

The twist ζ_a of n and l is defined as

$$\zeta_a = \frac{1}{2} < \nabla_3 e_4, e_a > . \tag{30}$$

Rizzi's angular momentum is defined as the following integral

$$L(\Omega_{(\alpha)}) = -\frac{1}{8\pi} \lim_{s\to\infty} \oint_{S_2} \Omega^a_{(\alpha)} \zeta_a dS_\gamma, \tag{31}$$

where $\Omega_{(\alpha)}$ are the SO(3) generators that satisfy the commutation relations $[\Omega_{(\alpha)}, \Omega_{(\beta)}]_{\rm L} = -\epsilon_{\alpha\beta}{}^{\gamma}\Omega_{(\gamma)}$, and s is an affine parameter of out-going null vector field l (Notice the change of sign from the Rizzi's original definition). The right hand side of (31) involves the limit that pulls the variables back to S_2, and dS_γ is the infinitesimal area element of S_2 with the standard metric γ. By using an identity

$$\zeta_a = \frac{1}{4}[n,\ l]_{{\rm L}a}. \tag{32}$$

one can straightforwardly show that

$$L(\Omega_{(\alpha)}) = \frac{1}{16\pi < n, l >} \lim_{s\to\infty} \oint_{S_2} \Omega^a_{(\alpha)} [n,\ l]_{{\rm L}a} dS_\gamma, \tag{33}$$

which is precisely the limit (25) of our quasilocal angular momentum, if the following identifications are made

$$dS_\gamma = d\mu, \quad \Omega^a_{(\alpha)} = \bar{\xi}^a_{(\alpha)}. \tag{34}$$

Therefore, the quasilocal angular momentum (17) is a quasilocal generalization of the angular momentum of gravitational fields proposed by A. Rizzi.

4 Summary

So far we have discussed the definition of the quasilocal angular momentum $L(\xi)$ and its noteworthy properties. In particular, the necessary and sufficient condition that ensures the gauge-independence of the quasilocal angular momentum is that ξ is divergence-free. And the Poisson algebra of the quasilocal angular momentum associated with an arbitrary tangent vector field ξ on N_2 is closed. Furthermore, the asymptotic limit of our quasilocal angular momentum is identical to the geometric angular momentum of A. Rizzi at null infinity, and its Poisson algebra reduces to the SO(3) commutation relations at null infinity.

Acknowledgment This work was supported by National Research Foundation of Korea (Grant 2015-R1D1A1A01-059407).

References

[1] J. H. Yoon and S. H. Oh, *Class. and Quant. Grav.* in press.
[2] J. H. Yoon, *Phys. Rev. D* **70**, 084037 (2004)
[3] Y. M. Cho, Q. H. Park, K. S. Soh and J. H. Yoon, *Phys. Lett. B* **286**, 251 (1992)
[4] R. A. d'Inverno and J. Stachel, *J. Math. Phys.* **19**, 2447 (1978)
[5] J. H. Yoon, *Class. and Quant. Grav.* **31**, 045005 (2014)
[6] A. Ashtekar and B. Krishnan, *Phys. Rev. Lett.* **89**, 261101 (2002)
[7] A. Rizzi, *Phys. Rev. Lett.* **81**, 1150 (1998)
[8] D. Christodoulou, *Phys. Rev. Lett.* **67**, 1486 (1991)

AUTHOR INDEX

AUTHOR INDEX